T0336024

STRING THEORY AND PARTICLE PHYSICS

String theory is one of the most active branches of theoretical physics, and has the potential to provide a unified description of all known particles and interactions. This book is a systematic introduction to the subject, focused on the detailed description of how string theory is connected to the real world of particle physics.

Aimed at graduate students and researchers working in high-energy physics, it provides explicit models of physics beyond the Standard Model. No prior knowledge of string theory is required as all necessary material is provided in the introductory chapters. The book provides particle phenomenologists with the information needed to understand string theory model building, and describes in detail several alternative approaches to model building, such as heterotic string compactifications, intersecting D-brane models, D-branes at singularities, and F-theory.

LUIS E. IBÁÑEZ is Professor of Theoretical Physics at the Universidad Autónoma de Madrid and member of the Instituto de Física Teórica-UAM/CSIC. One of the world leaders in physics beyond the Standard Model of particle physics, he has made important contributions to the construction of the supersymmetric Standard Model and superstring phenomenology.

ANGEL M. URANGA is Research Professor at the Consejo Superior de Investigaciones Científicas at the Instituto de Física Teórica-UAM/CSIC. He is one of the leading young string theorists working in the construction of models of particle physics, in particular due to his contribution on the use of D-branes to build realistic brane-world models.

Cover illustration: the authors thank Jorge Ibáñez-Albajar for his help with the design of the cover image for this book.

STRING THEORY AND PARTICLE PHYSICS: AN INTRODUCTION TO STRING PHENOMENOLOGY

LUIS E. IBÁÑEZ

Universidad Autónoma de Madrid and Instituto de Física Teórica IFT-UAM/CSIC

ANGEL M. URANGA

Instituto de Física Teórica IFT-UAM/CSIC

CAMBRIDGE
UNIVERSITY PRESS

CAMBRIDGE
UNIVERSITY PRESS

University Printing House, Cambridge CB2 8BS, United Kingdom

One Liberty Plaza, 20th Floor, New York, NY 10006, USA

477 Williamstown Road, Port Melbourne, VIC 3207, Australia

314-321, 3rd Floor, Plot 3, Splendor Forum, Jasola District Centre, New Delhi - 110025, India

79 Anson Road, #06-04/06, Singapore 079906

Cambridge University Press is part of the University of Cambridge.

It furthers the University's mission by disseminating knowledge in the pursuit of education, learning and research at the highest international levels of excellence.

www.cambridge.org
Information on this title: www.cambridge.org/9780521517522

First published 2012

A catalogue record for this publication is available from the British Library

Library of Congress Cataloging in Publication data
Ibáñez, Luis E., 1952–
String theory and particle physics : an introduction to string phenomenology / Luis E. Ibáñez, Angel M. Uranga.
p. cm.
Includes bibliographical references and index.
ISBN 978-0-521-51752-2 (Hardback)
1. String models. I. Uranga, A. (Angel) II. Title.
QC794.6.S85I23 2012
539.7′258–dc23

2011035562

ISBN 978-0-521-51752-2 Hardback

To our families

[. . .] vi el Aleph, desde todos los puntos, vi en el Aleph la tierra, y en la tierra otra vez el Aleph y en el Aleph la tierra [. . .] porque mis ojos habían visto ese objeto secreto y conjetural, cuyo nombre usurpan los hombres, pero que ningún hombre ha mirado: el inconcebible universo.

Contents

Preface

String theory is the leading candidate for a consistent quantum theory of gravity. It has also become a central area of research in mathematical physics, with different additional applications which range from heavy ion physics to condensed matter, cosmology or mathematics. Notwithstanding this, the excitement fostered in 1984 actually came from the coexistence of chiral anomaly free gauge theories and gravity in string theory, raising the expectation of an ultimate unification of Standard Model (SM) and gravitational interactions into a consistent string quantum theory. The enthusiasm was thus motivated by particle physics phenomenological goals.

Since then much effort has been dedicated to explore the possible embedding of the SM of particle physics in string theory, a field commonly known as string phenomenology. However, although there are by now several excellent books introducing the general field of string theory, there is no systematic and detailed coverage of the large body of knowledge accumulated in string phenomenology. This lack has become particularly acute after the duality revolution of 1995, when the advent of D-branes made the string engineering of non-trivial gauge theories more flexible, thus providing new avenues to realize the SM in string theory.

Consequently, and due to the seemingly imposing complexity of string theory, this field has not permeated much to many particle physics phenomenologists and model builders, who feel reluctant to struggle with a jungle of papers and reviews to extract the phenomenological aspects of string theory.

The main purpose of this book is to provide an elementary introduction to string theory, and to string phenomenology, in a systematic and self-contained way. It should be useful to particle phenomenologists and model builders, both senior and fresh. It will also be useful to string theorists interested in learning how (and how far) string theory may reproduce the observed SM physics.

The book has six chapters with introductory material. The first presents a brief summary of the SM structure, its puzzles, and several of its extensions, including Grand Unified Theories and extra dimensions. The second introduces the basic aspects of supersymmetry and its application to particle physics models, most notably the Minimal Supersymmetric

Standard Model (MSSM). These first two chapters serve to fix the notation and introduce concepts, appearing later when building string theory models of particle physics.

Chapters 3 to 6 constitute an introduction to the basics of string theory including the bosonic string (Chapter 3), and the heterotic, type II and type I superstrings (Chapter 4). The simplest toroidal compactification to four dimensions is described in Chapter 5, which also provides a first glimpse of D-branes. Chapter 6 describes D-branes and their role in string theory, as well as the different non-perturbative dualities in the theory. Our presentation in these chapters aims at getting the main physical results in the most comfortable way for the non-initiated, avoiding the machinery of conformal field theory (partly covered in an appendix). These four chapters are self-contained and constitute by themselves an introductory course on string theory, useful also to graduate students searching for a first contact with the formalism of string theory. String theorists acquainted with this material may safely jump over to Chapter 7.

Chapters 7 to 12 give a relatively detailed description of string compactifications giving rise to chiral theories in four dimensions, with emphasis on those with $\mathcal{N} = 1$ supersymmetry and a particle content close to the SM. They include different heterotic constructions, in Chapters 7 and 8, whose low-energy effective action is covered in Chapter 9, as well as type II orientifolds (and M- and F-theory related constructions), in Chapters 10 and 11, with their effective action discussed in Chapter 12. Detailed explicit examples of MSSM-like models are presented for the different compactification methods. The purpose is to enable the reader to obtain the massless spectrum and effective lagrangian of these string constructions, so as to grasp their contact to SM physics.

Chapters 13 and 14 introduce additional ingredients, most notably string instantons and closed string fluxes. Those ingredients give rise to extra contributions to the effective action relevant for aspects like Yukawa couplings, neutrino masses and moduli stabilization. Chapter 15 continues the study of moduli fixing and its interplay with supersymmetry breaking, reaching up to the generation of low-energy supersymmetry breaking masses in MSSM-like models. Further phenomenological issues are discussed in Chapter 16, and Chapter 17 contains a general discussion of the space of string vacua, in particular those resembling the SM or MSSM.

The optimum use of this book requires basic background of quantum field theory, group theory, and elementary notions of the SM of particle physics and general relativity. We have attempted to reduce the mathematics to a minimum, and to introduce the necessary definitions where required (including an appendix with the main geometrical and topological concepts used in the text).

We mark with an asterisk * those sections or subsections containing relevant material which may be skipped in a first reading of the book. Concerning the references, we have preferred not to insert citations in the main text and give a Bibliography for each chapter at the end of the book. These include some references to original literature, but mostly to reviews useful to the reader interested in further details. The list of references is (admittedly and necessarily) very incomplete and we apologize to many of our colleagues whose relevant work has not been cited. Finally, we have set up a webpage to publish corrections and errata for this book:

https://sites.google.com/site/stringtheoryandparticlephysics/

Many people and institutions have contributed to make this book possible. We thank our home institutions, the Departamento de Física Teórica of the Universidad Autónoma de Madrid (UAM), and the Instituto de Física Teórica IFT-UAM/CSIC of the Consejo Superior de Investigaciones Científicas and UAM. We thank our colleagues there, for creating a supportive and stimulating environment. A.M.U. also thanks the CERN TH group, for being "home" during the first half of this project. We are grateful to our colleagues and collaborators, for all the discussions during these years. In particular, we thank Luis Aparicio, Gerardo Aldazabal, Pablo G. Cámara, David G. Cerdeño, Anamaria Font, Iñaki Garcia-Etxebarria, Fernando Marchesano, Christoffer Petersson, Fernando Quevedo, Graham Ross, and Pablo Soler, for carefully reading selected chapters and making many improving suggestions. We also thank Bert Schellekens for discussions and for providing us with edited figures from his work. We are also grateful to the Cambridge University Press team, and especially to Simon Capelin, for suggesting the project, and for the gentle management throughout the process of writing. We finally thank our families, for giving the patience and support that is always required in such a demanding enterprise.

1

The Standard Model and beyond

In this chapter we overview the structure of the Standard Model (SM) of particle physics, its shortcomings, and different ideas for physics beyond the Standard Model (BSM) devised to address them. We discuss the three fine-tuning puzzles of the SM (the cosmological constant, strong CP and hierarchy problems) and several ideas put forward to solve them. Among the extensions of the SM, this chapter overviews Grand Unified Theories (GUTs), which explore the embedding of the SM gauge symmetry into a larger unified gauge group like $SU(5)$ or $SO(10)$, and models with extra spatial dimensions beyond the observed three. These ideas, along with supersymmetry, studied in Chapter 2, will reappear in the study of particle physics models from string theory in later chapters.

1.1 The Standard Model of particle physics

The Standard Model of particle physics constitutes an impressive success of twentieth century physics. It describes elementary particles and their electromagnetic, weak and strong interactions in a remarkable wide range of energies, and with unprecedented precision.

The SM is a quantum field theory based on a gauge group

$$G_{\rm SM} = SU(3) \times SU(2)_L \times U(1)_Y, \tag{1.1}$$

with $SU(3)$ describing strong interactions via Quantum Chromodynamics (QCD), and $SU(2)_L \times U(1)_Y$ describing electroweak (EW) interactions. The matter fields form three generations (or families) of quarks and leptons, described as Weyl 2-component spinors, with the EW structure

$$Q_L = \begin{pmatrix} U_L^i \\ D_L^i \end{pmatrix}, \ U_R^i, \ D_R^i, \ L = \begin{pmatrix} \nu_L^i \\ E_L^i \end{pmatrix}, \ E_R^i; \ i = 1, 2, 3. \tag{1.2}$$

We take all fields to be left-chirality Weyl spinors, independently of their L or R subindex, which merely denotes the $SU(2)_L$ transformation properties as doublets or singlets. In addition, quarks in Q_L transform as color triplets, while U_R, D_R transform as conjugate triplets. Gauge quantum numbers of the SM fermions are shown in Table 1.1.

The chirality of this fermionic spectrum is possibly one of the deepest properties of the SM. Describing particles in terms of Dirac spinors, it means that left- and right-chirality

Table 1.1 *Gauge quantum numbers of SM quarks, leptons and the Higgs scalars*

Field	$SU(3)$	$SU(2)_L$	$U(1)_Y$
$Q_L^i = (U^i, D^i)_L$	3	2	$1/6$
U_R^i	$\bar{3}$	1	$-2/3$
D_R^i	$\bar{3}$	1	$+1/3$
$L^i = (\nu^i, E^i)_L$	1	2	$-1/2$
E_R^i	1	1	$+1$
$H = (H^-, H^0)$	1	2	$-1/2$

components actually have *different* EW quantum numbers. In the above more suitable description of all fermions as 2-component left-handed Weyl spinors, chirality is the statement that they fall in a complex representation of the SM gauge group. In any picture, the implication is that explicit Dirac mass terms $m \overline{f_R} f_L$+h.c. are forbidden by gauge invariance, and fermions remain massless (until EW symmetry breaking, discussed below). The impossibility of mass terms makes chiral sets of fermions very robust, and easy to identify when deriving the light spectrum of particles from fundamental theories like string theory. Hence, the chirality of the fermion spectrum is a key guiding principle in building string models of particle physics.

In the SM the electroweak symmetry $SU(2)_L \times U(1)_Y$ is spontaneously broken down to the electromagnetic $U(1)_{EM}$ symmetry by a complex scalar Higgs field transforming as an $SU(2)_L$ doublet $H = (H^-, H^0)$ and with hypercharge $-1/2$. Its dynamics is parametrized in terms of a potential, devised to trigger a non-vanishing Higgs vacuum expectation value (vev) v

$$V = -\mu^2 |H|^2 + \lambda |H|^4 \quad \Rightarrow v^2 \equiv \langle |H| \rangle^2 = \mu^2/2\lambda. \qquad (1.3)$$

The vev defines the electrically neutral direction and is set to $\langle H^0 \rangle \simeq 170\,\text{GeV}$ in order to generate the $W^\pm$ and Z vector boson masses. Simultaneously it produces masses for quarks and leptons through the Yukawa couplings

$$\mathcal{L}_{\text{Yuk}} = Y_U^{ij} \, \overline{Q}_L^i \, U_R^j \, H^* + Y_D^{ij} \, \overline{Q}_L^i \, D_R^j \, H + Y_L^{ij} \, \overline{L}^i \, E_R^j \, H + \text{h.c.} \qquad (1.4)$$

These interactions are actually the most general consistent with gauge invariance and renormalizability, and accidentally are invariant under the global symmetries related to the baryon number B and the three family lepton numbers L_i. Regarding the SM as an effective theory, non-renormalizable operators violating these symmetries may, however, be present.

The hypercharges of the SM fermions in Table 1.1 are related to their usual electric charges by $Q_{EM} = Y + T_3$, where $T_3 = \text{diag} \left(\frac{1}{2}, -\frac{1}{2} \right)$ is an $SU(2)_L$ generator. They thus reproduce electric charge quantization, e.g. the equality in magnitude of the proton and

Table 1.2 *Masses of quarks and charged leptons at the EW scale in GeV*

U-quarks	u	c	t
	2×10^{-3}	5×10^{-1}	173
D-quarks	d	s	b
	4×10^{-3}	1×10^{-1}	3
Leptons	e	μ	τ
	0.51×10^{-3}	1.05×10^{-1}	1.7

electron charges. Although these hypercharge assignments look rather ad hoc, their values are dictated by quantum consistency of the theory. It is indeed easy to check that these are (modulo an irrelevant overall normalization) the only (family independent) assignments canceling all potential triangle gauge anomalies.

The above simple SM structure describes essentially all particle physics experimental data at present (with simple extensions to account for neutrino masses, see Section 1.2.6). Despite this character of a *Theory* of matter and interactions, it poses several intriguing questions, which probably hold the key to new physics at a more fundamental level, and thus motivate maintaining its status of *Model*. Most prominently, gravitational interactions are not included, in particular due to the difficulties in reconciling them with Quantum Mechanics. Gravity implies that the SM should be regarded as an effective theory with a cutoff at most the Planck scale

$$M_p = \frac{1}{\sqrt{8\pi}\, G_{\mathrm{N}}^{1/2}} = 1.2 \times 10^{19}\,\text{GeV}, \tag{1.5}$$

where G_{N} is Newton's constant. Other examples of hints for BSM physics arise from the fine tuning issues in Section 1.3.

Finally, another suggestive hint for an underlying structure is that the SM has many free parameters, which are *external* to the model, rather than predicted by it. There are three gauge coupling constants g_1, g_2, g_3, the QCD θ-parameter, the nine masses of quarks and leptons (plus those of neutrinos), as well as the quark CP violating phase (plus possible additional phases in the lepton sector). Finally, there are additional couplings in the Higgs sector. Most of the unknown parameters are related to the Higgs–Yukawa sector of the theory, which has not been fully tested experimentally, and is thus still poorly understood. One would hope for a more fundamental microscopic explanation of (at least some of) these parameters. Indeed, one of the most outstanding puzzles of the SM is the structure of fermion masses and mixing angles. The masses of quarks and leptons show a hierarchical structure, see Table 1.2, suggesting the possibility of some underlying pattern. The mass matrices m_{ij} obtained from the U, D, L Yukawa couplings are in general not diagonal, but may be diagonalized by bi-unitary transformations $V_L^{U,D,L}$, $V_R^{U,D,L}$, acting on the left- and right-handed fermions respectively. After these rotations, the $W^{\pm}$ gauge bosons couple to U- and D-quarks through the Cabibbo–Kobayashi–Maskawa (CKM) matrix, given by $U_{CKM} = V_L^U(V_L^D)^+$. Experimental measurements yield the following approximate structure for this matrix

$$|V_{CKM}| = \begin{pmatrix} 0.974 & 0.22 & 0.003 \\ 0.22 & 0.973 & 0.04 \\ 0.008 & 0.04 & 0.999 \end{pmatrix}. \tag{1.6}$$

Again this displays an interesting structure. It is close to the identity matrix, with small off-diagonal mixing entries, except for the Cabibbo (1-2) entry, which is somewhat larger. Experiments also show evidence for CP-violation in the quark sector, whose size may be measured in terms of the Jarlskog invariant $J = (3.0 \pm 0.3) \times 10^{-5}$.

The structure of neutrino masses and mixings turns out to be quite different from that of quarks. Neutrinos are much lighter than quarks and charged leptons. Solar and atmospheric oscillation experiments allow for the measurement of some squared-mass differences

$$\Delta m_{12}^2 = 8 \times 10^{-5} \, \text{eV}^2; \quad \Delta m_{23}^2 = 2.4 \times 10^{-3} \, \text{eV}^2, \tag{1.7}$$

which also suggest a possible hierarchical structure. On the other hand the neutrino Pontecorvo–Maki–Nakagawa–Sakata (PMNS) mixing matrix is quite different from the CKM one, since oscillation experiments reveal a structure of typically large mixing angles

$$|V_{PMNS}| = \begin{pmatrix} 0.77 - 0.86 & 0.50 - 0.63 & 0.0 - 0.22 \\ 0.22 - 0.56 & 0.44 - 0.73 & 0.57 - 0.80 \\ 0.21 - 0.55 & 0.40 - 0.71 & 0.59 - 0.82 \end{pmatrix}. \tag{1.8}$$

The search for a fundamental explanation of the fermion spectrum, gauge coupling constants, and other *external* SM parameters, has constituted a driving force to propose theories beyond the SM. A prototypical example is provided by Grand Unification Theories, reviewed next.

1.2 Grand Unified Theories

In the SM there are three gauge factors, and five matter multiplets Q_L, U_R, D_R, L, E_R per family. Grand Unified Theories (GUTs) propose that there is an underlying gauge group G_{GUT}, in general assumed to be simple, which at very high energies experiences spontaneous symmetry breaking down to the SM group. Thus the SM gauge factors are unified into a single gauge force, and the different SM matter fields are also (possibly partially) unified into multiplets of the larger symmetry G_{GUT}. This gauge group G_{GUT} must be at least of rank four and contain the SM group, $SU(3) \times SU(2)_L \times U(1)_Y$, and must admit complex representations to accommodate a chiral fermion spectrum. GUT theories can thus be classified according to the choice of G_{GUT} satisfying these conditions, see later sections.

1.2.1 Gauge coupling unification

A particularly compelling motivation for unification of the SM gauge groups into a *simple* group is the unification of gauge coupling constants into a single one. Indeed, there is convincing evidence for such unification, coming from the evolution of the three SM gauge couplings to high energies with the renormalization group equations (RGE). The one-loop evolution equations for the couplings $\alpha_a = g_a^2/(4\pi)$ are

$$\frac{1}{\alpha_a(Q^2)} = \frac{1}{\alpha_a(M^2)} + \frac{b_a}{4\pi} \log \frac{M^2}{Q^2}; \quad a = 1, 2, 3, \tag{1.9}$$

where Q, M are two energy scales, and b_a are the one-loop β-function coefficients, which for a $SU(N)$ gauge theory are

$$b = -\frac{11}{3}N + \frac{2}{3}T(R)n_f + \frac{1}{3}T(R)n_s, \tag{1.10}$$

with n_f, n_s the number of Weyl fermions and complex scalar fields in the representation R, and $T(R)$ is the corresponding quadratic Casimir, in the normalization $T(R) = 1/2$ for fundamentals. Expression (1.10) also holds for $U(1)$, by setting $N = 0$ and replacing $T(R) \to Y^2$. For the SM one has

$$\begin{pmatrix} b_1 \\ b_2 \\ b_3 \end{pmatrix} = \begin{pmatrix} 0 \\ -22/3 \\ -11 \end{pmatrix} + N_{\text{gen}} \begin{pmatrix} 4/3 \\ 4/3 \\ 4/3 \end{pmatrix} + N_{\text{Higgs}} \begin{pmatrix} 1/10 \\ 1/6 \\ 0 \end{pmatrix}. \tag{1.11}$$

where N_{gen}, N_{Higgs} denote the numbers of quark-lepton generations and Higgs doublets, and we have introduced an additional hypercharge normalization factor $Y^2 \to (3/5)Y^2$ to be justified later, just above (1.13).

Extrapolating the measured values of the three α_a from the EW scale up in energies, and assuming no further relevant degrees of freedom at intermediate scales (the so-called *desert hypothesis*), the three couplings tend to join around 10^{15} GeV into a single coupling α_{GUT}. An important general feature is that matter in full G_{GUT} multiplets does not modify the running of the *relative* gauge couplings at one loop, and hence does not spoil unification. The joining of the coupling constants using the minimal SM content is only in qualitative agreement with experiment, but becomes quite sharp in supersymmetric versions, as described in Section 2.6.2.

1.2.2 SU(5) GUTs

There is a unique compact simple Lie group of rank four admitting complex representations: $SU(5)$. It moreover contains the SM group as a maximal subgroup. $SU(5)$ GUTs illustrate many features of grand unification, like gauge coupling unification and proton decay, valid for other choices of simple GUT group.

The $SU(5)$ theory has $5^2 - 1 = 24$ gauge bosons, which include the 12 SM ones and 12 extra gauge bosons transforming as $SU(2)_L$ doublets and $SU(3)$ triplets, denoted by (X_r^+, Y_r^+), (X_r^-, Y_r^-), $r = 1, 2, 3$. The $SU(5)$ symmetry can be broken to the SM group by introducing the "GUT-Higgs" scalars Φ_{24} in the adjoint **24**. It acquires a large vev $\langle \Phi_{24} \rangle = \text{diag}\,(2v, 2v, 2v, -3v, -3v)$ commuting with the SM group, which is thus unbroken, whereas the X, Y gauge bosons get masses $M_{X,Y}^2 \simeq \alpha_{\text{GUT}} v^2$.

Each SM quark-lepton generation fits nicely into a reducible $SU(5)$ representation $\bar{\mathbf{5}} + \mathbf{10}$. For instance, fields in, e.g., the first generation can be written as

$$\bar{\mathbf{5}} = \begin{pmatrix} d_1^c \\ d_2^c \\ d_3^c \\ e^- \\ \nu_e \end{pmatrix}; \quad \mathbf{10} = \begin{pmatrix} 0 & u_3^c & u_2^c & u_1 & d_1 \\ & 0 & u_1^c & u_2 & d_2 \\ & & 0 & u_3 & d_3 \\ & & & 0 & e^+ \\ & & & & 0 \end{pmatrix}, \tag{1.12}$$

where the **10** is displayed as a 5×5 antisymmetric matrix, whose lower triangle entries are ignored for clarity. As usual, we have taken all fermion fields left-handed, and our notation for the first family relates to (1.2) and Table 1.1 by $d^c = D_R^1$, $(\nu_e, e^-) = L^1$, $u^c = U_R^1$, $(u, d) = Q_L^1$, $e^+ = E_R^1$. The assignment (1.12) reproduces the correct SM gauge quantum numbers upon breaking $SU(5)$. Note also that the combination $\bar{\mathbf{5}} + \mathbf{10}$ is free of $SU(5)$ anomalies.

$SU(5)$ unification has some interesting implications, holding also in other GUTs:

- *Charge quantization*: The electromagnetic $U(1)_{\text{EM}}$ generator belongs to $SU(5)$, and hence it must be traceless, $\text{Tr}\,Q_{\text{EM}} = 0$. This implies that, e.g., fermions in the $\bar{\mathbf{5}}$ in (1.12) must have electric charges adding to zero, hence $Q_{d^c} = -\frac{1}{3} Q_{e^-}$. This and a similar expression for u-quarks imply equality of the proton and electron charge, and thus that all charges are quantized, i.e. multiples of a basic unit.

- *Relationships among gauge couplings*: As mentioned, unification into a simple GUT group implies a unique gauge coupling constant, leading to two relations among SM gauge couplings. At the unification scale the non-abelian couplings satisfy $g_3 = g_2$. Standard hypercharge Y is realized in $SU(5)$ with an additional factor of $\sqrt{3/5}$, to comply with the normalization $\text{tr}\,T_{SU(5)}^2 = 1/2$ in the fundamental representation. This corresponds to the GUT relation $g_1^2 = (3/5) g_2^2$. For the weak mixing angle we have

$$\sin^2 \theta_W = \frac{\text{Tr}\,(T_3^2)}{\text{Tr}\,(Q_{\text{EM}}^2)} = \frac{1/2}{4/3} = \frac{3}{8}, \tag{1.13}$$

where in the second equality we have taken traces in the fundamental representation and $Q_{\text{EM}} = T_3 + Y$. These relationships apply at the unification scale M_{GUT}, and provide boundary conditions for gauge coupling running at lower energies according to

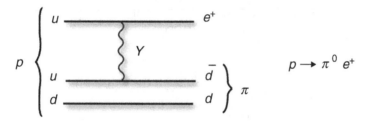

Figure 1.1 Dimension-6 contribution to the proton decay mode $p \to e^+\pi^0$.

the RGE (1.9). Combining the equations for the three couplings, the relations among couplings at the EW scale M_W are

$$\frac{1}{\alpha_3(M_W)} = \frac{3}{8}\left[\frac{1}{\alpha_{EM}(M_W)} - \frac{1}{2\pi}\left(b_1 + b_2 - \frac{8}{3}b_3\right)\log\frac{M_{GUT}}{M_W}\right],$$

$$\sin^2\theta_W(M_W) = \frac{3}{8} + \frac{5}{16\pi}\alpha_{EM}(M_W)\left(b_2 - \frac{3}{5}b_1\right)\log\frac{M_{GUT}}{M_W}. \qquad (1.14)$$

Using, e.g., the first equation, the unification scale can be computed to be $M_{GUT} \simeq 10^{15}$ GeV. The second then leads to the prediction $\sin^2\theta_W = 0.214 \pm 0.004$, only in qualitative agreement with the experimental result 0.2312 ± 0.0002. In the supersymmetric case in Section 2.6.2 the result gets much better, with an almost perfect quantitative agreement with the experimental value.

- *Baryon number violation and proton decay*: Since quarks and leptons live inside the same GUT multiplets, they can transform into each other by emission/absorption of the massive gauge bosons X, Y. This leads to baryon number violating transitions like the proton decay mode $p \to \pi^0 e^+$ in Figure 1.1. These transitions are suppressed by the large X, Y mass. On dimensional grounds, the amplitude scales as $1/M_X^2$, leading to a quite long proton lifetime $\tau_p \simeq M_X^4/m_p^5$, where m_p is the proton mass. A detailed computation yields $\tau_{p \to \pi^0 e^+} \simeq 4 \times 10^{29 \pm .7}$ years. This is actually well below the super-Kamiokande experiment bound $\tau_{p \to \pi^0 e^+} > 6.6 \times 10^{33}$ years, so the simplest $SU(5)$ GUT is excluded. However, in the supersymmetric version of this theory, proton decay through this kind of dimension-6 operator is more suppressed, as described in Section 2.6.2, so this variant of the theory is not ruled out.

- *Relationships among fermion masses*: In the SM the fermion masses arise from three independent kinds of Yukawa couplings (1.4). In $SU(5)$ the SM Higgs doublet sits in a GUT multiplet $H_{\bar{5}}$ in the representation $\bar{5}$ (along with a color triplet, to be made heavy as discussed in Section 2.6.2). There are only two kinds of Yukawa couplings, of the form

$$L_{Yuk}^{SU(5)} = Y_U^{ij}\,\bar{\psi}_{10}^i\,\psi_{10}^j\,H_{\bar{5}}^* + Y_{D,L}^{ij}\,\bar{\psi}_{\bar{5}}^i\,\psi_{10}^j\,H_{\bar{5}} + \text{h.c.} \qquad (1.15)$$

Hence there are relations between the Yukawas of charged leptons and D-quarks, $Y_D^{ij} = Y_L^{ij}$ at the GUT scale. At lower energies, the Yukawa couplings run according to the

RGEs, and in particular the QCD loop corrections enhance the quark Yukawas compared to the charged lepton ones. The leading QCD correction, e.g., for the third generation yields

$$\frac{m_b(M_W)}{m_\tau(M_W)} = \left(\frac{\alpha_3(M_W)}{\alpha_3(M_X)}\right)^{-\frac{\gamma}{2b_3}} \simeq 2, \qquad (1.16)$$

where $\gamma = 8$ is the quark QCD anomalous dimension coefficient. This result is in reasonable agreement with data for the third generation, but fails for the first two. The situation can be improved in more complicated GUT models involving additional Higgs multiplets, whose description is beyond this brief overview.

1.2.3 SO(10) GUTs

For rank five there are only two compact simple Lie groups admitting complex representations, $SU(6)$ and $SO(10)$. The use of $SU(6)$ for GUT model building is a simple extension of $SU(5)$ and does not lead to new remarkable features. On the other hand, $SO(10)$ GUTs extend $SU(5)$ in a more interesting way, leading to several genuinely new properties.

A remarkable property is that one SM family of quarks and leptons fits neatly into a single $SO(10)$ representation, the spinor representation **16**, which is the lowest-dimensional complex representation of $SO(10)$. For, e.g., the first family

$$\psi_{16} = \left(\nu_e,\ u_1,\ u_2,\ u_3;\ e^-,\ d_1,\ d_2,\ d_3;\ d_3^c,\ d_2^c,\ d_1^c,\ e^+;\ u_3^c,\ u_2^c,\ u_1^c,\ \nu_R\right).$$

The **16** actually contains an additional fermion, singlet under all SM gauge interactions, as is manifest in its decomposition under the $SU(5)$ subgroup, $\mathbf{16} = \mathbf{10} + \bar{\mathbf{5}} + \mathbf{1}$. The extension of the SM by these singlets is aesthetically pleasing, since they can be regarded as right-handed neutrinos, completing the right-handed spectrum of leptons in analogy with that of quarks. Hence $SO(10)$ GUTs predict the existence of a right-handed neutrino ν_R for each generation. Right-handed neutrinos may play a key role in the structure of neutrino masses, as described in Section 1.2.6.

The group $SO(10)$ has a maximal $SU(5) \times U(1)$ subgroup, with the adjoint decomposing as $\mathbf{45} = \mathbf{24} + \mathbf{10} + \overline{\mathbf{10}} + \mathbf{1}$. Unlike the $SU(5)$ case, breaking of the gauge symmetry down to the SM group G_{SM} may proceed via an intermediate stage of partial unification. Different patterns are

$$
\begin{aligned}
SO(10) &\longrightarrow & G_{\text{SM}} \\
SO(10) &\longrightarrow SU(3) \times SU(2)_L \times SU(2)_R \times U(1)_{B-L} \longrightarrow & G_{\text{SM}} \\
SO(10) &\longrightarrow \qquad\quad SU(5) \times U(1) \qquad\qquad \longrightarrow & G_{\text{SM}} \\
SO(10) &\longrightarrow \quad\ SU(4) \times SU(2)_L \times SU(2)_R \qquad\quad \longrightarrow & G_{\text{SM}}.
\end{aligned}
$$

The breaking of $SO(10)$ to the SM group requires not only adjoint scalars Φ_{45}, but also extra scalars ϕ_{16} transforming in the **16** in order to lower the rank. Models with direct $SO(10)$ breaking down to the SM group lead to predictions quite similar to those of $SU(5)$.

The intermediate step $SU(3) \times SU(2)_L \times SU(2)_R \times U(1)_{B-L}$, with $B - L$ denoting baryon minus lepton number, is often called the left–right symmetric model, and $SU(4) \times SU(2)_L \times SU(2)_R$ is known as the Pati–Salam model. These partial unifications are interesting even without resorting to an ultimate $SO(10)$ unification, although they fit nicely in the latter.

The minimal embedding of the EW Higgs doublet is in a multiplet of scalars H_{10} in the representation **10**. It decomposes as $5 + \bar{5}$ under $SU(5)$, and actually contains two ordinary Higgs doublets (plus additional triplets). They are often denoted H_u, H_d, since their vevs induce U- and D-quark masses. In the simplest $SO(10)$ GUT model, there is only one kind of Yukawa coupling

$$\mathcal{L}_{\text{Yuk}}^{SO(10)} = Y^{ij} \, \bar{\psi}_{\mathbf{16}}^i \, \psi_{\mathbf{16}}^j \, H_{10} + \text{h.c.} \tag{1.17}$$

This leads to a unification of Yukawas at the GUT scale,

$$Y_U^{ij} = Y_D^{ij} = Y_L^{ij} = Y_\nu^{ij}, \tag{1.18}$$

where Y_ν denotes the Yukawa coupling inducing neutrino Dirac masses, see Section 1.2.6. In particular, the relation for the third generation $Y_\tau = Y_b = Y_t$ may be consistent with experimental data for very large $\tan \beta = \langle H_u \rangle / \langle H_d \rangle \simeq 50 - 60$. However, this simplest scheme with a single Higgs H_{10} does not quite work, since it implies $Y_U^{ij} = Y_D^{ij}$ and hence aligned quark rotation matrices $V_L^U = V_L^D$, and there is no CKM mixing. This improves in models where the physical EW Higgs field involves further $SO(10)$ multiplets, like scalars in the **126**, or involving non-renormalizable couplings.

1.2.4 E_6 GUTs

At rank 6, the only compact simple Lie groups with complex representations are $SU(7)$ and E_6. Again $SU(7)$ adds no essentially new feature, whereas the exceptional group E_6 contains $SO(10) \times U(1)$ as a maximal subgroup, and does introduce some novelties.

The lowest-dimensional non-trivial representation in E_6 has dimension 27 and decomposes under the $SO(10)$ and $SU(5)$ subgroups as

$$\mathbf{27} = \mathbf{16} + \mathbf{10} + \mathbf{1} = (\mathbf{10} + \bar{\mathbf{5}} + \mathbf{1}) + (\mathbf{5} + \bar{\mathbf{5}}) + \mathbf{1}. \tag{1.19}$$

Thus the **27** contains one $SO(10)$ generation, but also some extra non-chiral D-like quarks and L-like leptons. This relative abundance of extra fields is a shortcoming of E_6 GUTs.

There are 78 gauge bosons decomposing under $SO(10)$ as $\mathbf{78} = \mathbf{45} + \mathbf{16} + \overline{\mathbf{16}} + \mathbf{1}$. The E_6 gauge symmetry may be broken down to the SM by Higgs fields in the adjoint and the **27**. The symmetry breaking may proceed also in various steps, for instance $E_6 \rightarrow SO(10) \times U(1)$, but also the breaking

$$E_6 \longrightarrow SU(3)_c \times SU(3)_L \times SU(3)_R \rightarrow G_{\text{SM}}. \tag{1.20}$$

This intermediate step, in which the QCD symmetry is on equal footing with left–right symmetries, is sometimes called *trinification*. An E_6 generation decomposes under (1.20) as $\mathbf{27} = (\mathbf{3}, \bar{\mathbf{3}}, \mathbf{1}) + (\bar{\mathbf{3}}, \mathbf{1}, \mathbf{3}) + (\mathbf{1}, \mathbf{3}, \bar{\mathbf{3}})$.

Although E_6 unification has in principle no compelling advantage over $SU(5)$ or $SO(10)$, it arises in a simple class of heterotic string compactifications, constructed in Section 7.3.2.

1.2.5 Flipped SU(5) unification

There are proposals in which the unification gauge group is not simple. The most prominent examples are "flipped" $SU(5)$ models, based on a $SU(5) \times U(1)$ gauge group. In the breaking $SU(5) \times U(1)_X \rightarrow SU(3) \times SU(2) \times U(1)_Y$, the hypercharge generator is identified with the linear combination

$$Y = \frac{1}{5}\left(Q_X - \frac{Q_{Y'}}{6}\right), \tag{1.21}$$

where Y' is the diagonal $SU(5)$ generator, given by $Q_{Y'} = \mathrm{diag}\,(-2, -2, -2, 3, 3)$ in the **5** representation. The flipped gauge group $SU(5) \times U(1)_X$ may be embedded into $SO(10)$.

One quark-lepton generation fits into the reducible structure

$$\mathbf{10}_1 = (Q_L, D_R, \nu_R), \quad \bar{\mathbf{5}}_{-3} = (U_R, L), \quad \mathbf{1}_5 = E_R, \tag{1.22}$$

where subindices denote $U(1)_X$ charges. The name "flipped" is due to the particle content of the representations, related to the standard $SU(5)$ by flipping $U_R \leftrightarrow D_R$, $E_R \leftrightarrow \nu_R$.

The symmetry breaking is obtained through scalars transforming as

$$\mathbf{5}_{-2} = (T, H_d), \quad \bar{\mathbf{5}}_2 = (\bar{T}, H_u); \quad \Phi_{\mathbf{10}_1}, \ \Phi_{\overline{\mathbf{10}}_{-1}}, \tag{1.23}$$

where $T, \bar{T}$ are color triplets with hypercharge $\pm 1/3$. Vevs for the **10**, $\overline{\mathbf{10}}$ along the hypercharge neutral components trigger the breaking down to the SM group. The **5**, $\bar{\mathbf{5}}$ contain the EW Higgs doublets, plus color triplets $T, \bar{T}$.

The gauge symmetries allow for three independent kinds of Yukawa couplings

$$L_{\mathrm{Yuk}}^{F.SU(5)} = Y_D^{ij}\, \overline{\psi}_{\mathbf{10}_1}^{i}\, \psi_{\mathbf{10}_1}^{j}\, H_{\mathbf{5}_{-2}} + Y_U^{ij}\, \overline{\psi}_{\bar{\mathbf{5}}_{-3}}^{i}\, \psi_{\mathbf{10}_1}^{j}\, H_{\bar{\mathbf{5}}_2} + Y_L^{ij}\, \overline{\psi}_{\bar{\mathbf{5}}_{-3}}^{i}\, E_R^{j}\, H_{\mathbf{5}_{-2}} + \mathrm{h.c.}$$

Hence flipped $SU(5)$ models imply no unification predictions for quark and lepton masses. Also, since the GUT group is not simple, there is in principle no unification of the hypercharge coupling with the remaining two. Despite these less appealing features, an advantage of flipped $SU(5)$ models is that they do not require large representations for GUT symmetry breaking. Since some large classes of string models cannot lead to adjoint scalars, flipped $SU(5)$ models provide an interesting possibility for unification in these string constructions.

1.2.6 Neutrino masses, seesaw mechanism, and GUTs

Non-zero neutrino masses require the introduction of additional ingredients beyond the minimal SM, namely right-handed neutrinos and/or a high scale of lepton number violation. At energies below the EW breaking scale, neutrinos are electrically neutral singlets.

Table 1.3 *Possible types of mass terms in the neutrino sector*

Type	Term	$|\Delta L|$	$|\Delta Y|$
Dirac	$m_D \, \bar{\nu}_L \nu_R$	0	1/2
LH-Majorana	$M_L \, \nu_L \nu_L$	2	1
RH-Majorana	$M_R \, \nu_R \nu_R$	2	0

There are three types of terms relevant for masses in the neutrino sector, classified according to their violation of lepton number L and hypercharge Y, and to their use or not of right-handed neutrinos, as summarized in Table 1.3.

In analogy with quark and charged lepton masses, it is possible to produce neutrino masses from Dirac mass terms, by introducing right-handed neutrinos, i.e. singlets under all SM gauge interactions, with Yukawa couplings $Y_\nu^{ij} \bar{\nu}_L^i \nu_R^j H$. This mechanism preserves lepton number, but provides no explanation for the smallness of observed neutrino masses, or their essentially left-handed character.

An alternative is to accept violations of lepton number at a relatively high scale M, leading to effective Majorana mass terms for left-handed neutrinos at lower scales. In particular the dimension-5 Weinberg operator

$$\frac{h_{ij}}{M} \, \nu_L^i \nu_L^j \, HH + \text{h.c.}, \tag{1.24}$$

which upon EW breaking produces left-handed neutrino Majorana masses of order

$$M_L^{ij} = \frac{h_{ij} \langle H \rangle^2}{M}. \tag{1.25}$$

Assuming, e.g., $h_{ij} \simeq 1$, and $M \simeq 10^{14} - 10^{15}$ GeV, neutrino masses fall in the ballpark of $M \simeq 0.01 - 0.1$ eV, consistent with neutrino oscillation experiments.

A natural implementation of the above idea is the *seesaw* mechanism. It arises when the theory includes right-handed neutrinos with a large Majorana mass M_R, and Yukawa couplings to left-handed neutrinos, leading to Dirac masses m_D. Diagonalization of the neutrino mass matrix leads to a predominantly left-handed neutrino eigenstate, with very small mass $\simeq m_D^2/M_R$, and a predominantly right-handed eigenstate, with very large mass $\simeq M_R$. The mechanism can be effectively regarded as generating an effective left-handed neutrino Majorana mass (1.25) with $M = M_R$, via the exchange of heavy right-handed neutrinos, as in Figure 1.2.

The seesaw mechanism can be implemented in many models of BSM physics, but fits well within the idea of GUTs, since the expected scale of right-handed neutrino Majorana mass is relatively close to M_{GUT}. In particular, $SO(10)$ GUTs naturally contain right-handed

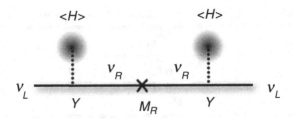

Figure 1.2 The exchange of a heavy right-handed neutrino gives rise to a seesaw contribution to the left-handed neutrino Majorana mass.

neutrinos, as explained in Section 1.2.3. In the simplest $SO(10)$ scheme, their masses come from the non-renormalizable coupling,

$$\frac{h_{ij}}{M_*} \nu_R^i \nu_R^j \langle \phi_{16}^* \phi_{16}^* \rangle. \tag{1.26}$$

Here M_* is some high scale, above the cutoff of the $SO(10)$ GUT theory (e.g. like the Planck scale), and ϕ_{16} is one of the GUT symmetry breaking scalars. Vevs $\langle \phi_{16} \rangle \simeq M_{\text{GUT}}$ induce right-handed neutrino Majorana masses $M_R \simeq M_{\text{GUT}}^2/M_*$. For $M_{\text{GUT}} \simeq 10^{16}$ GeV and, e.g., $M_* = M_p$, one gets the right scale $M_R \simeq 10^{14}$ GeV.

A further appealing implication of the neutrino mass seesaw mechanism is a possible explanation for baryogenesis, and thus the observed baryon–antibaryon asymmetry in the universe. Indeed, out-of-equilibrium decay of relic right-handed neutrinos could have produced a lepton asymmetry in the primordial universe. The latter is then transformed into a baryon asymmetry by $SU(2)_L$ instanton effects, which violate baryon and lepton number but preserve $B - L$. This mechanism for generating the observed baryon asymmetry in the universe is called *leptogenesis*.

1.3 The SM fine-tuning puzzles

Grand Unified Theories are interesting extensions of the SM, with many appealing features, like charge quantization, and the remarkable embeddings unifying quarks and leptons into simpler multiplets. They also nicely accommodate the seesaw mechanism to generate small neutrino masses. Finally, unification of gauge couplings works quite well in supersymmetric extensions of GUTs.

On the other hand GUTs do not address a number of legitimate questions. For instance, the framework does not implicate the gravitational interaction, and its possible interplay or even unification with the gauge interactions. It also does not explain other properties, like the existence of three generations, thus undermining any potential explanation for the full structure of fermion masses and mixings. Focusing on more quantitative aspects, the SM poses several questions, which remain unsolved in GUTs, related to the idea of *naturalness* or *fine-tuning*. Physical systems often contain small parameters, which are considered *natural* if the system develops a new symmetry when the parameter is set to zero. For instance, the smallness of fermion masses compared with any high-energy scale, like $M_{\rm GUT}$ or M_p, is *naturally explained* by chiral symmetry, i.e. independent rotations of left- and right-handed degrees of freedom. The SM however contains several parameters experimentally constrained to be extremely small, and for which no known symmetry exists within the SM, so they must be unnaturally *fine-tuned*, in the above sense. Their proper understanding/explanation may provide key hints of the nature of BSM physics, so we discuss them in turn.

1.3.1 The cosmological constant problem and the anthropic principle

In quantum field theory, vacuum energy is unphysical and can be shifted at will. When coupling to gravity, however, the energy-momentum tensor $T_{\mu\nu}$ gravitates and so does its vacuum value $\langle T_{\mu\nu} \rangle = -V_0 \, g_{\mu\nu}$. This introduces a cosmological constant $\Lambda_{\rm c.c.} = V_0$ in Einstein's equations

$$R_{\mu\nu} - \frac{1}{2} R g_{\mu\nu} = 8\pi G_N \langle T_{\mu\nu} \rangle_{\rm vac.} = -8\pi G_N V_0 g_{\mu\nu}, \tag{1.27}$$

where $R_{\mu\nu}$ is the Ricci tensor, and $R = R_\mu^\mu$ the scalar curvature. In the SM there are many physical contributions to the vacuum energy. For instance, in EW symmetry breaking the SM tree level Higgs scalar potential (1.3) has a value at its minimum $V_0 = -|\mu|^4/(4\lambda) \neq 0$. Similarly, there are other contributions from lower energies (e.g. the QCD condensates) or higher energies (e.g. possible GUT symmetry breaking phase transitions). In addition, there are loop corrections to the vacuum energy, which diverge quartically and lead to the theoretical expectation $\Lambda_{\rm c.c.} \simeq (M_{\rm cutoff})^4$. The cutoff scale could potentially be as large as the Planck scale, so the naive expectation would be $\Lambda_{\rm c.c.} \simeq M_p^4 \simeq 10^{112} \, ({\rm eV})^4$.

On the other hand, cosmological measurements from supernovae redshifts and the WMAP satellite indicate the existence of a dark energy density, which, barring alternative explanations, corresponds to a vacuum energy density

$$\Lambda_{\rm c.c.} \simeq (10^{-3} \, {\rm eV})^4, \tag{1.28}$$

around 124 orders of magnitude smaller than the expectation for $M_{\rm cutoff} = M_p$, and still 60 if $M_{\rm cutoff}$ is lowered down to the EW scale. Of course one can introduce a naked cosmological constant Λ_0, and adjust the total to agree with observational data. However, this seemingly harmless shift hides a dramatic fine-tuning of widely different contributions at very different scales, which constitutes the *cosmological constant problem*. A related question is the *coincidence problem*, i.e. explaining the similar relative densities of dark energy and matter density at present times. This is particularly surprising since vacuum energy density does not dilute in time, whereas matter density does, so that dark energy has become dynamically important only in a recent epoch.

The search for mechanisms to explain or avoid this fine-tuning is proving one of the most difficult challenges in theoretical physics, with many (unsuccessful) attempted solutions.

A possible explanation, attracting much attention in the last decade, is the proposal by S. Weinberg in 1987 of employing *anthropic* considerations to explain the smallness of the cosmological constant, as follows. Suppose Nature has the potentiality to scan through the possible values of the cosmological constant in different universes (or large regions thereof). Universes with cosmological constant of order the above naive expectations have an accelerated expansion too fast to develop structure formation by gravitational accretion. They thus have no galaxies, and not enough complexity to allow for the development of living observers to measure such cosmological constant. The correct expectation on the value of $\Lambda_{c.c.}$ is then that it would presumably have the largest possible value compatible with the existence of observers (us) to measure it. Interestingly, this upper bound comes relatively close to the observed value. This is particularly significant, since the argument and the prediction of $\Lambda_{c.c.}$ was proposed by Weinberg well before any serious observational evidence of a non-vanishing vacuum energy in the universe.

A possible dynamical realization of multiple regions of the universe scanning through different physical properties is the *multiverse* implemented by *eternal inflation* (see Section 16.6.1 for a brief discussion of inflation). In this type of inflationary model, large quantum fluctuations produced during inflation may locally change the value of the vacuum energy (and other physical properties) in some regions of the universe. These expand into *bubble universes* which can host new bubble nucleation processes. This eternal self-replication of the universe leads to multiple bubble universes populating different values for the vacuum energy.

Anthropic considerations are still controversial in the scientific community. In some ways they resemble Darwinian evolutionism, implying that some "fundamental" parameters in physics like $\Lambda_{c.c.}$ are environmental, i.e. have a historical rather than fundamental origin. This is not a new situation in the history of physics, where certain parameters considered fundamental at some point turned out to be environmental. An example is the solar system, where the number of planets and their orbits were once believed to be dictated by fundamental principles, e.g. Kepler's attempt to relate the planets known in his days to the platonic solids, the five regular polyhedra; they are now understood as defining a particular solution of celestial mechanics (among many possible planetary systems) chosen by historical accident in the formation of the solar system, and with an anthropically fixed parameter, the Sun–Earth distance. The seeming naivety, from our vantage point, of Kepler's approach should serve as reminder of the possibility of being incurring into analogous ones in present times, and as reassurance to contemplate a possible environmental explanation of $\Lambda_{c.c.}$

As discussed in Section 17.3, the anthropic solution to the cosmological constant problem seems particularly well suited to string theory, since it seemingly admits an extremely large number of solutions populating a large range of possible values of the vacuum energy. In the above analogy, this set of solutions of the underlying string theory would play the role of possible planetary systems consistent with the underlying celestial mechanics theory.

1.3.2 The strong CP problem and axions

The QCD lagrangian has a CP-violating piece of the form

$$\mathcal{L}_\theta = \frac{\theta}{32\pi^2} F_{\mu\nu} \tilde{F}^{\mu\nu}, \tag{1.29}$$

with $\tilde{F}^{\mu\nu} = \frac{1}{2}\epsilon^{\mu\nu\rho\sigma} F_{\rho\sigma}$. Although this term is a total derivative, it contributes non-trivially in the quantum theory through gauge instantons (see Section 13.1.1) and hence relates to the topological structure of the QCD vacuum. The CP-violating character of (1.29) is manifest by recasting it as $\vec{E}\cdot\vec{B}$, in terms of the chromo-electric and magnetic fields. Naively one would expect $\theta \simeq g_3 \simeq 1$. However, such value would induce too large contributions to the CP-violating electric dipole moment of the neutron. The latter is described by the effective operator

$$d_n \left(\bar{n}\gamma_5\sigma_{\mu\nu} F^{\mu\nu}_{\text{EM}} n\right), \tag{1.30}$$

where n denotes the neutron field. Its CP-violating character is clear in the non-relativistic limit, in which it reads $\vec{\sigma}\cdot\vec{E}$, with $\vec{\sigma}$ the neutron spin and $\vec{E}$ the electric field. The experimental bound on the neutron electric dipole moment is $|d_n| < 2.9 \times 10^{-26}$ e·cm, which in turn implies $\theta < 10^{-10}$. This is at least 10 orders of magnitude smaller than the naive order one value, thus leading to the so-called *strong CP problem*.

This fine-tuning problem hardly admits an explanation from anthropic considerations, since the value of the θ parameter does not seem to play a crucial role in the potential development of complex life and observers in the universe. Fortunately, there are explicit particle physics dynamical mechanisms which can explain the smallness of θ in a natural way.

The most attractive proposal in this direction is probably the introduction of a Peccei–Quinn *axion*. This is a very light pseudoscalar field $a(x)$ coupling to QCD as a θ-parameter

$$\mathcal{L}_a = \frac{a(x)}{32\pi^2 f_a} F_{\mu\nu} \tilde{F}^{\mu\nu}, \tag{1.31}$$

where f_a is the axion decay constant. A simple realization of the axion and its coupling (1.31) is as a Goldstone boson of a so-called Peccei–Quinn symmetry, a global $U(1)_{\text{PQ}}$ with a mixed $U(1)_{\text{PQ}} \times SU(3)^2$ anomaly, and spontaneously broken at a scale $\simeq f_a$. Although the minimal SM does not admit such a symmetry, since its only global symmetries are B and L_i, it can be implemented in simple extensions, e.g. including one additional singlet scalar. The Goldstone boson character of the axion forbids any perturbative contribution to its scalar potential. On the other hand, QCD instantons create a periodic potential of the form

$$V(a) \simeq 1 - \cos(\theta + a/f_a), \tag{1.32}$$

with a minimum at $\theta + a/f_a = 0$ (mod $2n\pi$), i.e. at a vanishing effective $\bar{\theta}$-parameter $\bar{\theta} = \theta + a/f_a$. The mechanism thus leads to a dynamical solution to the strong CP problem.

The axion field acquires a mass $m_a \simeq m_\pi f_\pi / f_a$, where m_π and $f_\pi \simeq 93$ MeV are the pion mass and decay constant, respectively. There are strong astrophysical and cosmological constraints on the value of f_a,

$$4 \times 10^9 \, \text{GeV} \lesssim f_a \lesssim 10^{12} \, \text{GeV}. \tag{1.33}$$

The lower bound is astrophysical, and arises from modifications of stellar evolution time scales due to energy loss by axion emission for smaller f_a. The upper bound is cosmological, and arises from the overclosing of the universe due to energy stored in axions for too large f_a. This latter bound is less reliable, and may be relaxed, e.g. by invoking the possible existence of regions in the universe with an atypically low axion density, selected on anthropic grounds. For axions in the allowed range, the axion is very light, $m_a < 10^{-2}$ eV, and couples very weakly to ordinary matter, with strength $\sim m_q / f_a$, with m_q some light quark mass. This type of axions are thus called *invisible*. These properties make invisible axions a possible candidate for cosmological dark matter particles.

Many extensions of the SM (including GUTs) can include a spontaneously broken $U(1)_{\text{PQ}}$ global symmetry, leading to invisible axion models. For instance, string theory compactifications produce natural axion candidates with the required couplings, as discussed in Section 16.2. In these and many other BSM scenarios, however, a non-trivial to realize requirement of the axion mechanism is that QCD must provide the dominant contribution to the axion potential.

1.3.3 The EW hierarchy problem

In simple terms, the *EW hierarchy problem* is the question of the origin of the hierarchy M_W / M_p between the EW and the Planck scale. There are two aspects to this question: there is a physical question of the mechanism generating the two widely different scales, and a *technical* question of keeping their hierarchy stable against quantum corrections. Indeed, the EW scale is fixed by the SM Higgs vev, related by (1.3) to its squared mass μ^2, but the latter parameter receives huge quadratically divergent corrections from graphs like those in Figure 1.3, dragging it to the cutoff scale of the theory, as

$$\delta \mu_{\text{Higgs}}^2 \simeq \frac{\alpha}{4\pi} \Lambda_{\text{cutoff}}^2. \tag{1.34}$$

The cutoff scale Λ_{cutoff} could be as large as the Planck scale M_p. Of course, it would be possible to choose the naked value of μ^2 such that the renormalized value is of order M_W^2. But this naively harmless shift actually hides a fine-tuning of many different loop contributions, with an incredible precision of, e.g., 10^{-34} for $\Lambda_{\text{cutoff}} = M_p$. Also, there are theories with a finite physical cutoff, like the string scale M_s in string models, in which this correction is actually finite and generically huge. The hierarchy problem is then the question of maintaining the Higgs sufficiently light to trigger EW symmetry breaking, given that its natural value would be an ultraviolet scale of the theory, such as M_{GUT}, M_p, or M_s.

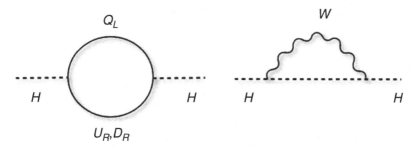

Figure 1.3 Loop diagrams contributing to the mass of the Higgs doublet in the SM.

The EW hierarchy problem lies at the heart of many proposals of BSM physics, since it points to the existence of new physics at the TeV scale to stabilize the hierarchy. Some proposed solutions are the following:

- *Dynamical generation by strongly coupled gauge sectors*: The SM already contains examples of dynamically generated mass scales stable against any large UV scale M of the theory. For instance, the smallness of the masses of QCD bound states, like protons, is explained by dimensional transmutation, as $m_p \simeq \Lambda_{QCD} \simeq \exp[1/(b_3 g_3^2)]M$, with b_3 the QCD one-loop β-function coefficient.

  The suggestion is to implement a similar mechanism at higher scales, and introduce a new gauge sector with strong dynamics generating the scale M_W, and reproducing the SM Higgs physics as its effective theory. A class of theories along this line are *technicolor* models, in which the EW scale is generated from a condensate of strongly interacting *techniquarks* which effectively plays the role of the Higgs field. This idea is quite attractive, although its explicit realization in specific models is somewhat contrived when trying to describe the observed quark and lepton masses and mixings.
- *Lowering the fundamental gravity scale*: In this approach the fundamental scale of gravity is taken only slightly above M_W, to avoid large radiative corrections. The weakness of gravitational interactions is then due to the existence of extra dimensions, with a large volume diluting the gravity field, or with a strong redshift (*warping*) weakening its effects, as described in the next section. The hierarchy problem then becomes the question of dynamically generating large volume or warping in the extra dimensions, in a quantum mechanically stable way. In string theory both extra dimensions and warping appear naturally, as shown in later chapters.
- *Low-energy supersymmetry*: In this approach the SM is extended to include one extra partner for each SM field, related to it by $\mathcal{N} = 1$ supersymmetry, a new symmetry relating bosons and fermions. These extra particles and new interactions precisely cancel the quadratically divergent loop corrections to the Higgs scalar, and stabilize the hierarchy. This approach has the nice property of being perturbative, and quite predictive, providing a benchmark scenario to be tested at the LHC. We describe $\mathcal{N} = 1$ supersymmetry and its implications as a solution of the hierarchy problem in Chapter 2. On the other hand,

supersymmetry is a basic theoretical physics tool, and a key ingredient in string compactifications, including many realizing particle physics models. Hence, supersymmetry will be pervasive in this book.

- *Anthropic approach*: One may contemplate the possibility that the ratio M_W/M_p is environmental rather than fundamental, and could receive an anthropic rather than a dynamical explanation. Indeed, it could be argued that if the Higgs scalar is not light enough, it cannot trigger EW symmetry breaking and the resulting set of interactions does not allow for the development of complex systems required for living observers. The argument is however more uncertain and less convincing than for the anthropic bound on the cosmological constant, and moreover many parameters enter into the EW hierarchy problem, making any quantitative statement difficult. On the other hand, there are several plausible dynamical mechanisms, like the three mentioned above, to address the hierarchy problem, making the anthropic approach to this question less compelling. The exploration of the TeV scale by the LHC will provide a more definite take on this issue, since physical mechanisms explaining the EW hierarchy require BSM physics in that range, whereas the anthropic explanation does not.

1.4 Extra dimensions

The world as we observe it is four-dimensional (4d), it has three space dimensions plus a time dimension. However, it is conceivable that there are extra space dimensions, with a finite size too small to be resolved with the presently accessible energies. The idea of a fifth dimension was proposed in 1921 by T. Kaluza, and further elaborated by O. Klein in 1926, in an attempt to unify gravitational and electromagnetic interactions. The idea of extra dimensions was retaken around 1970–80 in the context of supergravity theories. The maximally supersymmetric 4d gravity theory, $\mathcal{N} = 8$ supergravity, considered at the time a serious candidate for a unified theory of all interactions, is related to 11-dimensional (11d) $\mathcal{N} = 1$ supergravity by Kaluza–Klein compactification of seven dimensions. This unification proposal in terms of 11d supergravity, in its original form, was proven not viable by Witten in the early 1980s. The argument showed that higher-dimensional theories compactified on smooth manifolds cannot lead to 4d chiral fermions charged under non-abelian gauge interactions, unless both chirality and gauge interactions are already present in the higher-dimensional theory. Heuristically, the underlying reason is the fact that higher-dimensional spinor representations contain, when decomposed under the 4d Lorentz group, vector-like combinations of 4d Weyl spinors. The general no-go argument rules out smooth compactifications of 11d supergravity, and in general puts strong constraints on particle physics model building from higher-dimensional theories.

In the last decades the idea of extra dimensions has been reconsidered for two main reasons. First, it is natural in string theory, which leads to 10d and 11d theories, whose compactifications can however avoid the above no-go result, either due to the presence of higher-dimensional gauge bosons and chiral fermions (e.g. in heterotic string models), or by having them arise at special submanifolds in the compactification space (like D-branes or geometric singularities). Second, it has been realized that the size of the extra dimensions could be much larger than previously thought, leading to new phenomenological scenarios potentially testable at the LHC. In this section we review this second aspect, as the first is extensively covered in later chapters.

1.4.1 A fifth dimension

In this section we introduce the Kaluza–Klein (KK) compactification of five-dimensional (5d) theories on a circle, and explain the unobservability of the extra dimension at low energies. The examples generalize to KK compactification of more dimensions. We focus on the simplest bosonic fields, scalars and gravitons, which suffice to illustrate these points. The discussion of chiral fermions in compactifications is postponed to their appearance within string theory in later chapters.

Five-dimensional scalar field

Consider a 5d free massless scalar $\phi(x^M)$, $M = 0, \ldots, 4$, with action

$$S_{5d,\phi} = \int d^5 x \left(-\tfrac{1}{2} \partial_M \phi \partial^M \phi \right), \tag{1.35}$$

where we omit a dimensionful coefficient required by dimensional analysis. We consider the theory on 4d Minkowski space times a circle, $\mathbf{M}_4 \times \mathbf{S}^1$. This is described by letting the coordinate $y \equiv x^4$ have periodicity $y \simeq y + 2\pi R$, and so

$$\phi(x^\mu, y) = \phi(x^\mu, y + 2\pi R). \tag{1.36}$$

One can then expand the y-dependence in Fourier modes on the circle

$$\phi(x^\mu, y) = \sum_{k \in \mathbf{Z}} \phi_k(x^\mu) e^{iky/R}. \tag{1.37}$$

Substituting into (1.35), and integrating over y we obtain

$$S_{4d,\phi} = (2\pi R) \int d^4 x \left(-\frac{1}{2} \partial_\mu \phi_0 \partial^\mu \phi_0 \right) - (2\pi R) \sum_{k=1}^{\infty} \int d^4 x \left(\partial_\mu \phi_k \partial^\mu \phi_k^* + \frac{k^2}{R^2} \phi_k \phi_k^* \right),$$

where $\phi_k^* = \phi_{-k}$. This describes a 4d theory with a massless scalar ϕ_0 and an infinite tower of massive scalars, known as KK resonances, labeled by the KK momentum k, and with masses

$$m_k^2 = \frac{k^2}{R^2}; \quad k = 1, 2, 3, \ldots \tag{1.38}$$

At energies $E \ll 1/R$, only the *zero mode* ϕ_0 would be observable, so the effective theory is a 4d field theory with one massless scalar. The extra dimensions are unobservable at low energies, as their detection requires energies of order $1/R$ to access the tower of

Kaluza–Klein modes. It is easy to repeat the argument for a massive 5d scalar field with mass M, and show that the KK tower is shifted as $m_k^2 = M^2 + k^2/R^2$, so there is no zero mode.

Five-dimensional gravity

Things become even more interesting if we consider the KK compactification of 5d gravity, and its relation to 4d gravity. Recall the Einstein–Hilbert action for 4d gravity

$$S_{4d} = \frac{1}{2\kappa_4^2} \int d^4x \sqrt{-g}\, R_{4d}, \tag{1.39}$$

where $g = \det(g_{\mu\nu})$, $1/\kappa_4^2 = M_p^2/8\pi$ and R_{4d} is the 4d scalar curvature. Consider now the 5d gravity action

$$S_{5d} = \frac{M_5^3}{2} \int d^5x \sqrt{-G}\, R_{5d}, \tag{1.40}$$

where $G = \det(G_{MN})$, with $M, N = 0, 1, \ldots, 4$, and R_{5d} is the 5d scalar of curvature. Consider compactifying $y \equiv x^4$ into an S^1, and Fourier expand the metric,

$$G_{MN}(x^\mu, y) = \sum_{k\in\mathbf{Z}} G_{MN}^k(x^\mu)\, e^{iky/R}. \tag{1.41}$$

As for the scalar field above, the 4d theory contains a set of massless particles and an infinite tower of massive graviton KK modes. The massless states turn out to be a 4d graviton $g_{\mu\nu}$, a vector boson A_μ and a scalar σ, whose relationship with the zero mode metric is given by

$$(G_{MN}^0) = e^{\sigma/3} \left(\begin{array}{c|c} g_{\mu\nu}(x) + e^{-\sigma} A_\mu A_\nu & e^{-\sigma} A_\mu(x) \\ \hline e^{-\sigma} A_\mu(x) & e^{-\sigma} \end{array} \right). \tag{1.42}$$

The vector boson arises from the 5d metric mixed component $A_\mu \sim G_{\mu 4}$, and can be shown to have a $U(1)$ gauge invariance $A_\mu \rightarrow A_\mu - \partial_\mu \lambda$ inherited from local reparametrizations of S^1, $x^4 \rightarrow x^4 + \lambda(x^\mu)$. The scalar σ, known as the *radion*, is a *modulus* field, i.e. a scalar with vanishing potential, whose vev is arbitrary and parametrizes a microscopic parameter of the configuration, in this case the S^1 radius via $e^{-2\sigma/3} \sim G_{44}$. This model underlies the Kaluza–Klein attempt to unify gravity and electromagnetism through an extra dimension. Compactification of higher-dimensional theories on spaces with a (possibly non-abelian) isometry group G leads in a similar way to a 4d gauge group G. However, attempts to realize the SM along these lines fails since the appearance of charged chiral fermions requires extra ingredients, as already mentioned. These will be present in string theory, however, where the SM interactions are realized by alternative mechanisms.

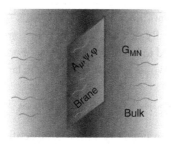

Figure 1.4 Brane-world matter localization.

Substituting (1.42) into the 5d action (1.40), one gets a 4d action for the zero modes

$$S_{KK}^0 = M_5^3 \pi R \int d^4x \sqrt{-g} \left(R_{4d}(g) - \frac{1}{6} \partial_\mu \sigma \partial^\mu \sigma - \frac{1}{4e^\sigma} F_{\mu\nu}^2 \right). \qquad (1.43)$$

This includes the 4d gravity action, radion kinetic term, and gauge field kinetic term with σ-dependent gauge coupling $g^2 = \exp \sigma$. Comparison with the 4d action (1.39) gives

$$M_p^2 = 16\pi^2 M_5^3 R. \qquad (1.44)$$

This suggests the possibility to have a low fundamental 5d gravity scale M_5 (even down to the EW scale), and generate a large 4d Planck scale (1.5) by taking R large enough. This is not possible in this simple setup, however, since 5d gauge/matter SM fields would produce too light KK replicas, which should have already been observed. The low fundamental scale is, however, phenomenologically viable in the *brane-world* scenario, and is reviewed next.

1.4.2 Branes and large extra dimensions

The *brane-world* idea is quite simple. In certain systems, non-gravitational degrees of freedom like, e.g., the SM fields may be forced to live in a subspace of a full higher-dimensional spacetime. Such submanifolds are called *branes*, as they generalize the notion of a *membrane* embedded in a larger space. In order to lead to 4d Poincaré invariance, the branes must span the observable 4d spacetime, but can be localized in (all or some of) the extra compact dimensions. On the other hand, gravitational interactions propagate throughout the full higher-dimensional spacetime, which is referred to as the *bulk*, see Figure 1.4.

It should be emphasized that localization of gauge and matter fields on branes must follow from *local* dynamics, and hence not by the compactification itself, which is a *global* property of spacetime. There are some field theory systems which realize localization of matter fields on certain submanifolds. However, localization of gauge and matter fields is particularly easy to describe in string theory, where it occurs on the volume of certain dynamical extended objects known as D-branes, introduced in Section 6.1.

There are branes of different dimensionality, so branes spanning p spatial dimensions (plus time) are more specifically denoted p-branes. In some phenomenological applications the SM fields are assumed to localize on a 3-brane. For such systems, the action splits into a 4d piece for the 3-brane volume, and a bulk piece in larger D-dimensional spacetime,

$$S_{\text{brane-world}} = S_{\text{brane}} + S_{\text{bulk}} = \int d^4 x \mathcal{L}_{\text{brane}} + \int d^D x \mathcal{L}_{\text{bulk}}. \tag{1.45}$$

As in the previous section, 4d gravity arises from compactification of D-dimensional gravity. However, since localized gauge and matter fields are already 4d in the present 3-brane case, they are insensitive to the compactification. This has at least two interesting implications: first, it is *in principle* easier to produce 4d chiral fermions, since they do not originate from the KK reduction of higher-dimensional fermions, which tend to lead to vector-like combinations as explained. Of course, the *actual* realization of chiral fermions is still a non-trivial requirement on the mechanism responsible for localization, and will be a useful guiding principle in brane model building in string theory. Second, there are no KK massive replicas for localized fields, e.g. the SM. Due to this key property, the brane-world scenario allows for very large extra dimensions, whose size is detectable only through their effects on gravity, and for which only mild experimental bounds exist.

These two general ideas underly the *large extra dimensions* (LED) scenario, proposed in 1998 by N. Arkani-Hamed, S. Dimopoulos, and G. Dvali as a possible solution to the hierarchy problem. Consider a $D = (4+n)$ dimensional space with the SM fields localized on a 3-brane, so that (1.45) becomes

$$S_{\text{LED}} = \frac{M^{2+n}}{2} \int d^4 x \int d^n y \sqrt{-G} R_{(4+n)d} + \int d^4 x \sqrt{-g} \mathcal{L}_{\text{SM}}, \tag{1.46}$$

where M and $R_{(4+n)d}$ are the D-dimensional Planck scale and scalar curvature, and G and g are the D- and four-dimensional metrics. The n extra dimensions parametrize a compactification space, whose detailed structure is not relevant at present, so we take the simplest case of compactifying all dimensions on circles of equal radius R. Performing the KK reduction of the gravity piece, we obtain a zero mode describing a 4d graviton, whose action picks up a volume factor from integration over the y-coordinates, generalizing (1.43)

$$\frac{1}{2} M^{2+n} (2\pi R)^n \int d^4 x \sqrt{g} R_4. \tag{1.47}$$

Hence the 4d Planck mass is

$$M_p^2 = 8\pi M^{2+n} (2\pi R)^n. \tag{1.48}$$

This equation shows that the fundamental scale M may be as small as desired, if the compactification volume controlled by R is large enough to reproduce the large 4d Planck scale. This leads to no observable effect on the localized SM sector. In a precise sense, gravity is diluted through the large internal volume, while gauge interactions are not, due to their localized nature.

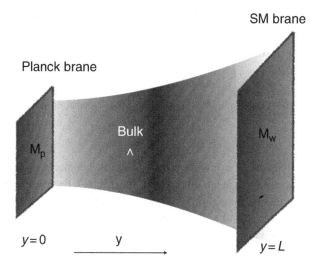

Figure 1.5 The Randall–Sundrum 5d model of warped dimensions.

In particular one may have $M \simeq$ TeV, so that there is no hierarchy between the EW and the fundamental gravitational scale, and thus no hierarchy problem in the traditional sense. The new avatar of the hierarchy problem is the need for a dynamical explanation of the large size R for the extra dimension, as compared to the fundamental length scale $1/M$. A fully satisfactory answer requires extra dynamical ingredients. As discussed in Section 16.5, the LED scenario can be easily realized in string theory, which moreover includes extra dynamics allowing the detailed discussion of the volume modulus stabilization.

1.4.3 Warped extra dimensions

The above scenarios of extra dimensions implicitly assumed a factorized ansatz of the background spacetime metric, $ds^2 = \eta_{\mu\nu}dx^\mu dx^\nu + g_{mn}(y)dy^m dy^n$. This is, however, not the most general ansatz consistent with 4d Poincaré invariance, since the 4d metric can be allowed to depend on the internal coordinates as $g_{\mu\nu} = f(y)\eta_{\mu\nu}$, with $f(y)$ known as *warp factor*. Warp factors actually arise in situations where branes are dynamical objects, with a finite tension T whose gravitational effect on the surrounding geometry may be non-negligible.

A simple toy model of this compactification with *warped dimensions* is the 5d configuration considered by L. Randall and R. Sundrum in 1999. It is described by a 5d spacetime with the extra dimension of finite extent $y \in [0, L]$, which can be obtained by quotienting an $\mathbf{S}^1$ of radius L/π by the $\mathbf{Z}_2$ action $y \to -y$. There are two 3-branes at the locations $y = 0, L$ as shown in Figure 1.5, with tensions $-T, T$, namely the localized brane lagrangians are $\mathcal{L}_{\text{brane}} = \pm\sqrt{-g}T$. They are termed the Planck and EW branes, because the graviton zero mode naturally localizes on the former, while the SM (or at least the SM Higgs) is assumed to localize on the latter.

One can show that in such a configuration the warped metric ansatz

$$ds^2 = e^{-2|y|/r} \eta_{\mu\nu} dx^\mu dx^\nu + dy^2 \tag{1.49}$$

solves the 5d Einstein's equations with a 5d bulk negative cosmological constant Λ

$$\Lambda = \frac{-24M^3}{r^2} \quad \text{with} \quad r = \frac{24M^3}{T}, \tag{1.50}$$

where M is the fundamental 5d gravity scale. The bulk geometry is a slice of 5d anti-de-Sitter (AdS$_5$) space, a fact enabling the use of the AdS/CFT correspondence (see Section 6.4) in the string theory realization of this scenario, see Section 14.1.3.

The crucial point is that the 4d metric at the two brane locations is different, due to the redshift effect of the warp factor

$$g_{\mu\nu}(y = L) = e^{-2L/r} g_{\mu\nu}(y = 0). \tag{1.51}$$

As a consequence, all mass scales on the brane at $y = L$ are exponentially suppressed with respect to the natural scale on the brane at $y = 0$. The latter corresponds to the 4d Planck scale M_p, since the 4d graviton is mainly localized around the $y = 0$ brane. Hence, if the SM (or at least the SM Higgs) is localized on the EW brane at $y = L$ the Higgs mass is naturally of order

$$m_{\text{Higgs}}^2 \simeq e^{-2L/r} M_p^2. \tag{1.52}$$

For moderate $L/r \simeq 16$, one gets $m_{\text{Higgs}} \simeq$ TeV, leading to a solution of the hierarchy problem, in terms of the exponential redshift in the extra dimensions. Interestingly, the above mentioned AdS/CFT correspondence translates this geometric origin of the hierarchy to the generation of scales by strong dynamics mentioned in Section 1.3.3, involving a 4d (approximately) conformal field theory sector.

Unlike the LED scenario, the Planck scale is of order of the fundamental scale M, since

$$M_p^2 = 8\pi M^3 \int_0^L dy e^{-2y/r} = 4\pi M^3 r (1 - e^{-2L/r}), \tag{1.53}$$

so typically $M_p \simeq M$. In the limit when $L \ll r$ we reproduce $M_p^2 \simeq 8\pi M^3 L$, the usual flat extra dimension result (1.44), (1.48), with the interval length L here playing the role of $2\pi R$ there. The phenomenology of warped compactifications, however, is very different from that of LED models. In particular, the KK replicas of the graviton have masses around the TeV, and may be produced at the LHC, e.g., à la Drell–Yan.

2

Supersymmetry

Supersymmetry (SUSY) is a symmetry which combines fermions and bosons into the same multiplets. It plays a crucial role in the structure of string theory, and in fact its first appearance as a symmetry in physics arose in trying to extend the bosonic string to include fermions. In string theory, SUSY guarantees the absence of divergences and of tachyonic scalars. Moreover, as discussed in Section 1.3.3, the simplest version of supersymmetry in four dimensions may provide a perturbative solution to the electroweak hierarchy problem, and thus is key to many proposals for physics beyond the Standard Model. For these reasons, most 4d string theory compactifications studied to date are supersymmetric, and lead to SUSY effective theories at low energies.

In this chapter we review general results for 4d $\mathcal{N} = 1$ SUSY and describe its possible role in stabilizing the electroweak scale against radiative corrections. We review local supersymmetry, which leads to the inclusion of gravitation and gives rise to 4d $\mathcal{N} = 1$ supergravity, which may play an important phenomenological role by mediating SUSY breaking. We also introduce the simplest SUSY extension of the SM, the Minimal Supersymmetric Standard Model (MSSM), for reference in future chapters. In addition, we provide a brief introduction to extended SUSY, which includes additional supersymmetry generators; although not of direct interest for particle physics, extended SUSY plays an important role in string theory, and actually appears in intermediate steps in explicit phenomenologically interesting string constructions.

2.1 Four-dimensional $\mathcal{N} = 1$ supersymmetry

Supersymmetry algebra

Supersymmetry is a symmetry which relates bosons to fermions, i.e. schematically a SUSY generator Q acts as

$$Q(\text{fermion}) = \text{boson}, \quad Q(\text{boson}) = \text{fermion}, \tag{2.1}$$

and so requires an equal number of fermionic and bosonic degrees of freedom. The simplest 4d system invariant under SUSY is a free theory with a Weyl fermion[1] ψ_α and a complex scalar Φ, whose action is

$$S = \int d^4x(-\partial^\mu\Phi^*\partial_\mu\Phi - i\bar\psi\bar\sigma^\mu\partial_\mu\psi). \tag{2.2}$$

This system has a current which is conserved *on-shell*, i.e. upon use of the equations of motion. This so-called supercurrent is

$$J_\alpha^\mu = (\partial_\nu\Phi^*\sigma^\nu\bar\sigma^\mu\psi)_\alpha, \quad \partial_\mu J_\alpha^\mu = 0, \tag{2.3}$$

which implies the conservation of the (super)charges

$$Q_\alpha = \int d^3x J_\alpha^0, \quad \bar Q_{\dot\alpha} = \int d^3x \bar J_{\dot\alpha}^0. \tag{2.4}$$

These charge generators have the unusual properties of being fermionic and transforming as Weyl spinors under the Lorentz group – rather than scalars, as more familiar symmetry generators. Also their algebra is generated by anticommutation relations, rather than by commutators. Since both Q and $\bar Q$ are conserved, their anticommutator should be a *bosonic* conserved quantity. The only candidate is the spacetime momentum P_μ, necessarily contracted with σ^μ to have the right spinorial structure. Indeed explicit computation leads to the (super)algebra

$$\{Q_\alpha, \bar Q_{\dot\alpha}\} = 2\sigma_{\alpha\dot\alpha}^\mu P_\mu, \tag{2.5}$$

with other (anti)commutators vanishing

$$\{Q_\alpha, Q_\beta\} = \{\bar Q_{\dot\alpha}, \bar Q_{\dot\beta}\} = [Q_\alpha, P_\mu] = [\bar Q_{\dot\alpha}, P_\mu] = 0. \tag{2.6}$$

Remarkably, Q_α, $\bar Q_{\dot\alpha}$ are not generators of an *internal* symmetry, rather they intertwine with the Poincaré algebra. In the following we focus on this 4d $\mathcal{N}=1$ supersymmetry algebra. Extended supersymmetry, with $\mathcal{N}>1$ sets of generators Q_α^I, $\bar Q_{\dot\alpha I}$, is briefly reviewed in Section 2.4.

Superfields and superspace: chiral superfield

In order to write down more general field theory actions, we are interested in realizing the supersymmetry algebra *off-shell*. This is achieved by using supermultiplets including additional auxiliary degrees of freedom, which eventually disappear *on-shell*, i.e. upon application of the equations of motion. The simplest supermultiplet is the *chiral multiplet*, which has the field content

$$(\Phi, \psi, F), \tag{2.7}$$

---

[1] We use here a metric signature $(-+++)$ and the Weyl spinor notation of Wess and Bagger. Thus we have 2-component spinors with undotted and dotted indices ψ_α, $\bar\psi^{\dot\alpha}$, transforming in representations $(1/2, 0)$ and $(0, 1/2)$ of the Lorentz group. A Dirac spinor contains two Weyl spinors, $\Psi_D = (\psi_\alpha, \bar\chi^{\dot\alpha})$, and a Dirac mass term reads $\psi^\alpha\chi_\alpha + \bar\psi_{\dot\alpha}\bar\chi^{\dot\alpha}$. Some useful identities are $\psi\chi \equiv \psi^\alpha\chi_\alpha = -\psi_\alpha\chi^\alpha = \chi^\alpha\psi_\alpha = \chi\psi$. One also defines $(\sigma_{\alpha\dot\alpha}^\mu) = (-\mathbf{1}, \vec\sigma)$ and $(\bar\sigma_{\dot\alpha\alpha}^\mu) = (-\mathbf{1}, -\vec\sigma)$.

where F is a complex scalar auxiliary field. A SUSY transformation rotates the components Φ, ψ, F into each other. The infinitesimal transformation parameters, denoted θ_α, $\bar{\theta}_{\dot{\alpha}}$, are anticommuting fermionic quantities, since they transform fermions into bosons and vice versa. Adding a term $|F|^2$ to the action (2.2), one can check its invariance under the SUSY transformations

$$\delta_\theta \Phi = \sqrt{2}\theta\psi,$$

$$\delta_\theta \psi_\alpha = i\sqrt{2}(\partial_\mu \Phi \sigma^\mu \bar{\theta})_\alpha + \sqrt{2}\theta_\alpha F,$$

$$\delta_\theta F = i\sqrt{2}(\bar{\theta}\bar{\sigma}^\mu \partial_\mu \psi). \tag{2.8}$$

With canonical mass dimensions $[\Phi] = 1$, $[\psi] = 3/2$, we have $[F] = 2$ and $[\theta] = -1/2$.

The systematic construction of 4d $\mathcal{N} = 1$ supersymmetric actions is made simpler by introducing the notion of superspace and superfields. *Superspace* is a generalization of 4d Minkowski space (x^0, x^1, x^2, x^3), on which Poincaré transformations act, by including additional anticommuting spinorial coordinates θ_α, $\bar{\theta}_{\dot{\alpha}}$, on which SUSY transformations act. Superspace is thus defined by the familiar four dimensions plus the *extra fermionic dimensions*, and is parametrized by coordinates $(x^\mu, \theta_\alpha, \bar{\theta}_{\dot{\alpha}})$. The anticommuting properties of the fermionic coordinates imply, e.g.,

$$\theta_1^2 = \theta_2^2 = 0, \quad \theta_1\theta_2 = -\theta_2\theta_1, \quad \int d\theta_\alpha = 0, \quad \int d\theta_\alpha \theta_\alpha = \frac{\partial}{\partial \theta_\alpha}\theta_\alpha = 1 \quad \text{(no sum).} \tag{2.9}$$

Superfields unify the different components of a supermultiplet into a single mathematical object. They are defined as fields depending on the superspace coordinates. From (2.9) it follows that superfields have a finite power-expansion in the fermionic coordinates, thus leading to a finite number of ordinary fields, filling out supermultiplets. Superfields may be endowed with Lorentz indices. The simplest superfield is however that with no such external indices, i.e. the scalar superfield

$$\Phi(x^\mu, \theta_\alpha, \bar{\theta}_{\dot{\alpha}}). \tag{2.10}$$

A general scalar superfield is reducible with respect to SUSY transformations, however. In order to extract its irreducible pieces, a useful strategy is to impose extra constraints. For instance, they can be introduced by using the SUSY covariant derivatives

$$D_\alpha = \frac{\partial}{\partial \theta^\alpha} + i\sigma^\mu_{\alpha\dot{\alpha}}\bar{\theta}^{\dot{\alpha}}\partial_\mu, \quad \bar{D}_{\dot{\alpha}} = -\frac{\partial}{\partial \bar{\theta}^{\dot{\alpha}}} - i\theta^\alpha \sigma^\mu_{\alpha\dot{\alpha}}\partial_\mu. \tag{2.11}$$

Chiral scalar superfields are characterized by the condition $\bar{D}_{\dot{\alpha}}\Phi = 0$, and turn out to contain the fields in the chiral multiplet (2.7). Concretely, the constraint is solved by the superfield structure

$$\Phi(x, \theta, \bar{\theta}) = \Phi(y) + \sqrt{2}\theta\psi(y) + \theta\theta F(y), \tag{2.12}$$

where $y = x + i\theta\sigma^\mu\bar{\theta}$. Here and in the following, we abuse notation and use the same symbol for the superfield Φ and its scalar component. Thus chiral superfields do not depend explicitly on $\bar{\theta}$, and (2.12) is just the superfield finite power expansion in θ. Antichiral superfields Φ^+ are similarly defined by the constraint $D_\alpha \Phi^+ = 0$, and lead to the conjugate field content $(\Phi^*, \bar{\psi}_{\dot\alpha}, F^*)$.

Other superfields are defined by constraints not involving SUSY derivatives. For instance, vector superfields $V(x, \theta, \bar\theta)$ are defined by the reality constraint $V = V^+$; as we review below, they are the simplest multiplets containing gauge bosons.

Supersymmetric actions for chiral multiplets

Our initial example of SUSY action (2.2) for the chiral multiplet is a free theory, but clearly the interest in SUSY relies on its implementation in interacting theories. The construction of SUSY invariant lagrangians is very easy in terms of superfields, as we now describe for chiral superfields. The key point is that the chiral superfield component with the highest θ power in (2.12) is an F auxiliary field, whose SUSY transformation (2.8) is a total derivative. Hence the 4d spacetime integral of the F-term (i.e. the θ^2 component) of any chiral superfield is invariant under SUSY. Since the product of chiral superfields (with the same chirality) yields another chiral superfield, the 4d spacetime integral of the F-term of an arbitrary polynomial of chiral superfields is SUSY invariant. In particular, the most general renormalizable supersymmetric couplings involving chiral superfields Φ_i have the form

$$\mathcal{L}_W = \int d^2\theta \, W(\Phi_i) + \text{h.c.} \equiv \int d^2\theta \left(\tfrac{1}{3} h^{ijk} \Phi_i \Phi_j \Phi_k + \tfrac{1}{2} m^{ij} \Phi_i \Phi_j + \lambda^i \Phi_i \right) + \text{h.c.} \tag{2.13}$$

The holomorphic function $W(\Phi_i)$ is called *superpotential*, and integration over $d^2\theta$ selects its F-term. Explicit integration over θ yields the interaction terms for the component fields, which include Yukawa couplings and fermion mass terms

$$\mathcal{L}_F = -\frac{1}{2} \frac{\partial W}{\partial \Phi_i \partial \Phi_j} \psi_i \psi_j + \text{h.c.} = -h^{ijk} \Phi_k \psi_i \psi_j - \frac{1}{2} m^{ij} \psi_i \psi_j + \text{h.c.} \tag{2.14}$$

SUSY interaction terms can be generalized to superpotentials given by arbitrary holomorphic functions of chiral superfields (defined by Taylor expansion).

Canonical kinetic terms for chiral multiplets are described in terms of superfields by

$$\int d^2\theta d^2\bar\theta \, \Phi_i{}^+ \Phi_i. \tag{2.15}$$

The SUSY invariance of this term is shown analogously. The product $\Phi_i^+ \Phi_i$ is a real superfield; its expansion in $\theta, \bar\theta$ has a $\theta^2 \bar\theta^2$ term (denoted D-term) whose SUSY variation is a total derivative, thus leading to a SUSY invariant term upon integration over 4d spacetime. Explicit integration over $\theta, \bar\theta$ in (2.15) produces the lagrangian for the component fields; it includes the kinetic terms for fermions and bosons, as well as terms $|F_i|^2$ for the non-propagating auxiliary fields. Using a coupling $F_i \partial W / \partial \Phi_i$ arising from (2.13), the equations of motion $F_i^* = -(\partial W)/(\partial \Phi_i)$ allow us to eliminate the auxiliary fields. This introduces a (so-called F-term) contribution to the scalar potential

$$V_F(\Phi_i) = \sum_i |F_i|^2 = \sum_i \left| \frac{\partial W}{\partial \Phi_i} \right|^2. \tag{2.16}$$

Altogether, the interactions for chiral superfields give rise to masses, Yukawa couplings, and a scalar potential, with coupling constants uniquely determined by the superpotential. This theory is known as the Wess–Zumino model.

Very often the lagrangian obtained from a given superpotential is invariant under $U(1)$ global symmetries acting differently on the fermionic and scalar components of chiral multiplets. This can be encoded as a $U(1)$ charge assignment to the superspace coordinates θ, i.e.

$$\theta \to e^{-i\gamma}\theta, \quad \Phi_i \to e^{a_i\gamma}\Phi_i, \quad \Phi_i^+ \to e^{-a_i\gamma}\Phi_i^+. \tag{2.17}$$

Such continuous $U(1)$ symmetries are called R-symmetries, denoted $U(1)_R$. Note that the $U(1)_R$ charge of fermions exceeds that of scalars by one unit. R-symmetry constrains the superpotential to couplings satisfying $\sum_i a_i = 2$.

Vector multiplets and SUSY gauge interactions

We now describe the introduction of gauge bosons and gauge interactions in SUSY theories. As already anticipated, gauge bosons are contained in *vector multiplets* defined by superfields satisfying the constraint $V = V^+$. Their (off-shell) field content is

$$V \to (\lambda_\alpha, A^\mu, D; C, \chi_\alpha, N), \tag{2.18}$$

where we have omitted gauge indices. This multiplet is subject to a generalized gauge invariance mentioned below, which in the so-called Wess–Zumino gauge fixing allows to gauge away the fields C, χ_α, N. The remaining fields are a gauge boson A^μ, a Weyl spinor λ_α in the adjoint representation, termed *gaugino*, and a real auxiliary scalar field D. Being in a real representation, the gauginos are often described as Majorana fermions. As mentioned later, these component fields are subject to standard gauge transformations. The vector superfield expansion in the Wess–Zumino gauge is

$$V(x, \theta, \bar{\theta}) = -\theta\sigma^\mu\bar{\theta}A_\mu(x) + i\theta\theta\bar{\theta}\bar{\lambda}(x) - i\bar{\theta}\bar{\theta}\theta\lambda(x) + \frac{1}{2}\theta\theta\bar{\theta}\bar{\theta}D(x). \tag{2.19}$$

In this gauge, $V^2 = -\frac{1}{2}\theta\theta\bar{\theta}\bar{\theta}A_\mu A^\mu$, and $V^n = 0$ for $n > 2$. The usual gauge invariant field strength $F_{\mu\nu}$ can be shown to belong to a spinorial chiral superfield W_α (i.e. $\bar{D}_{\dot{\alpha}}W_\alpha = 0$) defined in the abelian case as

$$W_\alpha = -\frac{1}{4}\bar{D}\bar{D}D_\alpha V, \quad \bar{W}_{\dot{\alpha}} = -\frac{1}{4}DD\bar{D}_{\dot{\alpha}}V, \tag{2.20}$$

and generalized in the non-abelian case to

$$W_\alpha = -\frac{1}{4}\bar{D}\bar{D}e^{-V}D_\alpha e^V, \quad \bar{W}_{\dot\alpha} = -\frac{1}{4}DDe^{-V}\bar{D}_{\dot\alpha}e^V, \tag{2.21}$$

where now $V = T_a V^a$, with T_a the gauge generators. The first term in the θ expansion of W_α is the gaugino, $W_\alpha = -i\lambda_\alpha + \cdots$.

The kinetic terms for gauge bosons and gauginos, and their gauge interactions, arise from

$$\mathcal{L} = \frac{1}{4}\mathrm{Tr}\int d^2\theta\, W^\alpha W_\alpha + \mathrm{h.c.} = \mathrm{Tr}\left(-\frac{1}{4}F_{\mu\nu}F^{\mu\nu} - i\lambda\sigma^\mu D_\mu\bar{\lambda} + \frac{1}{2}D^2\right). \tag{2.22}$$

The interactions of a gauge multiplet V with a chiral superfield Φ are described by

$$\int d^2\theta d^2\bar{\theta}\,\Phi^+ e^V\,\Phi. \tag{2.23}$$

The terms (2.22), (2.23) are invariant under the following generalized gauge transformations, with gauge parameters in chiral superfields $\Lambda(x,\theta) = T^a\Lambda_a(x,\theta)$

$$\Phi(x,\theta) \to e^{-i\Lambda}\Phi(x,\theta), \quad e^{V(x,\theta,\bar{\theta})} \to e^{-i\Lambda^+}e^V e^{i\Lambda}, \tag{2.24}$$

with Φ transforming in some representation of the gauge group. For $U(1)$ vector multiplets, the gauge transformation reads

$$V \to V + i(\Lambda - \Lambda^+) \tag{2.25}$$

As announced, expansion in components shows that generalized gauge transformations restricted to the Wess–Zumino gauge yield ordinary gauge transformations, and that the physical fields ψ, Φ, and A^μ, λ transform as usual under them. The term (2.23) includes kinetic terms for the chiral multiplet fermions and scalars, and their usual gauge invariant interactions with gauge bosons. In addition, it produces a linear term in the auxiliary fields D_a, which together with the $|D|^2$ term in (2.22) yields the equation of motion $D^a = -g\Phi_i^* T_{ij}^a \Phi_j$, where i, j run over gauge indices. Here we have rescaled $V \to gV$ in (2.23) to make the gauge coupling constant g manifest. Elimination of the auxiliary field leads to a (so-called D-term) contribution to the scalar potential

$$V_D = \sum_a \frac{1}{2}|D^a|^2 = \sum_a \frac{g^2}{2}|\Phi_i^* T_{ij}^a \Phi_j|^2, \tag{2.26}$$

In addition, (2.23) contains gaugino couplings to matter fields of the form

$$g(\Phi^* T^a \psi)\lambda_a + \mathrm{h.c.} \tag{2.27}$$

In some models the complete F- plus D-term scalar potential (2.16), (2.26) has flat directions, parametrized by vevs for some field or combination of fields. Moving along flat directions parametrized by chiral multiplets with gauge charges can have phenomenological advantages in model building, e.g. to break too large gauge symmetries, or to give masses to unwanted fields via Yukawa couplings. Flat directions associated to the singlet fields known as moduli are also ubiquitous in string compactifications, with their vevs encoding geometrical or other microscopic parameters of the model.

Finally, for a $U(1)$ gauge field there is an additional SUSY invariant term that one may add to the lagrangian, the Fayet–Illiopoulos (FI) term

$$\mathcal{L}_{\text{FI}} = \xi \int d^2\theta d^2\bar{\theta} V. \tag{2.28}$$

The integration selects the $\theta^2\bar{\theta}^2$ term, i.e. the auxiliary field D in (2.19), whose transformation is a total derivative, thus producing a SUSY invariant term upon 4d spacetime integration. The only effect of the FI term is to modify the $U(1)$ D-term scalar potential, which in a theory of chiral multiplets Φ_k with $U(1)$ charges q_k reads

$$V_{U(1)} = \frac{1}{2}|D|^2 = \frac{g^2}{2}\left|\sum_k q_k|\Phi_k|^2 + \xi\right|^2. \tag{2.29}$$

Field dependent couplings

In the above discussion we have considered the simplest SUSY actions for gauge and matter multiplets. These can be generalized to SUSY actions including additional functional dependence on chiral multiplets. This situation is ubiquitous in supergravity and string theory models, as discussed in later chapters. Usually these chiral multiplets are singlets under all gauge interactions, as we assume in the following.

For instance, it is possible to consider field dependent kinetic terms. Ignoring gauge interactions, they arise from the superspace term

$$\int d^2\theta d^2\bar{\theta} K(\Phi, \Phi^+), \tag{2.30}$$

where $K(\Phi, \Phi^+)$ is a real function of the chiral multiplets, known as the *Kähler potential*.[2] This provides a SUSY version of non-canonical kinetic terms of the form $K_{i\bar{j}}\partial_\mu \Phi^i \partial^\mu \overline{\Phi}^{\bar{j}}$, with

$$K_{i\bar{j}} = \frac{\partial^2 K}{\partial \Phi^i \partial \overline{\Phi}^{\bar{j}}}. \tag{2.31}$$

It is useful to regard the complex scalars Φ^i as coordinates parametrizing a complex manifold, and their kinetic term as a non-linear sigma model, analogous to that of pions in strong interactions. Manifolds admitting a metric which can be obtained from a Kähler potential ("Kähler metric" for short), as in (2.31), are mathematically known as Kähler manifolds. Canonical kinetic terms are obtained for $K = \Phi_i^+\Phi_i$, with the Kähler manifold being just flat space.

---

[2] In what follows, we use $\overline{\Phi}$ and Φ^* (indiscriminately) to denote the scalar component of the superfield Φ^+, or (abusing notation) even the whole superfield.

It is also possible to introduce field dependent gauge kinetic terms, with structure

$$\frac{1}{4}\int d^2\theta f(\Phi_i)\text{Tr}\,(W^\alpha W_\alpha) + \text{h.c.,} \tag{2.32}$$

where the *gauge kinetic function* $f(\Phi_i)$ is holomorphic. Its expansion, see (2.52) later, produces a gauge kinetic term $(\text{Re}\,f)(F_{\mu\nu}F^{\mu\nu})$ and an axion-like coupling of the form $(\text{Im}\,f)(F_{\mu\nu}\tilde{F}^{\mu\nu})$. Such couplings will turn out to be relevant to address the strong CP problem in string compactifications, as we will discuss in Section 16.2.

Finally, it is also possible to have field dependent FI terms, with structure

$$\int d^2\theta d^2\bar\theta\,[\xi(\Phi) + \bar\xi(\Phi^+)]V. \tag{2.33}$$

This concludes our summary of the structure of fields and interactions for $\mathcal{N} = 1$ gauge theories coupled to matter chiral multiplets.

Non-renormalization theorems

As already suggested, SUSY theories enjoy remarkable ultraviolet properties, with certain quantities protected from ultraviolet divergences. This in fact makes SUSY theories interesting to address the electroweak hierarchy problem.

It can be shown that loop corrections necessarily have a superspace structure of the form

$$\int d^2\theta d^2\bar\theta f(\Phi, \Phi^+). \tag{2.34}$$

Namely, they are integrals over the whole superspace, i.e. corrections to D-terms. Since superpotential couplings are F-terms, i.e. involve integration over only half of superspace, they do not receive loop corrections, and hence the tree level superpotential is exact in perturbation theory. On the other hand, D-terms like chiral multiplet kinetic terms do receive loop corrections, usually referred to as wave function renormalization.

The non-renormalization result would seem to also apply to gauge kinetic terms (2.22), (2.32) as well. However, such terms can be shown to admit an alternative expression as D-terms, and so can receive corrections. These are nevertheless constrained, and the holomorphic gauge kinetic function in (2.32) can be shown to receive only one-loop corrections; however, *physical* gauge couplings are defined by rescaling chiral multiplets to have canonical kinetic term, and this introduces non-trivial corrections to the physical gauge couplings to all loops.

2.2 SUSY breaking

SUSY implies that all fields in a supermultiplet must have the same mass. In applications to particle physics, since SUSY partners of ordinary particles have not been observed, supersymmetry should be broken at an energy above the electroweak scale or higher. Supersymmetry may be broken either spontaneously or explicitly. In analogy with other symmetries in particle physics, we may expect SUSY to be broken spontaneously, i.e. the lagrangian would be invariant under SUSY, but the vacuum state would not, that is

$$Q_\alpha |0\rangle \neq 0. \tag{2.35}$$

In theories with global SUSY (as opposed to supergravity theories, in which SUSY is a local symmetry, see Section 2.3), the order parameter for SUSY breaking is not the vev of a scalar but rather the groundstate energy, as follows. From the anticommutation relations (2.5) we have

$$H = P_0 = \frac{1}{4}(\bar{Q}_1 Q_1 + Q_1 \bar{Q}_1 + \bar{Q}_2 Q_2 + Q_2 \bar{Q}_2). \tag{2.36}$$

This is positive definite, leading to the bound for the groundstate energy

$$\langle 0|H|0\rangle \geq 0, \tag{2.37}$$

which is saturated if and only if the vacuum is invariant $Q_\alpha |0\rangle = 0$. Notice that this is consistent with the form of the scalar potential in the previous section, i.e.

$$\langle 0|H|0\rangle = \langle 0|V|0\rangle = \sum_i |F_i|^2 + \frac{1}{2}\sum_a |D^a|^2 \geq 0. \tag{2.38}$$

Hence SUSY breaking requires non-zero vevs for some of the auxiliary fields, i.e. $\langle 0|F_i|0\rangle \neq 0$ or/and $\langle 0|D^a|0\rangle \neq 0$ for some i, a.

There is a SUSY version of the Goldstone theorem which implies the existence of a massless particle with the transformation properties of the broken generator. In SUSY breaking, the broken generator Q_α is a fermionic spin-1/2 object, so the associated Goldstone particle is a Weyl spinor, the *goldstino* ψ_G, which is characterized by

$$\langle 0|\delta_\theta \psi_G|0\rangle \neq 0. \tag{2.39}$$

The goldstino is in general the fermionic partner of the non-vanishing F or D auxiliary fields breaking SUSY, or a combination thereof. We anticipate from Section 2.3 that in supergravity, i.e. local supersymmetry, there is a SUSY-Higgs mechanism in which the goldstino is eaten up by the spin-3/2 gravitino (SUSY partner of the graviton) and gets a mass. No massless particle is thus left.

It is fairly easy to construct theories with spontaneously broken SUSY. A simple class are the O'Raifeartaigh models, which involve chiral multiplets only and exhibit F-term SUSY breaking. Consider, e.g., a theory with three chiral superfields A, B, C with a superpotential

$$W = hAB^2 + MBC + fA. \tag{2.40}$$

The scalar potential is

$$V = |hB^2 + f|^2 + |2hAB + MC|^2 + |MB|^2. \tag{2.41}$$

For $M \neq 0$, the three terms cannot simultaneously vanish and SUSY is spontaneously broken. In this example the goldstino is a linear combination of the Weyl fermions in the model, the fermion partner of the non-vanishing F-term.

A different class of models involve a $U(1)$ gauge interaction with an FI term. Consider the Fayet model, with two chiral multiplets A, B with opposite charges under a $U(1)$ gauge symmetry, an FI parameter ξ and a superpotential, $W = MAB$. The scalar potential is

$$V = |MA|^2 + |MB|^2 + \frac{e^2}{2}(|A|^2 - |B|^2 + \xi)^2. \tag{2.42}$$

Again, for M, $\xi \neq 0$ the three terms in the potential cannot simultaneously vanish, and SUSY is broken. The model combines F- and D-term SUSY breaking, and the goldstino is a linear combination of the $U(1)$ gaugino and the Weyl fermions in the chiral multiplets.

These are examples of spontaneous SUSY breaking at the classical level. The scale of SUSY breaking in such models is given by the couplings in the superpotential and possibly the FI terms, and is hence put by hand in the system. In applications of SUSY to address the electroweak hierarchy problem, however, it is useful to consider mechanisms in which the scale of SUSY breaking M_{SB} is hierarchically small compared to some large cut-off scale M_* (e.g. the string scale), in a natural dynamical way. This is achieved by models of dynamical SUSY breaking, in which the scale of SUSY breaking is generated non-perturbatively, and is naturally small

$$M_{SB} \simeq e^{-c/g^2(M_*)} M_* \ll M_*. \tag{2.43}$$

Theories with dynamical SUSY breaking at their global minimum exist, but typically are either strongly coupled (and hence not amenable to perturbative analysis) or cumbersome. On the other hand, there are very simple theories with unbroken SUSY at their global minimum, but with dynamical SUSY breaking at a local but meta-stable minimum. The latter are physically acceptable for particle physics applications if the meta-stable vacuum lifetime is longer than the age of the universe. An example of such meta-stable vacua in supersymmetric gauge theories is described in Section 15.7.

A typical (although not completely general) problem of models with spontaneous breaking of global SUSY is the rather generic existence of an unbroken R-symmetry of the type described in (2.17). Such R-symmetries forbid gaugino masses, and although they can be generated in loops, the models are generically problematic for particle physics phenomenology. However, this problem can be avoided in theories with field dependent gauge kinetic functions, as in $\mathcal{N} = 1$ supergravity models derived from string compactifications.

A more direct way to break SUSY is to modify the action by terms breaking SUSY explicitly. Remarkably, this can be implemented without jeopardizing the nice ultraviolet properties of SUSY theories, through the so-called *soft* terms, which break SUSY explicitly, yet do not reintroduce quadratic ultraviolet divergences. In particle physics applications, this allows to break SUSY while retaining the solution to the electroweak hierarchy problem. There is a finite list of terms that are soft in this sense:

- Any dimension-2 operator, namely scalar mass terms of the following two types

$$m^2 \Phi^* \Phi, \quad m^2 \Phi\Phi + \text{h.c.} \tag{2.44}$$

- Gaugino Majorana mass terms

$$M\lambda^{\alpha}\lambda_{\alpha} + \text{h.c.} \tag{2.45}$$

- Trilinear scalar couplings

$$A'_{ijk}\Phi^i\Phi^j\Phi^k + \text{h.c.} \tag{2.46}$$

The proposal of breaking SUSY explicitly by soft terms is naively rather ad hoc, unmotivated, and unappealing. However, as reviewed in Section 2.6.5, this structure of SUSY breaking soft terms appears naturally as the low-energy limit of theories with spontaneously broken SUSY in local supersymmetry, i.e. supergravity theories, to which we turn in the next section.

2.3 $\mathcal{N} = 1$ Supergravity

Up to now we have considered SUSY as a *global* symmetry, with constant fermionic transformation parameters θ_{α}, $\bar{\theta}_{\dot{\alpha}}$. Fundamental symmetries of Nature are *local*, so it is reasonable to consider theories with local supersymmetry, i.e. invariant under SUSY transformations with spacetime dependent parameters $\theta_{\alpha}(x)$, $\bar{\theta}_{\dot{\alpha}}(x)$. Since the SUSY algebra involves the generator of spacetime translations P^{μ}, the local version of SUSY contains general spacetime coordinate transformations, the invariance group of general relativity. Thus local $\mathcal{N} = 1$ SUSY turns out to correspond to a supersymmetric version of gravity, known as $\mathcal{N} = 1$ supergravity. The spin-2 graviton $g_{\mu\nu}$ belongs to a gravity supermultiplet, which includes a spin-3/2 partner ψ_{α}^{μ}, the gravitino. The gravitino couples to the conserved supercurrent – such as (2.3) – so it is the *gauge particle* associated to local SUSY transformations.

The detailed construction of the supergravity action is beyond our scope, and we restrict to introducing a few relevant terms for reference in later chapters. Since 4d $\mathcal{N} = 1$ supergravity is non-renormalizable, it should be regarded as an effective theory, with a derivative expansion and a ultraviolet cutoff, below the Planck scale M_p. We review some ingredients of the two-derivative effective action of $\mathcal{N} = 1$ supergravity coupled to chiral multiplets and gauge fields. We use Φ^* or $\overline{\Phi}$ to denote the antichiral superfield Φ^+, and $\partial_i \equiv \partial/\partial\Phi_i$, $\partial_{\bar{i}} \equiv \partial/\partial\Phi_i^*$.

The couplings of chiral multiplets Φ_i are determined by a real Kähler function $G(\Phi_i, \Phi_i^*)$ defined in terms of the Kähler potential K and superpotential W as

$$G(\Phi_i, \Phi_i^*) = \kappa_4^2 K(\Phi_i, \Phi_i^*) + \log|W(\Phi_i)|^2, \tag{2.47}$$

where

$$\kappa_4^2 = 8\pi/M_p^2. \tag{2.48}$$

The complete supergravity action is invariant under Kähler transformations

$$K(\Phi_i, \Phi_i^*) \longrightarrow K(\Phi_i, \Phi_i^*) + F(\Phi_i) + \bar{F}(\Phi_i^*), \tag{2.49}$$

with $F(\Phi_i)$ an arbitrary holomorphic function of the chiral fields. This invariance is clear, e.g., for the Kähler metric (2.31). The F-term scalar potential is given by

$$V_F(\Phi_i, \Phi_i^*) = e^{\kappa_4^2 K}(K^{i\bar{j}} D_i W D_{\bar{j}} W^* - 3\kappa_4^2 |W|^2). \tag{2.50}$$

Also $D_i = \partial_i + \kappa_4^2(\partial_i K)$ is the "Kähler derivative," a covariant derivative with respect to (2.49), and $K^{i\bar{j}}$ is the inverse of the Kähler metric (2.31) $K_{i\bar{j}} = \partial_i \partial_{\bar{j}} K$. The auxiliary field F^i is given by

$$F^i = -e^{\kappa_4^2 K/2} K^{i\bar{j}} D_{\bar{j}} W^*. \tag{2.51}$$

The gauge kinetic terms in 4d $\mathcal{N} = 1$ supergravity include holomorphic gauge kinetic functions $f(\Phi_i)$ as in (2.32). The gauge bosons have kinetic and axionic couplings given by

$$-\frac{1}{4}(\text{Re } f)_{ab} F_{\mu\nu}^a F^{b\mu\nu} - \frac{1}{4}(\text{Im } f)_{ab} F_{\mu\nu}^a \tilde{F}^{b\mu\nu}, \tag{2.52}$$

with $\tilde{F}_{\mu\nu} = \frac{1}{2}\epsilon_{\mu\nu\rho\sigma} F^{\rho\sigma}$. Also a, b label gauge factors, and we allow a non-diagonal f_{ab} for $U(1)$ factors. Note that the $F\tilde{F}$ term is no longer a total derivative term, as it is for constant f. The gauge interactions with chiral matter fields are obtained by replacing $\Phi^+ \to \Phi^+ e^V$. Then the D-term scalar potential has the form

$$V_D = \frac{1}{2}[\text{Re } f^{-1}]^{ab}[K_i(T_a)_k^i \Phi^k][K_j(T_b)_k^j \Phi^k]^*, \tag{2.53}$$

where $K_i \equiv \partial_i K$, and i, j, k run over fields and their gauge indices. In the $\kappa_4^2 \to 0$ limit, gravitational effects decouple and the effective action reduces to the global SUSY one discussed in the previous section.

In principle one can consider the coupling to supergravity of a SUSY theory with a constant Fayet–Illiopoulos term. However, it turns out that this requires an exact unbroken global $U(1)$ symmetry. As explained in Section 17.1.4, global symmetries are expected not to exist in consistent quantum theories of gravity, so constant FI terms cannot be present in the supergravity actions arising from string theory compactifications. On the other hand, string compactifications do produce effective supergravity actions with field dependent FI terms, associated to Stückelberg masses for gauged $U(1)$'s, see Sections 9.5 and 12.4.

We now turn to the issue of spontaneous SUSY breaking in supergravity. Consider SUSY breaking by a non-zero F-term for a chiral superfield, whose fermion field is the goldstino η. The gravitino coupling to fermions in chiral multiplets has a term

$$\kappa_4 F \psi_\alpha^\mu \sigma_\mu^{\alpha\dot{\alpha}} \eta_{\dot{\alpha}}. \tag{2.54}$$

A non-zero vev for the auxiliary field F triggers the *super-Higgs mechanism*, by which the gravitino becomes massive by mixing with the goldstino η, which provides the two extra degrees of freedom to form a massive spin-3/2 particle. It is analogous to the standard Higgs mechanism, just replacing Goldstone bosons by goldstinos and scalar vevs by auxiliary field vevs. From (2.54) the gravitino mass is of order

$$m_{3/2} \simeq \sqrt{8\pi}\,\frac{F}{M_p}. \tag{2.55}$$

In supergravity, the order parameters of supersymmetry breaking are the vevs for F- and D-auxiliary fields. However, in sharp contrast with global SUSY, supersymmetry breaking is not directly equivalent to the vanishing of the vacuum energy (or 4d cosmological constant). This is already manifest from the structure of the scalar potential (2.50) which is not positive definite. In fact there are theories with supersymmetric vacua of negative vacuum energy, thus referred to as anti-de-Sitter (AdS) vacua.[3] Conversely, there exist simple models in which SUSY is spontaneously broken but the 4d cosmological constant vanishes. One such example is the Polonyi model, which contains a singlet chiral multiplet z with canonical kinetic term and a linear superpotential

$$W_P = m^2(\beta + z), \tag{2.56}$$

where m, β are constants. One can check that for $\beta = 2 - \sqrt{3}$ there is a minimum at $\langle z \rangle = \sqrt{3} - 1$, with vanishing vacuum energy but SUSY broken by $F_z \neq 0$. In models like this one, with canonical kinetic terms and vanishing vacuum energy, the condition $V = 0$ at the minimum relates the SUSY breaking scale and the vev of the superpotential, fixing the gravitino mass to be given specifically by

$$m_{3/2} = \sqrt{8\pi}\,\frac{F}{\sqrt{3}M_p}. \tag{2.57}$$

If SUSY breaking proceeds through a non-vanishing vev for an auxiliary field D, there is also a super-Higgs mechanism, with the gravitino becoming massive by eating up a gaugino.

2.4 Extended supersymmetry and supergravity*

In this section we introduce the basic ingredients of theories with extended supersymmetry in 4d. Although their systematic analysis usually includes similar results in other dimensions and their relations via dimensional reduction, we restrict to the 4d case, and in later chapters quote results in other dimensions as needed.

---

[3] This is consistent with the existence of SUSY algebras involving the AdS isometry group (instead of the Poincaré symmetry). We note in this respect that the de-Sitter isometry group is not compatible with SUSY, hence there are no supersymmetric de-Sitter vacua.

2.4.1 Extended SUSY and $\mathcal{N} = 2$, 4 supermultiplets

Extended supersymmetry is defined in terms of $\mathcal{N}$ sets of supercharges Q_α^I, $\bar{Q}_{\dot\alpha}^J$, with $I = 1, \ldots, \mathcal{N}$, satisfying an algebra generalizing (2.5)

$$\{Q_\alpha^I, \bar{Q}_{\dot\alpha J}\} = 2\sigma_{\alpha\dot\alpha}^\mu P_\mu \delta_J^I. \tag{2.58}$$

Hence $\mathcal{N}$ characterizes the number of supersymmetric charges, and the number of gravitinos when coupling to gravity. The R-symmetry group of the algebra contains an $SU(\mathcal{N})$ symmetry acting on them. Since each spinor supercharge contains four real components, $\mathcal{N}$-extended SUSY theories are said to have $4\mathcal{N}$ supersymmetries.

The construction of supermultiplets is analogous to the $\mathcal{N} = 1$ case, although *off-shell* superfield techniques are more involved or simply non-existing. Extended SUSY theories have larger supermultiplets, and more restricted lagrangians.

For 4d $\mathcal{N} = 2$ supersymmetry, the most familiar supermultiplets are:

- The *vector multiplet*, containing a gauge boson A_μ, a complex scalar Φ and two Majorana fermions λ^1, λ^2, all in the adjoint representation of the gauge group. It decomposes with respect to the 4d $\mathcal{N} = 1$ subalgebra as one vector multiplet and one adjoint chiral multiplet, for instance $V \equiv (A_\mu, \lambda^1)$, $\Phi \equiv (\Phi, \lambda^2)$ respectively. More generally, there is a family of $\mathcal{N} = 1$ subalgebras, determined by a phase φ, under which the precise components of these multiplets are (A_μ, λ^φ), (Φ, ψ^φ) with $\lambda^\varphi = \lambda^1 + \tan\varphi\lambda^2$, $\psi^\varphi = -\tan\varphi\lambda^1 + \lambda^2$. In the following we ignore such phases.
- The *hypermultiplet*, containing two complex scalars Q, $\tilde{Q}$ and two equal chirality Weyl fermions ψ_α, $\tilde{\psi}_\alpha$ (in conjugate representations of gauge and global symmetry groups). It decomposes as two 4d $\mathcal{N} = 1$ chiral multiplets (Q, ψ), $(\tilde{Q}, \tilde{\psi})$, in conjugate representations. For pseudoreal representations, it is consistent to take $(\tilde{Q}, \tilde{\psi})$ to be conjugates of (Q, ψ) to define a so-called half-hypermultiplet.
- Inclusion of gravity is achieved by coupling to a *gravity multiplet*, with a graviton $g_{\mu\nu}$, two opposite chirality gravitinos $\psi_{\mu,\alpha}$, $\psi_{\mu,\dot\alpha}$, and a gauge boson A_μ, termed the graviphoton.

The 4d $\mathcal{N} = 2$ effective action has completely decoupled kinetic terms for vector and hypermultiplets. The effective action for the vector multiplets is determined by the prepotential $\mathcal{F}(\Phi)$, a function holomorphic in the complex scalars in vector multiplets. The prepotential encodes the $\mathcal{N} = 1$ gauge kinetic functions and the Kähler potential for the adjoint chiral multiplet as

$$\text{Im}\left[\int d^2\theta d^2\bar{\theta}(\partial_i\mathcal{F})\overline{\Phi}_i + \int d^2\theta \frac{1}{2}(\partial_i\partial_j\mathcal{F})(W^\alpha)_i(W_\alpha)_j\right]. \tag{2.59}$$

The kinetic terms for hypermultiplets define a metric, constrained by supersymmetry to satisfy certain geometric properties, known as hyperkähler (for global supersymmetry) or quaternonic Kähler (for supergravity), whose structure is not necessary for our purposes.

Vector and hypermultiplets interact via $\mathcal{N} = 2$ gauge interactions, which in $\mathcal{N} = 1$ language correspond to couplings of the form (2.23), and to a superpotential necessarily of the following form

$$W_{\mathcal{N}=2} = \tilde{Q}\Phi Q, \tag{2.60}$$

with gauge indices contracted. In $\mathcal{N} = 2$ SUSY gauge field theories, without coupling to gravity, it can be shown that the one-loop correction to the prepotential is exact in perturbation theory, although it receives non-perturbative gauge instanton corrections; the latter are encoded in a beautiful geometric construct, the Seiberg-Witten curve, whose discussion is beyond our scope.

For 4d $\mathcal{N} = 4$ supersymmetry, the most familiar supermultiplets are:

- The *vector multiplet*, containing one gauge boson A_μ, four Majorana fermions λ_a, $a = 0, \ldots, 3$ and three complex scalars Φ_i, $i = 1, \ldots, 3$, all in the adjoint representation of the gauge group. Under 4d $\mathcal{N} = 2$ it decomposes as one vector multiplet $(A_\mu, \lambda_0, \lambda_1, \Phi_1)$ and one adjoint hypermultiplet $(\Phi_2, \Phi_3, \lambda_2, \lambda_3)$. Under 4d $\mathcal{N} = 1$ it decomposes as one vector multiplet e.g. (A_μ, λ_0), and three adjoint chiral multiplets $\Phi_i \equiv (\Phi_i, \lambda_i)$, $i = 1, \ldots, 3$.
- Gravity is described via the *gravity multiplet*, containing a graviton, four gravitinos, six $U(1)$ graviphotons, one complex scalar and four spin-1/2 fermions.

At the two-derivative level, the action for a 4d $\mathcal{N} = 4$ super-Yang–Mills gauge theory is completely determined by the gauge group. Under decomposition into $\mathcal{N} = 1$ notation, the form of the superpotential is

$$W_{\mathcal{N}=4} = \text{Tr}\,\Phi_1[\Phi_2, \Phi_3]. \tag{2.61}$$

Remarkably, gauge theories with global $\mathcal{N} = 4$ SUSY are exactly conformal interacting 4d field theories. This plays an important role in their S-duality, reviewed in Section 2.5.2, and in the simplest realizations of the AdS/CFT correspondence, in Section 6.4.

For 4d $\mathcal{N} = 8$ supersymmetry, the smallest supermultiplet already contains a graviton. The gravity multiplet contains a graviton, eight gravitinos, 28 $U(1)$ gauge bosons, 56 spin-1/2 particles, and 70 real scalars. Its action is most easily described as dimensional reduction of the 10d type II supergravities in Section 4.2.5, and we skip its discussion here.

2.4.2 Central charges and BPS states

Extended SUSY algebras admit generalizations of (2.58) including central charges Z^{IJ}, defined as commuting with all operators in the algebra, as follows

$$\{Q_\alpha^I, \bar{Q}_{\dot{\alpha}J}\} = 2\sigma^\mu_{\alpha\dot{\alpha}}P_\mu\delta^I_J, \quad \{Q_\alpha^I, Q_\beta^J\} = 2\epsilon_{\alpha\beta}Z^{IJ}, \quad \{\bar{Q}_{\dot{\alpha}I}, \bar{Q}_{\dot{\beta}J}\} = -2\epsilon_{\dot{\alpha}\dot{\beta}}Z^*_{IJ},$$

with other (anti)commutator relations unchanged. Here the Z^{IJ} are conserved charges, possibly depending on parameters of the theory, and antisymmetric in their indices. Hence there are no central charges in 4d $\mathcal{N} = 1$ supersymmetry. We also define $Z_{IJ} = -Z^{IJ}$.

Extended SUSY with central charges imply certain so-called Bogomol'nyi–Prasad–Sommerfield (BPS) bounds on the spectrum of the theory, as we now illustrate using the 4d $\mathcal{N} = 2$ case for simplicity. We choose a basis in which $Z^{IJ} = Z\epsilon^{IJ}$ with Z real. Consider a massive particle of mass m in a sector of central charge Z. In the rest frame of this particle, the algebra reads

$$\{Q_\alpha^I, \bar{Q}_{\dot\alpha J}\} = 2m\sigma_{\alpha\dot\alpha}^0\delta_J^I, \quad \{Q_\alpha^I, Q_\beta^J\} = 2\epsilon_{\alpha\beta}\epsilon^{IJ}Z, \quad \{\bar{Q}_{\dot\alpha I}, \bar{Q}_{\dot\beta J}\} = -2\epsilon_{\dot\alpha\dot\beta}\epsilon_{IJ}Z.$$

Consider the following linear combinations,

$$a_\alpha^\pm = \frac{1}{\sqrt{2}}[Q_\alpha^1 \pm \epsilon_{\alpha\beta}(\sigma^0)^{\beta\dot\beta}\bar{Q}_{\dot\beta 2}], \quad (a_\alpha^\pm)^\dagger = \frac{1}{\sqrt{2}}[\bar{Q}_{\dot\alpha 1} \pm \epsilon_{\dot\alpha\dot\beta}(\bar{\sigma}^0)^{\dot\beta\beta}Q_\beta^2].$$

They behave as (fermionic) creation and annihilator operators, satisfying

$$\{a_\alpha^\pm, (a_\beta^\pm)^\dagger\} = 2(m \pm Z)\delta_{\alpha\beta}. \tag{2.62}$$

They can be used to construct the particle supermultiplet by defining a lowest weight state $|0\rangle$ by $a_\alpha^\pm|0\rangle = 0$, and building other states by application of the raising operators $(a_\alpha^\pm)^\dagger$. Since the left-hand side of (2.62) is positive definite, so must be the eigenvalues of the right-hand side, and we can derive a (BPS) bound on the mass of states in a sector of given charge Z:

$$m \geq |Z|. \tag{2.63}$$

States saturating the bound are known as BPS states. They have the special property of being annihilated by certain linear combinations of the creation operators; for instance, a state $|\mathrm{BPS}_-\rangle$ with $m = Z$ is annihilated by $(a^-)^\dagger$, i.e.

$$\left(\bar{Q}_{\dot\alpha 1} - \epsilon_{\dot\alpha\dot\beta}(\bar{\sigma}^0)^{\dot\beta\beta}Q_\beta^2\right)|\mathrm{BPS}_-\rangle = 0. \tag{2.64}$$

In other words, BPS states preserve a subset (in this case, one half) of the supersymmetries. Since they are annihilated by certain creation operators, BPS multiplets have smaller number of states than generic non-BPS ones, and are known as "short multiplets." Since the number of degrees of freedom cannot jump discontinuously, such multiplets remain BPS under continuous changes of parameters of the theory, e.g. the coupling constants. Consequently, the BPS mass formula $m = |Z|$ is exact, and robustly survives the inclusion of, e.g., quantum (or other) corrections. Such quantities can be regarded as protected by non-renormalization theorems associated to the extended supersymmetry. The supersymmetry preserved by a BPS state depends on its BPS phase, i.e. the phase of its central charge Z. This can be rotated to zero (as above) for any single BPS state, but relative BPS phases between different BPS states cannot be removed and measure the mismatch of their preserved supersymmetries.

The above arguments can be generalized to supersymmetry algebras in other dimensions, containing tensorial central charges; their associated BPS states correspond to extended objects. These will play a crucial role in understanding the non-perturbative structure of string theory in Chapter 6, and in building type II orientifold compactifications in Chapters 10 and 11.

2.5 Non-perturbative dynamics in supersymmetric theories*

SUSY is a powerful symmetry which provides extra tools to analyze non-trivial dynamics of physically interesting systems, even beyond perturbation theory, as we briefly review for 4d $\mathcal{N} = 1, 4$ gauge theories.

2.5.1 Exact results for 4d $\mathcal{N} = 1$ SQCD and Seiberg duality

The constraints of holomorphy of the superpotential and the non-renormalization theorems in Section 2.1 have been successfully exploited to obtain non-trivial information on the low-energy dynamics of 4d $\mathcal{N} = 1$ theories. This is illustrated by supersymmetric QCD (SQCD) theories. These are $SU(N_c)$ gauge theories with N_f chiral multiplets (flavours) Q, $\tilde{Q}$ in the fundamental and anti-fundamental gauge representations, denoted by $\square, \overline{\square}$ respectively. We also take vanishing tree level superpotential. The non-anomalous global symmetry group is $SU(N_f)_L \times SU(N_f)_R \times U(1)_B \times U(1)_R$, where $U(1)_B$ is "baryon" number symmetry, and $U(1)_R$ is an R-symmetry, similar to (2.17). The gauge and global quantum numbers are given in the table below.

	$SU(N_c)$	$SU(N_f)_L$	$SU(N_f)_R$	$U(1)_B$	$U(1)_R$
Q	$\square$	$\overline{\square}$	1	1	$\frac{N_f - N_c}{N_f}$
$\tilde{Q}$	$\overline{\square}$	1	$\square$	-1	$\frac{N_f - N_c}{N_f}$

These theories have a rich pattern of non-trivial low-energy dynamics in different ranges of N_f, N_c. For instance, global symmetries and holomorphy can be used to show that for $0 \leq N_f < N_c$ there is a non-perturbative, so-called Affleck–Dine–Seiberg (ADS), superpotential

$$W_{\text{ADS}} = (N_c - N_f) \left(\frac{\Lambda^{3N_c - N_f}}{\det Q\tilde{Q}} \right)^{1/(N_c - N_f)}, \tag{2.65}$$

where Λ is the SQCD scale of strong dynamics (e.g. confinement). For $N_f = N_c - 1$ the superpotential can be understood as arising from a gauge instanton, see Section 13.1.1. For $N_f < N_c - 1$ the result can be indirectly derived from (2.65) by giving masses to a number of flavours and integrating them out. The superpotential determines the vacuum structure of the theories, as follows. For $0 < N_f \leq N_c - 1$ there is a runaway behavior for vevs of the "mesons" $M = Q\widetilde{Q}$, and there is no vacuum. In the $N_f = 0$ case, known as pure super-Yang–Mills (SYM), there are N_c supersymmetric vacua, labeled by an N_c-root of unity $\omega^{N_c} = 1$, which determines the value of the superpotential at the confined vacuum,

$$W_{\text{SYM}} = \omega N_c \Lambda^3. \tag{2.66}$$

This superpotential can be argued to produce a gaugino condensate $\langle \lambda\lambda \rangle \simeq \omega\Lambda^3$, so $\mathcal{N} = 1$ SYM strong dynamics is often termed "gaugino condensation."

For $N_f \geq N_c$ the symmetry and holomorphy arguments imply an exactly vanishing superpotential, even at non-perturbative level. The dynamics is very rich, but we focus on a particular property, known as Seiberg duality. This is a proposed exact equivalence of the infrared dynamics of two different ultraviolet theories. The above SQCD theory, referred to as "electric" description, is proposed to have an equivalent or dual "magnetic" description in the infrared, in terms of $SU(N_f - N_c)$ SQCD with N_f "dual" flavours $q, \widetilde{q}$, and singlets M, with gauge and global quantum numbers as given below:

	$SU(N_f - N_c)$	$SU(N_f)_L$	$SU(N_f)_R$	$U(1)_B$	$U(1)_R$
q	□	$\overline{\square}$	1	$\dfrac{N_c}{N_f - N_c}$	$\dfrac{N_c}{N_f}$
$\widetilde{q}$	$\overline{\square}$	1	□	$-\dfrac{N_c}{N_f - N_c}$	$\dfrac{N_c}{N_f}$
M	1	$\overline{\square}$	□	0	$2\dfrac{N_f - N_c}{N_f}$

and superpotential $W = Mqq'$, with color and flavour indices contracted. Although not proven, the proposed duality has a large body of indirect evidence supporting it.

An interesting application of Seiberg duality is to provide a weakly coupled dual description for the infrared dynamics of strongly coupled supersymmetric theories, as follows. For $N_c \leq N_f < \frac{3}{2}N_c$ the electric theory is strongly coupled in the infrared; in this so-called free magnetic phase, the magnetic theory is however weakly coupled in the infrared, and provides a perturbative description of the dynamics. The weakly coupled fields in the magnetic theory can be regarded as composites of the electric degrees of freedom, essentially "mesons" $M \sim Q\widetilde{Q}$ and "baryons" $q \sim Q^{N_c}, \widetilde{q} \sim \widetilde{Q}^{N_c}$. This is a supersymmetric version of QCD hadronic duality, in which mesons and baryons provide a low-energy description of quarks and gluons. It is easy to show that the Seiberg dual of the magnetic theory is the electric theory. Hence, in the complementary range $N_f \geq 3N_c$, dubbed "free electric phase," the electric theory is weakly coupled in the infrared (due to its too large number of flavors), and provides a weakly coupled dual description of the magnetic theory, which is strongly coupled in that regime. In the intermediate regime $3/2N_c \leq N_f \leq 3N_c$, known as the conformal window, both theories flow in the infrared to the same non-trivial interacting 4d conformal field theories, which does not admit a weakly coupled description. The dual theories remain just two complementary ultraviolet descriptions of this system.

2.5.2 S-duality in 4d $\mathcal{N} = 4$ super-Yang–Mills

The nomenclature "duality," "electric," and "magnetic," in the previous section is motivated by regarding the above equivalence as an $\mathcal{N} = 1$ analog (or remnant) of electric-magnetic dualities for theories with $\mathcal{N} = 2$, 4 extended supersymmetry. In this section we briefly review the S-duality of 4d $\mathcal{N} = 4$ SYM theories with $SU(N)$ group (ignoring its generalization to other gauge groups).

Gauge theories with 4d $\mathcal{N} = 4$ SUSY enjoy an electric-magnetic duality symmetry formally analogous to that of Maxwell vacuum equations of motion. Writing the latter in the differential form language of Appendix B, they read $dF = 0$, $d * F = 0$, and are invariant under linear transformations mixing F and $*F$. This symmetry is broken in the presence of electric charges since $dF = 0$ but $d * F = j_e$, with j_e the electric current 3-form. The symmetry would be restored by the inclusion of magnetic monopoles, which exist in Maxwell electromagnetism only as singular (so-called Dirac monopole) solutions. Smooth semi-classical solutions exist however in non-abelian gauge theories with scalars in the adjoint representation, when the latter acquire vevs and break the gauge symmetry down to an abelian subgroup $U(1)^r$. For instance, for $SU(2) \rightarrow U(1)$, the monopole solutions are classified by the second homotopy group of the broken generator coset $\Pi_2(SU(2)/U(1)) = \mathbf{Z}$, see section B.4.

Gauge theories with adjoint scalars occurred naturally in Section 2.4.1 in vector multiplets of extended $\mathcal{N} = 2$ or $\mathcal{N} = 4$ SUSY theories. In the presence of electric and magnetic charges, the equations of motion for the unbroken $U(1)$ are

$$dF = j_m, \quad d * F = j_e, \tag{2.67}$$

where j_e and j_m are the electric and magnetic current 3-forms. Equations (2.67) have a manifest $\mathbf{Z}_2$ symmetry $F \leftrightarrow *F$, $j_e \leftrightarrow j_m$, acting non-locally on the fundamental fields (i.e. the gauge potential) and exchanging electric and magnetic currents. It is actually a subgroup of a larger symmetry discussed later on.

As described in Section B.3, electric and magnetic charges obey the Dirac quantization condition $Q_e Q_m \in \mathbf{Z}$, where a factor of 2π relative to (B.29) has been absorbed in the charge definitions. Hence for electric charges quantized as $Q_e = gm$ with g being the gauge coupling and $m \in \mathbf{Z}$, magnetic charges are quantized as $Q_m = n/g$ with $n \in \mathbf{Z}$. Thus the $\mathbf{Z}_2$ transformation exchanging electric and magnetic charges must act non-trivially on the gauge coupling constant as $g \rightarrow 1/g$. It maps strong to weak coupling and vice versa, allowing to translate the strong coupling regime of the electric degrees of freedom into a weakly coupled dual, in terms of the monopoles. This remarkable duality can however work only for very special quantum field theories, because a consistent non-trivial transformation of the coupling constant requires that it should not run at quantum level. This is precisely the case for $\mathcal{N} = 4$ SUSY theories, which as mentioned in

Section 2.4.1 are exactly conformal. The coupling constant is a marginal coupling, a fixed number not subject to running, on which the duality transformation can act consistently at any scale. Hence only $\mathcal{N} = 4$ theories implement the strong–weak coupling duality in the form presented above. Dualities in $\mathcal{N} = 2, 1$ theories have a more involved structure, and must account for transmutation of the running gauge coupling into a strong dynamics scale Λ. A further bonus of $\mathcal{N} = 4$ theories is that they contain *BPS* electric and magnetic monopole states. Even though the latter are constructed as semiclassical solutions, they are guaranteed to exist in the full theory, by the robustness of BPS multiplets explained in Section 2.4.2.

Let us conclude by mentioning that in $\mathcal{N} = 4$ theories the above $\mathbf{Z}_2$ duality symmetry is actually a subgroup of a larger $SL(2, \mathbf{Z})$ symmetry. Introducing the θ angle in the complex coupling $\tau = \theta + i/g^2$, the duality acts as

$$\tau \to \frac{a\tau + b}{c\tau + d}, \quad \text{with } ad - bc = 1, \ a, b, c, d \in \mathbf{Z}, \tag{2.68}$$

with electric and magnetic charges transforming as an $SL(2, \mathbf{Z})$ doublet. Although the symmetry is classically $SL(2, \mathbf{R})$, only the discrete $SL(2, \mathbf{Z})$ subgroup is consistent with the general quantization condition (B.30), and survives in the quantum theory.

The $SL(2, \mathbf{Z})$ duality symmetry of 4d $\mathcal{N} = 4$ SYM theories is actually closely related to duality symmetries in string theory, e.g. in Section 6.4, with the complex coupling τ played by the complex dilaton modulus. The latter is often denoted S, hence motivating the name of S-duality for the duality of 4d $\mathcal{N} = 4$ SYM. Other usual terms are Montonen–Olive duality or strong–weak duality.

2.6 Low-energy supersymmetry and the MSSM
2.6.1 The MSSM fields and R-parity

The non-renormalization of the superpotential in 4d $\mathcal{N} = 1$ supersymmetric theories implies the ultraviolet stability of scalar masses. This suggests the application of SUSY to address the electroweak hierarchy problem, as anticipated in Section 1.3.3, for SUSY extensions of the SM. These must be $\mathcal{N} = 1$ theories, since the supermultiplets of extended SUSY in Section 2.4.1 contain vector-like combinations of fermions, and necessarily yield non-chiral theories. The Minimal Supersymmetric Standard Model (MSSM) is the minimal such $\mathcal{N} = 1$ extension, and includes the SM particles (gauge bosons, quarks, leptons, Higgs doublets) and their supersymmetric partners (called *gauginos, squarks, sleptons, higgsinos*). Gauge bosons and gauginos fill out vector supermultiplets, whereas the leptons, quarks, Higgses and their partners belong to chiral multiplets. The particle content is summarized in Table 2.1, with the SUSY partners conventionally denoted with a tilde. Each particle and its partner transform in the same way under $SU(3) \times SU(2)_L \times U(1)_Y$ since SUSY generators commute with the gauge generators. Note that there are two Higgs multiplets H_u, H_d with opposite hypercharges. This is in fact the minimal setting required to generate masses to both U- and D-quarks and leptons; indeed, the only way for a single

Table 2.1 *Particle content of the MSSM, classified by spin and supermultiplet type*

Vector $S = 1$	Multiplets $S = 1/2$	Chiral $S = 1/2$	Multiplets $S = 0$
g	$\tilde{g}$	q_L, q_R	$\tilde{q}_L, \tilde{q}_R$
$W^{\pm}, W^0$	$\tilde{W}^{\pm}, \tilde{W}^0$	l_L, l_R	$\tilde{l}_L, \tilde{l}_R$
B	$\tilde{B}$	$\tilde{H}_u, \tilde{H}_d$	H_u, H_d

Higgs doublet H to produce all masses is to appear conjugated in the Yukawa couplings, thus violating holomorphy of the superpotential. Moreover, the higgsino fermions in a single Higgs chiral multiplet would introduce $U(1)_Y$ anomalies, whereas in the MSSM they cancel among the two opposite hypercharge doublets. Thus the SUSY version of the SM essentially doubles the number of degrees of freedom, as dictated by the new symmetry; this can be regarded as amusingly analogous to the doubling of degrees of freedom by Dirac when introducing the notion of antiparticles!

The minimal phenomenologically consistent superpotential of a SUSY version of the SM is given by

$$W_{\mathrm{MSSM}} = Y_U^{ij} Q_i U_j H_u + Y_D^{ij} Q_i D_j H_d + Y_L^{ij} L_i E_j H_d + \mu H_u H_d, \tag{2.69}$$

where Q, U, D, L, E denote the chiral superfields containing the SM fermions in Table 1.1, with subindices L, R dropped from now on. The superpotential (2.69) also contains an explicit Higgs mass term, called the *μ-term*. It is in general required to obtain a phenomenologically satisfactory EW breaking and Higgs spectrum.

The above superpotential is *not* the most general one consistent with gauge invariance and renormalizability. There are other possible terms, of the form

$$W_R = \lambda^{ijk} U_i D_j D_k + \lambda^{ijk'} Q_i D_j L_k + \lambda^{ijk''} L_i L_j E_k + \mu_R^i L_i H_u. \tag{2.70}$$

The first of these terms violates baryon number in one unit, whereas the rest violate lepton number, also in one unit. The simultaneous presence of both types of terms leads to exceedingly fast proton decay, with lifetime of a few minutes. Note the sharp contrast with the non-SUSY SM, in which the most general renormalizable gauge invariant lagrangian automatically conserves both baryon and lepton numbers.

The SUSY SM thus requires additional symmetries forbidding the above dangerous couplings. The simplest possibility is a $\mathbf{Z}_2$ symmetry, called *R-parity*, under which the SM particles are even and their SUSY partners are odd. This symmetry forbids all terms in (2.70), and has additional phenomenological implications. For instance, SUSY particles must be produced pairwise in collider experiments. Furthermore, the lightest supersymmetric particle (LSP) is stable, since its decay into SM particles would violate R-parity. The LSP – typically a *neutralino*, linear combination of neutral gauginos and higgsinos, see Section 2.6.4 – turns out to be a natural candidate to explain the dark matter density in the universe. Reasonable choices of MSSM parameters give rise to LSP cosmological abundances in agreement with cosmological dark matter density measurements. However, it turns out that in the simplest unified schemes like the CMSSM (see Section 2.6.3) there is a tendency to obtain an excess of neutralino dark matter; this excess may be depleted in certain regions of the parameter space of sparticle masses, like the so-called coannihilation region, see Section 2.6.3.

It is worth mentioning that R-parity can be regarded as a discrete $\mathbf{Z}_2$ subgroup of the $U(1)_{B-L}$ symmetry. This viewpoint is natural in left–right symmetric extensions of the SM, already mentioned in Section 1.2.3, in which $U(1)_{B-L}$ is gauged. If the latter is spontaneously broken by the vev of a scalar with $B-L$ charge 2, there remains an unbroken (gauged) $\mathbf{Z}_2$ symmetry, playing the role of R-parity.

There are other possible choices of $\mathbf{Z}_N$ discrete symmetries ensuring proton stability, but allowing either B- or L-violating couplings. In these so-called R-parity violating models the LSP is no longer stable, and SUSY particles may be singly produced. Also, dark matter particles should be provided by other candidates (e.g. axions). One simple example is the $\mathbf{Z}_3$ *baryon triality* under which the chiral multiplets (Q, U, D, L, E, H_d, H_u) have charges $(1, \alpha^2, \alpha, \alpha^2, \alpha^2, \alpha^2, \alpha)$, $\alpha = \exp 2\pi i/3$. This symmetry forbids dimension-4 and 5 B-violating superpotential couplings but allows R-parity violating QDL, LLE, and LH_u terms. R-parity and this triality are the simplest discrete symmetries free of mixed anomalies with the SM gauge interactions.

2.6.2 The MSSM fields and unification

One can easily construct SUSY versions of the Grand Unified Theories of Section 1.2, dubbed SUSY-GUTs. For instance, the multiplet structure for $SU(5)$ SUSY-GUTs is shown in Table 2.2. The models require additional Higgs scalars Φ_{24} in the adjoint **24** (and their fermion partners $\tilde{\Phi}_{24}$) in order to break $SU(5)$ down to the SM. The usual Higgs doublets are contained in chiral multiplets H_5, $H_{\bar{5}}$ in the **5**, $\bar{\mathbf{5}}$ of $SU(5)$.

The SUSY models retain many qualitative properties of their non-SUSY counterparts, but with relevant quantitative differences. For instance, gauge coupling unification improves as depicted in Figure 2.1. Numerically, it differs from the non-SUSY case in the

Table 2.2 *Field content of SU(5) SUSY-GUTs*

Vector $S = 1$	Multiplets $S = 1/2$	Chiral $S = 1/2$	Multiplets $S = 0$
$A^\mu_{SU(5)}$	$\tilde{\lambda}_{SU(5)}$	$(\bar{\mathbf{5}} + \mathbf{10})$ $\tilde{H}_5, \tilde{H}_{\bar{5}}$	$(\tilde{\bar{\mathbf{5}}} + \tilde{\mathbf{10}})$ $H_5, H_{\bar{5}}$

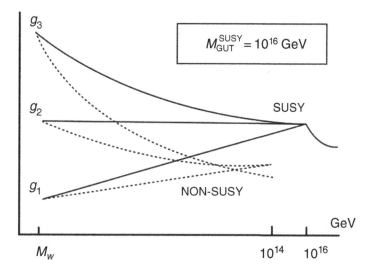

Figure 2.1 Qualitative picture of gauge coupling unification in $SU(5)$ Grand Unification, for the non-SUSY (dashed line) and SUSY case (continuous line).

extra SUSY particle contributions to the running of gauge couplings. This increases the one-loop β-functions, yielding

$$b_3 = -3; \quad b_2 = n_H; \quad b_1 = 6 + \frac{3}{5} n_H, \tag{2.71}$$

with n_H the number of $H_u + H_d$ Higgs sets ($n_H = 1$ in the minimal case). This produces two net effects, namely the increase of the unification scale to $M_{GUT} \simeq 10^{16}$ GeV (useful to suppress proton decay rates, see below), and the much better numerical unification of the three coupling constants (persisting at two loops with a few percent precision). The precise unification of gauge coupling constants when extrapolated from their observed low-energy values using SUSY extensions of the SM yields considerable support to the idea of low-energy supersymmetry.

It is worth mentioning that the improvement in gauge coupling unification is mainly due to the SUSY fermion particles (gauginos and higgsinos) rather than to the scalars (squarks and sleptons), since the latter fill complete $SU(5)$ representations and do not contribute at one loop. Incidentally, this has motivated the so-called *split SUSY* proposal, in which scalars have very large masses (up to 10^9 GeV), while fermions remain light (around the TeV scale) and lead to approximate gauge coupling unification. In this scheme, SUSY does not play any role in solving the electroweak hierarchy problem, which must be addressed by other means. One possible signature of this idea is the existence of quasi-stable gluinos, whose decay may be observable at the LHC.

Since the unification scale in SUSY-GUTs is larger than in non-SUSY ones, the proton decay through dimension-6 operators of Section 1.2.2 has a very small rate, with $\tau_{p \to \pi^0 e^+} \gtrsim 10^{36}$ years. There are, however, new dimension five operators $QQQL$ or $UUDE$

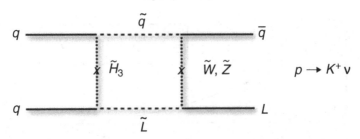

Figure 2.2 Diagram mediating proton decay via dimension five operators in $SU(5)$ SUSY-GUTs.

in the effective low-energy superpotential, which constitute new sources of proton decay. In the $SU(5)$ case they come from diagrams like the left half of Figure 2.2, involving exchange of a heavy color triplet higgsino ($SU(5)$ partner of doublet higgsinos). Such operators can be dressed by further exchanges of gauginos or higgsinos as in the right half of Figure 2.2 to contribute to proton decay. The dominant decay channel involves now kaons rather than pions, and for the minimal version of $SU(5)$ one gets $\tau_{p \to K^+ \bar{\nu}} \simeq (0.2 - 4) \times 10^{32}$ years, already excluded by the limits of order 6×10^{32} years from the super-Kamiokande experiment; however, there are many other SUSY-GUT models with slightly modified Higgs sectors which are consistent with proton decay experimental bounds.

The potential appearance of dimension five operators of the type discussed above is a general feature of many other SUSY extensions of the SM, besides SUSY-GUTs. A general lesson is that their role in proton decay can lead to strong constraints on specific models.

Concerning the prediction (1.16) for the m_b/m_τ ratio, there are two competing modifications in the SUSY case. On one hand, now $\beta_3 = -3$, $\gamma = 16/3$, so the exponent increases, while on the other, $\alpha_3(M_W)/\alpha_3(M_X)$ decreases; the net result is similar to the non-SUSY case. Detailed agreement with experiment, however, requires taking into account the top Yukawa contribution to the anomalous dimensions, and is only achieved for large $\tan \beta \equiv \langle H_u \rangle / \langle H_d \rangle$.

A generic problem of Grand Unification models which still persists in the SUSY case is *doublet–triplet splitting*, which is most simply illustrated in the $SU(5)$ case. Recall that the Higgs sector contains an adjoint field Φ_{24} to break the GUT group, and fields H_5, $H_{\bar{5}}$ in the $(5 + \bar{5})$. The latter contain the electroweak MSSM Higgs doublets H_u, H_d, but also color triplets H_3, $H_{\bar{3}}$. As described above, the latter can mediate proton decay, and hence must be very heavy. The model thus requires a large splitting between the light doublets and heavy triplets in the H_5, $H_{\bar{5}}$ Higgs multiplets. These masses are determined by the Higgs sector superpotential, which has the structure

$$W_{\text{SUSY-Higgs}} = M H_5 H_{\bar{5}} + \lambda_5 H_{\bar{5}} \Phi_{24} H_5. \tag{2.72}$$

Once the adjoint vev $\langle \Phi_{24} \rangle = v \, \text{diag} \, (1, 1, 1, -3/2, -3/2)$ breaks the GUT symmetry down to the SM one, the doublet and triplet get masses

$$m_{H_{\text{EW}}} = \left(M - \tfrac{3}{2} \lambda_5 v \right); \quad m_{H_3} = (M + \lambda_5 v). \tag{2.73}$$

The Higgs doublets can be made light at the price of fine-tuning $M = \frac{3}{2}\lambda_5 v + \mathcal{O}(M_W)$. In this SUSY version of the doublet–triplet splitting problem, such fine-tuned parameter relations are preserved by renormalization, and are thus technically natural, but remain rather ad hoc and unmotivated. There are several proposed explanations for the origin of such relations, e.g. based on $SO(10)$ GUT unification, with a particular choice of gauge symmetry breaking. In theories with extra dimensions, in particular string compactifications, there exist new mechanisms to address this issue.

2.6.3 Soft terms and radiative EW symmetry breaking

MSSM soft terms

As explained in Section 2.2, in particle physics models SUSY must be broken, at some scale M_{SUSY}, presumably in the TeV range. On general grounds, it is not surprising that precisely the SM fields remain light while their partners get heavy, since the SM gauge bosons and fermions are protected by gauge symmetry and chirality, whereas their partners gauginos and scalars are vector-like and can get masses of order M_{SUSY}.

Whatever the underlying source of SUSY breaking, its effects on the *visible* SUSY SM sector are efficiently parametrized as the soft terms introduced in Section 2.2. In the case of the MSSM, the soft terms are given by

$$
\begin{aligned}
\mathcal{L}_\lambda &= \frac{1}{2}\sum_a M_a \lambda_a \lambda_a + \text{h.c.},\\
\mathcal{L}_{m^2} &= -m^2_{H_d}|H_d|^2 - m^2_{H_u}|H_u|^2 - m^2_{Q_{ij}} Q_i Q_j^* - m^2_{U_{ij}} U_i U_j^*\\
&\quad - m^2_{D_{ij}} D_i D_j^* - m^2_{L_{ij}} L_i L_j^* - m^2_{E_{ij}} E_i E_j^*,\\
\mathcal{L}_{A,B} &= -A^{U\prime}_{ij} Q_i U_j H_u - A^{D\prime}_{ij} Q_i D_j H_d - A^{L\prime}_{ij} L_i E_j H_d - B' H_d H_u + \text{h.c.},
\end{aligned}
\tag{2.74}
$$

where Q, U, D, L, E, and H_u, H_d denote the scalar components of the chiral multiplets, and a, i label gauge factors and SM families, respectively. The above general lagrangian has many (concretely 107) new free parameters beyond those of the SM, seemingly spoiling the beauty of SUSY models. However, specific models of SUSY breaking have a drastically reduced number of free parameters, as discussed in Section 2.6.5. Furthermore, there are strong phenomenological constraints on them, in particular from the observed suppression of flavour changing neutral current (FCNC) transitions like $K^0 \leftrightarrow \bar{K}^0$. This forces the equal hypercharge squarks in the first two families to be almost degenerate in mass, $|\tilde{m}_1^2 - \tilde{m}_2^2| \leq 10\,\mathrm{GeV}^2$. The simplest way to satisfy FCNC constraints is to assume universal (i.e. family independent) SUSY masses for scalars. In Grand Unification Theories like $SU(5)$, one also expects unification of gaugino masses according to

$$
M_3 = M_2 = \frac{3}{5} M_1 \equiv M,
\tag{2.75}
$$

and also of masses for scalars in the same GUT multiplets; thus e.g. in an $SO(10)$ GUT
with flavour independence, we would have

$$m_Q^2 = m_U^2 = m_D^2 = m_L^2 = m_E^2 \equiv m^2 \times \mathbf{1}. \tag{2.76}$$

In many GUT models the trilinear scalar terms are also universal and proportional to the
Yukawa couplings,

$$A_{ij}^{U'} = A_{ij}^{D'} = A_{ij}^{L'} \equiv AY_{ij}^{U,D,L}. \tag{2.77}$$

One may also assume unification of the soft SUSY breaking Higgs mass parameters with
those of the squarks and sleptons

$$m_{H_u}^2 = m_{H_d}^2 = m^2, \tag{2.78}$$

although this is less motivated, even in the context of GUT models.

In models with these relations, the number of free parameters in the low-energy lagrangian is reduced from 107 to just five independent parameters, namely

$$M, \ m, \ A, \ B', \ \mu. \tag{2.79}$$

The MSSM model with these extra assumptions is known as the *constrained MSSM*
(CMSSM).

In theories with a large fundamental scale, like GUTs – or later, string compactifications –
relationships like (2.75), (2.76), (2.77), and (2.78) are high-scale boundary conditions. At
lower energies their values evolve according to the renormalization group equations. For
instance, gaugino masses evolve with energy as the ratio of coupling constants

$$M_i = \frac{\alpha_i(\mu)}{\alpha_{\text{GUT}}} M. \tag{2.80}$$

Radiative EW symmetry breaking

A particularly relevant question is whether the Higgs scalar potential obtained from soft
term SUSY breaking triggers appropriate EW symmetry breaking. The scalar potential for
the neutral Higgs fields $H_{u,d}^0$ from the superpotential and soft terms has the form

$$V_H = \frac{1}{8}(g_1^2 + g_2^2)(|H_d^0|^2 - |H_u^0|^2)^2 + \mu_d^2|H_d^0|^2 + \mu_u^2|H_u^0|^2 + (B'H_d^0H_u^0 + \text{h.c.}), \tag{2.81}$$

where $\mu_{u,d}^2 = m_{H_u,H_d}^2 + |\mu|^2$ at the unification scale. As it stands, this potential does not
trigger an appropriate $SU(2) \times U(1)$ breaking. Indeed, this requires a negative eigenvalue in
the matrix of Higgs squared masses, i.e. $\mu_d^2\mu_u^2 - |B'|^2 < 0$. However, since $\mu_d^2 = \mu_u^2 > 0$,
this may only happen if $\mu_d^4 = \mu_u^4 < |B'|^2$, in which case the scalar potential is unbounded
from below in the direction $\langle H_d \rangle = \langle H_u \rangle \to \infty$.

This is actually not a problem, since the existence of correct EW symmetry breaking
must be established using the Higgs potential at EW scales, i.e. including the quantum

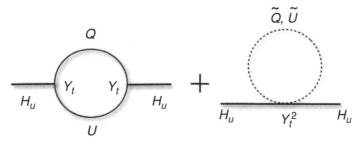

Figure 2.3 One-loop corrections to the H_u (mass)^2, controlled by the top Yukawa coupling y_t. After SUSY breaking the negative contribution from the first diagram wins over the positive contribution from the second.

corrections involved in running to low scales, as follows. Consider in particular the one-loop corrections to the masses of the Higgs fields. These include diagrams controlled by the Yukawa couplings of the Higgs field H_u to the U-quarks and squarks, see Figure 2.3. These corrections are clearly negligible except for those involving the top quark, which has a relatively large Yukawa coupling y_t. With unbroken SUSY, the first diagram in Figure 2.3 leads to a negative quadratically divergent contribution which is exactly canceled by the second diagram. Once SUSY is broken, the squarks get masses and the second diagram is suppressed compared to the first, leaving an overall uncanceled *negative* contribution

$$\delta\mu_u^2 \simeq -\frac{3}{16\pi^2} y_t^2 m_{\tilde{q}}^2 \log \frac{M_{\mathrm{GUT}}^2}{(Q_0^2 + m_{\tilde{q}}^2)}, \tag{2.82}$$

where the contribution is evaluated at a scale Q_0. For large enough y_t, i.e. a heavy enough top quark, this negative contribution may, at low energies, exceed the original positive contribution and trigger EW symmetry breaking. Similar diagrams exist for the Higgs field H_d but they are controlled by the bottom Yukawa coupling, and so give smaller contributions, which we ignore for simplicity.

One may worry that similar diagrams correcting the squared masses of squarks like $\tilde{t}$, $\tilde{b}$ could drive them to negative values, leading to minima breaking charge and color symmetries. However, for these colored scalars there are additional large and positive contributions from diagrams involving gluino loops, controlled by the large QCD coupling constant, which thus prevent $SU(3) \times U(1)_{\mathrm{EM}}$ breaking. Also for this reason, squarks are in general heavier than sleptons.

The resulting pattern of running masses for the Higgs scalar and the various SUSY particles of the MSSM is sketched in Figure 2.4. Hence the structure of the MSSM is such that quantum corrections produce the desired pattern of $SU(3) \times SU(2)_L \times U(1)_Y$ symmetry breaking in a natural and elegant way. This mechanism, known as *radiative electroweak symmetry breaking*, requires a heavy top quark, e.g. $m_t \simeq 70$–$190\,\mathrm{GeV}$ in the simplest constrained MSSM. The experimentally measured value of $m_t \simeq 170\,\mathrm{GeV}$ thus fits quite well with radiative EW symmetry breaking in the MSSM; this is a remarkable success of

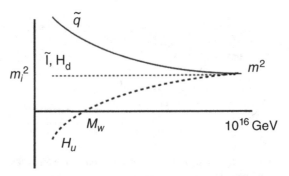

Figure 2.4 Qualitative view of the renormalization group evolution of the squared masses of MSSM scalar fields, from a universal value m^2 at a large scale. Squarks $\tilde{q}$ are heavier than sleptons $\tilde{l}$ due to large positive contributions from gluino loops. H_u is dragged to negative squared mass due to the large top contribution, and triggers EW symmetry breaking at low energies.

this mechanism, in particular given that at the time of its original formulation the existing theoretical prejudice was $m_t \simeq 30\text{–}40\,\text{GeV}$.

A more complete treatment of loop corrections makes use of the RGEs to run the soft terms down to low energies, producing a potential (2.81) for running couplings. In this description, $\mu_u^2 \neq \mu_d^2$ at low energies, and the potential can trigger correct EW symmetry breaking. Minimization of the potential can be shown to give

$$v^2 \equiv v_d^2 + v_u^2 = \frac{2[\mu_d^2 - \mu_u^2 - (\mu_d^2 + \mu_u^2)\cos 2\beta]}{(g_2^2 + g_1^2)\cos 2\beta}, \tag{2.83}$$

where $v_{d,u} = \langle H_{d,u}^0 \rangle$ and $\sin 2\beta \equiv 2|B'|/(\mu_d^2 + \mu_u^2)$, with $\tan\beta = v_u/v_d$. The non-zero B'-term forces the two vevs to align such that the electromagnetic $U(1)_{\text{EM}}$ remains unbroken. Using the W boson mass to fix $v^2 = 2M_W^2/g_2^2$, the Z boson mass can then be expressed as a polynomial in μ and the SUSY breaking parameters. For instance in the CMSSM one obtains a quadratic expression

$$M_{Z^0}^2 = c_1 M^2 + c_2 m^2 + c_3|A|^2 + c_4|\mu|^2 + c_5 MA + \cdots, \tag{2.84}$$

where $c_i = c_i(y_t, g_i)$ are calculable coefficients. Thus the soft parameters and top Yukawa y_t are constrained in order to obtain the correct experimental value for M_Z. There are wide regions of parameter space in which this works. However, the LEP and Tevatron bounds have restricted so much the SUSY parameter space that certain small amount of fine-tuning, at the few percent level, is required. This is sometimes called the *little hierarchy problem*.

We conclude this section by recalling some potential tuning problems of the MSSM with soft term SUSY breaking. There are several associated to *flavour physics*. For instance, as already mentioned, soft scalar masses of the first two generations should be family independent and diagonal to suppress large FCNC. Also, soft parameters can potentially lead to too large CP violation. Indeed, A, B', M, and μ are in general independent complex parameters, with in principle no symmetry (or other) relation among their phases, and generically lead to too large one-loop contributions to the electric dipole moment of the neutron. These constraints require such complex phases to be smaller than $\simeq 10^{-2}$, leading to important constraints on specific microscopic models of supersymmetry breaking.

Finally there is the following so-called μ *problem*. A phenomenologically correct low-energy Higgs potential requires the original superpotential (2.69) to contain a SUSY mass term μ of the same order of magnitude as the SUSY breaking soft terms. Although this can be set ad hoc, there is no obvious reason why a superpotential parameter like μ should have a scale similar to soft terms, which only appear after SUSY breaking. A number of solutions have been put forward to explain this relationship, e.g. the introduction of a singlet X, coupling as $W_H = X H_u H_d$, and getting a vev controlled by the SUSY breaking scale. This simple modification of the MSSM (which includes also an X^3 coupling) is called next-to-minimal SUSY SM (NMSSM). Another simple proposal is the so-called Giudice–Masiero mechanism, described in the context of gravity mediation models in Section 2.6.5.

2.6.4 The sparticle spectrum

For future reference we here list the supersymmetric particles (*sparticles*) appearing in the MSSM spectrum:

- *Gluinos*: The fermionic partners of gluons get their mass only from the SUSY breaking Majorana mass term M_3 in (2.74). As shown in (2.80) the running mass is proportional to the strong coupling α_3 so that the gluino is typically the heaviest gaugino in setups with universal soft terms.
- *Charginos*: The fermionic partners of the $W^{\pm}$ bosons and the charged higgsinos mix after EW symmetry breaking and have a mass matrix of the form

$$(\tilde{W}^+, \quad \tilde{H}_u^+) \begin{pmatrix} M_2 & g_2 \langle H_d \rangle \\ g_2 \langle H_u \rangle & \mu \end{pmatrix} \begin{pmatrix} \tilde{W}^- \\ \tilde{H}_d^- \end{pmatrix}. \qquad (2.85)$$

The off-diagonal terms are induced by (2.27) after replacing $H_{u,d}$ by their vevs. There are two massive eigenstates, the two charginos $\chi_1^{\pm}, \chi_2^{\pm}$, whose masses are simple functions of M_2, μ and $\tan \beta$.
- *Neutralinos*: The fermionic partners of W^0 and of the hypercharge gauge boson also mix with the two neutral higgsinos. There are four eigenstates $\chi_{1,2,3,4}^0$ which are linear combinations of those neutral states, and whose masses depend on the parameters M_2, M_1, μ, and $\tan \beta$. The lightest neutralino χ_1^0 is most frequently the lightest SUSY particle (LSP), and as we mentioned is a natural dark matter candidate.
- *Sfermions*: Squarks and sleptons (sfermions) come, for fixed flavour, in two varieties denoted left and right, $\tilde{q}_{L,R}, \tilde{l}_{L,R}$, according to the character of the corresponding SM fermion partner. They have four different sources for their masses. The leading one arises from the scalar soft terms in (2.74), which do not mix left and right sfermions. A second contribution comes from the trilinear couplings and mixes left and right scalars;

the A-parameters are usually proportional to the corresponding Yukawa coupling, so the mass contributions are non-negligible only for third generation sfermions. A third contribution is proportional to $M_{Z^0}^2$ and is obtained from the scalar potential V_D in (2.26), including sfermions and replacing the neutral Higgs scalars by their vevs. Finally, there is a SUSY contribution equal to the mass of the corresponding SM fermion, which again is non-negligible only for third generation sfermions. Most mass eigenstate sfermions are essentially of left or right type, except for stops, which experience large mixing from A-terms proportional to the top Yukawa coupling, and whose eigenstates are denoted $\tilde{t}_{1,2}$.

- *Higgs scalars*: The two complex Higgs doublets contain altogether eight scalars, three of which are the Goldstone bosons swallowed by the $W^\pm$, Z^0 gauge bosons. The other five degrees of freedom include one charged Higgs with a mass

$$m_{H^\pm}^2 = M_W^2 + \mu_u^2 + \mu_d^2, \tag{2.86}$$

and two neutral scalars h, H and a pseudoscalar A with masses

$$m_A^2 = \mu_u^2 + \mu_d^2, \tag{2.87}$$

$$m_{h,H}^2 = \frac{1}{2}\left(m_A^2 + M_{Z^0}^2 \mp \sqrt{m_A^4 + M_{Z^0}^4 - 4M_{Z^0}^2 m_A^2 \cos^2 2\beta}\right).$$

The spectrum of Higgses (and their couplings) are determined by two parameters, m_A and $\tan\beta$. The above tree-level formulae lead to the bounds

$$m_{H^\pm} \geq M_W; \quad m_h \leq m_A \leq m_H, \tag{2.88}$$

$$m_h \leq M_{Z^0}|\cos 2\beta| \leq m_H.$$

The upper bound on the lightest Higgs h comes from the fact that the Higgs quartic self-coupling in the MSSM is given by the EW coupling constants, see (2.81). Thus at tree level $m_h \leq M_{Z^0}$ in the MSSM, which is ruled out by LEP measurements; however, there are sizable loop corrections to these expressions from the large top Yukawa coupling, and diagrams like those in Figure 2.3 correct the lightest Higgs mass by

$$\delta m_h^2 \simeq \frac{3}{4\pi^2} v^2 y_t^4 \sin\beta^4 \log\frac{m_{\tilde{t}_1} m_{\tilde{t}_2}}{m_t^2}. \tag{2.89}$$

Including loop corrections, the upper bound on the lightest Higgs scalar mass becomes $m_h \leq 130\,\text{GeV}$, which is compatible with LEP limits. Thus the Higgs sector in the MSSM should be tested eventually at the LHC.

As already suggested by Figure 2.4, in unified models gluinos and squarks tend to be heavier than sleptons and charginos, and the lightest neutralino is often the LSP, see Figure 15.6 for an example of the SUSY spectrum. Squarks and gluinos should be copiously produced at the LHC. Squarks decay into quarks and charginos or neutralinos, $\tilde{q} \rightarrow q + \chi_{1,2}^\pm$, $\tilde{q} \rightarrow q + \chi_i^0$, or also into quark+gluino if the latter is lighter than the squark. Gluinos in turn decay into quark+antiquark and neutralinos (or charginos). In any event the

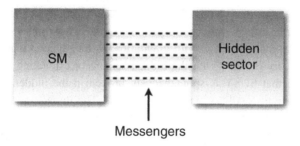

Figure 2.5 Structure of SUSY breaking in a hidden sector, transmitted to the SM sector by messenger degrees of freedom.

classical MSSM signature of SUSY particles at hadronic colliders is missing energy (from the unseen neutralinos), along with jets and leptons. The LHC at 14 TeV should eventually be able to search for gluinos and squarks with masses well above 1 TeV.

2.6.5 The origin of SUSY breaking soft terms

As described in Section 2.6.3, SUSY breaking in the MSSM can be efficiently *parametrized* using soft terms. However, we would like to understand their origin, and if possible to find predictive constraints and/or relations among them.

A first observation is that the SM fields cannot directly participate in a tree-level SUSY breaking model, of the kind introduced in Section 2.2, for the following reason. There is a general (tree level) constraint among boson and fermion masses, given by

$$\sum m_{J=0}^2 - 2 \sum m_{J=1/2}^2 + 3 \sum m_{J=1}^2 = 0. \tag{2.90}$$

This is in fact the equation ensuring the absence of quadratic divergences and hence is fulfilled even after soft SUSY breaking. Since SUSY commutes with electric charge, this expression holds separately for particles of a given charge. Thus, e.g., the sum of the D-squark squared masses cannot exceed $2m_b^2$, with m_b the b-quark mass, clearly contradicting experimental sparticle bounds.

A simple way to overcome this problem is to locate the source of SUSY breaking in a separate sector, termed *hidden sector*, and assume that SUSY breaking is transmitted to the SM *visible sector* by some mediating particles or interactions, the *messengers*, as sketched in Figure 2.5. Several mediation mechanisms have been proposed, the most popular being *gravity mediation*, *gauge mediation*, and *anomaly mediation*, reviewed next.

Gravity mediation

The minimal assumption for SUSY breaking mediation is that the gravitational interaction (plus possibly other Planck scale suppressed interactions) acts as the messenger, since gravity exists and must necessarily be present in any model.

To describe gravity mediation quantitatively, consider a prototypical structure, in which chiral multiplets split in two types, corresponding to the SM particles C^α and additional *hidden sector* chiral multiplets h_m, with gravitational strength couplings to the SM. This arises in string theory models, with the hidden sector corresponding to diverse moduli fields. We assume that SUSY is spontaneously broken in the hidden sector by a non-vanishing vev for some auxiliary field(s) $\langle F_m \rangle \neq 0$. The impact on the SM sector can be analyzed from the structure of the $\mathcal{N} = 1$ supergravity action. The superpotential and Kähler potential can be expanded in powers of the SM fields, with the general form

$$W = \hat{W}(h_m) + \frac{1}{2}\mu_{\alpha\beta}(h_m)C^\alpha C^\beta + \frac{1}{6}Y_{\alpha\beta\gamma}(h_m)C^\alpha C^\beta C^\gamma + \cdots, \tag{2.91}$$

$$K = \hat{K}(h_m, h_m^*) + K_{\bar{\alpha}\beta}(h_m, h_m^*)C^{*\bar{\alpha}}C^\beta + \left[\frac{1}{2}Z_{\alpha\beta}(h_m, h_m^*)C^\alpha C^\beta + \text{h.c.}\right] + \cdots.$$

The bilinear terms $\mu_{\alpha\beta}$ and $Z_{\alpha\beta}$ are often forbidden by gauge invariance in specific models, but are relevant for the Higgs μ-term. As indicated, the coefficients $K_{\bar{\alpha}\beta}$, $Y_{\alpha\beta\gamma}$, $\mu_{\alpha\beta}$, and $Z_{\alpha\beta}$ in (2.91) may in general depend on the hidden sector fields. Similarly, the SM gauge kinetic functions are considered to depend on these fields.

Spontaneous SUSY breaking produces a gravitino mass (2.55) $m_{3/2} \simeq \langle F_m \rangle / M_p$. Replacing h_m and F_m by their vevs in the supergravity lagrangian, and taking $M_p \to \infty$ keeping $m_{3/2}$ fixed, one obtains SUSY breaking soft terms in the effective SM lagrangian, with typical scale

$$m_{\text{soft}} \simeq m_{3/2} \simeq \frac{F_m}{M_p}. \tag{2.92}$$

For this to be in the TeV range, the hidden sector SUSY breaking scale is

$$F_m \simeq M_W M_p \simeq (10^{10} \, \text{GeV})^2. \tag{2.93}$$

It is now possible to obtain the specific expressions for all MSSM soft parameters in terms of the gauge kinetic functions f_a and the quantities in (2.91), as follows (for simplicity we set $\kappa_4^{-1} = M_p/\sqrt{8\pi} = 1$ in these formulae). The kinetic functions (2.32) produce masses for the canonically normalized gaugino fields

$$M_a = \frac{1}{2}(\text{Re } f_a)^{-1} F^m \partial_m f_a. \tag{2.94}$$

The (still unnormalized) Yukawa couplings and mass terms relate to the superpotential couplings in supergravity by

$$Y'_{\alpha\beta\gamma} = \frac{\hat{W}^*}{|\hat{W}|}e^{\hat{K}/2}Y_{\alpha\beta\gamma},$$

$$\mu'_{\alpha\beta} = \frac{\hat{W}^*}{|\hat{W}|}e^{\hat{K}/2}\mu_{\alpha\beta} + m_{3/2}Z_{\alpha\beta} - \bar{F}^{\bar{m}}\partial_{\bar{m}}Z_{\alpha\beta}. \tag{2.95}$$

Expanding the supergravity scalar potential (2.50) in the SM fields, and assuming vanishing cosmological constant, one obtains

$$V_{\text{soft}} = m'^2_{\overline{\alpha}\beta} C^{*\overline{\alpha}} C^\beta + \left(\frac{1}{6} A'_{\alpha\beta\gamma} C^\alpha C^\beta C^\gamma + \frac{1}{2} B'_{\alpha\beta} C^\alpha C^\beta + \text{h.c.} \right), \tag{2.96}$$

with

$$m'^2_{\overline{\alpha}\beta} = m^2_{3/2} K_{\overline{\alpha}\beta} - \overline{F^m}(\partial_{\overline{m}}\partial_n K_{\overline{\alpha}\beta} - \partial_{\overline{m}} K_{\overline{\alpha}\gamma} K^{\gamma\overline{\delta}}\partial_n K_{\overline{\delta}\beta}) F^n, \tag{2.97}$$

$$A'_{\alpha\beta\gamma} = \frac{\hat{W}^*}{|\hat{W}|} e^{\hat{K}/2} F^m \{ \hat{K}_m Y_{\alpha\beta\gamma} + \partial_m Y_{\alpha\beta\gamma}$$
$$- [K^{\delta\overline{\rho}}\partial_m K_{\overline{\rho}\alpha} Y_{\delta\beta\gamma} + (\alpha \leftrightarrow \beta) + (\alpha \leftrightarrow \gamma)]\}, \tag{2.98}$$

$$B'_{\alpha\beta} = \frac{\hat{W}^*}{|\hat{W}|} e^{\hat{K}/2} \{ F^m [\hat{K}_m \mu_{\alpha\beta} + \partial_m \mu_{\alpha\beta}$$
$$- (K^{\delta\overline{\rho}}\partial_m K_{\overline{\rho}\alpha}\mu_{\delta\beta} + (\alpha \leftrightarrow \beta))] - m_{3/2}\mu_{\alpha\beta} \}$$
$$+ 2m^2_{3/2} Z_{\alpha\beta} - m_{3/2}\overline{F^m}\partial_{\overline{m}} Z_{\alpha\beta}$$
$$+ m_{3/2} F^m [\partial_m Z_{\alpha\beta} - (K^{\delta\overline{\rho}}\partial_m K_{\overline{\rho}\alpha} Z_{\delta\beta} + (\alpha \leftrightarrow \beta))]$$
$$- \overline{F^m} F^n [\partial_{\overline{m}}\partial_n Z_{\alpha\beta} - (K^{\delta\overline{\rho}}\partial_n K_{\overline{\rho}\alpha}\partial_{\overline{m}} Z_{\delta\beta} + (\alpha \leftrightarrow \beta))]. \tag{2.99}$$

Here $K^{\alpha\overline{\beta}}$ is the inverse of the Kähler metric $K_{\alpha\overline{\beta}}$ for the visible sector matter fields. We thus see that spontaneous supersymmetry breaking in supergravity induces SUSY breaking soft terms of the type discussed in Section 2.2.

Note that the Kähler metrics $K_{\overline{\alpha}\beta}$, $\hat{K}_{\overline{m}n}$ of observable and hidden sectors may in general be non-diagonal. Upon normalization of the fields to get canonical kinetic terms, the first piece in (2.97) leads to universal diagonal soft masses, but the second piece generically induces *off-diagonal* contributions. These off-diagonal contributions can lead to too large, and thus phenomenologically problematic, FCNC transitions in the SM. Happily, as described in later chapters, string compactifications often lead to diagonal metrics $K_{\overline{\alpha}\beta}$, i.e.

$$K_{\overline{\alpha}\beta}(h_m, h_m^*) = \delta_{\overline{\alpha}\beta} K_\alpha(h_m, h_m^*). \tag{2.100}$$

Some simplifications take place in this diagonal metric case. In the MSSM, the Kähler potential and the superpotential have then the form

$$K = \hat{K}(h_m, h_m^*) + K_\alpha(h_m, h_m^*) C^{*\overline{\alpha}} C^\alpha + [Z(h_m, h_m^*) H_u H_d + \text{h.c.}], \tag{2.101}$$

$$W = \hat{W}(h_m) + \mu(h_m) H_u H_d$$
$$+ \sum_{\text{families}} [Y_U(h_m) Q H_u U + Y_D(h_m) Q H_d D + Y_E(h_m) L H_d E], \tag{2.102}$$

where $C^\alpha = Q, U, D, L, E, H_u, H_d$ in the first line, and for notational simplicity we have taken diagonal Yukawa couplings, $Y_{\alpha\beta\gamma} = Y_U, Y_D, Y_E$, in self-explanatory notation. The SUSY breaking soft terms in the effective action now read

$$\mathcal{L}_{\text{soft}} = \frac{1}{2}(M_a \widehat{\lambda}^a \widehat{\lambda}^a + \text{h.c.}) - m_\alpha^2 \widehat{C}^{*\overline{\alpha}} \widehat{C}^\alpha$$
$$- \left(\frac{1}{6} A_{\alpha\beta\gamma} \widehat{Y}_{\alpha\beta\gamma} \widehat{C}^\alpha \widehat{C}^\beta \widehat{C}^\gamma + B\widehat{\mu} \widehat{H}_u \widehat{H}_d + \text{h.c.} \right), \tag{2.103}$$

with

$$M_a = \frac{1}{2}(\text{Re } f_a)^{-1} F^m \partial_m f_a, \tag{2.104}$$

$$m_\alpha^2 = m_{3/2}^2 - \overline{F}^{\overline{m}} F^n \partial_{\overline{m}} \partial_n \log K_\alpha, \tag{2.105}$$

$$A_{\alpha\beta\gamma} = F^m [K_m + \partial_m \log Y_{\alpha\beta\gamma}^{(0)} - \partial_m \log(K_\alpha K_\beta K_\gamma)], \tag{2.106}$$

$$B = \widehat{\mu}^{-1}(K_{H_u} K_{H_d})^{-1/2} \left\{ \frac{W^*}{|W|} e^{K/2} \mu [F^m(\widehat{K}_m + \partial_m \log \mu \right.$$
$$- \partial_m \log(K_{H_u} K_{H_d})) - m_{3/2}]$$
$$+ 2m_{3/2}^2 Z - m_{3/2} \overline{F}^{\overline{m}} \partial_{\overline{m}} Z$$
$$+ m_{3/2} F^m [\partial_m Z - Z \partial_m \log(K_{H_u} K_{H_d})]$$
$$\left. - \overline{F}^{\overline{m}} F^n [\partial_{\overline{m}} \partial_n Z - \partial_{\overline{m}} Z \partial_n \log(K_{H_u} K_{H_d})] \right\}. \tag{2.107}$$

Here $\widehat{C}^\alpha$ and $\widehat{\lambda}^a$ are the *canonically normalized* scalars and gauginos respectively

$$\widehat{C}^\alpha = K_\alpha^{1/2} C^\alpha, \quad \widehat{\lambda}^a = (\text{Re} f_a)^{1/2} \lambda^a. \tag{2.108}$$

Also, we factored out the rescaled Yukawa coupling and μ parameter in the A and B terms

$$\widehat{Y}_{\alpha\beta\gamma} = Y_{\alpha\beta\gamma}^{(0)} \frac{W^*}{|W|} e^{K/2} (K_\alpha K_\beta K_\gamma)^{-1/2}, \tag{2.109}$$

$$\widehat{\mu} = \left(\frac{W^*}{|W|} e^{K/2} \mu + m_{3/2} Z - \overline{F}^{\overline{m}} \partial_{\overline{m}} Z \right) (K_{H_u} K_{H_d})^{-1/2}. \tag{2.110}$$

These expressions look rather complicated, but in specific models actually give rise to very compact expressions for the predicted soft terms. One example is the scheme often called *minimal supergravity*, in which the kinetic terms of matter fields are canonical ($K_{\alpha\bar{\beta}} = \delta_{\alpha\bar{\beta}}$). Assuming further that the Yukawas and μ-term are independent of hidden sector fields, and $Z = 0$, it yields the very simple boundary conditions

$$m_\alpha^2 = m_{3/2}^2, \quad B = A - m_{3/2}. \tag{2.111}$$

In this simple scheme the scalar masses are automatically universal. As shown in later chapters, in string compactifications the SM fields do not have canonical kinetic terms, and this kind of simple structure does not arise. Nevertheless, many specific models lead to other equally simple soft term relations, see Section 15.5.1.

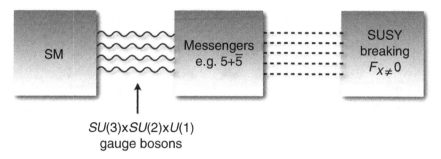

$SU(3) \times SU(2) \times U(1)$
gauge bosons

Figure 2.6 Structure of gauge mediated SUSY breaking.

One interesting aspect of gravity mediation is that, as observed by Giudice and Masiero, there is a simple solution to the μ-problem. Indeed, (2.110) shows that even if $\mu = 0$ in the original lagrangian, a physical $\widehat{\mu}$-term is generated if $Z \neq 0$. This $\widehat{\mu}$ term is naturally of the order of the other soft terms, thus solving the μ-problem.

Gauge mediation

In gauge mediated SUSY breaking (GMSB) the leading mediators are the SM gauge bosons and gauginos. To better illustrate some of the general features of this scenario, let us introduce a simple "minimal" model of GMSB, depicted in Figure 2.6. It contains a singlet chiral multiplet X, with $\langle X \rangle \neq 0$ and breaking SUSY by $\langle F_X \rangle \neq 0$. This singlet couples to new heavy vector-like messenger fields with SM quantum numbers, taken in complete $SU(5)$ representations to preserve one-loop gauge coupling unification, e.g. $\mathbf{5} + \bar{\mathbf{5}}$. These transmit SUSY breaking to the SM via their gauge interactions, so that SM gaugino masses appear at one loop, and scalar masses at two loops, via diagrams as in Figure 2.7. Their orders of magnitude are

$$M_a \simeq \frac{\alpha_a}{4\pi} \frac{F_X}{\langle X \rangle}; \quad m_{\tilde{q}}^2 \simeq \left(\frac{F_X}{\langle X \rangle}\right)^2 \sum_a C_a \left(\frac{\alpha_a}{4\pi}\right)^2, \quad (2.112)$$

with C_a the quadratic Casimirs of SM gauge factors. Hence, soft terms of the order of the electroweak scale require $F_X/\langle X \rangle \simeq 100 \, \text{TeV}$. In this case the LSP is a very light gravitino, with $m_{3/2} \simeq 10 - 100 \, \text{eV}$, and supergravity contributions to soft terms are negligible. Typically the next-to-lightest SUSY particle is a neutralino or a charged slepton. The standard experimental collider signatures for this scheme are processes with missing energy plus photons. In certain models the NLSP is long-lived and may decay inside the detector.

A rather generic problem of the simplest GMSB models is the difficulty in getting appropriate B and μ parameters. In some models they may arise through certain one-loop diagrams, but are two orders of magnitude larger than the soft scalar masses, which arise at two-loops and are more suppressed; the resulting scalar potential does not lead to consistent EW symmetry breaking. There are, however, generalized versions of gauge mediation in which these phenomenological problems are improved. On the other hand, the great

Supersymmetry

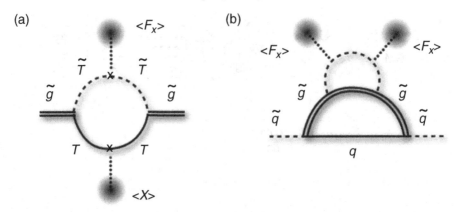

Figure 2.7 Diagrams generating SM soft terms in models of gauge mediation, with messengers denoted by T. (a) One-loop diagram producing gaugino masses; (b) two-loop diagram producing sfermion masses.

virtue of gauge mediation is that contributions to FCNC are very small, since gauge transmission is flavour-blind and renormalization effects are small due to the low microscopic SUSY breaking scale.

Anomaly mediation

In analogy with gravity mediation, this mechanism is based on SUSY breaking in supergravity, but assumes that the original action presents *sequestering*, i.e. it does not contain Planck mass suppressed couplings of hidden sector fields to the SM. The key observation is that despite sequestering, there always exists a class of one-loop soft terms appearing for all SM particles. They are due to a conformal anomaly, which arises because logarithmically divergent radiative corrections introduce a mass scale. The attractive aspect of anomaly mediation is that it is quite independent of the ultraviolet physics, and is determined by the one-loop properties of the low-energy effective theory, in particular the β-functions and anomalous dimensions γ_f of the SUSY SM. More explicitly, the soft terms have the structure

$$M_a = \frac{\beta_a}{g_a} m_{3/2},$$

$$m_f^2 = -\frac{1}{4} \left(\frac{\partial \gamma_f}{\partial g} \beta_g + \frac{\partial \gamma_f}{\partial Y} \beta_Y \right) m_{3/2}^2,$$

$$A_Y = -\frac{\beta_Y}{Y} m_{3/2}, \tag{2.113}$$

where Y denotes the corresponding Yukawa coupling. Note that soft terms are of order $(\alpha/4\pi) m_{3/2}$, and one finds that gaugino masses are not universal but rather satisfy ratios $M_1 : M_2 : M_3 = 2.8 : 1 : 7.1$ at the EW scale. This implies that the LSP is typically the neutral W gaugino (wino) and there is a relatively light chargino.

Besides its insensitivity to the ultraviolet, another appealing feature of anomaly mediation is its generic pattern of degenerate squark masses, thus suppressing unwanted FCNCs. On the other hand, although anomaly mediation contributions are always present, it is difficult to construct explicit supergravity/string scenarios in which they are dominant; it is indeed hard to construct models where all singlet chiral multiplets (e.g. the moduli in string models) are sufficiently decoupled from the observable sector. Finally, a phenomenological problem of anomaly mediation is that (2.113) leads to negative slepton squared masses, since their relevant β-functions are positive. This renders the simplest anomaly mediation models not viable. On the other hand, this problem may be overcome in simple extensions including additional (but more model-dependent) contributions, like D-terms from extra $U(1)$ symmetries, or contributions from moduli in string compactifications.

3

Introduction to string theory: the bosonic string

In the first two chapters we outlined the general structure of the SM and of several extensions envisaged to solve or understand some of its puzzles. It is clear that the SM or any of the extensions there described are at best just the low-energy effective description of some more fundamental theory. The SM itself contains interactions which are not asymptotically free and eventually lead to ultraviolet Landau poles, and so do its GUT extensions, whose scalar sector is not asymptotically free. These issues are not resolved in the previously mentioned extensions of the SM. In particular, models with extra dimensions even worsen the ultraviolet behavior of the theories, and the partial taming of the ultraviolet by SUSY bites back when its local version drives us into (super)gravity and its non-renormalizability. Finally, the explanation of certain SM properties like the family replication and its flavour physics, as well as other SM puzzles, seem to lie at a significantly more fundamental level.

Independently of the SM issues, there is the question of the quantum consistency of gravity. Einstein's gravity considered as a quantum field theory is not renormalizable, and should be regarded as an effective theory to be completed in the ultraviolet. String theory is arguably our best candidate to provide such completion and define a consistent quantum theory of gravity. String theory indeed provides an extension of Einstein's gravity, free of quantum divergences. It is also leading to important progress in other aspects of gravity, like accounting for the microscopic degrees of freedom of certain black holes, and explicitly realizing holography in terms of the AdS/CFT correspondence.

Most remarkably, string theory is not a purely gravitational theory, but contains the basic building blocks of the SM, naturally including non-abelian gauge interactions, charged chiral fermions in replicated families, fundamental scalars, Yukawa couplings, etc. String theories thus potentially constitute unified theories of all interactions including the SM and gravitation. It is a main purpose of this book to describe string theory models with low-energy particle content and effective action close to the SM of particle physics.

In this and the following three chapters we present an introduction to the basics of string theory. We start with a review of some basic features of perturbative string theory. Many general features are most simply exemplified in the context of the bosonic string theory, making it a good warm-up for the more interesting case of superstring theories in coming chapters. Our approach is down-to-earth and aimed to the computation of the massless string spectra; we use the light-cone quantization, and avoid the machinery of conformal field theory.

3.1 Generalities

We start with an overview of some general properties of string theory. Several points are mentioned only briefly for completeness, and will be developed for the bosonic string theory in later sections.

3.1.1 What are strings?

String theory proposes that elementary particles are *not* point-like, rather they are small one-dimensional extended objects, *strings*, of typical size $L_s = 1/M_s$, with M_s known as the string scale. They can be open or closed, as shown in Figure 3.1. At energies well below M_s, there is not enough resolution to perceive their spatial extent, and strings behave as point particles, with dynamics described by an effective theory of point particles, i.e. a quantum field theory. The string scale is very large compared with any experimentally probed energy scale, and in many string theory models is close to the 4d Planck scale. Certain string models, however, allow for lower values of the string scale, even down to the TeV range.

Strings can vibrate, see Figure 3.2, and different string oscillation states have different properties (masses, Lorentz and gauge quantum numbers, etc) and behave as different particles. The mass of the particle corresponding to a string oscillation state increases with the number of excited oscillators. Hence, the string vibration modes produce an infinite tower of particles, with masses increasing in steps of order M_s. At energies $E \ll M_s$, the only observable sector is the spectrum of massless particles (compared with the large scale M_s).

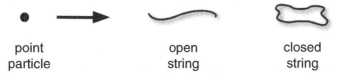

point
particle

open
string

closed
string

Figure 3.1 According to string theory, elementary particles are strings, which can be open or closed.

closed

open

Figure 3.2 Different oscillation modes of the same type of string correspond to different kinds of particles.

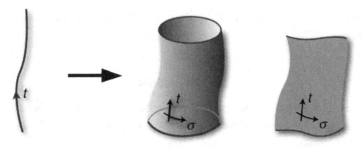

Figure 3.3 Worldsheets for closed and open strings. They reduce to worldlines in the point particle (low-energy) limit.

In all string theories, the massless sector of closed strings always contains a particle transforming under the Lorentz group as a 2-index traceless symmetric tensor G_{MN}, as shown in Section 3.2.1 for the bosonic string theory, and in Chapter 4 for the superstrings. Moreover, the string interaction rules (sketched below) can be used to show that this field interacts as a graviton, with dynamics invariant under general reparametrizations of the spacetime coordinates. Thus string theory automatically incorporates gravitational interactions, and contains a sector reducing to Einstein's general relativity. The remarkable fact that string theory necessarily implies gravity is one of the main successes of the theory.

3.1.2 The worldsheet

As a string evolves in time, it sweeps out a two-dimensional surface Σ in spacetime, known as the worldsheet, which is the analog of the spacetime worldline of a point particle. Closed strings correspond to worldsheets with no boundary, while open strings to worldsheets with boundaries. Any point in the worldsheet is labeled by two coordinates (t, σ), with t a "time" coordinate analogous to that in point particle worldlines, and with σ parametrizing the extended spatial dimension of the string at fixed t, see Figure 3.3.

A classical string configuration in D-dimensional Minkowski space $\mathbf{M}_D$ is given by a set of functions $X^M(t, \sigma)$ with $M = 0, \ldots, D-1$, which specify the spacetime position of the worldsheet point (t, σ). More formally, see Figure 3.4, the functions $X^M(t, \sigma)$ provide an embedding of a two-dimensional surface Σ (the abstract worldsheet) into D-dimensional spacetime $\mathbf{M}_D$ (called target space, in this view).

The string dynamics is defined by an action $S[X(t, \sigma)]$, a functional of the embeddings defining classical configurations. Although string actions are discussed in Section 3.2.1, it is useful to bring about some of their basic ingredients for the bosonic string theory. A natural proposal for the classical string action is the total area spanned by the worldsheet (in analogy with the point particle action given by the worldline interval)

$$S_{\text{NG}} = -\frac{1}{2\pi\alpha'} \int_{\Sigma} dA, \qquad (3.1)$$

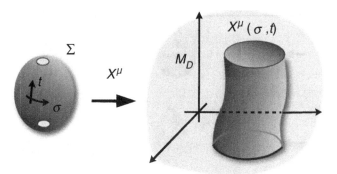

Figure 3.4 The functions $X^M(t, \sigma)$ define an embedding of a two-dimensional surface Σ into D-dimensional spacetime.

where $1/(2\pi\alpha') \simeq M_s^2$ is the string tension i.e. its energy per unit length. It can be expressed in terms of $X^M(t, \sigma)$ as

$$S_{\text{NG}} = -\frac{1}{2\pi\alpha'} \int_\Sigma \sqrt{-\det h} \, d\sigma \, dt, \tag{3.2}$$

where we have used the 2d worldsheet metric induced from the spacetime metric

$$h_{tt} = \partial_t X^M \, \partial_t X_M, \quad h_{\sigma\sigma} = \partial_\sigma X^M \, \partial_\sigma X_M, \quad h_{t\sigma} = \partial_t X^M \, \partial_\sigma X_M. \tag{3.3}$$

This is the so-called Nambu–Goto action, whose quantization is difficult due to the square root. Quantization of strings is actually carried out in terms of an alternative (but classically equivalent) action, known as the Polyakov action, which reads

$$S_{\text{P}} = -\frac{1}{4\pi\alpha'} \int_\Sigma d^2\xi \sqrt{-\det g} \, g^{ab}(t, \sigma) \, \partial_a X^M \, \partial_b X^N \eta_{MN}, \tag{3.4}$$

where $a, b = 0, 1$, and $\xi^0 = t$, $\xi^1 = \sigma$ are the worldsheet coordinates. In this description there is an additional function $g_{ab}(t, \sigma)$, which cannot be interpreted as an embedding. It rather should be regarded as a metric in the abstract worldsheet Σ. This abstract worldsheet viewpoint is useful in providing new physical intuitions on the system. For instance, from this perspective the action (3.4) describes a two-dimensional (2d) field theory coupled to 2d gravity. Many of the remarkable properties of string theory emerge from the subtle relation between the physics in this 2d world and physics in spacetime. As we will see in Chapter 4, different string theories are defined by different worldsheet structures (e.g. 2d field contents).

The action (3.4) has a number of local symmetries, i.e. gauge redundancies which should be removed by gauge fixing. For instance, the 2d metric is actually trivial, as there are no physical graviton polarization modes in two dimensions. Also, the only physical oscillation modes of the string are those transverse to the two directions of the worldsheet,

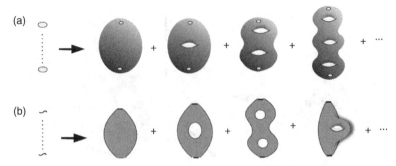

Figure 3.5 (a) The perturbative expansion for theories of closed strings. Higher orders are obtained by adding handles (closed string loops) to the worldsheet. (b) The perturbative expansion for theories of open and closed strings. Higher orders are obtained by adding boundaries (open string loops) and/or handles (closed string loops) to the worldsheet.

since oscillations along it can be removed by worldsheet coordinate reparametrizations. The dynamics thus reduces to a 2d quantum field theory of $D - 2$ free massless scalar fields on the worldsheet. Quantization of this system leads to an infinite set of decoupled harmonic oscillators, which correspond to the oscillation modes of the string. The Hilbert space of 2d quantum oscillation states corresponds to the spectrum of spacetime particles in the string theory. This will be computed for the closed bosonic string in Section 3.2.1 and for the open bosonic string in Section 3.3.1. For the moment, suffice it to quote that the massless closed string sector contains a spacetime field $G_{MN}(x)$ transforming as a two-index symmetric tensor of the spacetime Lorentz group (describing a graviton), a two-index antisymmetric tensor field $B_{MN}(x)$, and a real scalar $\phi(x)$, known as the dilaton. The massless sector of open strings contains a gauge boson $A_M(x)$.

3.1.3 String interactions

Although the 2d *worldsheet* theory is non-interacting, there are non-trivial interactions in the *spacetime* theory, which arise from the non-trivial worldsheet topologies, as we now describe. In perturbative string theory, scattering amplitudes between asymptotic states corresponding to different particles are computed as a quantum path integral, namely summing over all possible worldsheet geometries (2d surfaces) which interpolate between the string states in the external legs. In particular, see Figure 3.5, there is a sum over different worldsheet topologies, with increasing number of handles (and boundaries for theories with open strings). This is analogous to (and in fact at low energies reduces to) the sum over different Feynman diagrams in theories of point particles i.e. quantum field theory. Formally, the scattering amplitude has the structure

$$\langle \text{out}|\text{evolution}\,|\text{in}\rangle = \sum_{\text{worldsheets}} \int [\mathcal{D}X]\, e^{-S_{\text{P}}[X]} \mathcal{O}_{\text{in}}[X] \mathcal{O}_{\text{out}}[X], \tag{3.5}$$

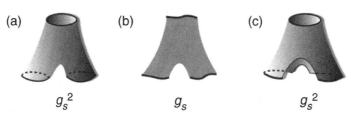

(a) g_s^2 (b) g_s (c) g_s^2

Figure 3.6 Basic interaction vertices and their strengths, joining three closed strings (a), three open strings (b), or two open strings into a closed string (c). The latter implies that any theory with open strings must contain also closed strings.

where the $\mathcal{O}[X]$ denote insertions of the so-called vertex operators, which implement the information about the "in" and "out" string states. They are further discussed in Appendix E, although the details are not essential in our present heuristic discussion.

Each worldsheet topology can be built by glueing a set of basic string interaction vertices, shown in Figure 3.6. The vertex in Figure 3.6(b) describes two open strings joining into one, and is weighted by a quantity g_s, the string coupling constant. A general property of string theory is that it does not have arbitrary external parameters. Indeed, as shown later in Section 3.1.5, the string coupling constant is not an external parameter in the theory, but rather is determined by the vev of the spacetime dilaton field $\phi(x)$, by

$$g_s = e^{\phi}. \tag{3.6}$$

The vertex in Figure 3.6(a) describes the interaction of three closed strings and is weighted by g_s^2. It is easy to show that worldsheets with g handles and n_b boundaries are weighted by $g_s^{-\chi}$, where χ is the genus or Euler characteristic, given by

$$\chi = 2 - 2g - n_b. \tag{3.7}$$

The sum over topologies, known as genus expansion, is a loop expansion with g_s as the loop counting parameter.

The above rules lead to an important consequence, related to the vertex in Figure 3.6(c); it describes two open strings combining into a closed string, by applying twice the interaction joining open string endpoints. This coupling exists in any theory of interacting open strings, so interacting theories of open strings must necessarily contain closed strings. On the other hand, it is consistent to have theories of just closed strings, with no open strings. The end result is that consistent string theories contain either closed strings only (i.e. worldsheets with handles, but no boundaries), or open and closed strings (i.e. worldsheets with handles and boundaries).

3.1.4 UV finiteness and critical dimension

A fundamental property of string theory is that the scattering amplitudes of the theory are unitary, and moreover finite, order by order in perturbation theory. It defines consistent

Figure 3.7 (a) A one-loop 4-point scattering amplitude in string theory, with external legs depicted as dashed lines. (b) As two vertices come close together, there is exchange of higher energy states, corresponding to longer strings. (c) The extreme UV regime involves exchange of very long strings for a very short distance, which can be reinterpreted as the IR of the dual channel, with very short strings across very long distances.

Figure 3.8 Conformal transformations rescale the small worldsheet region denoted with a dashed line and map a seemingly UV regime of coincident interaction vertices into an IR regime in the dual channel.

quantum theories, which are moreover free of the ultraviolet (UV) divergences of quantum field theory. String theory provides a regularization of quantum field theory, with an effective cutoff M_s, above which the amplitudes soften rather than diverge. This can be understood heuristically as follows. In quantum field theory, UV divergences occur when two interaction vertices coincide in spacetime. In string theory, vertices are delocalized in a region of size $L_s \simeq 1/M_s$, which acts as an effective position space cutoff for the amplitude.

A more precise description shows that in string theory the extended nature of strings turns UV regimes into infrared (IR) regimes, as illustrated in Figure 3.7. Consider an amplitude with large momentum/energy transfer in some internal leg. As this energy increases above M_s, it becomes possible to exchange states corresponding to long strings. The very high energy regime is dominated by exchange of very long strings for a very short distance, and leads to a worldsheet which, rather than a UV divergence, describes the IR limit of a new diagram involving a "dual" channel. A similar phenomenon occurs for amplitudes involving open strings. This mechanism underlies the properties of modular invariance and open–closed duality, see Sections 3.2.2 and 3.3.2 for these properties in bosonic string theories. This qualitative argument can be formalized in terms of "conformal invariance." As described in Section 3.2.1, this is a local symmetry of the worldsheet theory, which includes local rescalings expanding any small disk region in the worldsheet to a long capped tube. It thus relates a worldsheet with several almost coincident vertex operators (external leg insertions), to a worldsheet with finitely separated insertions at the end of a long capped tube (the dual channel), see Figure 3.8.

The UV behavior of string theory amplitudes depends on the tower of massive string oscillation modes, which effectively act as UV regulators. Their multiplicity and properties depend on the number of spacetime dimensions in which the string can oscillate. The condition that the UV divergences are appropriately cutoff actually constrains the number of spacetime dimensions D to a critical value, called *critical dimension*. The critical dimension can be understood in terms of the so-called conformal anomaly. The conformal invariance underlying the absence of UV divergences in string theory is potentially anomalous at the quantum level, so its anomaly must be canceled by choosing suitably the field content of the 2d worldsheet theory. Since the field content for, e.g., the bosonic theory is the set of embedding functions $X^M(t, \sigma)$, the only freedom available is their number D, which actually corresponds to the spacetime dimension.

In our light-cone quantization procedure in Section 3.2.1, conformal symmetry is eliminated by a gauge-fixing condition violating Lorentz invariance. The critical dimension will arise as a requirement to restore Lorentz invariance of physical quantities in the resulting quantum theory. There are thus different derivations of the critical dimension from requiring the quantum realization of different classical symmetries. This reflects the familiar property that in theories with several classical symmetries, the quantum anomaly can be shifted from a symmetry to another by using local counterterms. We will see that the critical dimension is $D = 26$ for bosonic string theories and $D = 10$ for the superstrings.

Quantitative computation of scattering amplitudes in string theory shows that the interactions of the massless spacetime fields is invariant under spacetime gauge symmetries – in upcoming computations of the spacetime spectrum in the light-cone gauge, these gauge invariances manifest as a reduced number of physical polarization states for the massless fields. In particular, the theory is invariant under general changes of coordinates, and the 2-index tensor G_{MN} behaves as a graviton. For theories including massless bosons A_M (like those including open strings), interactions of the latter are invariant under standard gauge transformations. Hence, string theory provides a unified description of gravitational and gauge interactions, consistent at the quantum level. The subject of *string phenomenology* or *string model building* is the development of tools to incorporate realistic gauge interactions and matter particles, and the construction of explicit particle physics models in string theory.

3.1.5 String theory in background fields

Given the interpretation of G_{MN} as the spacetime metric, it is natural to consider the dynamics of string theory in a curved spacetime, i.e. a general background $G_{MN}(X)$. This is described by a modified worldsheet action, given by a natural generalization of (3.4)

$$S_P^G = -\frac{1}{4\pi\alpha'} \int_\Sigma d\sigma \, dt \, \sqrt{-g} \, G_{MN}[X(t,\sigma)] \, g^{ab} \, \partial_a X^M(t,\sigma) \, \partial_b X^N(t,\sigma), \quad (3.8)$$

where $g = \det(g_{ab})$, and $G_{MN}[X]$ is a function(al) of $X(t, \sigma)$. At the classical level, this action reduces to a Nambu–Goto action providing the worldsheet area measured with the curved spacetime metric. The action (3.8) is known as the 2d non-linear sigma model, in analogy with the pion effective lagrangian in strong interactions, because it also involves scalar fields parametrizing a non-trivial curved space.

One may worry about the double role played by the spacetime graviton in string theory, as an oscillation mode of the underlying string, and as the quantum of the background spacetime metric field. In fact, the background metric can be regarded as a coherent superposition of a large number of gravitons (just like a macroscopic electromagnetic field can be regarded as a coherent superposition of photons); the worldsheet action of a string in a background metric field can be regarded as an efficient way of resumming the interaction of a string with all the graviton string states present in the background.

There are also other massless fields in the string spectrum, in particular the two-index antisymmetric tensor field B_{MN} and the dilaton ϕ in the closed string sector. It is natural to consider the string action coupled to general backgrounds for these fields. This is implemented by the additional terms in the action

$$S_{B,\phi} = \frac{1}{4\pi\alpha'} \int_{\Sigma} d\sigma \, dt \left[\epsilon^{ab} B_{MN} \, \partial_a X^M \, \partial_b X^N + \sqrt{-g} \, \alpha' R[g]\phi \right], \tag{3.9}$$

where $R[g]$ is the curvature scalar of the 2d metric $g_{ab}(t,\sigma)$.

These terms have a nice spacetime interpretation. Recasting B_{MN} as a differential form in spacetime $B_2 = \frac{1}{2} B_{MN} dX^M \wedge dX^N$ (see Appendix B), the corresponding term in (3.9) can be expressed as

$$S_B = \frac{1}{2\pi\alpha'} \int_{\Sigma} B_2. \tag{3.10}$$

This term is independent of the 2d metric, i.e. it is purely topological. It is a generalization, for a 2d worldsheet, of the minimal coupling $\oint_C A_1$ of a charged particle worldline C to an electromagnetic gauge potential 1-form $A_1 = A_M dx^M$. In the language of Section B.3, it means that strings are electrically charged under the field B_2, which behaves as a generalized gauge potential. The interaction (3.10) is indeed invariant under gauge transformations by 1-forms Λ_1

$$B_2 \rightarrow B_2 + d\Lambda_1, \tag{3.11}$$

as is easily shown using Stokes theorem (B.16).

The term involving ϕ is very special, as for constant ϕ it corresponds to the Einstein–Hilbert term for the 2d metric g_{ab}. However, 2d gravity is non-dynamical, and the integrated curvature scalar is actually a topological invariant, in fact the Euler characteristic (3.7),

$$S_\phi = \frac{1}{4\pi} \int_{\Sigma} R[g] = \chi = 2 - 2g - n_b. \tag{3.12}$$

(For theories with open strings, the contribution from the n_b worldsheet boundaries arises from boundary terms added to define the integral properly.) The coupling of the string to a constant background of ϕ thus weights each worldsheet diagram by a factor $\exp(-\phi\chi)$ in the euclidean path integral, justifying the relation (3.6) of the dilaton vev and the string coupling constant g_s. We again emphasize that the absence of external parameters is a general feature of string theory, and that all adimensional parameters are given by vevs of spacetime scalar fields (known as moduli).

Finally, for theories with open strings, the coupling of the string worldsheet to a background of the massless gauge boson A_M, is

$$S_{\partial \Sigma} = \int_{\partial \Sigma} d\xi^a \, A_M \, \partial_a X^M = \int_{\partial \Sigma} A_1, \qquad (3.13)$$

where $\partial \Sigma$ is the boundary of Σ, with length element $d\xi^a$. The last equality simply recasts the integral in the language of differential forms. For a freely propagating open string, spanning an infinite strip with boundaries at $\sigma = 0, \ell$, it reads

$$S_{\partial \Sigma} = \int dt \, A_M \, \partial_t X^M \big|_{\sigma=0} - \int dt \, A_M \, \partial_t X^M \big|_{\sigma=\ell}, \qquad (3.14)$$

with the relative sign arising from the opposite orientations of both boundaries. These couplings imply that the open string endpoints $\sigma = 0, \ell$ behave as charged particles, with charges ± 1 under the gauge potential. Incidentally, in a theory with open strings, the B_2 gauge transformation (3.11) requires a transformation of the gauge field strength, whose description we postpone to Section 6.1.2, see (6.11).

In the presence of non-trivial backgrounds, the worldsheet action describes an interacting 2d field theory. This is usually not exactly solvable, but may be studied perturbatively around the free theory. This is known as the α' expansion, because the expansion parameter is α'/R^2, with R a typical length scale of variation of any spacetime field, e.g. the curvature radius. Thus string theory in a general background has a double expansion, namely the genus expansion (i.e. spacetime loop expansion) controlled by g_s, and for each fixed worldsheet topology, the worldsheet loop expansion controlled by α'/R^2. The computation of g_s and α' corrections is in general involved, most often forcing the study of string theories to regimes where such corrections are negligible. The small α'/R^2 regime corresponds to a large radius or long distance regime, hence a low-energy limit rather insensitive to the extended nature of strings. Happily, the 2d worldsheet theory is free (or interacting but solvable) in some interesting classes of models, for which α' effects are computable and stringy effects manifest in spectacular ways, such as shown in Section 3.2.3. Concerning the small g_s approximation, in certain cases it is possible to go beyond the perturbative regime by using the dualities in Section 6.3.

3.1.6 Compactification

String theories have a good high-energy behavior only if spacetime has the critical dimension, $D = 26$ for bosonic strings or $D = 10$ for superstrings. In order to construct string theory models consistent with the observed four spacetime dimensions, we may exploit the mechanism of compactification introduced in Section 1.4.1, leading to string compactifications. Namely, the theory is defined on a spacetime $\mathbf{M}_4 \times \mathbf{X}_n$, where $\mathbf{X}_n$ is an n-dimensional compact manifold, with $n = 22, 6$ for bosonic strings and superstrings respectively. Denoting L the typical size of $\mathbf{X}_n$, the physics at energies $E \ll 1/L$ reduces to an effectively 4d theory.

String compactifications are described as in the previous section, regarding the compactification spacetime as a background for string propagation. Therefore, except for cases of solvable worldsheet theories, the system must be analyzed in the large volume approximation $\alpha'/L^2 \ll 1$, namely $L \gg 1/M_s$. In this description, the effective 4d physics regime $E \ll 1/L$ implies the point-particle regime $E \ll M_s$, and most features of the compactified theory and its low-energy physics can be described using the field theory KK compactification results in Section 1.4.1 (suitably generalized to higher-dimensional spaces).

Note that there are two kinds of towers of resonances in string compactifications. There is the tower of string resonances, corresponding to different oscillation modes, whose mass spectrum has spacing $\sim M_s$. In addition, there are towers of KK momentum resonances of the different higher-dimensional fields, with typical mass spacing of order $1/L$. In the large volume regime $1/L \ll M_s$ stringy excitations are much less relevant than KK resonances, justifying the field theory approximation.

3.2 Closed bosonic string

We leave the discussion of generalities of string theory, and move on to illustrate them quantitatively using the bosonic string theory. Although it cannot accommodate realistic physics (e.g. does not lead to spacetime fermions) and presents theoretical puzzles (e.g. its vacuum instability, signaled by a spacetime tachyon in its spectrum), it is a very useful playground for the purpose of fleshing out some of the above properties. Furthermore, the bosonic string theory plays an important role in the construction of heterotic superstrings, see Section 4.3.

3.2.1 Quantization and spectrum of the closed bosonic string

In this section we quantize the closed bosonic string and compute its spectrum of spacetime particles. The reader not interested in details is advised to go straight to the final recap in page 80.

Worldsheet action, symmetries and conformal invariance

The closed bosonic string theory corresponds to a genus expansion in terms of worldsheets with arbitrary number of handles, but without boundaries. The worldsheet Σ of a closed string propagating freely in spacetime is an infinite cylinder, which we parametrize by the two coordinates t, σ, collectively denoted ξ^a, $a = 0, 1$. The closed bosonic string propagating in flat D-dimensional Minkowski space is described by a set of worldsheet fields $X^M(t, \sigma)$, with $M = 0, \dots, D-1$, describing its embedding into spacetime. In the

Polyakov formulation there is also an additional 2d metric $g_{ab}(t, \sigma)$. The action is (3.4), namely

$$S_{\mathrm{P}}[X, g] = -\frac{1}{4\pi\alpha'} \int_{\Sigma} d^2\xi \, (-g)^{1/2} \, g^{ab} \, \partial_a X^M \, \partial_b X^N \eta_{MN}, \tag{3.15}$$

with $g = \det(g_{ab})$. It is easily shown to be classically equivalent to the Nambu–Goto action (3.2) as follows. Using $\delta g = -g \, g_{ab} \, \delta g^{ab}$, the classical equations of motion $\delta S / \delta g_{ab} = 0$ read

$$-\frac{1}{2} g_{ab} \, g^{cd} \partial_c X^M \, \partial_d X_M + \partial_a X^M \, \partial_b X_M = 0. \tag{3.16}$$

We define the metric induced on Σ from the ambient spacetime metric η_{MN}, by

$$h_{ab} = \eta_{MN} \, \partial_a X^M \, \partial_b X^N. \tag{3.17}$$

The equations of motion (3.16) give

$$h_{ab} = \frac{1}{2} g_{ab} \, g^{cd} \, h_{cd} \quad \Rightarrow \quad (-h)^{1/2} = \frac{1}{2} (-g)^{1/2} \, g^{cd} \, h_{cd}, \tag{3.18}$$

where the second equation follows from taking determinants in the first. Replacing it into (3.15) leads to the Nambu–Goto action (3.2)

$$S_{\mathrm{P}}[X(\xi), g_{\mathrm{clas}}(\xi)] = -\frac{1}{2\pi\alpha'} \int_{\Sigma} d^2\xi \, (-h)^{1/2} = S_{\mathrm{NG}}[X(\xi)]. \tag{3.19}$$

The Polyakov action allows a nice quantization of the 2d worldsheet theory, for which it is important to keep track of the following global and local symmetries:

- D-dimensional spacetime Poincaré invariance is a global symmetry of the worldsheet theory; it leaves the 2d coordinates ξ^a invariant, and acts on the 2d fields as

$$X'^M(\xi) = \Lambda^M_N X^N(\xi) + a^M,$$
$$g'_{ab}(\xi) = g_{ab}(\xi). \tag{3.20}$$

- Invariance under 2d local worldsheet coordinate reparametrizations, for which the fields $X^M(\xi)$ behave as scalars, while $g_{ab}(\xi)$ is a 2-index tensor,

$$\xi'^a = \xi'^a(\xi),$$
$$X'^M(\xi') = X^M(\xi),$$
$$g'_{ab}(\xi') = \frac{\partial \xi^c}{\partial \xi'^a} \frac{\partial \xi^d}{\partial \xi'^b} g_{cd}(\xi). \tag{3.21}$$

- Weyl invariance, i.e. invariance under local rescalings of the 2d metric

$$X'^M(\xi) = X^M(\xi),$$
$$g'_{ab}(\xi) = \Omega(\xi) \, g_{ab}(\xi). \tag{3.22}$$

In order to quantize the system, we need to gauge-fix the local invariances. The reparametrization of the two coordinates ξ^a and Weyl invariance can be used to remove the degrees of freedom of g_{ab}, by setting the so-called conformal gauge

$$g_{ab} = \eta_{ab}. \tag{3.23}$$

This is always possible locally on the worldsheet, and even globally for our infinite cylinder worldsheet topology. We thus have a flat system of orthogonal coordinates t, σ. We choose a reference line, running in the t direction, to correspond to $\sigma = 0$, and denote ℓ the total length of the string in the σ direction. The choice of reference line is arbitrary, and eventually we will need to impose invariance of the theory under its translation in σ, see later around (3.46).

The Polyakov action in this gauge reads

$$S_{\mathrm{P}} = -\frac{1}{4\pi\alpha'} \int_{\Sigma} d^2\xi \, \eta^{ab} \, \partial_a X^M \, \partial_b X^N \eta_{MN}. \tag{3.24}$$

The equations of motion are 2d wave equations

$$\left(\partial_t^2 - \partial_\sigma^2 \right) X^M = 0, \tag{3.25}$$

whose general solution is a superposition of left- and right-moving traveling waves

$$X^M(t, \sigma) = X_L^M(t + \sigma) + X_R^M(t - \sigma). \tag{3.26}$$

Even though the metric g_{ab} is removed, the equations of motion (3.16) must still be imposed as constraints in the remaining system. In components they read

$$\partial_t X^M \partial_t X_M + \partial_\sigma X^M \partial_\sigma X_M = 0, \quad \partial_t X^M \partial_\sigma X_M = 0. \tag{3.27}$$

Defining the left- and right-moving worldsheet coordinates $\xi_{L,R} = t \pm \sigma$; they can be rephrased as the so-called Virasoro constraints

$$\partial_{\xi_L} X^M \partial_{\xi_L} X_M = 0, \quad \partial_{\xi_R} X^M \partial_{\xi_R} X_M = 0. \tag{3.28}$$

The condition (3.23) still leaves a set of left-over local symmetries. Indeed, the metric (3.23) is $ds^2 = -d\xi_L d\xi_R$, which is unchanged under the combination of the change of coordinates

$$\xi_L \rightarrow f_L(\xi_L), \quad \xi_R \rightarrow f_R(\xi_R), \tag{3.29}$$

and a suitable Weyl rescaling. Reparametrizations (3.29) preserve the conformal structure of the 2d metric (i.e. the angles), and are known as conformal symmetries.

Theories invariant under this symmetry, like (3.24), are called conformal field theories (CFTs), and play a prominent role in worldsheet physics and string theory. The role of conformal symmetry in the UV properties of string theory, sketched in Section 3.1.4, can be understood from the geometric interpretation of conformal symmetry in the euclidean 2d theory. Defining $\tau = -it$, the worldsheet coordinates $\xi_{L,R} = i\tau \pm \sigma$ can be regarded as complex coordinates $z, \bar{z}$ on the worldsheet. In fact, they can be promoted to independent variables, for which the conformal transformations (3.29) read $z \to f(z)$, $\bar{z} \to \bar{f}(\bar{z})$, namely are holomorphic reparametrizations of the worldsheet. Mathematically, 2d surfaces related by holomorphic reparametrizations are said to have the same complex structure, and surfaces defined up to these transformations are known as Riemann surfaces. As mentioned in Section 3.1.3, conformal transformations can be regarded as including local rescalings on the worldsheet, e.g. the transformation $z' = e^{-iz}$ maps a small disk $0 < |z'| < \epsilon$ to a semi-infinite tube with Im $z \in (-\infty, \log \epsilon)$ and periodic Re z.

Light-cone quantization

Different quantization procedures are possible at this point. We proceed in light-cone quantization, which uses a simple gauge-fixing of the conformal symmetries, allowing to solve the Virasoro constraints explicitly, but at the price of giving up manifest Lorentz invariance. Other quantization procedures, like covariant BRST quantization, preserve manifest Lorentz invariance, but involve technical tools beyond the scope of this book.

We define the spacetime light-cone coordinates

$$X^{\pm} = \frac{1}{\sqrt{2}} (X^0 \pm X^1), \tag{3.30}$$

and use the index $i = 2, \ldots, D - 1$ for the remaining coordinates. The D-dimensional Minkowski spacetime metric is $\eta_{+-} = \eta_{-+} = -1$, $\eta_{ij} = \delta_{ij}$, so for a vector V^M we have $V_- = -V^+$, $V_+ = -V^-$, and $V_i = V^i$.

The solution (3.26) to the wave equation (3.25) for the light-cone coordinate X^+ reads $X^+(\xi) = X_L^+(\xi_L) + X_R^+(\xi_R)$. Since $t = (\xi_L + \xi_R)/2$, it transforms under (3.29) as $t \to [f_L(\xi_L) + f_R(\xi_R)]/2$, and we may use $f_{L,R} = 2X_{L,R}^+$ to set

$$X^+(t, \sigma) = t. \tag{3.31}$$

This gauge fixes the conformal symmetries, and breaks manifest Lorentz invariance. The Virasoro constraints (3.28) become linear in X^-

$$\partial_{\xi_L} X_L^- = \frac{1}{2} \left(\partial_{\xi_L} X_L^i \right)^2, \quad \partial_{\xi_R} X_R^- = \frac{1}{2} \left(\partial_{\xi_R} X_R^i \right)^2, \tag{3.32}$$

and can be solved by determining X^- in terms of X^i; the only degree of freedom of X^- remaining in (3.24) is the center of mass term,

$$x^-(t) = \frac{1}{\ell} \int_0^{\ell} d\sigma \, X^-(t, \sigma). \tag{3.33}$$

The lagrangian from (3.24) reads

$$\begin{aligned} L &= -\frac{1}{4\pi\alpha'} \int_0^{\ell} d\sigma \left(2 \, \partial_t X^- - \partial_t X^i \, \partial_t X^i + \partial_\sigma X^i \, \partial_\sigma X^i \right) \\ &= -\frac{\ell}{2\pi\alpha'} \partial_t x^-(t) - \frac{1}{4\pi\alpha'} \int_0^{\ell} d\sigma \left(-\partial_t X^i \, \partial_t X^i + \partial_\sigma X^i \, \partial_\sigma X^i \right). \end{aligned} \tag{3.34}$$

Actually $x^-(t)$ evolves linearly, with constant momentum

$$p_- = -p^+ = \frac{\partial L}{\partial(\partial_t x^-)} = -\frac{\ell}{2\pi\alpha'}; \tag{3.35}$$

hence the length of the string in light-cone gauge grows with the p^+ momentum

$$\ell = 2\pi\alpha' p^+. \tag{3.36}$$

The momenta conjugate to $X^i(t, \sigma)$ are

$$\Pi^i(t, \sigma) = \frac{\partial \mathcal{L}}{\partial(\partial_t X^i)} = \frac{1}{2\pi\alpha'} \partial_t X^i(t, \sigma), \tag{3.37}$$

and we obtain a hamiltonian describing $D - 2$ free bosons in two dimensions:

$$\begin{aligned} H &= p_- \partial_t x^-(t) + \int_0^\ell d\sigma\, \Pi_i(t, \sigma)\, \partial_t X^i(t, \sigma) - L \\ &= \frac{1}{4\pi\alpha'} \int_0^\ell d\sigma \left(\partial_t X^i\, \partial_t X^i + \partial_\sigma X^i\, \partial_\sigma X^i \right) \\ &= \frac{1}{2} \int_0^\ell d\sigma \left(2\pi\alpha'\, \Pi_i\, \Pi_i + \frac{1}{2\pi\alpha'} \partial_\sigma X^i\, \partial_\sigma X^i \right). \end{aligned} \tag{3.38}$$

Oscillator expansions and spectrum of states

It is convenient to expand the functions $X^i_{L,R}$ (3.26) in oscillation modes. For closed strings, the appropriate boundary conditions follow from periodicity in σ

$$X^i(t, \sigma + \ell) = X^i(t, \sigma). \tag{3.39}$$

The corresponding general form of X_L, X_R is the oscillator expansion

$$\begin{aligned} X^i_L(t + \sigma) &= \frac{x^i}{2} + \frac{p_i}{2p^+}(t + \sigma) + i\sqrt{\frac{\alpha'}{2}} \sum_{n \in \mathbf{Z}-\{0\}} \frac{\alpha^i_n}{n} e^{-2\pi i n(t+\sigma)/\ell}, \\ X^i_R(t - \sigma) &= \frac{x^i}{2} + \frac{p_i}{2p^+}(t - \sigma) + i\sqrt{\frac{\alpha'}{2}} \sum_{n \in \mathbf{Z}-\{0\}} \frac{\tilde{\alpha}^i_n}{n} e^{-2\pi i n(t-\sigma)/\ell}, \\ \to\ X^i(t, \sigma) &= x^i + \frac{p^i}{p^+} t + i\sqrt{\frac{\alpha'}{2}} \sum_{n \neq 0} \left[\frac{\alpha^i_n}{n} e^{-2\pi i n(t+\sigma)\ell} + \frac{\tilde{\alpha}^i_n}{n} e^{-2\pi i n(t-\sigma)\ell} \right]. \end{aligned} \tag{3.40}$$

The coefficients x^i, p_i define the center of mass coordinate and momentum, while the two infinite sets of coefficients α^i_n, $\tilde{\alpha}^i_n$ describe the amplitudes of the momentum n mode for left and right movers. Promoting the worldsheet degrees of freedom x^-, p^+, X^i, Π^i to operators, with canonical commutators, we obtain

$$[x^-, p^+] = -i, \quad [x^i, p_j] = i\delta^i_j,$$
$$[\alpha^i_m, \alpha^j_n] = [\tilde{\alpha}^i_m, \tilde{\alpha}^j_n] = m\,\delta^{ij}\,\delta_{m,-n}, \quad [\alpha^i_m, \tilde{\alpha}^j_n] = 0. \tag{3.41}$$

The hamiltonian (3.38) can be expressed in terms of the oscillators as

$$H = \sum_{i=2}^{D-1} \frac{p_i^2}{2p^+} + \frac{1}{\alpha' p^+}\left[\sum_i \sum_{n>0}\left(\alpha^i_{-n}\alpha^i_n + \tilde{\alpha}^i_{-n}\tilde{\alpha}^i_n\right) + E_0 + \tilde{E}_0\right]. \tag{3.42}$$

We get the quantum mechanics of the center of mass motion and two infinite sets of decoupled harmonic oscillators. Here we have normal-ordered the creation and annihilation modes, and E_0, $\tilde{E}_0$ are the zero point energies, to be discussed below.

The Hilbert space of string states is obtained by defining a groundstate, as a space-time momentum eigenstate annihilated by the positive modding oscillators. Ignoring the momentum label, we denote it by $|0\rangle$ and have

$$\alpha^i_n|0\rangle = \tilde{\alpha}^i_n|0\rangle = 0 \quad \forall n > 0. \tag{3.43}$$

Excited states are constructed by acting on $|0\rangle$ with creation operators α^i_{-n}, $\tilde{\alpha}^i_{-n}$, with $n > 0$. These operators can be applied in all possible ways, but there is an important additional constraint for physical states. Recall that our parametrization of the worldsheet had an implicit arbitrary choice of reference line $\sigma = 0$. The physical spectrum must be invariant under changes in this choice, i.e. under translations in σ. These are generated by the σ-momentum operator

$$P_\sigma = \int_0^\ell d\sigma\, \Pi_i\, \partial_\sigma X^i = \frac{2\pi}{\ell}\,(N - \tilde{N}), \tag{3.44}$$

where we have defined the oscillator number operators

$$N = \sum_i \sum_{n>0} \alpha^i_{-n}\alpha^i_n, \quad \tilde{N} = \sum_i \sum_{n>0} \tilde{\alpha}^i_{-n}\tilde{\alpha}^i_n. \tag{3.45}$$

Physical states must thus obey the so-called level matching constraint

$$N = \tilde{N}. \tag{3.46}$$

For future reference, we advance that the worldsheet theory is made up of two decoupled systems of left and right movers, whose quantum evolution is determined by independent hamiltonians. The only relation between the left and right sectors is in the level matching constraint satisfied by physical states.

Each string oscillation state corresponds to a particle from the spacetime viewpoint, whose mass is obtained from the mass shell condition

$$M^2 = -p^2 = 2p^+ p^- - p_i p_i. \tag{3.47}$$

Since $p^- = i \partial_{x^+} = i \partial_t = H$, we have $M^2 = 2p^+ H - p_i p_i$, and from (3.42) we get

$$\frac{\alpha' M^2}{2} = N + \tilde{N} + E_0 + \tilde{E}_0. \tag{3.48}$$

Hence the masses of spacetime particles increase with the number of oscillators in the corresponding string state. The mass formula also includes the zero point energies E_0, $\tilde{E}_0$. Noting in (3.41) that the modes α_n^i, $\tilde{\alpha}_n^i$ contribute n to the oscillator number, the zero point energy is naively

$$E_0 = \tilde{E}_0 = \sum_i \frac{1}{2} \sum_{n=1}^{\infty} n. \tag{3.49}$$

This is formally infinite, but its regularized value can be defined in the so-called zeta function regularization, as the limit $\epsilon \to 0$ of the non-singular part of

$$Z(\epsilon) = \frac{1}{2} \sum_{n=0}^{\infty} n e^{-n\epsilon} = -\frac{1}{2} \frac{d}{d\epsilon} \sum_{n=0}^{\infty} e^{-n\epsilon} = -\frac{1}{2} \frac{d}{d\epsilon} \left(\frac{1}{1 - e^{-\epsilon}} \right)$$
$$= -\frac{1}{2} \left(-\frac{1}{\epsilon^2} + \frac{1}{12} + \cdots \right). \tag{3.50}$$

This prescription produces the values

$$E_0 = \tilde{E}_0 = -\frac{D-2}{24}. \tag{3.51}$$

Alternatively, they could be obtained from the modular invariance in Section 3.2.2.

Light spectrum, critical dimension, and low-energy physics

The spectrum of light particles in the theory corresponds to the string states with smallest number of oscillators and satisfying (3.46). We have

$$N = \tilde{N} = 0 \qquad |k\rangle \qquad \alpha' M^2/2 = -(D-2)/12,$$
$$N = \tilde{N} = 1 \qquad \alpha_{-1}^i \tilde{\alpha}_{-1}^j |k\rangle \qquad \alpha' M^2/2 = 2 - (D-2)/12. \tag{3.52}$$

In the light-cone gauge there is manifest invariance only under an $SO(D-2)$ subgroup of the Lorentz group. It is possible to test the restoration of full Lorentz invariance, by constructing the Lorentz generators in terms of the oscillator operators, and checking their commutation relations. The computation shows that full Lorentz invariance is obtained only for a specific value of the spacetime dimension, the critical dimension $D = 26$. The appearance of the critical dimension from demanding full Lorentz invariance is consistent with its derivation from the conformal anomaly mentioned in Section 3.1.4 (see also

Section E.1); indeed, manifest Lorentz invariance is lost in gauge fixing the conformal symmetry with the light-cone condition (3.31).

It is actually possible to use a shortcut to recover the critical dimension, using only a *necessary* condition for restoration of full Lorentz invariance. Recall that in a D-dimensional Lorentz invariant theory, physical polarization states of particles belong to representations of the little group, i.e. the subgroup of Lorentz group preserving the particle D-momentum. For massive particles, the D-momentum can be brought to the form $P = (M, 0, \ldots, 0)$ in the particle rest frame, and the little group is $SO(D - 1)$. For massless particles, the D-momentum can be brought to the form $(E, E, 0, \ldots)$, and the little group is $SO(D - 2)$. The states in the second line in (3.52) transform as a two-index tensor with respect to $SO(D - 2)$ and cannot be arranged into a representation of $SO(D - 1)$. Hence these states must correspond to physical polarizations of *massless* particles, and this forces $D = 26$. Fixing this value from now on, the mass formula reads

$$\frac{\alpha' M^2}{2} = N + \tilde{N} - 2. \tag{3.53}$$

The light states, suitably decomposed in irreducible representations of $SO(24)$ are:

Sector	State	$\alpha'M^2$	$SO(24)$	26d field	
$N = \tilde{N} = 0$	$|k\rangle$	-4	1	T	(3.54)
$N = \tilde{N} = 1$	$\alpha^i_{-1}\alpha^j_{-1}|k\rangle$	0	$\square\!\square + \square\!\square + 1$	G_{MN}, B_{MN}, ϕ	

where the Young tableaux $\square$, $\square\!\square$ denote the two-index antisymmetric and traceless symmetric tensor representations, and the "$+1$" indicates the trace. It is possible to show that massive states appear in $SO(24)$ representations which can actually combine into representations of $SO(25)$, the little group for massive particles.

The string groundstate corresponds to a 26d spacetime tachyonic scalar field $T(x)$. This tachyon indicates that the theory just constructed is at a local maximum (rather than a minimum) of a putative spacetime scalar potential, and is therefore unstable. Despite much recent progress in understanding tachyons and their condensation in open string sectors, see Section 6.5.2, the fate of the theory upon closed string tachyon condensation is still unknown. Happily, these tachyonic instabilities are absent in the superstring theories, on which we focus in coming chapters.

The massless two-index tensor splits into irreducible representations of $SO(24)$. Its trace corresponds to a scalar field, the dilaton ϕ, whose vev fixes the string interaction coupling constant g_s by (3.6). The symmetric traceless part is the 26d graviton G_{MN}, while the antisymmetric part is the 26d 2-form field B_{MN}. As mentioned in Section 3.1.1 these fields have interactions invariant under spacetime changes of coordinates and gauge transformations (3.11), consistently with their reduced number of physical polarization modes.

For completeness, let us introduce the 26d effective theory for these massless fields (ignoring momentarily the tachyon) at energies below M_s. Its computation is beyond our scope, and we simply quote its structure

$$S_{26d}^{\text{s.f.}} = \frac{1}{2\kappa_0^2} \int d^{26}x \, (-G)^{1/2} e^{-2\phi} \left(R - \frac{1}{12} H_{MNP} H^{MNP} + 4\, \partial_M \phi \, \partial^M \phi \right) + \mathcal{O}(\alpha'),$$

where $G = \det(G_{MN})$, R is the 26d curvature scalar, and $H_{MNP} = \partial_{[M} B_{NP]}$, where the brackets indicate antisymmetrization of indices. The constant κ_0 is discussed below. As expected, the effective action is invariant under the spacetime local symmetries.

The above action is said to be in the string frame, since its fields are naturally associated with the above string oscillation Hilbert space states, or equivalently are those appearing in (3.8), (3.9). From the spacetime viewpoint, it is also convenient to redefine the fields as

$$\tilde{\phi} = \phi - \phi_0; \quad \tilde{G}_{MN} = e^{(\phi_0 - \phi)/6} G_{MN}, \tag{3.55}$$

where ϕ_0 is the dilaton vev. This produces the action in the so-called Einstein frame

$$S_{26d}^{\text{E.f.}} = \frac{1}{2\kappa^2} \int d^{26}X (-\tilde{G})^{1/2} \left(\tilde{R} - \frac{1}{12} e^{-\tilde{\phi}/3} H_{MNP} H^{MNP} - \frac{1}{6} \partial_M \tilde{\phi} \, \partial^M \tilde{\phi} \right) + \mathcal{O}(\alpha'),$$

with indices raised/lowered by $\tilde{G}$. It describes 26d gravity in the canonical Einstein–Hilbert form, with gravitational coupling constant $\kappa = \kappa_0 e^{\phi_0}$, coupled to a scalar and a 2-form gauge potential. Morally the 26d Planck scale is $M_{p,26d}^{24} = M_s^{24}/g_s^2$. The cutoff of the theory is M_s, above which full string theory takes over and completes the above effective theory.

Final recap

The results of the above quantization can be summarized as follows. The dynamics boils down to $(D-2)$ free 2d bosons $X^i(t, \sigma)$ describing string oscillations transverse to the worldsheet, with hamiltonian (3.38). They obey periodic boundary conditions (3.39) leading to an oscillator expansion (3.40) in terms of two infinite sets of decoupled harmonic oscillators α_n^i, $\tilde{\alpha}_n^i$. The spectrum is obtained by applying negative modding operators to a groundstate, subject to the level matching condition (3.46). Each state corresponds to a spacetime particle with mass (3.53), and the light spectrum is (3.54).

3.2.2 One-loop vacuum amplitude and modular invariance

Although the closed bosonic string theory cannot lead to realistic models of particle physics, it is a useful toy model for diverse features of string theory. We now focus on the remarkable UV properties of string theory, in the simplest diagram exhibiting them. This is the one-loop vacuum amplitude, also known as partition function, with a free closed string propagating and closing onto itself, defining a worldsheet with the topology of a two-torus with no external legs. From the spacetime viewpoint, the amplitude computes the one-loop correction to the vacuum energy, namely the spacetime cosmological constant.

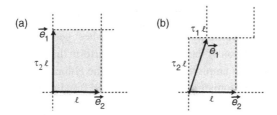

Figure 3.9 A two-torus can be constructed by modding out the two-dimensional plane by translations in a two-dimensional lattice. The unit cell is a parallelogram with sides identified. (a) The simplest case of rectangular two-torus: a closed string of length ℓ evolves for some time $t = \tau_2 \ell$ and closes back to the initial state. (b) The more general class, where the closed string is glued back to the original state modulo a shift by $\tau_1 \ell$ in the reference line in σ.

Considerations on this amplitude lead to a crucial consistency condition, *modular invariance* (which must be satisfied by all closed string theories), as follows. The amplitude requires summing over *all possible inequivalent* worldsheet geometries with two-torus topology. It is therefore crucial to incorporate all possible geometries, and to avoid possible overcountings. Indeed a given geometry can admit several different interpretations: a diagram corresponding to a two-torus with circle lengths ℓ_1 and ℓ_2 can be regarded either as a closed string of length ℓ_1 propagating over a distance ℓ_2, or as a closed string of length ℓ_2 propagating over a distance ℓ_1. These seemingly different processes are related by the exchange $\sigma \leftrightarrow t$, and so correspond to different reparametrizations of the same worldsheet geometry. Modular invariance is the invariance under these "large" coordinate transformations (i.e. not continuously connected to the identity). It controls the UV behavior of amplitudes, because it relates the UV regime of a given channel to the IR of the dual channel (amplitudes for highly energetic long strings propagating over short times turn into amplitudes for short strings propagating over long times).

The worldsheet geometry

We first characterize the possible geometries of the worldsheet Σ with two-torus topology. A two-torus can be described as the two-dimensional real plane, modded out by translations by vectors in a two-dimensional lattice, see Figure 3.9. The simplest such worldsheet geometries are rectangular two-tori, as shown in Figure 3.9(a), and correspond to a closed string (of σ-length ℓ) evolving for a time $\tau_2 \ell$ and closing back onto itself. Denoting $z = \sigma + i\,t$, the two-torus is defined by the identifications $z \equiv z + \ell, z \equiv z + i\tau_2 \ell$.

There is a more general possibility, however, which exploits the freedom to choose the reference line $\sigma = 0$. The geometry in Figure 3.9(b) corresponds to a closed string of length ℓ which evolves for a time $\tau_2 \ell$, and glues back to its original state, up to a shift by $\tau_1 \ell$ in the reference line. The two-torus is defined by the identifications $z \equiv z + \ell$ and $z \equiv z + \tau \ell$, with $\tau = \tau_1 + i\tau_2$. The parameter τ thus characterizes the different worldsheet geometries. Mathematically, it labels the complex structure of the two-torus, and characterizes genus one Riemann surfaces, i.e. two-torus worldsheet modulo conformal transformations.

Computation of the one-loop vacuum amplitude*

We wish to compute the amplitude $Z(\tau)$ for a worldsheet geometry with parameter τ. We use the euclidean formulation of the 2d theory, and perform the computation in the operator formalism. To compute the amplitude, we consider the Hilbert space $\mathcal{H}_{\rm cl.}$ of closed string states at $t = 0$, apply hamiltonian evolution until $t = \tau_2 \ell$ and glue the resulting state to the initial one, modulo a σ-translation by $\tau_1 \ell$. Namely

$$Z(\tau) = \sum_{|{\rm st.}\rangle \in \mathcal{H}_{\rm cl.}} \langle {\rm st.}| \, e^{-\tau_2 \ell H} \, e^{i \tau_1 \ell P_\sigma} \, |{\rm st.}\rangle. \tag{3.56}$$

For the closed bosonic string, H, P_σ are given by (3.42), (3.44), and we obtain

$$Z(\tau) = {\rm tr}\,_{\mathcal{H}_{\rm cl.}} \left(e^{-\tau_2 \ell H} e^{i \tau_1 \ell P} \right) = {\rm tr}\,_{\mathcal{H}_{\rm cl.}} \left[\exp\left(-\tau_2 \pi \alpha' \sum_i p_i^2\right) q^{N+E_0} \, \bar{q}^{\tilde{N}+\tilde{E}_0} \right], \tag{3.57}$$

where we have introduced

$$q = e^{2\pi i \tau}. \tag{3.58}$$

The above trace should run through the Hilbert space $\mathcal{H}_{\rm cl.}$ of physical states, constrained by the level matching condition (3.46). It is simpler, however, to trace over states in an extended Hilbert space $\mathcal{H}$ with unconstrained oscillator structure. Happily the final result is unchanged since contributions from unphysical states cancel out upon integration over geometries. In particular, integration over τ_1 automatically projects onto physical states

$$\int d\tau_1 \, e^{2\pi i \tau_1 (N - \tilde{N})} \simeq \delta_{N, \tilde{N}}. \tag{3.59}$$

We therefore evaluate $Z(\tau)$ as a trace over states characterized by the center of mass momentum, and the arbitrary oscillator occupation numbers.

$$\prod_{n,i} \left(\alpha^i_{-n}\right)^{K_{n,i}} \prod_{m,j} \left(\tilde{\alpha}^j_{-m}\right)^{\tilde{K}_{m,j}} |p\rangle. \tag{3.60}$$

Since the different pieces in these states are decoupled, the total Hilbert space splits as a product of Hilbert spaces for the center of mass and for each oscillator

$$\mathcal{H} = \mathcal{H}_{\rm c.m.} \otimes \left(\otimes_{i=2}^{25} \otimes_{n=1}^{\infty} \mathcal{H}_{i,n} \right)_L \otimes \left(\otimes_{i=2}^{25} \otimes_{n=1}^{\infty} \tilde{\mathcal{H}}_{i,n} \right)_R. \tag{3.61}$$

The trace over $\mathcal{H}$ splits as a product

$$Z(\tau) = Z_{\rm c.m.}(\tau) \left| q^{E_0} \prod_{i,n} Z_{i,n}(\tau) \right|^2, \tag{3.62}$$

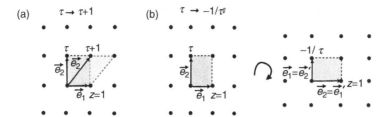

Figure 3.10 Modular transformations relating equivalent two-tori. (a) A transformation $\tau \to \tau + 1$ simply corresponds to changing the basis vector $\vec{e}_2$ in the lattice. (b) A transformation $\tau \to -1/\tau$ corresponds to an exchange of the basis vectors $\vec{e}_1, \vec{e}_2$, and an overall rescaling.

where

$$Z_{\text{c.m.}}(\tau) = \text{tr}\,_{\mathcal{H}_{\text{c.m.}}} \exp\left(-\tau_2 \pi \alpha' \sum_i p_i^2\right), \quad Z_{i,n}(\tau) = \text{tr}\,_{\mathcal{H}_{i,n}} q^{N_{i,n}}, \quad \text{with } N_{i,n} = \alpha_{-n}^i \alpha_n^i.$$

The trace over the center of mass degrees of freedom is

$$Z_{\text{c.m.}}(\tau) = \prod_i L_{X^i} \int \frac{dp_i}{2\pi} \langle p_i | e^{-\tau_2 \pi \alpha' \sum_i p_i^2} | p_i \rangle = V_{24} \left(\int \frac{dp}{2\pi} e^{-\tau_2 \pi \alpha' p^2} \right)^{24}$$

$$= V_{24} \,(4\pi^2 \alpha' \tau_2)^{-12}, \tag{3.63}$$

with L_{X^i} a regularized length along x^i, and $V_{24} = \prod_i L_{X^i}$ a regularized 24d volume.

Let us compute the oscillator factors. For a single oscillator, the trace over states $(\alpha_{-n}^i)^K |0\rangle$, for fixed i, n gives a contribution

$$Z_{i,n}(\tau) = \sum_{K=0}^{\infty} \langle 0 | \left(\alpha_n^i\right)^K q^{N_{i,n}} \left(\alpha_{-n}^i\right)^K |0\rangle = \sum_{K=0}^{\infty} q^{Kn} = (1 - q^n)^{-1}. \tag{3.64}$$

Replacing (3.63), (3.64) into (3.62), we get

$$Z(\tau) = V_{24}(4\pi^2 \alpha' \tau_2)^{-12} \left| q^{1/24} \prod_{n=1}^{\infty} (1 - q^n) \right|^{-48} = V_{24}(4\pi^2 \alpha' \tau_2)^{-12} |\eta(\tau)|^{-48},$$

where we have used the definition of the Dedekind η function (A.3). The infinite spacetime volume factor simply implies that the only well-defined quantity is the amplitude per unit volume, and will be ignored in similar computations in later chapters.

The $\mathbf{T}^2$ modular group

The complete partition function involves a sum over all inequivalent geometries, without overcounting. A crucial observation is therefore that the characterization of worldsheet geometries by τ is actually not one-to-one, but rather there are different values of τ corresponding to the *same* geometry. For instance, as shown in Figure 3.10, two geometries corresponding to τ and $\tau + 1$ are defined by the same two-dimensional lattice, hence are equivalent two-tori. In a slightly trickier way, geometries corresponding to τ and $-1/\tau$ are

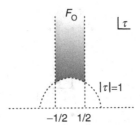

Figure 3.11 Fundamental domain F_0 of τ. Any point in the upper half plane can be mapped to some point in F_0 using the basic modular transformations $\tau \to \tau + 1, \tau \to -1/\tau$.

also equivalent, upon exchanging the roles of the two basis vectors (hence the roles of σ and t) and an overall rescaling.

It is possible to show that these two transformations $\tau \to \tau + 1$ and $\tau \to -1/\tau$ generate the whole set of equivalences of $\mathbf{T}^2$ geometries, called the modular group of the two-torus. Its elements have the general form

$$\tau \to \frac{a\tau + b}{c\tau + d} \quad \text{with } a, b, c, d \in \mathbf{Z} \quad \text{and } ad - bc = 1. \tag{3.65}$$

The modular group is $SL(2, \mathbf{Z})$, since the parameters a, b, c, d can be written as a 2×2 matrix of integer entries and unit determinant. This group appears in this book in very different contexts. We already encountered it in Section 2.5.2 when discussing S-duality in $\mathcal{N} = 4$ supersymmetric field theories, and will show up again in Section 6.3.1 when discussing type IIB string self-duality. It also arises when discussing T-duality symmetries of heterotic toroidal orbifold compactifications, see Section 9.6. In the present physical context, it corresponds to "large" changes of coordinates, under which the worldsheet torus geometry is invariant.

The set of inequivalent geometries is therefore characterized by the parameter τ in the upper half complex plane $\tau_2 > 0$, modulo $SL(2, \mathbf{Z})$ transformations. A choice of fundamental domain F_0 of τ, shown in Figure 3.11, is

$$-1/2 \leq \tau_1 < 1/2, \quad |\tau| \geq 1. \tag{3.66}$$

Modular invariance of the partition function

The closed bosonic string partition function $Z(\tau)$ should be the same for equivalent tori, since physics cannot change upon reparametrizations of the worldsheet. So $Z(\tau)$ should be modular invariant, i.e. invariant under the $SL(2, \mathbf{Z})$ modular transformations. This is not obvious a priori, since our gauge fixing and quantization dealt with "small" coordinate changes (continuously connected to the identity), whereas modular transformations imply "large" changes of coordinates. Invariance under the latter is however required to define a consistent sum over 2d geometries and hence a consistent string theory.

Happily, the above partition function is modular invariant, as is easily shown using the modular transformation properties of Dedekind's η function (A.4),

$$Z(\tau) \simeq \tau_2^{-12} |\eta(\tau)|^{-48} \overset{\tau \to \tau+1}{\longrightarrow} \tau_2^{-12} |\eta(\tau)|^{-48},$$

$$Z(\tau) \simeq \tau_2^{-12} |\eta(\tau)|^{-48} \overset{\tau \to -1/\tau}{\longrightarrow} \frac{(\tau_1^2 + \tau_2^2)^{12}}{\tau_2^{12}} \frac{1}{|\tau|^{24} |\eta(\tau)|^{48}} = \tau_2^{-12} |\eta(\tau)|^{-48}.$$

(3.67)

The complete vacuum amplitude (per unit spacetime volume) is obtained by summing over inequivalent geometries, i.e. integrating τ over the fundamental domain

$$Z = \int_{F_0} \frac{d^2\tau}{4\tau_2} (4\pi^2 \alpha' \tau_2)^{-12} |\eta(\tau)|^{-48},$$

(3.68)

where $d^2\tau/(4\tau_2)$ is an $SL(2, \mathbf{Z})$ invariant measure in τ parameter space.

UV behavior of the string amplitude

The restriction of the above integration to the fundamental region F_0 removes the region $\tau \simeq 0$, which corresponds to the UV regime of strings propagating for a very short time ($\tau_2 \simeq 0$). This regime is actually mapped by $\tau \to -1/\tau$ to the region around $\tau \to i\infty$, which is in F_0, but corresponds to an IR regime of string propagating for a very long time ($\tau_2 \simeq \infty$). This is a particular realization of the mechanism mentioned in Section 3.1.3. The lesson that UV regimes turn into dual IR channels follows from the geometry of the worldsheet, and holds in any string theory. In the bosonic theory, this IR channel is actually divergent due to an exponential contribution $e^{+\tau_2}$ from the spacetime tachyon. This is an artifact of bosonic string theory, however, absent in tachyon-free theories like the superstrings.

3.2.3 *Toroidal compactification of closed bosonic string and T-duality*

As discussed in Sections 1.4.1 and 3.1.6, compactification is a canonical mechanism to obtain low-energy 4d physics in a theory with extra dimensions. In bosonic string theory, we are interested in compactifications $\mathbf{M}_4 \times \mathbf{X}_{22}$, with $\mathbf{X}_{22}$ a 22d compact manifold. The simplest possibility is compactification on a torus, for which the most interesting phenomena are already present in the circle compactification of just one dimension. We thus focus on the circle compactification of the 26d closed bosonic string theory to a 25d theory.

Results in this section are useful in the construction of toroidal compactifications of superstrings (and orbifolds thereof). Also, they are relevant in the construction of ten-dimensional heterotic string theories, in Section 4.3.

Kaluza–Klein toroidal compactification of bosonic string theory

We start by applying the Kaluza–Klein field theory compactification to the massless spectrum of the closed bosonic string. This provides a good approximation to the large radius string theory compactification, when α' corrections are negligible, and can be used for comparison with the full string theory results derived later.

The KK compactification on an $\mathbf{S}^1$ of length $L = 2\pi R$ of the 26d massless fields G_{MN}, B_{MN}, ϕ can be carried out in close analogy with Section 1.4.1, leading to

$$
\begin{array}{ccl}
\text{26d field} & & \text{25d fields} \\
G_{MN} & \rightarrow & g_{\mu\nu}, A_\mu, \sigma, \\
B_{MN} & \rightarrow & b_{\mu\nu}, \hat{A}_\mu, \\
\phi & \rightarrow & \phi,
\end{array}
\tag{3.69}
$$

where M, N and μ, ν denote 26d and 25d indices, respectively. As in the 5d model in Section 1.4.1, the field A_μ arises from the mixed metric component $G_{\mu,25}$, and has a $U(1)$ gauge invariance inherited from local reparametrizations of the $\mathbf{S}^1$ coordinate, i.e. $x^{25} \rightarrow x^{25} + \lambda(x^\mu)$ implies $A_\mu \rightarrow A_\mu - \partial_\mu \lambda$. These transformations are generated by (local) translations in x^{25}, so the KK momentum mode k plays the role of $U(1)$ charge. Similarly, the 26d gauge invariance (3.11) of the 26d 2-form B_{MN} leads to a similar gauge invariance of the 25d 2-form $b_{\mu\nu}$, and a $U(1)$ gauge invariance of the mixed component $\hat{A}_\mu \sim B_{25,\mu}$. States charged under $\hat{A}_\mu$ are absent in field theory, but exist in string theory as winding states, see later.

The 26d effective action produces a 25d effective action for these fields. Skipping details, we just note the relation between their Planck scales by the analog of (1.44),

$$
M_{p,25d}^{23} = M_{p,26d}^{24} L \simeq \frac{M_s^{24} L}{g_s^2}.
\tag{3.70}
$$

The value of L is related to the vev of the "radion" field σ, which is a modulus field, i.e. has vanishing potential. It is a general feature that parameters of the compactification geometries correspond to massless moduli fields in the lower-dimensional theory. The presence of moduli in string compactifications poses phenomenological problems with experimental bounds on fifth forces, and with cosmological evolution. However, they can be stabilized to minima of dynamical scalar potentials in compactifications with additional ingredients, described in Chapter 14.

Concerning massive string states, recall that KK compactification on $\mathbf{S}^1$ of states of mass M give a shifted KK mass formula

$$
m_k^2 = M^2 + \left(\frac{k}{R}\right)^2.
\tag{3.71}
$$

Hence in the regime $E \ll 1/R \ll M_s$, string resonances are far heavier than KK replicas of massless fields, and can be neglected.

Toroidal compactification of closed bosonic string theory

We now consider the string theory description of the $\mathbf{S}^1$ compactification. From Section 3.1.5, the worldsheet action for a string propagating in $\mathbf{M}_{25} \times \mathbf{S}^1$ is given by (3.8)

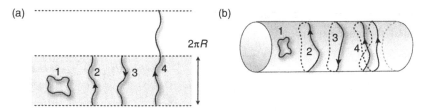

Figure 3.12 Closed string states in $\mathbf{S}^1$ compactifications. The closed string 1 is closed already in flat 26d space, and has no winding. Closed strings 2, 3 and 4 are closed only after the periodic identification defining the circle. States 2 and 3 have opposite winding numbers $w = \pm 1$, and state 4 has multiple winding $w = 2$.

with a metric in $\mathbf{M}_{25} \times \mathbf{S}^1$. In fact, this geometry is defined as a quotient of $\mathbf{M}_{26}$ by a translation, so it is locally flat and the worldsheet action is simply given by (3.4). This may seem puzzling, but simply means that the dynamics of the strings oscillations is *locally* identical for the theory on $\mathbf{M}_{25} \times \mathbf{S}^1$ and on $\mathbf{M}_{26}$. The difference between both geometries is indeed a global effect, i.e. the identification $x^{25} \simeq x^{25} + 2\pi R$. Consequently the effects of the compactification do not arise at the level of the local structure of the worldsheet, but in the boundary conditions for the 2d worldsheet fields.

Since the local worldsheet dynamics is exactly as in the uncompactified theory, the degrees of freedom are the 2d bosons $X^i(t, \sigma)$ with local hamiltonian as in (3.38)

$$H = \frac{1}{2} \int_0^\ell d\sigma \left(2\pi\alpha' \Pi_i \Pi_i + \frac{1}{2\pi\alpha'} \partial_\sigma X^i \partial_\sigma X^i \right). \tag{3.72}$$

In order to express it in terms of oscillator modes, we need to specify the boundary conditions. For X^i, $i = 2, \ldots, 24$, we have the usual periodic boundary conditions

$$X^i(t, \sigma + \ell) = X^i(t, \sigma) \quad \text{for} \quad i = 2, \ldots, 24. \tag{3.73}$$

For X^{25}, the periodicity in x^{25} allows a more general boundary condition

$$X^{25}(t, \sigma + \ell) = X^{25}(t, \sigma) + 2\pi R w, \quad w \in \mathbf{Z}. \tag{3.74}$$

For $w = 0$ it reduces to the standard periodic boundary condition, but for $w \neq 0$ it describes a closed string winding around $\mathbf{S}^1$ a number of times given by w, see Figure 3.12. The sign of the winding number w corresponds to the orientation of the string. It is easy to show that interactions between the different string states conserve the total winding number. Different values of w correspond to different closed string sectors, all coexisting in the theory; the complete spacetime 25d spectrum is given by the string states in all possible winding sectors.

Since the X^i, $i = 2, \ldots, 24$ behave as usual, their oscillator expansion is (3.40). On the other hand, the mode expansion for X^{25} has the structure

$$X^{25} = x^{25} + \frac{k/R}{p^+} t + \frac{2\pi R w}{\ell} \sigma + i\sqrt{\frac{\alpha'}{2}} \sum_{n \neq 0} \left[\frac{\alpha_n^{25}}{n} e^{-2\pi i n (t+\sigma)/\ell} + \frac{\tilde{\alpha}_n^{25}}{n} e^{-2\pi i n (t-\sigma)/\ell} \right].$$

As in field theory, the center of mass momentum is quantized $p_{25} = k/R$, $k \in \mathbf{Z}$.

The hamiltonian differs from that of the uncompactified theory only in new contributions to $\partial_\sigma X^{25}$ from winding (potential energy from string stretching). In fact

$$H = H_{w=0} + \frac{1}{2} \int_0^\ell d\sigma \frac{1}{2\pi\alpha'} \left(\frac{2\pi R w}{\ell} \right)^2$$

$$= \sum_{i=2}^{24} \frac{p_i^2}{2p^+} + \frac{(k/R)^2}{2p^+} + \frac{R^2 w^2}{2\alpha'^2 p^+} + \frac{1}{\alpha' p^+} (N + \tilde{N} - 2), \qquad (3.75)$$

where $H_{w=0}$ is given by (3.42) and we use the number operators (3.45). The Hilbert space is constructed by defining groundstates, labeled k, w, and applying oscillator creation operators. The level-matching constraint is $P_\sigma = 0$, with

$$P_\sigma = \int_0^\ell d\sigma \, \Pi_i \, \partial_\sigma X^i = \frac{2\pi}{\ell} (N - \tilde{N} + kw). \qquad (3.76)$$

Each state corresponds to a particle in the 25d spacetime, with mass

$$M_{25d}^2 = 2p^+ H - \sum_{i=2}^{24} p_i^2 = \frac{k^2}{R^2} + \frac{R^2}{\alpha'^2} w^2 + \frac{2}{\alpha'} (N + \tilde{N} - 2). \qquad (3.77)$$

Winding states arise from the extended nature of the string, so only the $w = 0$ sector can correspond to the point particle limit of the theory. Indeed, states with non-zero winding are very massive and decouple in the large radius limit, $\alpha'/R^2 \ll 1$. The Kaluza–Klein description should hold in this regime, and indeed the mass formula for $w = 0$ reproduces (3.71) with M the 26d mass in (3.53).

Forgetting the tachyon and its KK replicas, the massless modes are

$$\begin{aligned}
\alpha_{-1}^\mu \tilde{\alpha}_{-1}^\nu |0\rangle &\rightarrow & g_{\mu\nu}, b_{\mu\nu}, \varphi, \\
(\alpha_{-1}^\mu \tilde{\alpha}_{-1}^{25} + \alpha_{-1}^{25} \tilde{\alpha}_{-1}^\mu) |0\rangle &\rightarrow & A_\mu, \\
(\alpha_{-1}^\mu \tilde{\alpha}_{-1}^{25} - \alpha_{-1}^{25} \tilde{\alpha}_{-1}^\mu) |0\rangle &\rightarrow & \hat{A}_\mu, \\
\alpha_{-1}^{25} \tilde{\alpha}_{-1}^{25} |0\rangle &\rightarrow & \sigma,
\end{aligned} \qquad (3.78)$$

where μ, ν are 25d indices. The right-hand column describes the spacetime fields in a notation consistent with the KK analysis. As in field theory, massive states with internal momentum k are charged under the gauge boson $A_\mu \sim G_{\mu,25}$. In addition, in string theory there are states charged under $\hat{A}_\mu \sim B_{\mu,25}$, given by winding states. Indeed the coupling (3.10) leads to

$$\int_\Sigma B_{MN}\, \partial_a X^M\, \partial_b X^N \epsilon^{ab} \rightarrow \int dt \int_0^\ell d\sigma\, B_{\mu,25}\partial_\sigma X^{25}\, \partial_t X^\mu \simeq w \int dt\, \hat{A}_\mu \partial_t x^\mu, \quad (3.79)$$

where in the last step the string coordinate is reduced to its center of mass piece, since oscillators average to zero. Thus winding states couple to $\hat{A}_\mu$ with charge w.

Modular invariance of toroidally compactified closed bosonic string

Winding states are generically massive, and irrelevant at low energies. They can however become light at particular values of R, known as critical radii, as we discuss further in Section 5.2.2 in the more interesting setup of heterotic theories. Even at general radius, they play a crucial role in ensuring the good UV behavior of string amplitudes, as follows from the modular invariance of the one-loop partition function. Using arguments similar to Section 3.2.2, one can obtain

$$Z(\tau) = (2\pi\alpha'\tau_2)^{-\frac{23}{2}} |\eta(\tau)|^{-48} \sum_{k,w=-\infty}^{\infty} \exp\left[-\pi\tau_2\left(\frac{\alpha'k^2}{R^2} + \frac{R^2 w^2}{\alpha'} \right) + 2\pi i \tau_1\, kw \right].$$
$$(3.80)$$

The first factor is the trace over 25d center of mass momentum, the second is the trace over oscillators, and the third traces over internal momentum and winding degrees of freedom. This expression is clearly invariant under $\tau \to \tau + 1$, and invariance under $\tau \to -1/\tau$ can be shown by using the Poisson resummation (A.13) on both sums over k and w. In this process, momenta and winding exchange their roles under $\sigma \leftrightarrow t$, showing that winding states are crucial to obtain a modular invariant theory.

Left and right sectors in toroidal compactification

Closed string dynamics splits into decoupled left and right sectors. For future convenience, it is useful to recast some formulae making this split manifest. For instance, the X^{25} oscillator expansion gives

$$X_L^{25}(t+\sigma) = \frac{x^{25}}{2} + \sqrt{\frac{2}{\alpha'}}\frac{p_L}{2p^+}(t+\sigma) + i\sqrt{\frac{\alpha'}{2}} \sum_{n\in\mathbf{Z}-\{0\}} \frac{\alpha_n^{25}}{n} e^{-2\pi i n\,(t+\sigma)/\ell},$$

$$X_R^{25}(t-\sigma) = \frac{x^{25}}{2} + \sqrt{\frac{2}{\alpha'}}\frac{p_R}{2p^+}(t-\sigma) + i\sqrt{\frac{\alpha'}{2}} \sum_{n\in\mathbf{Z}-\{0\}} \frac{\tilde{\alpha}_n^{25}}{n} e^{-2\pi i n\,(t-\sigma)/\ell}, \quad (3.81)$$

with the left- and right-moving "momenta"

$$p_L = \sqrt{\frac{\alpha'}{2}}\left(\frac{k}{R} + \frac{wR}{\alpha'} \right); \quad p_R = \sqrt{\frac{\alpha'}{2}}\left(\frac{k}{R} - \frac{wR}{\alpha'} \right), \quad (3.82)$$

where the square root factors are introduced to simplify upcoming expressions. These sectors evolve with hamiltonians

$$
H_L = \frac{1}{4p^+} \sum_{i=1}^{24} p_i^2 + \frac{1}{\alpha' p^+} \left(\frac{p_L^2}{2} + N + E_0 \right),
$$

$$
H_R = \frac{1}{4p^+} \sum_{i=1}^{24} p_i^2 + \frac{1}{\alpha' p^+} \left(\frac{p_R^2}{2} + \tilde{N} + \tilde{E}_0 \right). \tag{3.83}
$$

They contribute to the mass of spacetime particles as $M^2 = M_L^2 + M_R^2$ with

$$
\frac{\alpha' M_L^2}{2} = \frac{p_L^2}{2} + N - 1, \qquad \frac{\alpha' M_R^2}{2} = \frac{p_R^2}{2} + \tilde{N} - 1. \tag{3.84}
$$

The level-matching constraint relating both sectors is just $M_L^2 = M_R^2$.

In this language, the partition function (3.80) can be written

$$
Z(\tau) = \mathrm{tr}_{\mathcal{H}_L}\, q^{\alpha' p^+ H_L}\, \mathrm{tr}_{\mathcal{H}_R}\, \bar{q}^{\alpha' p^+ H_R} = (2\pi \alpha' \tau_2)^{-23/2} |\eta(\tau)|^{-48} \sum_{(p_L, p_R) \in \Gamma_{1,1}} q^{\frac{p_L^2}{2}}\, \bar{q}^{\frac{p_R^2}{2}},
$$

where $\Gamma_{1,1}$ is the lattice of the vectors (p_L, p_R) in (3.82). In this language, modular invariance is a consequence of the lattice $\Gamma_{1,1}$ being *even* and *self-dual*, see Appendix A. Invariance under $\tau \to -1/\tau$ follows from self-duality of $\Gamma_{1,1}$, using the Poisson formula (A.14). Invariance under $\tau \to \tau + 1$ follows from $\Gamma_{1,1}$ being even: the square of any vector is an even integer, with respect to the Lorentzian scalar product

$$
(p_L, p_R) \cdot (p_L', p_R') = (p_L p_L' - p_R p_R') = kw' + k'w. \tag{3.85}
$$

T-duality

The existence of winding states in string theory leads to an amazing property, known as T-duality, which generalizes (in suitable sense) to other string theories. The mass formula (3.77) is invariant under the T-duality transformation

$$
R \to \frac{\alpha'}{R}, \quad k \leftrightarrow w. \tag{3.86}
$$

Namely, the complete spectrum of the theory at radius R is the same as the spectrum of the theory at radius $R' = \alpha'/R$, up to a relabeling of k and w. This extremely striking equivalence of two very different geometrical interpretations is possible because when the naive geometric intuitions hold in one theory, they fail in the other. Indeed if $R^2 \gg \alpha'$, so that stringy α' corrections are small, and classical geometry is a good approximation to the physics of the system, then $R'^2 \ll \alpha'$, and the T-dual system has strong stringy effects, invalidating familiar geometric intuitions.

An important consequence of T-duality is that the $R \to 0$ limit corresponds to a decompactification limit of the T-dual theory, $R' \to \infty$; the infinite tower of winding states becoming light in the $R \to 0$ limit are interpreted as an infinite tower of KK momentum states becoming light in the $R' \to \infty$ limit.

T-duality is not just an accidental property of the string spectrum, but rather an exact property of the fully fledged string theory. The two theories are equivalent, even at the interacting level, since they are described by equivalent worldsheet theories, as follows. Let us split the worldsheet theory into decoupled left and right sectors. Given the fields X_L^{25}, X_R^{25}, there are two ways to construct a physical spacetime coordinate, namely

$$X^{25} = X_L^{25} + X_R^{25}, \quad X'^{25} = X_L^{25} - X_R^{25}. \tag{3.87}$$

They are related by a flip of the right-moving coordinate, i.e. at the level of momenta

$$p_L \to p_L, \quad p_R \to -p_R. \tag{3.88}$$

Using (3.82) this corresponds to the T-duality transformation (3.86), so X^{25} and X'^{25} describe string compactifications on radii R and $R' = \alpha'/R$. T-duality can thus be described as a *parity operation on right movers*. This viewpoint is useful in the generalizations of the T-duality concept to superstrings.

A last important point is that T-duality also acts non-trivially on the dilaton. Using (3.6), the equality of the 25d Planck scale (3.70) of the dual theories requires

$$e^{-2\phi} R = e^{-2\phi'} R' \implies e^{-\phi'} = e^{-\phi} R \alpha'^{-1/2}. \tag{3.89}$$

T-duality is our first example of a non-trivial equivalence of different string theories. This and other dualities play an important role in the structure of the theory, in particular beyond the perturbative regime.

3.3 Open bosonic string

The open bosonic string provides an excellent playing ground to learn about general properties of open strings. The latter play a fundamental role in the construction of type I superstring, Section 4.4, and in the description of D-branes, Section 6.1.

3.3.1 Quantization and spectrum of the open bosonic string

Open strings are strings with endpoints, and upon propagation they span worldsheets with boundaries. Recall that theories with open strings necessarily include closed strings, so the total spacetime particle spectrum is given by the closed plus open string oscillation states. The fact that open strings couple to closed strings implies that they share the same local worldsheet structure, and both have identical local dynamics (with differences arising only as global boundary conditions).

Many of the properties of open strings are illustrated by the open bosonic string, whose closed string sector is the closed bosonic string, and so is defined in $D = 26$ dimensions; the worldsheet fields are the $X^M(t, \sigma)$ and the worldsheet metric $g_{ab}(t, \sigma)$, with the Polyakov action (3.15). The open string oscillation states are obtained by quantizing the theory on a worldsheet geometry given by an infinite strip, parametrized by coordinates t, σ, with $\sigma \in [0, \ell]$, so string endpoints are at $\sigma = 0, \ell$. In contrast, with the closed string, there is a natural choice of reference line $\sigma = 0$ at an open string endpoint, so there is no symmetry under σ-translations and no level matching constraint for physical states. Since the local dynamics on the open string worldsheet must be identical to the closed string one, the only physical excitations are the fields $X^i(t, \sigma)$, with light-cone hamiltonian (3.38)

$$H = \frac{1}{2} \int_0^\ell d\sigma \left(2\pi\alpha' \, \Pi_i \, \Pi_i + \frac{1}{2\pi\alpha'} \, \partial_\sigma X^i \, \partial_\sigma X^i \right). \tag{3.90}$$

This describes the freely oscillating fields $X^i(t, \sigma)$ obeying a 2d free wave equation, yielding a superposition of left- and right-moving waves. The general expansion is

$$X_L^i(t + \sigma) = \frac{x^i}{2} + \frac{p_i}{2p^+}(t + \sigma) + i\sqrt{\frac{\alpha'}{2}} \sum_\nu \frac{\alpha_\nu^i}{\nu} e^{-\pi i \, \nu \, (t+\sigma)/\ell},$$

$$X_R^i(t - \sigma) = \frac{x^i}{2} + \frac{p_i}{2p^+}(t - \sigma) + i\sqrt{\frac{\alpha'}{2}} \sum_\nu \frac{\tilde{\alpha}_\nu^i}{\nu} e^{-\pi i \, \nu \, (t-\sigma)/\ell}, \tag{3.91}$$

with ν a modding to be fixed by the boundary conditions. The above expansion conventionally differs from the closed string in a factor of 2 in the exponents. This is related to the *doubling trick*, which is used to describe the open string interval $\sigma \in [0, \ell]$ as a $\mathbf{Z}_2$ quotient of a circle of length 2ℓ.

Boundary conditions are determined by requiring the vanishing of boundary terms in the variational principle derivation of the equations of motion from (3.24)

$$\delta S_P = -\frac{1}{2\pi\alpha'} \int_\Sigma d^2\xi \, \eta^{ab} \, \partial_a X^M \, \partial_b \delta X_M$$

$$= -\frac{1}{2\pi\alpha'} \int_{-\infty}^\infty dt \, (\delta X^M \, \partial_\sigma X_M) \Big|_{\sigma=0}^{\sigma=\ell} + \frac{1}{2\pi\alpha'} \int_\Sigma d^2\xi \, \delta X^M \, \partial_a \partial^a X_M. \tag{3.92}$$

The second term leads to the same equations of motion as for the closed string, while the first corresponds to the boundary terms. In a Poincaré-invariant theory δX^M are unconstrained,[1] and the vanishing of the boundary terms requires

$$\partial_\sigma X^i \Big|_{\sigma=0,\ell} = 0. \tag{3.93}$$

---

[1] This is not the case for open string sectors describing lower-dimensional D-branes in Section 6.1. The breaking of Poincaré invariance is interpreted as the presence of a D-brane in the configuration.

These are Neumann boundary conditions on both open string endpoints, thus known as Neumann–Neumann or NN boundary conditions. They describe free endpoints, with no momentum flow, and link the left and right dynamics by

$$\partial_\sigma X_L^i + \partial_\sigma X_R^i = 0 \text{ at } \sigma = 0, \ell. \tag{3.94}$$

Namely a left-moving wave is reflected at a boundary into a right-moving one (and vice versa). Imposing the boundary condition at $\sigma = 0$ we obtain

$$\alpha_\nu^i = \tilde{\alpha}_\nu^i. \tag{3.95}$$

The boundary condition at $\sigma = \ell$ requires

$$\alpha_\nu^i \sin \pi \nu = 0, \text{ hence } \nu \in \mathbf{Z}. \tag{3.96}$$

The identification of left and right movers in (3.95) implies that (Poincaré invariant) open strings can couple only left–right symmetric closed strings. It also implies that the Hilbert space of an open string is exactly like one of the sides (say the left sector) of a closed string. Indeed, the hamiltonian (3.90) in terms of the oscillator modes becomes

$$H = \frac{\sum_i p_i^2}{2p^+} + \frac{1}{2\alpha' p^+} \left(\sum_i \sum_{n>0} \alpha_{-n}^i \alpha_n^i - 1 \right). \tag{3.97}$$

This is similar to the left-moving piece of the closed bosonic string hamiltonian, except for a factor of 2 tracing back to the oscillator expansion. The spacetime mass formula reads

$$\alpha' M^2 = N - 1, \text{ with } N = \sum_i \sum_{n>0} \alpha_{-n}^i \alpha_n^i. \tag{3.98}$$

The lightest modes are:

Sector	State	$\alpha' M^2$	$SO(24)$	26d field	
$N = 0$	$|0\rangle$	-1	$\mathbf{1}$	T	(3.99)
$N = 1$	$\alpha_{-1}^i |0\rangle$	0	$\square$	A_M	

where $\square$ is the Young tableaux notation for the vector representation. We obtain a 26d tachyonic scalar, and a $U(1)$ massless gauge boson. Note that we get the correct Lorentz little group for the massless particles. To this sector of open string states we have to add the closed string states (3.54).

In contrast with the closed string tachyon, the physical meaning of the open string tachyon is relatively well understood. The corresponding instability signals that the theory is actually sitting at a maximum of the potential for this field, and tends to roll down to a more stable minimum. The tachyon condensation process admits a spacetime picture similar to that in Section 6.5.2, in terms of decay of unstable spacetime-filling objects, the

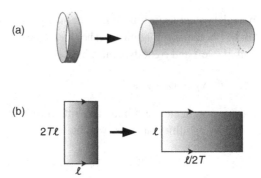

Figure 3.13 Open–closed duality: An open string propagating a time $2T\ell$ is geometrically the same as a closed string propagating a time $T'\ell$ with $T' = 1/(2T)$. (b) Displays an unfolding of (a) with the worldsheet given by a rectangle with two sides identified.

D-branes to be introduced in Section 6.1. In the present case, the final configuration surprisingly corresponds to a closed bosonic string theory, with no open string sector. Although open string sectors of the bosonic theory are unstable, they provide a useful warm-up for the more interesting (and stable) open string sectors in superstrings.

3.3.2 Open–closed duality*

We now consider the UV behavior of open string amplitudes, in the simplest one-loop open string diagram, namely the vacuum to vacuum amplitude given by the annulus or cylinder diagram. This corresponds to an open string evolving for a time $2T\ell$ and glueing back to itself, see Figure 3.13.

The amplitude is easily computed as a trace over the open string Hilbert space. An important difference with respect to the torus amplitude for the closed bosonic string is that there is a fixed reference line, and no freedom to shift it, thus no analog of τ_1; the only parameter of the geometry to integrate over is T. We have

$$Z = \int_0^\infty \frac{dT}{2T} Z(T), \tag{3.100}$$

where $1/(2T)$ is a suitable integration measure, and

$$Z(T) = \mathrm{Tr}_{\mathcal{H}_{op.}} e^{-2T\ell H_{op.}} = \mathrm{Tr}_{c.m.}\, e^{-2\pi\alpha'T \sum_i p_i^2}\, \mathrm{Tr}_{osc.}\, e^{-2\pi T(N-1)}. \tag{3.101}$$

Defining $q = e^{2\pi i(iT)}$, the traces organize in modular functions with parameter $\tau = iT$. Computing the traces in a by now familiar way we have

$$Z = \int_0^\infty \frac{dT}{2T} (8\pi^2\alpha'T)^{-12}\, \eta(iT)^{-24}. \tag{3.102}$$

The labeling of geometries in terms of T is one-to-one, and there is no modular invariance for the annulus. This implies that the UV region $T \to 0$ is not cut off from the amplitude, and leads to a divergence. However, there is an interpretation of this limit as an IR regime of a dual channel, which now involves closed strings, see Figure 3.13. This property, known as open–closed duality, is the fact that the annulus diagram can be regarded, in a dual channel, as a closed string diagram (in this case, a tree exchange between two boundaries).

We can indeed recast the amplitude (3.102) as a closed string one. Note that the exchange of the roles of σ and t in the annulus requires a rescaling to bring the new σ coordinate to the light-cone convention of having length ℓ, and hence the closed string propagates for a time $T'\ell$ with $T' = 1/(2T)$. Using the modular transformation properties

$$\eta(i/(2T')) = (2T')^{1/2} \eta(2iT'), \tag{3.103}$$

the amplitude (3.102) can be written

$$Z = \int_0^\infty \frac{dT'}{2T'} (8\pi^2\alpha')^{-12} \eta(2iT')^{-24}. \tag{3.104}$$

This has the structure of a sum over closed string states with some peculiarities: there is no power-like dependence on T', meaning that due to momentum conservation the closed states are created out of the vacuum with zero momentum; also, there is no analog of τ_1 since the closed string does not close onto itself; finally, due to the absence of integration over τ_1, the level matching on closed string states is imposed explicitly, leading to a doubling of the argument of the oscillator η functions.

3.3.3 Chan–Paton indices and non-abelian gauge symmetries

An essentially new feature of open strings is that they have endpoints. This allows to consider the introduction of degrees of freedom located at these preferred points, in a way consistent with the open string interaction vertices. It is indeed possible to introduce discrete degrees of freedom at the endpoints, known as Chan–Paton indices, described by labels $a = 1, \ldots, N$ taking N possible values. These labels will shortly be interpreted as indices under a suitable gauge group, and as D-branes later in Sections 5.3.4 and 6.1.2. The Chan–Paton indices have trivial dynamics, and just propagate unchanged, thus providing a label for each boundary of the worldsheet. Their behavior in open string interactions is shown in Figure 3.14. For bosonic strings, the value of N is arbitrary (but in superstrings it is eventually constrained, see Sections 4.4.1 and 4.4.3), and in principle each value of N defines a different theory.

The quantization of open strings with Chan–Paton factors is straightforward. Since Chan–Paton indices are non-dynamical, they do not enter in the hamiltonian, and simply label N^2 open string sectors, characterized by two independent indices at their endpoints. The quantization of each ab sector proceeds as in Section 3.3.1, resulting in just a multiplicity of N^2 for each state in the string spectrum (note that for oriented strings, the ab and

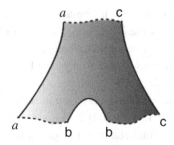

Figure 3.14 Open string interaction vertex with Chan–Paton factors.

ba open string sectors are different). Labeling the groundstate in each open string sector by its Chan–Paton indices, the lightest states from open strings are as follows:

Sector	State	$\alpha' M^2$	$SO(24)$	26d field
$N_{\text{osc}} = 0$	$\lvert ab \rangle$	-1	1	T_{ab}
$N_{\text{osc}} = 1$	$\alpha^i_{-1} \lvert ab \rangle$	0	$\square$	A^{ab}_M

where we have momentarily denoted the oscillator number operator by N_{osc}. The open string theory in Section 3.3.1 can be regarded as the particular case $N = 1$.

We obtain N^2 gauge bosons and N^2 scalar tachyons. The multiplicity of gauge bosons corresponds to a non-abelian enhancement of the gauge group to $U(N)$, as suggested by the $N \times N$ matrix structure of the states. A more detailed argument is as follows. The gauge boson A_{aa} in each aa open string sector generates a $U(1)$ gauge factor. Using the rules of open interactions in Figure 3.14, the gauge bosons A_{aa}, A_{bb} for $a \neq b$ do not interact among themselves, since they do not have common indices, and generate a $U(1)^N$ Cartan subalgebra. On the other hand, ab open string states interact with the A_{aa} and A_{bb} gauge bosons through the corresponding endpoint. From Figure 3.14 it can be shown that they carry charges $(+1, -1)$ under the generators $U(1)_a$, $U(1)_b$. Since charges of gauge bosons under Cartan generators correspond to roots of the gauge Lie algebra, the gauge group is enhanced by $N^2 - N$ non-zero roots of the form

$$(+1, -1, 0, \ldots, 0), \tag{3.105}$$

where underlining means permutation of the non-zero entries. These are the non-zero roots of $U(N)$, so the gauge group in a theory of open strings with N-valued Chan–Paton indices is $U(N)$. Similarly, the tachyonic scalars and other states in the open string sector transform in the adjoint representation of $U(N)$.

The fact that open string interactions, and in particular gauge interactions, take place through their endpoints, allows to assign gauge quantum numbers to the latter. The endpoints $\sigma = 0, \ell$ transform in the $\square_{+1}, \overline{\square}_{-1}$ representation under $U(N) = SU(N) \times U(1)$, in Young tableaux notation for the fundamental and anti-fundamental. The $U(1)$ charge, denoted by the subindices, is always correlated with the non-abelian representation, so it is often not made explicit. In this view, Chan–Paton indices label states in the non-trivial gauge representation carried by the string endpoints. The above assignment reproduces the quantum numbers of open strings, which carry two oppositely charged endpoints and transform in the $\square \times \overline{\square}$, which is the adjoint of $U(N)$. It is often convenient to package similar states in all the ab sectors into a single linear combination

$$\sum_{ab} \lambda_{ab}\, \mathcal{O}\, |ab\rangle, \qquad (3.106)$$

where $\mathcal{O}$ contains the oscillator part. The $N \times N$ matrix of coefficients λ is hermitian and provides a representation of $U(N)$ as the tensor product $\square \times \overline{\square}$, represented by the rows and columns of λ.

As already mentioned, the value of N is unconstrained in the bosonic theory. In particular, open–closed duality is obeyed trivially, as the cylinder amplitude is just (3.102), up to a multiplicity factor of N^2. The absence of constraints for N in the bosonic theory is related to the presence of open string tachyons. Indeed, the interpretation of open string sectors in Section 6.1 allows to regard N as the number of D-branes. In the bosonic theory these are unstable and can decay by tachyon condensation, thus interpolating among different values of N. In superstrings, open strings sectors describe stable D-branes, and their multiplicity N corresponds to a conserved charge, which is constrained by a topological consistency condition of the vacuum.

It is remarkable that the simple addition of non-dynamical Chan–Paton degrees of freedom leads to the rich dynamics of non-abelian gauge symmetry from the viewpoint of spacetime. This introduces an extremely simple way to obtain non-abelian gauge symmetries in string theory, exploited to build particle physics models in Chapters 10 and 11. We postpone the study of other properties of open string sectors, like their effective action or behavior under compactification and T-duality, to later chapters.

3.4 Unoriented bosonic string theory*

The theories we have been considering are oriented string theories, described by orientable worldsheets. We now turn to unoriented string theories, and illustrate their construction for closed and open bosonic strings. The main motivation to study unoriented string theories is their role in the construction of the type I superstring, Section 4.4, and other type II orientifolds. We advise the reader not interested in these constructions to skip this section in a first reading.

3.4.1 Generalities of unoriented string theories

Unoriented string theories can be defined by simply including unorientable worldsheets in the genus expansion of the theory. It is however convenient to present an equivalent but more intuitive construction, known as orientifold quotient or orientifolding. Consider a

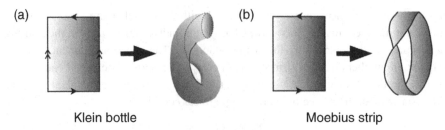

Klein bottle   Moebius strip

Figure 3.15 The Klein bottle (a) and Moebius strip (b) geometries as rectangles with suitable identifications of sides indicated by arrows.

string theory with identical left- and right-moving worldsheet sectors, like closed bosonic string theory.[2] We would like to construct a new string theory as a quotient by the symmetry Ω, called worldsheet parity, that exchanges left and right movers. In other words, the new theory is obtained by imposing that states related by left–right exchange are considered equivalent

$$|A\rangle_L \otimes |B\rangle_R \equiv |A\rangle_R \otimes |B\rangle_L. \tag{3.107}$$

Since Ω flips the orientation of the string, this quotient is known as orientifold.

This identification implies a drastic modification of the genus expansion, as already manifest for instance in the one-loop vacuum diagrams. The usual contribution is the torus, which corresponds to closed string states which evolve and are glued back to the original state. In the orientifold quotient theory, we can in addition consider processes in which the closed string state is glued back to the original, up to the action of Ω, i.e. orientation reversal. The diagram, shown in Figure 3.15(a), corresponds to a Klein bottle. A similar argument for open strings produces worldsheets with Moebius strip geometry, see Figure 3.15(b). The genus expansion of orientifold theories thus contains unorientable worldsheets, so they are unoriented string theories. The terms unoriented theories and orientifold theories are used interchangeably.

A general worldsheet for an unoriented string theory can be described as a sphere with an arbitrary number of handles and crosscaps (and boundaries if the theory includes open strings). A crosscap is a local surgery operation that introduces non-orientability in a surface. It corresponds to cutting a small disk in the surface and identifying antipodal points in the resulting boundary to close back the surface. For instance, a sphere with one crosscap corresponds to $\mathbf{RP}_2$, the real projective plane. Several non-orientable surfaces are displayed in this representation in Figure 3.16. In the genus expansion of an unoriented string theory, an amplitude mediated by a worldsheet with g handles, n_c crosscaps, and n_b boundaries is weighted by a factor of $g_s^{-\chi} = e^{-\chi\phi}$, where ϕ is the dilaton vev and χ is the general Euler characteristic

$$\chi = 2 - 2g - n_c - n_b. \tag{3.108}$$

---

[2] Or also type IIB superstring, Section 4.4.2. Giving up maximal Poincaré invariance, as in compactifications, allows the introduction of slightly modified orientifold quotients, which are valid for type IIA theories, see Sections 5.3.2 and 10.1.3.

RP^2 Klein bottle   Moebius strip

Figure 3.16 Several examples of non-orientable surfaces constructed by glueing crosscaps and boundaries to a sphere.

Unorientability is a global property of the surface, hence the orientifold quotient does not modify the local structure on the worldsheet. In particular, properties like the critical dimension of the orientifold theory are inherited from the parent oriented theory.

The spectrum of the unoriented theory is obtained from the spectrum of the parent oriented theory by truncating onto Ω invariant states. This also follows from the genus expansion: the spectrum of a theory can be obtained by extracting the states that contribute in the one-loop vacuum amplitude. The sum of one-loop vacuum contributions from the torus and the Klein bottle can be written in terms of the parent Hilbert space $\mathcal{H}_{\text{or.}}$ sketchily as

$$\text{Tr}_{\mathcal{H}_{\text{or.}}}\, q^{H_L}\bar{q}^{H_R} + \text{Tr}_{\mathcal{H}_{\text{or.}}}\, q^{H_L}\bar{q}^{H_R}\Omega = 2\,\text{Tr}_{\mathcal{H}_{\text{or.}}} \left[q^{H_L}\bar{q}^{H_R} \tfrac{1}{2}(1+\Omega) \right].$$

The piece $\tfrac{1}{2}(1+\Omega)$ is a projector that only keeps Ω invariant states. The sum over topologies thus projects out states non-invariant under Ω. A similar conclusion holds for open strings, by summing over the cylinder and Moebius strip diagrams.

3.4.2 Unoriented closed bosonic string

Let us focus on closed bosonic string theory to flesh out these ideas, and obtain the precise action of Ω on the string states. Since the local dynamics on the worldsheet is unchanged, it is enough to determine the Ω action for the degrees of freedom remaining after gauge fixing. The action Ω maps the worldsheet fields $X^i(t,\sigma)$ to $X_\Omega^i(t,\sigma)$ defined by the condition

$$X_\Omega^i(t,\sigma) = X^i(t,\ell-\sigma). \tag{3.109}$$

where the sign flip in σ encodes worldsheet orientation reversal. Using the oscillator expansion (3.40) we obtain

$$\Omega:\; x^i \to x^i;\quad p^i \to p^i;\quad \alpha_n^i \leftrightarrow \tilde{\alpha}_n^i, \tag{3.110}$$

which corresponds to an exchange of the left and right movers, as expected. Defining the groundstate as Ω-invariant, the spectrum of the quotient theory is obtained by applying Ω-invariant combinations of creation operators. The lightest states and their properties are:

Sector	State	$\alpha' M^2$	$SO(24)$	26d field
$N = \tilde{N} = 0$	$\lvert 0 \rangle$	-4	$\mathbf{1}$	T
$N = \tilde{N} = 1$	$\alpha_{-1}^{(i} \alpha_{-1}^{j)} \lvert 0 \rangle$	0	$\square + \mathbf{1}$	G_{MN}, ϕ

$$(3.111)$$

Besides the closed string tachyon, the theory contains a massless 26d graviton and a massless dilaton (whose vev controls the string coupling g_s). The massless 2-form B_{MN} has been projected out by the orientifold.

There is an interesting new ingredient in the UV behavior of orientifold theories, as compared to their parent oriented theories. Already at the one-loop level there is a new diagram, the Klein bottle, corresponding to a closed string propagating for a time $T\ell$ and closing onto itself up to the action of Ω. The UV regime corresponds to $T \to 0$, and can be mapped to the IR of a dual channel of small closed strings propagating between two crosscaps for a long time $T'\ell$ with $T' = 1/(4T) \to \infty$, see Figure 3.17(a). The computation of Klein bottle diagrams, in the physically more relevant context of superstrings, is further discussed in Appendix D.

3.4.3 Unoriented open bosonic string

As already suggested, it is also possible to perform an orientifold quotient of a theory with open string sectors. Here we describe the orientifold quotient of the open bosonic strings. Given that the main ideas are already familiar, we sketch the computation of the spectrum, emphasizing only the main novelties.

The action of Ω maps the fields $X^i(t, \sigma)$ to $X^i_\Omega(t, \sigma)$ satisfying (3.109). Using the mode expansions (3.91), we get

$$\Omega : \quad x^i \to x^i; \quad p^i \to p^i; \quad \alpha_n^i \leftrightarrow (-1)^n \tilde{\alpha}_n^i, \qquad (3.112)$$

which exchanges left and right sectors, up to a sign. The action of Ω on the oscillator groundstate can be taken to be trivial.

For an open string theory with N-valued Chan–Paton indices, there is a non-trivial action of Ω on the latter. Indeed, Ω flips the orientation of the open string, and exchanges the two endpoints, mapping the ab open string sector to the ba sector. In terms of the Chan–Paton matrix λ in (3.106), the action of Ω can be expressed in terms of a unitary matrix γ_Ω acting as

$$\Omega : \quad \lambda \to \gamma_\Omega \lambda^T \gamma_\Omega^{-1}. \qquad (3.113)$$

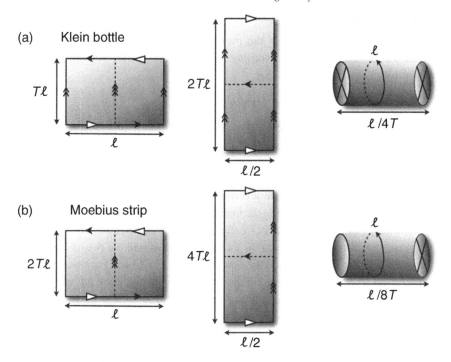

Figure 3.17 Dual channels for the Klein bottle (a) and Moebius strip (b). The topology of the dual channel becomes manifest by cutting the left-most diagrams along the dashed line, and glueing them as in the middle diagram. The right-most diagrams provide the dual channel interpretation as a tree-level closed string exchange between two crosscaps (a), or a crosscap and a boundary (b).

The transposition of λ exchanges rows and columns, and corresponds to exchange of string endpoints. Applying this twice one obtains

$$\lambda \to \gamma_\Omega \left(\gamma_\Omega^T \right)^{-1} \lambda \gamma_\Omega^T (\gamma_\Omega)^{-1}. \tag{3.114}$$

The condition $\Omega^2 = 1$ then requires

$$\gamma_\Omega = \pm \gamma_\Omega^T. \tag{3.115}$$

For the two sign choices, it can be brought without loss of generality to the form

$$O^- \text{projection}: \qquad\qquad \gamma_\Omega = \mathbf{1}_N,$$

$$O^+ \text{projection}: \gamma_\Omega = i\epsilon_{N/2} \equiv i \begin{pmatrix} 0 & \mathbf{1}_{N/2} \\ -\mathbf{1}_{N/2} & 0 \end{pmatrix} \quad \text{for } N \text{ even.} \tag{3.116}$$

The two projections lead to different unoriented open string theories, and are also known as SO and Sp projections for reasons to become clear shortly. Note that the O^+-projection is possible only for N even.

The spectrum of the unoriented string theory is obtained by keeping the states invariant under the combined action of Ω on the oscillators and Chan–Paton indices. The components of the groundstate $\lambda_{ab}|ab\rangle$ (or any state with even total oscillator number) are Ω-invariant for

$$\lambda_{ab}|ab\rangle \rightarrow \lambda = \gamma_\Omega \lambda^T \gamma_\Omega^{-1}. \tag{3.117}$$

Also the components of the massless state $\lambda_{ab}\alpha^i_{-1}|ab\rangle$ (or any other state with odd oscillator number) are invariant for

$$\lambda_{ab}\,\alpha^i_{-1}|ab\rangle \rightarrow \lambda = -\gamma_\Omega \lambda^T \gamma_\Omega^{-1}. \tag{3.118}$$

For the O^- projection the groundstate survives for $\lambda = \lambda^T$, yielding $N(N+1)/2$ surviving tachyons, out of the original N^2, while the first excited states survive for $\lambda = -\lambda^T$, yielding $N(N-1)/2$ gauge bosons. The matrix structure implies that the gauge bosons produce an $SO(N)$ gauge symmetry, under which the tachyons transform in the two-index symmetric representation. For the O^+ projection the conditions (3.117), (3.118) can be shown to lead to $USp(N)$ gauge bosons and tachyons in the two-index antisymmetric representation. The light states for the $O^\pm$ projections are thus:

Projection	Sector	State	$\alpha'M^2$	$SO(24)$	26d field	Gauge
O^-	$N_{osc}=0$	$\lambda\,|0\rangle$	-1	$\mathbf{1}$	T	$\Box\Box$ $+$ $\mathbf{1}$
	$N_{osc}=1$	$\alpha^i_{-1}\lambda\,|0\rangle$	0	$\Box$	A_M	$SO(N)$
O^+	$N_{osc}=0$	$\lambda\,|0\rangle$	-1	$\mathbf{1}$	T	$\Box\hspace{-0.4em}\Box$ $+$ $\mathbf{1}$
	$N_{osc}=1$	$\alpha^i_{-1}\lambda\,|0\rangle$	0	$\Box$	A_M	$USp(N)$

where we use Young tableaux notation for Lorentz and gauge representations, with $\Box\Box$ denoting the two-index symmetric traceless representation, and "$+1$" the trace piece. The above gauge representations are consistent with the gauge representation assignments for single open string endpoints, as follows. Since Ω flips the string orientation, exchanging the two endpoints, it identifies the $\Box$ at $\sigma=0$ with the $\overline{\Box}$ at $\sigma=\ell$. The open string thus transforms in a representation contained in the tensor product $\Box \times \Box = \Box\Box + \Box\hspace{-0.4em}\Box + \mathbf{1}$ (determined by the precise Ω action as described).

Similar to closed unoriented strings, the UV behavior of unoriented open string amplitudes is controlled by a new interesting dual channel in worldsheet diagrams. Indeed, consider the new one-loop vacuum diagram in open unoriented theories, namely the Moebius strip. It can be described as an open string propagating a time $2T\ell$ and closing onto itself up to the action of Ω. Its UV limit $T \rightarrow 0$ actually corresponds to an IR diagram in a dual channel of small closed strings propagating for a long time $T'\ell$ with $T' = 1/(8T) \rightarrow \infty$ between a boundary and a crosscap, see Figure 3.17(b). The computation of Moebius strip diagrams, in the physically more relevant context of superstrings, is further discussed in Appendix D.

4

Superstrings

In this chapter we construct the superstring theories in flat Minkowski spacetime. Their main advantage with respect to bosonic string theory is that they lead to spacetime fermions. In addition, they naturally lead to spacetime supersymmetry, which guarantees the absence of spacetime tachyons, and thus provide stable vacua. There are five such superstring theories, called the type IIA and IIB theories, the $E_8 \times E_8$ and $SO(32)$ heterotic theories, and type I theory, which we study in that order. These superstrings are the best studied string theories, and have been efficiently applied to the construction of realistic particle physics models, to be developed in later chapters.

4.1 Fermions on the worldsheet

We would like to construct new string theories which avoid the shortcomings of bosonic string theory, and in particular can lead to spacetime fermions. This requires modifying the worldsheet theory. However, the modification cannot amount to just changing the worldsheet action keeping the same 2d fields, since as discussed in Section 3.1.5 this just leads to the same bosonic string theory in a different spacetime background. Rather, we need to modify the worldsheet field content.

The most elegant modifications of the field content of a theory are those based on a symmetry principle; this is exploited by superstrings, whose worldsheet theory contains additional 2d fermionic fields as required by a 2d *super*symmetry on the worldsheet (hence the name). Concretely, there is a set of fermionic fields $\psi^M(\xi)$, partners of $X^M(\xi)$, and which are 2d spinors, but transform as vectors under the spacetime Lorentz group (since supersymmetry commutes with global symmetries). In addition, there is a worldsheet gravitino $\psi_a(\xi)$, related to $g_{ab}(\xi)$. It is important to emphasize that the presence of *worldsheet fermions* and *worldsheet supersymmetry* does not automatically imply the appearance of *spacetime fermions* and *spacetime supersymmetry*.[1] The subtle appearance of the latter is discussed in Section 4.2.3.

---

[1] Indeed there exist string theories, like the type 0 theories, with worldsheet fermions and supersymmetry, but with neither spacetime fermions nor spacetime supersymmetry. These theories have essentially no application to particle physics model building, and will not be discussed in this book.

Despite the new 2d field content, general results based on geometrical properties of the worldsheet extend to the superstrings. For instance, the description of the genus expansion, the reinterpretation of UV regimes as IR dual channels; or symmetry properties, like invariance under 2d diffeomorphisms, Weyl and conformal symmetries, their gauge-fixing, and the level matching conditions for closed strings. This facilitates the quantization procedure, since it suffices to supersymmetrize the worldsheet field content and hamiltonian of the bosonic theory after light-cone gauge fixing. Therefore, the propagation of superstrings on D-dimensional Minkowski space will involve $(D-2)$ free bosonic fields $X^i(t,\sigma)$ and free fermionic fields $\psi^i(t,\sigma)$, for $i = 2,\ldots,D-1$. Note, however, that since the field content differs from that in the bosonic theory, the critical dimension will be different. We therefore keep the spacetime dimension arbitrary, and eventually determine it from restoration of the appropriate symmetries in the quantum theory.

We now turn to the study of the different superstring theories, and from now drop the implicitly understood prefix "super."

4.2 Type II string theories

Type II string theory is a theory of closed strings, with worldsheet fields $X^i(t,\sigma)$, $\psi^i(t,\sigma)$, $i = 2,\ldots,D-1$, in the light-cone gauge. The theory describing the propagation of the string on Minkowski space is free, and the fields obey 2d wave equations. As for the closed bosonic string theory, it is conceptually useful to split the worldsheet theory into left and right sectors, with fields $X_L^i(t+\sigma)$, $\psi_L^i(t+\sigma)$, $X_R^i(t-\sigma)$, $\psi_R^i(t-\sigma)$. These are decoupled (except for a level matching condition in forming physical states), so we will describe most features for just, e.g., the left sector.

4.2.1 NS and R boundary conditions

Since we have closed strings, worldsheet bosonic fields obey periodic boundary conditions

$$X_L^i(t+\sigma+\ell) = X_L^i(t+\sigma), \tag{4.1}$$

leading to an oscillator expansion (3.40)

$$X_L^i(t+\sigma) = \frac{x^i}{2} + \frac{p_i}{2p^+}(t+\sigma) + i\sqrt{\frac{\alpha'}{2}}\sum_{n\in\mathbf{Z}-\{0\}}\frac{\alpha_n^i}{n}e^{-2\pi i n(t+\sigma)/\ell}. \tag{4.2}$$

These modes have commutation relations

$$[x^i, p_j] = i\delta_j^i; \quad \left[\alpha_m^i, \alpha_n^j\right] = m\,\delta_{ij}\,\delta_{m,-n} \tag{4.3}$$

The hamiltonian for bosons (thus labeled with B) is

$$H_{B,L} = \frac{\sum_i p_i^2}{4p^+} + \frac{1}{\alpha'p^+}\left(\sum_i\sum_{n>0}\alpha_{-n}^i\alpha_n^i + E_0^B\right), \quad \text{with } E_0^B \overset{\text{reg.}}{=} -\frac{1}{24}(D-2),$$

Concerning the fermionic fields, their local dynamics correspond to free oscillations, as expected from their 2d supersymmetry relation to the bosons. Since any observable of the 2d theory is quadratic in the fermions, periodicity of physical observables is consistent with periodic or antiperiodic fermion fields. Hence there are two consistent boundary conditions:

$$
\begin{aligned}
\text{Neveu–Schwarz NS} \quad & \psi_L^i(t + \sigma + \ell) = -\psi_L^i(t + \sigma), \\
\text{Ramond} \quad \text{R} \quad & \psi_L^i(t + \sigma + \ell) = \psi_L^i(t + \sigma).
\end{aligned}
\tag{4.4}
$$

These can be chosen independently for left and right sectors,[2] but the choice must be shared by all values of i due to Lorentz invariance. It would seem that this produces four different kinds of closed strings, namely NS-NS, NS-R, R-NS, and R-R strings, according to the boundary conditions obeyed by the left and right fermions. We will see in Section 4.2.2 that modular invariance actually forces these possibilities to coexist, and form just different *sectors* of the same theory.

We now construct the fermion Hilbert spaces for NS and R boundary conditions.

Neveu–Schwarz (NS) sector

Antiperiodic boundary conditions require the oscillator modding to be half-integer, leading to the oscillator expansion

$$
\psi_L^i(t + \sigma) = i\sqrt{\frac{\alpha'}{2}} \sum_{r \in \mathbf{Z}} \psi_{r+1/2}^i \, e^{-2\pi i(r+1/2)(t+\sigma)/\ell}.
\tag{4.5}
$$

Note that there are no zero modes in the expansion. The oscillator modes have anticommutators

$$
\left\{ \psi_{m+1/2}^i, \psi_{n+1/2}^j \right\} = \delta^{ij} \, \delta_{m+1/2, -(n+1/2)}.
\tag{4.6}
$$

The hamiltonian for the fermions in the NS sector (thus labeled with F_{NS}) is

$$
H_{F_{\text{NS}}, L} = \frac{1}{\alpha' p^+} \left[\sum_{r=0}^{\infty} \left(r + \tfrac{1}{2} \right) \psi_{-r-1/2}^i \psi_{r+1/2}^i + E_0^{F_{NS}} \right],
\tag{4.7}
$$

where the zero point energy for NS fermionic oscillators is

$$
E_0^{F_{\text{NS}}} = -\frac{1}{2}(D - 2) \sum_{n=0}^{\infty} \left(n + \tfrac{1}{2} \right) \overset{\text{reg.}}{\equiv} -\frac{1}{48}(D - 2).
\tag{4.8}
$$

This follows from a zeta-function regularization of the sum, generalizing (3.50)

$$
Z_\alpha = \frac{1}{2} \sum_{n=0}^{\infty} (n + \alpha) \overset{\text{reg.}}{\equiv} -\frac{1}{24} + \frac{1}{4}\alpha(1 - \alpha) \quad \text{for } 0 \le \alpha \le 1.
\tag{4.9}
$$

---

[2] To prevent confusions, we use the roman font R for "Ramond," and the italic *R* for "right."

The total bosonic and fermionic hamiltonian for the 2d theory in the NS sector is

$$H_{NS,L} = \frac{\sum_i p_i^2}{4p^+} + \frac{1}{\alpha' p^+} \left(N_{F_{NS}} + N_B - \frac{D-2}{16} \right),$$ (4.10)

where we have introduced the fermionic and bosonic number operators

$$N_{F_{NS}} = \sum_i \sum_{r=0}^{\infty} \left(r + \tfrac{1}{2} \right) \psi^i_{-r-1/2} \psi^i_{r+1/2}, \quad N_B = \sum_i \sum_{n>0} \alpha^i_{-n} \alpha^i_n.$$ (4.11)

The contribution of the left-moving sector to the spacetime mass is

$$M^2_{NS,L} = 2p^+ H_L - \frac{1}{2} \sum_i p_i^2,$$ (4.12)

namely

$$\frac{\alpha' M^2_{NS,L}}{2} = N_{F_{NS}} + N_B - \frac{D-2}{16}.$$ (4.13)

The spectrum in the NS sector is obtained by defining a groundstate $|0\rangle_{NS}$ annihilated by all positive modding oscillators

$$\psi^i_{r+1/2}|0\rangle_{NS} = 0 \quad \forall i, \forall r \geq 0; \quad \alpha^i_n|0\rangle_{NS} = 0 \quad \forall i, \forall n > 0,$$ (4.14)

and applying negative modding oscillators in all possible ways. Note that each fermion oscillator can be applied at most once, due to their anticommuting character. In Young tableaux notation, the left-moving states with smallest contribution to the mass are as follows:

State	$\left(\alpha' M^2_{NS,L} \right)/2$	$SO(D-2)$
$|0\rangle_{NS}$	$-\frac{(D-2)}{16}$	**1**
$\psi^i_{-1/2}|0\rangle_{NS}$	$\frac{1}{2} - \frac{(D-2)}{16}$	$\square$

(4.15)

It is already possible to compute the critical dimension from the proper implementation of Lorentz symmetry in the quantum theory. The above states will eventually combine with states in the right sector, satisfying the level matching condition $M_L^2 = M_R^2$, to form physical states corresponding to spacetime particles of mass $M^2 = M_L^2 + M_R^2 = 2M_L^2$. So already at this stage the left states must form representations of $SO(D-1)$ or $SO(D-2)$ little groups for massive or massless particles, i.e. $M_L \neq 0$ and $M_L = 0$, respectively. The states in the first excited level form a representation of $SO(D-2)$ and do not suffice to form a representation of $SO(D-1)$. Therefore they must be the left part of a massless particle, with $M_L = 0$. This fixes the value of the critical dimension $D = 10$. Using this value, the above table of light left states in an NS sector reads:

State	$\left(\alpha' M_{NS,L}^2\right)/2$	$SO(8)$
$\|0\rangle_{NS}$	$-\frac{1}{2}$	$\mathbf{1}$
$\psi_{-1/2}^i\|0\rangle_{NS}$	0	$\mathbf{8}_V$

$$(4.16)$$

where $\mathbf{8}_V$ denotes the vector representation of $SO(8)$.

Ramond (R) sector

Periodic boundary conditions require integer modding for fermionic oscillators,

$$\psi_L^i(t+\sigma) = i\sqrt{\frac{\alpha'}{2}} \sum_{r \in \mathbf{Z}} \psi_r^i\, e^{-2\pi i r(t+\sigma)/\ell}. \tag{4.17}$$

The anticommutation relations read

$$\left\{\psi_m^i, \psi_n^j\right\} = \delta^{ij}\,\delta_{m,-n}, \tag{4.18}$$

and the hamiltonian for the fermions in the R sector (thus labeled with F_R) is

$$H_{F_R,L} = \frac{1}{2\alpha' p^+}\left(\sum_{r=1}^{\infty} r\,\psi_{-r}^i\,\psi_r^i + E_0^{F_R}\right), \tag{4.19}$$

with

$$E_0^{F_R} = -\frac{1}{2}(D-2)\sum_{r=1}^{\infty} r \stackrel{\text{reg.}}{\equiv} \frac{1}{24}(D-2) \;\to\; E_0^{F_R} \stackrel{\text{reg.}}{\equiv} \frac{1}{3}, \tag{4.20}$$

where we anticipate that R and NS sectors coexist, and already use the critical dimension $D = 10$. The total hamiltonian in the R sector, and the left-moving contribution to the spacetime mass, are

$$H_{R,L} = \frac{\sum_i p_i^2}{4p^+} + \frac{1}{\alpha' p^+}\left(N_{F_R} + N_B\right),$$

$$\frac{\alpha' M_{R,L}^2}{2} - N_{F_R} + N_B. \tag{4.21}$$

Note the vanishing zero point energy due to 2d boson–fermion cancellation. We have introduced the oscillator number operators

$$N_{F_R} = \sum_{r=1}^{\infty} r\,\psi_{-r}^i\,\psi_r^i, \quad N_B = \sum_{n>0} \alpha_{-n}^i \alpha_n^i. \tag{4.22}$$

The groundstate is naively defined as annihilated by positive modding oscillators

$$\psi_r^i |0\rangle = 0 \quad \forall i, \forall r > 0; \quad \alpha_n^i |0\rangle = 0 \quad \forall i, \forall n > 0, \tag{4.23}$$

However, an important difference with respect to the NS sector is that there are fermion oscillator zero modes ψ_0^i, whose contribution to the string state energy is zero. This implies that the groundstate is degenerate, with the fermion zero mode operators relating different states. This action must be consistent with the anticommutation relations (4.18) restricted to the zero mode sector

$$\left\{ \psi_0^i, \psi_0^j \right\} = \delta^{ij}, \tag{4.24}$$

so the set of groundstates must form a representation of this Clifford algebra. Its construction is sketched below, although the result can be anticipated: the operators ψ_0^i behave as Dirac matrices of the $SO(8)$ symmetry group, acting on the set of groundstates which thus must transform as an $SO(8)$ spinor. To carry out the construction in more detail, we define the operators

$$A_a^{\pm} = \frac{1}{\sqrt{2}} \left(\psi_0^{2a} \pm i \psi_0^{2a+1} \right) \quad \text{for } a = 1, \dots, 4, \tag{4.25}$$

with anticommutation relations

$$\left\{ A_a^-, A_b^+ \right\} = \delta_{ab} \tag{4.26}$$

and all other anticommutators equal to zero. We define a "lowest weight state" by $A_a^- |0\rangle = 0$, and build the set of groundstates by application of the A_a^+ operators, i.e.

$$\begin{array}{cc} |0\rangle & A_{a_1}^+ |0\rangle \\ A_{a_1}^+ A_{a_2}^+ |0\rangle & A_{a_1}^+ A_{a_2}^+ A_{a_3}^+ |0\rangle \\ A_1^+ A_2^+ A_3^+ A_4^+ |0\rangle & \end{array} \tag{4.27}$$

They transform in a 16-component spinor representation of $SO(8)$. This representation describes a non-chiral spinor, and is reducible into two opposite-chirality representations. These are denoted $\mathbf{8}_S$, $\mathbf{8}_C$ and correspond to the two columns in (4.27), which have opposite eigenvalues of the parity operator $\Gamma = \psi_0^2 \cdots \psi_0^9$.

Since any excited state is massive, the table of light states in a R sector only includes the groundstates just discussed:

State	$\left(\alpha' M_{R,L}^2 \right)/2$	$SO(8)$
$|\mathbf{8}_C\rangle$	0	$\mathbf{8}_C$
$|\mathbf{8}_S\rangle$	0	$\mathbf{8}_S$

(4.28)

4.2.2 Modular invariant partition functions

To form physical states corresponding to spacetime particles, we must combine left and right states, each in the NS or R sector, obeying the level matching condition

$$M_L^2 = M_R^2. \tag{4.29}$$

The glueing is constrained by the eventual requirement of modular invariance of the resulting theory. This property is so restrictive that it is in fact more practical to first construct modular invariant partition functions, and later determine the corresponding glueing prescriptions. The reader not interested in these details may wish to skim through this and the next section, directly to Section 4.2.4.

The closed string partition functions splits into left and right sector partition functions

$$Z(\tau) = \text{Tr}_{\mathcal{H}_{\text{cl.}}} \left(e^{-\tau_2 \ell H} e^{i\tau_1 \ell P} \right) = (4\pi^2 \alpha' \tau_2)^{-4} \text{Tr}_{\mathcal{H}_L} q^{N+E_0} \text{Tr}_{\mathcal{H}_R} \bar{q}^{\tilde{N}+\tilde{E}_0}, \tag{4.30}$$

where the first factor is the center of mass momentum piece, and N, E_0 denote the total (bosonic and fermionic) left-moving oscillator number operator and zero point energy (and analogously for the right-moving $\tilde{N}$, $\tilde{E}_0$).

Since the partition function factorizes, we now consider the structure of the partition function for left movers, in the NS or R sector. As usual, the traces over the different degrees of freedom (fermions and bosons, for each spacetime direction, and each oscillator modding) factorize, and we get products of traces. The trace over the bosonic oscillators is computed just like in bosonic string theory

$$\text{Tr}_{\mathcal{H}_{\text{bos.}}} q^{N_B + E_0^B} = \eta(\tau)^{-8}. \tag{4.31}$$

To compute the partition function over the infinite set of fermionic oscillators, consider first the case of a single fermionic oscillator, ψ_ν^i, $\psi_{-\nu}^i$, with general modding $\nu > 0$. Its Hilbert space has just two states, the vacuum $|0\rangle$ and $\psi_{-\nu}^i|0\rangle$, so its contribution to the partition function is

$$\text{Tr}_{\mathcal{H}_{\nu,i}} q^{N_{F_\nu} + E_0^{F_\nu}} = q^{E_0^{F_\nu}} (1 + q^\nu). \tag{4.32}$$

For several decoupled fermionic harmonic oscillators, we simply get the product of partition functions for the individual ones. The results can be written compactly in terms of the modular functions defined in Appendix A. For instance, the partition function for eight NS fermionic coordinates is given by

$$\text{Tr}_{\mathcal{H}_{\text{NS}}} q^{N_{F_{\text{NS}}} + E_0^{F_{\text{NS}}}} = \left[q^{-1/48} \prod_{n=1}^{\infty} (1 + q^{n-1/2}) \right]^8 = \frac{\vartheta \begin{bmatrix} 0 \\ 0 \end{bmatrix}^4}{\eta^4}. \tag{4.33}$$

The partition function for eight R fermionic coordinates is given by

$$\mathrm{Tr}_{\mathcal{H}_R}\, q^{N_{F_R}+E_0^{F_R}} = 16 \left[q^{1/24} \prod_{n=1}^{\infty}(1+q^n) \right]^8 = \frac{\vartheta \begin{bmatrix} 1/2 \\ 0 \end{bmatrix}^4}{\eta^4}, \tag{4.34}$$

where we have taken into account the 16-fold degeneracy of the groundstate.

Using (A.12), it is easy to observe that modular transformations mix sectors with different boundary conditions. For instance, (4.33) and (4.34) are related by

$$-\frac{\vartheta \begin{bmatrix} 0 \\ 0 \end{bmatrix}^4}{\eta^4} \xrightarrow{\tau \to \tau+1} \frac{\vartheta \begin{bmatrix} 0 \\ 1/2 \end{bmatrix}^4}{\eta^4} \xrightarrow{\tau \to -1/\tau} \frac{\vartheta \begin{bmatrix} 1/2 \\ 0 \end{bmatrix}^4}{\eta^4}. \tag{4.35}$$

This implies that a modular invariant theory must contain coexisting string states in different sectors. The precise glueing prescription is most easily described by first proposing a modular invariant partition function, and later extracting its physical interpretation. Consider the two partition functions for left-moving fermions

$$Z_{\pm} = \frac{1}{2}\, \eta^{-4} \left(\vartheta \begin{bmatrix} 0 \\ 0 \end{bmatrix}^4 - \vartheta \begin{bmatrix} 0 \\ 1/2 \end{bmatrix}^4 - \vartheta \begin{bmatrix} 1/2 \\ 0 \end{bmatrix}^4 \pm \vartheta \begin{bmatrix} 1/2 \\ 1/2 \end{bmatrix}^4 \right), \tag{4.36}$$

where we have introduced a factor of $1/2$ for latter convenience. The terms added up in this partition functions are sometimes referred to as "spin structures," as they relate to the different periodicities of worldsheet fermions along the (t, σ) directions on the toroidal geometry.

The two choices $Z_{\pm}$ are invariant under $\tau \to \tau + 1$ and $\tau \to -1/\tau$, up to phase factors, hence we can construct two modular invariant theories with (left $\times$ right) partition functions

$$\text{type IIB}: \quad Z_+\overline{Z}_+, \quad \text{type IIA}: \quad Z_+\overline{Z}_-. \tag{4.37}$$

They define the type IIB and type IIA string theories. The theories defined by $Z_-\overline{Z}_+$, $Z_-\overline{Z}_-$ are actually related to the above by spacetime parity, as explained shortly, and do not lead to new theories.

Note that the theta function identity (A.8) implies the vanishing of the partition function. This signals the cancellation of the one-loop spacetime vacuum energy, due to spacetime supersymmetry of these theories. However, formal modular invariance of the above expressions is still a meaningful criterion, which ensures the finiteness of string amplitudes with external legs, which do not vanish identically. Similarly, the theta function with characteristic $(1/2, 1/2)$, although formally zero, must be present in (4.36), and its sign has implications for the physical spectrum.

4.2.3 GSO projection: the type IIA and type IIB theories

We now consider the physical interpretation of the above fermionic oscillator partition functions. The combination of theta functions with first characteristic 0 and 1/2 implies the coexistence of states in NS and R sectors. On the other hand, the presence of theta functions with different second characteristic implies a projection onto states with even or odd worldsheet fermion number, known as the Gliozzi–Scherk–Olive or GSO projection, as follows.

Let us introduce the operator $(-1)^F$, anticommuting with any fermionic oscillator

$$(-1)^F \psi_\nu^i = -\psi_\nu^i (-1)^F. \tag{4.38}$$

String states with even or odd number of fermionic oscillators have $(-1)^F$ eigenvalue $+1$ or -1. Using this, we can interpret the NS sector contribution to (4.36) as

$$\frac{1}{2} \eta^{-4} \left(\vartheta \begin{bmatrix} 0 \\ 0 \end{bmatrix}^4 - \vartheta \begin{bmatrix} 0 \\ 1/2 \end{bmatrix}^4 \right) = \frac{1}{2} \operatorname{Tr} \mathcal{H}_{NS} \, q^{N+E_0} - \frac{1}{2} \operatorname{Tr} \mathcal{H}_{NS} \, q^{N+E_0} (-1)^F$$

$$= \operatorname{Tr} \mathcal{H}_{NS} \left[q^{N+E_0} \frac{1 - (-1)^F}{2} \right], \tag{4.39}$$

where here N and E_0 refer to the fermion number operator and zero point energy in the NS sector. The operator $\frac{1}{2}[1 - (-1)^F]$ is a projector restricting the contribution to states with an odd number of fermionic oscillators. The GSO projection on the NS sector thus removes states with an even number of fermionic oscillators; for instance, it removes the tachyonic groundstate $|0\rangle_{NS}$ from the physical spectrum, while leaving the massless states $\psi_{-1/2}^i |0\rangle_{NS}$ in the physical spectrum.

Similarly, the R sector pieces of the partition functions (4.36) correspond to

$$Z_{\mp} : \frac{1}{2} \eta^{-4} \left(\vartheta \begin{bmatrix} 1/2 \\ 0 \end{bmatrix}^4 \pm \vartheta \begin{bmatrix} 1/2 \\ 1/2 \end{bmatrix}^4 \right) = \frac{1}{2} \operatorname{Tr} \mathcal{H}_R \, q^{N+E_0} \pm \frac{1}{2} \operatorname{Tr} \mathcal{H}_R \, q^{N+E_0} (-1)^F$$

$$= \operatorname{Tr} \mathcal{H}_R \left[q^{N+E_0} \frac{1 \pm (-1)^F}{2} \right], \tag{4.40}$$

where now N and E_0 are the fermion number operator and zero point energy in the R sector. The GSO projections implied by $Z_{\pm}$ differ in their action on the R sector. Focusing on the light states, the GSO projection associated to Z_+ projects out the R groundstates in the $\mathbf{8}_S$, while leaving those in the $\mathbf{8}_C$ in the spectrum; whereas the GSO projection associated to Z_- projects out the R groundstates in the $\mathbf{8}_C$ while leaving those in the $\mathbf{8}_S$ in the spectrum.

The two projections are related by spacetime parity, which exchanges $\mathbf{8}_S \leftrightarrow \mathbf{8}_C$. Therefore, the $Z_-\bar{Z}_-$ theory is just the parity reflected image of the $Z_+\bar{Z}_+$ theory, while the $Z_-\bar{Z}_+$ theory is the image of $Z_+\bar{Z}_-$. The two essentially different theories are the type IIB and IIA theories in (4.37), defined by having equal or opposite left and right GSO projections on the R sector. Their partition functions in terms of GSO-projected Hilbert space traces are schematically

$$
\begin{aligned}
Z_+ \overline{Z}_\mp &= (\operatorname{Tr} \mathcal{H}_{\mathrm{NS,GSO}} - \operatorname{Tr} \mathcal{H}_{\mathrm{R,GSO_-}}) \times (\operatorname{Tr} \mathcal{H}_{\mathrm{NS,GSO}} - \operatorname{Tr} \mathcal{H}_{\mathrm{R,GSO_\pm}})^* \\
&= \operatorname{Tr} \mathcal{H}_{\mathrm{NS,GSO}} \operatorname{Tr}^* \mathcal{H}_{\mathrm{NS,GSO}} - \operatorname{Tr} \mathcal{H}_{\mathrm{NS,GSO}} \operatorname{Tr}^* \mathcal{H}_{\mathrm{R,GSO_\pm}} \\
&\quad - \operatorname{Tr} \mathcal{H}_{\mathrm{R,GSO_-}} \operatorname{Tr}^* \mathcal{H}_{\mathrm{NS,GSO}} + \operatorname{Tr} \mathcal{H}_{\mathrm{R,GSO_-}} \operatorname{Tr}^* \mathcal{H}_{\mathrm{R,GSO_\pm}},
\end{aligned} \tag{4.41}
$$

in hopefully obvious notation. This determines the precise glueing of left and right sectors with NS and R fermions, into NS-NS, NS-R, R-NS and R-R states (often written unhyphenated, especially for NSNS, RR). We now turn to the computation of the spacetime spectrum. Incidentally, the negative sign in the contribution from NS-R and R-NS states anticipates that these states are spacetime fermions.

4.2.4 Massless spectrum of type IIB and IIA string theories

The modular invariant partition functions for type II theories implement a GSO projection, which in the NS sector removes the tachyonic groundstate, and leaves the state $|8_V\rangle$. On the R sector, the type IIB theory has equal GSO projections, which at the massless level remove the states $|8_S\rangle$ and leave the states $|8_C\rangle$, on both left and right sectors. The type IIA theory has different GSO projections, with the $|8_C\rangle$ surviving in the left sector, and the $|8_S\rangle$ surviving in the right sector.

The spacetime massless particle spectrum is obtained by glueing these GSO-surviving left and right states with $M_L^2 = M_R^2 = 0$, which therefore satisfy the level matching condition (4.29) automatically.

Type IIB string theory

The massless spectrum has the following structure:

Sector	$|\rangle_L \otimes |\rangle_R$	$SO(8)$	10d field
NS-NS	$8_V \otimes 8_V$	$1 + 28_V + 35_V$	ϕ, B_{MN}, G_{MN}
NS-R	$8_V \otimes 8_C$	$8_S + 56_S$	$\lambda_\alpha^1, \psi_{M\alpha}^1$
R-NS	$8_C \otimes 8_V$	$8_S + 56_S$	$\lambda_\alpha^2, \psi_{M\alpha}^2$
R-R	$8_C \otimes 8_C$	$1 + 28_C + 35_C$	a, C_{MN}, C_{MNPQ}

$$\tag{4.42}$$

where α is a chiral 10d spinor index. The NS-NS sector contains a dilaton scalar ϕ, a two-index antisymmetric tensor B_{MN} (a two-form B_2), and a 2-index traceless symmetric tensor G_{MN}, describing the graviton. The R-NS and NS-R sectors produce fermions, concretely two Rarita–Schwinger fields $\psi_{M\alpha}$, which are gravitinos of the spacetime supersymmetry of the theory (see below), and two spinors λ_α, known as dilatinos. The R-R sector contains several *p*-forms (*p*-index antisymmetric tensors), namely a scalar (or 0-form) a, often called axion, a 2-form C_{MN}, and a 4-form C_{MNPQ} with self-dual field strength, in the sense of Section B.3. These are denoted also as C_0, C_2, C_4.

Type IIA string theory

The massless spectrum has the following structure:

Sector	$|\rangle_L \otimes |\rangle_R$	$SO(8)$	10d field
NS-NS	$\mathbf{8}_V \otimes \mathbf{8}_V$	$1 + 28_V + 35_V$	ϕ, B_{MN}, G_{MN}
NS-R	$\mathbf{8}_V \otimes \mathbf{8}_S$	$\mathbf{8}_C + \mathbf{56}_C$	$\lambda^1_{\dot\alpha}, \psi^1_{M\dot\alpha}$
R-NS	$\mathbf{8}_C \otimes \mathbf{8}_V$	$\mathbf{8}_S + \mathbf{56}_S$	$\lambda^2_\alpha, \psi^2_{M\alpha}$
R-R	$\mathbf{8}_C \otimes \mathbf{8}_S$	$\mathbf{8}_V + \mathbf{56}_V$	C_M, C_{MNP}

$$(4.43)$$

where α, $\dot\alpha$ are opposite-chirality spinor indices. The NS-NS sector contains a dilaton ϕ, a 2-form B_2, and a graviton G_{MN}. The R-NS and NS-R sectors contain fermions, two opposite-chirality gravitinos $\psi^1_{M\dot\alpha}$, $\psi^2_{M\alpha}$, and two opposite-chirality dilatinos $\lambda^1_{\dot\alpha}$, λ^2_α. The R-R sector contains the 1- and 3-forms, C_1, C_3.

Type IIA and IIB string theories enjoy several important local symmetries in spacetime:

- Local changes of coordinates in spacetime, with G_{MN} as the graviton.
- Gauge transformations (B.21) of the *p*-forms,

$$C_p \to C_p + d\Lambda_{p-1}, \qquad (4.44)$$

for *p* even in type IIB and odd in type IIA theories. These fields behave as generalized gauge potentials. There are no states charged under them in the perturbative spectrum, but non-perturbative brane states couple to them, see Section 6.1.
- Gauge transformations (3.11) of the NS-NS 2-form B_2, just as for the bosonic theory. Similarly, the coupling (3.10) implies that strings are charged under this 2-form gauge field.
- Local supersymmetry, with 32 supercharges arranged in two 16-component spinors in 10d. The spectra correspond to the gravity multiplet of chiral or non-chiral 10d $\mathcal{N} = 2$ superalgebra, for type IIB (equal chirality supercharges) or type IIA (opposite chirality supercharges) theories, respectively.

Let us conclude by emphasizing once again that the appearance of fermions and spacetime supersymmetry is *not* an automatic consequence of the existence of worldsheet fermions and worldsheet supersymmetry. Rather these are features which arise from the structure of the GSO projection.

4.2.5 *Effective action and anomaly cancellation**

The high level of supersymmetry strongly constrains the effective actions for the above massless states. They correspond to the unique chiral or non-chiral 10d $\mathcal{N} = 2$ supergravity theories, whose bosonic part is described below. Due to the absence of UV divergences in the type II string theories, they provide a finite ultraviolet completion of these supergravity theories. Applications of type II theories to particle physics model building are based on the so-called orientifold compactifications, and are described in Chapters 10 and 11.

Anomaly cancellation

There is an important consequence of the well-defined quantum behavior of type IIB theory. The theory is chiral and it has potential gravitational anomalies from the contribution of the massless chiral fermions and the self-dual 4-form field. These anomalies would render the theory non-unitary and thus ill-defined. Anomaly cancellation in 10d imposes a highly constraining set of conditions, which are very remarkably satisfied by the massless field content of the theory. This cancellation, miraculous from the effective field theory perspective, is a low-energy manifestation of the good quantum behavior of type IIB string theory. Indeed, anomalies arise when there exists no regulator consistent with the classical symmetries of the theory, so they are ultimately not restored in the quantum theory. Type IIB theory is free of UV divergences, due to modular invariance, and acts as a regulator of the effective theory, preserving all its symmetries. Hence, modular invariance ensures that no anomaly can arise, or equivalently that all potential 10d anomalies cancel. Note that type IIA theory is non-chiral and thus trivially anomaly-free.

Type IIB effective action

Although the construction of the 10d effective actions for type II strings (and other upcoming string theories) is beyond the scope of the present book, we review their bosonic sector. For type IIB, the 10d effective action is

$$2\kappa_{10}^2 S_{\mathrm{IIB}} = \int d^{10}x \, (-G)^{\frac{1}{2}} \left[e^{-2\phi} \left(R + 4\partial_M \phi \, \partial^M \phi - \tfrac{1}{2}|H_3|^2 \right) - \tfrac{1}{2}|F_1|^2 - \tfrac{1}{2}|\tilde{F}_3| - \tfrac{1}{4}|\tilde{F}_5|^2 \right]$$
$$- \frac{1}{2} \int_{10d} C_4 \wedge H_3 \wedge F_3, \qquad (4.45)$$

where $2\kappa_{10}^2 = (2\pi)^7 \, \alpha'^4$. Also, $|F_p|^2$ is defined in (B.22), and

$$\tilde{F}_3 = F_3 - C_0 \, H_3, \quad \tilde{F}_5 = F_5 - \frac{1}{2}C_2 \wedge H_3 + \frac{1}{2}B_2 \wedge F_3. \qquad (4.46)$$

The action should be complemented with a self-duality constraint on the 5-form field strength. Note that the kinetic term of the NSNS 3-form field strength H_3 and the RR ones, F_3, F_5 differ in factors involving the dilaton.

Type IIA effective action

For type IIA, the 10d effective action is

$$2\kappa_{10}^2 S_{\mathrm{IIA}} = \int d^{10}x \, (-G)^{\frac{1}{2}} \left[e^{-2\phi} \left(R + 4\partial_\mu \phi \, \partial^\mu \phi - \tfrac{1}{2}|H_3|^2 \right) - \tfrac{1}{2}|F_2|^2 - \tfrac{1}{2}|\tilde{F}_4|^2 \right]$$
$$- \frac{1}{2} \int_{10d} B_2 \wedge F_4 \wedge F_4, \qquad (4.47)$$

where again $2\kappa_{10}^2 = (2\pi)^7 \alpha'^4$, $|F_p|^2$ is defined in (B.22), and

$$\tilde{F}_4 = dA_3 - A_1 \wedge H_3. \qquad (4.48)$$

As in the type IIB case the kinetic term of the NSNS 3-form field strength H_3 and the RR ones, F_2, F_4 differ in factors involving the dilaton.

For future reference in Chapter 14, we mention that type IIA supergravity admits a deformation by a mass parameter m, known as Romans mass, which may be regarded as a background field strength $F_0 = -m$. The resulting Romans theory contains additional contributions from its kinetic term plus additional Chern–Simons couplings

$$S'_{\mathrm{IIA}} = \tilde{S}_{\mathrm{IIA}} - \frac{1}{4\kappa_{10}^2} \int d^{10}x (-G)^{\frac{1}{2}} m^2 + \frac{1}{2\kappa_{10}^2} \int_{10d} m \, F_{10}, \qquad (4.49)$$

where $\tilde{S}_{\mathrm{IIA}}$ is obtained from (4.47) by the replacement

$$F_2 \to F_2 + mB_2, \quad F_4 \to F_4 + \frac{1}{2}mB_2 \wedge B_2, \quad \tilde{F}_4 \to \tilde{F}_4 + \frac{1}{2}mB_2 \wedge B_2. \qquad (4.50)$$

For completeness we also introduce the 11d $\mathcal{N} = 1$ supergravity action, which is the low-energy approximation to the dynamics of M-theory, see Section 6.3.2. The relevant fields are an 11d graviton $G_{\hat{M}\hat{N}}$, a 3-form C_3, and a gravitino $\Psi_{\hat{M}\alpha}$, with $\hat{M}$, α being 11d vector and spinor indices, respectively. These fields fill out the gravity multiplet of 11d $\mathcal{N} = 1$ supersymmetry. The bosonic piece of the action is

$$2\kappa_{11}^2 S_{11d} = \int d^{11}x \, (-G)^{\frac{1}{2}} \left(R - \tfrac{1}{2}|G_4|^2 \right) - \frac{1}{6} \int_{11d} C_3 \wedge G_4 \wedge G_4, \qquad (4.51)$$

where $G_4 = dC_3$ and κ_{11} is the 11d gravitational strength.

4.2.6 An aside: bosonization*

For future reference in the fermionic construction of the 10d heterotic string and its application to model building in Chapter 8, we introduce the concept of bosonization/ fermionization. This is a phenomenon relating different 2d field theories, by a (non-local) change of fermionic variables to bosonic ones or vice versa; this is consistent in 2d since all representations of the $SO(2)$ Lorentz group are one-dimensional, and there is no well-defined concept of spin. Although the idea is far more general, we restrict to the simple setup of free field theories, and only at the level of their partition functions, see Section E.1 for a complementary description.

Consider two left-moving free fermions with NS boundary conditions. This system is equivalent to a left-moving boson parametrizing a circle of the so-called critical radius $R = \sqrt{\alpha'}$, as follows. The partition function of the fermionic theory is

$$Z = \left[q^{-1/48} \prod_{n=1}^{\infty} \left(1 - q^{n-1/2}\right) \right]^2 = \frac{\vartheta \begin{bmatrix} 0 \\ 0 \end{bmatrix}(\tau)}{\eta(\tau)}. \tag{4.52}$$

Using the infinite sum expansion of theta functions (A.6), this can be recast as

$$Z = \frac{1}{\eta(\tau)} \sum_{n \in \mathbf{Z}} q^{n^2/2}, \tag{4.53}$$

which corresponds to the partition function of the bosonic theory, with η describing the oscillator contribution, and n labeling the discrete left momentum p_L. There is no center of mass momentum factor, since such degree of freedom is absent for purely left bosons with no right sector partner.

The idea can be used to bosonize the complete left-moving sector of a type II superstring. Using (A.6), the fermionic partition functions (4.36) can be recast as

$$Z_\pm = \frac{1}{2} \eta^{-4} \left[\sum_{n_1,n_2,n_3,n_4} q^{\sum_i n_i^2} - \sum_{n_1,n_2,n_3,n_4} q^{\sum_i n_i^2} e^{\pi i \sum_i n_i} \right.$$

$$\left. - \sum_{n_1,n_2,n_3,n_4} q^{\sum_i (n_i+1/2)^2} \pm \sum_{n_1,n_2,n_3,n_4} q^{\sum_i (n_i+1/2)^2} e^{\pi i \sum_i (n_i+1/2)} \right]. \tag{4.54}$$

By gathering terms we can write

$$Z_\pm = \eta^{-4} \left\{ \sum_{\vec{r}=(n_1,n_2,n_3,n_4)} q^{\vec{r}^2} \tfrac{1}{2} \left[1 - (-1)^{\sum_i n_i}\right] - \sum_{\vec{r}=\left(n_1+\frac{1}{2},\ldots,n_4+\frac{1}{2}\right)} q^{\vec{r}^2} \tfrac{1}{2} \left[1 \mp (-1)^{\sum_i n_i}\right] \right\}.$$

Defining the lattices $\Lambda^\pm$ of vectors of the form

$$(n_1, n_2, n_3, n_4), \qquad n_i \in \mathbf{Z}, \quad \sum_i n_i = \text{odd},$$

$$\left(n_1 + \tfrac{1}{2}, n_2 + \tfrac{1}{2}, n_3 + \tfrac{1}{2}, n_4 + \tfrac{1}{2}\right), \qquad n_i \in \mathbf{Z}, \quad \sum_i n_i = \text{even, odd for } \Lambda^-, \Lambda^+,$$

$$\tag{4.55}$$

we have

$$Z_\pm = \eta^{-4} \sum_{\vec{r} \in \Lambda^\pm} q^{\vec{r}^2} (-1)^{F_{\text{s.t.}}}, \tag{4.56}$$

where $F_{\text{s.t.}}$ is spacetime fermion number. This expression is a partition function for four left bosons parametrizing a four-torus, with left-moving momentum in the lattice $\Lambda^\pm$. In this bosonized language, the mass formula reads

$$\frac{\alpha' M_L^2}{2} = N_B + N_{F_B} + \frac{r^2}{2} - \frac{1}{2}, \tag{4.57}$$

where N_{F_B} is the number operator for the bosonized fermions, quantized in *integers*. Massless states have $N_B = N_{F_B} = 0$, and in the NS sector are of the form $r = (\pm 1, 0, 0, 0)$, and in the R sector (for e.g. Λ^+) are $r = \pm \left(\frac{1}{2}, \frac{1}{2}, \frac{1}{2}, -\frac{1}{2}\right)$, where underlining denotes permutations of entries. These are just the $SO(8)$ weights of the $\mathbf{8}_V$ and $\mathbf{8}_C$ representations associated to massless states.

4.3 Heterotic string theories

As already emphasized, left and right sectors in closed string theories provide well-defined independent 2d quantum field theories. The closed string theories we have constructed hitherto have left and right sectors described by similar degrees of freedom. However, it is conceivable to construct new string theories with different left and right degrees of freedom. Indeed, heterotic strings (from the greek *heteros*, meaning *different*) are constructed by taking, e.g., the left sector to be that of the bosonic theory and the right sector to be that of the superstring.

4.3.1 Quantization

Despite its asymmetric 2d field content, the construction leads to well-behaved string theories, with automatic 2d conformal invariance, and achieving modular invariance as described later on. The physical degrees of freedom in the light-cone gauge can be borrowed from previous constructions. In the right sector there are 8 bosons $X_R^i(t - \sigma)$ and 8 fermions $\psi_R^i(t - \sigma)$, for $i = 2, \ldots, 9$. In the left sector there are 24 bosons, split for convenience as $X_L^i(t + \sigma)$, $i = 2, \ldots, 9$ and $X_L^I(t + \sigma)$, $I = 1, \ldots, 16$. Note that we have already fixed the number of fields in each sector, a choice eventually confirmed by restoration of spacetime Lorentz invariance. Since the number of fields differs for left and right movers, an immediate question is the spacetime dimensionality of the theory. The heterotic string has ten physical spacetime dimensions

$$X^i(t, \sigma) = X_L^i(t + \sigma) + X_R^i(t - \sigma), \quad i = 2, \ldots, 9, \tag{4.58}$$

and there is only 10d Poincaré invariance for the flat space theory. The 16 extra left-moving bosons $X_L^I(t + \sigma)$ do not correspond to physical spacetime dimensions, but can be regarded as compactified on a 16d torus of stringy size $R = \sqrt{\alpha'}$, as made precise below.

Let us consider the quantization of heterotic strings propagating on flat 10d spacetime. The quantization of the right degrees of freedom X_R^i, ψ_R^i works as in Section 4.2.1. The bosons X_R^i satisfy periodic boundary conditions and have oscillator expansions given by the right sector version of (4.2)

$$X_R^i(t - \sigma) = \frac{x^i}{2} + \frac{p_i}{2p^+}(t - \sigma) + i\sqrt{\frac{\alpha'}{2}} \sum_{n \in \mathbf{Z} - \{0\}} \frac{\tilde{\alpha}_n^i}{n} e^{-2\pi i n(t-\sigma)/\ell}. \tag{4.59}$$

The fermions satisfy NS or R boundary conditions and have oscillator expansions given by the right sector version of (4.5), (4.17):

$$\psi_R^i(t - \sigma) = i\sqrt{\frac{\alpha'}{2}} \sum_{r \in \mathbf{Z}} \tilde{\psi}_{r+\nu}^i e^{-2\pi i(r+\nu)(t-\sigma)/\ell}, \tag{4.60}$$

with $\nu = 1/2, 0$ for the NS, R sectors. The oscillators have the familiar (anti)commutation relations, and the right sector hamiltonian is

$$H_R = \frac{\sum_i p_i^2}{4p^+} + \frac{1}{\alpha' p^+}(\tilde{N}_B + \tilde{N}_F + \tilde{E}_0)$$

$$\tilde{N}_B = \sum_i \sum_{n>0} \tilde{\alpha}_{-n}^i \tilde{\alpha}_n^i, \quad \tilde{N}_F = \sum_{r=0}^{\infty}(r+\nu)\,\psi_{-r-\nu}^i\,\psi_{r+\nu}^i, \quad \tilde{E}_0 = -2\nu(1-\nu). \tag{4.61}$$

The quantization of the left degrees of freedom X_L^i, $i = 2, \ldots, 9$, is also standard, with periodic boundary conditions and oscillator expansions (4.2). Following (4.58), they pair up with X_R^i to form physical coordinates of 10d spacetime. The left-moving degrees of freedom $X^I(t + \sigma)$ do not have right sector partners to pair up with; they can, however, be regarded as the left-moving part of coordinates whose right-moving part has been frozen to zero. This is possible for toroidally compactified dimensions at the critical radius $R = \sqrt{\alpha}$, as follows. Recall the mode expansions (3.81) for left- and right-moving bosons in a circle compactification of the bosonic theory,

$$X_L(t + \sigma) = \frac{x}{2} + \frac{p_L}{\sqrt{2\alpha'}\,p^+}(t + \sigma) + i\sqrt{\frac{\alpha'}{2}} \sum_{n \in \mathbf{Z} - \{0\}} \frac{\alpha_n}{n} e^{-2\pi i n(t+\sigma)/\ell},$$

$$X_R(t - \sigma) = \frac{x}{2} + \frac{p_R}{\sqrt{2\alpha'}\,p^+}(t - \sigma) + i\sqrt{\frac{\alpha'}{2}} \sum_{n \in \mathbf{Z} - \{0\}} \frac{\tilde{\alpha}_n}{n} e^{-2\pi i n(t-\sigma)/\ell}, \tag{4.62}$$

with left and right momenta (3.82)

$$p_L = \sqrt{\frac{\alpha'}{2}}\left(\frac{k}{R} + \frac{wR}{\alpha'}\right), \quad p_R = \sqrt{\frac{\alpha'}{2}}\left(\frac{k}{R} - \frac{wR}{\alpha'}\right). \tag{4.63}$$

The right sector dynamics can be made trivial, $X_R \equiv 0$, while still leaving non-trivial left sector dynamics, by imposing

$$x \equiv 0, \quad \tilde{\alpha}_n \equiv 0, \quad k \equiv w, \quad R \equiv \sqrt{\alpha'}. \tag{4.64}$$

The center of mass position is removed, the internal torus is frozen at the critical radius $\sqrt{\alpha'}$, and momentum and winding are related and can be described by just the left-moving momentum p_L. The purely left-moving bosons of the heterotic string are described by a 16d generalization of this mode expansion

$$X_L^I(t + \sigma) = \frac{P^I}{\sqrt{2\alpha'}p^+}(t + \sigma) + i\sqrt{\frac{\alpha'}{2}} \sum_{n \in \mathbf{Z} - \{0\}} \frac{\alpha_n^I}{n} e^{-2\pi i n (t+\sigma)/\ell}. \quad (4.65)$$

We point out that the factor $\sqrt{2\alpha'}$ makes momenta P^I adimensional; these momenta belong to a 16-dimensional lattice Λ_{16} discussed below.

The total left-moving hamiltonian is

$$H_L = \frac{\sum_i p_i^2}{4p^+} + \frac{\sum_I (P^I)^2}{2\alpha' p^+} + \frac{1}{\alpha' p^+}(N_B - 1),$$

$$\text{with } N_B = \sum_i \sum_{n>0} \alpha_{-n}^i \alpha_n^i + \sum_I \sum_{n>0} \alpha_{-n}^I \alpha_n^I. \quad (4.66)$$

The physical spectrum of spacetime particles is obtained by glueing left and right states, satisfying the level-matching condition $M_L^2 = M_R^2$, with

$$\frac{\alpha' M_R^2}{2} = \tilde{N}_B + \tilde{N}_F - 2\nu(1 - \nu),$$

$$\frac{\alpha' M_L^2}{2} = N_B + \frac{P^2}{2} - 1, \quad (4.67)$$

where $P^2 = \sum_I (P_I)^2$. As for type II theories, the construction of consistent theories is highly constrained by modular invariance, as studied next.

4.3.2 Modular invariance and the two heterotic string theories

To construct consistent theories, it is again simpler to first propose a modular invariant partition function, and later describe its implications for the physical spectrum. The partition function is factorized in a spacetime momentum piece, times the left and right sector partition functions, with a structure

$$Z(\tau) = (4\pi\alpha' \tau_2)^{-4} |\eta(\tau)|^{-16} \overline{Z}_\psi(\overline{\tau}) Z_{X^I}(\tau). \quad (4.68)$$

The first factor corresponds to the center of mass momentum piece, while the second is the trace over the oscillators of X_R^i, X_L^i. The factor $\overline{Z}_\psi$ is the trace over the right-moving fermionic oscillators in ψ_R^i. To obtain a nice modular behavior, the natural proposal for it is of the form (4.36). The two choices $Z_\pm$ lead to equivalent theories, related by 10d spacetime parity. For concreteness we take

$$\overline{Z}_\psi = \bar{\eta}^{-4} \left(\vartheta \begin{bmatrix} 0 \\ 0 \end{bmatrix}^4 - \vartheta \begin{bmatrix} 0 \\ 1/2 \end{bmatrix}^4 - \vartheta \begin{bmatrix} 1/2 \\ 0 \end{bmatrix}^4 + \vartheta \begin{bmatrix} 1/2 \\ 1/2 \end{bmatrix}^4 \right)^*. \quad (4.69)$$

This choice of partition function for ψ_R^i implies the coexistence of NS and R sectors, and a GSO projection removing the NS tachyon and selecting, at the massless level, the states $|8_V\rangle$ and $|8_C\rangle$ in the NS and R sectors, respectively.

Finally, the piece Z_{X^I} is a trace over the oscillators and left momentum for the 16 purely left-moving bosons X_L^I. It takes the form

$$Z_{X^I}(\tau) = \eta(\tau)^{-16} \sum_{P \in \Lambda_{16}} q^{P^2/2}. \tag{4.70}$$

Different choices of the lattice Λ_{16} define different heterotic theories. However, there are strong constraints on Λ_{16} arising from modular invariance of (4.68). Using the transformation properties of modular functions in Appendix A, invariance under $\tau \to \tau + 1$ boils down to the requirement

$$\sum_{P \in \Lambda_{16}} e^{2\pi i (\tau + 1) P^2/2} = \sum_{P \in \Lambda_{16}} e^{2\pi i \tau P^2/2}. \tag{4.71}$$

Invariance thus requires Λ_{16} to be an *even* lattice, namely $P^2 \in 2\mathbf{Z}$ for any $P \in \Lambda_{16}$, see Appendix A. Invariance under $\tau \to -1/\tau$ requires

$$Z_{X^I}(-1/\tau) = Z_{X^I}(\tau). \tag{4.72}$$

Using (A.4) and the Poisson resummation (A.14) we have

$$Z_{X^I}(-1/\tau) = (-i\tau)^{-8} \eta(\tau)^{-16} \sum_{P \in \Lambda_{16}} e^{2\pi i (-1/\tau) P^2/2}$$

$$= \eta(\tau)^{-16} \frac{1}{|\Lambda_{16}^*/\Lambda_{16}|} \sum_{P' \in \Lambda_{16}^*} e^{-2\pi i \tau P'^2/2}, \tag{4.73}$$

and invariance requires the lattice to be *self-dual*, $\Lambda_{16}^* = \Lambda_{16}$, see Appendix A.

Modular invariant heterotic theories are thus obtained for even and self-dual "compactification" lattices Λ_{16}.[3] Even and self-dual lattices (with euclidean signature scalar product) have been mathematically proven to be very rare, and for instance exist only in dimensions multiple of eight. Happily this includes our case of interest of 16-dimensional lattices, for which there exist only two inequivalent even self-dual lattices. They are defined as follows:

(i) The Spin(32)/$\mathbf{Z}_2$ lattice $\Lambda_{SO(32)}$, given by vectors of the form

$$\begin{matrix} (n_1, \ldots, n_{16}) \\ \left(n_1 + \frac{1}{2}, \ldots, n_{16} + \frac{1}{2}\right) \end{matrix} \qquad \text{with } n_I \in \mathbf{Z}, \text{ and } \sum_I n_I = \text{even}. \tag{4.74}$$

(ii) The $E_8 \times E_8$ lattice, which is the direct sum $\Lambda_{E_8 \times E_8} = \Gamma_{E_8} \oplus \Gamma_{E_8}$ of two 8d lattices Γ_{E_8}, each given by vectors

$$\begin{matrix} (n_1, \ldots, n_8) \\ \left(n_1 + \frac{1}{2}, \ldots, n_8 + \frac{1}{2}\right) \end{matrix} \qquad \text{with } n_I \in \mathbf{Z}, \text{ and } \sum_I n_I = \text{even}. \tag{4.75}$$

Namely, $\Lambda_{E_8 \times E_8}$ is given by vectors $(V; V')$, with $V, V' \in \Gamma_{E_8}$.

---

[3] This is similar to the requirement in Section 3.2.3, but for euclidean signature scalar product, as corresponds for purely left-moving momenta. The Narain lattice in Section 5.2.2 encompasses these two kinds of even self-dual lattices.

The two possible lattices define two different 10d heterotic string theories, whose 10d gauge groups are related to the above lattice names, as shown below.

4.3.3 Massless spectrum of heterotic string theories

The spectrum of these theories is obtained by glueing GSO-projected right states with left states, with level-matching $M_L^2 = M_R^2$. For light states, the GSO-projected right states can be taken from Section 4.2.1, and correspond to the $|8_V\rangle$ in the NS sector and $|8\rangle_C$ in the R sector, both with $M_R^2 = 0$. The lightest left-moving states, using (4.67) are given in the table below:

N_B, P	State	$(\alpha' M_L^2)/2$	$SO(8)$	
$N_B = 0,\ P = 0$	$\lvert 0 \rangle$	-1	$\mathbf{1}$	
$N_B = 1,\ P = 0$	$\alpha_{-1}^i \lvert 0 \rangle$	0	$\mathbf{8}_V$	(4.76)
$N_B = 1,\ P = 0$	$\alpha_{-1}^I \lvert 0 \rangle$	0	$\mathbf{1}$	
$N_B = 0,\ P^2 = 2$	$\lvert P \rangle$	0	$\mathbf{1}$	

The left groundstate has negative M_L^2 and cannot be level-matched to any right state. Therefore, there are no spacetime tachyon fields in the theory.

The left states in the last row have $P^2 = 2$, and can be regarded as additional massless states arising from the compactification on the stringy size torus. For the $\Lambda_{E_8 \times E_8}$ lattice, the momenta P with $P^2 = 2$ are of the form

$$(\pm, \pm, 0, 0, 0, 0, 0, 0; 0, 0, 0, 0, 0, 0, 0, 0)$$

$$\tfrac{1}{2}(\pm, \pm, \pm, \pm, \pm, \pm, \pm, \pm; 0, 0, 0, 0, 0, 0, 0, 0) \quad \text{with } \#- = \text{even},$$

$$(0, 0, 0, 0, 0, 0, 0, 0; \pm, \pm, 0, 0, 0, 0, 0, 0)$$

$$\tfrac{1}{2}(0, 0, 0, 0, 0, 0, 0, 0; \pm, \pm, \pm, \pm, \pm, \pm, \pm, \pm) \quad \text{with } \#- = \text{even}, \quad (4.77)$$

where $\pm$ denotes ± 1 and underlining means permutation over the corresponding entries. These vectors are the non-zero roots of the $E_8 \times E_8$ Lie algebra (hence the lattice name). For the $\Lambda_{SO(32)}$ lattice the momenta P with $P^2 = 2$ are

$$(\pm, \pm, 0, 0, 0, 0, 0, 0, 0, 0, 0, 0, 0, 0, 0, 0) , \quad (4.78)$$

which are the non-zero root vectors of $SO(32)$. Also note that momenta of the form $P = \tfrac{1}{2}(\pm, \pm, \ldots, \pm)$ have $P^2 = 4$ and give rise to massive states.

The massless spacetime particle spectrum has the following structure:

Sector	$\| \rangle_L \times \| \rangle_R$	$SO(8)$	10d field
NS	$\mathbf{8}_V \times \mathbf{8}_V$	$\mathbf{1} + \mathbf{28}_V + \mathbf{35}_V$	ϕ, B_{MN}, G_{MN}
R	$\mathbf{8}_V \times \mathbf{8}_C$	$\mathbf{8}_S + \mathbf{56}_S$	$\lambda_\alpha, \psi_{M\alpha}$
NS	$\alpha_{-1}^I \| 0 \rangle \times \mathbf{8}_V$	$\mathbf{8}_V$	$A_M^{(I)}$
NS	$\| P \rangle \times \mathbf{8}_V$	$\mathbf{8}_V$	$A_M^{(P)}$
R	$\alpha_{-1}^I \| 0 \rangle \times \mathbf{8}_C$	$\mathbf{8}_C$	$\lambda_{\dot\alpha}^{(I)}$
R	$\| P \rangle \times \mathbf{8}_C$	$\mathbf{8}_C$	$\lambda_{\dot\alpha}^{(P)}$

$$(4.79)$$

where $|P\rangle$ denotes the different states with $P^2 = 2$.

The massless fields in the upper part correspond to a scalar dilaton ϕ, a two-index anti-symmetric tensor field B_{MN} (a 2-form B_2), a graviton G_{MN}, a gravitino $\psi_{M\alpha}$, and a dilatino λ_α. The NS and R fields A_M, $\lambda_{\dot\alpha}$ in the lower part correspond to gauge bosons with respect to a spacetime non-abelian gauge symmetry, and the corresponding gauginos. The fields $A_M^{(I)}$ are gauge bosons of a $U(1)^{16}$ Cartan subalgebra, which heuristically can be thought of as arising from the Kaluza–Klein mechanism in the internal 16d toroidal compactification, in analogy with Section 1.4.1. The additional gauge bosons $A_M^{(P)}$ carry momentum P_I in the Ith direction, which gives their charge under the $U(1)_I$ gauge symmetry. As mentioned above, these momentum vectors (4.77), (4.78) correspond to roots of Lie algebras, and hence the corresponding gauge bosons in the string spectrum realize this algebra as a gauge symmetry in spacetime. The heterotic theory defined by $\Lambda_{E_8 \times E_8}$ has spacetime gauge bosons (and gauginos) of $E_8 \times E_8$, and is known as the $E_8 \times E_8$ heterotic string theory. The theory defined by $\Lambda_{SO(32)}$ has spacetime gauge bosons (and gauginos) of $SO(32)$, and is known as the $SO(32)$ heterotic string theory (or sometimes as the $Spin(32)/Z_2$ heterotic, since this is the actual global structure of its gauge group).

These theories enjoy several important local symmetries in spacetime, as follows:

- Local changes of coordinates in spacetime, with G_{MN} as the graviton.
- Gauge transformations (3.11) of the 2-form B_2, just as for the bosonic theory. Similarly, the coupling (3.10) implies that strings are charged under B_2.
- $E_8 \times E_8$ or $SO(32)$ gauge symmetry, under which the gauge bosons and their gauginos transform in the adjoint representation. States with momentum P in the 16d lattice transform in a representation with weight vectors P. It is worth emphasizing that the $E_8 \times E_8$ theory is the only string theory containing gauge symmetries with exceptional groups at the perturbative level, and consequently upon compactification can contain massless spinor representations $\mathbf{16}$ of $SO(10)$ at the perturbative level.

The appearance of non-abelian gauge symmetry in heterotic theories is intimately related to an underlying algebraic structure on the worldsheet theory, a Kac–Moody algebra, realized in terms of the internal bosons X^I as shown in Section E.3. This view-point is useful in the study of general properties of gauge groups and representations in heterotic compactifications.

- Local 10d $\mathcal{N} = 1$ supersymmetry, corresponding to a single gravitino field. There are 16 supercharges which can be arranged as a 10d spinor. The massless fields in the upper part of the table above correspond to the gravity multiplet, and those in the lower part to vector multiplets of the gauge group. As for type II superstrings, the appearance of spacetime supersymmetry is related to the GSO projection. There are indeed other choices of modular invariant partition functions which lead to non-supersymmetric heterotic string theories. These have not been exploited for phenomenological model building, and we omit their discussion.

4.3.4 Effective action and anomaly cancellation

The effective action for the above massless states is fixed by supersymmetry and describes the 10d $\mathcal{N} = 1$ supergravity multiplet, coupled to vector multiplets in the relevant gauge group. Given the good UV behavior of the theories, they provide a quantum theory including gauge and gravitational interactions. Applications of heterotic theory to particle physics model building are described in Chapters 7 and 8. The consistency of the theory also implies cancellation of 10d gauge and gravitational anomalies, in a very non-trivial way involving a novel contribution, known as the Green–Schwarz mechanism.

Effective action of heterotic theories

The bosonic part of the 10d effective action for the heterotic string theories is

$$S_{10} = \frac{1}{2\kappa_{10}^2} \int d^{10}x \, (-G)^{1/2} \, e^{-2\phi} \left(R + 4\partial_M \phi \partial^M \phi - \frac{1}{2}|H_3|^2 - \frac{\alpha'}{4}\mathrm{tr}\,_v|F|^2 \right)$$

(4.80)

where here and in what follows $\mathrm{tr}\,_v$ is the gauge trace normalized to the vector representation of $SO(32)$, or of the $SO(16)$ subgroup of E_8 in the $E_8 \times E_8$ theory. Note that $|F_2|^2 = \frac{1}{2}F_{MN}F^{MN}$, according to (B.22). Finally, H_3 is given by

$$H_3 = dB_2 - \frac{\alpha'}{4}\left(\omega_3 - \omega_3^{\mathrm{grav}} \right),$$

(4.81)

with gauge Chern–Simons terms

$$\omega_3 = \mathrm{tr}\,_v \left(A \wedge dA - \frac{i}{3}A \wedge A \wedge A \right), \quad d\omega_3 = \mathrm{tr}\,_v F^2,$$

(4.82)

and a similar expression (involving the spin connection) for the gravitational Chern–Simons ω_3^{grav}, with $d\omega_3^{\mathrm{grav}} = \mathrm{tr}\, R^2$, where R is the curvature 2-form, see Section B.3.

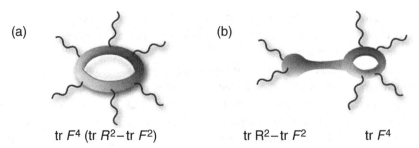

(a)   (b)

tr F^4 (tr R^2 − tr F^2)   tr R^2 − tr F^2 tr F^4

Figure 4.1 Infrared contributions to the one-loop six-leg amplitude in heterotic string theory. (a) The standard field theory hexagon anomaly, controlled by the chiral massless spectrum of the theory. (b) The Green–Schwarz diagram, which involves a tree-level coupling and a one-loop coupling, and a tree-level exchange of the 2-form.

Anomaly cancellation and Green–Schwarz mechanism

There is an important consequence of the well-defined quantum behavior of heterotic string theories; the theories are chiral, and have potential gravitational, gauge, and mixed anomalies from one-loop hexagon diagrams, see Figure 4.1(a). Their contribution to most anomalies cancel, but in contrast with type IIB theory there remains a mixed anomaly contribution to the six-point scattering amplitude, with factorized structure encoded in the anomaly polynomial

$$A_{\text{hex.}} \simeq \text{tr}\, F^4 \wedge (\,\text{tr}\, R^2 - \text{tr}\, F^2\,). \tag{4.83}$$

As described for type IIB, the UV finiteness of the theory implies a gauge invariant regularization of the theory, and hence no anomaly should be possible. Thus the above infrared contribution from the hexagon anomaly diagram should cancel against a different infrared contribution from a diagram corresponding to the same stringy topology. This anomaly cancellation mechanism, uncovered by Green and Schwarz in 1984 (and known under their name) is shown in Figure 4.1(b), and remarkably is only possible for $E_8 \times E_8$ and $SO(32)$. Its structure is closely related to the above 10d effective action (4.80), which contains a term of the form

$$-\int_{10d} \frac{e^{-2\phi}}{2}|H_3|^2 = -\int_{10d} \frac{e^{-2\phi}}{12} \left[\partial_{[M} B_{NP]} - \frac{\alpha'}{4} \left(\omega_{MNP} - \omega_{MNP}^{\text{grav}} \right) \right]^2, \tag{4.84}$$

involving the gauge and gravitational Chern–Simons terms. The cross term contains a coupling of the B field to two gauge bosons or two gravitons, which upon integration by parts can be written in differential form language as

$$\int_{10d} *_{10d} B_2 \wedge \left(\text{tr}\, R^2 - \text{tr}\, F^2 \right). \tag{4.85}$$

On the other hand, there is a term coupling the 2-form with four gauge bosons, which can be shown to arise as a one-loop correction

$$S_{B_2 F^4} = \int_{10d} B_2 \wedge \operatorname{tr} F^4. \tag{4.86}$$

The combination of these two terms produces a factorized contribution to the six-point amplitude, canceling the hexagon anomaly (4.83).

The definition (4.81) implies the following modified Bianchi identity for H_3,

$$dH_3 = \frac{\alpha'}{4} \left(\operatorname{tr} R^2 - \operatorname{tr} F^2 \right). \tag{4.87}$$

(For the $E_8 \times E_8$ heterotic string, there is a tr F^2 contribution from each E_8 gauge factor.) This modified Bianchi identity has important implications in the 4d heterotic compactifications on Calabi–Yau spaces in Section 7.3.

The Green–Schwarz anomaly cancellation mechanism has analogs in general 4d string compactifications, where certain $U(1)$ mixed triangle anomalies are canceled by diagrams involving the exchange of 2-form fields, as described in Section 9.5.

*4.3.5 Fermionic formulation**

Using the ideas in Section 4.2.6 it is possible to construct the heterotic string by replacing the 16 bosonic coordinates X^I by 32 left-moving fermions λ^A. In fact, we can use (A.6) to rewrite the partition functions Z_{X^I} in terms of theta functions, as

$$\Lambda_{E_8 \times E_8} : Z_{X^I}(\tau) = \eta(\tau)^{-16} \left(\vartheta \begin{bmatrix} 0 \\ 0 \end{bmatrix}^8 + \vartheta \begin{bmatrix} 0 \\ 1/2 \end{bmatrix}^8 + \vartheta \begin{bmatrix} 1/2 \\ 0 \end{bmatrix}^8 + \vartheta \begin{bmatrix} 1/2 \\ 1/2 \end{bmatrix}^8 \right)^2,$$

$$\Lambda_{SO(32)} : Z_{X^I}(\tau) = \eta(\tau)^{-16} \left(\vartheta \begin{bmatrix} 0 \\ 0 \end{bmatrix}^{16} + \vartheta \begin{bmatrix} 0 \\ 1/2 \end{bmatrix}^{16} + \vartheta \begin{bmatrix} 1/2 \\ 0 \end{bmatrix}^{16} + \vartheta \begin{bmatrix} 1/2 \\ 1/2 \end{bmatrix}^{16} \right).$$

$$\tag{4.88}$$

The modular invariance of the complete partition function can be easily checked also using this expression. The structure of these partition functions describes the analog of the GSO projection for the 32 fermions. For the $SO(32)$ theory we introduce $(-1)^{F_\lambda}$, anticommuting with the λ^A. The partition function describes the coexistence of NS and R sectors, and imposes a GSO projection onto $(-1)^{F_\lambda} = +1$ states. For the $E_8 \times E_8$ theory, the fermions split in two sets of 16, denoted $\lambda^{A_1}, \lambda^{A_2}$, with their respective operators $(-1)^{F_1}, (-1)^{F_2}$. For each set of 16 fermions there are NS and R sectors, and a GSO projection onto states with $(-1)^{F_i} = +1$. In other words, there is a sum over spin structures for the 32 internal fermions. We leave the reconstruction of the massless spectrum in terms of fermionic degrees of freedom as an exercise for the interested reader. For future reference we simply mention that left-moving NS fermion bilinears $\lambda^A_{-1/2} \lambda^B_{-1/2} |0\rangle$, $A, B = 1, \ldots, n$, lead to $SO(2n)$ gauge bosons. In the $E_8 \times E_8$ theory, the $SO(16)$ from each NS sector is enhanced to E_8 by additional gauge bosons arising from the degenerate Ramond groundstate, which transforms as a spinor **128** of $SO(16)$.

4.4 Type I string theory

In this section we describe the construction of type I string theory. This is a theory of unoriented open and closed superstrings, and requires background material from Sections 3.3 and 3.4. The construction is somewhat involved and proceeds in several steps. We show that a theory of open oriented superstrings is actually inconsistent due to an uncanceled charge (the "RR tadpole"). Next we show that the attempt to construct a theory of closed unoriented superstrings fails for a very similar reason. Type I string theory is then obtained by combining open and closed unoriented strings in a way such that their individual inconsistencies cancel.

With hindsight, we emphasize that the problems with oriented open superstrings are secretly related to demanding 10d Poincaré invariance. This requirement is natural in constructing the vacua of new 10d string theories. However, it can be relaxed in describing non-Poincaré invariant *states* in a theory, like the D-branes in Section 6.1, for which oriented open superstrings indeed play a crucial role.

4.4.1 Oriented open superstrings

Construction

It is natural to consider the construction of open superstring theories, i.e. open strings coupling to a closed superstring sector, and sharing its local worldsheet dynamics. Recalling that boundary conditions for open strings relate the left and right oscillators (by swapping them upon bouncing off the string endpoints), it is clearly not possible to construct open heterotic string theories. Concerning type II theories, a (10d Poincaré invariant) open string sector can potentially couple only to a completely left–right symmetric closed superstring, namely type IIB theory.

The worldsheet degrees of freedom for the open superstring are thus locally the same as for the type IIB string. In the light cone gauge they are given by $X_L^i(t + \sigma)$, $\psi_L^i(t + \sigma)$, $X_R^i(t - \sigma)$, $\psi_R^i(t - \sigma)$, with $i = 2, \ldots, 9$. Note that we have already fixed the spacetime dimension to the critical value. The quantization of the bosonic piece works exactly as for the open bosonic string in Section 3.3. For worldsheet fermions, as in closed superstrings, there are two possible choices of boundary conditions, also denoted NS and R

$$
\begin{array}{llll}
\text{NS:} & \psi_L^i = -\psi_R^i & \text{at } \sigma = 0\,, & \quad\quad \text{R:} & \psi_L^i = \psi_R^i & \text{at } \sigma = 0, \\
& \psi_L^i = \psi_R^i & \text{at } \sigma = \ell, & & \psi_L^i = \psi_R^i & \text{at } \sigma = \ell\,.
\end{array}
$$

This can be understood as (anti)periodicity along the circle present in the "doubling trick" mentioned in Section 3.3.1. Only the relative sign between $\sigma = 0, \ell$ is physical, and other choices of signs relate to these by field redefinitions. The general mode expansion for left and right fermions reads

$$\psi_L^i(t+\sigma) = i\sqrt{\frac{\alpha'}{2}} \sum_{r\in\mathbf{Z}} \psi_{r+\nu}^i \, e^{-\pi i(r+\nu)(t+\sigma)/\ell},$$

$$\psi_R^i(t-\sigma) = i\sqrt{\frac{\alpha'}{2}} \sum_{r\in\mathbf{Z}} \tilde\psi_{r+\nu}^i \, e^{-\pi i(r+\nu)(t-\sigma)/\ell}. \tag{4.89}$$

For NS boundary conditions, we have

$$\sigma = 0 \quad \sum_r \left(\psi_{r+\nu}^i + \tilde\psi_{r+\nu}^i\right) e^{-\pi i(r+\nu)t/\ell} = 0 \quad\to\quad \psi_{r+\nu}^i = -\tilde\psi_{r+\nu}^i,$$

$$\sigma = \ell \quad \sum_r \psi_{r+\nu}^i \cos\pi(r+\nu) \, e^{-\pi i(r+\nu)t/\ell} = 0 \quad\to\quad \nu = \frac{1}{2}. \tag{4.90}$$

For R boundary conditions, we have

$$\sigma = 0 \quad \sum_r \left(\psi_{r+\nu}^i - \tilde\psi_{r+\nu}^i\right) e^{-\pi i(r+\nu)t/\ell} = 0 \quad\to\quad \psi_{r+\nu}^i = \tilde\psi_{r+\nu}^i,$$

$$\sigma = \ell \quad \sum_r \psi_{r+\nu}^i \sin\pi(r+\nu) \, e^{-\pi i(r+\nu)t/\ell} = 0 \quad\to\quad \nu = 0. \tag{4.91}$$

As for bosons in (3.95), left and right fermionic oscillators are identified (up to a sign). Therefore the open superstring quantization, hamiltonian and Hilbert space are very similar to, e.g., the left sector of a type IIB string (except for factors of 2 from the exponent in the mode expansion). The hamiltonian and mass formula for spacetime particles are given by

$$H = \frac{\sum_i p_i^2}{2p^+} + \frac{1}{2\alpha'p^+}\left[N_B + N_{F_\nu} - 2\nu(1-\nu)\right],$$

$$\alpha'M^2 = N_B + N_{F_\nu} - 2\nu(1-\nu), \tag{4.92}$$

where N_B, and N_{F_ν} are the bosonic and fermionic oscillator number operators, and $\nu = 1/2, 0$ for NS, R sectors. In the NS sector, there is a groundstate annihilated by positive modding operators

$$\alpha_n^i|0\rangle = 0, \quad \psi_{n-1/2}^i|0\rangle = 0, \quad \forall n > 0, \, \forall i, \tag{4.93}$$

and the Hilbert space is built by applying negative modding oscillators. In the R sector, there are fermion zero modes, which satisfy a Clifford algebra (4.24). The groundstate is degenerate and forms an $\mathbf{8}_S + \mathbf{8}_C$ spinor representation of the spacetime $SO(8)$ Lorentz group. The light states in the NS and R sectors have a structure as in (4.16), (4.28).

At this point a natural question is whether consistency of the theory requires the coexistence of NS and R sectors, and a GSO projection on their spectra. This is indeed required for a consistent coupling to the sector of closed strings, as shown in Section D.1 from open-closed duality (recall Section 3.3.2). The required open string GSO projection is exactly as that on e.g. the left sector of a type IIB theory. For the light NS states, it eliminates the tachyonic groundstate and leaves the massless state $|\mathbf{8}_V\rangle$. On the R sector it projects out

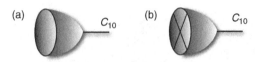

Figure 4.2 (a) Disk diagram leading to a tadpole for the RR 10-form C_{10} in theories with a 10d Poincaré invariant sector of open strings. (b) Crosscap diagram leading to a tadpole for the RR 10-form C_{10} in 10d Poincaré invariant unoriented theories.

the $|8_S\rangle$, and leaves the $|8_C\rangle$ groundstate. The massless spectrum of the open superstring is as follows:

Sector	State	$\alpha' M^2$	$SO(8)$	10d field
NS	$\psi^i_{-1/2}|0\rangle_{NS}$	0	8_V	A_M
R	$|8_C\rangle$	0	8_C	$\lambda_{\dot\alpha}$

$$(4.94)$$

The spacetime fields correspond to a 10d $U(1)$ gauge boson, and its gaugino superpartner, filling out a $U(1)$ vector multiplet of 10d $\mathcal{N} = 1$ supersymmetry. The complete massless spectrum of the theory also includes the closed type IIB string sector (4.42).

The addition of N-valued Chan–Paton indices to the open superstring is carried out as in Section 3.3.3, and is easily shown to be compatible with open–closed duality, see Section D.1. The N-valued Chan–Paton indices lead to an N^2 multiplicity of the above spectrum, corresponding to a $U(N)$ enhancement of the gauge symmetry.

The RR tadpole

The above construction would seem to produce an infinite family of consistent theories of oriented open and closed superstrings, labeled by the Chan–Paton index range N. Actually, there is an additional consistency condition, known as RR tadpole cancellation, which renders all these theories inconsistent. This inconsistency is in fact manifest already at the effective field theory level, as uncanceled gravitational and gauge anomalies of the 10d $\mathcal{N} = 1$ $U(N)$ vector multiplets from the open string sector (as mentioned in Section 4.2.4, the 10d $\mathcal{N} = 2$ gravity multiplet from the closed string sector is anomaly-free).

The microscopic inconsistency of the open string sector can be shown as follows. There is a disk diagram, shown in Figure 4.2(a), describing a tadpole; namely, the emission of a closed string out of the vacuum, or a one-point vertex for a spacetime field in the closed string sector. The computation of this diagram is carried out in Section D.1, but Poincaré invariance alone already suggests that the sourced fields can only be the graviton and dilaton in the NS-NS sector, and a 10-form field C_{10} in the RR sector. The latter is actually non-propagating – its field strength would be an 11-form, vanishing identically in a 10d spacetime – and does not appear in the physical spectrum of spacetime particles. However, the disk tadpole diagram implies that it does appear in the 10d spacetime action in a source term

$$S_{C_{10}} = N \, Q_{\text{disk}} \int_{M_{10}} C_{10}, \qquad (4.95)$$

where the factor of N arises from the Chan–Paton indices on the disk boundary, and Q_{disk} is a non-zero numerical coefficient. The RR 10-form does not appear anywhere else in the 10d action, so its equations of motion $\delta S_{C_{10}}/\delta C_{10} = 0$ read

$$N = 0. \qquad (4.96)$$

Rather than a condition on the field, we obtain a consistency condition on the theory, forcing the absence of open string sectors. It is thus not consistent to couple a *10d Poincaré invariant oriented open string* sector to type IIB theory.

This kind of RR tadpole cancellation conditions play a very important role in any model with open string sectors. For future reference, it is useful to reinterpret it as follows. The RR 10-form can be regarded as a generalized gauge potential, and $N Q_{\text{disk}}$ as a background charge in spacetime. A non-zero background charge is inconsistent with the gauge field equations of motion. This is analogous to the inconsistency of the equations of motion for an electromagnetic gauge field in a compact space sourced by a charge distribution with non-zero total charge. The analogy will become manifest in the analysis of RR tadpole cancellation conditions in compactifications, e.g. in Sections 5.3.4 and 10.3.1.

Incidentally, let us comment on tadpoles for NSNS fields. Since these fields are dynamical and have kinetic terms, the tadpole does not imply an inconsistency of the theory, but rather a linear source term in the 10d action. These signal that we are not expanding the theory around a minimum of its potential, but rather around a point with non-zero slope. For spacetime supersymmetric string theories, cancellation of RR tadpoles automatically implies the cancellation of NSNS tadpoles. This follows from "abstruse" identities of the theta functions appearing in their computation, see Appendix D. Equivalently, because the D-branes secretly described by the open string sector (see Section 6.1) are BPS, and so have equal tension and charge (recall Section 2.4.2). For non-supersymmetric theories, no such cancellation occurs, and the tadpoles contribute non-trivially to the dynamics of spacetime fields. For instance, the open bosonic string theories have disk tadpoles for the graviton and the dilaton.

4.4.2 Unoriented closed superstring

The Ω orientifold quotient

In Section 3.4.1 we introduced a general construction of (unoriented) string theories as quotients of left–right symmetric closed string theories by worldsheet orientation reversal Ω. It is natural to apply this recipe to the left–right symmetric type IIB theory, to produce an unoriented superstring theory. To construct this quotient, we need to obtain the action of Ω on the different mode operators, and truncate the Hilbert space of the type IIB theory to the Ω-invariant set.

The worldsheet theory is described by the bosonic and fermionic fields $X^i(t, \sigma)$, $\psi^i(t, \sigma)$. The action of Ω maps the bosons X^i to X^i_Ω, satisfying (3.109) and leading to (3.110) as in the bosonic theory. The action of Ω on the fermions ψ^i maps them to ψ^i_Ω, satisfying

$$\psi^i_\Omega(t, \sigma) = \psi^i(t, \ell - \sigma). \tag{4.97}$$

Recalling the oscillator expansions (4.5), (4.17), collectively written

$$\psi^i(t, \sigma) = i\sqrt{\frac{\alpha'}{2}} \sum_{r \in \mathbf{Z}} \left[\psi^i_{r+\nu} e^{-2\pi i (r+\nu)(t+\sigma)/\ell} + \tilde{\psi}^i_{r+\nu} e^{-2\pi i (r+\nu)(t-\sigma)/\ell} \right], \tag{4.98}$$

with $\nu = 1/2, 0$ for NS and R fermions, we have

$$\Omega: \ \psi^i_{r+\nu} \leftrightarrow (-1)^{2\nu} \tilde{\psi}^i_{r+\nu}. \tag{4.99}$$

In constructing Ω-invariant states, there is a subtlety in the RR sector. Since the left and right states in this sector are spacetime spinors, they anticommute, and Ω acts on physical states as $|A\rangle_L \times |B\rangle_R \to |A\rangle_R \times |B\rangle_L = - |B\rangle_L \times |A\rangle_R$; the Ω-invariant RR states are therefore of the form $|A\rangle_L \times |B\rangle_R - |B\rangle_L \times |A\rangle_R$. Note also that states in the NS-R sector must combine with states in the R-NS sector to form invariant combinations. The massless spectrum of the theory is given by:

Sector	State	$SO(8)$	10d field	
NS-NS	$[\,\lvert 8_V\rangle \otimes \lvert 8_V\rangle\,]_S$	$1 + 35_V$	ϕ, G_{MN}	(4.100)
NS-R + R-NS	$\lvert 8_V\rangle \otimes \lvert 8_C\rangle + \lvert 8_C\rangle \otimes \lvert 8_V\rangle$	$8_S + 56_S$	$\lambda_\alpha, \psi_{M\alpha}$	
R-R	$[\,\lvert 8_C\rangle \otimes \lvert 8_C\rangle\,]_A$	28_C	B_{MN}	

where the subindices A, S indicate that the (anti)symmetric combination should be kept. This spectrum corresponds to the gravity multiplet of 10d $\mathcal{N} = 1$ supersymmetry. Note that the orientifold projection has removed the NS-NS 2-form, and one linear combination of the two gravitinos of the original $\mathcal{N} = 2$ supersymmetric type IIB theory.

The RR tadpole

The theory as it stands is not consistent, as already manifest at the effective field theory level by the 10d anomalies of the 10d $\mathcal{N} = 1$ gravity multiplet. As in the oriented open string theories studied before, 10d Poincaré invariance suggests that the pathology of this theory also corresponds to an uncanceled tadpole for the RR 10-form C_{10}. It indeed arises from the crosscap diagram in Figure 4.2(b), which leads to a spacetime coupling

$$S_{C_{10}} = Q_{\text{crosscap}} \int_{M_{10}} C_{10}. \tag{4.101}$$

The computation in Section D.1 shows that the coefficient Q_{crosscap} is non-zero, thus leading to inconsistent equations of motion. As in the previous section, the inconsistency can be regarded as the presence of a spacetime filling background charge for the gauge field C_{10}. A *10 Poincaré invariant* theory of *unoriented closed* superstrings is therefore inconsistent due to an uncanceled RR tadpole.

4.4.3 Unoriented open and closed strings: type I string theory

The key idea underlying the construction of type I string theory is to consider a theory of unoriented open and closed strings, such that the RR tadpoles from the (simultaneously present) disk and crosscap diagrams cancel against each other. Namely, with both couplings (4.95) and (4.101) present, it is possible to obtain a consistent theory if there is a choice of N satisfying the constraints from the C_{10} equations of motion

$$Q_{\text{crosscap}} + N Q_{\text{disk}} = 0. \tag{4.102}$$

The closed string sector of the theory is the Ω-orientifold of type IIB theory in the previous section. The open string sector is given by unoriented open superstrings, discussed next.

Open unoriented superstrings

Unoriented open superstrings are constructed generalizing the unoriented open bosonic strings in Section 3.4.3. The action of Ω on worldsheet bosons is given by (3.112), while the action on fermions can be obtained from the relation

$$\psi^i_\Omega(t, \sigma) = \psi^i(t, \ell - \sigma), \tag{4.103}$$

and the expansion (4.89), namely

$$\psi^i(t, \sigma) = i\sqrt{\frac{\alpha'}{2}} \sum_{r \in \mathbf{Z}} \left[\psi^i_{r+\nu} e^{-\pi i(r+\nu)(t+\sigma)/\ell} + \tilde{\psi}^i_{r+\nu} e^{-\pi i(r+\nu)(t-\sigma)/\ell} \right], \tag{4.104}$$

with $\nu = 1/2, 0$ for NS and R fermions. The resulting Ω action on oscillators is

$$\Omega : \psi^i_{r+\nu} \longleftrightarrow e^{-\pi i(r+\nu)} \tilde{\psi}^i_{r+\nu}. \tag{4.105}$$

This is supplemented by $\tilde{\psi}^i_{r+\nu} = e^{2\pi i \nu} \psi^i_{r+\nu}$ from (4.90), (4.91). This introduces phases $e^{i\pi(r+3/2)}$ on NS oscillators $\psi_{-r-1/2}$, so in order to define a $\mathbf{Z}_2$ action on physical NS states, like $\psi_{-1/2}|0\rangle_{\text{NS}}$, there must be a non-trivial action of Ω on $|0\rangle_{\text{NS}}$, which can be shown must be given by

$$\Omega|0\rangle_{\text{NS}} = e^{-i\pi/2}|0\rangle_{\text{NS}}. \tag{4.106}$$

Finally, the action of Ω on the Chan–Paton indices is specified by a matrix γ_Ω as in (3.116), i.e. $\gamma_\Omega = \mathbf{1}_N$ or $\gamma_\Omega = i\epsilon_{N/2}$ for the SO or Sp projections, respectively.

The unoriented open string spectrum is obtained by considering the Ω-invariant states of the corresponding oriented open string spectrum, under the combined action of Ω on the groundstate, oscillator operators, and Chan–Paton indices. The light NS states in the parent oriented theory are $\lambda \psi^i_{-1/2}|0\rangle$, using the notation (3.106). The Ω-invariance condition is

$$\lambda = -\gamma_\Omega \lambda^T \gamma_\Omega^{-1}. \tag{4.107}$$

It can be shown that in the R sector the Ω action on the $|8_C\rangle$ groundstates is a minus sign, leading to the same Ω-invariance condition. This is consistent with spacetime supersymmetry of the string spectrum. Indeed the total massless spectrum is as follows:

Projection	Sector	State	$\alpha' M^2$	$SO(8)$	10d field	Gauge
SO	NS	$\lambda \psi^i_{-1/2}|0\rangle$	0	8_V	A_M	$SO(N)$
	R	$\lambda|8_C\rangle$	0	8_C	$\lambda_{\dot\alpha}$	⊟
Sp	NS	$\lambda \psi^i_{-1/2}|0\rangle$	0	8_V	A_M	$USp(N)$
	R	$\lambda|8_C\rangle$	0	8_C	$\lambda_{\dot\alpha}$	☐☐

It fills out $SO(N)$ or $USp(N)$ vector multiplets of 10d $\mathcal{N} = 1$ supersymmetry (gauge bosons and gauginos), for the SO and Sp projections, respectively.

Cancellation of RR tadpoles

As suggested above, the precise choice of projection and the value of N are constrained by cancellation of RR tadpoles. The computation in Section D.1 shows that the crosscap and disk tadpole coefficients are related as

$$Q^\pm_{\text{crosscap}} = \pm 32 Q_{\text{disk}}, \tag{4.108}$$

with the positive and negative signs for the Sp and SO projections, respectively. The RR tadpole cancellation condition reads $N \pm 32 = 0$, and so it is satisfied for an open string sector with $N = 32$ and SO orientifold projection. The resulting theory is known as type I string theory.

4.4.4 Massless spectrum of type I string theory

Type I string theory is thus a consistent theory of unoriented open and closed strings. Its massless spectrum is as follows:

Sector	State	$SO(8)$	10d field	Gauge
Closed NS-NS	$[\,|8_V\rangle \otimes |8_V\rangle\,]_S$	$1 + 35_V$	ϕ, G_{MN}	
NS-R + R-NS	$|8_V\rangle \otimes |8_C\rangle + |8_C\rangle \otimes |8_V\rangle$	$8_S + 56_S$	$\lambda_\alpha, \psi_{M\alpha}$	
R-R	$[\,|8_C\rangle \otimes |8_C\rangle|,]_A$	28_C	B_{MN}	
Open NS	$\lambda|8_V\rangle$	8_V	A_M	$SO(32)$
R	$\lambda|8_C\rangle$	8_C	$\lambda_{\dot\alpha}$	⊟

The massless fields in the closed string sector correspond to a scalar dilaton ϕ, a 2-form B_2 (often denoted C_2 in notation inherited from the parent IIB theory), a graviton G_{MN}, a gravitino $\psi_{M\alpha}$ and a dilatino λ_α. The massless fields in the open string sector A_M, $\lambda_{\dot{\alpha}}$ are gauge bosons and gauginos of the $SO(32)$ spacetime gauge symmetry. The theory enjoys the following spacetime local symmetries:

- Local changes of coordinates in spacetime, with G_{MN} as the graviton.
- Gauge transformations of the RR 2-form given by (4.44) or (B.21), for $p = 2$.
- $SO(32)$ gauge symmetry, under which gauge bosons and gauginos transform in the adjoint representation.
- Local 10d $\mathcal{N} = 1$ supersymmetry, corresponding to a single gravitino field.

There are 16 supercharges which can be arranged as a 10d spinor. The massless fields in the closed string sector correspond to the gravity multiplet, and massless fields in the open string sector to $SO(32)$ vector multiplets. As usual, the appearance of spacetime supersymmetry is related to the GSO and Ω projections. Indeed, there exist unoriented 10d open string theories with no spacetime supersymmetry.

4.4.5 Effective action and anomaly cancellation

The 10d effective action for the above massless states is fixed by supersymmetry, and describes the 10d, $\mathcal{N} = 1$ gravity multiplet, coupled to $SO(32)$ vector multiplets. Type I string theory provides an UV completion for this theory, and thus of its gauge and gravitational interactions. Applications of type I theory (and other generalized orientifold theories) to particle physics model building are described in Chapters 10 and 11. It is interesting to point out that the finiteness of amplitudes in type I theory is very closely related to RR tadpole cancellation. For instance, for the one loop vacuum-to-vacuum amplitude, the Klein bottle, Moebius strip, and cylinder diagrams each lead to UV divergences, which cancel out in the total sum due to RR tadpole cancellation.

Effective action of type I theory

The bosonic part of the 10d effective action of type I is

$$
S_{10} = \frac{1}{2\kappa_{10}^2} \int d^{10}x \, (-G)^{\frac{1}{2}} \left[e^{-2\phi} \left(R + 4 \partial_M \phi \, \partial^M \phi - \frac{1}{2} |\tilde{F}_3|^2 \right) \right]
$$
$$
- \frac{1}{2g_{10}^2} \int d^{10}x \, (-G)^{\frac{1}{2}} e^{-\phi} \mathrm{Tr}_v |F_2|^2, \tag{4.109}
$$

where

$$
g_{10}^2 = \kappa_{10} 2(2\pi)^{7/2} \alpha', \quad \tilde{F}_3 = dC_2 - \frac{\kappa_{10}^2}{g_{10}^2} (\omega_3 - \omega_{\mathrm{grav}}). \tag{4.110}
$$

Here ω_3, ω_{grav} are the gauge and gravitational Chern–Simons terms. Recall also that according to (B.22), $|F_2|^2 = \frac{1}{2} F_{MN} F^{MN}$.

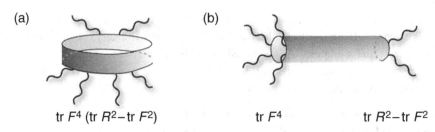

(a)   (b)

$$\mathrm{tr}\ F^4\ (\mathrm{tr}\ R^2 - \mathrm{tr}\ F^2) \qquad\qquad \mathrm{tr}\ F^4 \qquad\qquad \mathrm{tr}\ R^2 - \mathrm{tr}\ F^2$$

Figure 4.3 Infrared contributions to the one-loop six-leg amplitude in type I string theory. (a) Shows the standard field theory hexagon anomaly, controlled by the chiral massless spectrum of the theory. (b) Shows the Green–Schwarz diagram, which involves two disk couplings and a RR 2-form exchange.

Several terms of the above action can be understood as arising from the worldvolume action (6.6), (6.15) of D9-branes, certain spacetime filling objects present in the vacuum of type I theory, see Section 6.1.

Incidentally, the massless spectrum of type I and $SO(32)$ heterotic theories are identical, and their 10d effective actions are given by the same 10d $\mathcal{N} = 1$ supergravity theory (albeit in different variables). This is actually related to the non-perturbative duality between the two theories, see Section 6.3.5. Still, their perturbative expansions are very different, e.g. as reflected in the different dilaton dependences in type I and heterotic effective actions. This will be relevant in Section 16.1.1 for the discussion of gauge coupling unification in 4d compactifications of these theories.

Green–Schwarz anomaly cancellation

An important implication of the UV finiteness is the anomaly cancellation in type I theory. Since the massless spectrum of type I theory is exactly as for the $SO(32)$ heterotic, the anomaly cancellation discussion in effective field theory is essentially as in Section 4.3.4. Namely, there is a non-vanishing hexagon anomaly, with factorized structure (4.83), and which is canceled by a Green–Schwarz mechanism involving the Chern–Simons terms in $\tilde{F}_3$ and a coupling (4.86), now for the RR 2-form C_2. The microscopic diagrams involve worldsheets with annulus/cylinder topology, shown in Figure 4.3, and the Green–Schwarz couplings arise from disk diagrams. This Green–Schwarz anomaly cancellation mechanism has analogs in 4d orientifold compactifications of Chapters 10 and 11.

4.5 Summary

In this chapter we have introduced the five different 10d superstring theories. They contain fermion fields in their massless spectrum, and interestingly the heterotic and type I theories contain also non-abelian gauge symmetries. The five theories have a high degree of spacetime supersymmetry, with the same number of supercharges as 4d $\mathcal{N} = 8$ in type II theories and of 4d $\mathcal{N} = 4$ in heterotic and type I theories. This spacetime supersymmetry guarantees the absence of tachyons, and allows the detailed study of these theories and their compactifications. In 4d compactifications, the degree of supersymmetry can be reduced, leading to theories with 4d $\mathcal{N} = 1$ (or no) supersymmetry, with potential particle physics model building applications.

Although the ingredients in the construction of these theories are rather different, and they seem pretty much disconnected, there are several non-trivial relations among them. This occurs already at the perturbative level, by the T-dualities upon toroidal compactification in Chapter 5, and the mirror symmetry for compactifications on Calabi–Yau spaces in Section 10.1.2. But also most strikingly at the non-perturbative level, by the string dualities in Section 6.3. These relations lead to analogous dualities among their 4d compactifications, as will become manifest all along this book.

5

Toroidal compactification of superstrings

In this chapter we study the toroidal compactification of superstrings. The main motivation is to produce 4d theories at low energies, as a preliminary tool for toroidal orbifold constructions, and as a warm-up for more involved Calabi–Yau compactifications. Toroidal compactifications also frame interesting T-dualities among different superstring theories. We mostly focus on circle compactification of one dimension, which generalizes easily to compactifications to 4d.

5.1 Type II superstrings

We start with the circle compactification of type II theories, and the T-duality relation between type IIB and IIA compactifications.

5.1.1 Circle compactification

Consider type IIA or IIB theories compactified to 9d on a circle S^1 of radius R, namely $x^9 \simeq x^9 + 2\pi R$. As in Section 3.2.3 for closed bosonic string theory, the local 2d dynamics on the worldsheet is exactly as for the uncompactified theory; the only modifications arise from global considerations, and imply a quantized center of mass momentum in the compact dimension, and the inclusion of winding boundary conditions for the worldsheet bosons,

$$X^9(t, \sigma + \ell) = X^9(t, \sigma) + 2\pi R\, w. \tag{5.1}$$

In a sector of momentum k and winding w, the center of mass piece of the bosonic mode expansion is the analog of (3.81),

$$X_L^9 = \frac{x_0^9}{2} + \sqrt{\frac{2}{\alpha'}} \frac{p_L}{2p^+}(t + \sigma) + (\text{osc.}), \quad X_R^9 = \frac{x_0^9}{2} + \sqrt{\frac{2}{\alpha'}} \frac{p_R}{2p^+}(t - \sigma) + (\text{osc.}),$$

with

$$p_L = \sqrt{\frac{\alpha'}{2}} \left(\frac{k}{R} + \frac{wR}{\alpha'} \right), \quad p_R = \sqrt{\frac{\alpha'}{2}} \left(\frac{k}{R} - \frac{wR}{\alpha'} \right). \tag{5.2}$$

The fermionic degrees of freedom are described exactly as in non-compact 10d theory. The spacetime mass formulae are given by

$$\frac{\alpha' M_L^2}{2} = \frac{p_L^2}{2} + N_B + N_F + E_0, \quad \frac{\alpha' M_R^2}{2} = \frac{p_R^2}{2} + \tilde{N}_B + \tilde{N}_F + \tilde{E}_0, \quad (5.3)$$

with $E_0 = -1/2, 0$ in the NS, R sectors, and similarly for $\tilde{E}_0$. The level-matching constraint is simply $M_L^2 = M_R^2$.

These expressions provide the spectrum of 9d states at any radius R. For large R, the winding states are very heavy and the light spectrum reduces to the momentum excitations of states of the 10d theory, appropriately decomposed with respect to the 9d Lorentz symmetry. This precisely corresponds to the field theory KK compactification result, as expected in the large volume limit $\alpha'/R^2 \ll 1$. For illustration, consider the spectrum of 9d massless fields. The process of KK compactification and truncation to the zero mode amounts to decomposing the $SO(8)$ representations of 10d massless fields into representations of the 9d $SO(7)$ group. Working with left and right sectors independently, the required decompositions are

$$\mathbf{8}_V \to \mathbf{7} + \mathbf{1}, \quad \mathbf{8}_S \to \mathbf{8}, \quad \mathbf{8}_C \to \mathbf{8}, \quad (5.4)$$

where $\mathbf{7}$ and $\mathbf{8}$ are the vector and the unique spinor representation of $SO(7)$. Note that the 10d chirality is lost upon reducing to 9d, since there is no concept of chirality in odd dimensions. The glueing of left and right pieces can be carried out in the 10d theory, with subsequent decomposition into $SO(7)$ representations, or first decomposing and glueing subsequently. The resulting representations and 9d fields are shown in the following tables, where $M, N, \ldots$ and $\mu, \nu, \ldots$ are 10d and 9d indices respectively, and α is a spinor index. Generalization to compactification to lower dimensions is straightforward. For type IIB theory on $\mathbf{S}^1$ we have:

Type IIB on $\mathbf{S}^1$				
NSNS: $\mathbf{8}_V \otimes \mathbf{8}_V$	$\to$	$\mathbf{8}_V \otimes \mathbf{8}_V =$	$\mathbf{35}_V + \mathbf{28}_V + \mathbf{1}$	G_{MN}, B_{MN}, ϕ
$(\mathbf{7}+\mathbf{1}) \otimes (\mathbf{7}+\mathbf{1})$	$\to$	$\mathbf{7} \otimes \mathbf{7} =$	$\mathbf{27} + \mathbf{21} + \mathbf{1}$	$G_{\mu\nu}, B_{\mu\nu}, \phi$
		$(\mathbf{7} \otimes \mathbf{1}) + (\mathbf{1} \otimes \mathbf{7}) =$	$\mathbf{7} + \mathbf{7}$	$G_{9\mu}, B_{9\mu}$
		$\mathbf{1} \otimes \mathbf{1} =$	$\mathbf{1}$	G_{99}
NS-R: $\mathbf{8}_V \otimes \mathbf{8}_C$	$\to$	$\mathbf{8}_V \otimes \mathbf{8}_C =$	$\mathbf{56}_S + \mathbf{8}_S$	$\psi^1_{M\alpha}, \lambda^1_\alpha$
$(\mathbf{7}+\mathbf{1}) \otimes \mathbf{8}$	$\to$	$\mathbf{7} \otimes \mathbf{8} =$	$\mathbf{48} + \mathbf{8}$	$\psi^1_{\mu\alpha}, \lambda^1_\alpha$
	$\to$	$\mathbf{1} \otimes \mathbf{8} =$	$\mathbf{8}$	$\psi^1_{9\alpha}$

(continued)

R-NS: $8_C \otimes 8_V$	$\rightarrow$	$8_C \otimes 8_V =$	$56_S + 8_S$	$\psi^2_{M\alpha}, \lambda^2_{\dot\alpha}$
$8 \otimes (7+1)$	$\rightarrow$	$8 \otimes 7 =$	$48 \; + 8$	$\psi^2_{\mu\alpha}, \lambda^2_{\alpha}$
	$\rightarrow$	$8 \otimes 1 =$	8	$\psi^2_{9\alpha}$
R-R: $8_C \otimes 8_C$	$\rightarrow$	$8_C \otimes 8_C =$	$1 + 28_C + 35_C$	a, C_{MN}, C_{MNPQ}
$8 \otimes 8$	$\rightarrow$	$8 \otimes 8 =$	$1 + (7+21) + 35$	$a, C_{9\mu}, C_{\mu\nu}, C_{9\mu\nu\rho}$

Note that self-duality of the RR 4-form relates the component $C_{\mu\nu\rho\sigma}$ to $C_{9\mu\nu\rho}$, so it is not an independent field. The 9d field content of type IIA theory on $\mathbf{S}^1$ gives:

Type IIA on $\mathbf{S}^1$

NSNS: $8_V \otimes 8_V$	$\rightarrow$	$8_V \otimes 8_V =$	$35_V + 28_V + 1$	G_{MN}, B_{MN}, ϕ
$(7+1) \otimes (7+1)$	$\rightarrow$	$7 \otimes 7 =$	$27 \; + 21 \; + 1$	$G_{\mu\nu}, B_{\mu\nu}, \phi$
		$(7 \otimes 1) + (1 \otimes 7) =$	$7 \; + \; 7$	$G_{9\mu}, B_{9\mu}$
		$1 \otimes 1 =$	1	G_{99}
NS-R: $8_V \otimes 8_S$	$\rightarrow$	$8_V \otimes 8_S =$	$56_C + 8_C$	$\psi^1_{M\dot\alpha}, \lambda^1_{\dot\alpha}$
$(7+1) \otimes 8$	$\rightarrow$	$7 \otimes 8 =$	$48 \; + 8$	$\psi^1_{\mu\alpha}, \lambda^1_{\alpha}$
	$\rightarrow$	$1 \otimes 8 =$	8	$\psi^1_{9\alpha}$
R-NS: $8_C \otimes 8_V$	$\rightarrow$	$8_C \otimes 8_V =$	$56_S + 8_S$	$\psi^2_{M\alpha}, \lambda^2_{\dot\alpha}$
$8 \otimes (7+1)$	$\rightarrow$	$8 \otimes 7 =$	$48 \; + 8$	$\psi^2_{\mu\alpha}, \lambda^2_{\alpha}$
	$\rightarrow$	$8 \otimes 1 =$	8	$\psi^2_{9\alpha}$
R-R: $8_C \otimes 8_S$	$\rightarrow$	$8_C \otimes 8_S =$	$8_V \; + \; 56_V$	C_M, C_{MNP}
$8 \otimes 8$	$\rightarrow$	$8 \otimes 8 =$	$(1+7) + (21+35)$	$C_9, C_\mu, C_{9\mu\nu}, C_{\mu\nu\rho}$

The 9d theories from $\mathbf{S}^1$ compactification of type II theories have 32 supersymmetries, i.e. the equivalent of $\mathcal{N} = 8$ in 4d.

5.1.2 T-duality for type II theories

Circle compactifications of type IIA and IIB theories lead to the same 9d massless spectrum, as shown in the previous tables; rather than being an accident, this is a manifestation of the T-duality relation between the two theories, described next.

Recall from the bosonic theory, Section 3.2.3, that T-duality relates two theories compactified on circles of radii R and $R' = \alpha'/R$. A convenient derivation is to start with the theory compactified on a circle of radius R, decompose the worldsheet field parametrizing the $\mathbf{S}^1$ as $X(t, \sigma) = X_L + X_R$, and use the decoupled left and right degrees of freedom to form the coordinates of the T-dual theory $X'(t, \sigma) = X_L - X_R$. The T-dual describes a theory compactified on a circle of radius R'.

We can apply the same strategy in type II theory to construct the T-dual of, e.g., type IIB on a circle of radius R. As in the bosonic theory, the T-dual coordinate $X^{9'} = X_L - X_R$ describes a compactification on a circle of radius $R' = \alpha'/R$. However, an important novelty is that by worldsheet supersymmetry T-duality acts non-trivially also on the worldsheet fermion ψ^9, transforming the original $\psi^9(t, \sigma) = \psi_L^9 + \psi_R^9$ into the T-dual $\psi^{9'}(t, \sigma) = \psi_L^9 - \psi_R^9$. T-duality is effectively a spacetime parity operation on the right-movers, and hence turns the type IIB Ramond groundstate $|\mathbf{8}_C\rangle$ into an $|\mathbf{8}_S\rangle$ right-moving groundstate in the T-dual theory. The change of GSO projection for right-movers upon T-duality means that the T-dual of type IIB on a circle of radius R corresponds to type IIA theory compactified on a circle of radius $R' = \alpha'/R$.

The flip in the GSO projection can be derived more explicitly as follows. Recall the construction of Ramond groundstates in Section 4.2.1, in terms of the operators

$$A_a^{\pm} = \frac{1}{\sqrt{2}} \left(\psi_0^{2a} \pm i\psi_0^{2a+1} \right). \tag{5.5}$$

T-duality in direction 9 flips $\psi_0^9 \to -\psi_0^9$ on the right movers, and so $A_4^{\pm} \leftrightarrow A_4'^{\mp}$. Hence the groundstates $|0\rangle$ and $|0\rangle'$ of the original and T-dual theory, defined by $A_a^-|0\rangle = 0$ and $A_a'^-|0\rangle' = 0$, are not the same, but rather $|0\rangle' = A_4^+|0\rangle$; they have opposite eigenvalue of $(-1)^F$, and thus lead to different GSO projections in the original and T-dual theories.

It is easy to check the effect of T-duality on the 9d reduction of the massless 10d fields, by comparing their 9d spectra. For instance, for bosonic fields

$$
\begin{array}{ccc}
\text{IIA} & \overset{T}{\longleftrightarrow} & \text{IIB,} \\
G_{9\mu}, B_{9\mu} & \longleftrightarrow & B_{9\mu}, G_{9\mu}, \\
C_9, C_\mu & \longleftrightarrow & a, C_{9\mu}, \\
C_{9\mu\nu}, C_{\mu\nu\rho} & \longleftrightarrow & C_{\mu\nu}, C_{9\mu\nu\rho}.
\end{array}
\tag{5.6}
$$

In the NSNS sector, the mixed components of the metric and B-field are exchanged; in the RR sector, p-forms with an index along the T-dualized direction, lose it and become $(p-1)$-forms, and p-forms with no index along the T-dualized direction, gain it and become $(p+1)$-forms. Hence T-duality explains the equivalence of the massless 9d spectrum noticed in the previous section.

5.1.3 Compactification of several dimensions and background fields*

In $\mathbf{S}^1$ compactifications the only parameter associated to the internal space is the radius R. Upon compactification of several dimensions, there is a larger set of parameters describing the internal backgrounds, and it is interesting to describe the compactification for generic values of these moduli. This is nicely encoded in terms of the Narain lattice of left and right momenta generalizing (5.2). We consider compactification on a d-dimensional torus $\mathbf{T}^d$, defined by periodic coordinates $x^m \simeq x^m + 2\pi R$, and implement general torus geometries via an internal (flat) metric G_{mn}. We also allow a constant internal background NSNS field B_{mn}, with vanishing field strength. The parameters G_{mn}, B_{mn} are actually vevs for moduli fields of the resulting lower-dimensional theory. Concerning other massless fields, the dilaton controls the string coupling as usual, and we restrict to vanishing RR backgrounds, as there is no simple description of their coupling to the worldsheet theory.

The light-cone gauge-fixed action for the bosonic coordinates of a string propagating on such background is, from (3.8) and (3.9),

$$L = \frac{1}{2\pi\alpha'} \int_0^\infty d\sigma \left[\frac{1}{2} G_{mn} (\partial_t X^m \partial_t X^n - \partial_\sigma X^m \partial_\sigma X^n) + B_{mn} \partial_t X^m \partial_\sigma X^n \right]. \qquad (5.7)$$

The presence of the backgrounds and the periodicity of the coordinates x^i do not modify the worldsheet fermions or the oscillator piece for the worldsheet bosons. Thus it is enough to work with the center of mass piece in the mode expansion of the 2d bosons, which includes momentum and winding terms,

$$X^m(t, \sigma) = x_0^m + \dot{x}^m t + \frac{2\pi R}{\ell} w^m \sigma, \qquad (5.8)$$

where $\dot{x}^m$ are later related to quantized momenta k_m. Replacing into (5.7) gives

$$L = \frac{\ell}{2\pi\alpha'} \left\{ \frac{1}{2} G_{mn} \left[\dot{x}^m \dot{x}^n - \left(\frac{2\pi R}{\ell}\right)^2 w^m w^n \right] + \frac{2\pi R}{\ell} B_{mn} \dot{x}^m w^n \right\}. \qquad (5.9)$$

The canonical momentum conjugate to x^m is

$$p_m = \frac{\partial L}{\partial \dot{x}^m} = \frac{\ell}{2\pi\alpha'} \left(G_{mn} \dot{x}^n + \frac{2\pi R}{\ell} B_{mn} w^n \right), \qquad (5.10)$$

and must be quantized as $p_m = k_m/R$, with $k_m \in \mathbf{Z}$. This leads to

$$\dot{x}^m = \frac{G^{mn}}{p^+} \left(\frac{k_n}{R} - \frac{R}{\alpha'} B_{np} w^p \right). \qquad (5.11)$$

Replacing in (5.8) and splitting into left and right movers, we have left and right momenta generalizing (5.2),

$$p_{L,m} = \sqrt{\frac{\alpha'}{2}} \left[\frac{k_m}{R} + \frac{R}{\alpha'} (G_{mn} - B_{mn}) w^n \right],$$

$$p_{R,m} = \sqrt{\frac{\alpha'}{2}} \left[\frac{k_m}{R} + \frac{R}{\alpha'} (-G_{mn} - B_{mn}) w^n \right]. \qquad (5.12)$$

The mass formulae for particles in the lower-dimensional theory have the structure (5.3), also with the level matching condition $M_L^2 = M_R^2$.

The dimension $2d$ lattice $\Gamma_{d,d}$ of momenta (p_L, p_R) is known as the Narain lattice. It is self-dual, and even with respect to the Lorentzian (d, d) signature scalar product

$$(p_L, p_R) \cdot (p_L', p_R') = p_L^m p_{L,m}' - p_R^m p_{R,m}' = k^m w_m' + w^m k_m' \in \mathbf{Z}. \tag{5.13}$$

These two properties ensure modular invariance of the 1-loop partition function, which contains a piece

$$Z(\tau) = \dots \sum_{(k,w)} q^{\frac{p_L^2}{2}} \bar{q}^{\frac{p_R^2}{2}} = \dots \sum_{(p_L, p_R) \in \Gamma_{d,d}} q^{\frac{p_L^2}{2}} \bar{q}^{\frac{p_R^2}{2}}, \tag{5.14}$$

completely analogous to that of the bosonic string. Conversely, any even self-dual (d, d) signature lattice is of the form (5.12), so the latter parametrization exhausts the modular invariant toroidal compactifications of type II theory.

5.2 Heterotic superstrings

We now turn to the circle compactification of heterotic string theories, without and with backgrounds for the internal components of the 10d gauge fields (Wilson lines). These backgrounds allow to establish a T-duality between the $\mathbf{S}^1$ compactifications of the two heterotic theories.

5.2.1 Circle compactification without Wilson lines

Consider a heterotic string theory compactified to 9d on an $\mathbf{S}^1$ parametrized by a periodic coordinate, $x^9 \simeq x^9 + 2\pi R$. As in the bosonic and type II theories, the compactification only modifies the theory by the inclusion of winding sectors, and the restriction to quantized momenta in the compact direction. Therefore, different sectors of the theory will be labeled by left- and right-moving momenta

$$p_{L,R} = \sqrt{\frac{\alpha'}{2}} \left(\frac{k}{R} \pm \frac{R}{\alpha'} w \right), \tag{5.15}$$

as well as the internal 16d lattice left-moving momenta P^I in the $E_8 \times E_8$ or $\mathrm{Spin}(32)/\mathbf{Z}_2$ lattices. The mass formulae are given by

$$\frac{\alpha' M_L^2}{2} = \frac{P^2}{2} + \frac{p_L^2}{2} + N_B - 1, \qquad \frac{\alpha' M_R^2}{2} = \frac{p_R^2}{2} + \tilde{N}_B + \tilde{N}_F + \tilde{E}_0. \tag{5.16}$$

and the level-matching condition is $M_L^2 = M_R^2$. The spectrum of massless states at large radius corresponds to the field theory KK reduction of the massless fields of the 10d theory, as expected. We skip its detailed description, which can be easily worked out in analogy with the type II examples above. The resulting spectrum corresponds to a 9d theory with 16 supersymmetries, and containing the gravity multiplet coupled to vector multiplets of an $E_8 \times E_8$ or $SO(32)$ gauge group.

The toroidal compactification to lower dimensions is similarly obtained by simply decomposing the fields with respect to representations of the corresponding Lorentz group. Notice that, in any of these compactifications, chirality is lost. In particular, compactifications to 4d lead to theories with 4d $\mathcal{N} = 4$ supersymmetry, which are automatically non-chiral. Still, toroidal models are the starting point of orbifold compactifications, which do produce chiral theories, see Chapter 8.

5.2.2 Heterotic compactification with Wilson lines

The heterotic compactifications in the previous section are not the most general circle compactifications on which strings can be quantized exactly. In particular, it is possible to turn on a non-zero background for the internal components of the gauge fields A_9^a, where a denotes an index in the Lie algebra of the 10d gauge group. These so-called Wilson line backgrounds modify the spectrum of the lower-dimensional theory in interesting ways.

Field theory description of Wilson lines

The possibility to introduce Wilson lines in compactifications of extra dimensions is already present in field theory KK compactifications, as we now review in a toy model. Consider a 5d Yang–Mills gauge theory with gauge group G compactified to 4d on a spacetime $\mathbf{M}_4 \times \mathbf{S}^1$. We introduce the periodic coordinate $x^4 \simeq x^4 + 2\pi R$, and consider a constant background for A_4^a. Locally, this is pure gauge and could be removed, but globally the gauge parameter would not be well-defined in $\mathbf{S}^1$; for instance, for $G = U(1)$, the gauge background can be locally gauged away with a gauge transformation

$$A_\mu \to A_\mu + \partial_\mu \lambda \ \text{ with } \ \lambda = -\langle A_4 \rangle x^4, \tag{5.17}$$

but λ is not periodic and hence not globally well defined on $\mathbf{S}^1$. The Wilson line is thus physically observable; for a general group G, it can indeed be measured by the gauge-invariant quantity

$$W^a = \text{Tr} \exp\left(i \int_{\mathbf{S}^1} A^a \right) = \text{Tr} \exp\left(2\pi i R A_4^a \right). \tag{5.18}$$

The Wilson lines A_4^a are matrices in the adjoint representation of the 5d gauge group, and can be diagonalized without loss of generality. Equivalently one can restrict to Wilson line backgrounds for the Cartan generators of the 5d group, which we denote A_4^I, with $I = 1, \ldots, \text{rank}\,G$. From the analog of (5.18), the A_4^I are periodic with period $1/R$. For later use, it is convenient to rescale the Wilson lines into adimensional quantities, with period 1,

$$\tilde{A}_4^I = R A_4^I. \tag{5.19}$$

The presence of Wilson lines modifies the KK compactification to 4d, and changes the spectrum of light 4d fields. Indeed, consider the 5d Yang–Mills action

$$S_{5d} = \int_{M_4 \times \mathbf{S}^1} \text{Tr}\, \mathcal{F}_{MN} \mathcal{F}^{MN}, \tag{5.20}$$

with

$$\mathcal{F}_{MN} = \partial_{[M}\mathcal{A}_{N]} + [\mathcal{A}_M, \mathcal{A}_N]; \quad \mathcal{A}_M = \sum_a A_M^a T^a, \quad (5.21)$$

where T^a are generators of the gauge group G. Expanding this action around the background configuration, the terms $|[\mathcal{A}_M, \mathcal{A}_N]|^2$ lead upon compactification to terms of the form

$$\mathrm{Tr}\,|[\mathcal{A}_M, \mathcal{A}_N]|^2 \to \mathrm{Tr}\,|[\mathcal{A}_\mu, \mathcal{A}_4]|^2 \to \mathrm{Tr}\left(A_\mu^a A^{a\mu}\right), \text{ for } [T^a, \mathcal{A}_4] \neq 0. \quad (5.22)$$

Namely there are mass terms for the 4d vector bosons associated to generators not commuting with the Wilson line background. The unbroken 5d gauge group is broken in the compactification to a 4d group given by the commutant, i.e. the subgroup commuting with the internal gauge background. A generalization of this mechanism for discrete Wilson lines will be exploited in model building in heterotic compactifications in Sections 7.3.3, 7.4.2, and 8.3.

In the basis where Wilson lines are rotated to the Cartan subalgebra, we can give a more precise description of the surviving 4d gauge group, using the Cartan–Weyl basis $\{H_I, E_\alpha\}$ of the Lie algebra. Gauge bosons in the Cartan subalgebra commute with the Wilson lines, and thus remain massless in 4d, so the rank of the gauge group is preserved. For generators E_α associated with some non-zero root α, the corresponding gauge bosons survive in the massless sector if the commutator $[\mathcal{A}_4, E_\alpha] = \left[A_4^I H_I, E_\alpha\right] \propto \alpha_I A_4^I$ vanishes.

Wilson lines are described from the 4d perspective as vevs of scalars in the adjoint representation of G, and correspond to moduli of the compactification, i.e. have vanishing 4d scalar potential. The physical properties of the 4d theory depend on their vevs. For instance, for generic Wilson lines the surviving 4d gauge group is the Cartan subalgebra $U(1)^r$, with $r =$ rank G. At particular loci there are enhanced non-abelian gauge symmetries, e.g. for zero Wilson lines the 4d group is the whole G. Starting from a point of enhanced symmetry, turning on small Wilson lines corresponds from the 4d viewpoint to a Higgs mechanism by adjoint scalar vevs.

An alternative description of the KK compactification with Wilson lines, valid for fields in general representations of the gauge group, is as follows. The internal momentum of a KK mode in a gauge invariant theory is given by a covariant derivative; in the presence of Wilson lines, the momentum in the circle direction thus suffers a shift with respect to its naive quantized value, as

$$p_4 = \left(k + q_I \tilde{A}_4^I\right)\!/R \quad \text{with } k \in \mathbf{Z}, \quad (5.23)$$

where Wilson lines are normalized as in (5.19), and q^I denotes the charge of the corresponding field under the Ith $U(1)$ factor in the Cartan subalgebra. The KK states have masses shifted as $m^2 = p_4^2$, and there is a massless zero mode only if

$$q \cdot \tilde{A}_4 \in \mathbf{Z}. \quad (5.24)$$

For fields in the adjoint representation, this general rule reproduces the above breaking of the gauge group.

The generalization to $\mathbf{T}^d$ compactification of several dimensions is simple. It is possible to introduce gauge backgrounds A_m^a along all internal directions. In order to obtain a vacuum of the theory, the background should have vanishing non-abelian field strength, hence the backgrounds must be constant and commuting, $[\mathcal{A}_m, \mathcal{A}_n] = 0$; they can be diagonalized simultaneously, and parametrized in terms of Wilson line backgrounds A_m^I for the Cartan generators. Each Wilson line acts independently, and the modified KK compactification is defined by shifted internal momenta $p_m = k_m + q_I A_m^I$. There are massless modes for fields satisfying $q \cdot \tilde{A}_m \in \mathbf{Z}$, for all Wilson lines $\tilde{A}_m$ in $\mathbf{T}^d$. In the next section this effective field theory description will be recovered from the string theory description in the large radius regime, as expected.

*Heterotic string with Wilson lines and heterotic Narain lattice**

In this section we review the general toroidal $\mathbf{T}^d$ compactification of heterotic string theory in a general background for the metric G_{mn}, 2-form B_{mn}, and Wilson lines A_m^I, where $I = 1, \ldots, 16$ runs through the Cartan subalgebra of the 10d gauge group $E_8 \times E_8$ or $SO(32)$. The coupling of the worldsheet bosons to the metric and 2-form are as in (5.7), and the coupling to the Wilson lines is given by

$$S_{\text{WL}} = \frac{1}{4\pi\alpha'} \int d^2\xi \epsilon^{ab} \partial_a X^m \partial_b X_L^I A_m^I. \tag{5.25}$$

The quantization of the string in this background is slightly more involved than for type II, due to the purely left-moving character of the X^I, and we simply quote the result. The backgrounds modify neither the worldsheet fermions nor the bosonic oscillators, and their effect on the bosonic center of mass piece is encoded in suitable left and right momenta

$$P_L^I = P^I + R A_m^I w^m,$$

$$p_{L,m} = \sqrt{\frac{\alpha'}{2}} \left[\frac{k_m}{R} + \frac{R}{\alpha'}(G_{mn} - B_{mn})w^n - P^I A_m^I - \frac{R}{2} A_m^I A_n^I w^n \right],$$

$$p_{R,m} = \sqrt{\frac{\alpha'}{2}} \left[\frac{k_m}{R} + \frac{R}{\alpha'}(-G_{mn} - B_{mn})w^n - P^I A_m^I - \frac{R}{2} A_m^I A_n^I w^n \right]. \tag{5.26}$$

This momentum lattice is the (heterotic) Narain lattice. The mass formulae are

$$\frac{\alpha' M_L^2}{2} = \frac{P_L^2}{2} + \frac{p_L^2}{2} + N_B - 1, \qquad \frac{\alpha' M_R^2}{2} = \frac{p_R^2}{2} + \tilde{N}_B + \tilde{N}_F + \tilde{E}_0, \tag{5.27}$$

and the level matching condition $M_L^2 = M_R^2$. The lattice of momenta (5.26) is self-dual, and even with respect to the $(16 + d, d)$ signature Lorentzian scalar product

$$(p_L, p_R) \cdot (p_L', p_R') = \left(P_L^I P_L^{I'} + p_{L,m}^m p_{L,m}' - p_R^m p_{R,m}' \right). \tag{5.28}$$

This ensures that the partition function for these theories is modular invariant for any choice of background fields, so they define consistent vacua of the theory. Similarly, any even and self-dual $(16 + d, d)$-signature lattice can be expressed in the form (5.26). Note the parallelism with the type II discussion in Section 5.1.3.

At large R the string prescription reproduces the field theory KK reduction, as expected. Let us show it restricting for simplicity to square tori $G_{mn} = \delta_{mn}$, and vanishing B-field $B_{mn} = 0$. At large radius the winding states are heavy and the light spectrum arises from the $w = 0$ sector, for which the momenta have the form

$$P_L^I = P^I, \quad p_{L,m} = p_{R,m} = \sqrt{\tfrac{\alpha'}{2}}(k_m - P \cdot \tilde{A}_m)/R, \tag{5.29}$$

using the notation (5.19). Modulo an irrelevant sign, they reproduce the shifted momentum formula (5.23) of KK compactification with Wilson lines. Massless states of the lower-dimensional theory correspond to states with $P^2 = 2$ and $P \cdot \tilde{A}_m \in \mathbf{Z}$, namely KK zero modes of 10d gauge bosons and gauginos, associated to generators commuting with the Wilson lines. For generic Wilson lines the 4d gauge symmetry from the 10d gauge group is $U(1)^{16}$, and is enhanced at particular points in Wilson line moduli space.

In the stringy regime of small radii, the spectrum can differ from the KK approximation. In particular there can be more dramatic enhancements of the gauge group, even to groups larger than the 10d one, if states with non-zero momentum and/or winding become massless. This occurs if the Narain lattice (5.26) contains left-moving momenta satisfying

$$P_L^2 + p_L^2 = 2, \tag{5.30}$$

leading to additional massless left states which can pair up with the right groundstates to produce massless vector multiplets. For illustration, we consider an example leading to a 4d non-abelian gauge group $SO(44)$, saturating the maximal rank 22 for $\mathcal{N} = 4$ vector multiplets (with an extra $U(1)^6$ from graviphotons in the gravity multiplet, recall Section 2.4.1). It is obtained from an even self-dual lattice with vectors $(P_L, p_L; p_R)$ of the form

$$\left(p_1, \ldots, p_{16}; n_1, \ldots, n_6; n'_1, \ldots, n'_6\right), \quad \sum_I p_I + \sum_m n_m = \text{even}, \quad \sum_m n'_m = \text{even},$$

$$\left(p_1, \ldots, p_{16}; n_1, \ldots, n_6; n'_1, \ldots, n'_6\right), \quad \sum_I p_I + \sum_m n_m = \text{odd}, \quad \sum_m n'_m = \text{odd},$$

$$\left(p_1 + \tfrac{1}{2}, \ldots, p_{16} + \tfrac{1}{2}; n_1 + \tfrac{1}{2}, \ldots, n_6 + \tfrac{1}{2}; n'_1 + \tfrac{1}{2}, \ldots, n'_6 + \tfrac{1}{2}\right),$$

$$\sum_I p_I + \sum_m n_m = \text{even}, \quad \sum_m n'_m = \text{even},$$

$$\left(p_1 + \tfrac{1}{2}, \ldots, p_{16} + \tfrac{1}{2}; n_1 + \tfrac{1}{2}, \ldots, n_6 + \tfrac{1}{2}; n'_1 + \tfrac{1}{2}, \ldots, n'_6 + \tfrac{1}{2}\right),$$

$$\sum_I p_I + \sum_m n_m = \text{odd}, \quad \sum_m n'_m = \text{odd}. \tag{5.31}$$

It corresponds to compactification on a $\mathbf{T}^6$ given by the $SO(12)$ root lattice, at the critical radius, with a suitable B-field and Wilson line, see Section 8.4.1 for a similar example. Massless states corresponding to the non-zero roots of $SO(44)$ arise from vectors $(\pm, \pm, 0, \ldots; 0, \ldots, 0; 0, \ldots, 0)$, where underlining means permutation of entries.

T-duality of $E_8 \times E_8$ and $SO(32)$ circle compactification

In circle compactification of the $E_8 \times E_8$ and $SO(32)$ heterotic strings with general Wilson lines, they lead to the same gauge group $U(1)^{16}$, and the same spectrum of 9d massless fields. Rather than an accident, this is a manifestation of a T-duality relation between the two theories, acting non-trivially on the compactification radius and the Wilson line backgrounds. A detailed discussion is beyond our scope, and we merely state the result for a particular choice of Wilson lines, preserving an unbroken $SO(16)^2$, the maximal common subgroup of $E_8 \times E_8$ and $SO(32)$.

Consider the compactification of $E_8 \times E_8$ and $SO(32)$ heterotic theories on $\mathbf{S}^1$'s of radii R and R' respectively, with $G_{99} = 1$, $G'_{99} = 1$, and Wilson lines

$$(\tilde{A}^I) = \left(\tfrac{1}{2}, \tfrac{1}{2}, \tfrac{1}{2}, \tfrac{1}{2}, \tfrac{1}{2}, \tfrac{1}{2}, \tfrac{1}{2}, \tfrac{1}{2}; 0, 0, 0, 0, 0, 0, 0, 0\right) \quad \text{for } SO(32),$$

$$(\tilde{A}^{I'}) = (1, 0, 0, 0, 0, 0, 0, 0; 1, 0, 0, 0, 0, 0, 0, 0) \quad \text{for } E_8 \times E_8. \tag{5.32}$$

The Narain lattices (5.26) for these two compactifications are actually equivalent for $R' = \alpha'/(2R)$. This implies the two theories are described by the same worldsheet theory, and are therefore equivalent.

5.3 Type I toroidal compactification and D-branes

The circle compactification of the type I superstring provides a new illustration of the insights from T-duality; in the present context, it provides an easy first contact with D-branes.

5.3.1 Circle compactification without Wilson lines

We start by discussing the simplest case of compactification on a circle of radius R, with no Wilson lines. The compactification manifests differently in the closed and open string sectors of type I theory, as we now describe.

Closed string sector

The toroidal compactification of the closed sector of type I is simply the Ω projection of the toroidal compactification of type IIB theory in Section 5.1.1. In an $\mathbf{S}^1$ compactification, different sectors of the parent type IIB theory are characterized by the momentum k and winding w in (5.2). The effect of Ω on k, w, is obtained by recalling that Ω maps X^i to X^i_Ω defined by

$$X^i_\Omega(t, \sigma) = X^i(t, \ell - \sigma). \qquad (5.33)$$

Applied to X^9, we get that Ω acts as $x^9 \to x^9, k \to k, w \to -w$. Hence Ω-invariant states are linear combinations of opposite winding states, schematically of the form $|w\rangle + |-w\rangle$; note that the winding number is conserved only for modulo 2 in this theory, an observation which will prove relevant in interpreting the T-dual theory.

The $w = 0$ sector is mapped to itself under Ω. Hence, invariant states are obtained by applying to the groundstate sets of oscillators which are Ω-invariant by themselves; namely, symmetric combinations in the NSNS sector, antisymmetric combinations in the RR sector, and combinations of states in the NS-R and R-NS sectors. It is easy to check that the $w = 0$ sector gives the KK reduction of the massless fields in the original 10d theory, as expected since the $w = 0$ sector is the only relevant one in the large radius regime. The massless states correspond to the gravity multiplet of a 9d theory with 16 supersymmetries.

Open string sector

This is our first encounter with circle compactification of open strings. Although we could have considered this question in the context of the bosonic theory, the discussion is simple enough to carry it out directly for open superstrings.

Open strings with free endpoints cannot wrap around a compact dimension in a topologically non-trivial way, so there are no winding sectors. Indeed, it is easy to see that open strings with NN boundary conditions (3.93) do not allow for a term linear in σ in the oscillator expansion. The only effect of the circle compactification in the open string sector is the quantization of the internal momentum. The mode expansion for the worldsheet bosons is schematically

$$X^9(t, \sigma) = x^9 + \frac{k}{Rp^+} + \text{oscillators}, \qquad (5.34)$$

whereas that of worldsheet fermions is as usual. The spacetime mass formula is

$$M^2 = \frac{1}{\alpha'}(N_B + N_F + E_0) + \frac{k^2}{R^2}. \qquad (5.35)$$

Since there is no winding, this agrees with the mass formula (3.71) for field theory KK compactification. Massless states correspond to $k = 0$ and reproduce the $SO(32)$ gauge bosons, one real scalar in the adjoint representation, and fermion superpartners (with respect to the 16 supersymmetries in spacetime).

5.3.2 T-duality, orientifold compactifications and D-branes

Let us consider the T-dual of type I theory compactified on a circle of radius R. The T-dual configuration, sometimes dubbed type I$'$ theory, turns out to be best described in terms of new objects, *orientifold planes*, and *D-branes*, related to the closed and open sectors in the theory.

Closed string sector

The closed string sector of type I theory is type IIB modded out by Ω, compactified on a circle of radius R. The compact dimension is described by the worldsheet fields $X^9 = X_L^9 + X_R^9$, $\psi^9 = \psi_L^9 + \psi_R^9$. Since the orientifold quotient by Ω is felt only globally, via the introduction of non-orientable worldsheets in the genus expansion, it does not modify the local 2d worldsheet dynamics. Hence the T-dual theory is described by coordinates $X^{9\prime} = X_L^9 - X_R^9$ and fermions $\psi^{9\prime} = \psi_L^9 - \psi_R^9$. Using the results in Section 5.1.2, the local worldsheet dynamics of the T-dual theory is that of type IIA on a circle of radius $R' = \alpha'/R$. Actually, since the original theory is a quotient by Ω, the T-dual is a quotient of this type IIA compactification, by an action Ω'; the action of Ω' on X', ψ' can be obtained from the action of Ω on X, ψ. Since Ω acts as $X_L^9 \leftrightarrow X_R^9$, we find that Ω' acts as $X^{9\prime} \to -X^{9\prime}$. Hence the T-dual is an orientifold of type IIA (compactified on the T-dual radius) by $\Omega' = \Omega\mathcal{R}$, the combined action of worldsheet parity Ω and a spacetime reflection $\mathcal{R} : x^9 \to -x^9$.

The corresponding action on the worldsheet fermion, $\Omega' : \psi^{9\prime} \to -\psi^{9\prime}$, implies a flip of the GSO projection when exchanging the left and right sectors. It is then a consistent symmetry of type IIA theory, which has opposite GSO projections in the two sectors. This additional action in Ω' thus allows to relax the strict left–right symmetry requirement for unoriented theories imposed in Section 3.4.

The action of $\Omega\mathcal{R}$ on the modes of the oscillator expansion can be obtained from

$$X_{\Omega'}^9{}'(t, \sigma) = -X^{9\prime}(t, \ell - \sigma), \tag{5.36}$$

and similarly for the worldsheet fermions. The result, consistent with T-duality, is that $\Omega\mathcal{R}$ exchanges left and right oscillators, with suitable signs, and acts on center of mass quantum numbers as $x^9 \to -x^9$, $k^9 \to -k^9$, $w^9 \to w^9$.

Let us consider the geometric interpretation described by the T-dual closed string sector. The theory is essentially type IIA on a circle of radius R', modded out by a geometric action $\mathcal{R}$ (times Ω). This action has two fixed points at $x^9 = 0, \pi R'$, see Figure 5.1; the action $\mathcal{R}$ can be regarded as a reflection with respect to any of these points. For instance a closed string (with $w = 0$ winding for simplicity) with center of mass away from the fixed point is mapped to a closed string at the image point, and with opposite momentum. The existence of these special points violates translation invariance in the internal direction (but preserves 9d Poincaré invariance). Momentum is conserved only modulo 2, in agreement with the mod 2 conservation of winding number in the original type I compactification. Since $\Omega' = \Omega\mathcal{R}$ also flips the string orientation, the subspace fixed under $\mathcal{R}$ corresponds to a region where the orientation of a string can flip; heuristically, the image of a closed string at a fixed point is the same closed string with opposite orientation. These subspaces have

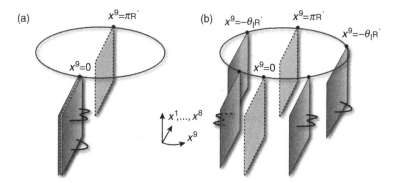

Figure 5.1 Structure of the T-dual of type I without (a) and with (b) Wilson lines. The closed string sector corresponds to a type IIA on $\mathbf{S}^1$ modded by an orientifold whose geometric action leaves two fixed points on $\mathbf{S}^1$, corresponding to two O8-planes at $x^9 = 0, \pi R'$. The open strings have their endpoints fixed at particular values in x^9, corresponding to D8-branes. In (a) they are all located at $x^9 = 0$, while in (b) their locations are controlled by the parameters θ_I.

eight space dimensions (parametrized by $(x^1, \ldots, x^8)$, with $x^9 = 0, \pi R'$, plus time), and are known as orientifold 8-planes, or O8-planes for short. Generalizations of these objects will be a crucial ingredient in the constructions known as orientifold compactifications in Chapters 10 and 11.

Open string sector

Consider now the open string sector in the T-dual theory. The local 2d dynamics of the T-dual open string sector should be that of (an orientifold of) type IIA theory. In particular, it implies that the interior of open string worldsheets propagates in 10d. However, since the original open string sector does not have winding number in x^9, the T-dual open string sector has no momentum in x^9, and leads to fields propagating only in 9d. This conundrum is clarified by computing the boundary conditions for the T-dual worldsheet fields $X^{9\prime}(t, \sigma)$, $\psi^{9\prime}(t, \sigma)$. Focusing on the bosonic fields, the NN boundary conditions (3.93) for $X^9(t, \sigma)$ read

$$\partial_\sigma X^9(t, \sigma) = \partial_\sigma X_L^9(t + \sigma) + \partial_\sigma X_R^9(t - \sigma) = 0 \quad \text{at } \sigma = 0, \ell. \tag{5.37}$$

Since $\partial_\sigma X_L = \partial_t X_L$ and $\partial_\sigma X_R = -\partial_t X_R$, in terms of $X^{\prime 9}$ we have

$$\partial_t X^{9\prime}(t, \sigma) = 0 \quad \text{at } \sigma = 0, \ell. \tag{5.38}$$

These are Dirichlet boundary conditions on both endpoints, or DD boundary conditions for short. They imply that, although the inside of the open string can move in 10d, the open string endpoints cannot move from a fixed value of the coordinate x^9. As will become clear in the next section, in the present situation (i.e. of vanishing Wilson lines in the original type I) the open string endpoints are fixed at $x^9 = 0$. Namely, an equivalent set of DD boundary conditions for our system is

$$X^{9\prime}(t, \sigma) = 0 \quad \text{at } \sigma = 0, \ell. \tag{5.39}$$

Concerning fermions, the only relevant information is that a DD boundary condition flips the GSO projection. Namely a left-moving Ramond groundstate $|\mathbf{8}_C\rangle$ bounces back at the open string boundary as a right-moving Ramond groundstate $|\mathbf{8}_S\rangle$. This is consistent with the coupling of this open string to a type IIA closed string sector. Needless to say, the remaining fields $X^i(\sigma, t)$, for $i = 2, \ldots, 8$, satisfy NN boundary conditions (and similarly for the associated worldsheet fermions).

The spacetime picture of the T-dual open string sector is more manifest by looking at the mode expansion satisfying DD boundary conditions,

$$X^{9\prime}(t, \sigma) = \frac{2\pi R'}{\ell} w'\sigma + \text{oscillators}. \tag{5.40}$$

The center of mass position and momentum are frozen to zero, so open string states are localized in the 9d plane $x^9 = 0$, see Figure 5.1(a). On the other hand, an arbitrary integer winding number is allowed; indeed open strings winding around the compact dimension cannot be unwound, since the endpoints must remain at a fixed location in x^9.

The locus where open string endpoints are located has eight space dimensions, plus time. It is usually referred to as Dirichlet 8-brane, or D8-brane for short. In the present picture, all open string endpoints are located at $x^9 = 0$, on top of an O8-plane, and it is not obvious that D8-branes and O8-planes are different objects. This additional intuition will become manifest upon the introduction of non-trivial Wilson lines in the next section.

The open string spectrum has the familiar structure, and a mass formula

$$M_{ab}^2 = \frac{R^2 w^2}{\alpha'^2} + \frac{1}{\alpha'}(N_B + N_F + E_0), \tag{5.41}$$

where ab denote the Chan–Paton indices and $E_0 = -1/2, 0$ in the NS, R sector. The massless sector is obtained for $w = 0$, oscillator structures as for massless 10d states, and any choice of Chan–Paton indices surviving the $\Omega\mathcal{R}$ orientifold projection (since the open strings are localized at an orientifold plane at $x^9 = 0$). In the NS sector we obtain 9d $SO(32)$ gauge bosons, and one real scalar in the adjoint representation. In the R sector we obtain the gauginos, i.e. one 9d fermion in the adjoint representation. As emphasized, these fields propagate only on the volume of the D8-brane defined by $x^9 = 0$. Therefore, the model is a simple (and still not realistic) realization of the brane-world scenario described in Section 1.4.2.

It is useful to describe the subset of supersymmetries of type IIA theory preserved by this orientifold compactification, in other words by the O8-planes and D8-branes. Let us denote the 32 supercharges of type IIA by two 16-component spinors Q_L, Q_R of $SO(10)$, arising from the left- or right-moving sector, respectively (namely, Q_L transforms left NS states into left R states, leaving the right part of the state invariant, and conversely for Q_R). They have opposite spacetime chirality, namely $\Gamma^{10}Q_L = -Q_L$ and $\Gamma^{10}Q_R = Q_R$, with $\Gamma^{10} = \Gamma^0 \cdots \Gamma^9$, and Γ^M being the 10d Dirac matrices. The orientifold and the open string boundary conditions exchange the left and right sectors, so they preserve a linear combination of the form

$$Q = \epsilon_L Q_L + \epsilon_R Q_R, \tag{5.42}$$

where the coefficient ϵ's are $SO(10)$ spinors with same chirality as the Q's. The model preserves 16 supersymmetries corresponding to the solutions of the condition

$$\epsilon_L = \Gamma^0 \cdots \Gamma^8 \epsilon_R. \tag{5.43}$$

This follows since both the orientifold action and the DD boundary conditions relate left and right sectors with a flip of the GSO projection, equivalently with the application of the Ramond fermion zero mode ψ_0^9, which corresponds to the Dirac matrix Γ^9 from the spacetime perspective. The orientifold projection and the open string boundary conditions reduce by half the supersymmetry of the underlying type IIA theory. The fact that D8-branes preserve some supersymmetries implies, in analogy with Section 2.4.2, that they are BPS states of the theory, as further developed in Section 6.1.

5.3.3 Circle compactification with Wilson lines

As in heterotic theories in Section 5.2.2, type I compactified on $\mathbf{S}^1$ contains 9d scalars in the adjoint representations of the gauge group. Their vevs parametrize the possibility of turning on Wilson lines A_9^a. Their effects on the compactification are remarkably simple, as we now describe.

The closed string sector is insensitive to the presence of Wilson lines, since its states are neutral under the gauge symmetry. In the open string sector, the coupling to the Wilson line backgrounds follows from the worldsheet term (3.14). Recalling from Section 3.3.3 that an open string with Chan–Paton indices ab has charge $+1$ and -1 under the ath and b^{th} $U(1)$ factors in the Cartan subalgebra, the worldsheet coupling to Wilson lines is

$$\Delta S = \int dt \left(A_9^a - A_9^b \right) \partial_t X^9. \tag{5.44}$$

In this expression, the Chan–Paton indices take values $a = 1, \ldots, 32$, subject to the orientifold action, which reduces the number of free Wilson line parameters to 16. These can be parametrized as

$$\left(A_9^1, \ldots, A_9^{32} \right) = -\frac{1}{2\pi R} (\theta_1, \theta_2, \ldots, \theta_{16}; -\theta_1, -\theta_2, \ldots, -\theta_{16}), \tag{5.45}$$

with the overall sign for later convenience. This can be regarded as the diagonalization (using complex variables) of the block-diagonal Wilson lines along the $SO(32)$ Cartan generators,

$$A_9 = \frac{1}{2\pi R} \text{diag} \left(\begin{pmatrix} 0 & \theta_1 \\ -\theta_1 & \end{pmatrix}, \ldots, \begin{pmatrix} 0 & \theta_{16} \\ -\theta_{16} & \end{pmatrix} \right). \tag{5.46}$$

In the following we often use an extended set θ_a, $a = 1, \ldots, 32$, with the understanding that $\theta_{16+a} = -\theta_a$, as in (5.45). Note that the periodicity of the Wilson line A_9^a implies a periodicity $\theta^a \simeq \theta^a + 2\pi$, a relevant fact for the T-dual interpretation.

The whole modification from (5.44) for an open string in the ab Chan–Paton sector amounts just to a shift of the center of mass momentum. Recalling the overall sign in (5.45), the shifted momentum is

$$p^9 = \frac{k}{R} + \frac{\theta_b - \theta_a}{2\pi R}. \tag{5.47}$$

Hence it reduces to an effect already familiar from the field theoretical analysis. The spacetime mass formula in the ab sector is

$$M_{ab}^2 = \left(\frac{k}{R} + \frac{\theta_b - \theta_a}{2\pi R} \right)^2 + \frac{1}{\alpha'}(N_B + N_F + E_0), \tag{5.48}$$

with $E_0 = -1/2, 0$ in the NS, R sector. For generic radius R and Wilson line parameters θ_a's, the gauge group is broken to $U(1)^{16}$, since only aa states produce massless modes. If N eigenvalues have equal value $\theta \neq 0, \pi$, there are additional massless fields, enhancing the gauge symmetry to $U(N)$. Finally, when N eigenvalues have equal value $\theta = 0$ or $\theta = \pi$, the gauge symmetry is enhanced to $SO(2N)$.

5.3.4 T-duality and D-branes

Many of the properties of this compactification are realized geometrically and become more manifest in the T-dual picture, to which we now turn.

Closed strings are insensitive to Wilson lines, so the T-dual closed string sector is as studied in Section 5.3.2, i.e. type IIA theory on a circle of radius R', modded out by $\Omega\mathcal{R}$. In the open string sector, the Wilson lines induce a non-trivial shift of the internal center of mass momentum. This turns into a shifted winding number in the T-dual. The mode expansion is

$$X^{9\prime}(t, \sigma) = \text{const.} + \frac{2\pi R'}{\ell} \left(w' + \frac{\theta_b - \theta_a}{2\pi} \right) \sigma + \text{oscillators.} \tag{5.49}$$

This implies that the open string endpoints of ab strings sit at different locations in x^9. In other words, the DD boundary conditions obeyed by an open string with Chan–Paton indices ab are

$$X^{9\prime}(t, \sigma)\big|_{\sigma=0} = \theta_a R', \quad X^{9\prime}(t, \sigma)\big|_{\sigma=\ell} = \theta_b R', \tag{5.50}$$

(so the σ-independent constant term in (5.49) is $\theta_a R'$). This implies that if θ_a, θ_b are different, the string has a minimal stretching proportional to $(\theta_b - \theta_a)$, equivalently a shift of the winding. The mass formula following from (5.49) is

$$M_{ab}^2 = \frac{R'^2}{\alpha'^2} \left(w' + \frac{\theta_b - \theta_a}{2\pi} \right)^2 + \frac{1}{\alpha'}(N_B + N_F + E_0). \tag{5.51}$$

These features are clearer using the already introduced concept of D8-branes, as the loci where open strings end. The configuration contains 32 D8-branes, labeled by an index a, corresponding to the hyperplanes $x^9 = \theta_a R'$, with $a = 1, \ldots, 32$ and the understanding that $\theta_{16+a} = -\theta_a$; there are thus 16 D8-branes at independent locations, and their 16 orientifold images making the configuration invariant under the symmetry $\mathcal{R}$, see Figure 5.1(b). In this language, the 2π periodicity of the type I Wilson line parameters θ_a is just the periodicity in the D8-brane position on the T-dual $\mathbf{S}^1$. An ab open string stretches from the ath to the bth D8-branes (in general winding w' times around the circle), thus providing a geometric interpretation of (5.51). This picture leads to the crucial insight, which permeates the construction of type II orientifold compactifications in Chapters 10 and 11, that Chan–Paton indices are labels for the different D-branes on which the open strings can end.

The massless open string spectrum is easily recovered in this picture. In the generic configuration of D8-branes at different locations, and away from the O8-planes (which always sit at $x^9 = 0, \pi R'$), the only massless states arise from open strings with both endpoints on the same D8-brane, and the gauge group is $U(1)^{16}$. When N D8-branes coincide, still away from O8-planes, there are additional ab massless states which enhance the group to $U(N)$. Finally, when N D8-branes (and their orientifold images) sit on top of an O8-plane, the group is enhanced to $SO(2N)$. In fact, the T-dual of the compactification without Wilson lines corresponds to taking all the 16 D8-branes (and their images) on top of the O8-plane at $x^9 = 0$. The localization of non-abelian gauge symmetries on the volume of coincident D8-branes is a realization of the brane-world scenario of Section 1.4.2, extensively exploited in type II orientifold compactifications in Chapters 10 and 11.

The spatial separation of D8-branes and O8-planes allows a geometric interpretation of the type I RR tadpole cancellation condition fixing the range of the Chan–Paton index. Under T-duality, the type I non-propagating RR 10-form loses its index along x^9 and turns into a RR 9-form field with indices along the non-compact dimensions. The crosscap and disk diagrams are localized on the O8-planes and D8-branes, and measure the charge of these objects under this RR 9-form gauge field. From the type I result, we obtain that an O8-plane has charge $Q_{O8} = -16Q_{D8}$. Cancellation of RR tadpole diagrams is thus equivalent to the requirement (by Gauss law) that in a compact space the total charge must be zero; the -32 units of charge from the two O8-planes is canceled by the introduction of 32 D8-branes.

The above ideas generalize easily to lower-dimensional compactification on further circles. This leads to more general orientifold compactifications, defined as toroidal compactifications of type IIA and IIB, quotiented by orientifold actions $\Omega\mathcal{R}$, with $\mathcal{R}$ a $\mathbf{Z}_2$ geometric action flipping d internal coordinates (with d even/odd for type IIB/IIA theory), and acting suitably on R groundstates. Geometrically, the compactifications contain 2^d Op-planes, with $p = 9 - d$, namely $(p + 1)$-dimensional planes fixed under $\mathcal{R}$. The open string sectors of these models have DD boundary conditions along d dimensions, and describe Dp-branes, i.e. $(p + 1)$-dimensional subspaces on which open strings endpoints are located. The crosscap and disk tadpoles imply that these objects are charged under the

RR $(p+1)$-form with

$$Q_{Op} = -2^{p-4}Q_{Dp}.$$ (5.52)

The Op-planes and Dp-branes preserve 16 of the 32 supersymmetries of the parent type II theory. Denoting Q_L, Q_R the generators of the latter, the preserved supersymmetries are combinations of the form (5.42) satisfying

$$\epsilon_L = \Gamma^0 \cdots \Gamma^p \epsilon_R.$$ (5.53)

In this language of D-branes and O-planes, type I theory can be described as containing one 10d spacetime filling O9-plane, whose -32 units of RR 10-form charge are canceled by 32 spacetime filling D9-branes, leading to open strings with NN boundary conditions and 32-valued Chan–Paton factors.

6

Branes and string duality

We have studied the main properties of string theory in the framework of perturbation theory. However, on physical grounds we expect the theory to have interesting non-perturbative dynamics; for instance, because the low-energy limit of string theory contains gauge theories, which do have non-perturbative effects. Although a complete definition of string theory at the non-perturbative level is lacking, there is a considerable body of knowledge on some of its properties. In this chapter we introduce certain non-perturbative extended states known as branes, and review their role in non-perturbative string theory and in string duality.

6.1 D-branes in string theory

We start by uncovering the existence in string theory of certain non-perturbative states which admit a remarkably simple description. They are the Dp-branes, extended objects of p spatial dimensions, which at weak coupling can be defined as $(p + 1)$-dimensional subspaces of spacetime on which open strings end. They are crucial in the study of non-perturbative dualities in string theory, but also in the construction of string compactifications, as exploited in Chapters 10 and 11. In fact, we have already encountered D-branes in the orientifold compactification of Section 5.3.

6.1.1 D-branes as non-perturbative states

In Section 5.3.4 we described compatifications of type IIA or IIB string theory with sectors of open strings ending on certain lower-dimensional planes, the D-branes. In fact, these D-branes already exist in the theory in flat 10d spacetime, as can be recovered by taking a suitable decompactification limit. In this section we interpret these D-branes in 10d as non-perturbative states of type II theory (or quotients thereof, like type I in Section 6.1.3).

Consider type IIA or IIB string theory in flat 10d spacetime, in the presence of a flat Dp-brane, i.e. a $(p + 1)$-dimensional plane with the property that the system contains sectors of open strings ending on it, see Figure 6.1. Since type II theory does not contain open strings in its vacuum, this suggests that the configuration at hand is actually an

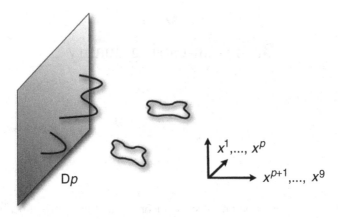

Figure 6.1 String theory in the presence of a Dp-brane, along the directions $x^0, \ldots, x^p$. The closed string sector describes the fluctuations of the theory around the vacuum, while the sector of open strings characterizes the fluctuations of the non-perturbative object.

excited state of type II theory. The Dp-brane should be regarded as a dynamical object, with the open string sector describing the excitations of the theory around it. This matches the fact that the open string sector does not preserve the 10d Poincaré invariance of the vacuum.

The properties of the D-brane can be studied in terms of the corresponding open string sector, or equivalently in terms of worldsheets with boundaries. For instance there are disk diagrams describing the emission of closed string states by the D-brane. For massless closed string states, these are identical to the tadpole diagrams of open string sectors in earlier chapters The corresponding emission of gravitons and components of the RR $(p + 1)$-form along the brane volume encode the D-brane tension (energy per unit volume) and RR charge. Using (3.7), the Euler characteristic of the disk is $\chi = 1$, so the D-brane tension scales with the string coupling as $T_p \simeq 1/g_s$, from arguments around (3.6). It diverges in the weak coupling limit, reflecting the non-perturbative nature of these states. It also implies its description as a formally rigid plane in the weak coupling regime, since it becomes static as $T_p \to \infty$.

Note that, except for the special case of D9-branes, which are actually spacetime filling, the Dp-brane has non-compact transverse dimensions. Hence the flux-lines of the corresponding RR field can escape to infinity, and there are no RR tadpole (i.e. charge) cancellation constraints. Consequently it is perfectly consistent to consider configurations of D-branes without orientifold planes. Moreover, the number of D-branes in the configuration is arbitrary;[1] configurations with different numbers of D-branes simply correspond to different states in the same type II theory, rather than to different theories. RR tadpole conditions only reappear in compactifications, when all dimensions transverse to a

---

[1] For D7- and D8-branes, the backreaction of the branes on the geometry modifies the asymptotic behavior of spacetime in the non-compact setup, see Section 6.2. This introduces additional subtleties, like forcing the compactification for too large number of D-branes, hence resulting in an effective bound even for non-compact setups.

Dp-brane sector are compact, and indeed impose constraints on the D-brane configuration, as in Section 5.3.4, or later in, e.g., Sections 10.3.1 and 11.4.2.

6.1.2 D-brane worldvolume spectrum and effective action

The spectrum of oscillations for open string sectors ending on Dp-branes encode the fluctuations of the theory around the D-brane, and can be used to describe its dynamics. We construct this spectrum for D-branes in type IIA and IIB theory. Most results are a reinterpretation of those for open string sectors in Section 5.3.4.

A single Dp-brane

Consider first a configuration with a single Dp-brane spanning $x^0, \ldots, x^p$, and transverse to the directions $x^{p+1}, \ldots, x^9$, see Figure 6.1. An open string with both endpoints on the Dp-brane has a worldsheet theory with 2d bosons $X^\mu(t, \sigma)$, $\mu = 2, \ldots, p$ (in the light-cone gauge) and $X^i(t, \sigma)$, $i = p+1, \ldots, 9$, and their 2d fermion partners; the worldsheet bosons X^μ satisfy Neumann boundary conditions, and X^i satisfy Dirichlet conditions (hence the name "D"-brane):

$$\text{NN}: \quad \partial_\sigma X^\mu(\sigma, t)|_{\sigma=0,\ell} = 0,$$
$$\text{DD}: \quad X^i(\sigma, t)|_{\sigma=0,\ell} = 0. \tag{6.1}$$

These indeed allow to drop the boundary terms in the variational principle derivation of the worldsheet equations of motion; this is already familiar for Neumann conditions, and for Dirichlet conditions it follows because in the variation

$$\delta S_P = -\frac{1}{2\pi\alpha'} \int_\Sigma d^2\xi \, \partial_a X^i \partial_a \delta X^i$$
$$= -\frac{1}{2\pi\alpha'} \int_{-\infty}^{\infty} dt \, (\delta X^i \partial_\sigma X^i)\Big|_{\sigma=0}^{\sigma=\ell} + \frac{1}{2\pi\alpha'} \int_\Sigma d^2\xi \, \delta X^i \, \partial_a \partial_a X^i, \tag{6.2}$$

the endpoints are fixed in the transverse directions, i.e. $\delta X^i = 0$ at $\sigma = 0, \ell$. The mode expansions for these boundary conditions are

$$X^\mu(\sigma, t) = x^\mu + \frac{p^\mu}{p^+} t + \sqrt{2\alpha'} \sum_{n \neq 0} \frac{\alpha_n^\mu}{n} \cos\left(\frac{\pi n \sigma}{\ell}\right) e^{-\pi i n t/\ell},$$
$$X^i(\sigma, t) = \sqrt{2\alpha'} \sum_{n \neq 0} \frac{\alpha_n^i}{n} \sin\left(\frac{\pi n \sigma}{\ell}\right) e^{-\pi i n t/\ell}. \tag{6.3}$$

There is a similar expansion for fermions, with moddings $n + \nu$ with $\nu = 1/2, 0$ in the NS and R sectors. Interestingly, each DD boundary condition in the R sector flips the sign between the left- and right-moving GSO projections. Therefore, type IIB theory is compatible with Dp-branes with odd p (even number of DD dimensions), while type IIA theory contains only Dp-branes with even p (odd number of DD dimensions).

From the structure of the oscillators, the spectrum of states is very similar to that of a purely NN open superstring sector. However, from (6.3), the open string center of mass is localized on the D-brane, so the corresponding particles propagate in its $(p+1)$-dimensional volume. Hence, the states must be expressed in terms of this lower-dimensional Lorentz group. In particular, the massless states are as follows:

Sector	State	$SO(p-1)$	$(p+1)$-dim field	
NS	$\psi^{\mu}_{-1/2}|0\rangle$	Vector	Gauge boson A_{μ}	(6.4)
	$\psi^{i}_{-1/2}|0\rangle$	Scalar	$9-p$ real scalars ϕ^i	
R	$|8_C\rangle$	spinor	fermions λ_{α}	

The multiplicity of fermions can be obtained from group-theoretical decomposition of the 8_C, and will be described explicitly in examples as needed. The above spectrum corresponds to a $U(1)$ vector supermultiplet with respect to 16 supersymmetries in $(p+1)$ dimensions. This is also often described as the dimensional reduction of the 10d $\mathcal{N}=1$ vector multiplet. A prototypical example is the D3-brane, whose massless open string states fill out a $U(1)$ vector multiplet of $\mathcal{N}=4$ supersymmetry in 4d, given by one gauge boson, six real scalars, and four Majorana fermions, recall Section 2.4.1.

•The supersymmetry of the open string spectrum reflects that the Dp-brane preserves half of the supersymmetries of type II theory, and is a BPS state, generalizing the discussion in Section 2.4.2. The supersymmetries preserved by a Dp-brane are combinations of left- and right-moving supersymmetries of the form (5.42), i.e. $Q = \epsilon_L Q_L + \epsilon_R Q_R$, satisfying

$$\epsilon_L = \Gamma^0 \cdots \Gamma^P \epsilon_R. \tag{6.5}$$

D-brane effective action

Open string modes on the D-brane describe the dynamics of the D-brane. This is particularly manifest for massless open string states; for instance, the $(9-p)$ massless real scalars ϕ^i on the Dp-brane are the Goldstone bosons of the translational symmetries of the vacuum, broken by the presence of the brane.[2] This implies that the vevs of these scalars parametrize the Dp-brane position in the transverse space $\mathbf{R}^{9-p}$. Going further, non-trivial profiles $\phi^i(x^{\mu})$ for these scalar fields describe fluctuations of the embedding of the Dp-brane *worldvolume* W_{p+1} on spacetime (analogous to the notion of worldline for point particles, or worldsheet for fundamental strings). The bottom line is that the $(p+1)$-dimensional effective action for the massless open string modes is the worldvolume action describing the dynamics of the Dp-brane, and is known as *D-brane action*.

---

[2] Similarly, the fermions are goldstinos of the supersymmetries of the vacuum broken by the presence of the D-brane.

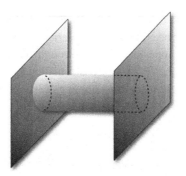

Figure 6.2 Closed string exchange between parallel Dp-branes.

There are several techniques to obtain this action, but we skip their description and merely present the result for the bosonic sector.[3] The D-brane action contains two pieces, the first being known as Dirac–Born–Infeld (DBI) action,

$$S_{\mathrm{DBI}} = -\mu_p \int_{W_{p+1}} d^{p+1}x \, e^{-\phi} \sqrt{-\det(P[G+B] - 2\pi\alpha'F)}. \tag{6.6}$$

The coefficient μ_p is given by

$$\mu_p = \frac{(\alpha')^{-(p+1)/2}}{(2\pi)^p}. \tag{6.7}$$

As mentioned below, it is related to the Dp-brane tension (and charge). Its computation, for which we refer to the Bibliography, involves the evaluation of the annulus/cylinder diagram in Figure 6.2, describing the interaction between two parallel D-branes by exchange of NSNS and RR closed strings.

We have also introduced $P[E]$, the "pullback" of a spacetime tensor E into the brane worldvolume, i.e. the induced worldvolume tensor. In particular, one has

$$P[G]_{\mu\nu} = G_{\mu\nu} + G_{\mu i}\,\partial_\nu\phi^i + \partial_\mu\phi^i\,G_{i\nu} + \partial_\mu\phi^i\partial_\nu\phi^j\,G_{ij}. \tag{6.8}$$

This introduces the dependence of the action on the embedding fields $\phi^i(x^\mu)$. Finally, $F_{\mu\nu}$ is the field strength of the worldvolume gauge field.

A heuristic interpretation of the DBI action is as follows. The DBI action describes the coupling of the D-brane to the NSNS fields, in particular to gravity. Ignoring the 2-form coupling and the gauge field strength dependent contributions, and taking constant dilaton, the coupling to the metric reproduces a generalized Nambu–Goto action of an extended p-dimensional brane with tension μ_p/g_s

$$S_{\mathrm{NG},Dp} = \frac{\mu_p}{g_s} \int d^{p+1}x \, (-g)^{1/2}, \tag{6.9}$$

---

[3] The action as described below suffices for our present purposes. A more general version for the non-abelian case is introduced and employed in Section 15.4.1.

with $g = \det P[G]$; note the analogy with (3.2). The coupling of B in combination with the metric can be understood because both fields arise microscopically from string states $\psi^M_{-1/2}\tilde{\psi}^N_{-1/2}|0\rangle$. Finally, the appearance of F through the combination $B - 2\pi\alpha' F$ is required by invariance under the B_2 gauge transformation (3.11) in the presence of world-sheets with boundaries; indeed, using the couplings (3.10), (3.13) the worldsheet action contains the terms

$$S_{\text{ws}} = \int_\Sigma B_2 - 2\pi\alpha' \int_{\partial\Sigma} A_1. \tag{6.10}$$

This is invariant under a B_2 gauge transformation, only if accompanied by a shift of A_1 to cancel the boundary term. Namely

$$B_2 \to B_2 + d\Lambda_1, \quad A_1 \to A_1 + \frac{1}{2\pi\alpha'}\Lambda_1. \tag{6.11}$$

Note that the second piece is not a gauge transformation for the D-brane worldvolume gauge field. The above invariance implies that all physical observables must depend on the gauge invariant combination

$$2\pi\alpha' \mathcal{F} = 2\pi\alpha' F - B_2, \tag{6.12}$$

where here and in what follows the pullback of B is understood implicitly.

Expanding the DBI action (6.6) in α', or in the field strength, the first term beyond (6.9) is a Yang–Mills action for the worldvolume gauge field

$$S_{\text{YM}} = \frac{\alpha'^{-(p-3)/2}}{4g_s(2\pi)^{p-2}} \int d^{p+1}x\,(-g)^{1/2}\text{tr}\,F_{\mu\nu}\,F^{\mu\nu}, \tag{6.13}$$

and its supersymmetric completion involving the scalar fields and fermions. We have introduced a trace with hindsight, anticipating the upcoming non-abelian generalization for coincident D-branes. From (6.13), the $(p+1)$-dimensional (dimensionful) gauge coupling constant is

$$g^2_{\text{YM}} = g_s\,\alpha'^{(p-3)/2}(2\pi)^{p-2}. \tag{6.14}$$

The second piece of the effective action describes topological couplings to the RR fields, known as Chern–Simons (CS) terms. It formally reads

$$S_{\text{CS}} = \mu_p \int_{W_{p+1}} P\left[\sum_q C_q\right] \wedge e^{2\pi\alpha' F - B_2} \wedge \hat{A}(R). \tag{6.15}$$

In the above expression it is implicit that the CS action picks up only the terms corresponding to $(p+1)$-forms, suitable for integration over the worldvolume W_{p+1}. The first factor corresponds to a formal sum of the spacetime RR q-form fields C_q, pulled back onto the brane volume. The third factor is the so-called A-roof polynomial $\hat{A}(R) = 1 - \frac{1}{24(8\pi^2)}\text{tr}\,R^2 + \cdots$, where R is the curvature 2-form defined in Section B.3, so is relevant only in the presence of spacetime curvature. Finally, the second factor includes a

contribution from the worldvolume gauge field – combined with the NSNS 2-form as in
(6.12). To display its physical meaning, consider the expansion of (6.15), for vanishing B_2
for simplicity,

$$
S_{CS} = \mu_p \left(\int_{W_{p+1}} C_{p+1} + (2\pi\alpha') \int_{W_{p+1}} C_{p-1} \wedge \mathrm{tr}\, F \right.
$$
$$
\left. + \frac{1}{2}(2\pi\alpha')^2 \int_{W_{p+1}} C_{p-3} \wedge \mathrm{tr}\, F^2 - \frac{1}{24(8\pi^2)} \int_{W_{p+1}} C_{p-3} \wedge \mathrm{tr}\, R^2 + \cdots \right), \tag{6.16}
$$

where gauge traces are introduced with hindsight for the upcoming non-abelian generalization. The topological nature of the CS action relates to the fact that it describes the
RR charges of D-branes. The first term in (6.16) is the Dp-brane coupling to the RR
$(p + 1)$-form. The subsequent terms describe the fact that worldvolume gauge field backgrounds (or spacetime curvature) induce lower-dimensional D-brane charges. For instance,
a Dp-brane carrying a non-trivial instanton background $\int F^2 = 8\pi^2 k$ (see (13.3) in Chapter 13) couples to C_{p-3} with coefficient

$$
\mu_p \frac{1}{2}(2\pi\alpha')^2(8\pi^2 k) \equiv k\mu_{p-4}. \tag{6.17}
$$

Namely there are k units of induced D$(p − 4)$-brane charge. Conversely, a D$(p − 4)$-brane
inside a Dp-brane can be regarded as a gauge instanton on the latter, see Section 13.1.2.
The induced D-brane charges from (6.16) will also be relevant in type IIB orientifold
models in Section 11.4.

We finally note that for D9-branes, the last line in (6.16) produces the gauge and gravitational Chern–Simons terms in the 10d type I effective action in Section 4.4.5 (save for a
gravitational O9-plane contribution whose discussion we skip).

Stack of coincident Dp-branes

It is possible to consider configurations of several parallel D-branes, either separated in
transverse space or in the limit of coincident D-branes. This simply amounts to considering several $(p + 1)$-dimensional parallel planes and considering all possible sectors of
open strings ending on them. From the perspective of D-branes as dynamical objects,
the stability of configurations of parallel branes is due to an exact cancellation of their
mutual interactions. These correspond to the cylinder diagram of Figure 6.2, which can be
computed as a one-loop open string diagram, which vanishes due to supersymmetry via
the ϑ-function abstruse identity – see (D.2) for a related computation. This supersymmetry underlies the BPS character of D-brane states, and implies a relationship between the
D-brane tension and charge, both controlled by μ_p as described above. This relation provides an alternative interpretation, in the closed string channel, for the zero-force among
parallel branes, since it entails an exact cancellation between exchange of NSNS fields (in
particular, gravitational attraction) and RR fields (in particular, Coulomb repulsion from
the $(p + 1)$-form interaction).

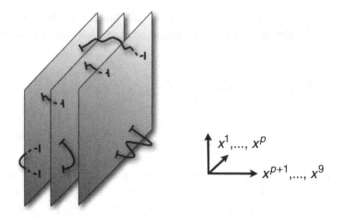

Figure 6.3 Configuration of multiple parallel Dp-branes.

We thus consider n Dp-branes, labeled $a = 1, \ldots, n$, spanning the directions $x^0, \ldots, x^p$, and sitting at the locations x_a^i in the $(9 - p)$ transverse directions, $i = p, \ldots, 9$, see Figure 6.3. The computation of the open string spectrum is a simple generalization of the single D-brane case, and analogous to systems in Section 5.3.4. There are n^2 open string sectors, labeled ab, corresponding to open strings stretching from the ath to the bth D-brane. Note that we are presently working with oriented strings, so ab and ba open strings are different. For each of these n^2 sectors, there are NN boundary conditions for the worldsheet bosons $X^\mu(t, \sigma)$ and DD for $X^i(t, \sigma)$ (and their respective fermion partners). Explicitly, for an ab string we have

$$\partial_\sigma X^\mu(\sigma, t)\big|_{\sigma=0,\ell} = 0,$$
$$X^i(\sigma, t)\big|_{\sigma=0} = x_a^i \quad ; \quad X^i(\sigma, t)\big|_{\sigma=\ell} = x_b^i. \tag{6.18}$$

The spacetime mass formula is the familiar one for open strings, including an additional contribution from stretching between the two D-branes

$$M_{ab}^2 = \sum_{i=p+1}^{9} \left(\frac{x_a^i - x_b^i}{2\pi\alpha'} \right)^2 + \frac{1}{\alpha'}(N_B + N_F + E_0), \tag{6.19}$$

with $E_0 = -1/2, 0$ in the NS, R sectors.

For n coincident D-branes, there are massless states in all n^2 sectors, leading to an enhancement of the gauge symmetry to a $U(n)$ group. The massless sector corresponds to a $(p+1)$-dimensional theory with $U(n)$ gauge bosons, $(9-p)$ real scalars in the adjoint representation, and a number of fermions in the adjoint. Note that all these open string sectors actually have endpoints on the same geometrical plane, but differ in the D-brane label on their endpoints; this is exactly the Chan–Paton degree of freedom introduced in Section 3.3.3, which we can now regard as labeling multiple underlying D-branes.

The D-brane interpretation of Chan–Paton labels is particularly advantageous for geometrically separated D-branes. In a generic such configuration, the aa open strings lead to $U(1)^n$ gauge bosons (plus scalar and fermion partners), while the lightest ab open strings are massive, with mass squared proportional to $\sum_i \left(x_a^i - x_b^i\right)^2$, and have charges $(+1, -1)$ under $U(1)_a \times U(1)_b$. The process of continuously moving away from coincident to slightly separated D-branes can be described as motion along a flat direction on the D-brane worldvolume theory, corresponding to a Higgs mechanism by the adjoint scalars. This is in agreement with the interpretation of the scalar ϕ^i, or rather their n independent eigenvalues upon diagonalization, as the D-brane positions x_a^i in transverse space.

We conclude by mentioning that the effective action for worldvolume massless fields in coincident D-branes is a non-abelian generalization of (6.6), (6.15). For our present purposes, it suffices to promote fields to the adjoint representation, and to take traces, as implicitly done in (6.13), (6.16). A more complete action is briefly touched upon in Section 15.4.1.

6.1.3 D-branes in type I string theory*

Type I theory is constructed as an orientifold quotient of type IIB theory, which leaves the RR 2-form (and its dual 6-form, in the sense of Section B.3) in the massless spectrum. The theory contains D1- and D5-branes coupling to those tensors. Their properties are described by suitable open string sectors, as above. The computations are however slightly more involved than in type II theories, since type I already contains open string sectors in its vacuum. These can be included in the D-brane language by regarding the type I vacuum as containing 32 spacetime filling D9-branes, on which open strings can end.

Type I D5-brane

The computation of the worldvolume spectrum for type I D5-branes is similar to that of type IIB D5-branes, with the additional ingredients of the D5-D9 open string sectors and the orientifold projection Ω.

We start working with oriented strings, postponing the Ω projection to a later stage. Consider a system of k coincident D5-branes, along the dimensions 012345, and N spacetime filling D9-branes. This D5/D9-brane configuration is a prototype of other $Dp/D(p+4)$-brane systems. The configuration preserves eight supersymmetries, given by the solutions of (6.5) for $p = 5, 9$ simultaneously. The theory lives on the 6d worldvolume of the D5-branes, and has 6d $\mathcal{N} = 1$ supersymmetry. This is the 6d analog of 4d $\mathcal{N} = 2$, to which it relates by dimensional reduction, so it shares its supermultiplet terminology (in Section 2.4.1); concretely, there are 6d $\mathcal{N} = 1$ vector multiplets (gauge boson plus one 6d Weyl fermion) and hypermultiplets (one 6d Weyl fermion plus two complex scalars).

The D5-D5 open string sector gives a $U(k)$ vector multiplet of the 16 supersymmetries preserved by the D5-brane alone. In 6d $\mathcal{N} = 1$ terms this correspond to a $U(k)$ vector multiplet, and an adjoint hypermultiplet. In the D5-D9 and D9-D5 sectors, open strings have the familiar NN boundary conditions on the direction 2345, and ND conditions on 6789, namely

$$\partial_\sigma X^i(t,\sigma)\big|_{\sigma=0}=0, \quad \partial_t X^i(t,\sigma)\big|_{\sigma=\ell}=0, \quad \text{for } i=6,7,8,9. \tag{6.20}$$

The ND boundary conditions restrict the string center of mass to the D5-brane location $x^i=0$, and forbid momentum and winding. Also oscillator moddings are shifted by $1/2$ with respect to their usual values, yielding bosonic oscillators $\alpha^i_{n+1/2}$ and fermionic oscillators $\psi^i_{n+\nu+1/2}$, with $\nu=1/2, 0$ for NS, R. The mass formula is

$$\alpha' M^2 = N_B + N_F, \tag{6.21}$$

with vanishing zero point energy $E_0 = 0$ in both the NS and R sectors. In the NS sector the DN boundary conditions shift the oscillator modding and lead to fermion zero modes in the directions 6789. The massless groundstate is degenerate and corresponds, after the GSO projection, to two 6d scalars (transforming as a Weyl spinor under the $SO(4)$ rotation group in 6789). In the R sector there are fermion zero modes along 2345, so the massless groundstate is degenerate, and after the GSO projection transforms as a 6d chiral spinor. Both sectors gather into one 6d $\mathcal{N}=1$ hypermultiplet, which due to Chan–Paton multiplicities transforms in the bi-fundamental representation $(\mathbf{k}, \overline{\mathbf{N}})$ under the D5- and D9-brane symmetries.

We now impose the Ω projection, which fixes the number of D9-branes to $N=32$ by RR tadpole cancellation, and acts on their Chan–Paton indices as the matrix $\gamma_{\Omega,9}$ introduced in Section 4.4.3. The Ω action on D5-brane Chan–Paton indices is given by a $k \times k$ matrix $\gamma_{\Omega,5}$, which, as can be shown, must be antisymmetric in order to yield an action satisfying $\Omega^2 = +1$ on D9-D5 and D5-D9 states. Without loss of generality,

$$\gamma_{\Omega,9} = \mathbf{1}_{32}; \quad \gamma_{\Omega,5} = \begin{pmatrix} 0 & \mathbf{1}_{k/2} \\ -\mathbf{1}_{k/2} & 0 \end{pmatrix}, \tag{6.22}$$

with k even for consistency. The Ω projection in the D5-D5 open strings acts with opposite signs on oscillators along DD and NN directions, as follows from imposing $X_\Omega(t,\sigma) = X(t,-\sigma)$ on the corresponding mode expansions. This implies different projection conditions for the 6d $\mathcal{N}=1$ hyper and vector multiplets, i.e.

$$\text{v.m.}: \lambda = -\gamma_{\Omega,5}\lambda^T \gamma^{-1}_{\Omega,5}, \quad \text{h.m.}: \lambda = \gamma_{\Omega,5}\lambda^T \gamma^{-1}_{\Omega,5}. \tag{6.23}$$

This yields a 6d $USp(k)$ vector multiplet, and one hypermultiplet in the two-index antisymmetric tensor representation (reducible into a singlet plus the rest).

The D5-D9 and D9-D5 sectors are swapped by Ω, so we may compute just the former and not perform any projection. It gives one 6d half-hypermultiplet[4] in the representation $(\mathbf{k}, \mathbf{32})$ under the $USp(k)_{55} \times SO(32)_{99}$ symmetry. Although 6d chiral theories are notoriously plagued with potential gauge anomalies, the D5-brane worldvolume theory is remarkably a chiral anomaly-free 6d theory, for any value of k.

---

[4] Recall from Section 2.4.1 that a half-hypermultiplet contains two real scalars and one Weyl fermion satisfying a reality condition, and exists for multiplets in pseudoreal gauge representations.

Type I D1-brane

The worldvolume theory on type I D1-branes can be worked out similarly, as we now sketch. A system of D1-branes along 01 in type I theory preserves eight supersymmetries, corresponding to solutions of (6.5) for $p = 1, 9$ simultaneously. All the supercharges have the same 2d chirality, so the supersymmetry on the D1-brane worldvolume is known as 2d $\mathcal{N} = (0, 8)$.

To construct the spectrum, consider k D1-branes and N D9-branes before the orientifold projection. The D1-D1 massless sector gives a $U(k)$ gauge symmetry (non-dynamical in 2d), and eight real scalars and eight fermions in the adjoint. In the D1-D9 and D9-D1 sectors, the eight light-cone directions have DN boundary conditions, and the zero point energy contribution to the spacetime mass is $E_0 = 1/2, 0$ for the NS, R sectors; all states in the NS sector are massive, but there are massless states from the (non-degenerate) R sector groundstate, which gives a 2d spinor transforming in the $(\mathbf{k}, \overline{\mathbf{N}})$ under the D1- and D9-brane gauge symmetries.

The Ω action on Chan–Paton factors is implemented by the matrices

$$\gamma_{\Omega,9} = \mathbf{1}_{32}; \quad \gamma_{\Omega,1} = \mathbf{1}_k. \tag{6.24}$$

The orientifold projection on D1-D1 states leads to a 2d theory with (non-dynamical) $SO(k)$ gauge symmetry, eight left chirality fermions in the adjoint, and eight real scalars and eight right chirality fermions in the two-index symmetric representation (reducible as a singlet plus the traceless part). The D1-D9 and D9-D1 sectors are swapped by Ω, so we keep the former and not perform any projection; it leads to one left chirality spinor, in the representation $(\mathbf{k}, \mathbf{32})$ under $SO(k)_{11} \times SO(32)_{99}$. The complete 2d field theory is free of anomalies. Incidentally, the $k = 1$ theory is closely related to a heterotic string worldsheet theory, for reasons explained in Section 6.3.5.

6.2 Supergravity description of non-perturbative states[*]

D-branes provide an example of non-perturbative states in string theory. Such states do not correspond to oscillation states of the string, but rather should be thought of as analogs of solitons in quantum field theory, like magnetic monopoles in certain gauge theories. Consequently it should be possible to construct non-perturbative states in string theory as collective excitations of the spacetime fields, described as classical solutions of the space-time equations of motion. In contrast to gauge field theory, in string theory there is no spacetime action for the complete theory, rather only the 10d supergravity effective actions for the massless fields, in Sections 4.2.5, 4.3.4, and 4.4.5. Classical solutions of these effective theories provide at best an approximate description of certain non-perturbative states of string theory. However, for BPS states, many properties are protected by supersymmetry against quantum and α' corrections, and can be reliably studied in the supergravity approximation.

The idea is illustrated for D-branes in Figure 6.4. The interaction of D-branes with closed strings leads to a non-trivial background for the metric and RR fields, which can be

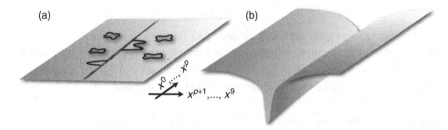

Figure 6.4 A Dp-brane interacts with closed strings via open strings (a), creating an effective super-gravity background (b) which describes the backreaction of the D-brane tension and charge on the configuration.

described as a solution to the supergravity equations of motion. The supergravity approach allows also the construction of other non-perturbative branes, for which no microscopic open string description exists.

In the following we just quote the supergravity solution for certain p-brane states in the different string theories, focusing on the simplest states, carrying charge under a single p-form field. In each string theory, for any $(p + 1)$-form gauge potential (including duals, in the sense of Section B.3), there exist BPS p-brane solutions (i.e. objects with p spatial dimensions, plus time) charged under it. These charges can be measured by computing the H_{8-p} flux around a $(8 - p)$-sphere surrounding the object in the transverse $(9 - p)$-dimensional space, as in (B.25) for $d = 10$. They are moreover quantized from the general arguments in Section B.3. We denote by x^μ, $\mu = 0, \dots, p$, the $(p + 1)$ dimensions along the brane worldvolume, and x^m, $m = p + 1, \dots, 9$, the $(9 - p)$ transverse dimensions.

The Dp-brane solution

These are states describing objects with p spatial dimensions plus time, and charged under the RR $(p+1)$-forms, so they exist for p odd in type IIB theory, p even in IIA, and $p = 1, 5$ in type I (and do not exist in heterotic theories). The supergravity solution for N coincident D-branes is, for $p \leq 6$,

$$ds^2 = Z(r)^{-1/2} \eta_{\mu\nu} dx^\mu dx^\nu + Z(r)^{1/2} dx^m dx^m,$$

$$e^{2\phi} = Z(r)^{(3-p)/2},$$

$$Z(r) = 1 + \frac{\rho^{7-p}}{r^{7-p}}; \quad \rho^{7-p} = g_s N \alpha'^{(7-p)/2} (4\pi)^{(5-p)/2} \Gamma\left(\tfrac{7-p}{2}\right),$$

$$H_{8-p} = \frac{N}{r^{8-p}} d(\text{vol})_{\mathbf{S}^{8-p}}, \qquad (6.25)$$

where $r = \sum_m |x^m|^2$ is the radial coordinate in the transverse space $\mathbf{R}^{9-p}$. Also $d(\text{vol})_{\mathbf{S}^{8-p}}$ is the volume form of a unit $(8 - p)$-sphere surrounding the Dp-brane location in $\mathbf{R}^{9-p}$, so that the solution describes a state with charge N under C_{p+1}, namely $\int_{\mathbf{S}^{8-p}} H_{8-p} = N$, where $H_{8-p} = *H_{p+2}$. The solution has Poincaré invariance along the coordinates x^μ, but only rotational invariance in the x^m, corresponding to a $(p + 1)$-dimensional object,

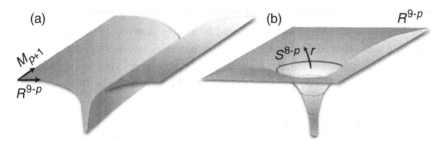

Figure 6.5 Two artistic views of the *p*-brane supergravity solutions. The second picture (b) shows only the transverse directions, where the *p*-brane looks point-like.

point-like in the transverse $\mathbf{R}^{9-p}$. The solution asymptotes to flat 10d spacetime at large r, and at smaller r develops a throat of characteristic size $\rho \simeq g_s N$, see Figure 6.5 for a pictorial view of the solution. The supergravity solution allows to recover other properties of D*p*-branes, like their tension and preserved supersymmetries.

The above solutions apply for $p \leq 6$, and although D7- and D8-branes exist in the corresponding theories, their supergravity description does not have asymptotically flat spacetime. Also, the D9-brane exists only as part of the vacuum of type I, and not as a state.

The NS5-brane solution

This solutions exists in type IIA, type IIB, and heterotic theories, but not in type I. It carries magnetic charge under the (NS) 2-form B_2 in these theories, or equivalently electric charge under its dual 6-form potential B_6. The solution only excites the metric, 2-form, and dilaton, and reads

$$ds^2 = \eta_{\mu\nu}dx^\mu dx^\nu + Z(r)dx^m dx^m,$$
$$e^{2\phi} = Z(r) = g_s^2 + \frac{N}{2\pi^2 r^2},$$
$$H_3 = *_{6789}d\phi. \tag{6.26}$$

The qualitative picture of the NS5-brane solution is as in Figure 6.5, for $p = 5$, but there are some important differences. First, the NS5-brane tension computed from the above solution scales as $T_{\text{NS5}} \sim 1/g_s^2$, so that NS branes are heavier than D*p*-branes at weak coupling. Second, in contrast with D*p*-branes, there is no microscopic description of these states even at weak coupling. A heuristic explanation from (6.26) is that the effective string coupling $e^{2\phi}$ blows up at the core $r \to 0$, even for arbitrarily small coupling g_s at asymptotic infinity $r \to \infty$.

The F1 solution

Last but not least, type IIA, IIB and heterotic string theories contain BPS solutions charged electrically under the (NS) 2-form B_2. They actually correspond to the fundamental string, described effectively via the supergravity background it creates (whose structure we skip). The fact that the fundamental string, denoted F1 in this context, can be described on an equal footing with other BPS extended objects of the theory has far reaching consequences, explored in the next section. Finally, note that since type I theory contains open strings, the fundamental string can break, and thus it carries no charges and is not BPS.

6.3 Strings at strong coupling and 10d string duality

Since string theory contains extended objects with different numbers of spatial dimensions, it is not a theory of strings only. Rather, the fact that the string appears to play the role of fundamental object is an artifact of the perturbative limit $g_s \ll 1$, in which all other extended states are heavy and decouple from the dynamics; at finite coupling g_s all branes are expected to be on equal footing, leading to a complicated structure for the theory, for which no appropriate formulation has been proposed yet.

This viewpoint suggests that there may exist other limits in parameter space in which the theory simplifies again, and admits a description in terms of other fundamental objects. In this section we explore the limit $g_s \to \infty$ of 10d string theories, and characterize the resulting theories, known as "dual" to the original string theory. In certain cases, the theory at infinite coupling turns out to be another string theory, a phenomenon known as string duality. In other situations, the infinite coupling limit leads to new theories without known microscopic description, like M-theory. Although there are no rigorous proofs for these duality statements, there is considerable body of evidence supporting the conjectured results.

A heuristic argument to propose candidate objects dominating the dynamics at strong coupling is to consider the brane states with lowest characteristic energy scale. From the Dp-brane tension $T_p \simeq \alpha'^{-(p+1)/2}/g_s$, we obtain the mass scale $M \simeq \alpha'^{-1/2}g_s^{-1/(p+1)}$, while for the NS5-brane $M \simeq \alpha'^{-1/2}g^{-1/3}$; the lowest energy scale is thus attained by the Dp-brane with lowest p in the theory.

6.3.1 The type IIB $SL(2, \mathbf{Z})$ self-duality

Consider the type IIB theory at strong coupling, $g_s \to \infty$. From the above argument the fundamental object governing the dynamics is the D1-brane. This is an extended one-dimensional object, a string, suggesting that the dual theory is a string theory. To constrain it further, we note that the set of massless states is protected by supersymmetry, and at strong coupling it is still given by the 10d $\mathcal{N} = 2$ chiral supergravity multiplet. This is compatible only with the dual theory being again a type IIB string theory, with the original D1-brane as the fundamental string.

This suggestion in fact fits nicely with a symmetry of the low-energy effective 10d supergravity theory. The action (4.45) once rewritten in the Einstein frame (see Section 14.1.1) is invariant under the transformation

$$\phi \to -\phi; \quad B_2 \to -C_2; \quad C_2 \to B_2, \tag{6.27}$$

with the RR scalar C_0, the 4-form C_4, and the Einstein frame metric invariant. This symmetry maps $g_s \to 1/g_s$, hence relates the strong and weak coupling regimes. It also simultaneously exchanges the NSNS and RR 2-forms (up to a sign, in order to leave the 10d CS couplings invariant), and hence the objects charged under them, the fundamental string and the D1-brane (and the magnetically charged NS5- and D5-branes). This so-called S-duality[5] is proposed to be a symmetry of the full string theory.

Actually, the Einstein frame type IIB supergravity action has a larger $SL(2, \mathbf{R})$ symmetry acting on the 2-forms and the complex string coupling $\tau = C_0 + i/g_s$ as

$$\tau \to \frac{a\tau + b}{c\tau + d}, \quad \begin{pmatrix} B_2 \\ C_2 \end{pmatrix} \to \begin{pmatrix} a & b \\ c & d \end{pmatrix} \begin{pmatrix} B_2 \\ C_2 \end{pmatrix}, \tag{6.28}$$

with $ad - bc = 1$. This continuous group cannot be a symmetry of the full string theory, as it violates charge quantization; however, this property is preserved by a discrete $SL(2, \mathbf{Z})$ subgroup, which is indeed promoted to a symmetry of the full type IIB string theory. It is generated by the $\tau \to -1/\tau$ action (6.27), and the action $\tau \to \tau + 1$, which is a discrete version of the perturbative axion shift symmetry, broken to a discrete subgroup by D-brane instanton effects, as in Chapter 13.

The $SL(2, \mathbf{Z})$ symmetry of type IIB organizes BPS branes in $SL(2, \mathbf{Z})$ multiplets. In particular, there are (p, q) strings, which can be regarded as bound states of p F1's (fundamental strings) and q D1-branes. Similarly, they have dual (p, q)-fivebranes. Of particular importance are the (p, q) 7-branes, which are relevant in the construction of generalized type IIB compactifications, known as F-theory models, to be developed in Section 11.5.

The type IIB $SL(2, \mathbf{Z})$ S-duality group is formally identical to the $\mathbf{T}^2$ modular group, described in Section 3.2.2 in a different context. This tantalizing hint for an underlying geometrical interpretation will be physically realized in Section 6.3.4.

6.3.2 Type IIA and M-theory on $\mathbf{S}^1$

Using the scaling argument in the introduction, the strong coupling dynamics of type IIA theory is dominated by the D0-branes. These are not string-like states, hence the dual theory is *not* a string theory. In order to gain insight into the nature of the dual theory, we focus on bound states of k D0-branes, whose mass is k/g_s in string units. At $g_s \to \infty$ there is an infinite tower of states, with equally spaced masses, becoming light simultaneously. This is tantalizingly similar to the spectrum of KK momentum states in a decompactification limit, and suggests that the strongly coupled type IIA theory is related to an 11d theory on $\mathbf{M}_{10} \times \mathbf{S}^1$, and that the IIA strong coupling limit is the $\mathbf{S}^1$ decompactification limit. The full 11d theory describing the infinite coupling limit of type IIA theory is known as M-theory. Although it is a quantum theory including 11d gravity, it is not a string theory, and it has no adimensional coupling constants; this absence of perturbative expansion

---

[5] The name is motivated by its formal analogy with the perturbative T-duality $R \to 1/R$, and by the fact that it acts on the dilaton, often denoted S in 4d compactifications.

parameters makes difficult a complete formulation for its microscopic structure, despite several partially satisfactory proposals.

There is, however, a precise formulation for the low-energy effective theory of 11d M-theory. It is given by 11d supergravity, with action (4.51). Its massless fields are an 11d metric $G_{\hat{M}\hat{N}}$, a 3-index antisymmetric tensor $C_{\hat{M}\hat{N}\hat{P}}$, and an 11d gravitino $\psi_{\hat{M}\hat{\alpha}}$, with hatted 11d indices. Indeed this is the unique 11d theory with the right supersymmetry structure, and moreover can be shown to reduce to the 10d IIA supergravity action (4.47) when compactified on $\mathbf{S}^1$. This leads to the relation between the IIA coupling, the string tension and the M-theory $\mathbf{S}^1$ radius

$$g_s = (M_{11}R)^{3/2}, \alpha' = \frac{1}{M_{11}^3 R}, \tag{6.29}$$

where M_{11} is the 11d Planck scale defined by $2\kappa_{11} = (2\pi)^8 M_{11}^{-9}$, and R is measured in the 11d metric. In particular the KK reduction of the massless fields produces the IIA massless sector, as given by the following list:

$$
\begin{aligned}
G_{\hat{M}\hat{N}} \longrightarrow \quad & G_{MN} && \text{graviton} \\
& G_{M,10} \to C_1 && \text{RR 1-form} \\
& G_{10,10} \to \phi && \text{dilaton} \\
C_{\hat{M}\hat{N}\hat{P}} \longrightarrow \quad & C_{MNP} \to C_3 && \text{RR 3-form} \\
& C_{MN10} \to B_2 && \text{NSNS 2-form} \\
\Psi_{\hat{M},\hat{\alpha}} \longrightarrow \quad & \psi_{M\alpha}, \psi_{M\dot{\alpha}} && \text{gravitinos} \\
& \psi_{10,\alpha}, \psi_{10,\dot{\alpha}} \to \psi_{\alpha}, \psi_{\dot{\alpha}} && \text{dilatinos}
\end{aligned}
$$

The M-theory lift of the different type IIA p-form fields implies an M-theory lift for the type IIA brane states charged under them, denoted Mp-branes. M-theory contains BPS M2- and M5-branes, which can be constructed as 11d supergravity solutions. We skip their detailed discussion, and merely describe their relation to type IIA branes upon $\mathbf{S}^1$ compactification, given by the table below:

IIA	$\longleftrightarrow$	M-theory on $\mathbf{S}^1$
D0-branes	$\longleftrightarrow$	KK momenta of 11d supergravitons
IIA string	$\longleftrightarrow$	M2 wrapped on $\mathbf{S}^1$
D2	$\longleftrightarrow$	unwrapped M2
D4	$\longleftrightarrow$	M5 wrapped on $\mathbf{S}^1$
NS5	$\longleftrightarrow$	unwrapped M5
D6	$\longleftrightarrow$	Kaluza-Klein monopole

The type IIA D8-brane has no well-defined lift to 11d, but this is partly expected since it does not have flat space asymptotics. The D6-brane lift is not an M-brane, but rather a purely geometrical background, known as a Kaluza–Klein monopole because it carries magnetic charges under the KK gauge field, and is further discussed in the next section.

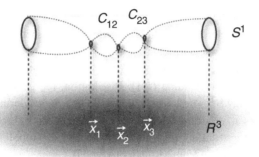

Figure 6.6 Pictorial view of the multi-center Taub-NUT geometry. It is an $\mathbf{S}^1$ fibration over $\mathbf{R}^3$, with fiber degenerating over the centers $\vec{x}_a$. The geometry contains non-trivial 2-cycles C_{ab} associated to pairs of centers, $\vec{x}_a, \vec{x}_b$.

6.3.3 Enhanced gauge symmetries from singularities in M-theory*

D6-branes will play an important role in a class of compactifications in Chapter 10. For future reference in Section 10.7.2, we introduce their description in M-theory and the mechanism leading to enhanced gauge symmetries by geometric singularities. This is also useful for the F-theory models in Section 11.5.

The D6-brane supergravity solution sources the type IIA metric, dilaton, and RR 1-form field. In the M-theory lift all these fields become components of the 11d metric, hence the M-theory lift of a D6-brane stack must be a purely metric background in 11d supergravity, i.e. not involving M2- or M5-branes. We thus have a geometry of the form $\mathbf{M}_7 \times \mathbf{X}_4$, with $\mathbf{X}_4$ a non-compact geometry asymptoting to $\mathbf{R}^3 \times \mathbf{S}^1$, with $\mathbf{R}^3$ the space transverse to the D6-branes and $\mathbf{S}^1$ the M-theory compactification circle. The space $\mathbf{X}_4$ must preserve half of the supersymmetries (in terms of the holonomy group introduced in Section 7.2.1, it must have $SU(2)$ holonomy), and this very restrictive condition, together with the asymptotics, fixes the metric to be of the form

$$ds^2 = \frac{V(\vec{x})}{4} d\vec{x}^2 + \frac{V(\vec{x})^{-1}}{4} \left(dx^{10} + \vec{\omega} \cdot d\vec{x}\right)^2,$$

$$V(\vec{x}) = 1 + \sum_{a=1}^{N} \frac{1}{|\vec{x} - \vec{x}_a|}, \qquad \vec{\nabla} \times \vec{\omega} = \vec{\nabla} V(\vec{x}). \tag{6.30}$$

This is known as the N-center Taub-NUT geometry, or Kaluza–Klein monopole. The coordinates $\vec{x}$ parametrize $\mathbf{R}^3$, and $\vec{\omega}$ is formally identical to the 3d vector potential for N Dirac magnetic monopoles at the locations $\vec{x}_a \in \mathbf{R}^3$. Indeed the N positions $\vec{x}_a$, correspond to the D6-branes locations in $\mathbf{R}^3$, and their magnetic charge N under the RR 1-form becomes a non-trivial twisting of the $\mathbf{S}^1$ fiber over the base $\mathbf{R}^3$. The points $\vec{x}_a$ are called centers and define points in $\mathbf{R}^3$ over which the $\mathbf{S}^1$ fiber shrinks to zero size. The full geometry is smooth, however, as long as the centers are not coincident. A sketch of the geometry is depicted in Figure 6.6.

Before considering the very interesting case of coincident centers, let us display some additional geometrical elements. Around each degenerate fiber over a center $\vec{x}_a$, the geometry supports a normalizable harmonic 2-form ω_a. The component of the M-theory 3-form along it, $C_3 = \omega_a \wedge A_1^a$, produces a $U(1)$ gauge boson, interpreted in the type IIA picture as the D6-brane worldvolume $U(1)$. In addition, the geometry contains non-trivial 2-cycles with topology $\mathbf{S}^2$. These can be constructed by considering a segment in $\mathbf{R}^3$ joining any two centers $\vec{x}_a$, $\vec{x}_b$, times the $\mathbf{S}^1$ fiber, see Figure 6.6; since the $\mathbf{S}^1$ shrinks to zero size at the centers, the resulting two-dimensional surface is topologically an $\mathbf{S}^2$, which we denote C_{ab}. In the reduction from M-theory to type IIA, states corresponding to M2-branes wrapped on the 2-spheres C_{ab} become fundamental strings stretching between the ath and bth D6-branes (the ba open string is obtained from M2-branes wrapping C_{ab} with opposite orientation, or equivalently from anti-M2-branes, see Section 6.5). The mass of the wrapped M2-branes is controlled by the volume of C_{ab}, which is proportional to $|\vec{x}_a - \vec{x}_b|$, in agreement with the stretching contribution to the type IIA open string mass formula. Also, the coupling of M2-branes to the M-theory 3-form implies that these states carry charges $(+1, -1)$ under the $U(1)_a \times U(1)_b$ gauge factors.

The above structure of 2-cycles and wrapped M2-branes provides insight into the physics of the Taub-NUT geometry for coincident centers. When k centers coincide, $\vec{x}_{a_1} = \ldots = \vec{x}_{a_k}$, the geometry (6.30) becomes singular. This is related to the fact that the corresponding $\mathbf{S}^2$'s shrink to zero size. A basis of collapsing 2-cycles is given by the $(k-1)$ 2-spheres $C_{a_p a_{p+1}}$. These singularities are called A_{k-1} singularities, because the intersection pattern for these independent $\mathbf{S}^2$'s reproduces the Dynkin diagram of A_{k-1}. This mathematical curiosity has a stunning physical interpretation in M-theory, because M2-branes wrapped on the $\mathbf{S}^2$'s collapsed at such singularity correspond to new charged massless states, which produce an enhanced non-abelian $U(k)$ gauge symmetry. In particular, M2-branes on the independent cycles $C_{a_p a_{p+1}}$ reproduce the gauge bosons of the simple roots of the $U(k) = A_{k-1}$ Lie algebra. This mechanism provides the M-theory picture of the gauge symmetry enhancement on coincident D6-branes.

In conclusion, M-theory at an A_{k-1} singularity develops an enhanced $U(k)$ gauge symmetry. This phenomenon generalizes to other simply-laced non-abelian symmetries, i.e. ADE Lie algebras, as follows. The A_{k-1} singularities can also be described as $\mathbf{C}^2/\mathbf{Z}_k$ orbifolds,[6] in the language of Section 8.1. These are quotients of $\mathbf{C}^2$ by a geometric rotation $(z_1, z_2) \rightarrow (e^{2\pi i/k} z_1, e^{-2\pi i/k} z_2)$, which belongs to $SU(2)$, implying that the geometry preserves half of the supersymmetries. This is true for other orbifolds $\mathbf{C}^2/\Gamma$, with $\Gamma \in SU(2)$, which actually have a mathematical classification in one-to-one correspondence to the A-D-E classification of Lie algebras, as determined by the Dynkin diagram intersection pattern of the collapsed $\mathbf{S}^2$'s. Extending our above arguments, this has the physical explanation that M-theory in an ADE singularity produces an enhanced ADE gauge symmetry.

---

[6] Incidentally, the process of continuous separation of coincident centers replaces the singular point by finite-size $\mathbf{S}^2$'s, and corresponds to the blowing-up mentioned in Section 8.2.6.

As described above, the A_{k-1} singularities are related to systems of k coincident D6-branes. Similarly, the D_k singularities produce $SO(2k)$ symmetries and are related to systems of $2k$ D6-branes on top of an O6-plane. Finally, E_6, E_7, and E_8 singularities cannot be related to any perturbative D6-brane system, and are genuinely M-theoretical. These singular geometries will play an important role in M- and F-theory model building in Sections 10.7.2 and 11.5.

6.3.4 M-theory on $\mathbf{T}^2$ *and type IIB on* $\mathbf{S}^1$

Upon compactification, non-perturbative dualities combine non-trivially with perturbative dualities like T-duality, eventually leading to an intricate duality web. We now describe an illustrative 9d example.

Recall from Section 5.1.2 that type IIA compactified to 9d on an $\mathbf{S}^1$ of radius R is T-dual to type IIB on a dual $\mathbf{S}^1$ of radius $R' = \alpha'/R$. This leads to a relation between M-theory compactified on $\mathbf{T}^2$ and type IIB compactified on $\mathbf{S}^1$. A careful analysis relates the parameters of both theories as follows: the type IIB complex coupling τ is identified with the complex structure parameter $\tau = R_2/R_1 e^{i\theta}$ of the M-theory $\mathbf{T}^2$, described in terms of two periodic coordinates of radii R_1, R_2 forming an angle θ. The type IIB $\mathbf{S}^1$ radius R is related to the M-theory $\mathbf{T}^2$ area A by $M_{11}^3 A \simeq 1/R$. The decompactified 10d type IIB string theory can be obtained by taking M-theory on a $\mathbf{T}^2$ in the limit of vanishing area, with the tower of light KK modes in this limit arising from M2-branes wrapped multiple times on $\mathbf{T}^2$.

This 9d duality provides a geometric interpretation for the type IIB $SL(2, \mathbf{Z})$ duality, in terms of the $SL(2, \mathbf{Z})$ modular group acting on the complex structure τ of the M-theory $\mathbf{T}^2$ (mathematically identical to that described in Section 3.2.2 in a different context). It is easy to relate the BPS branes in both theories, and we skip this detailed discussion. We merely point out that type IIB (p, q) 7-branes become geometric backgrounds in M-theory, similar to the above Taub-NUT geometries, in which a (p, q) 1-cycle of $\mathbf{T}^2$ shrinks to zero size. The geometrization of the type IIB $SL(2, \mathbf{Z})$ and of the (p, q) 7-branes are important ingredients in F-theory compactifications, whose construction and properties we postpone to Section 11.5.

6.3.5 Type I/$SO(32)$ heterotic duality

Let us turn to the analysis of the strong coupling limit of type I theory. From the heuristic scaling argument on page 168, the fundamental object dominating the strong coupling dynamics is the BPS D1-brane. This is a string-like object, suggesting that the dual theory is a string theory. Since the massless spectrum of the theory is protected by supersymmetry, and is given by the 10d $\mathcal{N} = 1$ gravity multiplet coupled to vector multiplets of $SO(32)$, the dual theory must be either again a type I theory or an $SO(32)$ heterotic theory. The criterion for the right choice is that the dual theory must have a BPS fundamental string, so it cannot be a type I theory (whose fundamental string can break and is not BPS); hence the dual theory is an $SO(32)$ heterotic theory.

This is also supported, in the low-energy approximation, by the fact that the effective actions (4.109), (4.80) for type I and the $SO(32)$ heterotic theories are equivalent up to a change of variables

$$\phi_{\text{typeI}} = -\phi_{\text{het.}}, \quad (G_{MN})_{\text{typeI}} = e^{-\phi_{\text{het.}}} (G_{MN})_{\text{het}}, \tag{6.31}$$

$$(\tilde{F}_3)_{\text{typeI}} = (H_3)_{\text{het.}}, \quad (A_{SO(32)})_{\text{typeI}} = (A_{SO(32)})_{\text{het.}},$$

which implies $(g_s)_{\text{het}} = 1/(g_s)_{\text{typeI}}$. This suggests that the type I theory at coupling g_s is equivalent to the $SO(32)$ heterotic at coupling $g_s' = 1/g_s$. Consequently the strong coupling limit of type I theory is described by a weakly coupled $SO(32)$ heterotic string theory, and vice versa. Additional supporting evidence comes from the type I D1-brane worldvolume spectrum in Section 6.1.3, which can be shown to be a particular (so-called Green–Schwarz) formulation of a heterotic worldsheet theory, in the fermionic version of Section 4.3.5.

This duality would seem to lead to some potential puzzles. For a start, heterotic theory contains massive spinor representations of $SO(32)$, naively absent in type I theory. Although these states are non-BPS, they are stable and should be present in the type I dual; they are actually realized as stable non-BPS D0-branes, discussed later in Section 6.5.3. A second potential problem is that the $SO(32)$ heterotic theory is T-dual to the $E_8 \times E_8$ theory, and hence can develop exceptional symmetries upon circle compactification. This is not possible for perturbative type I theory, but can occur in suitable strong coupling regimes, as discussed later in Section 6.3.7.

6.3.6 M-theory on $S^1/\mathbf{Z}_2$ and $E_8 \times E_8$ heterotic

For reasons that should become clear at the end of this section, the strong coupling limit of the $E_8 \times E_8$ heterotic is trickier to analyze directly, and we cannot resort to the heuristic scaling argument. Instead we start with the seemingly unrelated question of other possible M-theory compactifications to 10d, besides $\mathbf{S}^1$. The only such compactification is on the quotient $\mathbf{S}^1/\mathbf{Z}_2$, which a posteriori can be argued to provide the strong coupling dual of the $E_8 \times E_8$ heterotic.

Consider the compactification of M-theory on $\mathbf{S}^1$, modded out by a $\mathbf{Z}_2$ action with generator θ acting as

$$\theta : x^{10} \to -x^{10}, \quad C_3 \to -C_3. \tag{6.32}$$

This is a symmetry of the 11d supergravity action (4.51), with the flip of the 3-form required for the 11d Chern–Simons coupling $G_4 G_4 C_3$ to be invariant. The construction is a (geometrically simpler) M-theory analog of the string theory orbifold constructions in Section 8.1. The quotient space is an interval $\mathbf{I}$, so the M-theory spacetime $\mathbf{M}_{10} \times \mathbf{I}$ has two identical 10d boundaries at $x^{10} = 0, \pi R$, see Figure 6.7.

We may expect that for large R the system can be well described by taking the quotient of the supergravity effective theory. This corresponds to projecting the massless spectrum

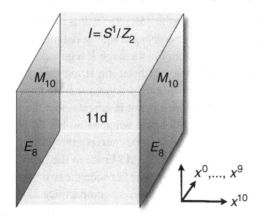

Figure 6.7 Spacetime geometry of the Hořava–Witten theory. M-theory on $\mathbf{M}_{10} \times \mathbf{S}^1/\mathbf{Z}_2$ has an 11d supergravity bulk with two 10d boundaries with localized E_8 vector multiplets.

of M-theory on $\mathbf{S}^1$ onto invariant states. The resulting fields in the notation of Section 6.3.2 are G_{MN}, ϕ, B_{MN}, $\psi_{M\alpha}$, and $\psi_{\dot{\alpha}}$. This corresponds to the gravity multiplet of 10d $\mathcal{N} = 1$ supersymmetry, reflecting the fact that the $\mathbf{Z}_2$ action breaks half of the supersymmetries.

This field content is chiral, and in fact anomalous, rendering the theory inconsistent. Hence, if M-theory is a consistent theory, it must manage to cancel the anomalies by producing additional massless fields beyond those in the supergravity prescription. Indeed, the expectation that the dynamics of M-theory on $\mathbf{S}^1/\mathbf{Z}_2$ can be approximated by supergravity fails, even for large R, near the orbifold fixed points; the key observation is that the action $x^{10} \to -x^{10}$ relates points which are arbitrarily close together as one approaches either of the two fixed points $x^{10} = 0, \pi R$. It is at these loci that the microscopic structure of M-theory can show up and produce effects beyond the supergravity approximation. This can (in fact for consistency *must*) include the appearance of additional massless states, in an M-theory generalization of twisted sectors in string theory orbifolds, see Chapter 8. Now recall from Section 4.3.4 that cancellation of gravity multiplet anomalies by the Green–Schwarz mechanism is only possible for $E_8 \times E_8$ or $SO(32)$ vector multiplets. In the present situation, the existence of two identical boundaries implies that for consistency there must appear one additional massless 10d $\mathcal{N} = 1$ vector multiplet of E_8 on each boundary.

The bottom line is that compactification of M-theory on the interval $\mathbf{S}^1/\mathbf{Z}_2$ leads to the 10d $\mathcal{N} = 1$ gravity multiplet, arising from the $\mathbf{Z}_2$ invariant fields of M-theory in the 11d "bulk," plus 10d $\mathcal{N} = 1$ $E_8 \times E_8$ vector multiplets, each factor propagating on each 10d boundary of spacetime. This system, depicted in Figure 6.7, is known as the Hořava–Witten theory.

The Hořava–Witten theory has precisely the same massless spectrum as the $E_8 \times E_8$ heterotic theory. Moreover, the supergravity effective action is determined by the matter content and supersymmetry, and is identical in both theories, with the M-theory $\mathbf{S}^1$ radius related to the heterotic coupling as $g_s = (M_{11}R)^{3/2}$. It is then natural to promote this to a full equivalence, and propose that the strong coupling regime of the $E_8 \times E_8$ heterotic is described by the Hořava–Witten theory in the large R regime.

It is easy to match the BPS states of heterotic string theory with M-brane states in $\mathbf{S}^1/\mathbf{Z}_2$. The M2-brane wrapped on $\mathbf{S}^1/\mathbf{Z}_2$ produces the fundamental string of heterotic theory, while the unwrapped M2-brane is not invariant under the $\mathbf{Z}_2$ action, and does not correspond to a BPS brane. The M5-brane wrapped on $\mathbf{S}^1/\mathbf{Z}_2$ is also not invariant, but the M5-brane sitting at a point in $\mathbf{S}^1/\mathbf{Z}_2$ does survive the projection and corresponds to a heterotic NS5-brane. The position of the M5-brane in the interval is parametrized by a worldvolume scalar field, and can interpolate between the two boundaries. The possibility to introduce such M5-branes will be exploited in compactifications in Section 7.5.

As a final remark, note that the $\mathbf{Z}_2$ quotient violates translational invariance in the M-theory $\mathbf{S}^1$, hence there is no conserved momentum quantum number, and so KK states are not BPS. This is why, in contrast to previous cases, the study of BPS states in the heterotic theory is not as insightful to identify the strong coupling limit of the $E_8 \times E_8$ heterotic as a decompactification limit.

6.3.7 Type I$'$ and Hořava–Witten theory[*]

The microscopic origin of the E_8 vector multiplets in Hořava–Witten theory is not manifest in the above construction, which just relies in the consistency of M-theory. However, it is possible to gain insight into the appearance of these massless degrees of freedom by using the duality web in 9d. In the following we re-derive the structure of Hořava–Witten theory starting from the 9d type I$'$ theory in Section 5.3, namely the T-dual to type I theory on $\mathbf{S}^1$. The construction also provides a partial answer to the M-theory lift of the type IIA D8-brane, an open issue in Section 6.3.2.

Consider type IIA on $\mathbf{S}^1$ modded out by the orientifold action $\Omega\mathcal{R}$, with the geometric action $\mathcal{R} : x^9 \to -x^9$ on the $\mathbf{S}^1$ introducing two O8-planes. Cancellation of RR tadpoles requires the introduction of 32 D8-branes, which we choose to distribute in two sets of 16, on top of the O8-planes, leading to $SO(16)^2$ gauge sectors. Regarding the $\mathbf{S}^1/\mathbf{Z}$ as an interval, the configuration contains type IIA theory in a 10d "bulk," with its spectrum truncated by a $\mathbf{Z}_2$ projection, and with $SO(16)$ vector multiplets located at each 9d boundary.

The construction is reminiscent of the Hořava–Witten theory, and can indeed be connected to it by a strong coupling limit, as follows. In the bulk we have type IIA theory, which at strong coupling lifts to 11d M-theory by growing an extra $\mathbf{S}^1$. We thus get M-theory on $\mathbf{S}^1/\mathbf{Z}_2 \times \mathbf{S}^1$, with the type IIA D0-brane states reproducing the KK momenta on $\mathbf{S}^1$. The 9d boundaries of type I$'$, supporting the $SO(16)$ gauge symmetries, become the 10d boundaries of Hořava–Witten theory, which should support E_8 gauge symmetries. This is actually not contradictory, since the 10d boundary actually contains the $\mathbf{S}^1$ factor, on which suitable Wilson lines can break the 10d E_8 group to the 9d $SO(16)$. From the type I$'$ perspective, the KK replicas of the $SO(16)$ gauge bosons arise from a tower of states of $2k$ D0-branes bound to the O8-plane, and transforming in the **120** of $SO(16)$; on the other

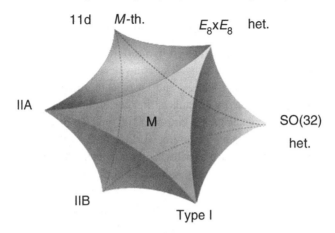

Figure 6.8 Artistic view of the different string theories and 11d supergravity as different limits of a unique theory underlying their duality relations.

hand, states of $2k + 1$ D0-branes bound to the O8-plane, and transforming in the **128** spinor representation, provide the tower (with no zero mode) of KK modes of the additional 10d E_8 gauge bosons. All these D0-brane towers become light in the strong coupling limit, signaling a decompactification limit, in which the boundary gauge symmetry is enhanced to E_8.

The above relation can be derived from a chain of previously studied dualities, as follows. The type I$'$ configuration is T-dual to type I with Wilson lines breaking the gauge symmetry to $SO(16)^2$. From Section 6.3.5, this is equivalent to the $SO(32)$ heterotic on $\mathbf{S}^1$ with $SO(16)^2$-preserving Wilson lines. The latter, using Section 5.2.2, is T-dual to the $E_8 \times E_8$ heterotic on $\mathbf{S}^1$ with $SO(16)^2$-preserving Wilson lines, which finally can be related to the Hořava–Witten theory, namely M-theory on $\mathbf{S}^1/\mathbf{Z}_2 \times \mathbf{S}^1$, with Wilson lines on $\mathbf{S}^1$ breaking each E_8 to $SO(16)$.

Final remarks

The study of strong coupling behavior of string theories adds a new layer in our understanding of string theory. In particular, it shows that all theories are related to each other, and to the novel M-theory, by a rich web of dualities, as depicted in Figure 6.8. Different string theories are actually regarded as different perturbative limits of a unique underlying theory, sometimes called also M-theory, in a qualitatively broader sense. Despite the lack of a proper formulation of the latter, this unified view of the different string theories is extremely appealing. Moreover, as discussed in the next section, there are further far-reaching dualities relating string theory to other physical theories, like quantum field theories. Interestingly, this rich web of dualities eventually manifests as relations among different constructions of phenomenologically interesting particle physics models in string theory.

6.4 AdS/CFT and gauge/gravity dualities

In Sections 6.1 and 6.2 we have introduced two different descriptions of the same kind of object, the Dp-branes in string theory. The equivalence of these two descriptions, in a particularly simple limit, leads to a new and far-reaching duality, known as AdS/CFT correspondence (also, gauge/gravity or gauge/string correspondence).

Consider a system of N coincident D3-branes in 10d type IIB theory. It can be described in terms of open strings with endpoints on its worldvolume, with N-valued Chan–Paton factors. In general, these strings have massless and massive modes on the 4d brane world-volume, which are moreover coupled to massless and massive states from the closed string sector in the 10d bulk. The system, however, simplifies enormously in the low-energy limit: first, all massive stringy states decouple; second, the massless closed strings decouple, since their (gravitational strength) interactions scale with the energy as E^2/M_p^2. The dynamics reduces to a 4d $\mathcal{N} = 4$ supersymmetric $U(N)$ gauge field theory with coupling $g_{YM}^2 \simeq g_s$, recall (6.14), and a decoupled 10d theory of free gravitons. The center of mass $U(1)$ in the 4d gauge theory is actually decoupled, and drops out of the coming discussion.

Consider now the D3-brane system as a supergravity solution. The metric in (6.25) for $p = 3$, in polar coordinates in $\mathbf{R}^6$, is

$$ds^2 = Z(r)^{-\frac{1}{2}}\eta_{\mu\nu}dx^\mu dx^\nu + Z(r)^{\frac{1}{2}}\left(dr^2 + r^2 ds_{S^5}^2\right), \quad Z = 1 + \frac{4\pi g_s N\alpha'^2}{r^4}. \tag{6.33}$$

The dilaton is constant and there are N units of RR 5-form flux on the $\mathbf{S}^5$ in the $\mathbf{R}^6$ transverse to the branes. The solution has a throat of characteristic size R with

$$R^4 = 4\pi g_s N\alpha'^2, \tag{6.34}$$

and asymptotes to flat 10d space. Type IIB string theory (including its full dynamics) on this background provides an alternative description of the D3-brane system, and we would like to consider a low-energy limit similar to that studied above. As before, there is a sector of weakly interacting low-energy gravitons, whose wavelength is too large to be sensitive to the throat size R, and so effectively propagate in flat 10d space. In addition, there is a further sector, localized in the throat region and hence decoupled from the above, which survives in the low-energy limit due to the following crucial effect. In a curved back-ground the energy is properly defined with respect to an asymptotic observer at infinity, and so excitations located at the position r in the radial coordinate have their energies red-shifted by $Z^{1/4}$; hence, in the low-energy limit the theory contains a sector corresponding to excitations of arbitrarily high energy but located sufficiently close to $r = 0$. This sector corresponds to full-fledged type IIB string theory in a background given by the so-called near-horizon limit $r \to 0$ of (6.33), i.e.

$$ds^2 = \frac{r^2}{R^2}\eta_{\mu\nu}dx^\mu dx^\nu + \frac{R^2}{r^2}dr^2 + R^2 ds_{S^5}^2. \tag{6.35}$$

The 4d coordinates and the radial coordinate combine together into a 5d anti-de-Sitter geometry, AdS$_5$, while the remaining coordinates describe a constant radius 5d sphere geometry, $\mathbf{S}^5$. Both 5d spaces have curvature length R.

Comparison of the low-energy limit of both descriptions thus leads to a surprising new equivalence. Type IIB string theory on AdS$_5 \times \mathbf{S}^5$ with radius R and N units of RR 5-form flux on $\mathbf{S}^5$ is equivalent to a 4d $\mathcal{N}=4$ $SU(N)$ gauge field theory. The equivalence surprisingly relates a string theory and a quantum field theory, in particular a theory containing gravity to a non-gravitational theory. There is however no obvious contradiction, since the equivalence is a "strong–weak" duality, in the following sense. The type IIB string theory description is under control when the curvature length is large compared with the string length $R^2 \gg \alpha'$; in this regime the α' expansion is a good approximation to the worldsheet dynamics, and the spacetime theory reduces to the familiar type IIB massless modes propagating on AdS$_5 \times \mathbf{S}^5$. On the other hand, the gauge theory description is under control in the perturbative regime, namely for small $\lambda = g_{\mathrm{YM}}^2 N$; the latter parameter is the 't Hooft coupling, which is the effective coupling in the gauge theory loop expansion for general number of colors. The relation (6.34) implies that the well-understood regime of string theory on low-curvature geometry is actually related to large N gauge theory in the large 't Hooft coupling regime; similarly the well-understood perturbative regime of the gauge theory maps to the poorly-understood regime of string theory on a space of string scale curvature.

The duality is also surprising in that it relates a 5d gravitational theory (once we perform the KK reduction on the compact $\mathbf{S}^5$) to a 4d non-gravitational theory. Actually, this is a very explicit example of *holography*, a concept originally introduced in the study of black hole physics. It states that a gravitational system in a $(d+1)$ dimensional region should admit a description in terms of a non-gravitational theory with degrees of freedom on the d-dimensional boundary. Indeed, the (conformal) boundary of AdS$_5$ is located at $r \to \infty$ and corresponds to a 4d Minkowski space (with coordinates x^μ), on which the holographic dual gauge theory description lives. Dualities of this kind are often referred to as holographic duality, or gauge/gravity or gauge/string correspondence.

The duality is also known as AdS/CFT correspondence, since, as mentioned in Section 2.4.1, the 4d $\mathcal{N}=4$ SYM theory is actually a 4d conformal field theory. This is nicely encoded in the dual AdS geometry, since AdS$_5$ has an $SO(2,4)$ isometry group, which is precisely the conformal group in four dimensions. In particular the rescaling isometry of AdS$_5$

$$x^\mu \to \Lambda x^\mu, \quad r \to \Lambda^{-1} r, \tag{6.36}$$

corresponds to a rescaling in the 4d gauge theory coordinates. The radial direction r (sometimes called the holographic direction) relates to the energy scale in the dual gauge theory side. Physics near $r \to \infty$ in the AdS side are dual to UV gauge theory phenomena, while physics at $r \to 0$ in the gravity side are dual to IR gauge theory phenomena.

There are other simple matchings of symmetries between the two dual theories. The $SO(6)$ isometry of $\mathbf{S}^5$ matches the R-symmetry of $\mathcal{N}=4$ theories, and the $SL(2,\mathbf{Z})$ S-duality of type IIB agrees with the S-duality of 4d $\mathcal{N}=4$ gauge field theories, described in Section 2.5.2. More significantly, there is a growing amount of non-trivial dynamical (rather than purely kinematical) evidence for the correctness of the AdS/CFT correspondence.

*Generalizations and the conifold throat**

The gauge/gravity correspondence generalizes in several interesting directions, and its applications provide a most active research area in string theory. We merely sketch a class of constructions for reference in later chapters. There is a class of gauge/gravity dual pairs, maintaining conformal invariance but with lower supersymmetry, based on running the above argument for D3-branes at a 6d conical singularity $\mathbf{Y}_6$ rather than in flat space. Denoting by $\mathbf{X}_5$ the 5d base of the 6d cone $\mathbf{Y}_6$, the upshot is that type IIB on $\mathrm{AdS}_5 \times \mathbf{X}_5$, with N units of RR 5-form flux on $\mathbf{X}_5$, is dual to a 4d gauge theory with conformal invariance, whose spectrum and interactions are determined by $\mathbf{Y}_6$.

A prototypical example is obtained by taking $\mathbf{Y}_6$ to be a singular conifold, described as the subspace of $\mathbf{C}^4$ defined by the equation

$$w_1^2 + w_2^2 + w_3^2 + w_4^2 = 0. \tag{6.37}$$

Incidentally, this is a simple example of a (non-compact) Calabi–Yau geometry, see Section 7.2. The gravity side of the dual pair is given by type IIB on AdS_5 times a 5d space called $T_{1,1}$, with $\mathbf{S}^2 \times \mathbf{S}^3$ topology. The dual gauge theory is given by the worldvolume theory on N D3-branes at a conifold geometry, see (11.72) in Section 11.3.4. It is a 4d $\mathcal{N} = 1$ $SU(N) \times SU(N)$ gauge theory, with chiral multiplets A_1, A_2 in the $(\square, \overline{\square})$ and B_1, B_2 in the $(\overline{\square}, \square)$, interacting through a superpotential $W = \epsilon_{ij}\epsilon_{kl}\mathrm{tr}\,(A_i B_k A_j B_l)$. This gauge theory admits an interesting non-conformal generalization obtained by taking different ranks for the two gauge factors; this theory has a non-trivial renormalization flow, and reproduces non-trivial infrared gauge phenomena, like confinement and chiral symmetry breaking. The gravity dual has a more involved structure and is only approximately AdS. The interior throat is capped at a finite value of the redshift factor, providing the gravity dual of the gauge theory dynamical scale. Such throats appear naturally in certain string compactifications, as described in Section 14.1.3, and provide a natural string theory embedding of the warped extra dimensions scenario of Section 1.4.3.

6.5 Brane–antibrane systems and non-BPS D-branes

In a relativistic quantum theory, for every particle or object in the theory there exists the corresponding anti-particle or anti-object. String theory thus contains antibranes, and in particular anti-Dp-branes, which we review in this section using open string techniques.

6.5.1 Anti D-branes

An anti-Dp-brane, denoted $\overline{\mathrm{D}p}$-brane for short, can be described as an object with p extended spatial dimensions and propagating in time, with the same coupling to NSNS closed string fields as a Dp-brane (in particular with the same tension) and opposite sign in the Chern–Simons action (6.15) coupling it to RR closed string fields (in particular opposite RR charges). It can also be defined as a Dp-brane with a change of worldvolume orientation, since the orientation flip changes the sign of the integrals in the Chern–Simons worldvolume action, while leaving the DBI action (6.6) invariant; the change of orientation can in particular be implemented as a time reversal, making this definition analogous to the description of anti-particles as particles traveling backwards in time.

From this definition it follows that a flat $\overline{Dp}$-brane in flat space is BPS, and preserves half of the supersymmetries. A $\overline{Dp}$-brane along the directions $0, 1, \ldots, p$ preserves the linear combinations of supercharges $Q = \epsilon_L Q_L + \epsilon_R Q_R$, with

$$\epsilon_L = -\Gamma^0 \cdots \Gamma^p \epsilon_R. \tag{6.38}$$

This condition is obtained from that of Dp-branes (6.5) with a sign flip due to change of orientation. The supersymmetries preserved by a $\overline{Dp}$-brane are precisely the supersymmetries broken by a Dp-brane along the same directions, and vice versa. This implies that a configuration containing parallel Dp- and $\overline{Dp}$-branes simultaneously, e.g. a brane–antibrane pair, breaks all the supersymmetries. Indeed, this is expected from the fact that branes and antibranes carry the same tension but opposite charge, so the overall state has non-vanishing energy and zero charge, and thus is not a BPS state. In analogy with electron–positron annihilation, branes and antibranes attract each other and eventually annihilate. Much of this process can be understood in terms of the open strings in the Dp-$\overline{Dp}$ brane system, as we describe in the following. For concreteness we focus on type II theories, and postpone the type I discussion to Section 6.5.3.

6.5.2 Open string spectrum and brane–antibrane annihilation

Consider N $\overline{Dp}$-branes along the directions $x^0, \ldots, x^p$ in flat 10d space. There is no absolute way to distinguish isolated $\overline{Dp}$-branes from Dp-branes, so at weak coupling they are described as a $(p + 1)$-dimensional plane on which open strings can end. The quantization of the open string sector, and the resulting spectrum, is exactly as for Dp-branes. The massless sector is a $(p + 1)$-dimensional $U(N)$ vector multiplet with respect to the 16 supersymmetries preserved by the $\overline{Dp}$-branes.

Consider now a configuration with a Dp brane–antibrane pair, both spanning $x^0, \ldots, x^p$ and sitting at different points in $\vec{x}_p$, $\vec{x}_{\bar{p}}$ in the transverse $\mathbf{R}^{9-p}$. The pp open strings with both endpoints on the Dp-brane are insensitive to the presence of the antibrane, and lead to a Dp-brane worldvolume $U(1)$ vector multiplet under the Dp-brane supersymmetries. Conversely, $\bar{p}\bar{p}$ open strings lead to a $\overline{Dp}$-brane worldvolume $U(1)$ vector multiplet under the $\overline{Dp}$-brane supersymmetries. Finally, there are $p\bar{p}$ and $\bar{p}p$ strings, stretching between both objects. These open strings have standard NN boundary conditions along $x^0, \ldots, x^p$ and DD along $x^{p+1}, \ldots, x^9$, respectively. However, they differ from pp or $\bar{p}\bar{p}$ strings in having the opposite GSO projection, as we now show from open–closed duality. Consider a cylinder diagram with one boundary on the Dp-brane and another on the $\overline{D}p$-brane. This diagram is closely related to a cylinder diagram obtained by replacing the $\overline{D}p$-brane by a putative second Dp-brane. For our present purposes, we focus on the worldsheet fermion contribution to these diagrams, which for the Dp-Dp cylinder is identical to that in eqs. (D.3) and (D.2), in the closed and open string channels respectively. The Dp-$\overline{D}p$ cylinder amplitude is obtained by simply flipping the sign of the last two contributions (with upper characteristic $1/2$) in (D.3); indeed this flips the sign of the RR exchange between the two objects, keeping the NSNS one unchanged. Translating into the open string channel, this flips the sign of the contributions of theta functions with lower characteristic $1/2$ in (D.2), effectively flipping the GSO projections in both the NS and R Dp-$\overline{D}p$ open string sectors, as announced.

Given the non-standard GSO projection, the resulting $p\bar{p} + \bar{p}p$ spectrum is non-supersymmetric, consistently with the fact that such open strings feel the breaking of all supersymmetries. The light spectrum is:

Sector	State	M^2	$(p+1)$-dim. field
NS	$|0\rangle_{NS}$	$-\frac{1}{2\alpha'} + \frac{|\vec{x}_p - \vec{x}_{\bar{p}}|^2}{(2\pi\alpha')^2}$	Complex scalar
R	$|8_S\rangle$	$\frac{|\vec{x}_p - \vec{x}_{\bar{p}}|^2}{(2\pi\alpha')^2}$	Fermions

All states have charges ± 1 under the $U(1)_p \times U(1)_{\bar{p}}$ symmetry. For stacks of multiple branes and antibranes, they transform in the $(\square, \overline{\square})$ of the non-abelian gauge symmetries.

For large separation between the Dp- and the $\overline{D}p$-branes, all such open string states are massive. However, the configuration is not stable, due to a non-vanishing attractive force due both to NSNS and RR exchange. The brane–antibrane pair is driven towards short separations, and upon reaching $|\vec{x}_p - \vec{x}_{\bar{p}}| < 2\pi^2\alpha'$ it develops a tachyonic mode. The spacetime interpretation of the tachyon is an instability against brane–antibrane annihilation, and the expected endpoint of tachyon condensation is the type II vacuum, with no open strings. The tachyon mass is of the order of the string scale, so a full microscopic description of its condensation requires string field theory tools and is well beyond our scope.

Antibranes and brane–antibrane systems are interesting ingredients in model building, since they introduce sources of non-supersymmetric dynamics under reasonable control. They will appear in Section 15.3.1 to construct de Sitter vacua, and in Section 16.6.2 to engineer models of cosmological inflation. They are also useful in the construction of stable non-BPS states in string theory, as described in the next section. Finally, more general non-supersymmetric D-brane systems provide an extra flexibility in D-brane model building, as used in Chapters 10 and 11.

6.5.3 Orientifolds and stable non-BPS D-branes*

Orientifold action on $\overline{D}p$-branes

We now describe the orientifold projection on open strings ending on $\overline{D}p$-branes. For open string in $\bar{p}\bar{p}$ sectors, the orientifold projection works exactly as in the corresponding pp open string sector, with the only modification of an additional minus sign for Ramond states. This follows from open–closed duality in the Moebius strip diagram, for Dp- or $\overline{D}p$-branes. Considering the closed channel diagram, describing the NSNS and RR exchange between the (anti)branes and the orientifold plane, there is a sign flip in the antibrane RR couplings. This translates into a sign flip in the orientifold action on $\bar{p}\bar{p}$ Ramond states, as compared with pp Ramond states.

As an example, consider N $\overline{D3}$'s on top of an O3$^{\pm}$-plane. In the O3^{-}/$\overline{D3}$ system, the NS sector produces 4d $SO(N)$ gauge bosons and six real scalars in the adjoint $\boxminus$, while the R sector produces four Majorana fermions, which due to the additional minus sign transform in the $\boxminus$. For the O3^{+}/$\overline{D3}$ system, there are $USp(N)$ gauge bosons, six real scalars in the $\boxminus$, and four Majorana fermions in the $\boxminus$. The spectra are completely non-supersymmetric, in agreement with the fact that Op-planes and $\overline{D}p$-branes preserve opposite supersymmetries. It is tachyon-free, however, since the antibrane and the orientifold plane cannot annihilate. Of course, the tree-level massless scalars receive loop corrections, which can jeopardize or improve the stability of these systems. In fact, there is an open string one-loop correction from the Moebius strip diagram, which equivalently describes the O3/$\overline{D3}$ interaction by closed string exchange. For the O3^{-}/$\overline{D3}$ system, the interaction is repulsive and the system at one-loop is unstable, with a runaway behavior pushing the $\overline{D3}$-branes infinitely away from the O3-plane. For the O3^{+}/$\overline{D3}$ system, however, the interaction is attractive and pulls the $\overline{D3}$-branes on top of the O3-plane, thus stabilizing the tree-level flat directions. This is a nice illustration that the use of $\overline{D}p$-branes can lead to non-supersymmetric systems under relatively good control.

The type I $\overline{D5}$-brane

It is interesting to point out that O-planes and anti-D-branes can preserve common supersymmetries. A simple example is the type I $\overline{D5}$-brane, with preserved supersymmetries given by

$$\epsilon_L = \Gamma^0 \cdots \Gamma^9 \epsilon_R, \quad \epsilon_L = -\Gamma^0 \cdots \Gamma^5 \epsilon_R. \tag{6.39}$$

This corresponds to 6d $\mathcal{N} = 1'$ supersymmetry, where the prime denotes that these supersymmetries are opposite to the 6d $\mathcal{N} = 1$ that would be preserved by D5-branes, recall Section 6.1.3. The computation of the $\overline{5}\overline{5}$ spectrum is similar to that of the type I D5-brane, with the additional sign in the Ω action on R states. We obtain 6d $\mathcal{N} = 1'$ vector multiplets of $USp(N)$ and one hypermultiplet in the $\boxminus$. The sign flip makes gauginos and hypermultiplet spinors exchange roles in going from the 55 case to the $\overline{5}\overline{5}$ case, thus flipping the $\mathcal{N} = 1$ supersymmetry to $\mathcal{N} = 1'$. The $\overline{5}9$ sector is mapped by Ω to the $9\overline{5}$, so we compute the former and impose no projection. The computation is similar to a 59 sector, up to the flipped GSO projection, and yields one bi-fundamental 6d $\mathcal{N} = 1'$ half-hypermultiplet.

The non-BPS D7-brane in type I theory

In orientifolds of type II theories, the orientifold projection may remove the tachyonic open string mode of brane–antibrane systems, leading to stable non-BPS brane states. There are several prototypical examples of stable non-BPS Dp-branes (denoted $\widehat{D p}$-branes) in type I theory, including the $\widehat{D0}$-brane, which produces non-BPS states in the $SO(32)$ spinor representation. For future reference, we focus on a different example, the type I $\widehat{D7}$-brane.

The type IIB RR 8-form (dual to the RR scalar, in the sense of section B.3) is odd under the orientifold action Ω, and is thus projected out in type I theory. Correspondingly, Ω maps a type IIB D7-brane to a $\overline{D7}$-brane, so the only invariant states are pairs of N D7- and $\overline{D7}$-branes, carrying no 8-form charge. These are non-BPS, and to assess their stability we need to carefully compute the Ω action on the D7-$\overline{D7}$ sector tachyon. Before the orientifold projection the $7\bar{7} + \bar{7}7$ sector contains a complex tachyon, and 8d spinors, both in the $(\square, \overline{\square})$ of $U(N)_7 \times U(N)_{\bar{7}}$. The Ω action exchanges the D7- and $\overline{D7}$-branes, leading to a single $U(N)$ gauge group; it also maps the $7\bar{7}$ sector to itself (and similarly for the $\bar{7}7$), leading to a truncation of the spectrum. Without entering the details, it projects the tachyon fields to the $\boxminus$ of $U(N)$. Therefore, for a single D7-$\overline{D7}$ pair, i.e. $N = 1$, there is no tachyon and it corresponds to a stable non-BPS D-brane, the type I $\widehat{D7}$-brane. For higher N, there are tachyon fields which can condense and lead to partial or total annihilation; for odd N, the antisymmetric matrix of tachyon vevs has necessarily one zero skew-eigenvalue, signaling that annihilation is not complete and there remains one left-over D7-$\overline{D7}$ pair, while for even N all skew-eigenvalues can be non-zero, leading to total annihilation. From the spacetime perspective, type I $\widehat{D7}$-branes carry a discrete $\mathbf{Z}_2$ charge, so that a single $\widehat{D7}$-brane is stable, but higher numbers of $\widehat{D7}$-branes can annihilate pairwise.

Actually there is a natural interpretation of this $\mathbf{Z}_2$-valued charge, in terms of the D9-brane gauge fields. In the previous discussion we have not taken into account the 79 and $\bar{7}9$ sectors, which actually contain a complex tachyon in the $(\mathbf{N}, \mathbf{32})$ of $U(N)_{\bar{7}} \times SO(32)_9$ surviving the Ω projection. This signals an instability of the $\widehat{D7}$-brane, even for $N = 1$, not towards annihilation into the vacuum, but rather towards dissolution of the $\widehat{D7}$-brane as a gauge field background in the D9-brane (forming a D9-$\widehat{D7}$-brane bound state). Such gauge field backgrounds must be non-trivial only in the transverse 2d, hence are characterized by embeddings of the $\mathbf{S}^1$ at infinity into the gauge group. Using Section B.4, they are classified by $\Pi_1(O(32)) = \mathbf{Z}_2$, where $O(32)$ is the actual global structure of the type I gauge group; this reproduces the $\mathbf{Z}_2$ valued $\widehat{D7}$-brane charge. Such discrete charges are known as K-theory charges, since the classification of gauge bundle topologies (modulo brane–antibrane annihilation) can be mathematically formulated in terms of the so-called K-theory groups. In the compactifications with D-branes and orientifolds in Sections 10.4.2 and 10.6, there are non-trivial consistency conditions from cancellation of D-brane K-theory charges, in addition to standard D-brane RR charge cancellation.

7

Calabi–Yau compactification of
heterotic superstrings

In Chapter 5 we showed that toroidal compactifications of superstrings can lead at low
energies to 4d theories, albeit with too much supersymmetry to allow for chirality, an
essential ingredient of particle physics. We will thus be interested in constructing more
general classes of compactifications leading to chiral theories in 4d, an enterprise starting
in this chapter and reaching until Chapter 12, and for which it is convenient to give an
aerial view.

7.1 A road map for string compactifications

There are different options to achieve the construction of chiral 4d compactifications in
string theory, summarized in Figure 7.1, and roughly classified according to the underlying 10d or 11d theory taken as starting point. In this book we will study in turn heterotic compactifications, type IIA orientifolds and type IIB orientifolds. In this chapter
and Chapter 8 we describe different varieties of heterotic string compactifications, whose
4d effective action is studied in Chapter 9. We consider models in the geometric realm,
using Calabi–Yau manifolds and toroidal orbifolds, and in the more abstract non-geometric
conformal field theory (CFT) framework, using asymmetric orbifolds, free fermions, and
Gepner constructions. In Chapter 10 we study type IIA orientifold constructions with gauge
fields localized on D6-branes, and matter fields at their intersections. We consider explicit
examples in toroidal orbifold compactifications, as well as Gepner constructions. Supersymmetric type IIA models with intersecting D6-branes are related to purely geometric
compactifications of M-theory on manifolds with G_2 holonomy, as we will also review.
Finally in Chapter 11 we describe different setups of type IIB orientifold compactifications. They include orientifolds of toroidal orbifolds, systems of D-branes at singularities,
and compactifications with magnetized D-branes. The non-perturbative regime of type IIB
orientifolds with 7-branes is better described in terms of F-theory compactifications, also
described there. The 4d effective action of these classes of models is studied in Chapter 12.
Within these classes of constructions we present explicit examples of models whose massless spectrum and couplings come close to the structure of the SM, usually referred to as
semi-realistic models.

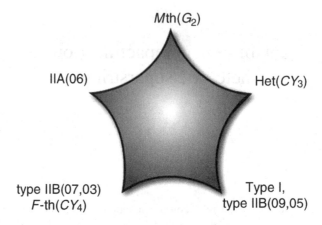

Figure 7.1 Five large classes of chiral 4d string constructions.

Many of these different classes of compactifications in the various underlying theories are related by dualities, and so must be considered as complementary tools to explore different corners of an underlying unique string/M-theory, as suggested in Figure 7.1.

There are some common general features of the compactification backgrounds considered in the different setups. In most constructions we consider compactifications with factorized background geometry $\mathbf{M}_4 \times \mathbf{X}_6$, where $\mathbf{X}_6$ is a compact space. This essentially assumes that the only non-trivial background turned on is the metric. Generalizations including additional fields, e.g. field strength backgrounds for p-form fields, lead to compactifications beyond the factorized ansatz, and are covered in Chapter 14.

For reasons explained in the next section, most constructions are also devised to produce 4d $\mathcal{N} = 1$ supersymmetric models, and are based on compactifications on Calabi–Yau (CY) manifolds (or their non-geometric analogs). Although specifically applied to heterotic compactification in this chapter, CY spaces are useful for all the different classes of compactifications summarized above, and their properties permeate the subject of 4d string model building. For instance, they play an important role in type II orientifold compactifications in Chapters 10 and 11, in which gauge sectors arise on the volume of $\mathrm{D}p$-branes wrapped on non-trivial cycles on the Calabi–Yau geometries.

Finally, in geometric compactifications on curved spaces, the background metric is not flat, and according to Section 3.1.5 the worldsheet 2d theory is interacting and in general not exactly solvable. The analysis is therefore carried out in the large volume regime $\alpha'/R^2 \ll 1$ (with R denoting a typical curvature radius of $\mathbf{X}_6$), where the 10d supergravity approximation is reliable. A full string theory description of these compactifications is available only in some special cases, like orbifolds or other abstract CFT constructions.

7.2 Generalities on Calabi–Yau compactification

We start by introducing Calabi–Yau manifolds, and motivating their use in string compactification. As mentioned above, these geometric results are valid for supersymmetry preserving heterotic and type II (orientifold) compactifications.

7.2.1 Supersymmetry and holonomy

There are several motivations to consider string compactifications preserving supersymmetry. For starters, they are automatically stable and do not contain tachyons. Also, they are simpler to study from the theoretical point of view; regarding the compactification backgrounds from the supergravity perspective, solutions to the (first-order) supersymmetry conditions can be shown to automatically be solutions to the (much more involved, second-order) supergravity equations of motion. From the phenomenological point of view, supersymmetric backgrounds may provide a good starting point to construct particle physics models, if supersymmetry is broken at low energies compared with the fundamental string scale, as reviewed in Chapter 2; this approach must be complemented by a discussion of supersymmetry breaking, which we postpone to Chapter 15. Finally, even if supersymmetry is not realized at the TeV scale, it is reasonable to consider supersymmetric compactifications as toy models to learn the behavior of more general string theory compactifications.

The condition that compactification on X_6 leads to some unbroken supersymmetry can be described as follows. Around each point P in $M_4 \times X_6$ the spacetime is locally 10d and we have a local set of 10d supercharges, which transform as spinors of $SO(10)$ (two for type II theories and one for type I and heterotic theories). The supersymmetries of the 4d theory correspond to supercharges which are well-defined globally on X_6. Since X_6 is curved, local supercharges at different points in X_6 are related by parallel transport[1] with the $SO(6)$ spin connection corresponding to the metric in X_6. In fact, a local supercharge at a point P parallel transported around a closed loop C will in general come back to P rotated by a non-trivial $SO(6)$ element R_C. Supercharges which get rotated upon parallel transport do not lead to globally well-defined supercharges; hence the condition that compactification on X_6 preserves some supersymmetry is that there exist non-trivial 6d spinors $\xi(x^m)$ in X_6, called Killing spinors, which are covariantly constant in X_6

$$\nabla_{X_6} \xi(x^m) = 0. \tag{7.1}$$

These define the internal profile of the supercharges surviving as supersymmetries of the 4d compactification. Equivalently, (7.1) is the equation defining the 6d wave function of the KK zero mode corresponding to the 4d gravitino, see Section 7.2.4.

The condition for the existence of solutions to (7.1) can be recast in terms of the *holonomy group* of X_6. The set of rotations R_C suffered by the spinor, for all possible closed loops C, is called the holonomy group of X_6 (see Figure 7.2), and depends on its metric via the spin connection. For a general 6d manifold, the holonomy group is $SO(6)$, and no component of a 10d spinor (transforming as 6d spinors **4** or **4̄** under it) remains unrotated; equivalently, there are no Killing spinors, so compactification on a generic holonomy space breaks all the supersymmetries. In order to preserve some supersymmetry, X_6 must be a

---

[1] Parallel transport of (e.g. a vector or spinor) χ along a path $x^m(s)$ parametrized by s, is defined by a one-parameter family $\chi(s)$, satisfying $(\partial x^m / \partial s) \nabla_m \chi(s) = 0$, with a suitable covariant derivative.

Figure 7.2 The holonomy group is given by the set of rotations R_C experienced under parallel transport around all possible closed loops in the manifold.

manifold of special holonomy, i.e. its holonomy group must be a subgroup of $SO(6)$ under which the decomposition of the **4** contains a singlet. This is satisfied by manifolds of $SU(3)$ holonomy, for which a chiral 10d spinor decomposes under the holonomy group times 4d Lorentz group as

$$SO(10) \quad \rightarrow \quad SO(6) \times SO(1,3) \quad \rightarrow \quad SU(3) \times SO(1,3),$$
$$\mathbf{16} \qquad\qquad (\mathbf{4}, \mathbf{2}) + (\bar{\mathbf{4}}, \mathbf{2}') \qquad\qquad (\mathbf{3}, \mathbf{2}) + (\bar{\mathbf{3}}, \mathbf{2}') + (\mathbf{1}, \mathbf{2}) + (\mathbf{1}, \mathbf{2}'),$$

where $\mathbf{2}, \mathbf{2}'$ denote the left- and right-handedness under the 4d Lorentz group. The $SU(3)$ singlet components in the last column lead to four unbroken 4d supercharges $Q_\alpha, \overline{Q}_{\dot\alpha}$ for each 10d spinor supercharge, while the others are broken by the compactification. Hence, compactifications of heterotic string theories on $SU(3)$ holonomy manifolds lead to 4d $\mathcal{N}=1$ supersymmetry, and have potential application to particle physics model building, as explored in later sections. Type II compactifications lead instead to 4d $\mathcal{N}=2$ supersymmetry, and are studied later in Section 10.1.1. Although non-chiral due to their high supersymmetry, their orientifolds (including type I models) preserve only $\mathcal{N}=1$, and their applications to model building are studied in Chapters 10, 11.

The above general argument using the holonomy group can be applied to determine the unbroken supersymmetries by different geometries in various dimensions. For instance, compactifications on manifolds with smaller holonomy groups lead to enhanced supersymmetry (since there are additional singlet spinor components). As discussed in Chapter 5, compactification on tori, which has trivial holonomy, preserves all supersymmetries, leading to 4d $\mathcal{N}=8$ for type II, and 4d $\mathcal{N}=4$ for heterotic and for type II orientifolds. Also, there exists a unique compact manifold (of dimension 4) with $SU(2)$ holonomy, called K3. Compactification on K3 $\times$ $\mathbf{T}^2$ leads to 4d $\mathcal{N}=4$ for type II and 4d $\mathcal{N}=2$ for heterotic and type II orientifolds. These theories are interesting from the theoretical viewpoint, but not relevant for particle physics model building. Hence in the following, we focus on spaces with strict $SU(3)$ holonomy. Other (larger) holonomy groups are useful in model building applications of M- and F-theory. For instance, compactification of M-theory on 7d spaces of G_2 holonomy produces 4d $\mathcal{N}=1$ models, see Section 10.7.2; compactification of M-theory on 8d manifolds of $SU(4)$ holonomy leads to 3d $\mathcal{N}=2$ supersymmetric theories, which can be lifted to 4d $\mathcal{N}=1$ models using the so-called F-theory, as described in Section 11.5.

7.2.2 Calabi–Yau manifolds

In principle it would seem difficult to determine explicitly whether a given $2N$-dimensional manifold **X** admits a metric of $SU(N)$ holonomy, since in principle it would require an explicit computation of the metric, and of its holonomy group. This is actually beyond present techniques, and in fact there are no analytic expressions of the metric of any (non-trivial) compact $SU(3)$ holonomy manifold (although they can be computed numerically). Happily there are powerful mathematical theorems which ensure the existence of $SU(N)$ holonomy metrics for manifolds which satisfy the so-called Calabi–Yau condition (and are hence called Calabi–Yau manifolds or CY N-folds), which are easy to verify in concrete examples. According to Calabi's conjecture, proved by Yau, an N-dimensional complex manifold which is Kähler and has vanishing first Chern class admits an $SU(N)$ holonomy metric. In the following we crack down on this jargon to get the essential ideas (see Appendix B for some useful background on geometry and topology).

An N-dimensional complex manifold is a $2N$-dimensional manifold on which one can introduce local complex coordinates which can be patched together globally in a consistent way (i.e. with holomorphic transition functions). More concretely, a $2N$-dimensional manifold is a complex manifold if it admits a globally defined *complex structure*, i.e. a mixed tensor I_m^n satisfying[2] $I_m^n I_n^l = -\delta_m^l$. This tensor can be used to define local complex coordinates (hence its name), as follows. Split the real coordinates in two sets x^j, y^j for $j = 1, \ldots, N$, so that locally $I = \begin{pmatrix} 0 & \mathbf{1}_N \\ -\mathbf{1}_N & 0 \end{pmatrix}$, and construct complex coordinates as $dz^j = dx^j + iI_k^j dy^k$ and $d\bar{z}^j = dx^j - iI_k^j dx^k$. Note that a given real differential manifold can admit many complex structures. A familiar example, is provided by the 2-torus, whose possible complex structures are parametrized by a complex number τ; the two real coordinates x, y can be combined to form complex coordinates via $dz = dx + \tau dy$ (as already encountered in Section 3.2.2 in the study of worldsheet geometries).

On a complex manifold, one can introduce a metric whose only non-zero components are mixed, namely $g_{i\bar{j}}$. Using this metric to lower one index from the complex structure we obtain a 2-form with mixed indices

$$J = g_{i\bar{j}} dz^i d\bar{z}^j, \tag{7.3}$$

(with implicit wedge products). A manifold is Kähler if this form is closed

$$dJ = 0, \tag{7.4}$$

in which case it is called the Kähler form. The relevance of the Kähler condition in our setup is that the condition (7.4) constrains the metric such that parallel transport does not

---

[2] There is an additional condition required for local complex coordinates to patch together holomorphically, given by the vanishing of the Niejenhuis tensor

$$N_{mn}^p = \partial_{[n} I_{m]}^p - I_{[m}^q I_{n]}^r \partial_r I_q^p = 0. \tag{7.2}$$

mix holomorphic and antiholomorphic indices, i.e. the holonomy group is at most $U(N) \simeq SU(N) \times U(1)$. Further reduction to $SU(N)$ requires the absence of holonomy in the overall $U(1)$ factor, and is related to the "vanishing of the first Chern class." This requires that the curvature 2-form $R_{i\bar{j}}$ defined in Section B.3 is cohomologically trivial, although we will not need further details.

Given a topological space satisfying the Calabi–Yau conditions, we would like to count the number of free parameters in the choice of its $SU(N)$ holonomy metric. This is the analog of the choice of compactification radius for a compactification on a space with the topology of a circle. This number of parameters is given in terms of certain topological invariants, the Hodge numbers, which refine for complex manifolds the Betti numbers in Section B.1. In a complex manifold, the classification of p-forms and their cohomology classes (and similarly p-chains, p-cycles, and their homology classes) can be refined according to their number of holomorphic and antiholomorphic indices. For instance, the 3-cohomology group in a complex threefold $\mathbf{X}_6$ splits as

$$H^3(\mathbf{X}_6) = H^{3,0}(\mathbf{X}_6) + H^{2,1}(\mathbf{X}_6) + H^{1,2}(\mathbf{X}_6) + H^{0,3}(\mathbf{X}_6), \tag{7.5}$$

where $H^{p,q}$ corresponds to forms with p holomorphic and q antiholomorphic indices (with basis $dz^{i_1} \wedge \cdots \wedge dz^{i_p} \wedge d\bar{z}^{\bar{j}_1} \wedge \cdots \wedge d\bar{z}^{\bar{j}_p}$). The integers $h_{p,q} = \dim H^{p,q}(\mathbf{X})$ are topological invariants, known as Hodge numbers. It is useful to also define the Euler characteristic, as the weighted sum of Hodge numbers

$$\chi(\mathbf{X}_6) = \sum_{p,q=0}^{3} (-1)^{p+q} h_{p,q}(\mathbf{X}_6) = 2[h_{1,1}(\mathbf{X}_6) - h_{2,1}(\mathbf{X}_6)]. \tag{7.6}$$

Hodge numbers are usually displayed in a diamond-shape diagram, known as Hodge diamond. For the case of CY threefolds $\mathbf{X}_6$ i.e. manifolds of $SU(3)$ holonomy (and not in a proper subgroup like $SU(2)$), the Hodge diamond has the structure

$$
\begin{array}{ccccccc}
 & & & h_{0,0} & & & \\
 & & h_{1,0} & & h_{0,1} & & \\
 & h_{2,0} & & h_{1,1} & & h_{0,2} & \\
h_{3,0} & & h_{2,1} & & h_{1,2} & & h_{0,3} \\
 & h_{3,1} & & h_{2,2} & & h_{3,1} & \\
 & & h_{3,2} & & h_{2,3} & & \\
 & & & h_{3,3} & & &
\end{array}
=
\begin{array}{ccccccc}
 & & & 1 & & & \\
 & & 0 & & 0 & & \\
 & 0 & & h_{1,1} & & 0 & \\
1 & & h_{2,1} & & h_{2,1} & & 1 \\
 & 0 & & h_{1,1} & & 0 & \\
 & & 0 & & 0 & & \\
 & & & 1 & & &
\end{array}
$$

The unique $(3,0)$ form, which we denote by Ω, is very important. In fact, a CY N-fold can be equivalently defined as a Kähler manifold admitting a nowhere vanishing $(N, 0)$ form Ω.

In the above discussion, the Kähler form is related to the complex structure via the metric. Turning this around, Yau's theorem guarantees that a choice of complex structure and of Kähler form (more precisely, of its cohomology class) determines the underlying

CY metric *uniquely*. We may use this argument to count the number of free parameters of the $SU(3)$ holonomy metric in a given CY manifold, as follows:

- The Kähler form is closed by (7.4). Moreover it cannot be exact, as (7.3) implies

$$\int_{\mathbf{X}_6} J \wedge J \wedge J \simeq \int_{\mathbf{X}_6} \sqrt{\det g} \, dz^1 d\bar{z}^1 \cdots dz^3 d\bar{z}^3 = \mathrm{Vol}(\mathbf{X}_6), \qquad (7.7)$$

while an exact $J = dA$ would give $\int J \cdots J = \int d(AJ \cdots J) = 0$. Thus J is cohomologically non-trivial and $(1, 1)$, so may be expanded in a basis $\{\omega_a\}$ of $H^{1,1}(\mathbf{X})$,

$$J = \sum_{a=1}^{h_{1,1}} t_a \omega_a. \qquad (7.8)$$

There are thus $h_{1,1}$ real parameters involved in the choice of the Kähler class. These are known as Kähler moduli of the CY, and control the sizes of even-dimensional $2n$-cycles of the internal space, whose volumes are measured by integrating the $2n$-form J^n over them, in analogy with (7.7).
- The choice of a complex structure tensor is equivalent, via contraction with Ω, to the choice of a $(2, 1)$ form $I_{ij\bar{k}} = \Omega_{ijl} I_{\bar{k}}^l$. Expanding on a basis, there are $h_{2,1}$ complex parameters involved in the choice of the complex structure. These are known as complex structure moduli of the CY, and control the sizes of 3-cycles of the internal space (whose volumes are measured by integrating Ω over them).

Thus the parameters required to specify a unique metric in a given CY compactification space are $h_{1,1}$ real Kähler moduli and $h_{2,1}$ complex structure moduli. In concrete examples, see next section, these numbers range from a few to the hundred.

We conclude by mentioning the direct connection between the geometric and the string theory viewpoints on the CY condition. In the string theory context, the original condition was the existence of a covariantly constant spinor ξ. Using it to sandwich products of Dirac matrices Γ_i, $\Gamma_{\bar{i}}$ one can construct the forms

$$J_{i\bar{j}} = -i \xi^\dagger \Gamma_i \Gamma_{\bar{j}} \xi, \quad \Omega_{ijk} = \xi^T \Gamma_i \Gamma_j \Gamma_k \xi, \qquad (7.9)$$

which can be shown to satisfy precisely the criteria of the Kähler form and the holomorphic 3-form appearing in the geometric viewpoint.

7.2.3 Examples

We now present several tools to construct explicit examples of CY threefolds.[3]

---

[3] To this list we could add the smooth CYs obtained upon blowing up the singularities of the toroidal orbifolds considered in Chapter 8.

Hypersurfaces in complex projective spaces and the quintic

A large class of CY manifolds can be constructed as hypersurfaces in (weighted) complex projective spaces. The complex projective space $\mathbf{P}_n$ or $\mathbf{CP}_n$ is defined as $\mathbf{C}^{n+1}$ with the origin removed, with the identification

$$(z_1, \ldots, z_{n+1}) \rightarrow (\lambda z_1, \ldots, \lambda z_{n+1}), \qquad (7.10)$$

where λ is a non-zero complex number, $\lambda \in \mathbf{C}^* \equiv \mathbf{C} - \{0\}$. The $n + 1$ coordinates z^i are called homogeneous coordinates, and points in $\mathbf{CP}_n$ are denoted $[z_1, \ldots, z_{n+1}]$. For example, when $z_{n+1} \neq 0$, we may use the $\mathbf{C}^*$ action (7.10) to set $z_{n+1} = 1$, and use $(z_1, \ldots, z_n)$ as local (so-called affine) complex coordinates. Complex projective spaces are compact Kähler manifolds, and its Kähler form is inherited from that of the parent $\mathbf{C}^{n+1}$

$$J = dz_1 d\bar{z}_1 + \cdots + dz_{n+1} d\bar{z}_{n+1}. \qquad (7.11)$$

The Kähler form on $\mathbf{CP}_n$ is further inherited to any subspace defined by holomorphic equations. This can be used e.g. to construct the so-called *quintic* CY, as a hypersurface of $\mathbf{CP}_4$ defined by $f_5(z_1, \ldots, z_5) = 0$, where $f_5(z_i)$ is a homogeneous polynomial of degree 5 in the homogeneous coordinates z_i. For instance

$$z_1^5 + z_2^5 + z_3^5 + z_4^5 + z_5^5 = 0. \qquad (7.12)$$

This condition together with the scalings in (7.10) reduce the number of independent real coordinates from ten to six. For general f_5, the hypersurface CY is denoted $\mathbf{CP}_4[5]$. The CY holomorphic 3-form is locally given by an expression of the form

$$\Omega = \frac{z_1 dz_2 dz_3 dz_4}{(\partial f_5 / \partial z_5)}, \qquad (7.13)$$

where again wedge product is understood. This is invariant under (7.10) precisely because the degree of f_5 cancels the $\mathbf{C}^*$ weights of the z_i in the numerator. The non-trivial Hodge numbers of the quintic are $h_{1,1} = 1$, $h_{2,1} = 101$. The Kähler modulus corresponds to an overall rescaling of the Kähler form (7.11), while the complex structure parameters correspond to the coefficients of monomials in f_5, of the form $z_1^{p_1} \cdots z_5^{p_5}$, with total degree 5, and $p_i \leq 3$ (monomials with some $p_i = 4$ may be removed by linear redefinitions of the z_i). The list of such independent monomials is

$$z_1 z_2 z_3 z_4 z_5, \ z_i z_j^2 z_k^2, \ z_i z_j z_k^3, \ z_i^2 z_j^3, \ z_i z_j z_k z_l^2 \qquad (7.14)$$

with different subindices within a monomial; this indeed gives $1 + 30 + 30 + 20 + 20 = 101$ independent deformations of (7.12). This CY admits a $\mathbf{Z}_5 \times \mathbf{Z}_5$ "freely acting" symmetry, i.e. without fixed points, generated by the actions $z_k \rightarrow z_{k+1}$ and $z_k \rightarrow \exp(2\pi i k/5) z_k$, respectively. The quotient by these symmetries is a new CY manifold with

$(h_{1,1}, h_{2,1}) = (1, 5)$, since the Kähler form is invariant and survives in the quotient, and there are five complex structure moduli, associated to the five invariant monomials

$$(z_1^3 z_2 z_5 + \cdots), (z_1^3 z_3 z_4 + \cdots), (z_1^2 z_2 z_3^2 + \cdots), (z_1^2 z_2^2 z_4 + \cdots), z_1 z_2 z_3 z_4 z_5, \qquad (7.15)$$

with the dots implying cyclic subindex permutations.

The above construction may be generalized using weighted complex projective spaces $\mathbf{WCP}_{k_1,\ldots,k_{n+1}}$ as ambient spaces. These are defined by taking $\mathbf{C}^{n+1}$ minus the origin, and imposing the identification

$$(z_1, \ldots, z_{n+1}) \to (\lambda^{k_1} z_1, \ldots, \lambda^{k_{n+1}} z_{n+1}). \qquad (7.16)$$

These spaces are compact and Kähler, with Kähler form essentially inherited from (7.11). The compact subspace $\mathbf{WCP}_{k_1,\ldots,k_5}[p]$ defined by the equation $f_p(z_1, \ldots, z_5) = 0$, with f_p a degree p homogeneous polynomial, is Calabi–Yau if $p = k_1 + \cdots + k_5$. Its holomorphic 3-form has the structure (7.13) with $f_5 \to f_p$. CYs in this class may have singularities, inherited from the ambient space; they are harmless in string theory, however, as will be clear from the study of orbifolds in Chapter 8.

Complete intersection CYs and the Tian–Yau manifold

It is possible to construct CY manifolds as subspaces of higher-dimensional complex projective spaces $\mathbf{CP}_{4+r}$. Using n independent equations $f_i = 0$, with f_i being a degree p_i polynomial, the resulting subspace, denoted $\mathbf{CP}_{4+r}[p_1, \ldots, p_r]$, is CY if $\sum p_i = r + 4$. The holomorphic 3-form has the local structure

$$\Omega = \frac{z_1 dz_2 dz_3 dz_4}{(\partial f_1 / \partial z_5) \cdots (\partial f_r / \partial z_{4+r})}. \qquad (7.17)$$

Since these CY spaces are defined as the intersection of the hypersurfaces $f_i = 0$, they are known as complete intersection Calabi–Yau spaces. The construction can be generalized to subspaces of weighted projective spaces $\mathbf{WCP}_{k_1,\ldots,k_{4+r}}[p_1, \ldots, p_r]$, with CY condition $\sum_j k_j = \sum_i p_i$, and holomorphic 3-form (7.17).

Finally, these constructions can be generalized by considering ambient spaces given by products of (possibly weighted) complex projective spaces. As should be familiar by now, the CY condition amounts to requiring that for the $\mathbf{C}^*$ action in each complex projective space, the sum of the degrees of the polynomial equations equals the sum of the weights of all the coordinates. Consider an illustrative example, with ambient space $\mathbf{CP}_3 \times \mathbf{CP}_3$, defined by three equations of degrees $(1, 1)$, $(3, 0)$ and $(0, 3)$ under the two $\mathbf{C}^*$ actions. For instance, denoting $x_i, y_i, i = 1, \ldots, 4$, the homogeneous coordinates of $\mathbf{CP}_3 \times \mathbf{CP}_3$, we consider the defining equations

$$f_1 \equiv \sum_i x_i y_i = 0, \quad f_2 \equiv \sum_i (x_i)^3 = 0, \quad f_3 \equiv \sum_i (y_i)^3 = 0. \qquad (7.18)$$

The information can be displayed in a table of weights under the $\mathbf{C}^*$ actions:

	x_1	x_2	x_3	x_4	y_1	y_2	y_3	y_4	f_1	f_2	f_3
$(\mathbf{C}^*)_1$	1	1	1	1	0	0	0	0	1	3	0
$(\mathbf{C}^*)_2$	0	0	0	0	1	1	1	1	1	0	3.

The Euler characteristic of this CY manifold is $\chi = -18$, with $h_{2,1} = 23$, $h_{1,1} = 14$. The 23 complex structure moduli correspond to the monomials

$$x_i x_j x_k, \; y_i y_j y_k \quad i \neq j \neq k \neq i; \quad x_i y_j \quad i, j \in (1, 2, 3, 4) \qquad (7.19)$$

with a monomial $x_1 y_1$ excluded, assuming it appears in the defining equation (7.18). This yields $4 + 4 + 15$ monomials. For future reference, we mention that this space has a freely acting $\mathbf{Z}_3$ symmetry, generated by

$$(x_1, x_2, x_3, x_4; y_1, y_2, y_3, y_4) \rightarrow (x_1, \alpha^2 x_2, \alpha x_3, \alpha x_4; y_1, \alpha y_2, \alpha^2 y_3, \alpha^2 y_4), \quad (7.20)$$

(where $\alpha = \exp(2\pi i/3)$). The quotient by this symmetry is a new CY, known as the Tian–Yau manifold, with Euler characteristic $\chi = -18/3 = -6$. The reader can check that only 9 of the 23 complex structure monomials are invariant under this symmetry so that $h_{2,1} = 9$ in this manifold and hence $h_{1,1} = 6$. This implies that when used for heterotic compactifications with "standard embedding," as in Section 7.3.2, it leads to a model with three net generations, studied in Section 7.3.3.

Toric CY manifolds*

The largest known class of CY manifolds is based on a simple generalization of the above ideas, known as toric geometry. The basic strategy is to generalize the above table of weights beyond the factorized structure. More specifically, we consider $3 + r + s$ complex coordinates z_i parametrizing $\mathbf{C}^{3+r+s}$, and introduce r $\mathbf{C}^*$ actions, labeled by $n = 1, \ldots, r$, under which z_i has non-negative weight $q_{i,n}$. The quotient of $\mathbf{C}^{3+r+s}$ (with a suitable set of points removed) by $(\mathbf{C}^*)^r$ leads to a compact Kähler space with $(3 + s)$ complex dimensions, that we use as ambient space for the CY space. The latter is defined as a subspace described by s independent equations $f_m(z_i) = 0, m = 1, \ldots, s$, of degrees $p_{m,n}$ under the nth $\mathbf{C}^*$ action. The CY condition is simply that for each n we have $\sum_i q_{i,n} = \sum_m p_{m,n}$.

These spaces are known as toric CYs, and their construction and properties (like Hodge numbers) are computable in terms of the "charge" matrices $q_{i,n}$, $p_{m,n}$. Systematic computer-based searches using these and related techniques have led to the construction of around 30 000 CY manifolds, with different Hodge numbers $(h_{1,1}, h_{2,1})$. Figure 7.3 shows the distribution of $h_{1,1} + h_{2,1}$ versus the Euler characteristic $\chi = 2(h_{1,1} - h_{2,1})$ of such CY manifolds for $h_{1,1} \leq h_{2,1}$. Typical values of Hodge numbers are in the hundreds, with a maximal value of $h_{1,1} + h_{2,1}$ obtained for CYs with $(h_{1,1}, h_{2,1}) = (491, 11), (251, 251)$. Only examples with $h_{1,1} \leq h_{2,1}$ are shown because the figure for $h_{1,1} \geq h_{2,1}$ would be symmetric under the exchange $h_{1,1} \leftrightarrow h_{2,1}$; for each CY manifold with Hodge numbers $(h_{1,1}, h_{2,1})$ there is a *mirror* manifold with Hodge numbers exchanged, $h_{1,1} \leftrightarrow h_{2,1}$. This is a numerical manifestation of the *mirror symmetry* described in Section 10.1.2.

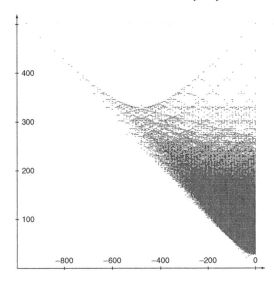

Figure 7.3 Distribution of $h_{1,1} + h_{2,1}$ versus $\chi = 2(h_{1,1} - h_{2,1})$ for CY toric constructions. Only CY's with $h_{1,1} \leq h_{2,1}$ are depicted, since others are related to these by mirror symmetry, described in Section 10.1.2. (Figure reproduced from Kreuzer and Skarke (2002)).

7.2.4 Massless modes in compactifications on curved spaces

The computation of the 4d massless spectrum in compactifications on CY manifolds is largely based on the Kaluza–Klein compactification approach, which provides a good approximation in the large volume regime. In fact, since massless fields are often counted by topological invariants of the compactification, their number is unchanged even if one moves away from the large volume regime (although other features, like their couplings, etc, do in general change). In the following we consider the KK compactification of different massless 10d fields to 4d on a compact 6d manifold $\mathbf{X}_6$. We keep the discussion general, and restrict to the CY case only if explicitly stated. Also, we use x^μ for the 4d non-compact coordinates, x^m for the coordinates in $\mathbf{X}^6$, and x^M for all coordinates collectively.

Scalar fields

Consider a free 10d scalar $\phi(x^M)$ on $\mathbf{M}_4 \times \mathbf{X}_6$. The basic strategy to compute the massless 4d fields is a clever generalization of the KK circle compactification in Section 1.4.1. The generalization of the Fourier decomposition (1.37) is an expansion

$$\phi(x^0, \ldots, x^9) = \sum_{\mathbf{k}} \phi_{6d}^{\mathbf{k}}(x^m) \phi_{4d}^{\mathbf{k}}(x^\mu), \tag{7.21}$$

where the internal profiles $\phi_{6d}^{\mathbf{k}}(x^m)$ are eigenfunctions of the laplacian $\Delta_{\mathbf{X}_6}$ in $\mathbf{X}_6$, characterized by a label (or set of labels) $\mathbf{k}$, and with eigenvalues $-\lambda^{(\mathbf{k})}$ (negative for compact $\mathbf{X}_6$). The above expansion is always possible, since eigenfunctions of $\Delta_{\mathbf{X}_6}$ form a basis of functions in $\mathbf{X}_6$. Moreover, since the 10d dalambertian factorizes as $\Box_{10d} = \Delta_{\mathbf{X}_6} + \Box_{4d}$, the 4d fields $\phi_{4d}^{\mathbf{k}}(x^\mu)$ obey the 4d equation of motion

$$\Box_{4d}\phi_{4d}^{\mathbf{k}}(x^\mu) - \lambda^{(\mathbf{k})}\phi_{4d}^{\mathbf{k}}(x^\mu) = 0. \tag{7.22}$$

Hence, the eigenvalue λ^k manifests as the squared mass of the corresponding 4d field. Massless 4d fields are associated to solutions of the zero mode equation in $\mathbf{X}_6$

$$\Delta_{\mathbf{X}_6}\phi_{6d}^{k}(x^m) = 0. \tag{7.23}$$

Although upcoming zero mode equations are in general difficult to solve, there are general results concerning the number of solutions. For the case at hand, it is a standard result (familiar from electromagnetism) that the Laplace equation in a compact space has a unique solution (given in fact by the constant function). We thus get a single massless 4d scalar field from the original 10d scalar field.

The basic idea generalizes to fields with other Lorentz index structures. An expansion ansatz into products of internal profiles times 4d fields leads to separation of variables for the kinetic term operator of the 10d field. Massless 4d fields are related to solutions of the zero mode equation in $\mathbf{X}_6$ for the internal part of the kinetic term operator. The number of solutions of such equations can be computed from general results, with no need to solve the equations themselves.

Antisymmetric tensor fields: p-forms

Let us apply this idea to the KK compactification of a 10d p-form field $C_p(x^M)$, i.e. an antisymmetric p-index tensor with gauge invariance (4.44), $C_p \to C_p + d\Lambda_{p-1}$. The ansatz leading to separation of variables will contain terms of the form

$$C_p(x^M) = c_q(x^m)C_{p-q}(x^\mu), \tag{7.24}$$

where the subindices denote the degree of the form, and wedge product is implicit. We have dropped the subindices denoting the 10d, 6d, or 4d nature of the fields (otherwise manifest in the coordinates they depend on), and the label $\mathbf{k}$. In short, a 10d p-form can lead to 4d $(p - q)$-forms by pointing q indices along a q-form in $\mathbf{X}_6$. From Section B.2, the kinetic term operator for 10d p-forms can be written

$$\Delta_{10d} = (dd^\dagger + d^\dagger d)_{\mathbf{X}_6} + \Delta_{4d}. \tag{7.25}$$

In order to get massless 4d fields, the zero mode equation for the q-forms on $\mathbf{X}_6$ is

$$(dd^\dagger + d^\dagger d)c_q(x^m) = 0, \tag{7.26}$$

where here and in similar expressions the subindex "6d" is implicit. Since this kinetic operator is positive definite, zero modes must satisfy $dc_q = 0$, $d^\dagger c_q = 0$, and are harmonic forms, see Section B.2. There is exactly one harmonic q-form per q-cohomology class, hence the number of zero mode q-forms in $\mathbf{X}_6$ is the Betti number $b_q(\mathbf{X}_6)$, which depends only on the topology of $\mathbf{X}_6$. Note that regarding scalar fields as 0-forms we recover our earlier result, since $b_0(\mathbf{X}_6) = 1$ for any compactification space $\mathbf{X}_6$. The classification of q-form zero modes by cohomology also follows from the fact that they must be closed $dc_q = 0$ (so that they have vanishing field strength, and hence internal kinetic energy), and gauge invariant, i.e. defined up to transformations $c_q \simeq c_q + df_{q-1}$. The bottom line is that the number of 4d massless $(p - q)$-forms arising from a 10d p-form is given by $b_q(\mathbf{X}_6)$. This will be widely used, e.g. in Sections 10.1.1 and 12.1.

The gauge invariance of the 10d forms descends to a gauge invariance of the 4d $(p - q)$-forms; this has in fact already appeared in the study of the $\mathbf{S}^1$ compactification of the bosonic string and the superstrings, in Sections 3.2.3 and 5.1.1.

The number of massless fields is a topological invariant, and so the counting cannot change by continuous deformations of the system, like the addition of interactions which modify the structure of the kinetic term operators. Therefore our results using the free field description are valid even when the corresponding fields are interacting (as is actually the case in supergravity theories).

Spinors

A similar approach can be followed to describe the KK compactification of 10d spinors ψ_{10d}. In the compactification ansatz there are components of the form

$$\psi_{10d}(x^M) = \chi_{6d}(x^m)\psi_{4d}(x^\mu), \tag{7.27}$$

where the quantum numbers of χ_{6d}, ψ_{4d} under the 6d and 4d Lorentz groups follow from the decomposition of the $SO(10)$ spinor representation under $SO(6) \times SO(1, 3)$,

$$\mathbf{16} = (\mathbf{4}, \mathbf{2}) + (\bar{\mathbf{4}}, \mathbf{2}'), \tag{7.28}$$

where $\mathbf{2}$, $\mathbf{2}'$ denote the left- and right-handedness under the 4d Lorentz group. The kinetic operator for the 10d spinor is the Dirac operator $\not{D}_{10d}$, covariantized with the spin connection (and the gauge connection in the presence of gauge backgrounds under which the spinor is charged). Acting on the $(\mathbf{4}, \mathbf{2})$ component it factorizes as $\not{D}_{10d} = \not{D}_{6d} + \not{D}_{4d}$, so the number of massless 4d left-handed Weyl fermions is given by the number of solutions of the zero mode equation on $\mathbf{X}_6$

$$\not{D}\chi_{6d}(x^m) = 0, \tag{7.29}$$

namely the dimension of the kernel of $\not{D}_{6d}$. Similarly, acting on the $(\bar{\mathbf{4}}, \mathbf{2}')$, the 10d Dirac operator factorizes as $\not{D}_{10d} = \not{D}_{6d}^\dagger + \not{D}_{4d}^\dagger$, and the number of massless 4d right-handed Weyl fermions is the dimension of the kernel of $\not{D}_{6d}^\dagger$. For spinors neutral under any gauge background, the zero mode condition is a rewriting of the Killing spinor condition (7.1), so in CY compactifications the $SU(3)$ holonomy dictates the existence of one massless 4d fermion per 10d fermion. For spinors carrying additional (e.g. gauge) quantum numbers introducing new terms in the covariantized Dirac operator, there are general results relating the number of zero modes to topological invariants of the configuration, see Sections 7.3 and 7.4. One particularly important quantity is the difference,

$$\text{ind}\,\slashed{D} = \dim\ker\,\slashed{D} - \dim\ker\,\slashed{D}^{\dagger}, \tag{7.30}$$

known as the index of the Dirac operator on $\mathbf{X}_6$; since it counts the difference of the numbers of left- and right-handed 4d spinors, it computes their net chirality. Powerful mathematical results, known as index theorems, state under which conditions such index can be non-zero, and provide formulae for its computation.

Metric

In KK compactifications the 10d metric can produce different kinds of 4d fields:

- The component $G_{\mu\nu}$ with both indices in the 4d non-compact directions has an internal structure that corresponds to a scalar field in $\mathbf{X}_6$. The zero mode equation is thus a scalar Laplace equation (7.23), which has a unique solution. Therefore the 10d graviton leads to a unique 4d graviton.

- The components $G_{\mu m}$ with one 4d index and one index in $\mathbf{X}_6$ leads to 4d vector bosons, with internal structure of a vector field in $\mathbf{X}_6$. The condition to obtain a massless 4d gauge boson is that the 6d vector is a Killing vector associated to a continuous isometry of $\mathbf{X}_6$. In other words, the 4d gauge group is the group of reparametrizations of x^m leaving $\mathbf{X}_6$ invariant, as already appeared in $\mathbf{S}^1$ compactifications in Sections 1.4.1, 3.2.3, and 5.1.1. For the case of CY spaces (with holonomy $SU(3)$, and not in a smaller subgroup), however, it can be mathematically shown that there are no continuous isometries, so there are no 4d gauge bosons arising from the KK compactification of the 10d metric.

- The components G_{mn} with both indices in $\mathbf{X}_6$ lead to 4d scalar fields. The massless 4d scalars correspond to moduli, whose vevs parametrize the geometry of the internal space. This is the generalization of the relation between the G_{99} component of the metric and the compactification radius in $\mathbf{S}^1$ compactifications. For the case of interest of CY compactifications, we have already described these parameters in Section 7.2.2: there are $h_{1,1}$ real scalars (Kähler moduli), and $h_{2,1}$ complex scalars (complex structure moduli).

Gravitinos

The KK compactification of gravitinos is closely related to the discussion of the number of preserved supersymmetries in Section 7.2.1. Consider the component where the vector index is along the non-compact 4d dimensions, while the spinor index splits as a 6d spinor times a 4d spinor; massless 4d fields from such components lead to 4d gravitinos. Their number provides the number of supersymmetries preserved by the compactification, namely the zero mode equation for the internal 6d field is the Killing spinor equation (7.1). For CY threefolds, there is a single Killing spinor on $\mathbf{X}_6$, so each 10d gravitino leads to one 4d gravitino.

Finally, the components where the vector index is along $\mathbf{X}_6$ give rise to 4d spin 1/2 fermions. Their internal structure is that of 6d spinors, with an additional $SO(6)$ vector index. Massless 4d fermions thus arise from zero modes of the Dirac operator $\not{D}_6$ for a 6d spinor transforming additionally in the **6** of the $SO(6)$ acting on the tangent space. Specializing (7.29) for this case, the zero mode equation is

$$\not{D}_6 \chi_{6d}(x^m) = 0. \tag{7.31}$$

For CY compactifications, explained later, they merely provide the superpartners of the Kähler and complex structure moduli. This fits the intuition that the 10d graviton and gravitinos, related by 10d supersymmetry, lead upon compactification to 4d massless modes which are related by the 4d supersymmetry.

It is time to put this technology to work and describe the massless spectrum of CY compactifications of heterotic strings. Compactifications for type II theories are studied in Section 10.1, and their orientifolds are discussed in Chapters 10 and 11.

7.3 Heterotic CY compactifications: standard embedding

We now initiate the study of compactifications of heterotic string theory on Calabi–Yau threefolds. They have potential phenomenological application to the construction of particle physics models, as we will emphasize. Note that in the supergravity approximation, the $SO(32)$ and $E_8 \times E_8$ heterotic compactifications are described in a very similar formalism, differing only in the group theory, and the resulting 4d gauge structures; for concreteness we focus on the $E_8 \times E_8$ compactifications, which are also more often considered for string phenomenology. We first consider the simplest compactification ansatz for internal gauge backgrounds, the so-called standard embedding, leaving its generalization for later sections. Also, in Chapter 8 we construct heterotic compactifications with exact worldsheet CFT description, including toroidal orbifolds, free fermion models, and Gepner models. They can be regarded as fully stringy versions of the geometric CY compactifications.

7.3.1 The gravity multiplet and compactification moduli

The 10d massless fields of the $E_8 \times E_8$ heterotic theory are the 10d $\mathcal{N} = 1$ gravity multiplet (whose bosonic content is the metric G_{MN}, the 2-form B_{MN}, the dilaton ϕ) and the 10d $\mathcal{N} = 1$ vector multiplets (with bosonic content given by the gauge bosons A_M^a). We consider the compactification of this theory on a CY threefold and for the moment focus on the KK reduction of the gravity multiplet. Since there is one 10d gravitino, the compactification leads to a 4d $\mathcal{N} = 1$ theory; this allows for chiral representations of the 4d gauge group, as will be explicitly discussed later on. In order to simplify the discussion, we carry out the KK compactification for the bosonic fields, and use the structure of 4d $\mathcal{N} = 1$ supermultiplets to complete the 4d massless fermion content. The supermultiplets we will need are the vector and chiral multiplets, described in Section 2.1. The KK compactification of the 10d $\mathcal{N}=1$ gravity multiplet is shown schematically in the table below:

Het	Gravity	$h_{1,1}$ Chiral	$h_{2,1}$ Chiral	Chiral	
$G \quad \rightarrow$	$g_{\mu\nu}$	$h_{1,1}$ Kähler	$2h_{2,1}$ Cmplx. Str.		(7.32)
$B \quad \rightarrow$		$B_{i\bar{j}}$		$B_{\mu\nu}$	
$\phi \quad \rightarrow$				ϕ	

Here we denote $B_{i\bar{j}}$ any of the $h_{1,1}$ scalars that arise from the 10d 2-form with 6d indices along a harmonic (1, 1)-form in $\mathbf{X}_6$. Note also that we need to dualize the 4d 2-form $B_{\mu\nu}$ into a 4d scalar in order to fill out a standard chiral multiplet, as mentioned in Section B.3, and described more explicitly in Section 9.1.1. The end result is the 4d $\mathcal{N}=1$ gravity multiplet, and $h_{1,1}+h_{2,1}+1$ chiral multiplets (which are neutral under the 4d gauge group to be discussed later).

For future reference, we mention the higher-dimensional origin of the massless fermions in these chiral multiplets. Those in the 4d dilaton chiral multiplet arise from the 10d dilatino, through zero modes satisfying (7.29). The fermion partners of Kähler and complex structure moduli arise from the 10d gravitino, with the vector index along the 6d internal space. Their zero modes are defined by a refined version of (7.31), in which the **6** of $SO(6)$ splits as $\mathbf{3}+\bar{\mathbf{3}}$ under the $SU(3)$ holonomy group. Namely massless fields arise from zero modes of $\not{D}_3$ and of $\not{D}_{\bar{3}} = \not{D}^\dagger$, whose multiplicities are determined by the CY Hodge numbers:

$$\dim \ker \not{D}_3 = h_{1,1}(\mathbf{X}_6), \quad \dim \ker \not{D}_3^\dagger = h_{2,1}(\mathbf{X}_6), \qquad (7.33)$$

as indeed required to complete the above chiral multiplets.

7.3.2 The vector multiplets: standard embedding

The discussion of the KK compactification of the 10d $\mathcal{N}=1$ vector multiplets requires taking into account an important ingredient in heterotic compactification on CY manifolds. In addition to the introduction of a non-trivial background for the internal components of the metric G_{mn}, the theory requires the introduction of a non-trivial background for the internal components of the 10d gauge fields A_m^a. This does not modify the KK reduction of the 10d gravity multiplet (since it is neutral under the 10d gauge group) but modifies that of the 10d vector multiplets.

The presence of such backgrounds is compatible with 4d Poincaré invariance, since only internal 6d components of the gauge fields are turned on. Also, they can be made compatible with 4d $\mathcal{N}=1$ supersymmetry, see later. However, the important point is that such backgrounds are required for consistency of the compactification. Recall from Section 4.3.4 that the 10d effective action for heterotic theories implies the modified Bianchi identity

(4.87) for B_2, which for the $E_8 \times E_8$ theory reads

$$dH_3 = \frac{\alpha'}{4}(\text{tr } R^2 - \text{tr } F^2 - \text{tr } F'^2), \tag{7.34}$$

where F, F', and R are the two E_8 gauge field strengths and the curvature 2-forms, respectively, see Section B.3. Since the left-hand side is an exact form in cohomology, so must the right-hand side. This requires to turn on non-trivial gauge field backgrounds satisfying the relation of cohomology classes

$$[\text{tr } F^2] + [\text{tr } F'^2] = [\text{tr } R^2]. \tag{7.35}$$

Note that in toroidal compactifications, the geometric curvature vanishes $R = 0$, and it is consistent to consider gauge backgrounds with zero field strength $F = F' = 0$, as in Section 5.2. Incidentally, the topological consistency condition for heterotic compactifications, and its derivation, is very analogous to the RR tadpole cancellation conditions for type II orientifolds in Chapters 10 and 11.

The upshot is that consistent heterotic CY compactifications require a non-trivial internal gauge background for (some of the) 10d gauge fields. In general, these lie inside a subgroup H of the 10d gauge group $G = E_8 \times E_8$. We momentarily focus on a simple choice suggested by the structure of (7.35), and take the gauge field strength to be locally identical to the curvature 2-form, namely $F = R$ point-wise; more general gauge backgrounds are discussed in later sections. Models with this "standard embedding" are known as (2, 2) compactifications, since their worldsheet CFT contains both left and right-moving $\mathcal{N} = 2$ superconformal algebras, see Section E.4. Models with non-standard embedding are known as (0, 2) compactifications.

In CY geometries, the metric connection, and thus the curvature, lies inside $SU(3)$, so the non-trivial gauge fields in standard embedding are in an $SU(3)$ subgroup of the 10d gauge group, in particular of one of the E_8 factors. The group E_8 has a maximal subgroup $E_6 \times SU(3)$, under which the adjoint decomposes as

$$
\begin{aligned}
E_8 &\rightarrow & E_6 \times SU(3), \\
\mathbf{248} &\rightarrow & (\mathbf{78}, \mathbf{1}) + (\mathbf{1}, \mathbf{8}) + (\mathbf{27}, \mathbf{3}) + (\overline{\mathbf{27}}, \overline{\mathbf{3}}).
\end{aligned} \tag{7.36}
$$

The standard embedding turns on a gauge field background in this $SU(3)$ factor. The 4d massless spectrum can be described as follows:

- There are vector multiplets of $E_6 \times E_8$. The 4d gauge bosons arise from the KK reduction of 10d gauge bosons transforming as singlets under $SU(3)$, and are associated to a constant 6d scalar zero mode, determined by a Laplace equation. The gauginos arise from the KK reduction of 10d gauginos transforming as singlets under $SU(3)$, and are associated to the unique zero mode satisfying the Dirac equation (7.29) for a neutral 6d spinor. The gauge group $E_6 \times E_8$ can heuristically be understood from arguments similar to Wilson line gauge symmetry breaking in page 143. Incidentally, we note that 10d gauge bosons singlets under $SU(3)$ cannot lead to massless scalars, since this would require aligning their Lorentz index along a harmonic 1-form in $\mathbf{X}_6$, and this is not possible because $b_1(\mathbf{X}_6) = 0$ for CY spaces.

- In addition, there are matter chiral multiplets transforming in the **27** and $\overline{\mathbf{27}}$ of E_6. Fermions in the **27** arise from 10d gauginos transforming in the **3** of $SU(3)$, with 6d profile given by zero modes of the Dirac operator $\not{D}_3$. Similarly fermions in the $\overline{\mathbf{27}}$ arise from zero modes of $\not{D}_{\bar{3}} = \not{D}_3^\dagger$. Since the gauge field strength is identified with the metric curvature, the zero mode multiplicities are identical to those in (7.33); thus the number of 4d $\mathcal{N} = 1$ chiral multiplets in the **27**, $\overline{\mathbf{27}}$ are given by

$$n_{\mathbf{27}} = h_{1,1}(\mathbf{X}_6), \quad n_{\overline{\mathbf{27}}} = h_{2,1}(\mathbf{X}_6). \tag{7.37}$$

The scalars in these chiral multiplets arise from components of the 10d gauge bosons with their polarization along internal directions of $\mathbf{X}_6$. Mathematically these can be described as 1-forms transforming in the representations **3**, $\bar{\mathbf{3}}$ under the $SU(3)$ gauge backgrounds, see Section B.1. Their zero modes are classified by the cohomology groups $H^1(\mathbf{X}_6, V)$, $H^1(\mathbf{X}_6, V^*)$, where V, V^* denote the vector spaces of the **3**, $\bar{\mathbf{3}}$, respectively; these groups can indeed be shown to have dimensions $h_{1,1}(\mathbf{X}_6)$, $h_{2,1}(\mathbf{X}_6)$, as required to complete the supermultiplets.

- Finally, there are also 4d massless singlet chiral multiplets, which are thus less relevant but which we discuss for completeness. Their fermions arise from 10d gauginos transforming in the **8**, and correspond to zero modes of the Dirac operator $\not{D}_8$. Their scalars arise from 10d gauge bosons with internal Lorentz index, and transforming in the **8** of $SU(3)$. These zero modes (and hence the singlet chiral multiplets) are mathematically classified by the cohomology group $H^1(\mathbf{X}_6, \mathrm{End}(V))$, where $\mathrm{End}(V)$ describes the vector space of the **8** – formally regarded as "endomorphisms" acting on the vector space V of the **3** i.e. as 3×3 matrices – since $\mathbf{3} \otimes \bar{\mathbf{3}} = \mathbf{8}(\oplus\mathbf{1})$. Although computable in explicit examples, the number of singlets cannot be expressed in terms of Hodge numbers of $\mathbf{X}_6$. These singlets are usually denoted vector bundle moduli, since their vevs describe continuous parameters of the choice of gauge field background, keeping it in $SU(3)$.

7.3.3 Phenomenological application: E_6 GUTs and discrete Wilson lines

The basic structure of the resulting spectrum is relatively close to a model of particle physics, as follows. The 4d gauge group E_6 is a possible GUT extension of the Standard Model, as reviewed in Section 1.2.4. The unbroken group E_8 is completely decoupled, and may play as a hidden sector possibly triggering supersymmetry breaking by gaugino condensation, see Section 15.2. The net number of families (not considering non-chiral pairs $\mathbf{27} + \overline{\mathbf{27}}$) is given by the index (7.30) of $\not{D}_3$, specifically

$$|n_{\mathbf{27}} - n_{\overline{\mathbf{27}}}| = |h_{1,1}(\mathbf{X}_6) - h_{2,1}(\mathbf{X}_6)| = \tfrac{1}{2}|\chi(\mathbf{X}_6)|, \tag{7.38}$$

where $\chi(\mathbf{X}_6)$ is the Euler characteristic (7.6). Non-chiral $\mathbf{27} + \overline{\mathbf{27}}$ pairs may in principle get large masses after further gauge and supersymmetry breaking and hence one expects that they can easily disappear from the massless spectrum. In addition, there are a number of singlets corresponding to (complex structure and Kähler) moduli of the compactification manifold and to vector bundle moduli.

Discrete Wilson lines and gauge symmetry breaking

Although the model has a GUT-like gauge group, it does not contain the GUT-Higgs adjoint multiplets necessary to break it to the Standard Model gauge group. In contrast with field theory, string theory is a rigid structure and we cannot choose to introduce this multiplet by hand. In fact, even modifying the compactification ansatz, the possibilities to get the required adjoint representations are severely limited in string theory, see Sections 9.8 and 17.1.2. However, for non-simply connected CY spaces (i.e. with $\Pi_1(\mathbf{X}_6) \neq 1$, see Section B.4), there exists the interesting alternative of breaking the E_6 group by turning on non-trivial Wilson lines, bringing it closer to the SM gauge group. These Wilson lines are analogous to those in toroidal compactifications in Section 5.2.2, but differ in not being associated to continuous parameters, since CY manifolds have $b_1(\mathbf{X}_6) = 0$ and there are no Wilson line moduli. Rather they are *discrete* Wilson lines, defined by a discrete set of gauge transformations along the non-trivial homotopy cycles in $\Pi_1(\mathbf{X}_6)$. CY manifolds constructed using the techniques in Section 7.2.3 tend to be simply connected, but an easy way to construct CY spaces with non-trivial $\Pi_1(\mathbf{X}_6) = G$ is to quotient a (simply connected) CY $\widetilde{\mathbf{X}}_6$ by a freely acting symmetry G, i.e. with no fixed points, as in the Tian–Yau example (7.20). This moreover leads to a potentially interesting reduction of the number of generations, since $\chi(\mathbf{X}_6) = \chi(\widetilde{\mathbf{X}}_6)/|G|$, with $|G|$ the number of elements of G.

Let γ_g denote the non-trivial homotopy cycle associated to an element $g \in G$. The discrete Wilson line is an internal gauge background A_m^a with vanishing field strength $F_{mn} = 0$, hence still preserving supersymmetry and the standard embedding condition. Its effect is encoded in the gauge invariant Wilson line operator

$$U_{\gamma_g} = P \exp\left(i \oint_{\gamma_g} A_m^a T_a dx^m\right), \tag{7.39}$$

where T_a are the gauge generators. Any field $\phi(x)$ defined on the quotient CY $\mathbf{X}_6$ suffers a gauge transformation upon moving around γ_g, inherited from the symmetry action on the covering CY space $\widetilde{\mathbf{X}}_6$

$$\phi(x) = U_{\gamma_g}\phi(gx). \tag{7.40}$$

Fields in the quotient are obtained by imposing invariance under the simultaneous action of the geometric symmetry G and its gauge embedding $\tilde{G}$ generated by the U_γ elements. The 4d gauge fields have a constant internal zero mode profile and are invariant under the G action, so they must be invariant under the $\tilde{G}$ group to remain massless. The 4d gauge group is thus the subgroup of E_6 commuting with the Wilson lines U_γ. At the group theory level, the effect is thus similar to that of an adjoint Higgs scalar vev.

To explore the possible structures of the unbroken gauge group, let us consider for notational purposes the $SU(3)_c \times SU(3)_L \times SU(3)_R$ maximal subgroup of E_6, mentioned in

Section 1.2.4, and for simplicity focus on $G = \mathbf{Z}_N$, with generator γ. Then the most general form for U_γ commuting with the SM gauge group is

$$U_\gamma = (\mathbf{1}_3)_c \otimes \mathrm{diag}\,(\alpha, \alpha, \alpha^{-2})_L \otimes \mathrm{diag}\,(\beta, \rho, \delta)_R, \qquad (7.41)$$

where the three matrices act on the three $SU(3)$ factors respectively, and $\alpha^N = \beta^N = \rho^N = \delta^N = 1$, $\beta\rho\delta = 1$. For $\alpha \neq \alpha^{-2}$ and β, ρ, δ all different (and $N > 6$), the unbroken group is $SU(3)_c \times SU(2)_L \times U(1)^3$, whereas larger non-abelian groups may be obtained for some of these phases being equal or $N \leq 6$. For abelian G the rank of the gauge group cannot be reduced, since the corresponding U_{γ_g} are all commuting and can be diagonalized simultaneously. However, there exist some examples of CY spaces with a non-abelian $\Pi_1(\mathbf{X}_6)$, which in principle allow for rank reduction through Wilson lines, although phenomenologically interesting examples have not been worked out.

Discrete Wilson lines and doublet–triplet splitting

The addition of Wilson lines may have another interesting application related to the doublet–triplet splitting problem, mentioned in Section 2.6.2. Wilson lines do not modify the standard embedding condition, and hence cannot change the net number of chiral fermions, still given by $\chi(\mathbf{X}_6)/2$. However, they can easily modify the non-chiral content included in pairs $(\mathbf{27} + \overline{\mathbf{27}})$. For illustration consider a compactification with $\Pi_1(\mathbf{X}_6) = \mathbf{Z}_N$ with $N > 4$, and with generator embedded as a Wilson line (7.41) of the form

$$U_\gamma = (\mathbf{1}_3)_c \otimes \mathrm{diag}\,(\alpha, \alpha, \alpha^{-2})_L \otimes \mathrm{diag}\,(\alpha, \alpha, \alpha^{-2})_R. \qquad (7.42)$$

The unbroken gauge group is $SU(3)_c \times SU(2)_L \times SU(2)_R \times U(1)^2$. Consider now a $\mathbf{27}$ in the massless spectrum, whose decomposition under the $SU(3)^3$ subgroup is $(\mathbf{3}, \bar{\mathbf{3}}, \mathbf{1}) + (\bar{\mathbf{3}}, \mathbf{1}, \bar{\mathbf{3}}) + (\mathbf{1}, \mathbf{3}, \bar{\mathbf{3}})$, which may be written as matrices

$$Q_L = \begin{pmatrix} u_1 & u_2 & u_3 \\ d_1 & d_2 & d_3 \\ D_1 & D_2 & D_3 \end{pmatrix}, \quad Q_R = \begin{pmatrix} u_1^c & u_2^c & u_3^c \\ d_1^c & d_2^c & d_3^c \\ D_1^c & D_2^c & D_3^c \end{pmatrix}, \quad L = \begin{pmatrix} H_u^0 & H_d^+ & e^- \\ H_u^- & H_d^0 & v_e \\ e^+ & v_R & N \end{pmatrix}.$$
$$(7.43)$$

Assuming for illustrative purposes that it is invariant under the geometric action G, it is easy to check that the only fields invariant under U_γ, are the EW Higgs doublets $H_{u,d}$; these are thus the only fields remaining in the massless spectrum. In particular their color triplets partners, which would mediate dangerous proton decay, are absent, i.e. get a large mass (of order the compactification scale) from the Wilson line. Incidentally, since the net number of chiral fermions cannot change, there should be an extra set of Higgs doublets from one of the $\overline{\mathbf{27}}$s.

Discrete Wilson lines thus provide a nice mechanism to split the doublets from the triplets. Its implementation in standard embedding compactifications does not lead to fully satisfactory E_6 GUTs, since there are colored E_6 partners of the SM families (of D, D^c type) which remain massless even after the introduction of Wilson lines. On the other hand, the introduction of discrete Wilson lines is a very general possibility, which may be implemented in non-standard embedding compactifications, as described in Section 7.4, leading to more satisfactory models. Discrete Wilson lines play also an important role in the construction of the toroidal orbifold models, see Section 8.3.

A 3-generation example from the Tian–Yau manifold

Consider the heterotic compactification on the Tian–Yau manifold constructed in Section 7.2.3, which has $\chi(\mathbf{X}_6) = -6$, $(h_{21}, h_{11}) = (9, 6)$, and hence produces three net E_6 generations and six vector-like $(\mathbf{27} + \overline{\mathbf{27}})$ pairs. Its construction as a quotient by the freely acting $\mathbf{Z}_3$ symmetry (7.20) implies that $\Pi_1(\mathbf{X}_6) = \mathbf{Z}_3$, and we may introduce a $\mathbf{Z}_3$ discrete Wilson line. Consider the choice

$$U_\gamma = (\mathbf{1}_3)_c \otimes (\alpha \mathbf{1}_3)_L \otimes (\alpha \mathbf{1}_3)_R. \tag{7.44}$$

The unbroken gauge group is $SU(3)_c \times SU(3)_L \times SU(3)_R$, incidentally the smallest possible with a $\mathbf{Z}_3$ symmetry. The computation of the massless matter fields surviving the Wilson line projection requires the information of the transformation properties of the original **27**s and $\overline{\mathbf{27}}$s under the geometric $\mathbf{Z}_3$ symmetry. This can be determined from their relation to the cohomology classes of the CY, e.g. the transformation properties of the monomials (7.19) associated to complex structure deformations. We skip this computation and the subsequent Wilson line projection, and simply quote the resulting massless spectrum, which in the notation in (7.43) reads

$$3Q_L + 3Q_R + 3L + 4(Q_L + \overline{Q_L}) + 4(Q_R + \overline{Q_R}) + 6(L + \overline{L}). \tag{7.45}$$

Note that the number of vector-like exotics has been substantially reduced.

Further reduction of the gauge group to the SM can be carried out by taking suitable D-flat directions in the 4d effective theory, giving vevs to SM singlet scalars transforming like ν_R. Skipping this analysis, we simply mention that two stages of symmetry breaking, assumed close to the string scale, can break $SU(3)^3$ down to the SM group and give large masses to all unwanted colored particles through Yukawa couplings. The final light spectrum is that of the MSSM, plus some additional extra vector-like charged and neutral leptons. A potential shortcoming of this class of symmetry breaking process is the difficulty to explain the absence of dimension-4 B and L violating operators, which generically appear once fields with quantum numbers of a ν_R scalar get vevs. This is a rather generic problem in many heterotic string constructions.

This 3-generation model was historically the first to be constructed and analyzed, and nicely illustrates the construction of SM-like particle physics models from (2, 2) heterotic string compactifications. There are other CY spaces with $|\chi(\mathbf{X}_6)| = 6$, which have been constructed more recently, but whose heterotic model building phenomenology has not been explored. Although most of them have a large number of vector-like $(\mathbf{27} + \overline{\mathbf{27}})$ pairs, making such analysis difficult, there is one example with $(h_{1,1}, h_{2,1}) = (1, 4)$. It is obtained as a quotient of a CY manifold defined by five monomial equations in $(\mathbf{CP}_2)^4$, and has $(h_{1,1}, h_{2,1}) = (8, 44)$.

7.4 Heterotic CY compactifications: non-standard embedding

It is possible to relax the compactification ansatz beyond the standard embedding, so that the starting 4d gauge group (before turning on Wilson lines) is already closer to the Standard Model; this provides a more flexible model building avenue. As mentioned above, these non-standard embedding models are also known as (0, 2) compactifications.

7.4.1 Generalities

The strategy is simply to choose a gauge background which lies in a larger subgroup H of $E_8 \times E_8$, so that the 4d gauge group is smaller. The gauge background is constrained in order to preserve 4d $\mathcal{N} = 1$ supersymmetry, as described later on.

Let us consider the computation of the 4d massless spectrum. Clearly, the KK compactification of the 10d gravity multiplet is unchanged, and leads to the 4d gravity multiplet plus the dilaton, complex structure and Kähler moduli chiral multiplets. The KK compactification of the 10d vector multiplets is a simple generalization of the standard embedding case, with differences on group theoretical decompositions, and thus in the topological invariants counting the numbers of zero modes. Consider for simplicity the gauge background in a subgroup H of E_8, so that the second E_8 factor can be ignored. Let us denote G_{4d} the 4d gauge group i.e. the commutant of H in E_8, and consider the decomposition of the E_8 adjoint

$$
\begin{aligned}
E_8 \quad &\to \quad H \times G_{4d}, \\
\mathbf{248} \quad &\to \quad (\mathbf{1}, \mathbf{adj}G_{4d}) + \sum_k (\mathbf{R}_{H,k}, \mathbf{R}_{G,k}) + (\mathbf{adj}H, \mathbf{1}), \qquad (7.46)
\end{aligned}
$$

where **adj** denotes the adjoint representation. For instance, for $H = SU(4)$ we have $G_{4d} = SO(10)$, and for $H = SU(5)$ we have $G_{4d} = SU(5)$. As in the standard embedding case, the 4d gauge bosons and gauginos arise from the 10d vector multiplets transforming as singlets under H, and their 6d zero modes are the unique solutions to the Laplace and Dirac equations for neutral fields on $\mathbf{X}_6$, respectively.

Charged matter chiral multiplets in the representation $\mathbf{R}_{G,k}$ of G_{4d} arise from the KK compactification of 10d gauge fields (with Lorentz index along $\mathbf{X}_6$) and 10d gauginos transforming in the non-trivial representation $\mathbf{R}_{H,k}$ of H. The number of 4d scalars is given by the dimension of the cohomology group $H^1(\mathbf{X}_6, V_{\mathbf{R}_{H,k}})$, while the number of 4d fermions is the dimension of the kernel of the Dirac operator $\not{D}_{\mathbf{R}_{H,k}}$; these are topological invariants of the configuration (i.e. $\mathbf{X}_6$ and the gauge background), which are equal for supersymmetry preserving backgrounds, but for which no closed expression exists in general. A more manageable yet interesting quantity is the *net* number of chiral multiplets in the representation $\mathbf{R}_{G,k}$, given by the index (7.30) of $\not{D}_{\mathbf{R}_{H,k}}$

$$
n_{\mathbf{R}_{G,k}} - n_{\mathbf{R}^*{}_{G,k}} = \mathrm{ind}\; \not{D}_{\mathbf{R}_{G,k}} = \int_{\mathbf{X}^6} \mathrm{Tr}\, {}_{\mathbf{R}_{G,k}} \left(\exp \frac{F}{2\pi} \right) \hat{A}(R) \qquad (7.47)
$$

where the last equality is the mathematical result known as the index theorem. In the last expression wedge products are implicit, the exponential is defined by formal Taylor expansion as in the Chern classes in Section B.3, $\hat{A}$ is the A-roof polynomial already appeared in a different context below (6.15), and integration over $\mathbf{X}_6$ picks the 6-form term in the expansion. These indices are non-zero in general, so the resulting spectrum is generically chiral. It is possible to use the above formula (and anomaly polynomials) to show that cubic non-abelian chiral anomalies cancel automatically, upon using (7.35).

Finally, uncharged chiral multiplets, i.e. vector bundle moduli, arise from 10d gauge fields and gauginos transforming as **adj** H, whose discussion we skip.

In order to provide explicit examples of this construction, one needs to take into account the requirement to preserve 4d $\mathcal{N} = 1$ supersymmetry. This imposes a set of conditions on the gauge background, the so-called Hermite–Yang–Mills equations

$$\text{(a) } F_{ij} = 0; \quad F_{\overline{i}\overline{j}} = 0; \quad \text{(b) } g^{i\overline{j}} F_{i\overline{j}} = 0. \tag{7.48}$$

From the perspective of the 4d effective action, these correspond to vanishing of F- and D-term contributions to the scalar potential, in the language of Chapter 2. Note that these are automatically satisfied in the standard embedding ansatz $F_{mn} = R_{mn}$; the first equation is satisfied since the curvature is a $(1, 1)$-form, while the second follows from the vanishing of the first Chern class. For future comparison with similar conditions in type IIB orientifolds in Sections 11.4.2 and 11.4.3, we recast (7.48) in differential form language as

$$\text{(a) } F_{2,0} = F_{0,2} = 0; \quad \text{(b) } F \wedge J \wedge J = 0, \tag{7.49}$$

where $F_{2,0}$, $F_{0,2}$ are the $(2, 0)$, $(0, 2)$ components of the field strength 2-form. Also, we have used $*_{6d} J \simeq J \wedge J$, which follows from (7.7).

In analogy with the $\mathcal{N} = 1$ supersymmetry conditions for the metric, these equations are difficult to solve explicitly. Fortunately, there is also a theorem (by Donaldson (1985) and Uhlenbeck and Yau (1986)) which guarantees the existence of a solution for these equations for gauge backgrounds satisfying two (almost topological) conditions. A detailed discussion of the conditions is beyond our scope, and we merely quote them for completeness:

- The complexified gauge bundle is holomorphic. This is equivalent to (7.49a).
- For such a holomorphic gauge bundle V, a solution to (7.49b) exists if the bundle satisfies a so-called stability condition. Mathematically, the stability condition is that the slope of the bundle, defined as

$$\mu(V) = \frac{\int_{\mathbf{X}_6} J \wedge J \wedge F}{\text{rank}(V)}, \tag{7.50}$$

is smaller than the slope of any sub-bundle, $\mu(V') \geq \mu(V)$ for any sub-bundle V' (roughly speaking, for any gauge background in a subgroup of the structure group of V). Physically, it means that the gauge background does not split dynamically into lower-energy gauge configurations, corresponding to the sub-bundles.

The classification or even the construction of stable holomorphic bundles over Calabi–Yau manifold is a developing field in mathematics, with partial results which have already been applied to the physics of heterotic string compactifications. Although these techniques are beyond our scope, in the next section we sketch a couple of examples to give a flavour of the main ingredients. Note also that giving vevs to $\mathbf{27} + \overline{\mathbf{27}}$ pairs along true F- and D-flat directions in (2, 2) models, leads also to (0, 2) vacua; thus some classes of (0, 2) models may be understood as deformations of (2, 2) compactifications.

7.4.2 Examples*

An SO(10)-based example

This example is based on a CY manifold $\widetilde{\mathbf{X}}_6$, described mathematically as a double elliptic fibration over $\mathbf{CP}_1$. In other words, it is obtained by erecting the product of two $\mathbf{T}^2$'s at each point in a base $\mathbf{CP}_1$, with complex structure parameters varying holomorphically along this base. More explicitly, it is defined by the equations

$$
\begin{aligned}
y^2 &= x^3 + f_6(z)x + g_4(z), \\
y'^2 &= x'^3 + f_6'(z)x' + g_4'(z),
\end{aligned}
\tag{7.51}
$$

where z is an affine coordinate in $\mathbf{CP}_1$ and the subindices denote the degree of the corresponding polynomial. In the language of toric geometry of Section 7.2.3, it can be described by the weight assignments

	z	z'	x	y	u	x'	y'	u'	f_1	f_2
$(\mathbf{C}^*)_1$	0	0	2	3	1	0	0	0	6	0
$(\mathbf{C}^*)_2$	0	0	0	0	0	2	3	1	0	6
$(\mathbf{C}^*)_3$	1	1	2	3	0	2	3	0	6	6,

where f_1, f_2 are related to the two equations in (7.51). Here $[z, z']$ are homogeneous coordinates in the $\mathbf{CP}_1$ base, and in a local patch z' can be eliminated using $(\mathbf{C}^*)_3$ to set $z' = 1$. The coordinates (x, y, u) and (x', y', u') describe the two $\mathbf{T}^2$ fibers, and we can set $u = u' = 1$ using $(\mathbf{C}^*)_1$ and $(\mathbf{C}^*)_2$ to recover (7.51).

We consider the locus in which $\widetilde{\mathbf{X}}_6$ admits a $\mathbf{Z}_3 \times \mathbf{Z}_3$ symmetry generated by $(z, x, y, x', y') \rightarrow (\alpha^2 z, \alpha x, y, x', y')$ and $(z, x, y, x', y') \rightarrow (\alpha^2 z, x, y, \alpha x', y')$, with $\alpha = e^{2\pi i/3}$, by suitably restricting the defining polynomials f_4, g_6, f_4', g_6'. The $\mathbf{Z}_3 \times \mathbf{Z}_3$ acts by shifts in the $\mathbf{T}^2$ fibers, and so is freely acting. The quotient $\mathbf{X}_6 = \widetilde{\mathbf{X}}_6/(\mathbf{Z}_3 \times \mathbf{Z}_3)$ is a smooth CY, with $\Pi_1(\mathbf{X}_6) = \mathbf{Z}_3 \times \mathbf{Z}_3$, later used to introduce discrete Wilson lines. The numbers of geometric moduli of $\mathbf{X}_6$ can be computed to be $(h_{1,1}, h_{2,1}) = (3, 3)$.

We continue the discussion in terms of the covering space $\widetilde{\mathbf{X}}_6$. The gauge background on one E_8 is associated to a bundle V, with structure group $SU(4)$, so that $G_{4d} = SO(10)$

in the absence of Wilson lines, and we may use the GUT language of Section 1.2.3 at this intermediate stage of the construction. Using the decomposition

$$E_8 \rightarrow SU(4) \times SO(10),$$
$$\mathbf{248} \rightarrow (\mathbf{1}, \mathbf{45}) + (\mathbf{4}, \mathbf{16}) + (\bar{\mathbf{4}}, \overline{\mathbf{16}}) + (\mathbf{6}, \mathbf{10}) + (\mathbf{15}, \mathbf{1}), \tag{7.52}$$

and recalling the previous section, the number of 4d massless $SO(10)$ families $\mathbf{16}$, conjugate families $\overline{\mathbf{16}}$, Higgses $\mathbf{10}$, and vector bundle moduli are

$$n_{16} = \dim H^1(\tilde{\mathbf{X}}, V), \qquad n_{\overline{16}} = \dim H^1(\tilde{\mathbf{X}}_6, V^*),$$
$$n_{10} = \dim H^1(\tilde{\mathbf{X}}_6, \Lambda^2 V), \qquad n_1 = \dim H^1(\tilde{\mathbf{X}}_6, \mathrm{End}(V)), \tag{7.53}$$

where $V, V^*, \Lambda^2 V, \mathrm{End}(V)$ denote the vector spaces of the $\mathbf{4}, \bar{\mathbf{4}}, \mathbf{6}$, and $\mathbf{15}$ of $SU(4)$.

This matter content is modified by the introduction of non-trivial Wilson lines. As in Section 7.3.3, we introduce Wilson lines along the non-trivial 1-cycles γ_g, γ_h, associated to the generators g, h of $\Pi_1(\mathbf{X}) = \mathbf{Z}_3 \times \mathbf{Z}_3$, embedded into the Cartan subgroup $U(1)^5 \subset SO(10)$ as

$$U_{\gamma_g} = \mathrm{diag}\,(\alpha^2 \mathbf{1}_3, \mathbf{1}_2), \qquad U_{\gamma_h} = \alpha^2 \mathbf{1}_5. \tag{7.54}$$

They lead to the breaking

$$SO(10) \xrightarrow{\gamma_h} \xrightarrow{\gamma_h} SU(5) \xrightarrow{\gamma_g} SU(3) \times SU(2) \times U(1)_Y \times U(1)_{B-L}. \tag{7.55}$$

For the matter multiplets, surviving 4d states must be invariant under the combined Wilson line plus geometric action of $\mathbf{Z}_3 \times \mathbf{Z}_3$. Working out the group theory, the Wilson line phases picked up by fields in the original $SO(10)$ multiplets are

$$\mathbf{16} \rightarrow [(\mathbf{3}, \mathbf{2}; 1, 1)_{\alpha, \alpha^2} + (\bar{\mathbf{3}}, \mathbf{1}; -4, -1)_{\alpha^2, \alpha^2} + (\mathbf{1}, \mathbf{1}; 6, 3)_{1, \alpha^2}]$$
$$+ [(\bar{\mathbf{3}}, \mathbf{1}; 2, -1)_{\alpha^2, 1} + (\mathbf{1}, \mathbf{2}; -3, -3)_{1, 1}] + (\mathbf{1}, \mathbf{1}; 0, 3)_{1, \alpha},$$
$$\mathbf{10} \rightarrow [(\mathbf{3}, \mathbf{1}; -2, -2)_{\alpha^2, \alpha^2} + (\mathbf{1}, \mathbf{2}; 3, 0)_{1, \alpha^2}] + \text{c.c.}, \tag{7.56}$$

where the entries in parentheses denote the representations and charges under the gauge group (7.55) and subindices denote the phases under $U_{\gamma_g}, U_{\gamma_h}$. The computation of the geometric action requires the detailed construction of the gauge bundles, and information of the $\mathbf{Z}_3 \times \mathbf{Z}_3$ action on the corresponding cohomology groups. Referring to the literature for the details, we simply quote the existence of a stable holomorphic $SU(4)$ bundle with

$$\dim H^1(\tilde{\mathbf{X}}_6, V) = 27, \quad \dim H^1(\tilde{\mathbf{X}}_6, V^*) = 0, \quad \dim H^1(\tilde{\mathbf{X}}_6, \Lambda^2 V) = 4, \tag{7.57}$$

and with the corresponding harmonic forms having g, h eigenvalues

$$R_{g, H^1(V)} = \mathbf{1}_3 \otimes \mathrm{diag}(1, 1, 1, \alpha, \alpha, \alpha, \alpha^2, \alpha^2, \alpha^2), \quad R_{g, H^1(\Lambda^2 V)} = \mathrm{diag}(1, 1, \alpha, \alpha^2),$$
$$R_{h, H^1(V)} = \mathbf{1}_3 \otimes \mathrm{diag}(1, \alpha, \alpha^2, 1, \alpha, \alpha^2, 1, \alpha, \alpha^2), \quad R_{h, H^1(\Lambda^2 V)} = \mathrm{diag}(\alpha, \alpha^2, \alpha^2, \alpha). \tag{7.58}$$

The massless spectrum is given by the states invariant under the combined $\mathbf{Z}_3 \times \mathbf{Z}_3$ actions in (7.56), (7.58). As the reader can check, it is given by three quark and lepton generations (including right-handed neutrinos), and one Higgs doublet pair. The complete definition of the heterotic compactification actually requires specifying the gauge background on the second E_8, such that the consistency condition (7.35) is satisfied. Since these additional choices do not modify the structure of the visible sector, we skip their discussion, and rather turn to some phenomenological properties of the above MSSM-like sector.

The gauge group is $SU(3) \times SU(2)_L \times U(1)_Y \times U(1)_{B-L}$ and there are three families of SM quarks and leptons (plus right-handed neutrinos), and one pair of Higgs doublets. In addition, there are 13 vector bundle moduli. The massless spectrum is relatively simple compared to other heterotic constructions like the orbifolds or fermionic models in the next chapter. In particular there are no massless exotic fractionally charged states, which often appear in other heterotic string constructions. Since the only SM singlets charged under $U(1)_{B-L}$ are the ν_R sneutrinos, they must acquire vevs to break the $U(1)_{B-L}$ symmetry, leading to R-parity violation. In order to avoid the generation of large R-parity B/L-violating dimension-4 couplings, the $U(1)_{B-L}$ breaking should occur not much above the EW scale. Finally, in this particular model the leading Yukawa couplings can be shown to yield one massless generation and two generations with EW scale masses.

SU(5)-based examples

Another class of models is based on a quotient of the CY $\widetilde{\mathbf{X}}_6$ in (7.51) by a freely acting $\mathbf{Z}_2$ symmetry. Introducing an $SU(5)$ bundle, the decomposition

$$
\begin{aligned}
E_8 \quad &\rightarrow \qquad\qquad\qquad SU(5) \times SU(5), \\
\mathbf{248} \quad &\rightarrow \quad (\mathbf{24}, \mathbf{1}) + (\mathbf{5}, \mathbf{10}) + (\bar{\mathbf{5}}, \overline{\mathbf{10}}) + (\mathbf{10}, \bar{\mathbf{5}}) + (\overline{\mathbf{10}}, \mathbf{5}) + (\mathbf{1}, \mathbf{24}),
\end{aligned} \qquad (7.59)
$$

shows that, before Wilson lines, the 4d gauge group is $SU(5)$ and there are matter multiplets in representations $\mathbf{5}, \bar{\mathbf{5}}, \mathbf{10}, \overline{\mathbf{10}}$, with multiplicities computed by suitable cohomology groups. The introduction of $\mathbf{Z}_2$ Wilson lines can break the group down to the SM $SU(3) \times SU(2) \times U(1)_Y$. There exist models leading to three quark-lepton families, plus a varying number of Higgs doublet pairs. Models based in $SU(5)$ have the generic problem that it is difficult to ensure the absence of B/L-violating dimension-4 Yukawa couplings. The most explored example in this class has the nice feature of not producing such R-parity violating couplings, but also the drawback of yielding D-quarks and leptons massless at leading order.

7.4.3 Models with U(1) bundles

In the previous discussions we have taken the internal gauge background to lie in a *simple* subgroup of the 10d gauge group. The extension to semi-simple subgroups is straightforward. A more non-trivial generalization is the inclusion of non-trivial gauge backgrounds on $U(1)$ subgroups, referred to as $U(1)$ bundles. They lead to a more general class of constructions, with novel properties; for instance, they can lead to 4d $U(1)$ gauge factors without the use of CY spaces with non-trivial $\Pi_1(\mathbf{X}_6)$ to support Wilson lines.

One relevant comment in this respect is the following. A natural suggestion to reproduce the structure of the SM is to consider gauge backgrounds on $SU(5) \times U(1)_Y$. Here the first $SU(5)$ gauge background leads to the breaking $E_8 \to SU(5)_{\text{GUT}}$, and the $U(1)_Y$ background is along the hypercharge generator in $SU(5)_{\text{GUT}}$, thus breaking the group down to $SU(3) \times SU(2) \times U(1)_Y$. However, in such a construction the hypercharge gauge boson becomes massive due to the 10d gauge Chern–Simons described in Section 4.3.4. Indeed, the 2-form kinetic term (4.84) produces upon compactification a 4d term with the structure

$$|H_3|^2 \to \left(\partial_\mu B_{mn} - \frac{\alpha'}{4} \omega^Y_{\mu mn} \right)^2 = \left(\partial_\mu B_{mn} - \frac{\alpha'}{4} A^Y_\mu F^Y_{mn} \right)^2, \qquad (7.60)$$

where the superindex Y stands for the hypercharge generator. For a non-vanishing hypercharge flux $F^Y_{ij} \neq 0$, the cross term in this expression corresponds to a Stückelberg mass term (see Section 9.5.2) for the hypercharge gauge boson. It becomes massive, by swallowing a combination of the scalars $B_{i\bar{j}}$, which lie in Kähler moduli, recall (7.32). Hence the suggestion is not phenomenologically viable in the heterotic setup. It will however be revisited in F-theory model building in Section 11.5.3.

Nevertheless, heterotic compactifications with $U(1)$ bundles display several interesting features, in particular concerning cancellation of (multiple) $U(1)$ mixed anomalies, by a 4d version of the Green–Schwarz mechanism, and the generation of $U(1)$ masses (both for anomalous and non-anomalous $U(1)$'s). However, their study is very analogous to the magnetized D-brane models in Section 11.4, as expected from heterotic-type I duality, and we skip their detailed discussion.

7.5 CY compactifications of Hořava–Witten theory

As described in Section 6.3.6 the 10d $E_8 \times E_8$ heterotic theory can be realized as Hořava–Witten theory, 11d M-theory on the interval $\mathbf{I} = \mathbf{S}^1/\mathbf{Z}_2$. In the large interval radius regime, the configuration describes the strong coupling regime of the heterotic theory, in terms of 11d supergravity on $\mathbf{M}_{10} \times \mathbf{I}$, coupled to 10d $\mathcal{N} = 1$ E_8 vector multiplets on the boundaries. In this section we explore the CY compactification of this theory.

7.5.1 Strong coupling description of heterotic CY compactification

We consider 11d supergravity on $\mathbf{M}_4 \times \mathbf{X}_6 \times \mathbf{I}$, coupled to the boundary E_8 vector multiplets propagating on $\mathbf{M}_4 \times \mathbf{X}_6$, in the presence of non-trivial gauge backgrounds in subgroups H_1, H_2 of the corresponding E_8's, see Figure 7.4; these compactifications are known as Hořava–Witten compactifications or heterotic M-theory models. Many of its properties are identical to the corresponding heterotic CY compactification, for instance the consistency condition (7.35), or the computation of the 4d massless spectrum. Therefore, any smooth CY compactification of the $E_8 \times E_8$ heterotic theory can be automatically lifted to a smooth CY compactification of Hořava–Witten theory. The system provides a particular realization

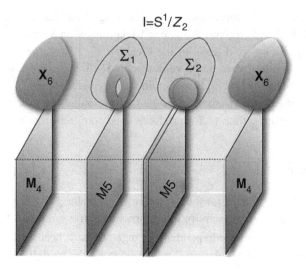

Figure 7.4 Compactification of Hořava–Witten theory on a CY manifold $\mathbf{X}_6$. The figure also depicts the possibility of introducing M5-branes wrapped on holomorphic 2-cycles Σ_i and located at points in the interval.

of the brane-world scenario introduced in Section 1.4.2. Some aspects of the effective action are discussed in Section 9.2, and some implications for gauge coupling unification are reviewed in Section 16.1.1.

There is, however, a novel feature arising in this 11d lift. From Section 6.3.6, the heterotic 2-form B_2 and its field strength H_3 are lifted to the M-theory 3-form C_3 and its field strength G_4. Since the Bianchi identity (7.34) for H_3 is related to the Green–Schwarz cancellation of anomalies from the E_8 vector multiplets, localization of the latter at the boundaries leads to localized contributions to the Bianchi identify for G_4 in the 11d picture. Namely

$$(dG_4)_{\hat{L}\hat{M}\hat{N}\hat{P}10} \simeq (Q_4)_{\hat{L}\hat{M}\hat{N}\hat{P}}\delta(x^{10}) - (Q_4')_{\hat{L}\hat{M}\hat{N}\hat{P}}\delta(x^{10} - \pi R), \qquad (7.61)$$

where the relative minus sign arises from the boundary orientations in the 11d picture, and

$$Q_4 = \operatorname{tr} F^2 - \frac{1}{2}\operatorname{tr} R^2, \qquad Q_4' = \operatorname{tr} F'^2 - \frac{1}{2}\operatorname{tr} R^2. \qquad (7.62)$$

In general it is not possible to have $\operatorname{tr} F^2 = (1/2)\operatorname{tr} R^2$ locally, due to quantization of the Chern and Pontryagin classes, see Section B.3. Consequently, the G_4 background necessarily varies along the interval. Although a detailed study of the complete supergravity background is beyond our scope, we explain an interesting effect. Since the flux G_4 gravitates, it induces a variation of the CY metric along x^{10}, which can be averaged as a variation of the Kähler moduli t_a in (7.8), sourced by the 4-forms (7.62). At leading order this is governed by a Laplace equation,

$$\partial_{x^{10}}^2 t_a \simeq \epsilon_S[\beta_a\delta(x^{10}) + \beta_a'\delta(x^{10} - \pi R)], \qquad (7.63)$$

with the expansion parameter, and coefficients

$$\epsilon_S = \left(\frac{\kappa_{11}}{4\pi}\right)^{\frac{2}{3}} \frac{2\pi\rho}{v^{\frac{2}{3}}}, \quad \beta_a = \int_{\mathbf{X}_6} Q_4 \wedge \omega_a, \quad \beta'_a = \int_{\mathbf{X}_6} Q'_4 \wedge \omega_a, \tag{7.64}$$

where ρ, v are the interval length and CY volume, and $\beta_a = -\beta'_a$ due to (7.35). The solutions are linear $t_a(y) \simeq t_a(0) + \epsilon_S\beta_a|y|$. Restricting to the overall CY size modulus t, we obtain a linear variation with slope $\epsilon_S\beta$, with $\beta = \int_{\mathbf{X}_6} Q_4 \wedge J$. This leads to a modification of the gauge kinetic functions, involving the Kähler moduli, as we describe in Section 9.2.

7.5.2 Models with fivebranes*

Given the interesting dynamics of the heterotic 2-form B_2 (or its Hořava–Witten lift) it is worthwhile to explore a further generalization of its Bianchi identity. This corresponds to the introduction of NS5-branes, described in Section 6.2 as non-perturbative states with $(5 + 1)$ extended dimensions, coupling magnetically to B_2. They can be introduced in heterotic CY compactifications, preserving 4d Poincaré invariance, by letting them span $\mathbf{M}_4 \times \Sigma$, where Σ is a non-trivial homology 2-cycle in $\mathbf{X}_6$, which must be holomorphic for the NS5-brane to preserve supersymmetry. In a configuration with N_a NS5-branes on 2-cycles Σ_a, (7.34) is modified to

$$dH_3 \simeq \frac{1}{8\pi^2}(\mathrm{tr}\, R^2 - \mathrm{tr}\, F^2 - \mathrm{tr}\, F'^2) - \sum_a N_a\delta_4(\Sigma_a), \tag{7.65}$$

(and similarly for the $SO(32)$ heterotic case). Here $\delta_4(\Sigma_a)$ is the Poincaré dual of the 2-cycle Σ_a (see Section B.1), namely a delta 4-form supported on Σ_a and with legs in the CY directions transverse to it. Incidentally, note that a NS5-brane on a 2-cycle contributes as a gauge instanton background (on the dual 4-cycle), see Section 13.1. In fact a NS5-brane can be regarded as a gauge instanton in the zero size limit, sometimes denoted "small instanton," which can in fact continuously turn into a semiclassical gauge instanton background.

Since NS5-branes are non-perturbative, and the effective string coupling diverges at their core, it is more natural to describe these configurations in the strong coupling dual theory. For the $SO(32)$ heterotic, this corresponds to type I compactifications with D5-branes, thus closely related to Chapter 11. For the $E_8 \times E_8$ heterotic, this corresponds to compactification of Hořava–Witten theory, with M5-branes wrapped on the CY 2-cycles Σ_a, and localized at points in the M-theory interval. The M5-branes need not be located on the interval boundaries, rather can be positioned at arbitrary points y_a (with image M5-branes at $-y_a$ as required by the $\mathbf{Z}_2$ symmetry). Lack of a proper microscopic description of M5-branes has relegated their role to just an extra ingredient to satisfy the consistency condition of the compactification. From (7.65) we can derive the generalization of (7.35)

$$[\mathrm{tr}\, F^2] + [\mathrm{tr}\, F'^2] - [\mathrm{tr}\, R^2] + \sum_a N_a[\Sigma_a] = 0, \tag{7.66}$$

where the factor of $8\pi^2$ has been rescaled into the gauge and gravitational curvature terms, making them integer-valued cohomology classes, see Section B.3. The presence of the M5-branes backreacts on the CY metric leading to a piecewise linear evolution of the CY Kähler moduli along x^{10}, with slope changing in crossing the M5-brane locations, see the references for details.

8

Heterotic string orbifolds and other exact
CFT constructions

In Chapter 7 we have studied the construction of 4d $\mathcal{N} = 1$ heterotic string vacua by compactification on smooth CY manifolds. The analysis of such compactifications is however limited because their worldsheet theory is not exactly solvable. Therefore, we must rely on the Kaluza–Klein reduction of the 10d supergravity action, which provides a good approximation to the 4d physics only in the large volume regime. In this chapter we study 4d $\mathcal{N} = 1$ heterotic string vacua obtained from toroidal orbifold compactifications and other exact CFT constructions, which overcome this limitation. These are simple α'-exact compactifications, which share many properties with more general CY compactifications (and in fact are often closely related to them), and which allow a very explicit construction of phenomenologically interesting particle physics models. We describe toroidal orbifolds, which provide a free CFT description of geometries which can be regarded as singular limits of CY spaces. We also introduce asymmetric orbifolds and free fermionic models, which are also described by free worldsheet CFTs but in general do not admit a geometric interpretation. Finally we describe Gepner models (and orbifolds thereof), defined by interacting but solvable CFTs and which can often be regarded as compactifications on CY spaces of stringy size. We focus on the $E_8 \times E_8$ heterotic string theory, although the basic rules apply in complete analogy to the $SO(32)$ theory.

8.1 Toroidal orbifolds

Let us start by describing the geometry of toroidal orbifolds. The main ingredients are useful for heterotic string compactifications, as well as for other string theories, like type II orientifolds. Toroidal orbifolds are defined by quotients of tori by a discrete group of symmetries. The resulting geometries are locally flat, except at a set of singular points, which can be regarded as concentrating the curvature of the compact space. These points are singular from the geometrical viewpoint, but are well behaved in string theory, as described in coming sections.

Let us start with a simple example, with just two real dimensions. Define a 2-torus as the quotient $\mathbf{T}^2 = \mathbf{R}^2 / \Lambda_2$, where Λ_2 is a lattice generated by two vectors $\{\vec{e}_1, \vec{e}_2\}$. Namely, points in $\mathbf{R}^2$ differing by a translation in Λ_2 are considered equivalent

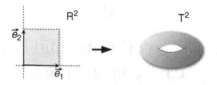

Figure 8.1 The 2-torus geometry as a quotient of $\mathbf{R}^2$ by translations in a two-dimensional lattice. The identification of opposite sides in the unit cell (shaded) yields a $\mathbf{T}^2$ topology.

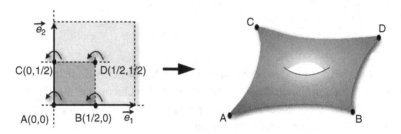

Figure 8.2 Geometry of a $\mathbf{T}^2/\mathbf{Z}_2$ orbifold. The orbifold action on the $\mathbf{T}^2$ acts as a rotation by π, and has four fixed points. Any point in the torus (light shaded rectangle) can be mapped inside the orbifold unit cell (dark shaded). The picture on the right shows an artistic view of the resulting space.

$$\vec{x} \equiv \vec{x} + n_1 \vec{e}_1 + n_2 \vec{e}_2, \quad n_1, n_2 \in \mathbf{Z}. \tag{8.1}$$

The resulting space has the topology of a 2-torus, see Figure 8.1.

Let us now construct a $\mathbf{Z}_2$ orbifold of this 2-torus. Introduce the operation θ defined as a reflection with respect to the origin, $\theta(\vec{x}) = -\vec{x}$, and denote by P the $\mathbf{Z}_2$ symmetry group $\{1, \theta\}$. The orbifold $\mathbf{T}^2/P$ is defined by identification of points of $\mathbf{T}^2$ related by the action θ. The resulting geometry is flat except at points fixed under the orbifold action; there are four points (A, B, C, D) in $\mathbf{T}^2$ fixed under θ, equivalently points in $\mathbf{R}^2$ fixed under θ modulo translations in Λ_2

$$\vec{x}_f = \theta \vec{x}_f + \sum_a m_a^f \vec{e}_a \quad \text{for some } m_a^f \in \mathbf{Z}. \tag{8.2}$$

The four fixed points $\vec{x}_f$ have coordinates $(0, 0)$, $(\frac{1}{2}, 0)$, $(0, \frac{1}{2})$ or $(\frac{1}{2}, \frac{1}{2})$ in the basis $\{\vec{e}_1, \vec{e}_2\}$. The point $(0, 0)$ is fixed already in $\mathbf{R}^2$, while the other three are fixed in $\mathbf{T}^2$ but not in $\mathbf{R}^2$ (as they require non-trivial translations, i.e. some $m_a^f \neq 0$). Points in $\mathbf{T}^2$ fixed under an orbifold action become conical singularities in the quotient space, as shown in Figure 8.2. The toroidal orbifold can thus be defined as a quotient of $\mathbf{R}^2$ by the so-called "space group" S, whose elements $(\theta^k; \vec{u})$ are given by rotations in the orbifold group P (called the "point group") and translations in Λ_2. Although possible, we will not consider here cases in which the orbifold rotation is accompanied by a translation in $\mathbf{R}^2$.

These ideas easily generalize to orbifolds of $\mathbf{T}^6$, which are suitable for string compactifications. Such orbifolds can be defined as quotients $\mathbf{R}^6/S$, where S is the orbifold space group, generated by translations in a six-dimensional lattice Λ_6, and rotations in a symmetry group P of Λ_6; equivalently, as a quotient $\mathbf{T}^6/P$, where the $\mathbf{T}^6$ is a quotient $\mathbf{R}^6/\Lambda_6$. For the identification of points under the orbifold twist to be well defined, the point group P must act crystallographically, namely its elements θ must be symmetries of the torus lattice, i.e. $\theta \vec{e}_i \in \Lambda_6$. A simple class of orbifolds have Λ_6 given by a rank-6 Lie algebra root lattice, and P generated by elements of its Weyl group. For simplicity, we restrict our discussion to abelian orbifold groups (defining abelian toroidal orbifolds), see later for explicit examples.

It is convenient to define complex coordinates for the six extra dimensions as $z^1 = \frac{1}{\sqrt{2}}(x^1 + ix^2)$, and similarly for z^2, z^3. The generator θ of the $\mathbf{Z}_N$ orbifold group can be written as an $SO(6)$ rotation

$$\theta = \exp[\,2\pi i\,(v_1 J_{12} + v_2 J_{34} + v_3 J_{45})\,], \tag{8.3}$$

where $J_{i,i+1}$ are the SO(6) Cartan generators, acting as rotations on the ith complex plane. The above action, in terms of the complex bosonic coordinates, reads

$$z^i \to e^{2\pi i v_i} z^i. \tag{8.4}$$

The rotation must be of order N, hence $N v_i \in \mathbf{Z}$. Also, for θ^N to act trivially on spinors, we must require $\sum_i N v_i$ = even. We fix our convention for twist vectors (v_1, v_2, v_3) such that $-1 < v_i < 1$.

In general, non-supersymmetric orbifolds lead to tachyons in the spectrum, localized at the singular points. In order to avoid such instabilities, and also for phenomenological particle physics applications, we focus on compactifications preserving 4d $\mathcal{N} = 1$ supersymmetry. Using our intuitions from CY compactifications, recall Section 7.2.1, this requires the holonomy group of the internal space to be in $SU(3)$. Since the underlying $\mathbf{T}^6$ is flat, the holonomy group is precisely the point group $\mathbf{Z}_N$; the supersymmetry condition is thus that the generator (8.3) is in $SU(3)$ i.e. $\pm v_1 \pm v_2 \pm v_3 = 0$ mod 1, for some choice of signs. The same condition can be recovered by explicit computation of the string spectrum, in coming sections. Without loss of generality we take the twist vectors to satisfy the condition with all positive signs

$$v_1 + v_2 + v_3 = 0. \tag{8.5}$$

Abelian subgroups of $SU(3)$ are of the form $\mathbf{Z}_N$ or $\mathbf{Z}_N \times \mathbf{Z}_M$. Groups $\mathbf{Z}_N$ are generated by an order N twist θ, so its elements are θ^k with $k = 0, \ldots, N - 1$. The groups $\mathbf{Z}_N \times \mathbf{Z}_M$ have two generators θ, ω, of orders N, M, with M a multiple of N; the elements have the structure $\theta^k \omega^l$, with $k = 0, \ldots, N - 1, l = 0, \ldots, M - 1$. Products of more $\mathbf{Z}_N$'s are not contained in $SU(3)$ (or can be brought to the earlier forms). The structure of abelian orbifold twists is very constrained, and there is a finite set of twists, given in Table 8.1, compatible with $\mathcal{N} = 1$ SUSY, and with crystallographic symmetries of 6-tori. Note that the order of twists preserving precisely 4d $\mathcal{N} = 1$ is constrained to six possible values $N = 3, 4, 6, 7, 8, 12$. For $N = 2, 3, 4, 6$ there are twists which lie in $SU(2)$ and thus preserve an enhanced 4d $\mathcal{N} = 2$. However, they are interesting since they can appear as a subset of larger orbifold groups preserving only $\mathcal{N} = 1$. Note also that twists of the

Table 8.1 *Possible abelian orbifold actions preserving 4d $\mathcal{N} = 1$ or $\mathcal{N} = 2$ supersymmetry and acting crystallographically. We denote them by the orbifold group they generate, and use a prime to distinguish groups of the same order. Note, however, that some twists can appear as elements in a larger orbifold group*

$\mathcal{N} = 2$	$\mathbf{Z}_2$	$\frac{1}{2}(1,-1,0)$	$\mathbf{Z}_3$	$\frac{1}{3}(1,-1,0)$	$\mathbf{Z}_4$	$\frac{1}{4}(1,-1,0)$	$\mathbf{Z}_6$	$\frac{1}{6}(1,-1,0)$
$\mathcal{N} = 1$	$\mathbf{Z}_3$	$\frac{1}{3}(1,1,-2)$	$\mathbf{Z}_4$	$\frac{1}{4}(1,1,-2)$	$\mathbf{Z}_6$	$\frac{1}{6}(1,1,-2)$	$\mathbf{Z}_6'$	$\frac{1}{6}(1,2,-3)$
	$\mathbf{Z}_7$	$\frac{1}{7}(1,2,-3)$	$\mathbf{Z}_8$	$\frac{1}{8}(1,2,-3)$	$\mathbf{Z}_8'$	$\frac{1}{8}(1,3,-4)$		
	$\mathbf{Z}_{12}$	$\frac{1}{12}(1,4,-5)$	$\mathbf{Z}_{12}'$	$\frac{1}{12}(1,5,-6)$				

same order can have different geometric action on the $\mathbf{T}^6$ complex coordinates, as happens for $N = 6, 8, 12$. Finally, even for a given set of v_a's, the orbifold may be compatible with different $\mathbf{T}^6$ geometries; for instance, the $\mathbf{Z}_3$ orbifold action is compatible with different 6-tori, e.g. defined by the root lattice of $SU(3)^3$ or that of E_6. Some properties of orbifolds are independent of the underlying torus lattice, and depend only on the orbifold twist eigenvalues. For instance, the Hodge numbers are independent of the lattice choice as long as it defines a factorizable $\mathbf{T}^6 = \mathbf{T}^2 \times \mathbf{T}^2 \times \mathbf{T}^2$. Other properties instead depend on the whole space group structure, for instance those related to the presence of Wilson lines, see Section 8.3, or the Hodge numbers of orbifolds with non-factorized 6-tori.

8.2 Heterotic compactification on toroidal orbifolds

We are now ready to describe the construction of orbifold compactifications of the heterotic string. However, many ideas are also useful in the construction of orbifold compactifications in other string theories, like type II orientifolds. We use a Lorentz index notation adapted to compactification, analogous to that used in Chapter 7, with indices $m, n, \ldots = 4, \ldots, 9$ and $i, j, \ldots = 1, \ldots, 3$ for real and complex compact dimensions respectively, and use $\mu = 2, 3$ for non-compact coordinates transverse to the light-cone (or sometimes abusing notation, $\mu = 0, \ldots, 3$).

8.2.1 Untwisted and twisted sectors

Recall from Section 4.3 the worldsheet degrees of freedom of the heterotic string in the light-cone gauge, in the bosonic formulation; there are right-moving bosons and fermions $X_R^\mu, X_R^m, \psi_R^\mu, \psi_R^m$, with $\mu = 2, 3$, $m = 4, \ldots 9$, and left-moving bosons X_L^μ, X_L^m, X_L^I, with $I = 1, \ldots, 16$. It is convenient to combine the coordinates $X^m(t, \sigma)$ into three complex bosonic coordinates $Z^i(t, \sigma)$ (and conjugates), and similarly right-moving fermions $\psi_R^m(t - \sigma)$ are complexified into $\Psi_R^i(t - \sigma)$. In compactifications on a toroidal orbifold

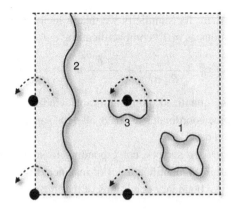

Figure 8.3 Three types of closed strings in a $\mathbf{Z}_2$ orbifold: string **1** is closed already in $\mathbf{R}^2$, string **2** is a winding mode, closed in $\mathbf{T}^2$ but not on $\mathbf{R}^2$; both are in the untwisted sector of the orbifold. Finally, string **3** is not closed in either $\mathbf{R}^2$ or $\mathbf{T}^2$, but is closed in the orbifold quotient, and belongs to the twisted sector. Note that twisted strings must be centered around the orbifold fixed points.

$\mathbf{T}^6/P$, the orbifold group P acts on the Z^i, Ψ_R^i as inherited from the geometric action (8.4). In addition, we may specify a non-trivial action on the internal coordinates X^I, namely an embedding of the orbifold action in the gauge degrees of freedom. Denoting the group of such actions by G, we have the structure

$$\frac{\mathbf{T}_{R+L}^6}{P} \otimes \frac{\mathbf{T}_L^{16,\text{Het}}}{G}, \tag{8.6}$$

where $\mathbf{T}_L^{16,\text{Het}}$ denotes the internal 16d "torus." A large class of such orbifold actions on the X^I can be represented by letting the $\mathbf{Z}_N$ generator θ act as a shift

$$\theta : X^I \rightarrow X^I + V^I. \tag{8.7}$$

For this gauge action to represent the $\mathbf{Z}_N$ orbifold action, we must have

$$N V \in \Lambda_{16d}, \tag{8.8}$$

where Λ_{16d} is the $E_8 \times E_8$ or $Spin(32)/\mathbb{Z}_2$ lattice, recall Section 4.3.2. The shift action can be regarded as a non-trivial gauge transformation associated to the twist, felt locally around the orbifold fixed point, and morally due to a gauge field strength background F_{mn}^a hidden at the singular points.

The spectrum of closed strings in the resulting orbifold falls in two sectors, see Figure 8.3. The *untwisted* sector corresponds to strings which are closed already on the underlying torus, before the orbifold identification. Thus they behave as strings on a $\mathbf{T}^6$ compactification, which can be studied by simple generalization of Section 5.2.1. The $\mathbf{T}^6$ simply introduces the possibility of winding for worldsheet bosons,

$$X^m(t, \sigma + \ell) = X^m(t, \sigma) + 2\pi R_m w^m, \quad \text{no sum}, \tag{8.9}$$

with no sum in m, and where for simplicity we take a factorized torus. These untwisted strings have mode expansions as in $\mathbf{T}^6$ compactification, e.g. for bosons

$$X^m(t, \sigma) = x_0^m + \frac{k_m}{R_m p^+} t + \frac{2\pi R_m}{\ell} w^m \sigma + \text{(oscillators)}, \qquad (8.10)$$

where p^m and w^m are the quantized momentum and winding vectors, respectively. The left-handed internal bosonic coordinates X^I have also the same boundary conditions as in the toroidally compactified case.

There are in addition *twisted* sectors, corresponding to strings which are closed only taking into account the orbifold identification. For instance, in a θ-twisted sector, internal 6d bosonic coordinates have boundary conditions of the form

$$X(t, \sigma + \ell) = \theta X(t, \sigma) + 2\pi R w. \qquad (8.11)$$

Using the complexified bosonic and fermionic fields along $\mathbf{T}^6$, and recalling the geometric and internal orbifold actions (8.4), (8.7), the boundary conditions in a θ-twisted sector are

$$Z^i(t, \sigma + \ell) = e^{2\pi i v_i} Z^i(t, \sigma) + 2\pi R \mathbf{w}^i,$$

$$\Psi_R^i(t - \sigma - \ell) = \pm e^{2\pi i v_i} \Psi_R^i(t - \sigma),$$

$$X^I(t + \sigma + \ell) = X^I(t + \sigma) + V^I, \qquad (8.12)$$

where $\mathbf{w}^i$ represents a complexified winding vector, and the $-/+$ sign for fermions corresponds to NS/R boundary conditions. The boundary conditions for Z^i are incompatible with momentum and winding contributions, and give a mode expansion

$$Z^i(t, \sigma) = Z_f^i + i\sqrt{\frac{\alpha'}{2}} \left[\sum_{n \neq 0} \frac{\alpha_{n-v_i}^i}{n - v_i} e^{-2\pi i (n-v_i)(t+\sigma)/\ell} + \frac{\tilde{\alpha}_{n+v_i}^i}{n + v_i} e^{-2\pi i (n+v_i)(t-\sigma)/\ell} \right].$$

$$(8.13)$$

The twisted boundary conditions fix the center of mass at the fixed points, defined by (8.2), whose complex coordinates we have denoted Z_f^i, leading to an extra multiplicity for each orbifold twist. Twisted sectors are then labeled by the element $g \in P$, and by the corresponding fixed point; in other words, sectors in an orbifold are in one-to-one correspondence with the elements of the space group. The orbifold phase rotation also induces a shift in the modding of the oscillators $\alpha_{n-v_i}^i$, $\tilde{\alpha}_{n+v_i}^i$, and similarly for the right-moving fermionic oscillators $\Psi_{n-v_i}^i$, $\tilde{\Psi}_{n+v_i}^i$. In addition, the conjugate fields $\bar{Z}^i$, $\bar{\Psi}_R^i$, lead to oscillators $\bar{\alpha}_{n+v_i}^i$, $\bar{\tilde{\alpha}}_{n-v_i}^i$, $\bar{\Psi}_{n+v_i}^i$, $\bar{\tilde{\Psi}}_{n-v_i}^i$.

The mode expansion for the internal bosonic coordinates has the structure

$$X^I(t + \sigma) = \frac{(P^I + V^I)}{\sqrt{2\alpha'} p^+} (t + \sigma) + i\sqrt{\frac{\alpha'}{2}} \sum_{n \neq 0} \frac{\alpha_n^I}{n} e^{-2\pi i n (t+\sigma)/\ell}. \qquad (8.14)$$

We now proceed to the construction of the untwisted and twisted massless spectrum.

8.2.2 The untwisted massless spectrum

In the untwisted sector, the mass formulae are as in a $\mathbf{T}^6$ compactification. By generalizing (5.16), we have

$$\frac{\alpha' M_R^2}{2} = \frac{p_R^2}{2} + \tilde{N}_B + \tilde{N}_{F_B} + \frac{r^2}{2} - \frac{1}{2},$$

$$\frac{\alpha' M_L^2}{2} = \frac{p_L^2}{2} + N_B + \frac{P^2}{2} - 1, \tag{8.15}$$

where right-moving fermions have been bosonized as in Section 4.2.6. Specifically, $r = (r_1, r_2, r_3, r_0)$ is a weight of the $SO(8)$ Lorentz symmetry manifest in the light-cone gauge, satisfying the GSO projection $\sum_{i=0}^{3} r^i = $ odd, equivalently $r \in \Lambda^+$ in (4.55). Also N_{F_B} is the number operator for the bosonized fermions, and is quantized in *integers*. Massless right-moving states have $\tilde{N}_B = \tilde{N}_{F_B} = 0$, $p_R = 0$; in the NS sector they are of the form $r = (\pm 1, 0, 0, 0)$, and in the R sector they are $r = \pm(\frac{1}{2}, \frac{1}{2}, \frac{1}{2}, -\frac{1}{2})$, where underlining denotes possible permutations of entries. These are just the $SO(8)$ weights of the $\mathbf{8}_V$ and $\mathbf{8}_C$ representations associated to massless states. In our convention the last entry encodes the 4d spacetime Lorentz quantum numbers. The level-matching condition is $M_R^2 = M_L^2$, hence these massless right-moving states must combine with massless left-moving states. These arise from states with $p_L = 0$, $N_B = 1$, $P = 0$, or with $p_L = 0$, $N_B = 0$, $P^2 = 2$. Since the actual space is a quotient of $\mathbf{T}^6$, the actual spectrum is given by states invariant under the orbifold action. At the massless level, we have the following 4d fields:

- The 4d graviton $g_{\mu\nu}$, a two-index antisymmetric tensor $B_{\mu\nu}$, and the dilaton ϕ, plus fermionic partners, from states

$$|(0, 0, 0, \pm 1)\rangle_R \otimes \alpha_{-1}^{\mu} |0\rangle_L, \quad |\pm(\tfrac{1}{2}, \tfrac{1}{2}, \tfrac{1}{2}, -\tfrac{1}{2})\rangle_R \otimes \alpha_{-1}^{\mu} |0\rangle_L. \tag{8.16}$$

As described in detail in Section 9.1.1, the 4d two-index antisymmetric tensor is equivalent to a massless pseudoscalar, which combines with the dilaton into a complex scalar denoted S.

- There are 4d $\mathcal{N} = 1$ vector multiplets, i.e. gauge bosons and gauginos, from

$$|(0, 0, 0, \pm 1)\rangle_R \otimes \alpha_{-1}^{I} |0\rangle_L, \quad |\pm(\tfrac{1}{2}, \tfrac{1}{2}, \tfrac{1}{2}, -\tfrac{1}{2})\rangle_R \otimes \alpha_{-1}^{I} |0\rangle_L,$$

$$|(0, 0, 0, \pm 1)\rangle_R \otimes |P^I\rangle_{P^2=2}, \quad |\pm(\tfrac{1}{2}, \tfrac{1}{2}, \tfrac{1}{2}, -\tfrac{1}{2})\rangle_R \otimes |P^I\rangle_{P^2=2}. \tag{8.17}$$

The former define the Cartan subalgebra of the 4d gauge group, while the latter provide the non-zero roots. Since these are charged, they pick up phases under the gauge shift V, and the orbifold projection onto G-invariant states requires $\exp(2\pi i\, P \cdot V) = 1$; the 4d gauge group is thus smaller than the original 10d one.

- There are 4d complex scalar fields, which can be interpreted as zero modes of the metric and 2-form field internal components, g_{mn}, B_{mn}, where m, n run over all $\mathbf{T}^6$ coordinates.

$$|(\pm 1, 0, 0, 0)\rangle_R \otimes \alpha_{-1}^{m} |0\rangle_L, \quad |\pm(-\tfrac{1}{2}, \tfrac{1}{2}, \tfrac{1}{2}, \tfrac{1}{2})\rangle_R \otimes \alpha_{-1}^{m} |0\rangle_L. \tag{8.18}$$

Using complex coordinates, the generic orbifold invariant states correspond to fields $g_{i\bar{j}}$, $B_{i\bar{j}}$, with $\exp[2\pi i(v_i - v_j)] = 1$ (and their fermionic partners), which combine into Kähler moduli chiral multiplets $T_{i\bar{j}}$. In certain orbifolds, e.g. even order $\mathbf{Z}_N$ models, there can also be some surviving modes g_{ij}, $g_{\bar{i}\bar{j}}$, etc., which give rise to complex structure moduli fields U_{ij}.

- Finally there are charged matter chiral multiplets C_j^P, $j = 1, 2, 3$, from states

$$| (\underline{\pm 1, 0, 0}, 0) \rangle_R \otimes |P^I\rangle_{P^2=2}, \quad |\pm \left(-\tfrac{1}{2}, \underline{\tfrac{1}{2}, \tfrac{1}{2}, \tfrac{1}{2}}\right) \rangle_R \otimes |P^I\rangle_{P^2=2}, \qquad (8.19)$$

where j corresponds to the location of the different entry among the underlined ones. The projection onto states invariant under $P \times G$ amounts to the condition $\exp 2\pi i$ $(r \cdot v - P \cdot V) = 1$, where we have defined the enlarged twist vector $v = (v_1, v_2, v_3, 0)$ to also include the 4d spacetime.

Note that the above states fill out multiplets of 4d $\mathcal{N} = 1$ supersymmetry; the reader may easily check that this follows from the $SU(3)$ condition $v_1 + v_2 + v_3 = 0$.

The massless untwisted spectrum is thus a simple truncation of a purely toroidal $\mathbf{T}^6$ compactification onto states invariant under the discrete symmetry $P \times G$. The resulting massless spectrum is in general chiral, and anomalous if considered by itself. The complete massless spectrum including twisted sector states is, however, anomaly-free. Since the latter sectors are in fact required by modular invariance of the theory (see Section 8.2.5), this is a particular instance of a general link between modular invariance and anomaly cancellation, whose discussion in full generality is, however, beyond our scope.

8.2.3 The twisted massless spectrum

Strings in twisted sectors also give rise to massless states, localized around orbifold fixed points; the number of fixed points for a given twist provides an overall multiplicity for the corresponding twisted sector states. This is modified in the presence of Wilson lines, which make different fixed points inequivalent; the present analysis however extends to this case with only minor changes, see Section 8.3.2.

Consider a θ^n-twisted sector in a $\mathbf{Z}_N$ orbifold. Using the mode expansions (8.13) (8.14), the mass formulae in bosonized language can be shown to be

$$\frac{\alpha' M_R^2}{2} = \tilde{N}_B + \tilde{N}_{F_B} + \frac{(r + nv)^2}{2} + E_0 - \frac{1}{2},$$

$$\frac{\alpha' M_L^2}{2} = N_B + \frac{(P + nV)^2}{2} + E_0 - 1, \qquad (8.20)$$

where $r \in \Lambda^+$ in (4.55). As already mentioned, there are no momentum or winding contributions. Note also that the fractional modding of the twisted oscillators shifts the normal ordering zero point energy. From (4.9) one gets

$$E_0 = \sum_{i=1}^{3} \tfrac{1}{2} |nv_i| \, (1 - |nv_i|), \tag{8.21}$$

for $-1 < nv_i < 1$, which can always be achieved by using shifts in the $SO(8)$ root lattice if necessary (simultaneous 2π rotations in two complex planes). Twisted sector massless states are explicitly constructed in the examples in upcoming sections.

Twisted sector states must also be projected requiring invariance under the orbifold action $P \times G$. The precise conditions turn out to be slightly trickier than in the untwisted sector, but can be derived from modular invariance, see Section 8.2.5. A second consequence of modular invariance is a consistency condition on the shift vector V, which is already manifest at this stage; indeed, using (8.20), the level matching condition $M_R^2 = M_L^2$ is possible only if

$$N \, (V^2 - v^2) = 0 \quad \mathrm{mod}\ 2. \tag{8.22}$$

A simple check is to focus, e.g., on the sector of (massive) states with $r^2 = 1$, $r \cdot v = 0$, $P^2 = 2$, $P \cdot V = 0$; the condition $M_R^2 = M_L^2$ then requires the quantity $(V^2 - v^2)/2$ to be a multiple of $1/N$, so that it can be balanced to achieve left–right mass equality.

A simple solution of (8.22), in both the $E_8 \times E_8$ and $SO(32)$ heterotics, is

$$V = (v_1, v_2, v_3, 0, \ldots, 0; 0, \ldots, 0), \tag{8.23}$$

where the 16 entries have been split in two sets of 8, for possible application to the $E_8 \times E_8$ heterotic. The above gauge shift satisfies (8.8), as follows from the conditions below (8.4). The choice (8.23) corresponds to identifying the orbifold action on the gauge degrees of freedom with the geometric action, and provides the orbifold version of the standard embedding $F = R$ of heterotic CY compactifications in Section 7.3. For non-standard embeddings, (8.22) turns out to be a very strong constraint on possible gauge embeddings of a given $\mathbf{Z}_N$ geometric orbifold action.

8.2.4 The $\mathbf{Z}_3$ orbifold of $E_8 \times E_8$ heterotic with standard embedding

We now flesh out the above definitions with the example of the $\mathbf{Z}_3$ orbifold with standard embedding, which has a particularly simple structure. Specifically all massless twisted sector states survive the orbifold projection, so we need not deal with the general projection conditions, which we postpone until Section 8.2.5. It is possible to carry out the construction for both the $E_8 \times E_8$ and $SO(32)$ heterotic theories. We focus on the former case, which is illustrative enough, and moreover provides a good starting point to construct interesting particle physics models, see Section 8.3.3.

We take the underlying 6-torus to be $\mathbf{T}^6 = \mathbf{R}^6 / \Lambda_{SU(3)^3}$, where $\Lambda_{SU(3)^3}$ is the root lattice of the $SU(3)^3$ algebra. The point group $P = \mathbf{Z}_3$ has a generator θ acting as a simultaneous rotation of the three 2-planes by $2\pi/3$, see Figure 8.4. The rotation of the complex coordinates Z^i by $\alpha = \exp(2\pi i/3)$ corresponds to $v = (1, 1, -2)/3$, recalling the constraints below (8.4), and also (8.5). The action θ leaves three fixed points on each $\mathbf{T}^2$, denoted

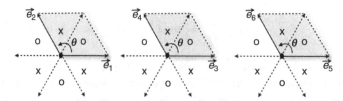

Figure 8.4 Geometry of the $\mathbf{T}^6/\mathbf{Z}_3$ orbifold. The vectors $\vec{e}_m$ define the $SU(3)^3$ root lattice of the underlying $\mathbf{T}^6$. The orbifold generator θ rotates all complex planes simultaneously by $2\pi/3$, and leaves 27 fixed points. The three fixed coordinates on each $\mathbf{T}^2$ are indicated as a cross ($\times$), a black circle ($\bullet$), and an empty circle ($\circ$).

with crosses, and black and empty circles in Figure 8.4. Each such point is actually fixed up to a specific shift in the $SU(3)^3$ lattice; for instance the point ($\bullet$, $\times$, $\circ$) needs a shift by $\vec{e}_3 + \vec{e}_4 + \vec{e}_5$ to regain its initial position after a θ rotation. Its associated space group element is thus $(\theta, \vec{e}_3 + \vec{e}_4 + \vec{e}_5)$.

For the action G of the orbifold on the gauge degrees of freedom, we choose the standard embedding (8.23), namely $V = \left(\frac{1}{3}, \frac{1}{3}, -\frac{2}{3}, 0, \ldots, 0\right)$. Notice that $3V$ belongs to the E_8 lattice, as required by (8.8).

We now turn to the computation of the untwisted and twisted massless spectrum. In the untwisted sector, besides the gravity and dilaton multiplets, there are nine Kähler moduli $T_{i\bar{j}}$ of the form (8.18). There are also 4d gauge bosons and gauginos (8.17), as follows. The 16 Cartan generators are unaffected by the orbifold action and survive in the spectrum. For states leading to non-zero roots we need to impose the orbifold projection $P \cdot V \in \mathbf{Z}$, which is satisfied by the following sets of internal momenta P. A first set is given by

$$(0, 0, 0, \pm 1, \pm 1, 0, 0, 0) \qquad \longrightarrow \qquad SO(10),$$

$$\pm\left(-\frac{1}{2}, -\frac{1}{2}, -\frac{1}{2}, \pm\frac{1}{2}, \pm\frac{1}{2}, \ldots, \pm\frac{1}{2}\right) \qquad \longrightarrow \qquad \mathbf{16} + \overline{\mathbf{16}}, \qquad (8.24)$$

(with the second line states containing an even number of minus signs). They complete the non-zero roots of E_6. The additional states

$$(1, -1, 0, 0, 0, 0, 0, 0), \qquad (8.25)$$

are the non-zero roots of $SU(3)$ in the embedding $E_6 \times SU(3) \subset E_8$. Finally, there are states corresponding to the second E_8 factor, which remains unbroken. The complete gauge group is

$$E_6 \times SU(3) \times E_8'. \qquad (8.26)$$

Note that the $E_6 \times E_8'$ structure agrees with the gauge group unbroken by the CY compactifications with standard embedding in Section 7.3.2.

The untwisted sector also contains chiral matter fields (8.19). The three right-moving states with eigenvalue α^2 under θ can form invariant states by combining with left-moving states transforming as α, namely $P \cdot V = 1/3$ mod $\mathbf{Z}$. This is the case for the following sets of states

$$\left(-\tfrac{1}{2}, -\tfrac{1}{2}, \tfrac{1}{2}, \pm\tfrac{1}{2}, \ldots, \pm\tfrac{1}{2}\right) \quad \longrightarrow \quad (\mathbf{3}, \mathbf{16}),$$

$$(1, 0, 0, \pm 1, 0, 0, 0, 0) \quad \longrightarrow \quad (\mathbf{3}, \mathbf{10}),$$

$$(-1, -1, 0, 0, 0, 0, 0, 0) \quad \longrightarrow \quad (\mathbf{3}, \mathbf{1}). \tag{8.27}$$

States in the first line exist for an even total number of minus signs. They correspond to weight vectors of a state transforming in a $\mathbf{3}$ of $SU(3)$ and a $\mathbf{16}$ of the $SO(10)$ subgroup of E_6. States in the second and third lines transform in a $(\mathbf{3}, \mathbf{10})$ and $(\mathbf{3}, \mathbf{1})$, respectively; the triplet structure can be checked by noting that states are related by shifts of non-zero roots (8.25) of $SU(3)$. These states combine into a $(\mathbf{3}, \mathbf{27})$ of the 4d gauge group $SU(3) \times E_6$. There are in fact three such chiral multiplets, from the three possible choices of right-moving state. The antiparticles of these chiral multiplets arise from right-moving states transforming with α under θ, combined with left-moving states with $P \cdot V = -1/3$ mod $\mathbf{Z}$, which are given by the above vectors with opposite sign. Altogether the untwisted sector of the $\mathbf{Z}_3$ orbifold gives rise to nine E_6 families, transforming as $3(\mathbf{3}, \mathbf{27})$ under $SU(3) \times E_6$.

Let us move on to the massless spectrum from the twisted sectors. In the $\mathbf{Z}_3$ orbifold, the θ^2-twisted sector leads to the antiparticles of states in the θ-twisted sector, hence we can just focus on the latter. There are 27 fixed points, which in the present case are all equivalent and lead to a 27-fold degeneracy of the spectrum. Using (8.20) and the zero point energy $E_0 = 1/3$ from (8.21), the massless twisted strings at any of these points must satisfy

$$\frac{\alpha' M_R^2}{2} = \tilde{N}_B + \tilde{N}_{F_B} + \frac{(r+v)^2}{2} - \frac{1}{6} = 0,$$

$$\frac{\alpha' M_L^2}{2} = N_B + \frac{(P+V)^2}{2} - \frac{2}{3} = 0. \tag{8.28}$$

For $\tilde{N}_B = \tilde{N}_{F_B} = 0$ there are right-moving massless NS states with $r + v = \left(\tfrac{1}{3}, \tfrac{1}{3}, \tfrac{1}{3}, 0\right)$ and R states with $r + v = \left(-\tfrac{1}{6}, -\tfrac{1}{6}, -\tfrac{1}{6}, -\tfrac{1}{2}\right)$. In the left-moving sector there are 27 massless states with $N_B = 0$ and gauge vector $P + V$ given by

$$\left(-\tfrac{1}{6}, -\tfrac{1}{6}, -\tfrac{1}{6}, \pm\tfrac{1}{2}, \ldots, \pm\tfrac{1}{2}\right) \rightarrow (\mathbf{1}, \mathbf{16}),$$

$$\left(\tfrac{1}{3}, \tfrac{1}{3}, \tfrac{1}{3}, \pm 1, 0, 0, 0, 0\right) \rightarrow (\mathbf{1}, \mathbf{10}),$$

$$\left(-\tfrac{2}{3}, -\tfrac{2}{3}, -\tfrac{2}{3}, 0, 0, 0, 0, 0, 0\right) \rightarrow (\mathbf{1}, \mathbf{1}). \tag{8.29}$$

Combining the left and right pieces, we obtain a chiral multiplet in the $(\mathbf{1}, \mathbf{27})$ of $SU(3) \times E_6$. These states survive the orbifold projection because the left- and right-moving states are θ-invariant, i.e. $\exp[2\pi i (r+v) \cdot v] = 1$ and $\exp[2\pi i (P+V) \cdot V] = 1$.

There are further massless left-moving states with one twisted oscillator, $N_B = 1/3$. They arise from internal momentum vectors with $(P + V)^2 = 2/3$, given by

$$\left(\tfrac{1}{3}, \tfrac{1}{3}, -\tfrac{2}{3}, 0, 0, 0, 0, 0\right). \tag{8.30}$$

They transform in the $(\bar{\mathbf{3}}, \mathbf{1})$ of $SU(3) \times E_6$. These chiral multiplets have an extra multiplicity from the three possible choices of twisted oscillator $\alpha^i_{-1/3}$. These states also survive the orbifold projection since the oscillator phase transformation by α is compensated by the gauge shift phase $\exp[2\pi i (P + V) \cdot V] = \alpha^2$.

The complete gauge group and chiral matter spectrum of this $\mathbf{T}^6/\mathbf{Z}_3$ orbifold is

$$SU(3) \times E_6 \times E_8'$$
$$3(\mathbf{3}, \mathbf{27}) + 27(\mathbf{1}, \mathbf{27}) + 81(\bar{\mathbf{3}}, \mathbf{1}). \tag{8.31}$$

We emphasize the intricate cancellation of $SU(3)$ anomalies among untwisted and twisted sector states, involving multiplicity factors from diverse sources, like fixed point degeneracies and oscillator number multiplicities.

The above orbifold model can be regarded as a heterotic CY compactification with standard embedding. Indeed $\mathbf{T}^6/\mathbf{Z}_3$ can be regarded mathematically as a singular limit of a smooth CY space with Hodge numbers $(h_{1,1}, h_{2,1}) = (36, 0)$, as described below; the spectrum of families (7.37) in CY compactification indeed agrees with the above orbifold spectrum, which contains 36 families in the $\mathbf{27}$ of E_6, with no conjugate families. The smooth CY is obtained from the orbifold by a mathematical procedure known as blowing-up, which replaces the singular points by smooth subspaces (in this case, with $\mathbf{CP}_2$ topology). The Kähler moduli parametrizing the sizes of the latter are localized around the singularities in the orbifold limit; the heterotic orbifold model encodes the blowing-up procedure as flat directions of the effective theory, which are parametrized by a D-flat combination of the three twisted oscillator states $(\bar{\mathbf{3}}, \mathbf{1})$ at each fixed point. These 27 moduli, together with the nine untwisted Kähler moduli, reconstruct the Hodge numbers $(h_{1,1}, h_{2,1}) = (36, 0)$ of the smooth CY geometry. Note that along this blow-up direction, the enhanced $SU(3)$ gauge factor is broken, leaving $E_6 \times E_8'$ as the generic 4d gauge group of the theory, again in agreement with CY compactifications with standard embedding.

8.2.5 Modular invariance and orbifold projection

As discussed in Section 3.2.2, modular invariance of the one-loop partition function is an important consistency condition of closed string theories; this is also the case for the abelian orbifold models at hand. In fact, modular invariance underlies important properties of the low-energy effective theory, like anomaly cancellation, by requiring the presence of twisted sectors whose massless states cancel potential gauge anomalies of the untwisted sector. Modular invariance is also an important ingredient in the definition of an appropriate orbifold projection on twisted sectors in general orbifold models, the so-called generalized GSO projection.

Twisted sectors from modular invariance

Consider the one-loop vacuum diagram, described by a toroidal worldsheet characterized by the parameter τ, namely $z = (it + \sigma)/\ell$ with identifications $z \simeq z + 1$, $z \simeq z + \tau$. The amplitude in the path integral formalism involves an integration over all worldsheet fields compatible with the worldsheet geometry. Focusing, e.g., on the worldsheet bosons, the theory in flat 10d spacetime only allows for periodic boundary conditions around the two non-trivial cycles in the directions (σ, t)

$$X^i(z + 1) = X^i(z), \quad X^i(z + \tau) = X^i(z). \tag{8.32}$$

In a toroidal orbifold, however, the bosons describing the 6d orbifold compactification admit twisted boundary conditions

$$X^i(z + 1) = g \, X^i(z), \quad X^i(z + \tau) = h \, X^i(z). \tag{8.33}$$

Generalizing to all worldsheet fields, the twists g, h are elements of $S \times G$, i.e. the space group and its gauge embedding. The amplitude is a sum over different contributions $Z(g, h)$ for the different boundary conditions (g, h) – note that for abelian orbifolds $[g, h] = 0$, and both twists are compatible. The contributions $Z(1, h)$ have a trivial twist in the σ direction, and correspond to a one-loop diagram of untwisted states; this untwisted partition function has the form

$$Z_U = \sum_{h \in S \times G} Z(1, h). \tag{8.34}$$

As emphasized below, the sum over group elements can be recast as a projection operator truncating the physical spectrum onto $S \times G$ invariant states in the untwisted sector; this is analogous to the GSO projection in Section 4.2.3.

The untwisted partition function is clearly not modular invariant, since it involved an artificial choice of σ, which is not preserved by reparametrizations of the worldsheet. In other words, modular transformations mix sectors with different boundary conditions (g, h). A general $SL(2, \mathbf{Z})$ modular transformation (3.65)

$$\tau \to \tau' = \frac{a\tau + b}{c\tau + d}, \quad \text{with } a, b, c, d \in \mathbf{Z} \text{ and } ad - bc = 1, \tag{8.35}$$

maps the (g, h) sector to $(g^a h^b, g^c h^d)$. For example, for $a = d = 0, b = -c = -1$, we have $\tau \to -1/\tau$ and the boundary conditions (g, h) are changed to (h^{-1}, g). These transformations relate pieces of the untwisted partition function to twisted sectors

$$Z(1, h) \longrightarrow Z(h^{-1}, 1). \tag{8.36}$$

A modular invariant partition function is thus obtained by summing over all possible boundary conditions, and includes untwisted and twisted sector contributions

$$Z = Z_U + Z_T = \sum_h Z(1, h) + \sum_{g \neq 1} Z(g, h). \tag{8.37}$$

Modular invariance thus requires the presence of twisted sectors, as announced.

Explicit partition functions and orbifold projection

Let us now display the explicit structure of these partition functions for the case of $\mathbf{Z}_N$ orbifolds, with factorized toroidal lattices for simplicity; they will provide an explicit expression for the orbifold projection in general $\mathbf{Z}_N$ orbifold twisted sectors. Although our earlier models were restricted to the $E_8 \times E_8$ heterotic theory, with standard embedding $V = (v_1, v_2, v_3, 0, \ldots, 0)$, $V' = 0$, the upcoming results are derived for general gauge embedding, and for either heterotic theory.

The partition function has the general structure

$$Z = \int \frac{d^2\tau}{\tau_2^2} \, (4\pi^2\alpha'\tau_2)^{-1} \, \frac{1}{N} \sum_{n,m=0}^{N-1} \eta(n,m) \, Z_{(\theta^n,\theta^m)}(\tau) \, \overline{Z}_{(\theta^n,\theta^m)}(\bar{\tau}). \qquad (8.38)$$

The first factor arises from the 4d momentum integration, the $\eta(n,m)$ are phases to be fixed by modular invariance, and $Z_{(\theta^n,\theta^m)}$, $\overline{Z}_{(\theta^n,\theta^m)}(\bar{\tau})$ are the left- and right-moving partition function in the sector twisted by (θ^n, θ^m).

Consider first the left-moving partition function. Using techniques similar to those of the 10d case in Sections 3.2.2, 4.2.2, and 4.3.2, it can be computed to be

$$Z_{(\theta^n,\theta^m)} = \frac{C(n,m)\,\eta}{\vartheta\begin{bmatrix} 1/2 + nv_1 \\ 1/2 + mv_1 \end{bmatrix} \vartheta\begin{bmatrix} 1/2 + nv_2 \\ 1/2 + mv_2 \end{bmatrix} \vartheta\begin{bmatrix} 1/2 + nv_3 \\ 1/2 + mv_3 \end{bmatrix}} \sum_{P\in\Lambda_{16}} q^{\frac{(P+nV)^2}{2}} e^{(P+nV)\cdot mV},$$

$$(8.39)$$

where η and ϑ are the Dedekind and theta functions, respectively; see Appendix A. The first factor corresponds to the contributions from the 10d spacetime bosonic oscillators, and includes the effect of the orbifold twist. The second factor corresponds to the 16d internal bosons, and includes the orbifold gauge shift. We have not included constant phases which cancel between left and right sectors, e.g. for bosonic oscillator contributions. The coefficients $C(n,m)$ are also associated with the oscillators and are given by

$$C(n,m) = \prod_i \left[-2\sin(\pi m v_i) \right], \text{ for } m \neq 0,$$

$$= \prod_i \left[-2\sin(\pi n v_i) \right], \text{ for } m = 0, \qquad (8.40)$$

with the product running only over non-zero v_i's. In sectors with some $nv_i \in \mathbf{Z}$, the twist θ^n has a fixed plane, and we should include an additional factor $\exp\left(2\pi i \tau p_L^2/2\right)$ from the sum over discrete momentum and winding. We skip them for simplicity, and because they generically involve massive states and thus do not influence our main result, the derivation of the orbifold projection for massless states.

The right-moving partition function is given by

$$
\overline{Z}_{(\theta^n,\theta^m)}(\bar{\tau}) = C(n,m)^* \sum_{\alpha,\beta} \overline{\eta}_{\alpha\beta} \frac{\overline{\vartheta}\begin{bmatrix} \alpha+nv_1 \\ \beta+mv_1 \end{bmatrix} \overline{\vartheta}\begin{bmatrix} \alpha+nv_2 \\ \beta+mv_2 \end{bmatrix} \overline{\vartheta}\begin{bmatrix} \alpha+nv_3 \\ \beta+mv_3 \end{bmatrix} \overline{\vartheta}\begin{bmatrix} \alpha \\ \beta \end{bmatrix}}{\overline{\vartheta}\begin{bmatrix} 1/2+nv_1 \\ 1/2+mv_1 \end{bmatrix} \overline{\vartheta}\begin{bmatrix} 1/2+nv_2 \\ 1/2+mv_2 \end{bmatrix} \overline{\vartheta}\begin{bmatrix} 1/2+nv_3 \\ 1/2+mv_3 \end{bmatrix} \overline{\eta}^3}
$$

$$
= \frac{C(n,m)^* \, \overline{\eta}}{\overline{\vartheta}\begin{bmatrix} 1/2+nv_1 \\ 1/2+mv_1 \end{bmatrix} \overline{\vartheta}\begin{bmatrix} 1/2+nv_2 \\ 1/2+mv_2 \end{bmatrix} \overline{\vartheta}\begin{bmatrix} 1/2+nv_3 \\ 1/2+mv_3 \end{bmatrix}} \sum_{r\in\Lambda^+} \overline{q}^{\frac{(r+nv)^2}{2}} e^{(r+nv)\cdot mv} .
$$

$$(8.41)$$

Again we have not included right-moving momentum contributions for possible fixed planes, and have also ignored constant phases canceling against the left-moving contribution. In the first line, the numerator and denominator correspond to the fermionic and bosonic contributions. In the second line, we have bosonized the fermions, whose contribution is the lattice sum.

The overall multiplicities $|C(n,m)|^2$ in the total partition function are in general associated to the number of fixed points in the twisted sectors; for the untwisted sector, it cancels against a similar factor from the oscillator pieces in the denominator, see (A.10), so no extra multiplicity is generated.

We now use these partition functions to derive the orbifold projection in the different sectors, first in the simplest case of the $\mathbf{Z}_3$ orbifold, and subsequently in the general $\mathbf{Z}_N$ case. Consider the left-moving $(1,\theta)$ partition function of the $\mathbf{Z}_3$ orbifold of the $E_8 \times E_8$ theory,

$$
Z_{(1,\theta)}(\tau) = \frac{\sum_{P\in\Lambda_{16}} q^{P^2/2} \, e^{2\pi i \, P\cdot V}}{q^{1/3} \, \eta^8 \, \prod_k (1-q^k)^2 \, (1-\alpha q^k)^3 \, (1-\alpha^2 q^k)^3} , \tag{8.42}
$$

where we recall that $q = e^{2\pi i \tau}$ and $\alpha = e^{2\pi i/3}$. The $(1,\theta)$ right-handed partition function is given by (the conjugate of) a similar expression, with P and V replaced by r and v, respectively. The partition functions in the $(1,\theta^m)$ sectors can be obtained by replacing $V \to mV$ and $v \to mv$. The total untwisted partition function (8.34) is obtained by summing over $m = 0, 1, 2$. The different terms can be gathered into a single one, with an effective projection, which for untwisted massless states is

$$
P_U = \frac{1}{3} (1 + \Delta + \Delta^2) \quad \text{with } \Delta = \exp\left[2\pi i \, (P \cdot V - r \cdot v)\right]. \tag{8.43}
$$

Such projectors are sometimes referred to as generalized GSO projectors. This reproduces the orbifold projection for untwisted states used in Section 8.2.4; gauge bosons have $r \cdot v = 0$, so the orbifold projection requires $P \cdot V \in \mathbf{Z}$, while for charged matter we have the more general condition $P \cdot V = r \cdot v \mod \mathbf{Z}$.

For twisted sectors we can obtain the orbifold projection using a similar logic. Consider the left-moving (θ, θ) partition function for the $\mathbf{Z}_3$ orbifold,

$$Z_{(\theta,\theta)}(\tau) = -3\sqrt{3} \, \frac{\sum_{P \in \Lambda_{16}} q^{(P+V)^2/2} \, e^{2\pi i \, (P+V) \cdot V}}{q^{1/3} \, \eta^8 \, \prod_m \left[(1 - q^m)^2 (1 - \lambda^2 q^{m-1/3})^3 (1 - \lambda q^{m-2/3})^3 \right]}.$$

The corresponding right-handed partition function is given by (the conjugate of) a similar expression, with P and V replaced by r and v, respectively. The overall factor $(3\sqrt{3})^2 = 27$ reproduces the number of fixed points, as announced. The partition functions in the (θ, θ^m) sectors can be easily obtained using modular transformations $\tau \to \tau + 1$. Summing over m, the different terms can be gathered into a single one, with an orbifold projector, which for massless states with no oscillators is

$$P_\theta = \frac{1}{3} \left(1 + \Delta_\theta + \Delta_\theta^2\right) \quad \text{with } \Delta_\theta = \exp\{2\pi i \, [(P+V) \cdot V - (r+v) \cdot v]\}. \quad (8.44)$$

For massless states involving oscillators, the projector includes an additional phase in Δ_θ. This reproduces the orbifold projection for θ-twisted states used in Section 8.2.4. Namely, massless twisted states with no oscillators have $(r+v) \cdot v = 0$, so the orbifold projection requires $(P+V) \cdot V \in \mathbf{Z}$; this is in fact automatically satisfied by all massless θ-twisted states in the $\mathbf{Z}_3$ orbifold, since the masslessness condition $(P+V)^2 = 4/3$ implies that $(P+V) \cdot V = 1/2(P+V)^2 - 1/2 P^2 + 1/2 V^2$ is integer and then $\Delta_\theta = 1$. For the massless oscillator states with $N_B = 1/3$ the modified projector has an additional phase and reads $P'_\theta = (1/3) \left(1 + \alpha \Delta_\theta + \alpha^2 \Delta_\theta^2\right)$; again the masslessness condition, $(P+V)^2 = 2/3$, implies that all states survive the projection. The fact that all massless states in twisted sectors survive the orbifold projection is a general feature of the $\mathbf{Z}_3$ orbifold (and more generally for $\mathbf{Z}_N$ orbifold with prime N), and also holds for non-standard embeddings and in the presence of non-trivial Wilson lines, see Section 8.3.

For $\mathbf{Z}_N$ orbifolds with N not prime the orbifold projection on massless states is non-trivial, and can remove some massless states from the physical spectrum. We construct the corresponding projectors by the logic employed in the above $\mathbf{Z}_3$ example. In a general θ^n-twisted sector, the orbifold projector for massless states is

$$P_{\theta^n} = \frac{1}{N} \sum_{m=0}^{N-1} e^{-i\pi nm(V^2 - v^2)} \, \tilde{\chi}(\theta^n, \theta^m) \, \Delta_{\theta^n}^m, \quad (8.45)$$

where

$$\Delta_{\theta^n} = \exp\{2\pi i \, [N_B + (P+nV) \cdot V - (r+nv) \cdot v]\}, \quad (8.46)$$

and the coefficients $\tilde{\chi}(\theta^n, \theta^m)$ are

$$\tilde{\chi}(1, \theta^m) = 1,$$
$$\tilde{\chi}(\theta^n, \theta^m) = \chi(\theta^n, \theta^m), \quad \text{if } \theta^n \text{ has no fixed planes},$$
$$= \hat{\chi}(\theta^n, \theta^m), \quad \text{if } \theta^n \text{ has fixed planes}. \quad (8.47)$$

Here $\chi(\theta^n, \theta^m)$ is the number of points left fixed simultaneously by θ^n and θ^m. If θ^n leaves fixed tori, we must use $\hat{\chi}(\theta^n, \theta^m)$, which is the number of simultaneous fixed points in the subspace actually rotated by θ^n.

The above structure of the orbifold projectors allows us to recover the consistency condition (8.22); it follows from requiring identical phase factor in the sectors $(\theta, 1)$ and (θ, θ^N), and equivalently, by invariance of the partition function under the modular transformation $\tau \rightarrow \tau + N$.

As mentioned, the above projectors apply for general gauge embedding shifts V, and also for models with Wilson lines. It also generalizes easily to abelian $\mathbf{Z}_N \times \mathbf{Z}_M$ orbifolds, with an extra subtlety, the possible presence of extra phases due to the "discrete torsion" described in the next section.

8.2.6 More abelian orbifolds

We now describe the geometric structure of more general abelian $\mathbf{Z}_N$ and $\mathbf{Z}_N \times \mathbf{Z}_M$ orbifolds, and construct new heterotic string compactifications with standard embedding. However, the information is also useful for compactification of other string theories, e.g. type II orientifolds. In later sections they are also applied to heterotic models with non-standard embeddings.

Z_N orbifolds

Orbifolds $\mathbf{T}^6/\mathbf{Z}_N$ are defined by a choice of underlying toroidal lattice, on which the generator θ acts crystallographically. Table 8.2 provides a list of possible $\mathbf{T}^6/\mathbf{Z}_N$ orbifolds, with their twist vector v. The lattices are described as root lattices of a Lie algebra, with θ an element of its Weyl group.

A simple class of modular invariant heterotic models is obtained for the standard embedding (8.23). As already mentioned, this is an orbifold version of the identification of gauge and geometric curvatures in the standard embedding CY compactifications in Section 7.3.2. Such orbifolds of the $E_8 \times E_8$ theory lead to a 4d gauge group $E_6 \times G_0 \times E_8'$, with G_0 some subgroup of $SU(3)$; specifically, $G_0 = SU(3)$ for $\mathbf{Z}_3$, $G_0 = SU(2) \times U(1)$ for $\mathbf{Z}_4$, $\mathbf{Z}_6$ and $G_0 = U(1)^2$ for the remaining cases. The $E_6 \times E_8'$ structure agrees with that of heterotic CY compactifications with standard embedding. In fact, as described for the $\mathbf{Z}_3$ case, the orbifold spaces can be related to smooth CY manifolds by blowing-up, i.e. replacing the singular points by finite size subspaces. The moduli controlling the size of the latter parametrize flat directions of the low-energy effective theory involving twisted sector fields of the orbifold model. In general these twisted sector fields are charged under the accidentally enhanced gauge factor G_0, which is thus generically broken as the orbifold is blown-up to a smooth CY.

The spectrum of heterotic orbifolds with standard embedding contains E_6 families and conjugate families, in the $\mathbf{27}$ and $\overline{\mathbf{27}}$ respectively. This allows to define an orbifold version of the Hodge numbers $(h_{1,1}, h_{2,1})$, and the Euler characteristic $\chi = 2(h_{11} - h_{12})$, which

Table 8.2 $\mathbf{T}^6/\mathbf{Z}_N$ orbifolds, with their twist vector v, and Lie algebras defining examples of possible underlying torus lattices. The last column provides the Hodge numbers of the smooth CY obtained by blowing up the singular points. They provide the numbers of E_6 27s and $\overline{27}$s in the corresponding heterotic orbifolds with standard embedding

$\mathbf{Z}_N$	(v_1, v_2, v_3)	Lattice	$(h_{1,1}, h_{12})$
$\mathbf{Z}_3$	$\frac{1}{3}(1, 1, -2)$	$SU(3)^3, E_6$	(36,0)
$\mathbf{Z}_4$	$\frac{1}{4}(1, 1, -2)$	$SO(4)^3$	(31,7)
		$SU(4)^2$	(25, 1)
$\mathbf{Z}_6$	$\frac{1}{6}(1, 1, -2)$	$SU(3) \times G_2^2$	(29,5)
$\mathbf{Z}_6'$	$\frac{1}{6}(1, 2, -3)$	$SU(2)^2 \times SU(3) \times G_2$	(35,11)
$\mathbf{Z}_7$	$\frac{1}{7}(1, 2, -3)$	$SU(7)$	(24, 0)
$\mathbf{Z}_8$	$\frac{1}{8}(1, 2, -3)$	$SO(5) \times SO(9)$	(27,3)
$\mathbf{Z}_8'$	$\frac{1}{8}(1, 3, -4)$	$SO(9) \times SO(4)$	(31,7)
$\mathbf{Z}_{12}$	$\frac{1}{12}(1, 4, -5)$	E_6	(25,1)
$\mathbf{Z}_{12}'$	$\frac{1}{12}(1, 5, -6)$	$SO(4) \times F_4$	(31,7)

in fact agree with those of the corresponding smooth CY space. Using the formulae discussed for the orbifold projection, one can derive a simple expression for the orbifold Euler characteristic. In slightly more general notation, it reads

$$\chi = \frac{1}{|P|} \sum_{g,h \in P} \chi(g, h), \tag{8.48}$$

where $|P|$ is the order of the point group P, i.e. N for a $\mathbf{Z}_N$ orbifold, and $\chi(g, h)$ is the number of points fixed under g, h simultaneously (taken zero when there is a common fixed torus). The above formula holds for the $\mathbf{Z}_N \times \mathbf{Z}_M$ orbifolds discussed below, and even for non-abelian orbifolds, by restricting the sum to commuting elements $[g, h] = 0$ of the orbifold group. A closed formula for the number of fixed points for a twist g with vector $v^{(g)}$ is provided by the Lefschetz fixed point theorem

$$n_g = \det'(1 - g) = \left| \prod_i 2 \sin \pi v_i^{(g)} \right|^2, \tag{8.49}$$

where the prime restricts the product to the subspace actually rotated by g. These fixed point multiplicities allow for a purely geometrical computation of the orbifold Hodge numbers. Their values are provided in the last column of Table 8.2, and in general depend on the choice of underlying toroidal lattice. Note that $\chi/2$ provides the net number of E_6 generations for these heterotic models with standard embedding, and is 36 for $\mathbf{Z}_3$ and 24 in all other cases. This is not very promising from the viewpoint of particle physics model building. However, the same geometric orbifolds can lead to much more interesting models, upon introduction of non-standard embeddings and Wilson lines.

$Z_N \times Z_M$ orbifolds and discrete torsion

Abelian discrete subgroups of $SU(3)$ can also be of the form $\mathbf{Z}_N \times \mathbf{Z}_M$, with M being a multiple of N. These can be generated by elements θ, ω, with the structure of the twist vectors given by the $\mathcal{N} = 1$ or $\mathcal{N} = 2$ preserving elements in Table 8.1. With two $\mathcal{N} = 2$ twists one has, e.g., $v = (1, 0, -1)/N$, $w = (0, 1, -1)/M$ with $N, M = 2, 3, 4, 6$. Since the twists are chosen to have different fixed planes, the complete orbifold preserves only 4d $\mathcal{N} = 1$ supersymmetry. It turns out that certain pairs (N, M) are not possible choices, because the corresponding $\mathbf{Z}_N$, $\mathbf{Z}_M$ cannot be realized simultaneously as symmetries of a six-dimensional lattice; an easy-to-check necessary condition is that, for any element of the group, the putative number of fixed points (8.49) must be integer. The list of consistent $\mathbf{Z}_N \times \mathbf{Z}_M$ twists is displayed in Table 8.3, mostly for the simplest choice of factorized toroidal lattices. Table 8.3 also provides the orbifold Hodge numbers, which can be obtained by computing the spectrum of E_6 families in heterotic models with standard embedding. The spectra of general orbifold compactifications can be obtained in close analogy with the previous $\mathbf{Z}_N$ models. We therefore skip its discussion, and instead focus on an important new ingredient, the choice of discrete torsion, which produces different models with the same underlying orbifold and gauge shift.

The one-loop partition function for a $\mathbf{Z}_N \times \mathbf{Z}_M$ orbifold has the general form

$$Z = \frac{1}{NM} \sum_{g,h} \epsilon(g, h) \, Z(g, h), \tag{8.50}$$

with $g, h \in \mathbf{Z}_N \times \mathbf{Z}_M$. For $\mathbf{Z}_N$ orbifolds, the modular group relates all non-trivial (θ^n, θ^m) sectors, so phases like $\epsilon(\theta^n, \theta^m)$ are fixed unambiguously by modular invariance; indeed, a $\tau \to 1/\tau$ transformation connects, e.g., the untwisted $(1, \theta^n)$ sectors to the twisted sectors $(\theta^n, 1)$, which can subsequently be related to (θ^n, θ^{nm}) by $\tau \to \tau + m$. For a $\mathbf{Z}_N \times \mathbf{Z}_M$ orbifold, however, there are disjoint orbits of the modular group; for instance, the (θ, ω) sector is not related by modular transformations to the untwisted $(1, \theta^n \omega^m)$ sectors. There is thus the freedom to assign a non-trivial phase $\epsilon(\theta, \omega)$ to the pieces of the one-loop partition function in the orbit of the (θ, ω) sector. Such phases are subject to additional constraints from two-loop modular invariance, etc., whose derivation lies beyond our scope. The bottom line is that the most general phase assignment is of the form

$$\epsilon(\theta^k \omega^l, \theta^t \omega^s) = \epsilon^{(ks - lt)}, \tag{8.51}$$

where the phase $\epsilon = \epsilon(\theta, \omega)$, known as discrete torsion, is an Nth root of unity (with our convention that M is a multiple of N). The impact of this phase on the spectrum is that the ω orbifold projector acts on the θ-twisted states with an additional phase ϵ (and vice versa). This extra phase can be regarded as arising from non-trivial discrete 2-form field background $B_{i\bar{i}}$ on orbifold fixed tori. For standard embedding models, the Hodge numbers for the different choices of discrete torsion are shown in Table 8.3.

Table 8.3 $\mathbf{Z}_N \times \mathbf{Z}_M$ orbifolds, with their twist vectors v, w, and Hodge numbers for different choices of discrete torsion

$\mathbf{Z}_N \times \mathbf{Z}_M$	$\theta = (v_1, v_2, v_3)$	$\omega = (w_1, w_2, w_3)$	(h_{11}, h_{12})
$\mathbf{Z}_2 \times \mathbf{Z}_2 \ (SO(4)^3)$	$1/2(1, 0, -1)$	$1/2(0, 1, -1)$	$(51,3),(3,51)$
$\mathbf{Z}_2 \times \mathbf{Z}_2 \ (SO(12))$	$1/2(1, 0, -1)$	$1/2(0, 1, -1)$	$(27,3),(3,27)$
$\mathbf{Z}_3 \times \mathbf{Z}_3$	$1/3(1, 0, -1)$	$1/3(0, 1, -1)$	$(84,0),(3,27)$
$\mathbf{Z}_2 \times \mathbf{Z}_4$	$1/2(1, 0, -1)$	$1/4(0, 1, -1)$	$(61,1),(21,9)$
$\mathbf{Z}_2 \times \mathbf{Z}_6'$	$1/2(1, 0, -1)$	$1/6(1, 1, -2)$	$(36,0),(15,15)$
$\mathbf{Z}_2 \times \mathbf{Z}_6$	$1/2(1, 0, -1)$	$1/6(0, 1, -1)$	$(51,3),(19,19)$
$\mathbf{Z}_4 \times \mathbf{Z}_4$	$1/4(1, 0, -1)$	$1/4(0, 1, -1)$	$(90,0),(42,0),(6,12)$
$\mathbf{Z}_3 \times \mathbf{Z}_6$	$1/3(1, 0, -1)$	$1/6(0, 1, -1)$	$(73,1),(13,13)$
$\mathbf{Z}_6 \times \mathbf{Z}_6$	$1/6(1, 0, -1)$	$1/6(0, 1, -1)$	$(84,0),(51,3),(27,3),(9,9)$

For illustration consider the simplest orbifold allowing for discrete torsion, the $\mathbf{Z}_2 \times \mathbf{Z}_2$ orbifold. The generators are $v = (1, 0, -1)/2$, $w = (0, 1, -1)/2$, and we choose the underlying torus to be rectangular, i.e. defined by an $SO(4)^3$ root lattice. Let us compute the orbifold Euler characteristic using the generalization of (8.48). Each of the elements θ, ω, and $\theta\omega$ leave some 2-torus invariant, hence

$$\chi(\theta, 1) = \chi(\omega, 1) = \chi(\theta\omega, 1) = 0. \tag{8.52}$$

On the other hand, we have

$$\chi(\theta, \omega) = \chi(\theta, \theta\omega) = \chi(\omega, \theta\omega) = 4 \times 4 \times 4 = 64. \tag{8.53}$$

Using $\chi(g, h) = \chi(h, g)$, we can add up the contributions to the Euler characteristic to get $\chi = (1/4) \times 2 \times (64 + 64 + 64) = 96$. A detailed computation of the orbifold spectrum shows that this arises from a contribution of 16 **27**s for each twisted sector (corresponding to the 16 fixed tori). In addition, there are three vector-like copies $(\mathbf{27} + \overline{\mathbf{27}})$ in the untwisted sector. Hence the Hodge numbers are $(h_{1,1}, h_{2,1}) = (51, 3)$, as shown in Table 8.3.

This result corresponds to a particular choice of discrete torsion, given by a trivial phase $\epsilon = 1$, implicit in the orbifold projector leading to (8.48). It is straightforward to compute the Hodge numbers of the model with discrete torsion $\epsilon = -1$; this simply flips the orbifold projections in twisted sectors, leading to 16 $\overline{\mathbf{27}}$s instead of the **27**s. Hence $(h_{1,1}, h_{2,1}) = (3, 51)$, as shown also in Table 8.3.

Note that in this case the change in discrete torsion exchanges h_{11} and h_{12}. Although this is not a general feature of orbifolds with discrete torsion, in this case it follows because the two $\mathbf{Z}_2 \times \mathbf{Z}_2$ happen to be related by mirror symmetry (see Section 10.1.2). Moreover, in this case mirror symmetry admits a simple realization as three T-dualities, along one circle direction in each 2-torus, making this example possibly the simplest case of a mirror CY pair. The two versions of this orbifold will also provide a fruitful arena for model building of type IIA and IIB orientifolds.

Table 8.4 *The four modular invariant embeddings of the* $E_8 \times E_8$ *heterotic* $\mathbf{Z}_3$ *orbifold*

V^I	Gauge group	Untwisted	Twisted
$\frac{1}{3}(1,1,2,0,\ldots,0)\times$ $(0,\ldots,0)$	$[E_6 \times SU(3)] \times E'_8$	$3(\mathbf{27},\mathbf{3};\mathbf{1'})$	$27\,(\mathbf{27},\mathbf{1};\mathbf{1'})$ $+81\,(\mathbf{1},\mathbf{\bar3};\mathbf{1'})$
$\frac{1}{3}(1,1,2,0,\ldots,0)\times$ $\frac{1}{3}(1,1,2,0,\ldots,0)$	$[E_6 \times SU(3)]^2$	$3(\mathbf{27},\mathbf{3};\mathbf{1'},\mathbf{1'})$ $+(\mathbf{1},\mathbf{1};\mathbf{27'},\mathbf{3'})$	$27(\mathbf{1},\mathbf{3};\mathbf{1'},\mathbf{\bar3'})]$
$\frac{1}{3}(1,1,0,\ldots,0)\times$ $\frac{1}{3}(2,0,\ldots,0)$	$[E_7 \times U(1)]\times$ $[SO(14) \times U(1)]'$	$3(\mathbf{56},\mathbf{1'}) + 3(\mathbf{1},\mathbf{1'})$ $+3(\mathbf{1},\mathbf{14'}) + 3(\mathbf{1},\mathbf{64'})$	$27[(\mathbf{1},\mathbf{1'}) + (\mathbf{1},\mathbf{14'})]$ $+81(\mathbf{1},\mathbf{1'})$
$\frac{1}{3}(1,1,1,1,2,0,\ldots,0)\times$ $\frac{1}{3}(2,0,\ldots,0)$	$SU(9)\times$ $SO(14)' \times U(1)'$	$3(\mathbf{84},\mathbf{1'}) + 3(\mathbf{1},\mathbf{14'})$ $+3(\mathbf{1},\mathbf{64'})$	$27(\mathbf{\bar9},\mathbf{1'})$

8.3 Non-standard embeddings and Wilson lines

We are now ready to construct more general orbifold models, by relaxing the standard embedding condition and allowing for the introduction of Wilson lines. These are orbifold analogs of the more general compactifications on smooth CY spaces in Sections 7.4 and 7.3.3.

8.3.1 Non-standard embeddings

The standard embedding of the point group P into the $E_8 \times E_8$ degrees of freedom is not the only consistent possibility in orbifold compactification; any other choice of shift V^I satisfying (8.8) and (8.22) leads to a consistent model. For each given $\mathbf{Z}_N$ or $\mathbf{Z}_N \times \mathbf{Z}_M$ orbifold, there is a finite list of such shifts V^I (up to translations in the 16d internal lattice). For the $E_8 \times E_8$ heterotic, there are four possible choices for $\mathbf{Z}_3$ and 12 for $\mathbf{Z}_4$, but the number increases to $\simeq 3000$ for the $\mathbf{Z}_{12}$ orbifolds. As an illustration we consider the $\mathbf{Z}_3$ orbifold, whose consistent gauge embeddings are shown in Table 8.4, along with the corresponding gauge group, and untwisted and twisted particle content.

A point worth remarking is that the $U(1)$ gauge factors in these models have non-zero triangle anomalies; they are however canceled by a 4d analog of the Green–Schwarz mechanism, which moreover renders the corresponding gauge bosons massive, and triggers further gauge symmetry breaking. We postpone the description of this phenomenon and its implications until Section 9.5.

From the phenomenological viewpoint of particle physics model building, neither of these four possible $\mathbf{Z}_3$ orbifold embeddings seems very exciting: the second model in Table 8.4 has nine families for each of the two E_6 factors; the third model is non-chiral with respect to the non-abelian groups; the fourth model contains an $SU(9)$ group which would also lead to nine families if decomposed into $SU(5)$ representations.

A similar classification of models can be carried out for other $\mathbf{Z}_N$ and $\mathbf{Z}_N \times Z_M$ models. The resulting theories are, however, similarly uninteresting from the viewpoint of particle physics model building. There is, moreover, a general pattern that the higher the order of the orbifold, the smaller the size of the 4d non-abelian gauge group. As we discuss in the next section, however, the situation and model building prospects of heterotic orbifolds is much improved through the addition of Wilson lines.

8.3.2 Introduction of Wilson lines

In all previous models, we have considered the underlying toroidal compactification to have a rather simple choice of background fields, essentially the metric moduli determining the six-dimensional toroidal lattice. In particular, we have not considered turning on additional background fields, like the internal components of the 2-index antisymmetric tensor field B_{mn}, or internal components of the $E_8 \times E_8$ or $SO(32)$ gauge fields A_m^a. The former imply an asymmetric choice of compactification torus for left- and right-moving degrees of freedom, and are naturally discussed in the framework of asymmetric orbifolds, see Section 8.4. The latter correspond to Wilson lines, already discussed in toroidal compactifications in Section 5.2.2, and in Section 7.3.3 for smooth CY compactifications. Here we describe their effect on the spectrum of toroidal orbifolds.

We start with a few basic observations. In $\mathbf{T}^6$ compactifications it is possible to have Wilson lines along each of the six non-trivial 1-cycles. These must be mutually commuting, so that the non-abelian commutator terms in the field strength tensor vanish. By simultaneously rotating them into the Cartan generators, we can label the background gauge fields A_m^I with $m = 1, \ldots, 6$ and $I = 1, \ldots, 16$. Denoting the $\mathbf{T}^6$ lattice vectors by $\vec{e}_r$, $r = 1, \ldots, 6$, we also define

$$a_r^I = \int_{\vec{e}_r} dx^m \, A_m^I, \tag{8.54}$$

which are the analogs of $\tilde{A}$ in (5.19). In toroidal orbifolds, the orbifold action may relate some of these Wilson lines, leaving a smaller set of independent choices.

The Wilson lines along the directions $\vec{e} = \sum_r n_r \vec{e}_r$ implement gauge transformations $\exp\left(2\pi i \sum_r n_r a_r^I P_I\right)$ on states with internal momenta P_I, hence we can regard Wilson lines as embedding the lattice translations into the gauge degrees of freedom. In the orbifold case, the orbifold gauge embedding and the Wilson lines thus combine to provide an embedding of the complete space group S into the gauge degrees of freedom. In this language, Wilson lines commute because the lattice translations are commuting, $[(1, \vec{e}_r), (1, \vec{e}_s)] = 0$. On the other hand, since orbifold twists and Wilson lines need not commute, Wilson lines and orbifold gauge shifts may be non-commuting. This implies that Wilson lines may not only break the gauge symmetry, but also lower its rank. For simplicity we consider the commuting case, so that both orbifold twists and Wilson lines gauge actions can be represented by shifts V^I and a_r^I along the Cartan generators. In space group embedding language,

$$\text{Twist}: (\theta, 0; V^I); \quad \text{Wilson line}: \left(1, \vec{e}_r; a_r^I\right), \tag{8.55}$$

where the first two entries denote the space group element, and the last denotes its gauge embedding. Other elements are given by group multiplication, e.g. $(\theta, \vec{e}_r; V^I + a_r^I)$.

The orbifold symmetry implies that the Wilson lines corresponding to vectors $\vec{e}_r$ rotated by the twist θ are discrete; i.e. since $\sum_n \theta^n \vec{e}_r = 0$, we have $(\theta, \vec{e}_r)^N = (1, 0)$ in the space group S, and therefore the gauge embedding must obey

$$(\theta, \vec{e}_r; V + a_r)^N = (1, 0; 0) \implies N(V + a_r) \in \Lambda_{16d} \implies N a_r \in \Lambda_{16d}. \tag{8.56}$$

Note also that, as mentioned above, in general not all Wilson lines are independent; if $\theta \vec{e}_r = \sum_s n_{rs} \vec{e}_s$, then $a_r = \sum_s n_{rs} a_s \bmod \Lambda_{16d}$. This reduces the number of independent Wilson lines. For example, in the $\mathbf{Z}_3$ orbifold with $SU(3)^3$ root lattice, there is only one independent Wilson line for each 2-torus, e.g. $\vec{e}_2 = \theta \vec{e}_1$ and hence $a_1 = a_2 \bmod \Lambda_{16d}$.

Let us consider the effect of the Wilson lines on the orbifold massless spectrum. In the untwisted sector, strings essentially propagate as in the underlying torus. As described in Section 5.2.2, a state with gauge charge vector P has its 6d momentum shifted as $p_r \to p_r + P \cdot a_r$, and this extra shift contributes to its 4d mass. The effect of the Wilson line on massless states is therefore a projection,

$$P^I a_r^I \in \mathbf{Z}, \tag{8.57}$$

similar to (5.24). Consequently both the gauge group and untwisted matter fields in the orbifold are reduced, as compared with models with no Wilson lines.

Wilson lines have a non-trivial effect on twisted sectors as well. A twisted sector string at a fixed point f associated to an element of the space group $\left(\theta^n, \sum_r k_r^f \vec{e}_r\right)$ has boundary conditions for, e.g., the $\mathbf{T}^6$ bosonic coordinates

$$X^m(t, \sigma + \ell) = \left(\theta^n X(t, \sigma)\right)^m + \sum_r k_r^f e_r^m, \tag{8.58}$$

where e_r^m are the components of $\vec{e}_r$. The corresponding boundary conditions on the left-handed internal coordinates are

$$X^I(t + \sigma + \ell) = X^I(t + \sigma) + n V^I + \sum_r k_r^f a_r^I. \tag{8.59}$$

That is, in this twisted sector and fixed point f, the gauge momenta P are shifted to $P + n V + \sum_r k_r^f a_r$. Hence, in a sector fixed under $\left(\theta^n, \sum_r k_r^f \vec{e}_r\right)$, all formulae in previous sections hold with the replacement $nV \to nV + \sum_r k_r^f a_r$. For instance, the left-moving mass formula in this sector takes the form

$$\frac{\alpha' M_L^2}{2} = N_B + \frac{\left(P + nV + \sum_r k_r^f a_r\right)^2}{2} + E_0 - 1. \tag{8.60}$$

The punchline is that Wilson lines make different fixed points feel different (in general non-standard) orbifold gauge shifts, and lead to different twisted massless spectra. This significantly enriches the model building prospects of toroidal orbifolds. The introduction of Wilson lines can reduce the size of the gauge group, bringing it closer to the SM one, as well as the number of families, usually too large in simpler orbifold models. Explicit examples are described in Section 8.3.3.

As expected, there are non-trivial modular invariance constraints on Wilson lines. They are given by the replacement $nV \rightarrow nV + \sum_r k_r^f a_r$ in (8.22) for all possible fixed points, associated to space group elements $\left(\theta^n, \sum_r k_r^f \vec{e}_r\right)$, i.e.

$$N \left[\left(nV + \sum_r k_r^f a_r \right)^2 - n^2 v^2 \right] = 0 \mod 2. \qquad (8.61)$$

Including Wilson lines increases enormously the number of orbifold vacua based on a given $\mathbf{Z}_N$ or $\mathbf{Z}_N \times \mathbf{Z}_M$ orbifold geometry.

8.3.3 A first step in semi-realistic model building

In this section we construct some simple examples of heterotic toroidal orbifold compactifications leading to interesting models of particle physics. Orbifolds with standard embedding do not lead to phenomenologically interesting models. The use of non-standard embeddings and Wilson lines, however, allows the construction of interesting semi-realistic particle physics models, close to (supersymmetric versions of) the Standard Model. This has been explored in the literature for several orbifold groups, including the $\mathbf{Z}_3$, $\mathbf{Z}_2 \times \mathbf{Z}_2$, $\mathbf{Z}_6$, $\mathbf{Z}_6'$, $\mathbf{Z}_3 \times \mathbf{Z}_3$, $\mathbf{Z}_{12}, \ldots$, but a systematic review of these explorations is beyond the scope of this book. We rather provide two explicit examples, and emphasize some of their general properties. Further analysis of models in this class from the viewpoint of the 4d effective action leads to an improvement in their model building applications, described in Section 9.7.

A Z_3 orbifold with two Wilson lines

Our first example is based on the $\mathbf{Z}_3$ orbifold, for which the orbifold projection in twisted states is particularly simple, as all massless states survive it. It is remarkable that despite the simplicity of the setup, it is possible to obtain models surprisingly close to the MSSM. An interesting observation is that in this and similar $\mathbf{Z}_3$-based models, the three SM generations are related to the three complex planes in the internal compactification space, nicely geometrizing the SM family triplication.

As in Section 8.2.4, we consider the $E_8 \times E_8$ heterotic on $\mathbf{T}^6/\mathbf{Z}_3$ with the underlying toroidal geometry given by the $SU(3)^3$ root lattice, see Figure 8.4. The first torus has three fixed points ($\bullet, \circ, \times$), which are invariant under θ modulo a translation by $m_1 \vec{e}_1 + m_2 \vec{e}_2$ with $(m_1, m_2) = (0, 0), (1, 0), (1, 1)$ respectively, and analogously for the other two tori. The orbifold generator must be embedded as one of the four gauge shifts satisfying the modular invariance constraint. Concerning the Wilson lines, the $\mathbf{Z}_3$ rotations leave only three independent Wilson lines, one per 2-torus, i.e. $a_1 = a_2, a_3 = a_4, a_5 = a_6$; they are also subject to modular invariance constraints. We consider the following choice of gauge embeddings

$$V = \tfrac{1}{3}(1, 1, 1, 1, 2, 0, 0, 0) \times \tfrac{1}{3}(2, 0, \dots, 0),$$
$$a_1 = a_2 = \tfrac{1}{3}(0, 0, \dots, 0, 2) \times \tfrac{1}{3}(0, 1, 1, 0, \dots, 0). \tag{8.62}$$

They satisfy (8.56), namely $3a_1 \in \Lambda_{E_8 \times E_8}$, and the constraints (8.61), since the fixed points $(1, 0)$ and $(1, 1)$ in the first 2-torus have local gauge shifts $V + a_1$ and $V - a_1$ mod $\Lambda_{E_8 \times E_8}$, which satisfy $3[(V \pm a_1)^2 - v^2] =$ even.

The model with shift V and no Wilson line would give rise to the fourth model in Table 8.4, with gauge group $SU(9) \times SO(14) \times U(1)$. The Wilson line induces the projection $P \cdot a_1 \in \mathbf{Z}$ for untwisted states, triggering the breaking

$$SU(9) \to SU(5) \times SU(2)^2 \times U(1)^2. \tag{8.63}$$

With hindsight we can tentatively identify this $SU(5)$ with a GUT symmetry, Section 1.2.2. As the reader may easily check, the untwisted chiral matter from the first E_8 gets reduced to

$$3\,[\,(\overline{\mathbf{10}}, \mathbf{1}, \mathbf{1}) + (\mathbf{5}, \mathbf{2}, \mathbf{1}) + (\mathbf{5}, \mathbf{1}, \mathbf{2})\,]. \tag{8.64}$$

There are three $\overline{\mathbf{10}}$s and twelve $\mathbf{5}$s; the anomalies of this untwisted spectrum will be canceled by additional massless states in the twisted sectors, to which we turn.

The 27 fixed points split into three sets with nine points each, according to the local gauge shifts they feel. The nine fixed points corresponding to $(0, 0)$ in the first two-torus are fixed under θ with no additional translation in the first complex plane; they are insensitive to the Wilson line, so the gauge embedding shift is V. The corresponding massless twisted sector fields are as in the case with no Wilson line, but with gauge representations decomposed according to (8.63). This gives

$$9\,[\,(\overline{\mathbf{5}}, \mathbf{1}, \mathbf{1}) + (\mathbf{1}, \mathbf{2}, \mathbf{1}) + (\mathbf{1}, \mathbf{1}, \mathbf{2})\,]. \tag{8.65}$$

Note how the $SU(5)$ anomalies from the untwisted sector spectrum are neatly canceled by these twisted fields.

The nine fixed points corresponding to $(1, 0)$ in the first two-torus feel a local gauge shift $V + a_1$. This shift is equivalent, modulo $\Lambda_{E_8 \times E_8}$, to that of the second model in Table 8.4, and would lead by itself to an $[E_6 \times SU(3)]^2$ gauge symmetry. We can thus borrow its twisted sector massless fields, and decompose it with respect to the actual 4d gauge symmetry; the $SU(3)$ triplets decompose under the $SU(2)$'s as singlets and doublets. The last set of nine fixed points, corresponding to $(1, 1)$ in the first two-torus, feel a shift $V + 2a_1$. This shift is again equivalent to that of the second model in Table 8.4, modulo $\Lambda_{E_8 \times E_8}$ and reordering of entries leading to a slightly different embedding inside E_8. It locally produces an $[E_6 \times SU(3)]^2$ structure, and its twisted sectors lead to additional doublets and singlets. The gauge symmetry structure of this model is depicted in Figure 8.5.

All in all, this model has the $SU(5)$ GUT content

$$3\,(\overline{\mathbf{10}} + \mathbf{5}) + 9\,(\mathbf{5} + \overline{\mathbf{5}}). \tag{8.66}$$

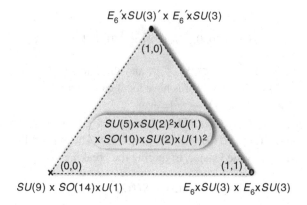

Figure 8.5 The three different sets of fixed points have different local gauge shifts breaking to different local gauge symmetries. The actual 4d gauge symmetry of the model is the overlap of the gauge symmetries on the fixed points.

Namely, three standard families plus nine extra vector-like $\mathbf{5} + \bar{\mathbf{5}}$-plets. This simple model illustrates how the addition of Wilson lines and non-standard embeddings in $\mathbf{Z}_N$ orbifolds can lead to potentially realistic particle physics models. The model is however not a really viable particle physics model, since it lacks the massless adjoints to trigger higgsing down to the SM group. This actually follows from very general constraints on the possible representations in the massless spectrum of heterotic compactifications, as described in Section 17.1.2, but may be overcome by realizing the gauge group at higher Kac–Moody level, see Section 9.8.

Given this situation, the simplest solution to construct realistic particle physics models is instead to directly produce a gauge symmetry of the form $SU(3) \times SU(2) \times U(1)^n \times G_h$, containing an SM gauge factor. This can be achieved for the above $\mathbf{Z}_3$ model with the addition of a second Wilson line $a_3 = a_4$ in the second two-torus. Consider the choice

$$a_3 = a_4 = \tfrac{1}{3} (1, 1, 1, 2, 1, 0, 1, 1) \times (0, 0, \ldots, 0). \tag{8.67}$$

In the untwisted sector, the Wilson line projection $P \cdot a_3 \in \mathbf{Z}$ breaks further the gauge group down to $SU(3) \times SU(2) \times U(1)^8 \times SO(10)'$, which contains the SM gauge symmetry as proposed above. It also truncates the untwisted matter spectrum (8.64) to

$$3 \left[(\mathbf{3}, \mathbf{2}) + (\bar{\mathbf{3}}, \mathbf{1}) + (\mathbf{1}, \mathbf{2}) \right]. \tag{8.68}$$

These fields transform under the non-abelian SM gauge factor as three left-handed quarks Q_L, three right-handed quarks, and three Higgs multiplets or lepton doublets.

The 27 fixed points split now into nine different sectors, each with multiplicity three, for which the local orbifold gauge shift is

$$V + m_1 \, a_1 + m_3 \, a_3, \tag{8.69}$$

with $m_1, m_3 = 0, 1, -1$ corresponding respectively to the fixed point position •, ○, ×, in the first two two-tori. The massless fields in these twisted sectors are

$$3\,[\,5\,(\bar{\mathbf{3}}, \mathbf{1}) + 4\,(\mathbf{3}, \mathbf{1}) + 12\,(\mathbf{1}, \mathbf{2}) + 57\,(\mathbf{1}, \mathbf{1})\,]. \tag{8.70}$$

Note how these twisted sector fields neatly cancel the $SU(3)$ anomalies from the untwisted sector spectrum. The twisted sector produces a net number of three triplets, which can play additional right-handed quarks of the SM gauge sector. In addition there are extra doublets which can play the role of lepton/Higgs multiplets, and singlets which can include the right-handed leptons. The full massless spectrum of charged chiral multiplets is given in Appendix C. It is worth emphasizing that in this simple orbifold example the doublet–triplet splitting is not an issue since, as we will see in Section 9.7, all extra triplets beyond quarks become massive. The complete study of the massless spectrum requires a careful analysis of the $U(1)$ symmetries of the model, and of the particular linear combination leading to the hypercharge $U(1)$. We postpone this analysis to Section 9.7.

In any event, it is remarkable that a simple symmetry breaking structure based on the simplest abelian orbifold can already provide models unexpectedly close to the SM spectrum. Still the model seems to contain a large number of additional $U(1)$'s and additional matter beyond the minimal SM. This is a rather generic feature of toroidal orbifolds containing the SM spectrum, which however is greatly improved under a more careful analysis. Indeed, as further developed in Section 9.5, heterotic string compactifications have one-loop corrections, controlled by the structure of $U(1)$ anomalies. These corrections trigger breaking of gauge symmetries along flat directions of the scalar potential, in which all $U(1)$ factors except the hypercharge $U(1)$ are broken, and most exotic matter fields beyond the SM get a large mass. This is a spectacular improvement of the model building prospects of heterotic orbifolds beyond their tree level structure.

A Z'_6 orbifold model

Our second example is based on the $\mathbf{Z}'_6$ orbifold, with shift $v = \frac{1}{6}(1, 2, -3)$, and illustrates the possibility of building realistic particle physics models using even order orbifolds. It also enjoys the property of exploiting an underlying $SO(10)$ GUT symmetry, useful to produce models with a canonical $SU(5)$-like normalization for hypercharge. The underlying torus is defined by the root lattice of $G_2 \times SU(3) \times SO(4)$, and we choose the gauge embedding of the generator θ to be

$$V = \tfrac{1}{6}(3, 3, 2, 0, \ldots, 0) \times \tfrac{1}{6}(2, 0, \ldots, 0). \tag{8.71}$$

This breaks the first E_8 symmetry down to $SO(10) \times SU(2) \times U(1)^2$. The symmetry is further broken by the addition of two Wilson lines a_2, a_3 along the second and third 2-tori, given by

$$\begin{aligned}
a_2 &= \tfrac{1}{2}(1, 0, 1, 1, 1, 0, 0, 0) \times \tfrac{1}{4}(-3, 1, 1, -1, 1, 1, 1, -1), \\
a_3 &= \tfrac{1}{3}(1, 0, 0, 1, 1, 1, 1, 1) \times \tfrac{1}{3}(3, 1, 1, 1, 0, 0, 0, 0).
\end{aligned} \tag{8.72}$$

The unbroken gauge symmetry is then $SU(3) \times SU(2) \times U(1)^9 \times [SU(4) \times SU(2)']$. Using the techniques described in previous sections, a straightforward but somewhat tedious computation leads to the following massless particle spectrum

$$3(\mathbf{3}, \mathbf{2})_{1/6} + 3(\mathbf{\bar{3}}, \mathbf{1})_{-2/3} + 7(\mathbf{\bar{3}}, \mathbf{1})_{1/3} + 4(\mathbf{3}, \mathbf{1})_{-1/3} + 8(\mathbf{1}, \mathbf{2})_{-1/2}$$
$$+ 5(\mathbf{1}, \mathbf{2})_{1/2} + 8(\mathbf{1}, \mathbf{2})_0 + 3(\mathbf{1}, \mathbf{1})_1 + 16[(\mathbf{1}, \mathbf{1})_{1/2} + \text{h.c.}] + \text{SM singlets},$$

where the subindex is the charge under a $U(1)$ subgroup of $U(1)^9$, eventually identified with hypercharge.

The model contains an SM gauge symmetry factor, and the matter contains three quark and lepton families. As in the above $\mathbf{Z}_3$ example, the model as it stands contains additional gauge factors and exotic matter multiplets, in particular with fractional charges. This is a very generic feature of heterotic orbifolds with non-standard embedding, see Section 17.1.3. However, as already mentioned for the $\mathbf{Z}_3$ orbifold example, there are flat directions in which all these exotics become naturally massive. Again, a detailed analysis of this effect requires a systematic study of the low-energy effective action for heterotic string compactifications, to which we turn in Chapter 9.

8.4 Asymmetric orbifolds*

The heterotic string is clearly an asymmetric construction in which left- and right-movers are treated differently. This left–right asymmetric treatment may be extended to the internal coordinates in a compactification to 4d. In the resulting string constructions this internal sector does not admit a geometrical interpretation, since the left and right bosonic fields cannot in general be combined into well-defined spacetime coordinates. The constructions are nevertheless perfectly consistent and can be made modular invariant, leading to consistent 4d string vacua. Despite their non-geometric character, we abuse language and still refer to them as asymmetric "compactifications." Left–right asymmetric compactifications can in principle be constructed within different frameworks, including the fermionic models and Gepner model constructions in upcoming sections. Here we focus on their realization as asymmetric abelian toroidal orbifolds, or asymmetric orbifolds for short.

8.4.1 General construction

The idea is to consider an underlying toroidal compactification, and mod out by orbifold groups G_L, G_R acting differently on the left and right sectors, so that (8.6) is replaced by

$$\frac{\mathbf{T}_R^6}{G_R} \otimes \frac{\mathbf{T}_L^{22}}{G_L}. \tag{8.73}$$

The condition to preserve 4d $\mathcal{N} = 1$ supersymmetry is that the action of G_R on $\mathbf{T}_R^6$ is a subgroup of $SU(3)$, so that right-moving states in NS and R sectors pair up into spacetime supermultiplets.

The underlying toroidal compactification is described in terms of the Narain lattice construction in Section 5.2.2, namely through an even self-dual lattice $\Gamma_{22,6}$ of left and right momenta (P_L, P_R). Left–right symmetric orbifolds can be obtained e.g. from lattices of the form $\Gamma_{22,6} = \Gamma_{16,0} \oplus 6\,\Gamma_{1,1}$, with identical orbifold twists for left and right sectors, with $\Gamma_{16,0} = \Lambda_{E_8 \times E_8}$ or $\Lambda_{SO(32)}$, and $\Gamma_{1,1}$ is the lattice (3.82) of winding/momentum on $\mathbf{S}^1$.

Asymmetric orbifolds are obtained by considering more general orbifold twists, $g = (\theta_L, V_L; \theta_R, V_R)$ acting on left/right momenta as

$$g|P_L, P_R\rangle = e^{2\pi i\,(P_L \cdot V_L - P_R \cdot V_R)}\,|\,\theta_L P_L;\theta_R P_R\,\rangle, \tag{8.74}$$

where θ_L, θ_R, are rotations of $\mathbf{T}_L^{22}$, $\mathbf{T}_R^6$, and V_L, V_R are shifts on these tori. Note that V_L includes the action on the gauge degrees of freedom through a shift.

This type of construction can be carried out following the same idea used in defining the 10d heterotic string in Section 4.3.1. One formally introduces auxiliary worldsheet fields to obtain a left–right symmetric structure, on which the orbifold twists can act as usual, and then imposes the auxiliary coordinates to be frozen to zero. The vanishing of the momenta of the auxiliary bosonic coordinates usually requires the parameters of the underlying torus, like the metric or the B_{mn} background, to be frozen at particular values.

To illustrate the idea, consider the description of a purely left-moving toroidal compactification on a factorized torus $\mathbf{T}^2 \times \mathbf{T}^2 \times \mathbf{T}^2$, with each 2-torus defined by an $SU(3)$ root lattice. This can be defined by introducing auxiliary right-moving coordinates, so the full system describes a geometric toroidal compactification, for which left/right momenta $(p_L, p_R) \in \Gamma_{d,d}$ are as in (5.12). With hindsight we take the following backgrounds for the metric G_{mn} and B_{mn} fields for each $SU(3)$ torus

$$(G_{mn}) = \begin{pmatrix} 1 & -\frac{1}{2} \\ -\frac{1}{2} & 1 \end{pmatrix}, \quad (B_{mn}) = \begin{pmatrix} 0 & -\frac{1}{2} \\ \frac{1}{2} & 0 \end{pmatrix}, \tag{8.75}$$

where $G_{mn} = \vec{e}_m \cdot \vec{e}_n$, with $\vec{e}_1, \vec{e}_2$ generating the $SU(3)$ root lattice. Demanding that the auxiliary right-handed momenta p_R vanish identically, the radius is frozen at $R = \sqrt{\alpha'}$ and the left-moving momenta on each $\mathbf{T}^2$ have the structure

$$p_{L,m} = \sqrt{2}\,G_{mn} w^n. \tag{8.76}$$

When used as momenta (5.26) in heterotic string compactifications (with no Wilson lines), states $(w^1, w^2) = \pm(1, 0), \pm(0, 1), \pm(1, 1)$ have $(p_L)^2 = G^{mn} p_{L,m} p_{L,n} = 2$ and give rise to massless gauge bosons. They correspond to the non-zero roots of $SU(3)$, and combine with the generic KK $U(1)^2$ gauge bosons to produce an enhanced $SU(3)$ gauge group. These enhanced gauge symmetries are stringy in nature, since they involve winding states, and have already been mentioned in Section 5.2.2. The bottom line is that the freezing of the auxiliary degrees of freedom required to define left–right asymmetric models leads to freezing of moduli, at loci of enhanced gauge symmetries. This general property still holds when including the (asymmetric) orbifold twists.

Let us now describe the computation of the spectrum in an asymmetric orbifold. The untwisted sector is obtained as usual by performing a projection on states invariant under the asymmetric twist (8.74). In addition there are twisted sectors, required by modular invariance as usual. In contrast to the symmetric orbifold case, they do not admit a simple geometric interpretation, as there is no well-defined geometric notion of orbifold fixed points. It is possible, however, to obtain and interpret the multiplicity of twisted sectors from the auxiliary left–right symmetric construction. In a left–right symmetric orbifold with a sector twisted by $\theta = \theta_R = \theta_L$, the fixed point multiplicity is $\det'(1 - \theta)$ by the Lefschetz formula (8.49), with the prime indicating that the determinant is only over θ-eigenvalues not equal to 1. For asymmetric orbifolds, the result produced from modular invariance is analogous, modulo the following differences. Denoting I the sublattice of $\Gamma_{22,6}$ invariant under $\theta = (\theta_L, \theta_R)$ and I^* its dual, the degeneracy factor can be shown to be

$$D = \sqrt{\frac{\det'(1 - \theta_L) \det'(1 - \theta_R)}{|I^*/I|}}, \tag{8.77}$$

where $|I^*/I|$ is the index of I on I^*, see Appendix A. This multiplicity is derived as in Section 8.2.5, as the prefactor of twisted sector partition functions, obtained by modular transformations from the untwisted sector partition functions; e.g. the denominator arises from the Poisson resummation (A.14) required after a modular transformation $\tau \to -1/\tau$ from the untwisted sector. For symmetric orbifolds $I = I^*$, and one recovers the geometric fixed point multiplicity. In general, modular invariance guarantees that D is always an integer number.

As usual, modular invariance implies constraints on the allowed orbifold actions. Let us define the shift vector $V = (V_L, V_R)$ and denote by V^* its projection onto the invariant sublattice I, i.e. $V = V^* + b$ with $V^* \in I$ and b orthogonal to I. Let us focus on $\mathbf{Z}_N$ orbifolds, and denote by $e^{2\pi i\, r_i/N}$, $e^{2\pi i\, \tilde{r}_s/N}$, the complex eigenvalues of θ_L, θ_R, respectively. For odd N, modular invariance implies the constraint

$$(NV^*)^2 - \sum_i r_i^2 = 0 \mod N. \tag{8.78}$$

For even N the above condition is mod $2N$, and there are the additional constraints

$$\sum_s \tilde{r}_s = 0 \mod 2, \quad L \cdot \theta^{N/2} \cdot L = 0 \mod 2, \quad \forall L \in \Gamma_{22,6}. \tag{8.79}$$

In the presence of Wilson lines in the underlying torus there are additional constraints, see later for a partial discussion.

8.4.2 An asymmetric $\mathbf{Z}_3$ orbifold example

For illustration we construct an example of an asymmetric $\mathbf{Z}_3$ orbifold of the $E_8 \times E_8$ theory, based on the $SU(3)^3$ root lattice toroidal compactification with the above background metric and B-field (8.75) on each $\mathbf{T}^2$; this leads to a lattice $\Gamma_{2,2}$ of left and right momenta.

We take the Narain lattice to be $\Gamma_{22,6} = \Gamma_{E_8} \oplus \Gamma_{E_8} \oplus 3\,\Gamma_{2,2}$, and mod out by an asymmetric twist Θ of the form

$$\Theta | P_L, P_R \rangle = e^{2\pi i\, P_L \cdot V_L} | P_L; \theta_R\, P_R \rangle, \qquad (8.80)$$

acting as a rotation θ_R on right-moving momenta, and as a shift V_L on the left-movers. We consider the simple example of a right-moving twist identical to that in the symmetric $\mathbf{Z}_3$ orbifold, hence $\tilde{r} = (1, 1, -2)$, and a left-moving shift of "standard embedding" form, $V_L = \frac{1}{3}(1, 1, -2, 0, \ldots, 0)$ in one of the Γ_{E_8} lattices; this satisfies the modular invariance constraint (8.78). States in the untwisted sector have momenta in the invariant sublattice I, which has the form

$$I = \Gamma_{E_8} \oplus \Gamma_{E_8} \oplus 3(\Lambda_{SU(3)}, 0), \qquad (8.81)$$

with $\Lambda_{SU(3)}$ the $SU(3)$ *root* lattice. Modular transformations of the untwisted partition function imply that Θ-twisted sector states have momenta in the dual lattice I^* shifted by V^*, hence lie in the lattice

$$I^* + V^* = \Gamma_{E_8} \oplus (\Gamma_{E_8} + V^*) \oplus 3(\Lambda_W, 0), \qquad (8.82)$$

where Λ_W is the $SU(3)$ *weight* lattice.

Let us turn to the computation of the 4d massless spectrum in this model. The untwisted sector produces the 4d graviton, the antisymmetric tensor $B_{\mu\nu}$, the dilaton ϕ, gauge bosons of $E_8 \times E_6 \times SU(3) \times SU(3)^3$, and their 4d $\mathcal{N} = 1$ partners. There are also charged chiral multiples transforming as three copies of $(\mathbf{27}, \mathbf{3}; \mathbf{1}, \mathbf{1}, \mathbf{1})$ under $E_6 \times SU(3) \times SU(3)^3$, as in the symmetric $\mathbf{Z}_3$ orbifold with standard embedding in Section 8.2.4.

Concerning the twisted sector, in this case $|I^*/I| = 3 \times 3 \times 3$ and $\det(1 - \theta_R) = 27$, so the twisted sector degeneracy (8.77) is just $D = 1$. The right-moving mass formula in the Θ-twisted sector is the same as in the symmetric $\mathbf{Z}_3$ orbifold (8.28), with $v = \tilde{r}/N$. The left-moving mass formula is

$$\frac{\alpha' M_L^2}{2} = \frac{(P + V^*)^2}{2} + \frac{p_L^2}{2} + N_B - 1. \qquad (8.83)$$

with $P \in \Lambda_{E_8 \times E_8}$ and p_L in the weight lattice of $SU(3)^3$; since the left-moving sector is not rotated, it differs from the symmetric $\mathbf{Z}_3$ orbifold expression (8.28) by the fact that there are six independent components of the momentum p_L, and that the normal ordering constant is -1 instead of $-2/3$. Note also that in our example $V^* = V_L$. There are massless states transforming as

$$p_L^2 = 0,\ (P + V^*)^2 - 2,\ (\overline{\mathbf{27}}, \mathbf{3}; \mathbf{1}, \mathbf{1}, \mathbf{1}),$$
$$p_L^2 = 4/3,\ (P + V)^2 = 4/3,\ (\mathbf{27}, \mathbf{1}; \underline{\mathbf{3}, \mathbf{1}, \mathbf{1}}) + (\mathbf{27}, \mathbf{1}; \underline{\bar{\mathbf{3}}, \mathbf{1}, \mathbf{1}}), \qquad (8.84)$$
$$p_L^2 = 8/3,\ (P + V)^2 = 2/3,\ (\mathbf{1}, \bar{\mathbf{3}}; \underline{\mathbf{3}, \mathbf{3}, \mathbf{1}}) + (\mathbf{1}, \bar{\mathbf{3}}; \underline{\bar{\mathbf{3}}, \bar{\mathbf{3}}, \mathbf{1}}) + (\mathbf{1}, \bar{\mathbf{3}}; \underline{\bar{\mathbf{3}}, \mathbf{3}, \mathbf{1}}),$$

all with $N_B = 0$. The number of net E_6 generations in this model is 24, compared to 36 in the symmetric $\mathbf{Z}_3$ orbifold, and unlike the latter there are $\mathbf{27} + \overline{\mathbf{27}}$ pairs.

It is possible to build models including Wilson lines in the underlying toroidal compactification. These can be constructed starting with a parent model with no Wilson lines and introducing an extra shift action on the lattice $\Gamma_{22,6}$, generalizing the embedding of the space group as a gauge shift in Section 8.3.2. A useful necessary condition for an asymmetric orbifold to admit Wilson lines is that the twisted sector degeneracy in the parent model must be bigger than one. Indeed, an order N Wilson line a changes the invariant lattice I to $I_N = \{L \in I \mid L \cdot a \in \mathbf{Z}\}$, its dual to $I_N^* = \{P^* + ka \mid P^* \in I^*, k = 0, \ldots, N-1\}$, and the index to $\left|I_N^*/I_N\right| = N^2\left|I^*/I\right|$, so the twisted sector degeneracy is reduced by a factor of N. For instance, the above asymmetric $\mathbf{Z}_3$ orbifold does not allow the introduction of Wilson lines, since $D = 1$. On the other hand, the analogous asymmetric $\mathbf{Z}_3$ orbifold with the underlying $\mathbf{T}^6$ defined instead by an E_6 root lattice has non-trivial twisted sector degeneracy $D = 3$, since $|I^*/I| = 3$ and $\det(1 - \theta_R) = 27$. Such models admit the introduction of one Wilson line of order 3.

The exploration of particle physics model building using asymmetric toroidal orbifolds has been rather limited, since they offer no obvious advantages over symmetric orbifolds. However, the fermionic models described in the next section may be regarded as asymmetric orbifolds (by the bosonization in Section 4.2.6). They provide a formulation more amenable to systematic model search, which indeed leads to phenomenologically interesting models.

One interesting general property of asymmetric orbifolds, already manifest in the above $\mathbf{Z}_3$ example, is the generic absence of untwisted moduli, e.g. $T_{i\bar{j}}$, which are quite generically projected out by the asymmetric orbifold action (equivalently, frozen at particular values to allow for the asymmetry). This suggests that non-geometric compactifications like asymmetric orbifolds may be useful in fixing the moduli in string compactifications. Indeed, certain asymmetric orbifolds can be related to the non-geometric compactifications in Section 14.4, which contain non-geometric fluxes fixing the moduli.

8.5 The fermionic construction

The bosonization/fermionization in Section 4.2.6 states that a left-moving real bosonic coordinate X compactified on a circle at the critical radius $R = \sqrt{\alpha'}$ is equivalent to a complex fermion; this was already exploited in Section 4.3.5 to replace the 16 purely left-moving bosonic coordinates X^I by 32 real fermions. It can also be used to define a new construction technique for 4d heterotic string models, which fermionizes all left and right worldsheet bosons, except those corresponding to the non-compact 4d. This "compactification" framework is known as the fermionic construction, and can be regarded as providing a very tractable version of the earlier (symmetric or asymmetric) toroidal orbifolds with compactification torus at the critical radius. The worldsheet theories are defined in terms of free fermions, with different possible boundary conditions, which allow for a α'-exact CFT computation of the 4d spectrum. The fermionic construction leads to a large class of 4d compactifications, including semi-realistic 4d $\mathcal{N} = 1$ models with a spectrum close to the MSSM. In this section we briefly review the general properties of these constructions.

8.5.1 Boundary conditions and spectrum

Let us fix the notation for the worldsheet fermions obtained after fermionization of the compact bosonic coordinates. In the right sector there are 20 real fermions, as follows:

- Two real fermions ψ^μ corresponding to the 4d spacetime (in light-cone gauge).
- For each internal coordinate $m = 1, \ldots, 6$ there is a triplet of real fermions (x^m, y^m, ω^m), two of which arise from fermionization of the bosonic coordinate.

In the left sector there are 44 real fermions, as follows:
- For each internal coordinate $m = 1, \ldots, 6$ there is a pair of real fermions $(\bar{y}^m, \bar{\omega}^m)$.
- The 16 complex fermions of the 10d fermionic formulation, are split for later convenience into three sets of $16 = 5 + 3 + 8$, denoted $(\bar{\psi}^1, \ldots, \bar{\psi}^5, \bar{\eta}^1, \bar{\eta}^2, \bar{\eta}^3, \bar{\phi}^1, \ldots, \bar{\phi}^8)$.

The 20 right- and 44 left-moving real fermions can be jointly treated as a vector $\mathbf{f}$ with 64 components f^i. These satisfy 2d boundary conditions, which we write as

$$f_i(\sigma + \ell) = -e^{i\pi\alpha_i} f_i(\sigma),\tag{8.85}$$

so that, e.g., antiperiodic (NS) and periodic (R) boundary conditions correspond to $\alpha_i = 0, 1$. The phases α_i may in general be fractional, although in most models in the literature this is the case only for the 16 left-moving complex fermions.

A model is defined by a set of p 64-component "basis vectors" $\mathbf{V}_a$, $a = 1, \ldots, p$, of order N_a, i.e. $N_a \mathbf{V}_a = 0 \pmod 2$. Their linear combinations with integer coefficients generate possible vectors of boundary conditions $\boldsymbol{\alpha}$ for the 64 fermions,

$$\boldsymbol{\alpha} = \sum_{a=1}^{p} \sum_{n_a=1}^{N_a-1} n_a \mathbf{V}_a.\tag{8.86}$$

Any pair of, e.g., left-handed fermions whose boundary conditions match in all such sectors can be combined into a complex fermion. Their bilinears can be used to construct massless left-moving states leading to spacetime gauge bosons, in analogy with the realization of gauge bosons at the end of Section 4.3.5. On the other hand, real fermions not combining into complex ones do not lead to such gauge bosons; hence models with unpaired real fermions lead to reduction of the gauge group rank. Boundary conditions for real fermions must be periodic or antiperiodic, but those for complex fermions admit fractional phases.

The one-loop partition function includes a sum over all possible boundary conditions generated by the basis vectors,

$$Z = \sum_{\boldsymbol{\alpha}, \boldsymbol{\beta}} C(\boldsymbol{\alpha}, \boldsymbol{\beta}) \, Z(\boldsymbol{\alpha}, \boldsymbol{\beta}).\tag{8.87}$$

The terms $Z(\boldsymbol{\alpha}, \boldsymbol{\beta})$ contain the product of all the individual fermion partition functions. The coefficients $C(\boldsymbol{\alpha}, \boldsymbol{\beta})$ are complex phases constrained by modular invariance conditions, in analogy with the orbifold compactifications in previous sections. The freedom in the choice of these phases can be parametrized by a $p \times p$ matrix k_{ab} of rational numbers with $-1 < k_{ab} \leq 1$. In terms of this matrix, we have

$$C(V_a, V_b) = (-1)^{s_a + s_b} \exp\left(i\pi \, k_{ba} - \frac{1}{2} V_a \cdot V_b\right), \qquad (8.88)$$

where s_a is the 4d component of V_a and the dot product is Lorentzian, namely $V \cdot W = V_L \cdot W_L - V_R \cdot W_R$ in terms of the left- and right-moving parts. For two sectors $\alpha = \sum_{a=1}^p n_a V_a$, $\beta = \sum_{a=1}^p m_a V_a$ the phases in (8.87) are

$$C(\alpha, \beta) = (-1)^{s_\alpha + s_\beta} \exp\left[i\pi \sum_{a,b} m_a \left(k_{ab} - \frac{1}{2} V_a \cdot V_b\right) n_b\right]. \qquad (8.89)$$

One loop modular invariance of the partition function implies the following conditions on the basis vectors V_a and the matrix k_{ab}

$$k_{ab} + k_{ba} = \frac{1}{2} V_a \cdot V_b \bmod 2,$$

$$N_a \, k_{ab} = 0 \bmod 2,$$

$$k_{aa} + k_{a1} = -s_a + \frac{1}{4} V_a \cdot V_a \bmod 2, \qquad (8.90)$$

where the subindex 1 in the last line denotes the vector $V_a = \mathbf{1}$, with all the 64 entries equal to 1, which is required (by modular invariance) to be included in the basis. This basis vector generates the "untwisted" sector of the theory, which includes the gravity multiplet and part of gauge bosons, as shown in the examples in Section 8.5.2. The explicit dependence of the conditions (8.90) on k_{ab} may be removed by multiplying by appropriate integers, so that the constraints may be recast in terms of the V_a's and their orders N_a,

$$N_{[ab]} V_a \cdot V_b = 0 \bmod 4, \qquad N_a V_a \cdot V_a = 0 \bmod 8, \qquad (8.91)$$

where $N_{[ab]}$ is the least common multiple of N_a and N_b. From these conditions it can be shown, e.g., that the net number of real fermions simultaneously periodic for any three basis vectors must be even.

The massless spectrum is obtained by acting on the vacuum $|0\rangle_\alpha$ of each sector of boundary conditions α with the relevant fermionic creation operators ψ^i_{-r}. Sectors with fermion zero modes lead to degenerate groundstates, in particular N complex fermions with periodic boundary conditions lead to a degenerate vacuum transforming as a spinor of the corresponding $SO(2N)$, in close analogy with the Ramond fermions in Section 4.2.1.

To each complex fermion g^i one can associate a $U(1)_i$ charge, which may be realized as a gauge boson. Each state in any sector α of the Hilbert space has a definite charge under these $U(1)$ symmetries, denoted Q^α_i; these charges are directly related to the boundary conditions α_i and the oscillator numbers by

$$Q^\alpha_i = \frac{1}{2} \alpha_i + F^\alpha_i, \qquad (8.92)$$

where F_i^α is a fermion number operator counting $+1$ for a negative modding fermionic oscillator ψ_{-r}^i and -1 for its positive modding counterpart. One can also define analogous (pseudo)charges for real fermions, in which case F_i^α counts $(+1)$ for each oscillator. Concerning sectors with zero modes, the fermion number is defined by $F_i^\alpha = 0, -1$ for the two vacuum states associated to each complex fermion with periodic boundary conditions. The charges (8.92) form charge (or "pseudo-charge" for real fermions) vectors, $\mathbf{Q}^\alpha = \frac{1}{2}\alpha + \mathbf{F}^\alpha$, associated with states in a sector α. They are useful to compute the massless spectrum, and to express the generalized GSO conditions implied by modular invariance. In the present case the projections for states in a sector $\alpha = \sum_{a=1}^p a_a V_a$ can be shown to impose the condition

$$V_b \cdot \mathbf{Q}^\alpha = \sum_a a_a k_{ba} + s_b \mod 2 , \quad \forall V_b. \tag{8.93}$$

The mass formulae may also be written in terms of the charges as

$$\frac{\alpha' M_R^2}{2} = -\frac{1}{2} + \frac{1}{2}\mathbf{Q}_R^\alpha \cdot \mathbf{Q}_R^\alpha,$$

$$\frac{\alpha' M_L^2}{2} = -1 + \frac{1}{2}\mathbf{Q}_L^\alpha \cdot \mathbf{Q}_L^\alpha. \tag{8.94}$$

Level matching requires $M_R = M_L$ for each state, as usual.

8.5.2 Examples

We illustrate the model building possibilities of the present framework with a couple of examples.

An SO(10) example

Let us sketch the structure of a prototypical 4d model with semi-realistic spectrum. We introduce the five basis vectors $\mathbf{V}_a = (\mathbf{1}, \mathbf{S}, \mathbf{b_1}, \mathbf{b_2}, \mathbf{b_3})$ in Table 8.5, where 1(0) denote R(NS) boundary conditions, and furthermore choose the phases

$$C(\mathbf{b_i}, \mathbf{b_j}) = C(\mathbf{b_i}, \mathbf{S}) = C(\mathbf{1}, \mathbf{1}) = -1, \quad i, j = 1, 2, 3, \tag{8.95}$$

with the remaining phases being determined by the consistency constraints. These determine the GSO projection, whose discussion we skip.

In the absence of the basis vectors $\mathbf{b_i}$, the model corresponds to a toroidal 4d $\mathcal{N} = 4$ compactification of the $SO(32)$ heterotic theory, on a $\mathbf{T}^6$ at a point of maximal enhanced $SO(44)$ gauge symmetry, already described in terms of the Narain lattice in Section 5.2.2. The bosonic states arise from the sector $\alpha = 2 \times \mathbf{1} = \mathbf{0}$, for which all worldsheet fermions have NS boundary conditions. The massless states

$$\tilde{\psi}_{-1/2}^\mu \alpha_{-1}^\nu |0\rangle_\mathbf{0}, \tag{8.96}$$

Table 8.5 *Boundary conditions for right and left fermions defining a 4d $SO(10)$ model*

	ψ^μ	x^1, x^2	x^3, x^4	x^5, x^6	$\bar\psi^1, \ldots, \bar\psi^5$	$\bar\eta^1, \bar\eta^2, \bar\eta^3$	$\bar\phi^1, \ldots, \bar\phi^8$
1	1	1	1	1	$1, \ldots, 1$	$1, 1, 1$	$1, \ldots, 1$
S	1	1	1	1	$0, \ldots, 0$	$0, 0, 0$	$0, \ldots, 0$
$\mathbf{b_1}$	1	1	0	0	$1, \ldots, 1$	$1, 0, 0$	$0, \ldots, 0$
$\mathbf{b_2}$	1	0	1	0	$1, \ldots, 1$	$0, 1, 0$	$0, \ldots, 0$
$\mathbf{b_3}$	1	0	0	1	$1, \ldots, 1$	$0, 0, 1$	$0, \ldots, 0$

	$y^3, \ldots, y^6$	$\bar{y}^3, \ldots, \bar{y}^6$	$y^1, y^2, \omega^5, \omega^6$	$\bar{y}^1, \bar{y}^2, \bar\omega^5, \bar\omega^6$	$\omega^1, \ldots, \omega^4$	$\bar\omega^1, \ldots, \bar\omega^4$
1	$1, \ldots, 1$	$1, \ldots, 1$	$1, \ldots, 1$	$1, \ldots, 1$	$1, \ldots, 1$	$1, \ldots, 1$
S	$0, \ldots, 0$	$0, \ldots, 0$	$0, \ldots, 0$	$0, \ldots, 0$	$0, \ldots, 0$	$0, \ldots, 0$
$\mathbf{b_1}$	$1, \ldots, 1$	$1, \ldots, 1$	$0, \ldots, 0$	$0, \ldots, 0$	$0, \ldots, 0$	$0, \ldots, 0$
$\mathbf{b_2}$	$0, \ldots, 0$	$0, \ldots, 0$	$1, \ldots, 1$	$1, \ldots, 1$	$0, \ldots, 0$	$0, \ldots, 0$
$\mathbf{b_3}$	$0, \ldots, 0$	$0, \ldots, 0$	$0, \ldots, 0$	$0, \ldots, 0$	$1, \ldots, 1$	$1, \ldots, 1$

correspond to the 4d graviton, dilaton, and two-index antisymmetric field $B_{\mu\nu}$. The $SO(44)$ gauge bosons correspond to states

$$\tilde\psi^\mu_{-1/2} b^i_{-1/2} b^j_{-1/2}|0\rangle, \quad \tilde\psi^\mu_{-1/2} b^{i\,*}_{-1/2} b^{j\,*}_{-1/2}|0\rangle, \quad \tilde\psi^\mu_{-1/2} b^i_{-1/2} b^{j\,*}_{-1/2}|0\rangle, \tag{8.97}$$

where b, b^* are oscillator operators corresponding to the 22 complex left-moving fermions. The $\mathcal{N} = 4$ vector multiplet scalar partners are obtained by replacing $\tilde\psi^\mu$ by the oscillators of any of the other six right-moving fermions, while the fermion partners are obtained from the **S** sector. There is also an extra $U(1)^6$ from right-moving fermions tensored with the light-cone bosonic oscillators, corresponding to the $\mathcal{N} = 4$ graviphotons, c.f. Section 2.4.1.

The presence of the additional three basis vectors $\mathbf{b_i}$ implements an effective $\mathbf{Z}_2 \times \mathbf{Z}_2$ orbifold projection, preserving only 4d $\mathcal{N} = 1$ supersymmetry. In addition, the gauge symmetry is broken to $SO(10) \times SO(6)^3 \times E_8$. The gauge symmetry with Cartan generators associated to the fermions $\bar\psi^2, \ldots, \bar\psi^5, \bar\eta^i, \bar{y}^j, \bar\omega^j$ is reduced by the GSO projection, while that associated to $\bar\phi^1, \ldots, \bar\phi^8$ is enhanced from $SO(16)$ to E_8. The computation of the massless spectrum shows that each of the $\mathbf{b_i}$ sectors leads to 16 $SO(10)$ families; these spinors **16** arise from the fermion zero modes of the five complex fermions $\bar\phi^1, \ldots, \bar\phi^5$ with periodic boundary conditions, while the multiplicity 16 comes from the additional $6+6$ real fermions with periodic boundary conditions, truncated by the different GSO projections. In total the model has 48 generations. There are also six multiplets transforming as **10**s of $SO(10)$, as well as a number of singlets.

A 3-generation MSSM-like example

The above model is not a realistic $SO(10)$ GUT, since it has too many generations and no adjoint scalars to break the symmetry down to the SM. An improved example can be

Table 8.6 *Additional boundary conditions to obtain a MSSM-like model*

	ψ^μ	x^1, x^2	x^3, x^4	x^5, x^6	$\bar\psi^1,\ldots,\bar\psi^5$	$\bar\eta^1$	$\bar\eta^2$	$\bar\eta^3$	$\bar\phi^1,\ldots,\bar\phi^8$
c_1	1	1	0	0	1, 1, 1, 1, 1	1	0	0	0, 0, 0, 0, 0, 0, 0, 0
c_2	1	0	0	1	1, 1, 1, 0, 0	1	0	1	1, 1, 1, 1, 0, 0, 0, 0
c_3	1	0	1	0	$\frac{1}{2},\frac{1}{2},\frac{1}{2},\frac{1}{2},\frac{1}{2}$	$\frac{1}{2}$	$\frac{1}{2}$	$\frac{1}{2}$	$\frac{1}{2}, 0, 1, 1, \frac{1}{2}, \frac{1}{2}, \frac{1}{2}, 1$

	y^3, y^6	$y^4, \bar y^4$	$y^5, \bar y^5$	$\bar y^3, \bar y^6$	y^1, ω^6	$y^2, \bar y^2$	$\omega^5, \bar\omega^5$	$\bar y^1, \bar\omega^6$	ω^1, ω^3	$\omega^2, \bar\omega^2$	$\omega^4, \bar\omega^4$	$\bar\omega^1, \bar\omega^3$
c_1	1	0	0	1	0	0	1	0	0	0	1	0
c_2	0	0	0	1	0	1	0	1	1	0	1	0
c_3	0	0	1	1	1	0	0	1	0	1	0	0

obtained by adding the extra basis vectors c_i in Table 8.6, to break the symmetry and reduce the number of generations.

We simply quote the existence of an example with a modular invariant choice of phases $C(\alpha, \beta)$, in which the number of generations is reduced to three, and the gauge group is $SU(3) \times SU(2) \times U(1)^{12} \times (SO(4) \times SU(3))$, i.e. containing the SM and an extra (not completely hidden) sector. Note that the gauge group has rank 19, arising from the 16 complex fermions $\bar\phi^a$, $\bar\psi^1,\ldots,\bar\psi^5$, $\bar\eta^1, \bar\eta^2, \bar\eta^3$ and the three pairs $(\bar y^3, \bar y^6)$, $(\bar y^1, \bar\omega^6)$ and $(\bar\omega^1, \bar\omega^3)$, which have equal boundary conditions and thus combine into complex fermions. The remaining left-moving fermions are unpaired and do not produce extra gauge bosons. A linear combination of the 12 $U(1)$'s may be identified with hypercharge, while another combination is anomalous and gets massive through the Green–Schwarz mechanism, described in Section 9.5. Besides the SM chiral multiplets, there are four sets of MSSM Higgs doublets, and further multiplets involving the visible and the non-abelian "hidden" sector $SO(4) \times SU(3)$. There are also extra $SU(3) \times SU(2)$ singlets with fractional $Q = \pm 1/2$ charges, which nevertheless become very massive in the gauge symmetry breaking induced by the one-loop FI term associated with an anomalous $U(1)$, see Section 9.5.

There exist many other models in the literature, constructed with these techniques, including also models based on the flipped $SU(5)$ unification of Section 1.2.5. The structure of their 4d spectrum is quite similar to the models obtained from orbifold compactifications; in fact, the above example is related to a $\mathbf{Z}_2 \times \mathbf{Z}_2$ symmetric orbifold. Hence, their 4d effective action and phenomenology also follow similar lines, and are not further pursued here. An interesting new possibility within the fermionic construction is the use of sets of unpaired real fermions for rank reduction of the gauge group. In particular this can be used to construct models where the gauge symmetry is realized at higher Kac–Moody level, see Section 9.8; this allows the construction of heterotic string models with $SU(5)$ or $SO(10)$ GUT symmetry with massless adjoint scalars to play the role of GUT-Higgses. A detailed discussion is however beyond our scope.

Let us finally remark that the 2d fermion-boson equivalence may also be used in the opposite direction, i.e. one can bosonize all worldsheet fermions and develop a bosonic formalism to construct 4d compactifications. This is the path followed by the so-called *covariant lattice* approach (furthermore based on a Lorentz covariant formalism, as suggested by its name), which also allows the construction of a large number of 4d $\mathcal{N} = 1$ string vacua. We will not discuss it here, however, since this formalism has not been much used in phenomenological string model building.

8.6 Gepner models

Asymmetric orbifolds or fermionic models are simple examples of the use of abstract 2d CFTs to describe the internal "compactification" sector, without resorting to an explicit geometric interpretation. This can be generalized to the construction of 4d string theory models in which the internal sector is described by interacting (but solvable) 2d CFTs. In this section we illustrate this possibility in the so-called Gepner model "compactifications" of the $E_8 \times E_8$ heterotic string. Since interactions on the 2d worldsheet are related to curved backgrounds, such compactifications are related to compactifications on curved CY manifolds of stringy size, as we will also explain.

8.6.1 General construction

Gepner models are based on a tensor product of $\mathcal{N} = 2$ superconformal minimal models. The construction requires a number of CFT ingredients, for which we refer to Appendix E.

Minimal models

Conformal symmetry in 2d is powerful enough to constrain the dynamics of non-trivial interacting 2d CFTs. In particular, there is a complete characterization of $\mathcal{N} = 2$ superconformal field theories (SCFTs) with $c < 3$, which are moreover solvable, e.g. the exact spectrum of primary fields is known. These so-called *minimal models* are the simplest interacting $\mathcal{N} = 2$ SCFTs, and can be constructed as cosets using the Sugawara construction in Section E.3. We now review their main properties relevant to our present purposes.

Minimal models are chiral (namely purely, e.g., left-moving) 2d $\mathcal{N} = 2$ SCFTs. They are denoted by $\mathcal{A}_k$, with a positive integer label k, known as the level, which is related to the central charge by

$$c_k = \frac{3k}{k+2}. \tag{8.98}$$

The minimal model $\mathcal{A}_k$ has a finite number of unitary highest weight representations, i.e. primary states. They are denoted $|l, q, s\rangle$, where the quantum numbers are integers in the ranges

$$0 \le l \le k, \ \ 0 \le |q - s| \le l, \ \ l + q + s = 0 \bmod 2, \tag{8.99}$$

with $s = 0, 2$ for the NS sector and $s = \pm 1$ in the R sector. Here q is defined modulo $2(k + 2)$ and s is defined modulo 4. Recalling the $\mathcal{N} = 2$ superconformal algebra in section E.4, each state $|l, q, s\rangle$ has an associated conformal weight and superconformal $U(1)$ charge. They are given by

$$h = \frac{l(l+2) - q^2}{4(k+2)} + \frac{s^2}{8}, \quad Q = \frac{-q}{k+2} + \frac{s}{2}. \tag{8.100}$$

Left and right minimal models can be tensored into $\mathcal{A}_k \otimes \tilde{\mathcal{A}}_k$. There is an automorphism implying an equivalence of states

$$|l, q, s\rangle \otimes |\bar{l}, \bar{q}, \bar{s}\rangle \equiv |k - l, q + k + 2, s + 2\rangle \otimes |k - \bar{l}, \bar{q} + k + 2, \bar{s} + 2\rangle, \tag{8.101}$$

where right-moving states are denoted with overlines. These transformations are useful to bring states with arbitrary quantum number into the standard range (8.99). The partition function for $\mathcal{A}_k \otimes \tilde{\mathcal{A}}_k$ reads

$$Z = \sum_{l\bar{l}, q, \bar{q}, s, \bar{s}} Z_{l, \bar{l}} \, \chi_q^{ls} \, \chi_{\bar{q}}^{\bar{l}\bar{s}*}, \tag{8.102}$$

where the χ's are $\mathcal{N} = 2$ characters, and the $Z_{l, \bar{l}}$ are integer multiplicities as in (E.30). They are given by the so-called $SU(2)$ affine modular invariants, which admit an ADE classification. The A-series is the diagonal invariant $Z_{l\bar{l}} = \delta_{l\bar{l}}$ and exists for all levels. The D-series exists for all even levels, while the exceptional invariants exist only for $k = 10, 16, 28$.

Tensoring $\mathcal{N} = 2$ minimal models and projections

Minimal models can be used to construct string compactifications. For type II strings the total central charge for right-movers must be $\tilde{c} = 15$ and the Minkowski degrees of freedom contribute $\tilde{c} = 6$, so the internal CFT must have central charge $\tilde{c} = 9$. Since each minimal model has $c_k < 3$, we must tensor r $\mathcal{N} = 2$ minimal models into $\mathcal{A}_{k_1 \cdots k_r} = \mathcal{A}_{k_1} \otimes \cdots \otimes \mathcal{A}_{k_r}$, with

$$\sum_{i=1}^{r} \frac{3k_i}{k_i + 2} = 9, \tag{8.103}$$

and use the resulting CFT $\mathcal{A}_{k_1 \cdots k_r}$ for the left and right sectors. For heterotic string compactifications, we need $\tilde{c} = 9$, $c = 22$, hence we take $\tilde{\mathcal{A}}_{k_1 \cdots k_r}$ as the internal right sector CFT, and in the left sector we take $\mathcal{A}_{k_1 \cdots k_r}$, supplemented with a level one $SO(10) \times E_8$ Kac–Moody algebra, with $c_{KM} = 13$. These heterotic models with left and right $\mathcal{N} = 2$ SCFT factors are called $(2, 2)$, and are closely related to CY (or orbifold) compactifications with standard embedding, as already mentioned in Sections 7.3.2 and 8.2.3. As described shortly, the left-moving superconformal $U(1)$ current enhances the $SO(10)$ symmetry into an E_6, showing up as a 4d gauge symmetry. Modifications not preserving the left-moving SCFT structure, thus called $(0, 2)$ models, generically lead to smaller gauge groups (as in non-standard embedding geometric compactifications), as studied later on. The extra E_8

will not play an important role at present, and is momentarily ignored. States are therefore denoted as

$$V_L \otimes V_R = (w; q_1, \ldots, q_r; s_1, \ldots, s_r) \otimes (\bar{w}; \bar{q}_1, \ldots, \bar{q}_r; \bar{s}_1, \ldots, \bar{s}_r), \quad (8.104)$$

where $\bar{w}$ is an $SO(2)$ weight, associated to the 4d spacetime coordinates (in light-cone gauge), and w is an $SO(10)$ weight.

The tensor product $\mathcal{A}_{k_1 \cdots k_r}$ actually does not give rise to a consistent $(2, 2)$ compactification, and requires certain projections to obtain the correct spin-statistics in all sectors, and a GSO projection to produce spacetime 4d $\mathcal{N} = 1$ supersymmetry; the resulting theories are known as Gepner models. These projections can be understood as extensions of the CFT algebra by simple currents, see Section E.5.3. Here we provide a more pragmatic implementation.

The generalized GSO projection amounts to truncating to states with odd total superconformal $U(1)$ charge Q_{tot}. To provide a useful expression for Q_{tot}, we introduce a scalar product of states in $\mathcal{A}_{k_1 \cdots k_r}$

$$V \cdot V' = \frac{1}{4} w \cdot w' + \frac{1}{4} \sum_{i=1}^{r} s_i s_i' - \frac{1}{2} \sum_{i=1}^{r} \frac{q_i q_i'}{k_i + 2}. \quad (8.105)$$

In addition we define the vector

$$\beta_0 = (c; 1, \ldots, 1; 1, \ldots, 1), \quad (8.106)$$

where c is an $SO(2)$ conjugate spinor weight. Then the total $U(1)$ superconformal charge of a state V is given by $Q_{\text{tot}}(V) = 2\beta_0 \cdot V$. The above mentioned generalized GSO projection is therefore

$$Q_{\text{tot}}(V) = 2\beta_0 \cdot V = \text{odd}. \quad (8.107)$$

In the right sector this implies spacetime 4d $\mathcal{N} = 1$ supersymmetry, while for the left sector it guarantees the presence of 4d gauge bosons transforming as $SO(10)$ spinors, enhancing $SO(10) \times U(1)_{\text{tot}}$ to an E_6 spacetime gauge symmetry. Additional projections are required to ensure that the fermions in the various sectors have aligned (NS or R) boundary conditions. To implement them we introduce r vectors

$$\beta_i = (v; 0, \ldots, 0; \underbrace{0, \ldots, 2, \ldots, 0}_{\text{entry "2" at } i\text{th position}}), \quad i = 1, \ldots, r, \quad (8.108)$$

with v an $SO(2)$ vector weight. The alignment projection is then

$$2\beta_i \cdot V = \text{even}. \quad (8.109)$$

The above projections are performed independently on the right and left states denoted by V_R and V_L. Modular invariance requires these two vectors to differ by an integer linear combination of β_0 and β_i, i.e.

$$V_L - V_R = m_0 \beta_0 + \sum_{i=1}^{r} m_i \beta_i, \quad \text{with } m_0, m_i \in \mathbf{Z}. \tag{8.110}$$

The sectors with $(m_0, m_1) \neq (0, 0)$ can be regarded as twisted sectors of the orbifold quotient implicit in the projections (8.107), (8.109), see also Section 8.6.4.

Finally, to compensate for the different modular transformations of the right-moving $SO(2)$ and left-moving $SO(10)$, the state V_L must be shifted by a vector $\mu = (v; 0, \ldots, 0; 0, \ldots, 0)$ – however, not to be included when computing the total charge Q_{tot} of states. With the above conditions and projections, the overall Gepner model invariant partition function has the structure

$$Z = \frac{1}{2^r} \sum_{l_L, V_L, l_R, V_R} Z_{l_L l_R} \, \chi_{V_L}^{l_L} \, \chi_{V_R}^{l_R \, *}. \tag{8.111}$$

Here $Z_{l_L l_R}$ is the product of the multiplicities of the r affine invariants. Unless stated otherwise, we take the diagonal invariant $Z_{l_L l_R} = \delta_{l_L l_R}$, available for all k.

8.6.2 Massless spectrum and examples

The mass formulae for the string states in heterotic Gepner models read

$$\frac{\alpha' M_R^2}{2} = \tilde{N} + \frac{\bar{w}^2}{2} + \bar{h}_{\text{int}} - \frac{1}{2},$$

$$\frac{\alpha' M_L^2}{2} = N + \frac{w^2}{2} + \frac{P^2}{2} + h_{\text{int}} - 1, \tag{8.112}$$

where h_{int}, $\bar{h}_{\text{int}}$ are the internal CFT conformal weight contributions to the mass, as in (E.14). Also $\bar{w}$ and w are $SO(2)$, $SO(10)$ weights, and P is an E_8 lattice vector. As usual $\tilde{N}$, N are oscillator numbers. It is easy to show that at the massless level, these theories contain the 4d $\mathcal{N} = 1$ gravity multiplet, neutral chiral multiplets describing the dilaton and other moduli, and vector multiplets of $E_6 \times E_8 \times U(1)^{r-1}$.

In addition there are a number of charged chiral multiplets in the **27**s and $\overline{\mathbf{27}}$s of E_6. They are most easily tagged by focusing on the right-moving NS sector and the left-moving weights of the **10** of $SO(10)$; the states completing the multiplets are obtained by SUSY transformations and E_6 symmetry, which correspond to translation of the vectors V_R, V_L by the β_0 generator. Such right-moving massless states are obtained for $\bar{w} = 0$ and $\bar{h}_{\text{int}} = 1/2$, while left-moving massless states in the **10** of $SO(10)$ are obtained for $P^2 = 0$, $w^2 = 1$, $h_{\text{int}} = 1/2$. We thus use (8.100) to look for combinations of r primary states satisfying $\sum_{i=1}^{r} h_i = 1/2$, and with V_R surviving the different projections (8.107), (8.109). Such states can be checked to be combinations of primary fields of the form $|\bar{l}, \bar{l}, 0\rangle$, which can be shown to have $\tilde{Q}_{\text{tot}} = \sum_i (-\bar{l}_i)/(k_i + 2) = -1$ for a massless state; antiparticles arise from similar product states with the replacement $|\bar{l}, \bar{l}, 0\rangle \rightarrow |\bar{l}, -\bar{l}, 0\rangle$. The same kind of states can then be used in the left-moving sector; tensor products of states $|l, l, 0\rangle$ have $Q_{\text{tot}} = -1$, and are eventually enhanced to **27**s of E_6, while products of states $|l, -l, 0\rangle$ have $Q_{\text{tot}} = +1$, and lead to $\overline{\textbf{27}}$s. As mentioned, the additional states beyond the **10** arise by shifting V_L, V_R by the β_0 vector.[1]

The E_6 singlets in the massless spectrum are obtained for $P = w = 0$, $h_{\text{int}} = 1$ and $Q_{\text{tot}} = -1$ (or $Q'_{\text{tot}} = 0$ after addition of the μ shift). These states can be seen to have either all $s_i = 0$ or only one of them equal to 2. The latter are simply obtained by making the replacement $s_i \rightarrow s_i + 2$ in each of the states associated to the **27**s, $\overline{\textbf{27}}$s. In the related geometric CY compactifications, see Section 8.6.3, they are associated with CY Kähler and complex structure moduli. To obtain the former there is no simple rule and one must scan for massless states verifying the projections and with $V_R - V_L = m_0 \beta_0 + m_i \beta_i$.

Examples

As a first example let us consider the model $(1)^9$, i.e. the tensor product of nine copies of the $k = 1$ minimal model, with $c_{k=1} = 1$ and hence total central charge $c = 9$, as required. In this case the 16+16 gauge bosons enhancing $SO(10) \times U(1)$ to E_6 come from massless states (8.104) of the form $(v; 0^9, 0^9)_R \otimes (s; 1^9, 1^9)_L$ and $(v; 0^9, 0^9)_R \otimes (c; -1^9, -1^9)_L$. Concerning charged matter, the only primaries $|l, q, s\rangle$ of the form $|l, l, 0\rangle$ with $l \leq k$ are $|1, 1, 0\rangle$ or $|0, 0, 0\rangle$, and hence the states with $h = 1/2$ and satisfying the projections are the 84 states

$$|1, 1, 0\rangle^3 \, |0, 0, 0\rangle^6, \tag{8.113}$$

where tensor products are implicit, and underlining means including all permutations. Using these states in both the left and right sectors we obtain 84 massless **27**s and no $\overline{\textbf{27}}$s. In addition there are 252 singlets coming from states $|1, 1, 2\rangle|1, 1, 0\rangle^2|0, 0, 0\rangle^6$. This spectrum is very reminiscent of the $\mathbf{Z}_3 \times \mathbf{Z}_3$ symmetric orbifold (with no discrete torsion) in Section 8.2.6, differing only in the additional presence of a $U(1)^6$ gauge symmetry and six singlets. The exact Gepner model spectrum is, however, recovered in the $\mathbf{Z}_3 \times \mathbf{Z}_3$ orbifold by choosing the underlying torus at a critical radius and specific backgrounds, as in (8.75); the extra $U(1)^6$ arises as the subgroup of the enhanced $SU(3)^3$ surviving the orbifold projection. Conversely, the generic orbifold spectrum is recovered from the Gepner model upon higgsing the $U(1)$'s using the six extra singlets. This simple example nicely illustrates the relation between Gepner models and geometric compactifications, further discussed in the next section.

As a second example, we consider the $(3)^5$ model, obtained by tensoring five copies of the $k = 3$ minimal model, with $c_{k=5} = 9/5$ and hence adding up to $c = 9$. The primaries

---

[1] Incidentally, note that the roles of the **27**s and $\overline{\textbf{27}}$s are exchanged by flipping the relative sign of left and right total superconformal charge. This is the CFT realization of the mirror symmetry in the corresponding geometric compactifications, see Section 10.1.2. The exchange of E_6 representations reflects, in $(2, 2)$ compactifications, the exchange of h_{11} and h_{21}.

of type $|l, l, 0\rangle$ are $|0, 0, 0\rangle$, $|1, 1, 0\rangle$, $|2, 2, 0\rangle$ or $|3, 3, 0\rangle$. The list of product states with total $h_{\text{int}} = 1/2$ is given by:

$$
\begin{array}{cc}
\text{States} & \# \\
|1, 1, 0\rangle^5 & 1 \\
\underline{|2, 2, 0\rangle^2 \, |1, 1, 0\rangle \, |0, 0, 0\rangle^2} & 30 \\
\underline{|2, 2, 0\rangle \, |1, 1, 0\rangle^3 \, |0, 0, 0\rangle} & 20 \\
\underline{|3, 3, 0\rangle \, |2, 2, 0\rangle \, |0, 0, 0\rangle^3} & 20 \\
\underline{|3, 3, 0\rangle \, |1, 1, 0\rangle^2 \, |0, 0, 0\rangle^2} & 30 \\
|1, -1, 0\rangle^5 & 1
\end{array}
\tag{8.114}
$$

where the underlining means permutation of the factors, and the right column gives the number of such permutations. The last state leads to one $\overline{27}$, while the rest produce 101 27s. Interestingly, these numbers correspond to the Hodge numbers $(h_{11}, h_{21}) = (1, 101)$ of the quintic CY in Section 7.2.3, which upon compactification with standard embedding thus reproduces the spectrum of the Gepner model (modulo enhancements). As we will see momentarily this is not a coincidence, as $(2, 2)$ Gepner models can be regarded as CY compactifications at particular points of their moduli space, corresponding to stringy size.

There is a finite set of $(2, 2)$ models obtained by tensoring minimal models as described above, whose construction algorithm is well suited for computer scans. There are 168 models based on the diagonal and exceptional modular invariants, and a total of 1176 using all possible modular invariants. The net number of E_6 generations $n_g = |n_{27} - n_{\overline{27}}|$ ranges between 0 and 480. Among these, there is only one example which can lead to a $(2, 2)$ 3-generation model. It is based on the tensor product $1_A \times 16_E \times 16_E \times 16_E$, of one $k = 1$ and three $k = 16$ minimal models, with the subindices A, E indicating the diagonal or exceptional modular invariants. It has 27 net generations, and is actually related to the Tian–Yau manifold in Section 7.2.3. The model can be turned into a 3-generation one by quotienting by a suitable $\mathbf{Z}_3 \times \mathbf{Z}_3$ symmetry, in analogy with the geometric counterpart in Section 7.3.3.

8.6.3 Connection to CY compactification*

The seeming connections between Gepner models and geometric CY manifolds mentioned above are examples of a general correspondence. Gepner models correspond to compactifications on certain CYs at particular points in moduli space, away from the large volume regime. This can be elegantly shown in the so-called linear sigma model description for the worldsheet theory, see the references. Here we present a less rigorous heuristic argument requiring less machinery on 2d $(2, 2)$ supersymmetric field theory. We also restrict to the case of diagonal modular invariants.

A first ingredient is the realization of $\mathcal{N} = 2$ minimal models as infrared fixed points of $\mathcal{N} = 2$ Landau–Ginzburg (LG) models. A 2d $\mathcal{N} = 2$ LG theory is described by a single chiral multiplet X, whose action reads

$$S = \int d^2z \, d^4\theta \, K(X, \bar{X}) + \left(\int d^2z \, d^2\theta \, W(X) + \text{h.c.} \right), \tag{8.115}$$

in 2d $(2, 2)$ superspace language–which is analogous to the 4d $\mathcal{N} = 1$ superspace, to which it relates via dimensional reduction. In particular, W is a polynomial superpotential, which we take to be of the form $W = X^{k+2}$ with k integer. These theories can be shown to flow to infrared fixed points corresponding to $\mathcal{N} = 2$ SCFTs with central charge $c = 3k/(k + 2)$, and hence must correspond to the $\mathcal{A}_k \otimes \tilde{\mathcal{A}}_k \, \mathcal{N} = 2$ minimal models. Hence, LG models with $W = X^{k+2}$ provide a lagrangian formulation of the level k minimal model.

Consider now the tensor product of r minimal models used in Gepner models. Their corresponding LG theory has chiral multiplets X_i, $i = 1, \ldots, r$, and action

$$S = \int d^2z \, d^4\theta \sum_{i=1}^{r} K_i(X_i, \bar{X}_i) + \left(\int d^2z \, d^2\theta \sum_{i=1}^{r} X_i^{k_i+2} + \text{h.c.} \right). \tag{8.116}$$

Let us concentrate for concreteness on the case $r = 5$, although the results are more general. In that case the condition $c = \sum_i [3k_i/(k_i + 2)] = 9$ implies

$$\sum_{i=1}^{5} \omega_i = 1, \quad \text{with } \omega_i = \frac{1}{k_i + 2}. \tag{8.117}$$

The correspondence associates this $r = 5$ Gepner model with a CY manifold, defined as a hypersurface in a complex projective space, recall Section 7.2.3. The ambient space is the weighted projective space $\mathbf{WCP}_{D\omega_1, \ldots, D\omega_r}$ defined by the identification

$$(X_1, \ldots, X_5) \sim (\lambda^{D\omega_1} X_1, \ldots, \lambda^{D\omega_5} X_5) \text{ with } \lambda \in \mathbf{C}^*, \tag{8.118}$$

where D is the least common multiple of the $(k_i + 2)$. The defining equation is

$$W = \sum_{i=1}^{5} X_i^{k_i+2} = 0, \tag{8.119}$$

which is indeed homogeneous of degree D.

The connection between the $(3)^5$ Gepner model and the quintic in $\mathbf{CP}_4$ is an example of this correspondence. There is indeed a formal analogy between the quintic LG superpotential $W = \sum_i X_i^5$ and the defining equation $\sum_i X^5 = 0$ of the quintic CY, although the condition $W = 0$ may not seem obvious from the LG viewpoint. A heuristic argument for this condition in the LG theory is as follows. In the infrared fixed point, the kinetic term in (8.116) is irrelevant (in the renormalization group sense) and can be dropped. The path integral of the theory thus has the structure

$$\int [\mathcal{D}X_1] \cdots [\mathcal{D}X_5] \exp \left[i \int d^2z \, d^2\theta \, W(X_1, \ldots, X_5) + \text{h.c.}) \right]. \tag{8.120}$$

Let us choose a path in which, e.g., $X_1 \neq 0$ and define

$$\xi_1^{\omega_1} = X_1, \quad \xi_i = \frac{X_i}{\xi_1^{\omega_i}}, \tag{8.121}$$

so that $W(X_1, \ldots, X_5) = \xi_1 W(1, \xi_2, \ldots, \xi_5)$. Then the path integral becomes

$$\int [\mathcal{D}\xi_1] \cdots [\mathcal{D}\xi_5] \exp\left[i \int d^2z \, d^2\theta \, \xi_1 \, W(1, \xi_2, \ldots, \xi_5) + \text{h.c.} \right]. \tag{8.122}$$

Here we have used that the jacobian of the change of variables is actually trivial,

$$J = \xi_1^{1 - \sum_i \omega_i} = 1, \tag{8.123}$$

where the second equality follows from the $c = 9$ condition (8.117). The ξ_1 integral then yields a delta function enforcing $W(1, \xi_2, \ldots, \xi_5) = 0$; proceeding analogously for the other four fields X_i, we cover the field space with patches obtaining a similar result. Regarding the scalar components of the X_i superfields as coordinates, the string target space is indeed the submanifold $W = 0$. The ambient space is projective since the above auxiliary coordinates like $\xi_2, \ldots, \xi_5$ play the role of affine coordinates in particular patches.

More precise derivations allow to deal with additional subtleties, like the appearance of orbifold actions $X_i \to e^{2\pi i \omega_i} X_i$ in the LG models, which relate to the β_0 twist. We leave these subtleties, as well as diverse generalizations to cases $r \neq 5$, and to more general CY manifolds, for the references. We hope the above discussion suffices to give a flavour of connection between Gepner models and CY manifolds.

8.6.4 Orbifolding Gepner models

Starting from a given $(2, 2)$ Gepner construction one may obtain a large class of models by modding out by different discrete symmetries of the theory. In particular, the level k minimal model with diagonal modular invariant has a $\mathbf{Z}_{k+2}$ symmetry, acting on primaries as

$$|l, q, s\rangle \otimes |\bar{l}, \bar{q}, \bar{s}\rangle \to e^{\frac{2\pi i Q}{k+2}} |l, q, s\rangle \otimes |\bar{l}, \bar{q}, \bar{s}\rangle, \tag{8.124}$$

where $Q = (q + \bar{q})/2$. Let us define the element $\theta = \exp[2\pi i \gamma Q/(k+2)]$, with $\gamma \in \mathbf{Z}$, which generates a $\mathbf{Z}_M$ symmetry of the minimal model, with M being the smallest integer for which $M\gamma = 0 \mod (k+2)$. It is possible to construct a quotient of the original minimal model by this $\mathbf{Z}_M$ symmetry, in close analogy with the worldsheet description of the orbifolding procedure in Section 8.2.5. For instance, the partition function of the orbifolded theory has the structure

$$Z = \frac{1}{M} \sum_{n,m=0}^{M-1} Z(n, m), \tag{8.125}$$

where (n, m) refers to boundary conditions closed up to θ^n and θ^m in the σ and t worldsheet torus directions, respectively. The $Z(0, 0)$ partition function has the form (8.102), and for generic twisted sectors one obtains

$$Z(n, m) = \frac{1}{2} \sum_{l, \bar{l}, q, \bar{q}, s, \bar{s}} \exp\left(2\pi i \, \frac{m\gamma(Q + n\gamma)}{k + 2}\right) Z_{l\bar{l}} \, \chi_{q+2n\gamma}^{ls} \, \chi_{\bar{q}}^{\bar{l}\bar{s}*}. \tag{8.126}$$

Equations (8.125) and (8.126) show the appearance of twisted sectors. In a fixed θ^n-twisted sector, the sum over m implements a projection onto invariant states; since the characters in (8.126) are independent of m, the generalized GSO projector is

$$P(\theta^n) = \frac{1}{M}\left[1 + e^{2\pi i \, \Delta(n)} + \cdots + e^{2\pi i \, (M-1)\Delta(n)}\right], \quad \Delta(n) = \frac{\gamma(Q + n\gamma)}{k + 2},$$

and the invariance condition is $\Delta(\theta^n) \in \mathbf{Z}$.

Consider a heterotic Gepner model compactification based on the tensor product of r minimal models, $\mathcal{A}_{k_1} \otimes \cdots \otimes \mathcal{A}_{k_r}$. There is a discrete symmetry group

$$G = \mathbf{Z}_{k_1+2} \times \cdots \times \mathbf{Z}_{k_r+2}. \tag{8.127}$$

The full symmetry group also includes permutations of identical blocks, if present; however, we restrict to the construction of quotients by a subgroup of G. Consider a set of integers $\gamma_i \in \mathbf{Z}$, $i = 1, \ldots, r$, and define the element θ acting as $\exp[2\pi i \gamma_i Q_i/(k_i + 2)]$ on $\mathcal{A}_{k_i}$. It generates a $\mathbf{Z}_M$ subgroup of G, with M being the smallest integer for which $M(\gamma_1, \ldots, \gamma_r) = (0, \ldots, 0) \mod (k_1 + 2, \ldots, k_r + 2)$. For the quotient to preserve supersymmetry, it must be compatible with the β_0 projection, corresponding to the condition

$$2\beta_0 \cdot \gamma = -\sum_{i=1}^{r} \frac{\gamma_i}{k_i + 2} \in \mathbf{Z}. \tag{8.128}$$

where $\gamma = (0; \gamma_1, \ldots, \gamma_r; 0, \ldots, 0)$, and the scalar product is (8.105).

The $\mathbf{Z}_M$ orbifold quotient of the Gepner model is constructed in analogy with the above single minimal model case. The new feature is the appearance of θ^n-twisted sectors in which $V_L - V_R$ lies on an enlarged lattice given by

$$V_L - V_R = n(2\gamma) + m_0\beta_0 + m_i\beta_i, \text{ with } n = 0, \ldots, M - 1. \tag{8.129}$$

The invariance condition on states in the θ^n-twisted sector is

$$\Delta(n) = -\gamma \cdot (V_R + V_L) \in \mathbf{Z}, \tag{8.130}$$

with V_R and V_L related as in (8.129). As usual in orbifold constructions, there is a twofold effect on the massless spectrum of the original Gepner model; on one hand the additional projections (8.130) reduce the number of states (e.g. of E_6 generations), while on the other there appear new twisted states associated to the enlarged lattice (8.129).

The systematic construction of orbifolds of the original Gepner models leads to a large number of new (2, 2) models. For illustration, let us consider the example of the $(1)^9$ model orbifolded by the $(\mathbf{Z}_3)^3$ group generated by the following twist vectors, satisfying (8.128).

$$\gamma_1 = (0; 1, 1, 1, 0, 0, 0, -1, -1, -1; 0, \ldots, 0),$$
$$\gamma_2 = (0; 1, 1, 1, -1, -1, -1, 0, 0, 0; 0, \ldots, 0),$$
$$\gamma_3 = (0; 0, 0, 0, 1, 1, 1, -1, -1, -1; 0, \ldots, 0). \tag{8.131}$$

Let us describe the structure of E_6 families in the orbifold theory. To avoid clutter, we simplify notation and describe primaries by just their q-labels. Hence the states (8.113) associated to the E_6 generations in the original $(1)^9$ model are denoted by $(\underline{111000000})$. The reader can check that the 27 states $(\underline{100}, \underline{100}, \underline{100})$ and the three states $(111, 000, 000)$, $(000, 111, 000)$, $(000, 000, 111)$ survive the projections. In addition, there are six extra twisted states of the form $(111, 000, 000)_R \otimes (000, 111, 000)_L$, and permutations keeping the "111" block at different positions in the left and right sectors, which are allowed now in the enlarged lattice. The resulting spectrum has 36 E_6 **27**s (and no $\overline{\mathbf{27}}$s), and in fact agrees with the massless spectrum of the toroidal $\mathbf{Z}_3$ orbifold in Section 8.2.4, modulo an additional $U(1)^6$ and singlets, for already familiar reasons.

As a second example we consider the $(3)^5$ Gepner model modded out by the $(\mathbf{Z}_5)^3$ group generated by the following twist vectors, satisfying (8.128):

$$\gamma_1 = (0; 0, 1, 2, 3, 4; 0, \ldots, 0),$$
$$\gamma_2 = (0; 0, 1, 1, 4, 4; 0, \ldots, 0),$$
$$\gamma_3 = (0; 0, 0, 0, 1, 4; 0, \ldots, 0). \tag{8.132}$$

One can check that the massless spectrum of the orbifold construction has 101 $\overline{\mathbf{27}}$s and one **27**, i.e. their multiplicities are reversed compared to the original $(3)^5$ model, providing the CFT counterpart of the compactification on the mirror manifold of the quintic. This is actually a general lesson, and mirrors of Gepner model compactifications can be generated by taking appropriate orbifold quotients.

Let us finally comment that one can further generalize these constructions by allowing for discrete torsion, i.e. allowing for different relative phases among disconnected modular orbits in the partition function, in close analogy with the toroidal orbifold discrete torsion in Section 8.2.6. A further possible generalization would be the construction of left–right asymmetric quotients, with independent twists γ_R, γ_L, in analogy with asymmetric toroidal orbifolds in Section 8.4; however, such asymmetric quotients turn out to be equivalent to left–right symmetric quotients with appropriate discrete torsion.

8.6.5 (0,2) constructions and model building

It is easy to generalize the above rules to construct $(0, 2)$ models. The simplest such modification is to give up the projection (8.107) $2\beta_0 \cdot V_L = $ odd for left-movers, while maintaining it for right-movers. This still preserves 4d $\mathcal{N} = 1$ supersymmetry, but prevents the enhancement of $SO(10) \times U(1)^r$ to $E_6 \times U(1)^{r-1}$.

More generally, it is possible to construct $\mathbf{Z}_M$ orbifold quotients with a non-trivial embedding into the $E_8 \times SO(10)$ gauge degrees of freedom, through a shift vector A

such that MA is in the $SO(10) \times E_8$ root lattice. This is quite analogous to the addition of discrete Wilson lines and/or non-standard embeddings in orbifold compactifications that we studied in Section 8.3. Using similar techniques, the term $Z(n, m)$ in the left-handed partition function associated to the $SO(10) \times E_8$ is

$$e^{-\pi i \, nm \, A^2} \, \eta^{-13} \sum_{P \in \Gamma^*} q^{\frac{(P+nA)^2}{2}} \, e^{2\pi i \, (P+nA) \cdot mA}. \tag{8.133}$$

where from now on P denotes a vector in Γ^*, the dual of the $SO(10) \times E_8$ root lattice. Note the similarity with the 16d momentum contributions in, e.g., (8.39). The modular invariance constraint is

$$MA^2 = 0 \bmod 2. \tag{8.134}$$

From (8.133), the left-handed mass formula is similar to (8.112) with the $SO(10) \times E_8$ term replaced by $(P + nA)^2/2$. The orbifold projection (8.130) also becomes

$$\Delta(n) = -\gamma \cdot (V_R + V_L) - \left[(P + nA) \cdot A - \frac{1}{2}nA^2 \right] \in \mathbf{Z}, \tag{8.135}$$

with V_R, V_L related as in (8.129).

These modified rules, which may be combined with discrete torsion, allow the construction of a large class of $(0, 2)$ Gepner model orbifolds, in which the $SO(10) \times E_8$ gauge symmetry is generically broken to a smaller gauge group. Note that for this purpose the number of generations of the parent $(2, 2)$ models is not relevant, since the twists significantly modify the massless spectrum. These constructions can be exploited for heterotic string particle physics model building. Since some of the Gepner models are closely related to certain toroidal orbifolds, it is for instance possible to recover the 3-generation $SU(3) \times SU(2) \times U(1)^n$ $\mathbf{Z}_3$ orbifold model in Section 8.3.3; it is in fact reproduced by the $(1)^9$ Gepner model orbifolded by the three twists γ_i in (8.131), and adding two further twists $\gamma_4 \otimes A_4$, $\gamma_5 \otimes A_5$, the counterparts of the two Wilson lines in the orbifold case.

The most systematic searches for MSSM-like particle physics models based on Gepner models lean on their description in terms of simple currents, see Section E.5. There exist computer scans, within the class of such $(0, 2)$ Gepner models and with the SM arising from the $SO(10)$ lattice, leading to some 600 MSSM-like 3-generation models (including left–right symmetric and trinification models), incidentally all coming from the $1 \times 16_E \times 16_E \times 16_E$ tensor product. An even larger class of $(0, 2)$ models can be obtained by replacing one or more of the left-moving $\mathcal{N} = 2$ SCFT factors and the E_8' by a non-supersymmetric CFT with the same modular transformation properties; within this class, there are many other 3-generation MSSM-like models, now based on other tensor product structures. We direct the reader to the references for details.

Generically, MSSM-like models from heterotic compactifications based on Gepner models contain an enlarged gauge group (due to the $U(1)$'s from the $\mathcal{N} = 2$ minimal models) and extra multiplets (often including fractionally charged exotics) beyond the minimal content, as also usual in other heterotic constructions. The latter may be phenomenologically problematic, see Section 17.1.3, unless they are vector-like (as is often the case) and get very large, e.g. string scale, masses. This may happen by moving off the Gepner point in the moduli space of the corresponding CY compactification. Also, many models contain pseudo-anomalous $U(1)$'s, which as described in Section 9.5 lead to one-loop generation of a Fayet–Illiopoulos term, whose cancellation triggers further gauge symmetry breaking and in general makes many of these exotics massive, see Section 9.7 for this phenomenon in the context of toroidal orbifolds.

9

Heterotic string compactifications: effective action

In this chapter we consider different aspects of the low-energy 4d effective action of heterotic compactifications. After discussing the structure of the effective action for the general CY compactifications of Chapter 7, we concentrate on the case of the abelian orbifold compactifications of Chapter 8, in which several additional aspects can be described in an explicit quantitative fashion. We discuss the form of the Kähler potential and the superpotential, as well as the gauge kinetic function and their one-loop corrections. In toroidal orbifolds the low-energy effective action must be modular invariant under T-duality transformations, and this leads to strong constraints on its structure. A general feature of heterotic compactifications is that they often have a non-trivial cancellation of mixed $U(1)$ anomalies, involving a generalized Green–Schwarz mechanism. This renders some $U(1)$ gauge bosons massive, and introduces a one-loop re-stabilization of the vacuum, triggering the breaking of extra gauge symmetries and the disappearance of charged matter multiplets. This phenomenon can be efficiently exploited in the construction of improved models of particle physics, as we describe using explicit examples.

9.1 A first look at the heterotic 4d $\mathcal{N} = 1$ effective action

We start by motivating the general structure of the effective action of heterotic string geometric (CY or orbifold) compactifications by using an intuitive description of the truncation to the zero mode sector. The resulting structure is quite general, as can be shown to largely follow from symmetries of the system. We finally describe more explicit results for CY compactifications with standard embedding.

9.1.1 A toy description from dimensional reduction

At energies well below the string and compactification scales, the 4d physics can be described in terms of a low-energy effective field theory. We are mostly interested in compactifications with unbroken SUSY, so (recalling Section 2.3) the 4d $\mathcal{N} = 1$ supergravity effective action is given in terms of the gauge kinetic functions f_a, the Kähler potential K and the superpotential W, all of them functions of the massless chiral multiplets in the compactification. In order to get a first intuitive view of the structure of the effective action, we consider these quantities for the untwisted sector of the $\mathbf{Z}_3$ orbifold of the $E_8 \times E_8$ heterotic string. As suggested from its construction in Section 8.2, the KK compactification may be obtained by simply making a $\mathbf{Z}_3$ orbifold projection of the underlying toroidal compactification. Despite the simplified setup, it nicely illustrates several key aspects of more general compactifications.

Recall the bosonic sector of the 10d heterotic action (4.80)

$$S_{10} = \frac{1}{2\kappa_{10}^2} \int d^{10}x \, (-G)^{1/2} \, e^{-2\phi} \left(R + 4\partial_M \phi \partial^M \phi - \frac{1}{2} |H_3|^2 - \frac{\alpha'}{4} \mathrm{tr}_v |F|^2 \right),$$

$$(9.1)$$

where H_3 is given by

$$H_3 = d B_2 - \frac{\alpha'}{4} \mathrm{tr}_v \left(A \wedge dA - \frac{i}{3} A \wedge A \wedge A \right),$$

$$(9.2)$$

with traces normalized to the vector representation of the $SO(16)$ subgroup of E_8, and we have ignored the gravitational Chern–Simons term. As described in Section 8.2.4, the massless bosonic fields in the untwisted sector include the graviton $g_{\mu\nu}$, the 2-form $B_{\mu\nu}$, the dilaton ϕ, and the scalars $g_{i\bar{j}}$, $B_{i\bar{j}}$ from the internal metric and antisymmetric tensor. In addition there are three massless charged scalars C_j, $j = 1, 2, 3$, transforming as representations $(\mathbf{3}, \mathbf{27})$ under $SU(3) \times E_6$. Ignoring momentarily these charged fields and the gauge bosons, one can consider the toroidal reduction of (9.1) restricting to the neutral bosonic fields. The computation can be shown to give

$$S_4 = \frac{1}{2\kappa_4^2} \int d^4x \, (-g)^{1/2} \left[R - 2 \, \partial_\mu \phi_4 \, \partial^\mu \phi_4 - \frac{1}{2} e^{-4\phi_4} |H_3|^2 \right.$$
$$\left. - \frac{1}{2} g^{i\bar{j}} g^{k\bar{l}} \left(\partial_\mu g_{i\bar{l}} \, \partial^\mu g_{\bar{j}k} + \partial_\mu B_{i\bar{l}} \, \partial^\mu B_{\bar{j}k} \right) \right].$$

$$(9.3)$$

Here we have defined

$$\kappa_4^2 = \kappa_{10}^2 e^{2\phi} / V_6, \quad \phi_4 = \phi - \frac{1}{2} \log(V_6/\alpha'^3),$$

$$(9.4)$$

where $V_6 = [\det(g_{mn})]^{1/2}$ is the compact volume and $\kappa_{10}^2 = \frac{1}{2}(2\pi)^7 \alpha'^4$. The action is written in the Einstein frame, by rescaling $g_{\mu\nu} = \exp(-2\phi_4)G_{\mu\nu}$ to get the pure gravity action in the Einstein–Hilbert form, c.f. the discussion on page 80.

In 4d a 2-index antisymmetric field $B_{\mu\nu}$ may be equivalently described in terms of a real scalar $a(x)$, as follows (see also Section B.3). Writing the 2-form action in terms of H_3, with a Lagrange multiplier $a(x)$ to enforce $d H_3 = 0$, and hence $H_3 = d B_2$, one has

$$-\frac{1}{2} \int d^4x \, (-g)^{1/2} e^{-4\phi_4} |H_3|^2 + \int_{4d} a \, d H_3,$$

$$(9.5)$$

where we use differential form language, see Appendix B. Note the invariance $a(x) \rightarrow a(x) + c$, with constant c, to be exploited in the next section. Imposing the equations of motion for H_3 one obtains

$$H^{\mu\nu\rho} = -(-g)^{-\frac{1}{2}} \exp(4\phi_4) \, \epsilon^{\mu\nu\rho\tau} \, \partial_\tau a, \tag{9.6}$$

which upon replacement in (9.5) produces the kinetic term for the field $a(x)$

$$-\frac{1}{2} \int d^4x \, (-g)^{1/2} \, e^{4\phi_4} \partial_\mu a \partial^\mu a. \tag{9.7}$$

Then the action (9.3) may expressed as

$$S_4 = \frac{1}{2\kappa_4^2} \int d^4x \, (-g)^{1/2} \left[R - \frac{2 \partial_\mu S \partial^\mu S^*}{(S + S^*)^2} - \frac{1}{2} g^{i\bar{j}} \, g^{k\bar{l}} \left(\partial_\mu T_{i\bar{l}} \, \partial^\mu T_{k\bar{j}}^* \right) \right], \tag{9.8}$$

where we have defined the 4d complex dilaton and complexified Kähler moduli

$$S = e^{-2\phi_4} + ia = e^{-2\phi}(V_6/\alpha'^3) + ia, \quad T_{i\bar{j}} = g_{i\bar{j}} + i B_{i\bar{j}}. \tag{9.9}$$

The field S is a complexified version of the 4d dilaton, with an imaginary part given by the universal scalar field a. The T-fields are nine Kähler moduli whose real parts govern the compact geometry. It is easy to check that the structure of the kinetic terms corresponds to a Kähler potential

$$\kappa_4^2 K = -\log(S + S^*) - \log \det \left(T_{i\bar{j}} + T_{i\bar{j}}^* \right). \tag{9.10}$$

The scalar $a(x)$ is an axion-like field because of its coupling to the gauge field strengths. Indeed, including the gauge bosons, the action (9.5) becomes

$$-\frac{1}{2} \int d^4x \, (-g)^{1/2} \, e^{-4\phi_4} \, |H_3|^2 + \int a \left[dH_3 + \frac{\alpha'}{4} \operatorname{tr}_v(F \wedge F) \right], \tag{9.11}$$

which now implements the modified Bianchi identity discussed in Section 7.3.2. Proceeding as above, one obtains the couplings

$$-\frac{\alpha'}{8\kappa_4^2} \int e^{-2\phi_4} \operatorname{tr}_v |F|^2 - i \frac{\alpha'}{8\kappa_4^2} \int a \operatorname{tr}_v(F \wedge F). \tag{9.12}$$

Reabsorbing $\langle \exp(2\phi_4) \rangle = 4\kappa_4^2/\alpha'$ (and recalling that the definition (B.22) implies $|F|^2 = \frac{1}{2} F_{\mu\nu} F^{\mu\nu}$), the gauge kinetic function reads

$$f_{ab} = S \, \delta_{ab}, \tag{9.13}$$

where a, b run through the different 4d gauge factors. Inclusion of the gravitational Chern–Simons term ω_3^{grav} would similarly lead to a coupling $a\operatorname{tr}(R \wedge R)$, which plays a role in the anomaly cancellation mechanism in Section 9.5.1.

Let us consider now the terms containing the charged scalar fields C_j arising from the internal components A_j of the E_8 gauge generators transforming in the $(\mathbf{3}, \mathbf{27})$ under

$SU(3) \times E_6$. It is possible to show that the 10d action produces kinetic terms for these charged fields, given by a modified Kähler potential

$$\kappa_4^2 K = -\log(S + S^*) - \log \det \left[T_{i\bar{j}} + T_{ij}^* - \alpha' \operatorname{tr}_v \left(C_i C_j^* \right) \right]. \tag{9.14}$$

The correct supergravity fields associated to the Kähler moduli are then defined as

$$T_{i\bar{j}} = g_{i\bar{j}} + i B_{i\bar{j}} + \frac{\alpha'}{2} \operatorname{tr}_v \left(C_i C_j^* \right). \tag{9.15}$$

In what follows we will often reabsorb the α' factor into the definition of the matter fields C_i. There is also a superpotential for the charged fields arising from the Chern–Simons piece $(A \wedge A \wedge A)$ inside H_3, which reads

$$W = \epsilon^{ijk} \epsilon_{pqr} d_{PQR} C_i^{pP} C_j^{qQ} C_k^{rR}, \tag{9.16}$$

where p, q, r run through the **3** of $SU(3)$, P, Q, R run through the **27** of E_6, and d_{PQR} is the cubic E_6 invariant tensor associated with the singlet term in the tensor product $\mathbf{27}^3$. The corresponding F-term and the D-term scalar potentials stem from the dimensional reduction of the non-abelian piece of $F_{mn} F^{mn}$.

The Kähler potential dependence on the $T_{i\bar{j}}$ has an interesting structure, known as "no-scale." It can already be discussed in a simplified supergravity model with Kähler potential

$$\kappa_4^2 K = -\log(S + S^*) - 3 \log (T + T^*). \tag{9.17}$$

This is in fact equivalent to studying the dependence of the $\mathbf{Z}_3$ orbifold untwisted sector in the overall modulus direction, $T_{i\bar{i}} \equiv T$, with the off-diagonal $T_{i\bar{j}}$ set to zero. For models with superpotential W independent of T, we have

$$D_T W = -\frac{3W}{T + T^*}. \tag{9.18}$$

Recall now the supergravity F-term scalar potential (2.50)

$$V = e^{\kappa_4^2 K} \left(K^{a\bar{b}} D_a W D_{\bar{b}} W^* - 3 \kappa_4^2 |W|^2 \right), \tag{9.19}$$

where $D_a W = \partial_a W + \kappa_4^2 W \partial_a K$, and the sum goes over all the chiral fields. Since $\kappa_4^{-2} K^{TT^*} = \frac{1}{3}(T + T^*)^2$, the $-3\kappa_4^2 |W|^2$ piece is exactly canceled by the contribution of the T modulus, and the remaining scalar potential is positive definite. If this residual potential has a minimum at zero energy, the potential has a flat direction in T with spontaneous SUSY breaking by the T-modulus. Since the value of T is undetermined, this class of supergravity potentials are referred as "no-scale." Such potentials arise in all toroidal orbifold compactifications, and in the large volume limit of CY compactifications, both in heterotic and in type II orientifolds. The no-scale structure of heterotic effective actions will be further discussed in Section 15.2, in relation to moduli stabilization and SUSY breaking.

The above analysis involves only the untwisted moduli and matter fields of the theory. The effective action for twisted sector fields is discussed in Section 9.3.

9.1.2 General structure of the effective action

The orbifold example in the previous section is illustrative in many respects, but usually the effective action for a general heterotic compactification is more involved. There are, however, some general properties of the effective action which apply to any 4d $\mathcal{N} = 1$ heterotic compactification, as we now review. These general properties rely on the existence of certain symmetries of the compactification, in the leading α' and g_s approximation. In this general discussion we denote the Kähler and complex structure moduli fields by T^i, U^n, and the charged chiral multiplets by C^I.

The tree level dependence of the effective action on the complex dilaton S is universal. At leading order in the string loop expansion, the Kähler potential $K(S, T^i, U^n, C^I)$ dependence on the complex dilaton field S is given by

$$\kappa_4^2 \, K(S, T^i, U^n, C^I) = -\log(S + S^*) + K(T^i, U^n, C^I). \qquad (9.20)$$

The gauge kinetic function is holomorphic, and in heterotic compactifications is, at tree level, universal for the different 4d gauge factors and given by

$$f_{ab} = S \, \delta_{ab}, \qquad (9.21)$$

as already encountered in (9.13). Note that, from (9.12), in the heterotic case the string scale $M_s^2 = 1/\alpha'$ and the Planck scale $M_p^2 = 8\pi/\kappa_4^2$ are related to the 4d gauge coupling constant $g^2 = \langle \exp(2\phi_4) \rangle$ by

$$\frac{M_p^2}{M_s^2} = 16 \frac{4\pi}{g^2} = \frac{16}{\alpha_G}, \qquad (9.22)$$

where we now change to the standard particle physics normalization with $\mathrm{tr}\,(T^a T^b) = \frac{1}{2}\delta^{ab}$ for $SU(n)$ generators in the fundamental, which introduces an extra factor 2. Here α_G is a unified fine structure constant, i.e. $\alpha_G \simeq 1/24$ in a GUT-like setting. Thus the string scale is only slightly below the Planck scale $M_p = 1.2 \times 10^{19}$ GeV, namely $M_s \simeq 6.1 \times 10^{17}$ GeV.

The reason underlying the universality properties of the dilaton is that the definition of ϕ_4 and the axion a is independent of the compactification details. Consequently, expressions derived in the orbifold example in Section 9.1.1 are valid in general. Note in particular that, as seen in (9.6), the axion has only derivative couplings and so the action is (as already mentioned) invariant under

$$S \longrightarrow S + ic, \qquad (9.23)$$

where c is an arbitrary real constant. Thus S must appear in the Kähler potential in the combination $S + S^*$. The shift symmetry (9.23) is also preserved by the gauge kinetic term, since $c \, F \wedge F$ is a total derivative, but it is broken to a discrete subgroup by non-perturbative gauge instantons, see Section 13.1.2.

The log dependence of the Kähler potential on $(S + S^*)$ may also be understood in terms of classical symmetries of the theory, for geometric (CY or orbifold) compactifications.

In particular, it is easy to check (e.g. in the orbifold example above) that under the rescalings

$$S \rightarrow \lambda S, g_{\mu\nu} \rightarrow \lambda g_{\mu\nu}, \tag{9.24}$$

the effective action scales like $S_4 \rightarrow \lambda^{-1} S_4$. This rescaling of the action is irrelevant at the classical level, and just shows that the power of the dilaton counts the loop order. This scaling behavior for, e.g., the supergravity scalar potential, requires that $e^K \rightarrow \lambda^{-1} e^K$, implying $K \propto -\log(S + S^*)$.

The shift symmetry (9.23) also guarantees that the superpotential W receives no loop corrections; it directly implies that the superpotential is independent of the axion a, hence, by holomorphy in S, it must be also independent of the 4d dilaton Re S, which is the loop counting parameter. An analogous 4d $\mathcal{N}=1$ non-renormalization theorem applies to the gauge kinetic function which can only receive one-loop corrections, which are S-independent and hence consistent with the classical shift symmetry.

Let us finally recast other pieces of the effective action, in language useful for application to phenomenological models. Concerning the general Kähler potential in (9.20), we are mainly interested in terms at most quadratic on the matter fields C^I. We thus perform an expansion

$$\begin{aligned} K(T^i, U^n, C^I) = & \hat{K}(T^i, T^{i*}, U^n, U^{n*}) + K_{I\bar{J}}^C(T^i, T^{i*}, U^n, U^{n*}) C^I C^{J*} \\ & + Z_{IJ}^C(T^i, T^{i*}, U^n, U^{n*})(C^I C^J + \text{h.c.}), \end{aligned} \tag{9.25}$$

valid for small vevs for matter fields C^I, as compared to moduli vevs. Similarly, the general structure of the superpotential, in an expansion in the matter fields, is

$$W = h_{IJK}(T^i, U^n) C^I C^J C^K + \mathcal{O}(C^4) + \cdots. \tag{9.26}$$

The functions $h_{IJK}(T^i, U^n)$ are holomorphic in the moduli fields and independent of the dilaton. Note also that there are no terms with a dimension of less than four, and no superpotential for the moduli; the latter is however generated by introducing diverse fluxes, see Section 14.5, and by instanton effects, Sections 13.1.2 and 15.2.1.

9.1.3 Effective action of (2,2) compactifications

In solvable compactifications like orbifolds one can explicitly compute the Kähler potential and the metric of the matter fields. On the other hand, such computations are intractable for general CY compactifications. Still, in certain classes of compactifications, there exist additional symmetries providing further control on the effective action. This is the case for heterotic compactifications with standard embedding, i.e. (2, 2) models, studied in Section 7.3.2. The CY moduli sector of such compactifications is identical to that in compactifications of type II theories, which have an enhanced 4d $\mathcal{N}=2$ SUSY. Recalling Section 2.4.1, this leads to a more restricted Kähler potential, which in fact extends to the heterotic models as follows. The Kähler and complex structure moduli spaces are decoupled, namely

$$\hat{K}(T^i, T^{i*}, U^n, U^{n*}) = K^{(T)}(T^i, T^{i*}) + K^{(U)}(U^n, U^{n*}). \qquad (9.27)$$

These Kähler potentials are moreover determined by the prepotentials, two holomorphic functions $\mathcal{F}(T_i)$ and $\mathcal{G}(U_n)$, through

$$K^{(T)} = -\log\left[\sum_i (\mathcal{F}_i + \mathcal{F}_i^*) \, (T^i + T^{i*}) - 2(\mathcal{F} + \mathcal{F}^*)\right], \qquad (9.28)$$

where the sum runs through all the Kähler moduli. An analogous formula defines K_U from the complex structure prepotential $\mathcal{G}(U_n)$. In large volume CY compactifications, the Kähler potentials are given by

$$K^{(T)} = -\log \int_{\mathbf{X}_6} J \wedge J \wedge J, \quad K^{(U)} = -\log\left(-i \int_{\mathbf{X}_6} \Omega \wedge \bar{\Omega}\right), \qquad (9.29)$$

where J and Ω are the CY Kähler 2-form and holomorphic 3-form introduced in Section 7.2.2. Note that the first of these integrals is essentially the CY volume V_6, recall (7.7). We may define an overall CY Kähler modulus $T = V_6^{1/3} + i\eta$, with $\eta \simeq \int_{\mathbf{X}_6} B_2 \wedge J \wedge J$, whose large volume Kähler potential is

$$K(T, T^*) = -3\log(T + T^*). \qquad (9.30)$$

At finite volume, these expressions in general receive corrections from worldsheet instanton processes in which a string worldsheet wraps a compact 2-cycle Σ_i in the CY $\mathbf{X}_6$. Such effects are proportional to e^{-T_i}, where $\mathrm{Re}\, T_i = \int_{\Sigma_i} J$ is the 2-cycle volume, controlled by the Kähler moduli; hence, by the decoupling (9.27), they do not correct the Kähler potential for complex structure moduli. These effects are independent of the string coupling, hence they appear at tree level in spacetime. Their contribution to the Kähler potential for Kähler moduli is however sub-leading in the large volume regime, on which we focus. Analogous effects for the superpotential however generate interesting contributions to Yukawa couplings in orbifold models, see Section 9.3.2.

The above formulae can be applied to the untwisted Kähler moduli of the $\mathbf{Z}_3$ orbifold (or any other orbifold), leading to expressions in agreement with the direct computations in the orbifold model in the previous section. In this case, the untwisted moduli Kähler potential receives no worldsheet instanton corrections due to the higher supersymmetry of the underlying toroidal compactification.

In standard embedding CY compactifications, the internal zero mode for the 4d matter fields in the $\mathbf{27}, \overline{\mathbf{27}}$ are given by harmonic forms in the CY, and are in one-to-one correspondence with Kähler and complex structure moduli. This provides an expression for the charged fields Kähler metrics, as follows

$$K_{i\bar{j}}^{\mathbf{27}} = e^{-\frac{1}{3}(K^{(T)} - K^{(U)})} K_{i\bar{j}}^{(T)} (T^i, T^{i*}),$$
$$K_{i\bar{j}}^{\overline{\mathbf{27}}} = e^{-\frac{1}{3}(K^{(U)} - K^{(T)})} K_{i\bar{j}}^{(U)} (U^n, U^{n*}), \qquad (9.31)$$

where $K^{(T)}_{i\bar{j}}$, $K^{(U)}_{i\bar{j}}$ are the Kähler metrics for Kähler and complex structure moduli, respectively. From (9.30), the dependence on the overall Kähler modulus T is

$$K^{27} \propto K^{\overline{27}} \propto \frac{1}{T + T^*}. \tag{9.32}$$

This also reproduces the orbifold Kähler potential (9.14) to first order in $|C|^2$. The underlying $\mathcal{N} = 2$ structure also restricts the form of the cubic superpotentials $\mathbf{27}^3$, $\overline{\mathbf{27}}^3$. They are given in terms of the prepotentials by

$$W = W_{ijk}(T_i)\, C^i_{\mathbf{27}}\, C^j_{\mathbf{27}}\, C^k_{\mathbf{27}} + W_{nmp}(U_n)\, C^n_{\overline{\mathbf{27}}}\, C^m_{\overline{\mathbf{27}}}\, C^p_{\overline{\mathbf{27}}}, \tag{9.33}$$

with

$$W_{ijk}(T_i) = \partial_i \partial_j \partial_k\, \mathcal{F}(T_i), \quad W_{nmp}(U_n) = \partial_n \partial_m \partial_p\, \mathcal{G}(U_n). \tag{9.34}$$

The moduli dependence of the cubic couplings also factorizes. Since the strength of world-sheet instanton corrections is determined by Kähler moduli, the $\overline{\mathbf{27}}^3$ Yukawa couplings are exact in α' (and in g_s as usual). This property, together with the mirror symmetry exchanging the T and U moduli of mirror CY spaces (see Section 10.1.2), allows in certain CYs the computation of the α'-exact K_T.

Let us sketch the microscopic computation of the above superpotential terms in large volume CY compactifications, using Kaluza–Klein reduction of the 10d lagrangian. Yukawa couplings arise from the coupling of the 10d gauginos λ to internal components of the gauge fields A_m

$$\int d^{10}x\, \mathrm{Tr}\,(\bar{\lambda}\, \Gamma^m\, [A_m, \lambda]). \tag{9.35}$$

The 4d Yukawa coupling constants are then given by overlap integrals of the internal zero mode profiles for these fields. In compactifications with standard embedding, these are harmonic forms on the CY $\mathbf{X}_6$. In particular, the 6d profiles for the $\mathbf{27}$s are harmonic $(1,1)$-forms $\left\{\omega_i^{(1,1)}\right\}$, $i = 1, \ldots, h_{1,1}$, so the cubic superpotential is

$$W_C = d^{PQR}\, C_i^P\, C_j^Q\, C_k^R \int_{\mathbf{X}_6} \omega_i^{(1,1)} \wedge \omega_j^{(1,1)} \wedge \omega_k^{(1,1)}, \tag{9.36}$$

where P, Q, R run through the $\mathbf{27}$ of E_6, and d_{PQR} is the cubic E_6 invariant tensor. The overlap integrals are independent of the metric and define a set of topological integers, known as CY triple intersection numbers,

$$D_{ijk} = \int_{\mathbf{X}_6} \omega_i^{(1,1)} \wedge \omega_j^{(1,1)} \wedge \omega_k^{(1,1)}. \tag{9.37}$$

The $\overline{\mathbf{27}}$s are related to the $(2, 1)$-forms $\left\{\omega_n^{(2,1)}\right\}$, $n = 1, \ldots, h_{2,1}$, so the cubic superpotential is schematically

$$W_C = d_{PQR} \, C_m^P \, C_n^N \, C_p^Q \int_{\mathbf{X}_6} \Omega \cdot \left(\omega_m^{(2,1)} \wedge \omega_n^{(2,1)} \wedge \omega_p^{(2,1)}\right), \tag{9.38}$$

where Ω is the CY holomorphic 3-form, and the dot denotes its contraction with one anti-holomorphic index from each $(2, 1)$ form. Since the overlap integral involves Ω, it involves data from the CY metric and is thus not topological in this case.

These general results for CY compactifications refer only to $(2, 2)$ models, so the gauge group is $E_6 \times E_8$. In particular, formulae (9.36), (9.38) for the Yukawa couplings have a limited practical interest for particle physics model building, as they do not apply directly to the phenomenologically more interesting case of general compactifications with non-standard embedding. Unfortunately, there is little information about the general structure of the effective action for this more general class of heterotic compactifications. The Yukawa couplings are in principle given by overlap integrals analogous to (9.36), (9.38), using the relevant internal zero mode wave functions. They can be explicitly computed by solving the Dirac and Klein–Gordon equations in simple setups, like toroidal models with $U(1)$ bundles, as we study for type IIB and F-theory models in Sections 12.5.4 and 12.7. For certain curved CY spaces, the exact wave functions are not known, but the integrals can be computed by using forms in the appropriate cohomology classes, i.e. in the same equivalence class in respect to integration. Finally, another class of models in which Yukawa couplings are explicitly computable, even for $(0, 2)$ compactifications, is abelian toroidal orbifolds, as we explain in Section 9.3.

9.2 Heterotic M-theory effective action*

In this section we detour from perturbative heterotic vacua and consider aspects of the strongly coupled regime of these compactifications. For the $SO(32)$ theory this is given by type I compactifications, whose effective action fits within the analysis in Chapter 12. Here we focus on the $E_8 \times E_8$ heterotic compactifications on a CY manifold $\mathbf{X}_6$, for simplicity with standard embedding. The strong coupling regime is described by the compactification of Hořava–Witten theory, i.e. M-theory on $\mathbf{S}^1/\mathbf{Z}_2 \times \mathbf{X}_6$, recall Section 7.5.

We denote ρ the radius of the underlying $\mathbf{S}^1$ and V_6 the CY volume. The regime in which the compactification scales lie below the 11d Planck scale admits a description in terms of higher dimensional supergravity. Namely, at energies below the 11d Planck scale but above the compactification scales, the theory is described by 11d supergravity on $\mathbf{S}^1/\mathbf{Z}_2 \times \mathbf{X}_6$ coupled to 10d $\mathcal{N} = 1$ E_8 vector multiplets localized at the 10d boundaries at $x^{10} = 0, \pi\rho$. The bosonic piece of the effective action for massless fields has the structure

$$S_{\mathrm{HW}} = -\frac{1}{2\kappa_{11}^2} \int d^{11}x \, (-G)^{\frac{1}{2}} \, R \; - \; \frac{1}{8\pi \left(4\pi\kappa_{11}^2\right)^{2/3}} \sum_{i=1,2} \int_{\mathbf{M}_{10,i}} d^{10}x \, (-G_{10})^{\frac{1}{2}} \mathrm{tr} \, F_i^2 + \cdots,$$

where κ_{11} is the 11d gravitational constant, G denotes the determinant of the 11d metric and G_{10} its restriction to the boundaries $\mathbf{M}_{10,i}$, $i = 1, 2$. The first integral extends over the 11d bulk spacetime whereas the second is only 10d.

The effective 4d theory at energies below the compactification scales can be obtained by KK reduction, as in previous sections. The result, to leading order in κ_{11}, and focusing on its dependence on V_6 and ρ, is

$$S_{HW,4} = -\frac{1}{2\kappa_{11}^2} V_6 \rho \int d^4x (-g)^{\frac{1}{2}} R - \sum_{i=1,2} \frac{1}{8\pi \left(4\pi\kappa_{11}^2\right)^{2/3}} V_6 \int d^4x (-g)^{\frac{1}{2}} \operatorname{tr} F_i^2 + \cdots,$$

where $g_{\mu\nu}$ is the 4d metric. The 4d Newton's constant G_N and the (unified) gauge fine structure constant α_G are given by

$$G_N = \frac{\kappa_{11}^2}{16\pi^2 V_6 \rho}; \qquad \alpha_G = \frac{\left(4\pi\kappa_{11}^2\right)^{2/3}}{2V_6}. \tag{9.39}$$

The impact of these relations on the scale of gauge coupling unification is discussed in Section 16.1.1.

The 4d effective action may be described in 4d $\mathcal{N} = 1$ supergravity language, by introducing the overall moduli S, T, in analogy with the previous sections but with slightly different definitions

$$S = V_6 + ia; \qquad T = \rho V_6^{1/3} + i\eta + \frac{1}{2}|C|^2, \tag{9.40}$$

where C denotes charged matter multiplets, e.g. in the **27** of E_6. The leading Kähler potential and gauge kinetic function turn out to be identical to those of the perturbative compactifications, but there are corrections at higher order in κ_{11}, concretely

$$
\begin{aligned}
K &= -\log(S + S^*) + \epsilon \frac{|C|^2}{(S + S^*)} - 3\log(T + T^* - |C|^2), \\
f_{E_6} &= S + \epsilon T; \qquad f_{E_8} = S - \epsilon T,
\end{aligned} \tag{9.41}
$$

with ϵ given by

$$\epsilon = -\frac{\rho}{16V_6} \left(\frac{\kappa_{11}}{4\pi}\right)^{2/3} \int_{\mathbf{X}_6} J \wedge \operatorname{tr}(R \wedge R). \tag{9.42}$$

These corrections arise from the variation of the CY moduli along the interval, mentioned in Section 7.5.1. The corrections have a structure similar to the perturbative loop corrections described in upcoming sections, although in the present case ϵ need not be parametrically small.

9.3 Effective action of orbifold models

In compactifications described by a solvable worldsheet CFT, it is in principle possible to compute the effective action. For instance, for toroidal orbifolds there are very explicit results for the effective action and its dependence on untwisted moduli, as we describe in this section.

9.3.1 Kähler potential and metrics in abelian orbifolds

Recall that for (2, 2) orbifold models, i.e. with standard embedding, the numbers of Kähler and complex structure moduli are given by the Hodge numbers (h_{11}, h_{21}) in Tables 8.2 and 8.3 in Section 8.2.6. For non-standard embedding, i.e. general (0, 2) orbifold models, the untwisted Kähler and complex structure moduli are as in (2, 2) compactifications, while the presence of extra (typically Kähler) moduli from twisted sectors is model-dependent. In this section we restrict to the dependence of the effective action on the untwisted moduli T_i, U_n, which are always present. Their number and properties however differ for the various abelian orbifolds. For instance, the Kähler potentials for the different orbifolds can be described according to their untwisted Hodge numbers $\left(h_{11}^U, h_{21}^U\right)$, as follows:

(i) $h_{11}^U = 9, \, h_{21}^U = 0 \implies K = -\log \det \left(T_{i\bar{j}} + T_{i\bar{j}}^*\right)$,

(ii)

$$h_{11}^U = 5, \, h_{21}^U = 1, 0 \implies K = -\log \left(T_1 + T_1^*\right) - \log \det \left(T_{i\bar{j}} + T_{i\bar{j}}^*\right)$$

$$-\sum_{n=1}^{h_{21}^U} \log \left(U_n + U_n^*\right), \, i, j = 2, 3, \tag{9.43}$$

(iii) $h_{11}^U = 3, \, h_{21}^U = 0, 1, 3 \implies K = -\sum_{i=1}^{3} \log \left(T_i + T_i^*\right) - \sum_{n=1}^{h_{21}^U} \log \left(U_n + U_n^*\right)$,

where $T_i \equiv T_{i\bar{i}}$. The first class corresponds to the $\mathbf{Z}_3$ orbifold, whereas the second includes the $\mathbf{Z}_4$ (with $h_{21}^U = 1$) and the $\mathbf{Z}_6$ (with $h_{21}^U = 0$); the third group describes the $\mathbf{Z}_7$, $\mathbf{Z}_8$, and $\mathbf{Z}_{12}$ (with $h_{21}^U = 0$), the $\mathbf{Z}_6'$ and $\mathbf{Z}_8$ (with $h_{21}^U = 1$), the $\mathbf{Z}_2 \times \mathbf{Z}_2$ (with $h_{11}^U = h_{21}^U = 3$), and the remaining $\mathbf{Z}_N \times \mathbf{Z}_M$ orbifolds. These results may be obtained directly by an orbifold projection of the corresponding toroidal compactification, as described above for the $\mathbf{Z}_3$ case. It may also be obtained by computing four-point amplitudes in string theory and proposing a Kähler potential reproducing the result. Note that, setting $T_1 = T_2 = T_3 \equiv T$ and $T_{i\bar{j}} = 0$ ($i \neq j$), the Kähler potential for the overall volume modulus field is $K = -3\log(T + T^*)$ for all models, as we already encountered in (9.17) for the $\mathbf{Z}_3$ orbifold.

The Kähler potential for twisted moduli is more model dependent, and in general more difficult to obtain. In certain cases, it has been computed as an expansion in the corresponding fields, in analogy with the charged matter fields in twisted sectors in the discussion below.

Let us thus turn to the Kähler potential for the charged matter fields, both from untwisted and twisted sectors. The phenomenologically most relevant piece is given by the terms quadratic in the charged chiral multiplets, which can be computed as functions of the untwisted moduli. Restricting ourselves for simplicity to the dependence on U_n and the diagonal moduli T_i, the matter field Kähler metrics in the notation of (9.25) have the structure

$$K_{I\bar{J}} = \delta_{I\bar{J}} \prod_{i=1}^{3} (T_i + \overline{T}_i)^{n_I^i} \prod_{n=1}^{h_{21}^U} (U_n + \overline{U}_n)^{l_I^n}. \qquad (9.44)$$

The kinetic terms for each matter field C^I are determined by the numbers n_I^i, l_I^n; these are rational numbers, known as *modular weights*, for reasons to be explained in Section 9.6. For untwisted matter fields C_j associated to the jth complex plane the modular weights are

$$n_j^i = -\delta_j^i, \quad l_j^i = -\delta_j^i. \qquad (9.45)$$

For instance, for the $\mathbf{Z}_3$ orbifold, the Kähler potential for an untwisted field corresponding to, e.g., the first complex plane is

$$K_{CC^*} = \frac{1}{T_1 + \overline{T}_1}, \qquad (9.46)$$

in agreement with the expansion of (9.14) to first order in $|C|^2$. This is in fact a general rule for abelian orbifolds, for which the untwisted matter field Kähler potential may be obtained from (9.43) by the replacement

$$T_i + \overline{T}_i \longrightarrow T_i + \overline{T}_i + \alpha'|C_i|^2. \qquad (9.47)$$

For abelian orbifolds this yields a closed expression for the untwisted field Kähler metric to all orders in the matter fields. This general result can also be derived from Kaluza–Klein reduction from the 10d action.

Analogous results are obtained for the dependence of untwisted matter field Kähler metrics on the complex structure moduli U_n. These are present in orbifolds with some two-torus rotated only by order two elements, like the third complex plane in, e.g., the $\mathbf{Z}_4$, $\mathbf{Z}_6'$ orbifolds, or any plane in the $\mathbf{Z}_2 \times \mathbf{Z}_2$. The piece of the Kähler potential describing the untwisted moduli T_3, U_3 and their associated untwisted matter fields C_3^T, C_3^U can be shown to have the structure

$$K = -\log\left[(T_3 + T_3^*)(U_3 + U_3^*) - \alpha'\left(C_3^T + C_3^{U*} \right)\left(C_3^{T*} + C_3^U \right) \right]$$

$$\simeq -K(T_3, U_3) + \alpha' \frac{|C_3^T|^2 + |C_3^U|^2 + \left(C_3^T C_3^U + \text{h.c.} \right)}{(T_3 + T_3^*)(U_3 + U_3^*)}. \qquad (9.48)$$

The diagonal metrics for C_3^T, C_3^U are as expected from (9.44), (9.45), but there is an interesting mixing $C_3^T C_3^U$ + h.c., i.e. a coupling $\mathbf{27} \cdot \overline{\mathbf{27}}$ for $(2, 2)$ compactifications. This has the structure (2.91) in Section 2.6.5, which as mentioned there, is the kind of coupling necessary for the Giudice–Masiero mechanism to induce mass terms after SUSY-breaking, recall Section 2.6.5.

For twisted sector matter fields, the Kähler potential is known only in an expansion on the charged fields. The computation requires advanced CFT techniques, and we merely quote the result for the quadratic piece, i.e. the kinetic terms. They also have a structure

(9.44), but with fractional modular weights. Consider a twisted state with no oscillator operators, in a sector with twist $v = (v_1, v_2, v_3)$, with negative v_3. Defining $(\eta_1, \eta_2, \eta_3) = (v_1, v_2, v_3 + 1)$, one has

$$\text{For } \eta_i \neq 0 \quad \rightarrow \quad n_I^i = -(1 - \eta^i), \quad l_I^i = -(1 - \eta^i),$$
$$\text{For } \eta_i = 0 \quad \rightarrow \quad n_I^i = 0, \tag{9.49}$$

For twisted sector states with oscillator operators

$$\text{For } \eta_i \neq 0 \quad \rightarrow \quad n_I^i = -(1 - \eta^i + p^i - q^i), \quad l_I^i = -(1 - \eta^i + q^i - p^i),$$
$$\text{For } \eta_i = 0 \quad \rightarrow \quad n_I^i = 0, \tag{9.50}$$

where p^i and q^i denote the numbers of $\alpha^i, \bar{\alpha}^i$ left-moving oscillators, respectively.

In order to keep track of the dependence of kinetic terms on the overall Kähler modulus $T_1 = T_2 = T_3 \equiv T$, we define overall modular weights for matter fields. They are given by $n_I = \sum_{i=1}^{3} n_I^i$, and take particularly simple forms

$$n = -1 \qquad \text{(untwisted)},$$
$$n = -2 - p + q \quad \text{(twisted, all planes rotated)},$$
$$n = -1 - p + q \quad \text{(twisted, only two planes rotated)}, \tag{9.51}$$

where $p = \sum_i p^i$ and $q = \sum_i q^i$. Note that overall modular weights are always integer. In specific models, including phenomenologically interesting ones, the possible values of modular weights turn out to be very restricted. The overall weights are typically $n = 0, -1, -2$, with rare appearances of more negative values. This is because more negative modular weights arise only for oscillator states, which tend to be massive because of the oscillator contribution to the mass formula.

The structure of the kinetic terms, in particular their moduli dependence, plays a key role in several phenomenologically relevant quantities, like SUSY-breaking soft terms in Section 15.5.2 or physical Yukawa couplings in the next section.

9.3.2 Superpotential couplings in abelian orbifolds

In abelian toroidal orbifolds it is possible to compute explicitly the superpotential couplings. The superpotential arises as a tree level amplitude, i.e. from a genus 0 worldsheet $\Sigma = \mathbf{S}^2$. It may be extracted from the calculation of CFT correlators on $\mathbf{S}^2$ involving two fermions and arbitrary numbers of scalars, see Figure 9.1:

$$\langle V_F \, V_F \, V_B \, \ldots, V_B \rangle_{\mathbf{S}^2}, \tag{9.52}$$

where $V_{F,B}$ are vertex operators for fermions and bosons respectively, see Appendix E for more details. We will not describe the rather technical computation of these correlators, but simply derive certain selection rules determining the allowed couplings, which suffice for many model building applications. The selection rules follow easily from the structure of the vertex operators and the symmetry of the orbifold.

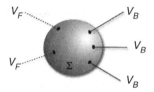

Figure 9.1 CFT correlator computing a superpotential coupling for fields associated to the vertex operators inserted.

Vertex operators in heterotic theory are products $V_L \otimes V_R$, where V_L, V_R are vertex operators of conformal weight $h = \tilde{h} = 1$ of the purely bosonic left sector and the supersymmetric right sector, respectively. In the right-moving sector, untwisted state vertex operators have the structure (E.20), (E.19)

$$\text{NS}: \quad V_{-1}^B = e^{-\tilde{\phi}} e^{i\alpha_v \cdot \tilde{H}} e^{ik_\mu \tilde{X}^\mu},$$

$$\text{R}: \quad V_{-1/2}^F = e^{-\frac{\tilde{\phi}}{2}} e^{i\alpha_s \cdot \tilde{H}} e^{ik_\mu \tilde{X}^\mu}, \tag{9.53}$$

in the (-1)- and $(-1/2)$-pictures, respectively. Here k^μ is the 4d momentum, and 4d scalars arise from $SO(10)$ vector weights on the internal 6d, i.e. $\alpha_v = (\pm 1, 0, 0, 0, 0)$, where underlining means permutation of entries. Also, $\alpha_s = (\pm\frac{1}{2}, \pm\frac{1}{2}, \dots, \pm\frac{1}{2})$ are $SO(10)$ spinor weights, with an even number of '+' signs. The fields $\tilde{H}$ are the bosonized right-moving fermions, and $\tilde{\phi}$ is the bosonized superconformal ghost.

For states in sectors twisted by the element $h = (\theta^m, \vec{w})$ of the orbifold space group, the right-moving vertex operators are a generalization of the above

$$\text{NS}: \quad V_{-1}^{B,h} = e^{-\tilde{\phi}} e^{i(\alpha_v + mv) \cdot \tilde{H}} e^{ik_\mu \tilde{X}^\mu} \tilde{\sigma}_h,$$

$$\text{R}: \quad V_{-1/2}^{F,h} = e^{-\frac{\tilde{\phi}}{2}} e^{i(\alpha_s + mv) \cdot \tilde{H}} e^{ik_\mu \tilde{X}^\mu} \tilde{\sigma}_h, \tag{9.54}$$

where $v = (v_1, v_2, v_3, 0, 0)$ is the twist vector of θ, and $\tilde{\sigma}_h$ is a "twist field," i.e. an operator which creates the vacuum in the h-twisted sector out of the CFT groundstate (the untwisted vacuum).

As explained in Section E.2, our amplitudes of interest have the form

$$\langle V_{-\frac{1}{2}}(\bar{z}_1) V_{-\frac{1}{2}}(\bar{z}_2) V_{-1}(\bar{z}_3) V_0(\bar{z}_4) \cdots V_0(\bar{z}_n) \rangle_{S^2}, \tag{9.55}$$

with total superconformal ghost charge -2, as required for a sphere amplitude. Note that the computation of non-renormalizable superpotential couplings requires the use of vertex operators in the 0-picture, like (E.18) or its twisted sector generalizations. We will not need their detailed expressions.

The left-moving sector of the heterotic theory is purely bosonic, so there are no superconformal ghost charges. One must just fix the positions of three vertices, using $c\tilde{c}$ insertions, and integrate only over any additional vertices. Ignoring this for notational simplicity, the relevant vertex operators for untwisted states are

State	Operator	State	Operator
$\alpha^i\lvert 0\rangle$	$V_L = \partial Z^i\, e^{ik_\mu X^\mu}$	$\alpha^I_{-1}\lvert 0\rangle$	$V_L = \partial X^I\, e^{ik_\mu X^\mu}$
$\bar\alpha^i\lvert 0\rangle$	$V_L = \partial \bar{Z}^i\, e^{ik_\mu X^\mu}$	$\lvert P\rangle_{P^2=2}$	$V_L = e^{i P_I X^I}\, e^{ik_\mu X^\mu}$

where Z^i denotes the complex coordinate along the ith $\mathbf{T}^2$. The left- and right-moving vertex operators can be easily paired (in orbifold invariant combinations) to obtain the untwisted matter fermions and bosons, and untwisted moduli.

Finally, the vertex operators for states in a $(\theta^m, \vec{w})$-twisted sector are

$$V_L = e^{i(P+mV)\cdot X}\, \sigma_h\, e^{ikx}, \tag{9.56}$$

where V is the embedding of the orbifold twist in the gauge degrees of freedom and σ_h is a left-moving twist operator for $h = (\theta^m, \vec{w})$.

We can now exploit the symmetries of correlators involving these vertex operators to extract useful selection rules, on top of the obvious ones like 4d momentum conservation and gauge invariance (arising from worldsheet conservation of the currents $i\partial X^\mu$, $i\bar\partial X^\mu$, $i\bar\partial X^I$). There are essentially three additional selection rules:

- *Space and point group selection rule*: Each vertex operator has an associated space group element $h = (\theta^m, \vec{w})$, determining the twisted sector and the fixed point at which the corresponding state lies. The operators create branch cuts on the worldsheet, across which the 2d worldsheet fields are acted upon by h; the jump of the 2d fields around a vertex operator turns, by the the state-operator map in Figure E.2, into a twisted boundary condition. Consider now contours surrounding each vertex operator, and deform them into a single contour surrounding all vertex operators. Since the latter can be shrunk to zero (on the other side of the worldsheet $\mathbf{S}^2$), it must have trivial space group twist. Since the total space group jump is the product of the individual ones, the product of all the space group elements h for the fields in the amplitude must be the identity, otherwise the coupling vanishes. Restricting this rule to the $\mathbf{Z}_N$ point group, implies that couplings of untwisted fields ϕ^U_i and θ^q-twisted fields ϕ^T_q of the form

$$\prod_{i,q} \left(\phi^U_i\right)^{k_i} \left(\phi^T_q\right)^{k_q} \tag{9.57}$$

vanish unless $\sum_q (q k_q) = 0 \bmod N$. The translation part of the space group selection rule is a constraint on the location of the fixed points associated to the twisted fields in the correlators. For example, in the $\mathbf{Z}_3$ orbifold, cubic couplings among twisted sector fields are allowed only if the fixed points are all equal or all different. Note that untwisted fields are not restricted by this rule.
- *Right-moving symmetries*: The right-moving vertex operators contain a piece carrying momentum along the bosonized fermions $\tilde{H}$, dubbed H-momentum. Momentum conservation, i.e. conservation of the right-moving worldsheet current $i\partial H$, provides a non-trivial selection rule, which in fact forbids many couplings for both twisted and

untwisted fields. By fermionization it can be recast as conservation of 2d fermion number. From the 4d spacetime viewpoint, it may be understood in terms of R-symmetries in the effective 4d field theory.

- When the underlying $\mathbf{T}^6$ factorizes as a product of several $\mathbf{T}^2$'s, the correlators must be invariant under independent phase rotations of each of the complex coordinates. This rule is particularly restrictive for states involving left-moving oscillator modes, i.e. vertex operators involving $\bar{\partial} Z^i$, $\overline{\partial Z}^{\bar{i}}$.

To put the above rules to use, consider the example of a Yukawa coupling among three untwisted chiral multiplets in a $\mathbf{Z}_N$ orbifold. As already discussed, there are three kinds of untwisted matter chiral multiplets C_i, with $i = 1, 2, 3$, associated with the three complex planes. Their Yukawa couplings have the structure (9.53)

$$Y^U_{ijk} \propto \left\langle V^i_F V^j_F V^k_B \right\rangle_{L \times R}, \tag{9.58}$$

with $L \times R$ indicating that operators and the correlator are taken in the full left times right worldsheet theory. Such a correlator is obviously invariant under the space group rule and independent phase rotations. On the other hand, H-momentum conservation requires the $SO(10)$ weights of the three vertices to add up to zero,

$$(0,0,0,0,0) = \alpha^i_s + \alpha^j_s + \alpha^k_v = \tfrac{1}{2}(\underbrace{+--}_{i}; +, -) + \tfrac{1}{2}(\underbrace{-+-}_{j}; -, +) + (\underbrace{0,0,+1}_{k}; 0, 0),$$

where the underbrace label indicates that the "+" entry position is the ith, jth, and kth, respectively. The H-momentum rule thus requires $i \neq j \neq k \neq i$, consistently with the dimensional reduction result in (9.16) $Y^U_{ijk} \sim \epsilon_{ijk}$. It is indeed a general (and expected) fact that couplings for untwisted fields can be obtained from KK reduction of the 10d theory.

Yukawa couplings involving three twisted fields are computable, with technical subtleties associated to the non-trivial correlators of twist operators, and we simply review the structure of the result. Consider two states associated with the fixed points f_a, f_b in sectors twisted by $\mathbf{Z}_N$ elements with shifts v, w respectively. They couple to a third state living at a fixed point f_c in the twisted sector with shift $-v - w$. The string theory result for the Yukawa coupling (restricting to odd N for simplicity) is

$$\tilde{Y}^{vw}_{f_a f_b f_c} = (2\pi)^{3/2} V^{1/2}_{\Lambda_6} \left(\prod_{j=1}^3 \Gamma_{v_j w_j} \right)^{1/2} \sum_{V \in f_a - f_b + \Lambda_6} e^{-2\pi V^t (G + iB) L V}, \tag{9.59}$$

where Λ_6 is the compactification torus lattice, V_{Λ_6} is its volume in string units, and G, B are the internal metric and 2-form background. Also $L = \prod_j L_j$ with

$$L_j = \frac{1}{|\lambda_j|} \left[\mathbf{1_2} + i \operatorname{sign}(\lambda_j) \epsilon \right], \quad \epsilon = \begin{pmatrix} 0 & 1 \\ -1 & 0 \end{pmatrix}, \tag{9.60}$$

and $\lambda_j = \cot(\pi v_j) + \cot(\pi w_j)$, $j = 1, 2, 3$. Finally

$$\Gamma_{v_j w_j} = \frac{\Gamma(1 - v_j) \Gamma(1 - w_j) \Gamma(v_j + w_j)}{\Gamma(v_j) \Gamma(w_j) \Gamma(1 - v_j - w_j)}, \tag{9.61}$$

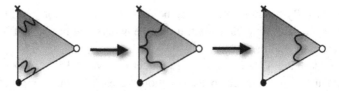

Figure 9.2 Yukawa coupling among three twisted sector fields. A pair of closed strings twisted around two fixed points stretch and join, giving rise to a recombined closed string twisted around a third fixed point.

where shifts are assumed in the range $0 < v_j + w_j < 1$. The exponential factor describes a sum over worldsheet instantons, processes in which the string worldsheet wraps a 2-cycle in the internal space passing through the three fixed points (see Figure 9.2), with generically a multiple wrapping characterized by the vectors V. Similar results are obtained for the case of even N, for which the range of the vectors V is slightly more complicated. The last factor is the semiclassical contribution to the amplitude given by $\exp(-S_{\text{clas}})$, the exponential of the worldsheet instanton action, i.e. the area swept by the worldsheet. This amplitude, for three different fixed points, is exponentially suppressed as e^{-R^2}, since the V's are measured in units of the lattice size; in principle this might be an interesting property to reproduce the observed hierarchies in fermion masses, although in practice it is not so easy to construct semi-realistic models achieving this goal for all SM fermions.

The above Yukawa couplings are *physical*, i.e. for fields with canonical kinetic terms, recall (2.109). The semiclassical piece can be recast in terms of holomorphic ϑ-functions depending on the untwisted Kähler moduli, and corresponds to the holomorphic Yukawa couplings in the superpotential. The remaining factors, which can be regarded as arising from fluctuations around the semiclassical worldsheet instanton (sometimes called the quantum contribution), are related to the wave function normalization of the participating fields as in (2.109). Incidentally, Yukawa couplings in type II orientifolds in Section 12.5.3 have an analogous structure.

In general CY compactifications, we do not have the CFT tools allowing for the explicit computations as above. However the discrete symmetries often present in these models give rise to selection rules strongly constraining the structure of Yukawa couplings. This is for example the case of the Tian–Yau 3-generation model in Section 7.3.2, and the models with non-standard embedding in Section 7.4. For Gepner models, one may in principle compute the Yukawa couplings from CFT correlators, but this has not been worked out in semi-realistic models, except for the Gepner version of the Tian–Yau example. We direct the reader to the Bibliography for further details.

9.4 Gauge couplings and Kac–Moody level

In general heterotic string compactifications, and in particular those described by exact CFTs, there is a possible subtlety in the relationship (9.22) between the Planck and string scales with the gauge coupling. The precise relation depends on the level of the Kac–Moody algebra underlying the corresponding gauge symmetries, see Section E.3. In general, the 4d gauge bosons arise from states with vertex operator (in the (-1)-picture for the right sector part, see Section E.2)

$$V_a^\mu \simeq e^{-\tilde\phi}\,\tilde\psi^\mu(\bar z)\,J_a(z)\,e^{ik\cdot X}, \tag{9.62}$$

where the operators $J_a(z)$ are Kac–Moody (KM) currents satisfying the OPE (E.21)

$$J_a(z)J_b(w) \sim \frac{(k/2)\delta^{ab}}{(z-w)^2} + \frac{if^{abc}J_c(w)}{(z-w)} + \cdots, \tag{9.63}$$

with f_{abc} being the gauge group structure constants. We use the standard particle physics normalization $\mathrm{tr}\,(T^aT^b)=\frac12\delta^{ab}$ of $SU(n)$ generators in the fundamental.

In the 10d heterotic theory the KM algebra is realized at level 1, by the operators (E.24). In standard geometric compactifications, this realization is inherited by the 4d non-abelian gauge factors, which are hence realized at level 1. However, it is possible to construct other vacua realizing gauge symmetries at higher level, as discussed later in Section 9.8. Also, 4d $U(1)$ gauge factors are in general given by linear combinations $J_Q = i Q_I \partial X^I$ of the Cartan currents in (E.24), with OPE's

$$J_Q(z)J_Q(w) \sim \frac{Q^2\,\delta^{ab}}{(z-w)^2} + \cdots. \tag{9.64}$$

Comparison with (9.63) shows that $U(1)$s have an associated KM level $k_1 = 2Q^2$.

The KM level of a 4d gauge symmetry modifies the gauge coupling unification relations. The generalized version of (9.22) for a gauge factor at level k may be computed explicitly from vertex correlators, and reads

$$\frac{M_p^2}{M_s^2} = 16\,\frac{4\pi}{g^2 k} = \frac{16}{\alpha k}, \tag{9.65}$$

with g the YM gauge coupling. For a heterotic compactification with a low-energy sector including the SM gauge group $SU(3) \times SU(2) \times U(1)$, with factors realized at levels k_3, k_2, k_1, we thus have

$$k_3\alpha_3 = k_2\alpha_2 = k_1\alpha_1. \tag{9.66}$$

Most compactifications, certainly including standard geometric compactifications, have $k_3 = k_2$ and thus unification of $SU(3)$ and $SU(2)$ couplings at the string scale. On the other hand, k_1 is determined by the more model-dependent hypercharge embedding, and yields the successful $SU(5)$ value $k_1 = 5/3$ only in very special circumstances. For instance, this value is obtained in orbifold models in which the SM is embedded into an $SO(10)$ subgroup of E_8. In general, GUT-like gauge coupling boundary conditions are natural in heterotic string compactifications with an underlying GUT symmetry, even in the absence of a full low-energy GUT theory. We postpone a more detailed review of gauge coupling unification to Section 16.1.1.

9.5 Anomalous $U(1)$s and Fayet–Illiopoulos terms

One of the prominent features of 4d string vacua is the ubiquitous presence of "pseudo-anomalous" $U(1)$s, i.e. gauged $U(1)$ symmetries with non-zero triangle contributions to the mixed non-abelian (and gravitational) anomalies. These are, however, canceled by a 4d version of the Green–Schwarz mechanism mentioned in Section 4.3.4. Although the total anomaly cancels and the models are consistent, we often refer to these $U(1)$s as "anomalous." In this section we describe this mechanism and its implications for the 4d effective action of heterotic string vacua.

9.5.1 A Z_3 orbifold example

We start with an illustrative warm-up example within the Z_3 orbifolds in Section 8.3.1. Consider the last model in Table 8.4, with gauge group $SU(9) \times SO(14) \times U(1)$. The $U(1)$ corresponds to the generator $Q = (0, \ldots, 0) \times (1, 0, \ldots, 0)$ and the matter fields transform as $3(\mathbf{84}, \mathbf{1})_0 + 3(\mathbf{1}, \mathbf{14})_{-1} + 3(\mathbf{1}, \mathbf{64})_{1/2} + 27(\overline{\mathbf{9}}, \mathbf{1})_{2/3}$ under $SU(9) \times SO(14) \times U(1)$. The subindices indicate the $U(1)$ charges, which are given by the scalar product $Q \cdot P$ for untwisted fields and $Q \cdot (P + V)$ for twisted fields. The one-loop triangle contributions to mixed $U(1) \times SU(9)^2$, $U(1) \times SO(14)^2$ anomalies, as well as the cubic $U(1)^3$ and mixed gravitational anomaly, give anomaly coefficients

$$A_{SU(9)} = 9; \quad A_{SO(14)} = 9; \quad A_{U(1)^3} = 54; \quad A_{\mathrm{grav}} = 24 \times 9. \tag{9.67}$$

The anomalies are related by

$$\frac{A_{SU(9)}}{k_9} = \frac{A_{SO(14)}}{k_{14}} = \frac{A_{U(1)^3}}{3k_1} = \frac{A_{\mathrm{grav}}}{24}, \tag{9.68}$$

where $k_9 = k_{14} = 1$ and $k_1 = 2Q^2 = 2$ are the $SU(9)$, $SO(14)$, and $U(1)$ KM levels. This universality is actually crucial in the anomaly cancellation mechanism. Recall the coupling (9.12) of the axion-like field $a = \mathrm{Im}\, S$ to gauge fields, schematically

$$k\, a\mathrm{tr}\, (F \wedge F), \tag{9.69}$$

where we have included general Kac–Moody levels. The above gauge anomalies are canceled if the axion transforms with a shift under $U(1)$ gauge transformations

$$A_\mu \rightarrow A_\mu + \partial_\mu \Lambda(x); \quad a \rightarrow a - c\Lambda(x), \tag{9.70}$$

where in our case $c = 9/(8\pi^2)$ for the anomalies to cancel. Diagrammatically the cancellation mechanism is depicted in Figure 9.3. The anomaly generated by the standard triangle graph is canceled by a diagram in which the anomalous $U(1)_X$ mixes with the axion-like field a, which in turn couples to $\mathrm{tr}\, (F \wedge F)$ for any gauge group in the compactification. Note that the cubic $U(1)^3$ anomaly has an extra factor $1/3$ due to permutation symmetry. Similarly, the mixed gravitational anomaly cancellation involves a curvature coupling of the axion $a\mathrm{tr}\, (R \wedge R)$, which has an extra normalization factor 24.

Figure 9.3 Cancellation of mixed $U(1)_X$ anomalies by the 4d Green–Schwarz mechanism.

9.5.2 General description of U(1) Stückelberg masses

The above 4d Green–Schwarz mechanism is very general and occurs in most classes of 4d string vacua. In perturbative heterotic vacua like orbifolds, or CY compactifications (without $U(1)$ bundles), only the axion field $a = \text{Im } S$ participates in the anomaly cancellation mechanism, and hence there is at most one anomalous $U(1)$, with universal mixed anomaly coefficients. In other 4d string vacua, like type II orientifolds or CYs with $U(1)$ bundles, there are in general several anomalous $U(1)$s with anomalies canceled via exchange of several axion-like fields, see Sections 10.3.1, 11.2.2, 11.3.2, and 11.4.2. Given the generality of the phenomenon, it is useful to devise a lagrangian for the relevant fields, in order to describe the properties of the scalar and the $U(1)$ gauge boson. This can be done in a general form, easily generalized to any 4d string vacuum, using the following 4d Stückelberg lagrangian, which involves a $U(1)$ boson A_μ and a two-index antisymmetric tensor $B_{\mu\nu}$,

$$\mathcal{L} = -\frac{1}{12}h^{\mu\nu\rho}h_{\mu\nu\rho} - \frac{1}{4g^2}F^{\mu\nu}F_{\mu\nu} + \frac{c}{4}\epsilon^{\mu\nu\rho\sigma}B_{\mu\nu}F_{\rho\sigma}, \qquad (9.71)$$

where

$$h_{\mu\nu\rho} = \partial_{[\mu}B_{\nu\rho]}, \quad F_{\mu\nu} = \partial_{[\mu}A_{\nu]}. \qquad (9.72)$$

Here we have not made explicit the dilaton dependence, which varies in the different 4d vacua. The above lagrangian contains just standard kinetic terms plus an additional crucial term of the form $B \wedge F$. In order to dualize the 2-index tensor to a scalar, we proceed as in Section 9.1.1 and consider $h_{\mu\nu\rho}$ as an arbitrary tensor, enforcing the constraint $dh = 0$ via a Lagrange multiplier field $\eta(x)$, as follows

$$\mathcal{L}_0 = -\frac{1}{12}h^{\mu\nu\rho}h_{\mu\nu\rho} - \frac{1}{4g^2}F^{\mu\nu}F_{\mu\nu} - \frac{c}{6}\epsilon^{\mu\nu\rho\sigma}h_{\mu\nu\rho}A_\sigma - \frac{1}{6}\eta\,\epsilon^{\mu\nu\rho\sigma}\partial_\mu h_{\nu\rho\sigma}. \qquad (9.73)$$

Integrating out η, we get $dh = 0$, which in 4d flat space implies $h = dB$ and gives back (9.71). On the other hand, integrating by parts the last term in (9.73) produces an action quadratic in $h_{\mu\nu\rho}$, which can be solved to obtain

$$h^{\mu\nu\rho} = -\epsilon^{\mu\nu\rho\sigma}(cA_\sigma + \partial_\sigma\eta). \qquad (9.74)$$

Inserting this back in (9.73) one obtains

$$\mathcal{L}_A = -\frac{1}{4g^2} F^{\mu\nu} F_{\mu\nu} - \frac{1}{2} \left(cA_\sigma + \partial_\sigma \eta\right)^2, \tag{9.75}$$

which describes a massive $U(1)$ gauge boson, whose longitudinal component is the pseudoscalar η. Note that the mass term in (9.75) is gauge invariant only if the η shifts as $\eta \to \eta - c\Lambda(x)$ under $U(1)$ gauge transformations, exactly as for the axion-like field a in (9.70). This transformation for the axionic field a is thus a direct consequence of the existence of a $B \wedge F$ coupling. We conclude that the axion field participating in anomaly cancellation in heterotic models gets eaten up into the massive $U(1)$ gauge boson and disappears from the low-energy spectrum.

Conversely, any $U(1)$ gauge boson with a $B \wedge F$ coupling acquires a mass, whether the $U(1)$ is anomalous or not. In fact, mixed $U(1)$ triangle anomalies are correlated with the existence of additional couplings $\eta F \wedge F$ to other gauge bosons, as required by the Green–Schwarz mechanism. This is always the case in the heterotic vacua under consideration, but not necessarily in type II orientifolds, in which there may exist Stückelberg couplings for non-anomalous $U(1)$s, see Sections 10.3.1 and 11.5.3.

In 4d $\mathcal{N} = 1$ models, SUSY has additional implications. We introduce the chiral superfield Φ, whose complex scalar component is $\Phi = \varphi + i\eta$, and promote the $U(1)$ boson to a vector multiplet V. The Kähler potential for Φ must have a structure

$$K(\Phi + \Phi^* + cV), \tag{9.76}$$

consistent with the shift of η, i.e. $\Phi \to \Phi - ic\Lambda$ under $U(1)$ gauge transformations (2.25) $V \to +i(\Lambda - \Lambda^*)$, with gauge parameter in a chiral multiplet Λ. Expanding (9.76) in components, the term linear in V produces a Fayet–Illiopoulos term

$$\xi = c \left.\frac{\partial K(\Phi + \Phi^*)}{\partial \Phi}\right|_{V=0}, \tag{9.77}$$

controlled by $\mathrm{Re}\,\Phi = \varphi$. Thus supersymmetry relates the $B \wedge F$ and the FI couplings.

Let us finally mention that the $B \wedge F$ coupling appears at one-loop order in heterotic compactifications. It may be seen as arising after dimensional reduction of the 10d one-loop coupling (4.86), i.e. $\int_{10d} B_2 \wedge F^4 \to \int_{4d} B_2 \wedge \mathrm{tr}\, F \int_{X_6} F^3$. In the type II orientifold compactifications in Chapters 10 and 11, this coupling arises at tree level from the D-brane action.

9.5.3 *U(1) mass and FI terms in heterotic vacua*

Coming back to the heterotic case, the role of η and Φ are played by the axion a and the dilaton S. Using the Kähler potential in (9.77), the effective action contains a field-dependent FI term for the anomalous $U(1)_X$:

$$\xi_X = \partial_S K(S, S^*) c = -\frac{c}{\kappa_4^2 (S + S^*)} = -4cM_s^2, \tag{9.78}$$

where we have used (9.22). Furthermore, expanding the Kähler potential (9.76) to second order in V one finds that the mass of the $U(1)$ boson with canonical kinetic term is
$M_{U(1)_X} = cg M_s$. Since Re $S = 1/g^2$ the FI term is, as explained, a one-loop effect. The
expression for the FI terms follows from anomaly cancellation and supersymmetry, so it is
not renormalized, i.e. receives no higher-loop corrections. The coefficient c can be computed for general heterotic compactifications and depends only on the massless charged
chiral fields of the theory. It is given by $c = \mathrm{Tr}\, Q_X/[48(2\pi)^2]$, with the trace running
through all the massless fermions charged under $U(1)_X$. The D-term scalar potential of
the theory has then the form

$$V_X = \frac{1}{S + S^*} \left| \frac{\mathrm{Tr}\, Q_X}{48(2\pi)^2} \frac{1}{\kappa_4^2 (S + S^*)} + \sum_i q_{X,i} |\Phi_i|^2 \right|^2, \tag{9.79}$$

where $q_{X,i}$ is the $U(1)_X$ charge of the field Φ_i. Since the FI term is controlled by the
string coupling, it cannot be taken to zero. In principle this term could be a source of
SUSY-breaking appearing at one-loop, because for all $\Phi_i = 0$ (and assuming the vev of S
has been fixed by another sector of the lagrangian), there is a non-zero D-term; however,
in all known 4d heterotic vacua, the FI term in the potential can always be canceled by
vevs for some suitable charged fields Φ_i, with the appropriate charge sign. The implication is that heterotic vacua with one anomalous $U(1)$ are re-stabilized at one-loop. The
re-stabilization by Φ_i vevs necessarily implies spontaneous breaking of the anomalous
$U(1)_X$, and possibly of other gauge symmetries. As we remark later in Section 9.7, this
turns out to be a phenomenologically welcome property. Note also that the actual scalar
eaten by the anomalous $U(1)_X$ is thus a linear combination of Im S and phases of Φ_i fields.

Incidentally, due to the constraint (9.66), in a putative KM level 1 heterotic vacuum
containing the SM gauge group and an anomalous $U(1)_X$, one can directly compute the
weak angle θ_W in terms of the anomaly coefficients, i.e.

$$\sin^2 \theta_W = \frac{k_2}{k_1 + k_2} = \frac{A_2}{A_1 + A_2}, \tag{9.80}$$

where A_2, A_1 are the $U(1)_X$-$SU(2)^2$ and $U(1)_X$-$U(1)_Y^2$ mixed anomaly coefficients. This
property is interesting because it depends only on low-energy data, i.e. the $U(1)_X$ charges
of massless chiral fields transforming under the SM group. One can get, e.g., the canonical $\sin^2 \theta_W = 3/8$ value for suitable $U(1)_X$ charges, with no need of any underlying $SU(5)$
symmetry. Phenomenological (field theory) models have been constructed in which $U(1)_X$
is a flavour symmetry with different charges for different generations; the corresponding FI
term is assumed to trigger vevs for SM singlets breaking the $U(1)_X$ and producing hierarchical Yukawa couplings through the Froggatt–Nielsen mechanism. Although this is potentially an attractive scenario for the generation of the observed structure of fermion masses,
no concrete heterotic compactification with the required properties has been produced
as yet.

9.6 T-duality and the effective action

The dualities of string theory in higher dimensions have interesting consequences for 4d vacua. This is nicely illustrated by the implications of T-dualities (generalizing that in Section 3.2.3) for the effective action of 4d heterotic orbifold vacua.

Consider first heterotic strings on a single $\mathbf{T}^2$, constructed as $\mathbf{R}^2$ modded out by a lattice of translations generated by $\vec{e}_1, \vec{e}_2$. There are four real scalar moduli, which can be grouped into the $\mathbf{T}^2$ Kähler modulus T and complex structure U

$$T = \sqrt{G} + iB; \quad U = \frac{\sqrt{G}}{G_{22}} + i\,\frac{G_{12}}{G_{22}}. \tag{9.81}$$

Here $G = \det(G_{ij})$, with $G_{ij} = \vec{e}_i \cdot \vec{e}_j$ defining the metric, and $B_{ij} = B\epsilon_{ij}$ is a constant 2-form background. Restricting ourselves to states with neither oscillators nor 16d gauge momenta, $\tilde{N} = N = P^I = 0$, we are left with the $\mathbf{T}^2$ momenta and winding states. The mass formula for this sector can be recast as

$$M^2 = \frac{|\,m_1 + n_1 TU + i(Um_2 + Tn_2)\,|^2}{\alpha'(T + T^*)(U + U^*)}, \tag{9.82}$$

where m_i and n_i are $\mathbf{T}^2$ momenta and windings. This expression is invariant under

$$
\begin{aligned}
T \to T + i\,, &\quad m_1 \to m_1 + n_2\,,\ m_2 \to m_2 - n_1, \\
T \to 1/T\,, &\quad m_1 \to n_2\,,\ n_1 \to m_2\,,\ m_2 \to -n_1\,,\ n_2 \to -m_1, \\
T \leftrightarrow U\,, &\quad m_2 \leftrightarrow n_2.
\end{aligned}
\tag{9.83}
$$

The first two transformations generate a modular group $SL(2, \mathbf{Z})_T$ acting on T. It involves exchange of winding and momenta, so it generalizes the T-duality for $\mathbf{S}^1$ compactifications. The general such $SL(2, \mathbf{Z})_T$ T-duality transformation is

$$T \to \frac{aT - ib}{icT + d}, \quad a, b, c, d \in \mathbf{Z}, \quad ad - bc = 1, \tag{9.84}$$

which indeed has the structure (A.1) by taking $\tau = iT$. This symmetry restricts the moduli space of T to the fundamental domain $-\frac{1}{2} \leq \operatorname{Im} T < \frac{1}{2}, |T| \geq 1$.

The third transformation in (9.83) exchanges the Kähler and complex structure moduli, and is the simplest example of mirror symmetry, see Section 10.1.2. This implies an $SL(2, \mathbf{Z})_U$ symmetry acting on U, whose interpretation is just the geometric (non-stringy) modular group associated to reparametrizations of $\mathbf{T}^2$.

These results extend to the abelian toroidal orbifolds, which inherit such T-duality symmetries from the underlying $\mathbf{T}^6$. We consider toroidal geometries which factorize as products of three $\mathbf{T}^2$. In the absence of additional backgrounds (e.g. discrete Wilson lines), the orbifold projection respects the underlying T-duality symmetries, and the orbifolds have (at least) an $SL(2, \mathbf{Z})^3$ invariance acting on the three diagonal T_i moduli, as

$$T_i \to \frac{a_i T_i - ib_i}{ic_i T_i + d_i}\,, \quad a_i, b_i, c_i d_i \in \mathbf{Z}\,, \quad a_i d_i - b_i c_i = 1. \tag{9.85}$$

(For non-factorized toroidal geometries, the T-duality group is smaller but still contains an $SL(2, \mathbf{Z})$ for the overall modulus T.) Invariance under the duality group has interesting implications for the effective action. Recalling (9.44), the Kähler potential for matter fields, to quadratic order, is given by

$$K_{\text{matter}} = \delta_{I\bar{J}} \prod_{i=1}^{3} \left(T_i + T_i^*\right)^{n_I^i} \prod_{n=1}^{h_{21}^U} \left(U_n + U_n^*\right)^{l_I^n} C_I \bar{C}_J. \tag{9.86}$$

Invariance under the modular group $\Gamma = [SL(2, \mathbf{Z})]_T^3 \times [SL(2, \mathbf{Z})]_U^{h_{21}^U}$ requires that the matter fields transform (up to constant matrices) as

$$C_I \rightarrow C_I \prod_{i=1}^{3} \left(i c_i T_i + d_i\right)^{n_I^i} \prod_{m=1}^{h_{21}} \left(i c_m U_n + d_m\right)^{l_I^m}, \tag{9.87}$$

hence the name "modular weights" for n_I^i, l_I^m. Consider now the S, T_i dependent piece of the supergravity Kähler function (2.47) (setting $\kappa_4 = 1$ for clarity)

$$G = -\log(S + S^*) - \sum_i \log\left(T_i + T_i^*\right) + \log|W|^2. \tag{9.88}$$

Under the duality transformations (9.85), we have

$$-\log\left(T_i + T_i^*\right) \rightarrow -\log\left(T_i + T_i^*\right) + \log|i c_i T_i + d_i|^2, \tag{9.89}$$

so that invariance requires that the superpotential transforms as

$$W \rightarrow e^{i\delta(a_i, b_i, c_i, d_i)} \frac{W}{i c_i T_i + d_i}, \tag{9.90}$$

where δ are possible moduli-independent phases. Using the matter field transformations (9.87), the Yukawa coupling constant h_{IJK} must transform as

$$h_{IJK} \rightarrow h_{IJK} \prod_{i=1}^{3} \left(i c_i T_i + d_i\right)^{\left(-1 - n_I^i - n_J^i - n_K^i\right)}. \tag{9.91}$$

Note that this is consistent with a moduli-independent Yukawa coupling constant for three untwisted matter fields, since they have modular weights $n_i = (-1, 0, 0)$, $(0, -1, 0)$, $(0, 0, -1)$. Yukawa couplings involving twisted fields are, however, constrained by the above transformation properties. We emphasize that duality invariance applies not only to cubic couplings but also to higher-dimensional and non-perturbative pieces of the superpotential, as we exploit in the study of SUSY breaking in Section 15.2.2. The duality invariance of heterotic orbifold compactifications is thus a strong constraint on the structure of their effective action.

The existence of infinite discrete duality groups acting on moduli is not restricted to toroidal compactifications and orbifolds thereof. They also arise in certain CY compactifications, although they are harder to characterize and have not been systematically exploited in phenomenological applications.

9.6.1 One-loop corrections and duality invariance

One-loop effects in the effective action are expected to be small in heterotic compactifications at weak coupling. However, there are some situations in which they may be relevant because the corresponding physical quantities are known with good precision. This is the case for the one-loop threshold corrections to the gauge coupling constants. These corrections are relevant to the issue of non-perturbative SUSY breaking and moduli stabilization in heterotic vacua, see Section 15.2.2.

From (9.21), the tree level gauge couplings have a unified value g. Analogously to (1.9), at a low-energy scale Q the one-loop corrected couplings for the different gauge factors, labeled with an index a, are

$$\frac{16\pi^2}{g_a^2(Q^2)} = k_a \frac{16\pi^2}{g^2} + b_a \log \frac{M_G^2}{Q^2} + \Delta_a + k_a \Delta^{\text{univ}}, \tag{9.92}$$

where $M_G \simeq M_s$ – determined more precisely later, see (9.96) – and b_a are the one-loop β-function coefficients, see (2.71). The group dependent threshold corrections are contained in Δ_a, and there is also a universal correction Δ^{univ} which may be reabsorbed in the definition of g. Both depend on the moduli but not on the dilaton S, and include contributions from massive string and Kaluza–Klein states. The group dependent threshold correction can be obtained from a one-loop string diagram, for whose computation we refer to the Bibliography. For any perturbative heterotic compactification it is formally given by the expression

$$\Delta_a = \int_{F_0} \frac{d^2\tau}{\tau_2} \left[B_a(\tau, \bar{\tau}) - b_a \right], \tag{9.93}$$

where F_0 is the fundamental domain for the worldsheet torus parameter τ, c.f. Section 3.2.2. Here B_a is given by the general expression

$$B_a(\tau, \bar{\tau}) = \frac{2}{|\eta(\tau)|^4} \sum_{\substack{(s_1, s_2) \\ \text{even } s}} (-1)^{s_1 + s_2} \frac{1}{2\pi i} \frac{d\bar{Z}_\Psi(s, \bar{\tau})}{d\bar{\tau}}$$

$$\times \operatorname{Tr}_{s_1} \left[q^{L_0 - \frac{11}{12}} \bar{q}^{\tilde{L}_0 - \frac{3}{8}} (-1)^{s_2 \tilde{F}} \left(T_a^2 - \frac{k_a}{8\pi \tau_2} \right) \right]_{\text{int}}. \tag{9.94}$$

Here the s_1, s_2 sum (over spin structures) runs over values 0, 1 corresponding to NS or R boundary conditions, and "even s" excludes the case $(s_1, s_2) = (1, 1)$. The first line contains universal factors independent of the specific compactification, and arise from the 4d coordinates. The $|\eta|^{-4}$ arises from bosonic oscillators, and Z_Ψ is the partition function for one complex fermion, with the derivative ultimately due to the insertion of (the right-moving piece of) two gauge boson vertex operators. The second line traces over the internal compactification CFT, as indicated by the subscript "int." The exponents of $q = e^{2\pi i \tau}$ (and $\bar{q}$) are essentially the internal CFT hamiltonian, written conventionally in terms of the internal CFT Virasoro generators L_0, $\tilde{L}_0$ and zero point energies – see Appendix E, e.g. (E.30) – and $(-1)^{s_2 \tilde{F}}$ implements the right-moving GSO projection. Finally, the generators T_a of the 4d gauge factor G_a arise from the insertion of (the left-moving piece of) two gauge boson vertex operators.

As expected on physical grounds, B_a reduces at low energies to the standard β-function coefficients, e.g. (1.10). Namely their infrared limit is

$$\lim_{\tau_2 \to \infty} B_a = -\frac{11}{3} \text{tr}_V \left(T_a^2 \right) + \frac{2}{3} \text{tr}_F \left(T_a^2 \right) + \frac{1}{3} \text{tr}_S \left(T_a^2 \right) = b_a, \qquad (9.95)$$

where the indices V, F, S stand for vector, fermion, and scalar massless fields; hence the difference in (9.93) contains just the contribution from massive modes. The above threshold correction expression is valid in the $\overline{\text{DR}}$ renormalization scheme with the definition

$$M_G = \frac{2e^{(1-\gamma)/2} 3^{-3/4}}{\sqrt{2\pi\alpha'}} = \frac{e^{(1-\gamma)/2} 3^{-3/4}}{4\pi} g\, M_P = 5.27 \times g \times 10^{17}\, \text{GeV}, \qquad (9.96)$$

where $\gamma \simeq 0.57722$ is Euler's constant. Hence M_G is of order 3.8×10^{17} GeV for g the standard value in GUTs; this is about a factor 20 larger than the unification scale M_{GUT} generated from extrapolation of the low-energy couplings in the MSSM, as we further discuss in Section 16.1.1.

In order to get a more specific result, one needs the explicit form of the one-loop amplitude of the internal theory. In the case of toroidal orbifolds there are explicit expressions for these partition functions. The resulting threshold corrections generically depend on the Kähler and complex structure moduli, and hence may be substantial for large or moderate moduli vevs. For simplicity let us consider abelian orbifolds with factorized $\mathbf{T}^6$. The internal part of (9.94) decomposes into sectors with boundary conditions (g, h) along the (σ, t) coordinates on the genus-one worldsheet, so the threshold correction is obtained by replacing the second line of (9.94) by

$$\frac{1}{|P|} \sum_{g,h \in P} \text{Tr}_{(g,s_1)} \left[q^{L_0 - \frac{11}{12}} \, \bar{q}^{\tilde{L}_0 - \frac{3}{8}} \, (-1)^{s_2 \tilde{F}} \, h \left(T_a^2 - \frac{k_a}{8\pi\tau_2} \right) \right], \qquad (9.97)$$

where P is the orbifold point group and $|P|$ its order. Each (g, h) sector preserves some unbroken SUSY, either $\mathcal{N} = 4$, $\mathcal{N} = 2$ or $\mathcal{N} = 1$, when g and h have three, one or no simultaneous fixed planes, respectively. The only $\mathcal{N} = 4$ sector is the untwisted $(g, h) = (1, 1)$, and contributes to neither the β-function nor to the threshold corrections. Sectors (g, h) preserving only $\mathcal{N} = 1$ have neither winding nor momentum states, and yield no moduli dependent threshold corrections, although in general produce moduli independent corrections c_a to Δ_a. Finally, sectors preserving $\mathcal{N} = 2$, with g and h leaving a common fixed plane, allow for winding and momenta on the unrotated $\mathbf{T}^2$, and produce threshold corrections depending on the corresponding moduli. The explicit evaluation of the one-loop diagram leads to a closed expression for the group dependent threshold corrections in abelian orbifolds

$$\Delta_a = -\sum_i \frac{|P^{(i)}|}{|P|} \, b_a^{i \, \mathcal{N}=2} \left[\log(\,|\eta(T_i)|^4 \operatorname{Re} T_i\,) + \log(\,|\eta(U_i)|^4 \operatorname{Re} U_i\,) \right] + c_a, \qquad (9.98)$$

where $P^{(i)}$ is the subgroup of P leaving the ith complex plane fixed, $b_a^{i \, \mathcal{N}=2}$ are the β-function coefficients of the corresponding $\mathcal{N}=2$ subsector, and η is Dedekind's modular function. In some cases certain complex structure fields U_i are not moduli of the full $\mathcal{N}=1$ orbifold, rather they are fixed by other orbifold elements, and must consequently be replaced by constants in the above expression.

Concerning the universal piece Δ^{univ} in (9.92), it is given by an expression

$$\Delta^{\text{univ}} = -\sum_{i=1,2,3} \delta_{\text{GS}}^i \left[\log\left(T_i + T_i^*\right) + \log\left(U_i + U_i^*\right) \right] + \Omega\left(T_i, U_i, T_i^*, U_i^*\right) + \text{const},$$

$$(9.99)$$

where the δ_{GS}^i are constants whose meaning we discuss shortly, and Ω is a function whose most relevant property is to be invariant under modular transformations of the moduli T_i, U_i.

Under a T-duality transformation (9.85) we have

$$\operatorname{Re} T_i \ \to \ \frac{\operatorname{Re} T_i}{|ic_i T_i + d_i|^2}, \qquad \eta(T_i)^2 \to \eta(T_i)^2(ic_i T_i + d_i), \qquad (9.100)$$

and similarly for the U_i moduli. Therefore the group dependent threshold corrections Δ_a are invariant under duality transformations. On the other hand, the universal piece Δ^{univ} is not duality invariant; however, recalling (9.21), invariance can be restored by proposing a (Green–Schwarz-like) transformation of the dilaton

$$S \to S - \frac{1}{8\pi^2} \delta_{\text{GS}}^i \, \log(ic_i T_i + d_i). \qquad (9.101)$$

Then, in order to render the dilaton Kähler potential invariant, we need to include an appropriate one-loop correction as

$$K(S, S^*) = -\log\left(S + S^* - \frac{\delta_{\text{GS}}^i}{8\pi^2} \log\left(T_i + T_i^*\right)\right). \qquad (9.102)$$

This is reminiscent of the modification of Kähler potentials in Section 9.5 due to the Green–Schwarz $U(1)$ anomaly cancellation mechanism. This is not a coincidence, as we discuss in the next section.

9.6.2 T-duality anomalies

It is instructive to also analyze the one-loop corrections to gauge couplings in terms of the effective low-energy lagrangian, without explicit reference to the effect of heavy string states. Consider the one-loop effective lagrangian for gauge bosons, including the dependence on all massless moduli, collectively denoted M_i. In this language, the moduli dependent one-loop correction to gauge couplings arises from triangle diagrams of charged fields coupling to two gauge bosons, and to one moduli background insertion. The latter is described by a composite Kähler connection, proportional to $K\left(M_i, M_i^*\right)$, which couples both to gauginos and charged matter fields C^I; it is associated with the invariance of the theory under Kähler transformations (2.49). There is a further coupling of the moduli to the matter fields C^I, arising from their non-canonical moduli dependent kinetic term K_{IJ}, as in (9.32), (9.44). These two types of couplings, together with the tree level part, give rise to the following (non-local) effective supersymmetric lagrangian

$$\sum_a \frac{1}{4} \int d^2\theta \, W^a W^a \left\{ k_a S - \frac{1}{16\pi^2} \frac{\Box^{-1}}{16} \overline{DD} DD \left[\left(C(G_a) - \sum_I T(R_I) \right) K \left(M_i, M_i^* \right) \right. \right.$$

$$\left. \left. + 2 \sum_I T(R_I) \log \det K_{IJ} \left(M_i, M_i^* \right) \right] \right\} + \text{h.c.}, \tag{9.103}$$

where W^a is the field strength superfield (2.20) and D is the SUSY covariant derivative (2.11). Also $C(G_a)$ and $T(R_I)$ are the quadratic Casimirs in the adjoint and in the representation R_I under which C^I transforms. The expansion of this expression in components leads to a non-local contribution to the term $F \wedge F$ and a local contribution to the inverse coupling constant.

Let us now concentrate again on the case of abelian toroidal orbifolds with factorized underlying $\mathbf{T}^6$. Replacing (9.43) and (9.44) into (9.103), we obtain

$$\sum_a \frac{1}{4} \int d^2\theta \, W^a W^a \left\{ k_a S - \frac{1}{16\pi^2} \frac{\Box^{-1}}{16} \overline{DD} DD \left[\sum_{i=1}^{3} b_a^{i\,\prime} \log \left(T_i + T_i^* \right) \right. \right.$$

$$\left. \left. + \sum_{m=1}^{h_{21}} b_a^{m\,\prime} \log \left(U_m + U_m^* \right) \right] \right\} + \text{h.c.}, \tag{9.104}$$

with

$$b_a^{i\,\prime} = -C(G_a) + \sum_I T(R_I) \left(1 + 2n_I^i \right),$$

$$b_a^{m\,\prime} = -C(G_a) + \sum_I T(R_I) \left(1 + 2l_I^m \right). \tag{9.105}$$

This effective action is clearly not invariant under modular transformations of the T_i and U_i moduli, despite these being duality symmetries of the theory. In particular, the non-invariance of the coefficient of $F \wedge F$ term signals the presence of a so-called *duality anomaly*. This is in fact not surprising, since we need to take into account some additional ingredients considered in the previous section. First, the transformation of the complex dilaton under the T-duality transformations, as in (9.101), provides a (in general) partial cancellation of this duality anomaly. The remaining anomaly is finally completely canceled by the one-loop T_i-dependent holomorphic correction to the gauge kinetic function, which may be written as

$$f_a^{\text{tree}} + f_a^{\text{one-loop}} = k_a S - \frac{1}{16\pi^2} \sum_{i=1}^{3} \left(b_a^{i\,\prime} - k_a \delta_{\text{GS}}^i \right) \log \eta(T_i)^4, \qquad (9.106)$$

where δ_{GS}^i is the universal coefficient in (9.99), and we have skipped an analogous U_i-dependent piece playing no role in T-duality. The above coefficient of $\log \eta(T_i)$ should be equal to the $\mathcal{N} = 2$ β-function coefficients $b_a^{i\,N=2}$ in (9.98), while the $b_a^{i\,\prime}$ may be computed from (9.105) with the knowledge of just the charged massless spectrum. One can then derive the value of the universal (gauge group independent) coefficients δ_{GS}^i for the compactification at hand; this is easily done in specific examples, indeed producing universal values δ_{GS}^i. Finally, note that the above expression for the gauge kinetic function does not receive higher-loop perturbative corrections, as they would violate the shift invariance discussed in Section 9.1.2.

The above T-duality anomaly cancellation gives important constraints in specific models. In particular, for orbifolds with the ith complex plane rotated by all orbifold group elements, there are no threshold corrections depending on the moduli T_i, U_i; hence the coefficients of the log term in (9.106) must vanish for all gauge group factors, so $\delta_{\text{GS}}^i = b_a^{i\,\prime}/k_a$ for all a, implying the universality relation

$$\frac{b_1^{i\,\prime}}{k_1} = \frac{b_2^{i\,\prime}}{k_2} = \cdots = \frac{b_n^{i\,\prime}}{k_n}, \qquad (9.107)$$

where n denotes the number of gauge factors. We illustrate the power of these constraints for, e.g., $\mathbf{Z}_3$ orbifold model building. Consider models with gauge group $SU(3) \times SU(2) \times G'$, and three quark-lepton generations; for instance, the explicit model in Section 8.3.3, although the argument is completely general and independent of the detailed construction. The $\mathbf{Z}_3$ actions have no fixed planes, so there are no T_i-dependent threshold corrections and the duality anomaly must cancel just through the Green–Schwarz mechanism. For present purposes, it suffices to restrict to duality anomalies with respect to the overall Kähler modulus T. Recall that modular weights (n^1, n^2, n^3) for untwisted fields are $(-1, 0, 0)$, $(0, -1, 0)$, $(0, 0, -1)$, while for twisted states with no oscillators they are $\left(-\frac{2}{3}, -\frac{2}{3}, -\frac{2}{3}\right)$, so the overall modular weights are $n = \sum_i n^i = -1, -2$ for untwisted and twisted fields, respectively. The $SU(3)^2$ and $SU(2)^2$ mixed duality anomalies have coefficients

$$b'_{SU(3)} = \sum_i b^{i\,\prime}_{SU(3)} = -9 + \tfrac{1}{2} \left(N_3^U - N_3^T \right);$$

$$b'_{SU(2)} = \sum_i b^{i\,\prime}_{SU(2)} = -6 + \tfrac{1}{2} \left(N_2^U - N_2^T \right),$$

where $N_3^{U,T}$ and $N_2^{U,T}$ are the net number of $SU(3)$ triplets and $SU(2)$ doublets in untwisted and twisted sectors. Assuming the usual situation of equal KM levels, (9.107) implies $b'_{SU(3)} = b'_{SU(2)}$. Using also that in these models $N_2^U - N_3^U = 3$, one concludes that $N_2^T - N_3^T = 9$ without needing to perform the (usually lengthy) computation of the full twisted spectrum. We have hence proven that models based on the $\mathbf{Z}_3$ orbifold have 12 more doublets than triplets. This argument provides the underlying reason for this pattern, common to all semi-realistic $\mathbf{Z}_3$ models with two Wilson lines; it also shows that this class of orbifolds cannot lead to the minimal content of the MSSM, at least before vacuum re-stabilization.

Additional constraints may follow from the cancellation of mixed gravitational duality anomalies. Their computation parallels the above, except that all massless fermions contribute, including the gravitino, dilatino, and modulinos. The analogous anomaly coefficient can be computed to be

$$b^{i'}{}_{\text{grav.}} = 21 + 1 + \delta^i_{(M)} - \sum_a \dim G_a + \sum_I \left(1 + 2n^i_I\right), \qquad (9.108)$$

where the sum in a runs through 4d gauge factors and accounts for the gaugino contributions, while the sum in I runs through massless matter chiral multiplets; the gravitino contributes 21, the dilatino contributes 1 (since it has zero modular weight), and $\delta^i_{(M)}$ is the contribution from the modulinos, which depends on the orbifold considered. In models with the ith complex plane rotated by all orbifold elements, there is no T_i-dependent threshold correction, and we obtain a universality relation between the gravitational and gauge T-duality anomaly coefficients

$$b^{i'}{}_{\text{grav.}} = 24 \, \frac{b^{i'}{}_{\text{gauge}}}{k}. \qquad (9.109)$$

This relation yields further powerful constraints, which are in fact sensitive to the whole massless spectrum, including neutral particles and possible hidden sectors.

9.7 Orbifold model building revisited

In Section 8.3.3 we presented two explicit semi-realistic models based on the $\mathbf{Z}_3$ and $\mathbf{Z}'_6$ orbifolds. Both contain additional $U(1)$ gauge symmetries, as well as extra chiral matter beyond the MSSM content. The new results described in this chapter naturally lead to a significant removal of additional gauge symmetries and extra charged matter. For concreteness we focus on the $\mathbf{Z}_3$ case, due to its relative simplicity, but a similar analysis applies to other semi-realistic orbifold models.

The full spectrum of charged chiral fields in this $\mathbf{Z}_3$ model is shown in Appendix C. The gauge group is $SU(3) \times SU(2) \times U(1)^8 \times SO(10)$, with $U(1)$ charge generators Q_r in (C.1). The reader may easily check that the $U(1)_X$ generated by Q_X has mixed gauge and gravitational triangle anomalies, with

$$A_{SU(3)} = A_{SU(2)} = A_{SO(10)} = \frac{A_{U(1)_r}}{k_r} = \frac{A_{U(1)^3_X}}{3k_X} = \frac{A_{\text{grav}}}{24} = 54. \qquad (9.110)$$

Here r runs through the seven non-anomalous orthogonal $U(1)$s, and $k_r = 2Q_r \cdot Q_r$ are the $U(1)$ normalizations, as in Section 9.4. The above is the universality relation expected from the $U(1)_X$ Green–Schwarz anomaly cancellation. The D-term potential (9.79) reads

$$V_X = \frac{1}{S + S^*} \left| \frac{27}{(2\pi)^2} \frac{1}{\kappa_4^2 (S + S^*)} + \sum_i q_X^i |\phi_i|^2 \right|^2, \tag{9.111}$$

where the sum goes over all the scalar fields charged under $U(1)_X$. As already mentioned, this potential induces vevs for some fields charged under $U(1)_X$, to minimize the D-term energy, and re-stabilize onto a SUSY preserving minimum. In particular the singlets S_i, Y_i in Appendix C are charged under $U(1)_X$ (and other $U(1)$ generators), and can acquire vevs consistently with D- and F-flatness. The F-flatness may be checked by using the selection rules summarized in Section 9.3.2. This can break all the $U(1)$ gauge symmetries except one, to be identified with hypercharge. An explicit choice of vevs fulfilling these properties is given by $\langle S_1 \rangle = \langle S_2 \rangle = \langle S_3 \rangle = \langle S_6 \rangle = \langle S_8 \rangle = \langle Y_1 \rangle = v$. It parametrizes a flat direction in which the Y generator defined by

$$6Y = \frac{1}{3} Q_1 - \frac{1}{2} Q_2 + Q_4, \tag{9.112}$$

remains unbroken and plays the role of hypercharge, while the remaining $U(1)^7$ symmetry is broken. The re-stabilization process yields additional byproducts. The S_i and Y_i singlet fields can have Yukawa couplings with most of the extra vector-like colored and charged particles in the model, i.e. they are allowed by the selection rules in Section 9.3.2; their vevs give large masses to all exotics with fractional charges (all the fields labeled A in Appendix C). Also, these vevs imply that two physical U-quarks are mixtures of untwisted and twisted sector fields. The right-handed D-quarks and leptons and the Higgs H_d remain purely in the twisted sector, whereas left-handed quarks Q_L and the Higgs H_u remain purely untwisted fields. These facts will be relevant to the Yukawa textures, see below. When the dust settles, the remaining low-energy spectrum of the model is shown in Table 9.1. The spectrum corresponds to the MSSM content, plus three extra copies of vector-like leptons, and additional singlets. Comparison with the (pretty involved) original spectrum in Appendix C illustrates the remarkable improvement in particle physics model building applications of heterotic string orbifolds, once one-loop effects are consistently included.

We refrain from entering a detailed phenomenological analysis of this particular model, but rather summarize a few properties, emphasizing those common to other models in this class of compactifications:

- *Yukawa couplings*: The selection rules in Section 9.3.2 imply that two of the U-quarks have renormalizable Yukawa couplings $\epsilon_{ijk} \left(Q_L^i u_1^j \overline{G}_1^k \right)$ as in (9.58), whereas the third remains massless at this order. The two massive ones may be hierarchical due to their mixing with twisted states (see Table 9.1). Renormalizable Yukawa couplings for the D-quarks are forbidden by point group selection rules, since fields Q_L and D belong to

Table 9.1 *Spectrum of the* $\mathbf{Z}_3$ *example after vacuum re-stabilization, in the field notation of Appendix C. Here* $\sim$ *indicates "linear combination of the fields involved"*

SM field	string state	#	Y
Q_L	Q_L^i	3	1/6
U	$u_1; \left(u_1^1 \sim u_2\right); \left(u_1^2 \sim u_2\right)$	3	$-2/3$
D	$(d_1 - d_2)$	3	1/3
L	G_5	3	$-1/2$
E	$(l_1 + l_2) - (l_3 + l_4)$	3	1
H_d	$(G_2 - G_3)$	1	$-1/2$
H_u	$\overline{G}_1^3$	1	1/2
L'	$(G_1 - G_4)$	3	$-1/2$
$\overline{L}'$	$\overline{G}_2$	3	1/2
E'	l_5	3	1
$\overline{E}'$	$(\bar{l}_1 + \bar{l}_2) - (\bar{l}_3 + \bar{l}_4)$	3	-1

the untwisted and twisted sectors, respectively. For one D-quark they can, however, arise at the non-renormalizable level. Concerning charged leptons they all have cubic Yukawa couplings of the type (9.59). The bottom line is that a qualitatively reasonable Yukawa coupling hierarchy can be achieved.

- *R-parity and proton decay*: There are R-parity violating couplings of (DDU) type from non-renormalizable operators, which violate baryon number; however, lepton-violating couplings are absent, so fast proton decay is not induced.
- *Gauge couplings constants*: The hypercharge normalization is $k_1 = 2Q_Y^2 = 11/3$, instead of the canonical $SU(5)$ GUT value $k_1 = 5/3$. This fact, combined with the contribution to the running from the extra vector-like leptons, make the agreement with gauge coupling unification possible only at the cost of using approximately (not exactly) flat directions to induce intermediate scale masses for these exotics.

Some of the phenomenological properties of this $\mathbf{Z}_3$ model may be improved in more elaborate orbifolds with several twisted sectors. Explicit semi-realistic MSSM-like models have been constructed using, e.g., the $\mathbf{Z}_2 \times \mathbf{Z}_2$, $\mathbf{Z}_6$, $\mathbf{Z}_6'$ and $\mathbf{Z}_{12}'$ orbifolds. In particular, models like the $\mathbf{Z}_6'$ orbifold in Section 8.3.3 embed the SM gauge group into an underlying $SO(10)$, producing the standard hypercharge normalization $k_1 = 5/3$. Their Yukawa coupling structure also has a simple hierarchical structure. The phenomenological analysis of these models is nevertheless quite similar to the above $\mathbf{Z}_3$ example. There is typically one anomalous $U(1)_X$, whose one-loop FI term triggers further gauge symmetry breaking down to the SM group, by vevs for a number of hypercharge neutral fields. In the process, most or all exotics get large masses, so that the low-energy spectrum is, at least in the most successful constructions, just the MSSM one. A few families of quarks and leptons have renormalizable Yukawa couplings, while others (eventually the lighter ones) acquire them from non-renormalizable couplings involving some of the singlet vevs.

The least attractive aspect of these models is the large number of particles beyond the MSSM in the initial (i.e. tree-level) massless spectrum, as exemplified in the table in Appendix C. This also implies that the low-energy physics depends very sensitively on the particular choice of flat direction used to cancel the FI term. This is the case, e.g., for the existence or not of R-parity in the final model, and also for the Yukawa coupling textures of the lighter generations. On the other hand, a nice feature of these constructions is the very natural appearance of fermion mass hierarchies, according to their origin from renormalizable Yukawa couplings or higher-dimensional operators. This can be regarded to some extent as analogous to the Froggatt–Nielsen mechanism widely used in flavour physics model building.

In summary, heterotic toroidal abelian orbifolds allow a simple construction of very explicit compactifications reproducing models remarkably close to the structure of the MSSM, or simple extensions thereof. Their general structure moreover provides insights into particle physics properties, like the existence of several families, gauge coupling unification, or the possible origin of Yukawa coupling textures. This is a remarkable achievement of heterotic string theory.

9.8 Higher Kac–Moody level models and string GUTs*

In Section 1.2 we introduced GUTs as an attractive idea to explain the unification of coupling constants and other features of the SM. However, the heterotic compactifications presented so far do not lead to satisfactory 4d GUTs. Actually, GUT groups like $SU(5)$, $SO(10)$ are easily obtained in heterotic compactifications; for instance, the CY compactification examples in Sections 7.3.2 and 7.4.2, before introduction of discrete Wilson lines, or the hidden sector in the $\mathbf{Z}_3$ orbifold model in the previous section, which is a 3-generation $SO(10)$ model. However, a satisfactory GUT model must contain adjoint Higgs multiplets in the 4d effective theory to trigger GUT symmetry breaking, and they are not present in these compactifications. As explained in Section 17.1.2, the absence of adjoint multiplets in the massless spectrum is a general property of gauge symmetries realized at Kac–Moody level $k = 1$ (see Sections 9.4 and E.3), as all constructions hitherto. Hence their application to particle physics must follow the alternative route of realizing directly the SM, or non-GUT extensions thereof, like $SU(3)^3$ or flipped $SU(5)$.

It is however possible to construct string GUT models with adjoint Higgs breaking by realizing the GUT symmetry at higher KM level $k > 1$. Such heterotic models have been constructed using both symmetric and asymmetric orbifolds as well as the fermionic construction. The simplest strategy to obtain higher-level models in orbifolds is to start from a model with a level 1 realization of k copies of a GUT group G, namely $G^k = G_{(1)} \times \cdots \times G_{(k)}$. In a subsequent stage, one embeds some $\mathbf{Z}_k$ space group action (like an orbifold twist or a discrete Wilson line) in the gauge group, as a cyclic permutation $G_{(A)} \to G_{(A+1)}$ (with A understood mod k). This leads to a breaking of the gauge group to the diagonal subgroup as

$$G_{(1)} \times \cdots \times G_{(k)} \longrightarrow G_{\text{diag}}. \qquad (9.113)$$

The generators of G_{diag} are $T_{\text{diag}}^a \sim \sum_{A=1}^N T_{(A)}^a$, and similarly for the corresponding Kac–Moody operators, which realize a KM algebra at level k, see section E.3. One simple way to understand the change in the unification relations (9.65) is that, denoting g the coupling for the $G_{(A)}$ factors (taken equal for $\mathbf{Z}_k$ symmetry), the gauge coupling of the diagonal group G_{diag} is given by $g/\sqrt{k}$.

As a simple application of these ideas, a $k = 2$ $SU(5)$ model may be obtained starting from a $SU(5) \times SU(5)$ theory, and modding out by a $\mathbf{Z}_2$ permutation of the two factors. If the original theory contains massless chiral multiplets in the "bi-fundamental" $(\mathbf{5}, \bar{\mathbf{5}})$ representation, it ends up as a representation $\mathbf{5} \times \bar{\mathbf{5}} = \mathbf{24} + \mathbf{1}$ of $SU(5)_{\text{diag}}$; this nicely agrees with the above mentioned fact that massless adjoints are possible for higher-level KM realizations. Incidentally, this kind of permutation twists can thus lead to substantial rank reduction. Similar constructions have also been carried out in the fermionic models of Section 8.5, where the rank reduction and change in normalization follows from the use of real fermions which cannot combine into complex pairs; although rank reduction by real fermions does not guarantee the realization of higher-level KM algebras, there exist appropriate choices of fermion boundary conditions which achieve this goal. Finally, permutation of gauge factors could be implemented in geometric CY compactifications, e.g., by discrete Wilson lines, although this possibility has not been exploited as yet.

In the literature there are several 3-generation $SO(10)$ and $SU(5)$ GUTs with adjoint chiral multiplets, see the references. A rather generic drawback of these models is that the adjoint chiral multiplets turn out to be moduli, in particular they have no self-couplings ϕ^3 or ϕ^2 in the superpotential; the adjoints thus produce left-over additional massless chiral multiplets beyond the MSSM, which for, e.g., $SU(5)$ GUTs of this kind transform under the SM group as

$$(\mathbf{8}, \mathbf{1})_0 + (\mathbf{1}, \mathbf{3})_0 + (\mathbf{1}, \mathbf{1})_0. \qquad (9.114)$$

These multiplets may acquire masses after supersymmetry breaking, i.e. around or slightly above the EW scale. But unfortunately their contribution to the running of the gauge couplings spoil their unification, one of the motivations of the GUT idea, and of the potential interest of string GUTs. A further problem is the doublet–triplet splitting problem mentioned in Section 2.6.2, which in this case becomes more acute since explicit mass terms for the adjoints are absent. These issues are however avoided in string models with an *underlying* GUT structure, broken by Wilson lines (as in CY or orbifolds in Chapters 7 and 8, or F-theory/type IIB models in Section 11.5.3).

10

Type IIA orientifolds: intersecting brane worlds

We now move to a different corner in the pentagon in Figure 7.1 and turn to the construction of type IIA orientifold compactifications down to four dimensions. The use of D-branes and their properties is crucial in these constructions, and provides a beautiful geometrical picture of the resulting 4d theories; gauge symmetries arise from overlapping D-branes, while charged chiral fermions arise from intersections among D-branes. The models are dubbed intersecting brane models and provide simple realizations of particle physics models in string theory.

10.1 Type II on CY and orientifolding

We now step back to Section 7.2 and consider the 4d compactification of type II theories on CY manifolds. They lead to 4d $\mathcal{N} = 2$ theories and are not suitable for particle physics model building, although some of their properties, like mirror symmetry, permeate the analysis of more realistic setups. We describe the introduction of orientifold projections truncating the supersymmetry to 4d $\mathcal{N} = 1$. These CY orientifolds provide the closed string sector of the compactifications in this and the next chapter, which contain sectors of D-branes (to achieve RR tadpole cancellation) realizing non-trivial chiral gauge theories on their worldvolume.

10.1.1 Calabi–Yau compactification of type II string theories

Recall the general properties of KK compactification on CY manifolds from Section 7.2. Type IIA and IIB theories have two 10d gravitinos, so their CY compactifications contain two 4d gravitinos and correspond to 4d $\mathcal{N} = 2$ supersymmetric vacua. This amount of supersymmetry does not allow for chirality, and such models are not directly suitable for particle physics model building; however, they provide the starting point of CY orientifolds, see Section 10.1.3. In order to describe the massless spectrum of type II CY compactifications, we perform the KK compactification for the bosonic fields, and complete them to 4d $\mathcal{N} = 2$ supermultiplets, whose structure we recall from Section 2.4.1:

- The gravity multiplet, containing a graviton, a gauge boson (graviphoton), and two gravitinos of opposite chiralities.
- The vector multiplet, containing a gauge boson, a complex scalar, and two Majorana fermions, all in the adjoint representation of the gauge group.
- The hypermultiplet, containing two complex scalars and two left-handed Weyl fermions (in conjugate representations).

Type IIA on CY

Recall that the bosonic fields for 10d type IIA are the graviton G, the NSNS 2-form B, the dilaton ϕ, and the RR 1-form A_1 and 3-form C_3. Their KK reduction, recalling Section 7.2.4, is encoded in the following table:

IIA		Gravity	$h_{1,1}$ Vector	$h_{2,1}$ Hyper	Hyper
G	$\rightarrow$	$G_{\mu\nu}$	$h_{1,1}$ Kähler	$2h_{2,1}$ Cmplx. Str.	
B	$\rightarrow$		$B_{i\bar{j}}$		$B_{\mu\nu}$
ϕ	$\rightarrow$				ϕ
A_1	$\rightarrow$	A_μ			
C_3	$\rightarrow$		$C_{i\bar{j}\mu}$	$C_{ij\bar{k}}, C_{\bar{i}\bar{j}k}$	$C_{ijk}, C_{\bar{i}\bar{j}\bar{k}}$

The organization in columns describes which 4d $\mathcal{N}=2$ multiplet the fields fill in. We need to dualize the 4d 2-form $B_{\mu\nu}$ into a 4d scalar (as in Section 9.1.1, see also Section B.3) in order to fill out a standard hypermultiplet. Also, we have simplified the notation for zero modes of p-forms, by denoting as, e.g., $B_{i\bar{j}}$ any of the $h_{1,1}$ scalars that arise from a 10d 2-form with 6d indices along a harmonic $(1, 1)$-form in $\mathbf{X}_6$.

The complete massless spectrum of type IIA on a CY is thus given by the 4d $\mathcal{N}=2$ supergravity multiplet, $h_{1,1}$ vector multiplets (with abelian group) and $h_{2,1}+1$ hypermultiplets (neutral under the gauge group).

Type IIB on CY

The bosonic fields for 10d type IIB are the graviton G, the NSNS 2-form B, the dilaton ϕ, and the RR 0-form a, 2-form C_2, and 4-form C_4 (with self-dual field strength). Their KK reduction is encoded in the following table:

IIB		Gravity	$h_{2,1}$ Vector	$h_{1,1}$ Hyper	Hyper
G	$\rightarrow$	$G_{\mu\nu}$	$2h_{2,1}$ Cmplx. Str.	$h_{1,1}$ Kähler	
B	$\rightarrow$			$B_{i\bar{j}}$	$B_{\mu\nu}$
ϕ	$\rightarrow$				ϕ
a					a
$\tilde{B}_2$	$\rightarrow$			$C_{i\bar{j}}$	$C_{\mu\nu}$
C_4	$\rightarrow$	$C_{ijk\mu}$	$C_{ij\bar{k}\mu}$	$C_{i\bar{j}\mu\nu}$	

In obtaining this spectrum, the self-duality of C_4 relates components associated to dual forms in $\mathbf{X}_6$, e.g. like $C_{ijk\,\mu}$ and $C_{\overline{ijk}\,\mu}$. Also we have again used the 4d duality between 2-forms and scalars to fill out standard hypermultiplets.

The massless spectrum is then given by the 4d $\mathcal{N} = 2$ supergravity multiplet, $h_{2,1}$ vector multiplets (with abelian gauge group) and $(h_{1,1} + 1)$ neutral hypermultiplets.

10.1.2 Mirror symmetry

The massless spectra of type IIA and IIB CY compactifications have a very similar structure, up to exchange of the roles of $h_{1,1}$ and $h_{2,1}$. This is not an accident, but rather the reflection of a much deeper relationship, *mirror symmetry*, already mentioned in, e.g., Sections 7.2.3, 8.6.2, 9.1.3, and 9.6. Roughly speaking,[1] mirror symmetry is a duality establishing that for each CY manifold $\mathbf{X}_6$ there exists another (mirror) CY manifold $\mathbf{Y}_6$, such that compactification of type IIA string theory on $\mathbf{X}_6$ is exactly equivalent to compactification of type IIB string theory on $\mathbf{Y}_6$. The simplest consequence of this equivalence is that their Hodge numbers are related by

$$h_{1,1}(\mathbf{X}_6) = h_{2,1}(\mathbf{Y}_6), \quad h_{2,1}(\mathbf{X}_6) = h_{1,1}(\mathbf{Y}_6), \tag{10.1}$$

so their Hodge diamonds are related by a mirror reflection along a diagonal (hence the name). For many CY spaces there are systematic tools to construct the mirror manifold, leading to a large class of dual pairs. Also, there is a heuristic (yet rigorous in a particular regime in CY moduli space[2]) description of mirror symmetry as T-duality in three directions in the CY geometry. An example is the (triple) application of the $T \leftrightarrow U$ duality of Section 9.6 in a factorized $\mathbf{T}^6 = \mathbf{T}^2 \times \mathbf{T}^2 \times \mathbf{T}^2$. The T-duality realization of mirror symmetry allows its generalization to type II orientifolds, i.e. models with orientifold planes and D-branes, since the T-duality action on the latter is very simple. For instance, it relates the intersecting brane models in this chapter to magnetized brane models in Section 11.4.

Mirror symmetry relates two different microscopic descriptions of the same 4d physical quantities. In particular, quantities which have a purely field theoretical description in one picture may arise from stringy physics in the mirror. For instance, momentum states (which are present already in the field theoretical approximation) are mapped to (stringy) winding states in the T-dual. Also, as described in Section 9.1.3, the kinetic term for complex structure moduli (describing vector multiplets in type IIB compactifications, or $\overline{\mathbf{27}}$s in (2, 2) heterotic compactifications), which has no α' corrections, is mapped to the kinetic term for Kähler moduli in the mirror CY compactification (describing vector multiplets in type IIA, or $\mathbf{27}$s in (2, 2) heterotic models), which receives worldsheet instanton corrections. This has an analog in the superpotential for charged chiral multiplets in models with D-branes, see Sections 12.5.3 and 12.5.4. This quantity has no α' corrections in type IIB

---

[1] The precise statement is more subtle, as the mirror dual of a geometric CY compactification may correspond to an abstract non-geometric CFT, without a large volume description.

[2] The so-called large complex structure limit, in which the CY splits as a three-dimensional base space, and a $\mathbf{T}^3$ fiber, along which one performs three T-dualities to obtain the mirror CY.

orientifolds, and can be obtained in a field theory computation, while in the mirror type IIA orientifold, the same quantity is recovered from a sum over α' corrections from (open) worldsheet instantons.

10.1.3 Type II orientifolds

Orientifold quotients of type II CY compactifications reduce the supersymmetry to 4d $\mathcal{N} = 1$, and provide a formidable playground for phenomenological model building. These models eventually include D-branes and can lead to a string theory realization of the brane world scenario, described in Section 1.4.2.

The construction generalizes the orientifolding procedure in Section 5.3.4. We consider type II on a CY $\mathbf{X}_6$, and mod out by the orientifold action $\Omega\mathcal{R}$, where Ω is (essentially) worldsheet parity[3] and $\mathcal{R}$ is a $\mathbf{Z}_2$ geometric symmetry of $\mathbf{X}_6$. The set of points fixed under $\mathcal{R}$ defines the orientifold planes in the model, denoted Op-planes, where p denotes their number of spatial dimensions. The Op-planes span the 4d Minkowski spacetime and wrap a compact cycle on $\mathbf{X}_6$, of dimension $(p - 3)$. They are sources of RR charge, which must be canceled by introducing Dp-branes, also spanning the 4d Minkowski spacetime, and wrapping cycles on $\mathbf{X}_6$. A novel feature, which will prove crucial in obtaining chiral theories, is that in general D-branes and O-planes can wrap different cycles in $\mathbf{X}_6$, still consistently with RR tadpole cancellation. The requirement that the orientifold preserves 4d $\mathcal{N} = 1$ supersymmetry restricts the possible forms of the geometric action $\mathcal{R}$, which are different in type IIA and type IIB orientifolds. In this chapter we focus on type IIA orientifolds, leaving the type IIB case for Chapter 11.

Generalities on type IIA orientifolds

For type IIA orientifolds, the orientifold action is of the form $\Omega\mathcal{R}(-1)^{F_L}$, where F_L is left-moving spacetime fermion number (acting as -1 on R-NS and R-R states), and $\mathcal{R}$ is a $\mathbf{Z}_2$ symmetry of $\mathbf{X}_6$ acting antiholomorphically on its complex coordinates, and hence as $J \to -J$, $\Omega_3 \to \overline{\Omega}_3$. The $(-1)^{F_L}$ operation is needed for the orientifold action to square to the identity operator.

An illustrative example is the generalization of the construction in Section 5.3.4, related to type I toroidal compactification by three T-dualities (along the directions 5,7,9 introduced below). Consider type IIA compactified on a (for simplicity, square) torus with complex coordinates $z_i = x^i + iy^i$, with the identifications $z_i \simeq z_i + R$, $z_i \simeq z_i + iR$. We then quotient by $\Omega\mathcal{R}(-1)^{F_L}$, with

$$\mathcal{R} : (z_1, z_2, z_3) \to (\bar{z}_1, \bar{z}_2, \bar{z}_3), \tag{10.2}$$

namely $y^i \to -y^i$ with x^i invariant. The orientifold introduces O6-planes at the fixed points of $\mathcal{R}$, namely there are eight O6-planes along 4d Minkowski spacetime plus the

---

[3] To avoid notation clashes, in this section we denote the CY holomorphic 3-form by Ω_3.

directions x^i, and sitting at $y^i = 0, R/2$. Recalling Section 5.3.4, each O6-plane generates a RR tadpole for the component of the 7-form C_7 along the x^i's, corresponding to its RR charge (5.52), $Q_{O6} = -4$ (in units of D6-brane charge in the covering $\mathbf{T}^6$). This RR tadpole/charge is most simply canceled by introducing 32 D6-branes, wrapped along the x^i's, and localized (in a $\mathbf{Z}_2$ invariant fashion) in the y^i coordinates. For instance, all 32 D6-branes on top of an O6-plane, without loss of generality taken at $y^1 = y^2 = y^3 = 0$, lead to a massless open string spectrum of $SO(32)$ vector multiplets of the 4d $\mathcal{N} = 4$ supersymmetry of the model. Other D6-brane distributions produce different gauge symmetry patterns (corresponding to turning on Wilson lines in the T-dual type I), still with 4d $\mathcal{N} = 4$. For instance, a symmetric distribution with 4 D6-branes on top of each O6-plane produces an $SO(4)$ gauge factor for each such set. This construction admits a very natural generalization to arbitrary CY spaces $\mathbf{X}_6$, useful in Section 10.7.2.

Simple, toroidal models with parallel D6-branes are not very interesting, as they preserve too much supersymmetry and are always non-chiral. Fortunately, there are more general configurations, with D6-branes wrapping intersecting cycles, which do produce chiral spectra, as discussed in detail in the next sections.

Closed string spectrum of type IIA orientifolds

Before entering the systematic study of the D-brane configurations and their physics, we conclude with the discussion of the closed string spectrum in type IIA CY orientifolds. This is obtained by considering the closed string spectrum of type IIA on $\mathbf{X}_6$ in Section 10.1.1, and truncating to states invariant under the orientifold action. For instance, the 4d $\mathcal{N} = 2$ gravity multiplet reduces to the 4d $\mathcal{N} = 1$ gravity multiplet, and the complex structure hypermultiplets truncate to $h_{2,1} + 1$ complex structure chiral multiplets. Finally, the $h_{1,1}$ 4d $\mathcal{N} = 2$ vector multiplets project down to $h_{1,1}^+$ 4d $\mathcal{N} = 1$ vector multiplets and $h_{1,1}^-$ chiral multiplets, where $h_{1,1}^{\pm}$ denote the numbers of $(1, 1)$-forms even and odd under the geometric action $\mathcal{R}$ in the orientifold operation. The microscopic description of these multiplets in terms of 10d fields is described in Section 12.1.1.

10.2 Intersecting D6-branes in flat 10d space

The physically most interesting ingredients in type IIA CY orientifold models are the D6-branes, and in particular their intersections. It proves useful to first become acquainted with configurations of D6-branes wrapped on 3-cycles on CY manifolds, and subsequently discuss the introduction of O6-planes. In fact, many important lessons about the physics at the intersection of D6-branes can be already drawn from configurations of intersecting D6-branes in flat 10d space.

10.2.1 Local geometry and spectrum

The basic configuration of intersecting D-branes leading to chiral 4d fermions at their intersection corresponds to two stacks of D6-branes in flat 10d, intersecting over a 4d

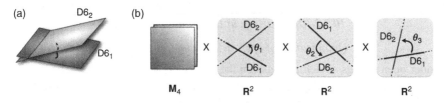

Figure 10.1 Two pictures of the configuration of two D6-branes intersecting over a 4d subspace of their volumes.

subspace of their volumes. Consider flat 10d space, decomposed as $\mathbf{M}_4 \times \mathbf{R}^2 \times \mathbf{R}^2 \times \mathbf{R}^2$, and two stacks of D6-branes, spanning $\mathbf{M}_4$ times a line in each of the three 2-planes. Figure 10.1 provides two depictions of the configuration. The local geometry is fully specified by the three angles θ_i which define the rotation between the two stacks of D6-branes. As we discuss below, the chiral fermions are localized at the intersection of the brane volumes. Incidentally, the angles $(\theta_1, \theta_2, \theta_3)$ bear some formal analogy with the orbifold twist (v_1, v_2, v_3) of Chapter 8.

The appearance of chirality, namely violation of parity, can be understood from the fact that the geometry of the two D-branes introduces a preferred orientation in the transverse 6d space; the rotation from the first to the second D6-brane defines an orientation by a 6d version of the "right-hand rule." This also explains why one should choose configurations of D6-branes, as for example two sets of type IIB D5-branes intersecting over 4d do not lead to 4d chiral fermions, since they do not have enough dimensions to define an orientation in the extra 6d space.

Let us describe the open string spectrum on a system with one stack of N_1 coincident D6-branes (denoted $D6_1$'s) intersecting a second stack of N_2 D6-branes (denoted $D6_2$'s). The open strings fall into three classes or sectors:

- $6_1 6_1$: Strings stretching among $D6_1$-branes produce 7d $U(N_1)$ gauge bosons, three real adjoint scalars and fermion superpartners, propagating over the 7d worldvolume of the $D6_1$-branes.
- $6_2 6_2$: Similarly, strings stretching among $D6_2$-branes give 7d $U(N_2)$ gauge bosons, three real adjoint scalars and fermion superpartners, on the 7d $D6_2$-brane worldvolume.
- $6_1 6_2 + 6_2 6_1$: Open strings between both kinds of D6-branes are naturally localized at their intersection, to minimize their stretching, and lead to fields charged in the bi-fundamental representation $(\mathbf{N_1}, \overline{\mathbf{N}_2})$ of $U(N_1) \times U(N_2)$ (or its conjugate). The detailed computation of their spectrum is carried out in the next section, but we advance the punchline. The intersecting configuration leads to a 4d chiral fermion in the $(\mathbf{N_1}, \overline{\mathbf{N}_2})$.[4] In addition to the chiral fermion, the intersection also leads to several potentially light scalars, whose detailed structure is given later on.

---

[4] Or its conjugate, as encoded in the orientation defined by the intersection; this will be implicitly taken into account in our discussions. We systematically follow the convention of using 4d left-handed Weyl spinors.

10.2.2 Open strings at D6-brane intersections

Quantization of open strings and light spectrum

Consider open strings stretching between the $D6_1$ and $D6_2$-branes. The boundary conditions for the coordinates along $\mathbf{M}_4$ are of the NN kind, and lead to the oscillators α_n^μ, Ψ_{n+r}^μ, as in Section 6.1.2. For the directions where the branes form non-trivial angles, e.g. in the (x^4, x^5) 2-plane, we have boundary conditions

$$\partial_\sigma X^4\Big|_{\sigma=0} = 0 \qquad \left(\cos\theta_1\, \partial_\sigma X^4 + \sin\theta_1\, \partial_\sigma X^5\right)\Big|_{\sigma=\ell} = 0,$$

$$\partial_t X^5\Big|_{\sigma=0} = 0 \qquad \left(-\sin\theta_1\, \partial_t X^4 + \cos\theta_1\, \partial_t X^5\right)\Big|_{\sigma=\ell} = 0, \tag{10.3}$$

and similarly for the remaining 2-planes, with angles θ_2, θ_3. Defining complex coordinates $Z^i = X^{2i+2} + i X^{2i+3}$, $i = 1, 2, 3$, the boundary conditions read

$$\partial_\sigma\left(\operatorname{Re} Z^i\right)\Big|_{\sigma=0} = 0; \quad \partial_t\left(\operatorname{Im} Z^i\right)\Big|_{\sigma=0} = 0,$$

$$\partial_\sigma\left[\operatorname{Re}\left(e^{i\theta_i} Z^i\right)\right]\Big|_{\sigma=\ell} = 0; \quad \partial_t\left[\operatorname{Im}\left(e^{i\theta_i} Z^i\right)\right]\Big|_{\sigma=\ell} = 0, \tag{10.4}$$

Using mode expansions for these coordinates, these boundary conditions shift the oscillator moddings by an amount $\pm\nu_i = \pm\theta^i/\pi$. The oscillator operators (which are now associated to complex coordinates) are $\alpha_{n-\nu_i}^i$, $\bar{\alpha}_{n+\nu_i}^i$, $\Psi_{n+r-\nu_i}^i$, $\bar{\Psi}_{n+r+\nu_i}^i$, with $r = \frac{1}{2}, 0$ in the NS, R sectors. Also, the centre of mass degrees of freedom for the bosonic coordinates are frozen in these directions, so that the open strings are localized at the D6-brane intersection.

The structure of oscillators is identical to (the right-moving piece in) an orbifold twisted sector with twist vector $\nu = (\nu_1, \nu_2, \nu_3)$, except for these being continuous parameters, rather than Nth roots of unity. Still, the construction of the oscillation states and their mass formula is identical to the orbifold case. In particular, and upon bosonization of the worldsheet fermions, the spacetime mass formula is

$$\alpha' M^2 = N_B + N_{F_B} + \frac{(r+\nu)^2}{2} + \sum_i \tfrac{1}{2} |\nu_i| \left(1 - |\nu_i|\right) - \frac{1}{2} \tag{10.5}$$

where r belongs to the lattice Λ^+ in (4.55), due to the GSO projection, and we assume $-1 \le \nu_i \le 1$. This is identical to the right-moving mass in (8.20), save for the factor of 2 related to the "doubling trick" in Section 3.3.1. These states are localized at the 4d intersection and transform in the bi-fundamental $(\mathbf{N}_1, \overline{\mathbf{N}}_2)_{1,-1}$ of the $U(N_1) \times U(N_2)$ gauge factor, with the subindices denoting the $U(1)_1 \times U(1)_2$ charges. Hereafter we will omit the $U(1)$ charges, with the implicit understanding that fundamental/anti-fundamental representations of $SU(N)$ carry charge ± 1 under the corresponding $U(1)$. The antiparticles arise from the $D6_2$-$D6_1$ open string sector. The light spectrum contains, in the NS sector, a set of light (massive, massless or tachyonic) scalars with masses depending on the θ_i, while the R sector always contains a massless 4d chiral fermion. Before bosonization, the latter arises from the R groundstate, which is degenerate due to two fermion zero modes along the

M_4 directions. The chirality of the 4d fermion is due to the GSO projection, and its hand-edness is determined by the orientation of the intersection. As will shortly become clear, we consider intersection angles in a range allowing for $\sum_i \theta_i = 0$, for instance $\theta_1, \theta_2 \geq 0$, $\theta_3 \leq 0$. In this range, some of the light states are as follows:

Sector	State	$\alpha' M^2$	4d field
NS	$(\nu_1, \nu_2, 1 + \nu_3, 0)$	$\frac{1}{2\pi}(\theta_1 + \theta_2 + \theta_3)$	Scalar
	$(-1 + \nu_1, \nu_2, \nu_3, 0)$	$\frac{1}{2\pi}(-\theta_1 + \theta_2 - \theta_3)$	Scalar
	$(\nu_1, -1 + \nu_2, \nu_3, 0)$	$\frac{1}{2\pi}(\theta_1 - \theta_2 - \theta_3)$	Scalar
	$(1 - \nu_1, 1 - \nu_2, 1 + \nu_3, 0)$	$1 - \frac{1}{2\pi}(-\theta_1 - \theta_2 + \theta_3)$	Scalar
R	$\left(-\frac{1}{2} + \nu_1, -\frac{1}{2} + \nu_2, \frac{1}{2} + \nu_3, -\frac{1}{2}\right)$	0	Ch. fermion

$$(10.6)$$

Supersymmetry and tachyons

It is interesting to consider the conditions under which there is some supersymmetry pre-served by the combined system of two D6-brane stacks. Using (6.5), this requires the existence of spinors ϵ_L, ϵ_R, satisfying

$$\epsilon_L = \Gamma_6 \epsilon_R \quad \text{with} \quad \Gamma_6 = \Gamma^0 \cdots \Gamma^3 \Gamma^4 \Gamma^6 \Gamma^8,$$
$$\epsilon_L = \Gamma_{6'} \epsilon_R \quad \text{with} \quad \Gamma_{6'} = \Gamma^0 \cdots \Gamma^3 \Gamma^{4'} \Gamma^{6'} \Gamma^{8'}, \quad (10.7)$$

where 4, 6, 8 and $4', 6', 8'$ denote the directions along the two D6-branes in the six dimen-sions 4, 5, 6, 7, 8, and 9. Introducing the $SO(6)$ rotation Θ that rotates the D6$_1$- to the D6$_2$-brane stack, we have $\Gamma_{6'} = \Theta \Gamma_6 \Theta^{-1}$. Supersymmetries preserved by the D6$_1$-branes are also solutions of the second equation in (10.7) if and only if they are invariant under Θ; hence, Θ has invariant spinors and, recalling a similar discussion in Section 7.2.1, it must belong to an $SU(3)$ subgroup of $SO(6)$. This constraint can be spelled out explicitly by rewriting Θ in the $SO(6)$ vector representation (in complex coordinates) as

$$\Theta = \text{diag}\,(e^{i\theta_1}, e^{-i\theta_1}, e^{i\theta_2}, e^{-i\theta_2}, e^{i\theta_3}, e^{-i\theta_3}). \quad (10.8)$$

The condition that the rotation is within $SU(3)$ is

$$\theta_1 \pm \theta_2 \pm \theta_3 = 0 \mod 2\pi, \text{ for some choice of signs.} \quad (10.9)$$

Configurations satisfying this kind of condition generically preserve 4d $\mathcal{N} = 1$ supersym-metry. We will eventually be interested in constructing supersymmetric models with the sign choice corresponding to

$$\theta_1 + \theta_2 + \theta_3 = 0 \pmod{2\pi} \quad \text{i.e.} \quad \nu_1 + \nu_2 + \nu_3 = 0 \pmod{2} \quad (10.10)$$

which is formally analogous to the condition for orbifold twists in Chapter 8. The pre-viously computed open string spectrum reflects the preserved supersymmetry, since the above condition ensures that the first complex scalar in Table 10.6 becomes massless and

(a)   (b)

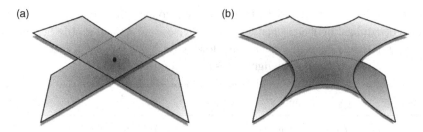

Figure 10.2 Recombination of two intersecting D6-branes into a single smooth one, corresponding to a vev for a scalar at the intersection.

combines with the chiral fermion into a 4d $\mathcal{N} = 1$ chiral multiplet (one can show that massive states also gather into suitable supermultiplets). In the special case of one vanishing angle, e.g. $\theta_1 = 0$, supersymmetric systems have $\theta_2 + \theta_3 = 0$ and preserve 4d $\mathcal{N} = 2$, while for $\theta_1 = \theta_2 = \theta_3 = 0$ we have a 4d $\mathcal{N} = 4$ system of parallel branes. The additional massless scalars in these cases suitably arise from the earlier open string spectrum.

For generic (not necessarily supersymmetric) angles, light scalars at intersections may be massless, massive or tachyonic. The massless case corresponds to a supersymmetric system, with the massless scalar corresponding to a modulus. Its vev parametrizes the possibility of recombining the two intersecting D-branes into a single smooth one, as pictorially shown in Figure 10.2. Geometrically, the intersecting configuration belongs to a family of supersymmetry preserving configurations of (generically recombined) D-branes; a geometrically simpler analog is the recombination of two intersecting 2-planes in $\mathbf{C}^2$, described by the holomorphic equation $uv = 0$, to the smooth 2-cycle $uv = \epsilon$, with ϵ the recombination modulus. General supersymmetric 3-cycles are mathematically known as special lagrangian submanifolds, and will be useful in Section 10.3.2.

In configurations with tachyonic scalars this recombination is triggered dynamically, by condensation of the tachyon at the intersection. Geometrically, the recombined 3-cycle has (suitably regularized) volume smaller than the sum of the volumes of the intersecting 3-cycles. In the very special case in which branes are parallel but with opposite orientation, e.g. $\theta_1 = \theta_2 = 0$, $\theta_3 = -1$, they become brane–antibrane pairs, recall Section 6.5.1, and the tachyon condensation describes their annihilation, as in Section 6.5.2. Configurations in which all light scalars have positive squared masses are non-supersymmetric, yet dynamically stable against recombination. Geometrically, the recombined 3-cycle has volume bigger than the sum of the volumes of the intersecting 3-cycles.

10.3 Compactification and an example of a toroidal model

We have succeeded in describing configurations of D-branes leading to charged chiral fermions, and can now employ them to build 4d compactifications of type IIA models with D6-branes. We consider type IIA string theory compactified on a CY $\mathbf{X}_6$, so that we have supersymmetry in the closed string sector. In addition, we introduce D6-branes;

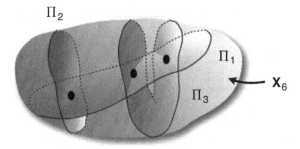

Figure 10.3 Compactification with intersecting D6-branes wrapped on 3-cycles.

specifically we introduce stacks (labeled by an index a) of N_a D6$_a$-branes spanning 4d Minkowski space and wrapped on 3-cycles Π_a in $\mathbf{X}_6$, see Figure 10.3. Introduction of orientifold planes will be discussed in Section 10.4. Each D6-brane stack leads to a 4d gauge factor, while each intersection between D6-brane stacks leads to charged chiral fermions. A key novelty in compact models is that 3-cycles in a 6d compact space generically intersect at multiple points, leading to replication of the charged chiral fermions. In the particle physics models constructed in upcoming sections, this is a beautiful mechanism to explain/reproduce the appearance of replicated generations of chiral fermions in Nature.

10.3.1 Toroidal models

Many features of general compactifications with intersecting D6-branes can be illustrated in the simpler setup of toroidal compactifications. Compactification on more general spaces is discussed in Section 10.3.2.

Construction

Consider type IIA compactified on a factorized 6-torus $\mathbf{T}^6 = \mathbf{T}^2 \times \mathbf{T}^2 \times \mathbf{T}^2$. A simple way to introduce intersecting D6-branes is to consider stacks of N_a D6$_a$-branes spanning 4d spacetime and wrapping a 1-cycle $\left(n_a^i, m_a^i\right)$ in the ith 2-torus; namely, the D6$_a$-branes wrap n_a^i, m_a^i times along the horizontal and vertical directions in the ith two-torus, see Figure 10.4 for examples. The 3-cycles Π_a are thus the product of three 1-cycles in the three 2-tori of $\mathbf{T}^6$. Note that, for each i, the integers (n^i, m^i) must be coprime, as we implicitly take in the following; otherwise, the system describes r D-branes with wrapping numbers $(n/r, m/r)$, where $r = \gcd(n, m)$. There exist more general non-factorizable 3-cycles, some of which can be obtained by recombination of intersecting factorized 3-cycles, or effectively by considering D8-branes with worldvolume magnetic fields, see Section 11.4.4. In this chapter, however, we restrict ourselves to the above factorizable 3-cycles for simplicity.

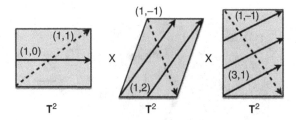

Figure 10.4 Examples of intersecting 3-cycles and their wrapping numbers in $\mathbf{T}^6$.

Near each intersection between D6-brane stacks, the configuration clearly reduces to the intersecting D6-branes in flat 10d space studied in Section 10.2. One important novelty in compact models is that the angles between branes are derived quantities, and depend on the closed string geometric moduli; for instance, for a rectangular $\mathbf{T}^2$ of radii R_1, R_2 along the horizontal and vertical directions, the angle between the 1-cycles $(1, 0)$ and (n, m) is

$$\tan \theta = \frac{m R_2}{n R_1} = \frac{m}{n} \operatorname{Re} U, \tag{10.11}$$

where U is the $\mathbf{T}^2$ complex structure modulus, e.g. as defined in (9.81) (modulo the replacement $R_2 \leftrightarrow R_1$).

As mentioned above, two 3-cycles in general intersect at multiple points. In the toroidal case, the intersection number is given by the product of the number of intersections in each 2-torus, and reads

$$I_{ab} = \left(n_a^1 m_b^1 - m_a^1 n_b^1 \right) \times \left(n_a^2 m_b^2 - m_a^2 n_b^2 \right) \times \left(n_a^3 m_b^3 - m_a^3 n_b^3 \right). \tag{10.12}$$

For example, the two branes in Figure 10.4 have intersection number $12 = (1) \times (-3) \times (-4)$. It is useful to exploit the topological tools in Section B.1 and introduce the 3-homology class $[\Pi_a]$ of the 3-cycle Π_a. Introducing the $\mathbf{T}^2$ basic homology 1-cycles $[a]$, $[b]$, the 1-cycle (n, m) has an associated 1-homology class $n[a] + m[b]$. Then a 3-cycle with wrapping numbers $\left(n_a^i, m_a^i \right)$ has 3-homology class

$$[\Pi_a] = \otimes_{i=1}^3 \left(n_a^i [a_i] + m_a^i [b_i] \right). \tag{10.13}$$

The multiplicity (10.12) is the homological intersection number, $I_{ab} = [\Pi_a] \cdot [\Pi_b]$, a simple generalization of (B.19). The result follows easily from $[a_i] \cdot [b_j] = \delta_{ij}$ and linearity and antisymmetry of the intersection pairing. It is worthwhile to mention that the homological intersection number weights intersections with a sign, depending on their orientation; as expected on physical grounds, it is an index measuring the *net* number of chiral fermions in a given bi-fundamental representation.

These basic data, i.e. the multiplicities N_a, and the 3-cycle intersection numbers I_{ab}, are sufficient to compute the gauge symmetry and chiral matter content of the 4d compactification, as follows. The closed string sector is just a toroidal compactification and produces

Figure 10.5 RR tadpole cancellation as a Gauss law. In a compact space, RR p-form potential flux-lines cannot escape and the total RR charge must vanish.

4d $\mathcal{N} = 8$ supergravity. This can be reduced to 4d $\mathcal{N} = 1$ in more general orbifold or CY orientifold models, see later. In addition there exist different open string sectors:

- $6_a 6_a$: Strings stretched among D6-branes in the ath stack produce 4d $U(N_a)$ gauge bosons, 6 real adjoint scalars and 4 adjoint Majorana fermions, filling out a vector multiplet of the 4d $\mathcal{N} = 4$ supersymmetry preserved by the corresponding brane. As described in the references, this can be reduced to just the $\mathcal{N} = 1$ vector multiplet in certain orbifold models; this corresponds geometrically to wrapping D6-branes on "rigid" 3-cycles, with $b_1(\Pi_a) = 0$.
- $6_a 6_b + 6_b 6_a$: Strings stretched between the ath and bth stack lead to I_{ab} replicated left-handed chiral fermions in the bi-fundamental representation $(\mathbf{N_a}, \overline{\mathbf{N_b}})$. Negative intersection numbers indicate a positive number of right-handed chiral fermions. Additional light scalars may be present, with masses as in (10.6), in terms of angles fixed by the $\mathbf{T}^2$ moduli and wrapping numbers, recall (10.11).

The gauge symmetry and chiral matter content of the models are

$$\begin{array}{ll} \text{Gauge} & \otimes_a U(N_a), \\ \text{Ch. fermions} & \sum_{a,b} I_{ab} (\mathbf{N_a}, \overline{\mathbf{N_b}}). \end{array} \tag{10.14}$$

Intersecting brane models naturally lead to four-dimensional theories with interesting non-abelian gauge symmetries and replicated charged chiral fermions. They provide a rich playground for particle physics model building in string theory.

RR tadpole cancellation

String theories with open string sectors must satisfy the crucial consistency condition of RR tadpole cancellation, recall Sections 4.4.3 and 5.3.4. Equivalently, since D6-branes are sources of charge under the RR 7-form, Gauss law implies that in a compact space their total charges must add up to zero (since RR field flux-lines have nowhere to escape, see Figure 10.5). In our setup, the vector of RR charges is encoded in the D6-brane 3-cycle homology, so the condition reads

$$[\Pi_{\text{tot}}] \equiv \sum_a N_a \, [\Pi_a] = 0. \tag{10.15}$$

In more technical detail, RR tadpole cancellation is the requirement of consistency of the equations of motion for RR fields. In our situation, the spacetime action for the RR 7-form C_7 is roughly of the form (see Appendix B for notation)

$$S_{C_7} = \int_{\mathbf{M_4} \times \mathbf{X_6}} F_8 \wedge *F_8 + \sum_a N_a \int_{\mathbf{M_4} \times \Pi_a} C_7$$

$$= \int_{\mathbf{M_4} \times \mathbf{X_6}} C_7 \wedge dF_2 + \sum_a N_a \int_{\mathbf{M_4} \times \mathbf{X_6}} C_7 \wedge \delta_3(\Pi_a), \qquad (10.16)$$

where F_8 is the 8-form field strength, F_2 its Hodge dual, and $\delta_3(\Pi_a)$ is the Poincaré dual of Π_a, i.e. a 3-form delta function localized on Π_a. The first term comes from the 10d action (4.47) (rewriting the F_2 kinetic piece in terms of the dual F_8) and the second from the D6-brane CS action (6.16). The equation of motion for C_7 corresponds to the modified Bianchi identity

$$dF_2 = \sum_a N_a \delta(\Pi_a). \qquad (10.17)$$

The left-hand side is an exact form and vanishes in cohomology. Taking this equation in homology thus leads to condition (10.15).

In the toroidal setup, using (10.13) the RR tadpole conditions (10.15) become

$$\sum_a N_a n_a^1 n_a^2 n_a^3 = 0,$$

$$\sum_a N_a n_a^1 n_a^2 m_a^3 = 0, \text{ and permutations,}$$

$$\sum_a N_a n_a^1 m_a^2 m_a^3 = 0, \text{ and permutations,}$$

$$\sum_a N_a m_a^1 m_a^2 m_a^3 = 0. \qquad (10.18)$$

Since the total D6-brane charge adds up to zero, models without O6-planes implicitly contain brane and antibrane charges; they are non-supersymmetric, and may have potential instabilities. However, their study is a useful warm-up for the supersymmetric models in upcoming sections, achieved by the introduction of O6-planes.

Anomaly cancellation

Cancellation of RR tadpoles in the underlying string theory configuration implies cancellation of anomalies in the 4d effective field theory with spectrum (10.14), as follows.

Cubic non-abelian anomalies The $SU(N_a)^3$ cubic anomaly is proportional to the number of fundamental minus anti-fundamental representations of $SU(N_a)$,

$$A_a = \sum_b I_{ab} N_b. \qquad (10.19)$$

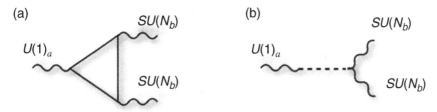

Figure 10.6 Triangle and Green–Schwarz diagrams contributing to the mixed $U(1)$–non-abelian anomalies.

This vanishes as a consequence of RR tadpole cancellation,[5] as easily shown by taking the intersection of $[\Pi_{\text{tot}}]$ in (10.15) with any $[\Pi]$ to get

$$0 = [\Pi_a] \cdot \sum_b N_b \, [\Pi_b] = \sum_b N_b I_{ab}. \tag{10.20}$$

Cancellation of mixed anomalies The $U(1)_a\text{-}SU(N_b)^2$ mixed anomalies cancel, but involving, as discussed shortly, a 4d Green–Schwarz mechanism mediated by closed string RR fields, see Figure 10.6; this is analogous to the heterotic case discussed in Section 9.5. On the other hand, mixed gravitational triangle anomalies cancel automatically, without Green–Schwarz contributions, since the sum of $U(1)_a$ charges is $N_a \sum_b I_{ab} N_b$, which vanishes from (10.15).

The triangle diagrams for $U(1)_a\text{-}SU(N_b)^2$ anomalies give a contribution which, even after using RR tadpole conditions, is non-zero and proportional to

$$A_{ab} \simeq N_a \, I_{ab}. \tag{10.21}$$

In addition, the theory contains contributions from Green–Schwarz diagrams, in which the $U(1)_a$ gauge boson mixes with a 2-form which subsequently couples to two $SU(N_b)$ gauge bosons. These couplings arise from the KK reduction of the D6-brane couplings $N_a \int_{D6_a} C_5 \wedge \text{tr} \, F_a$ and $\int_{D6_b} C_3 \wedge \text{tr} \, F_b^2$ in the CS action (6.16), as follows. Let us introduce a basis $\{[\alpha_k]\}$ of 3-cycles, and its dual basis $\{[\beta^k]\}$, i.e. $[\alpha_k] \cdot [\beta^l] = \delta_k^l$. We can define the KK reduced 4d 2-forms and scalar fields

$$B_2^k = \int_{[\alpha_k]} C_5, \quad a_l = \int_{[\beta^l]} C_3, \quad \text{with } \partial_\mu B_{\nu\rho}^k = -\delta^{kl} \, \epsilon_{\mu\nu\rho\sigma} \, \partial^\sigma a_l, \tag{10.22}$$

where the 4d duality relation follows from the 10d duality between C_5 and C_3, see Section B.3. The KK reduced 4d couplings read

---

[5] Interestingly RR tadpole cancellation is slightly stronger than cancellation of cubic non-abelian anomalies; the former requires that the number of fundamental minus anti-fundamentals vanishes even for the cases $N_a = 1, 2$, in which no gauge theory anomaly exists. This observation will prove relevant in phenomenological model building in Section 10.5.1.

$$N_a \int_{D6_a} C_5 \wedge \text{tr } F_a \;\to\; N_a Q_{ak} \int_{4d} B_2^k \text{ tr } F_a,$$

$$\int_{D6_b} C_3 \wedge \text{tr } F_b^2 \;\to\; q_b^l \int_{4d} a_l \text{ tr } F_b^2, \tag{10.23}$$

with $Q_{ak} = [\Pi_a] \cdot [\alpha_k]$, $q_b^l = [\Pi_b] \cdot [\beta^l]$. The total amplitude is proportional to

$$A_{ab}^{GS} = -N_a \sum_k Q_{ak} q_b^l \delta_l^k = \cdots = -N_a I_{ab}, \tag{10.24}$$

where the dots stand for some simple algebraic massaging. These contributions have the precise structure required to cancel the triangle piece (10.21).

This mechanism is analogous to that operating in the heterotic compactifications in Section 9.5, with some important differences. First, in D-brane models there are multiple RR fields participating in the anomaly cancellation, so no universality relation holds among the different anomalies. Second, these fields are complex structure moduli, unrelated to the complex dilaton, so the FI term is not controlled by the string coupling; it rather arises at tree-level and is a combination of geometric moduli, so it can vanish with no need of vacuum re-stabilization. Finally, the coupling to the RR fields renders the $U(1)$ gauge bosons massive, with a mass scale related to the string scale (with some volume factor suppressions, but not directly related to the string coupling). These points are further elaborated in Section 12.4.

A relevant observation at this point is that not only anomalous but also some non-anomalous $U(1)$s may have $B \wedge F$ couplings (10.23) and get massive by the Stückelberg mechanism of Section 9.5.2. Denoting the $U(1)_a$ generator by Q_a, the condition for a $U(1)$ linear combination $\sum_a c_a Q_a$ to remain massless is

$$\sum_a N_a Q_{ak} c_a = 0, \quad \text{for all } k. \tag{10.25}$$

Such $U(1)$ factors remain as gauge symmetries of the effective theory. Indeed, the $B \wedge F$ couplings (10.23) imply that a_k shift under $U(1)$ gauge transformations as

$$A_\mu^a \to A_\mu^a + \partial_\mu \Lambda_a, \quad a_k \to a_k + \sum_a N_a Q_{ak} \Lambda_a, \tag{10.26}$$

so that linear combinations satisfying (10.25) imply no shift of the RR scalars. The orthogonal $U(1)$ linear combinations are massive due to their couplings to RR fields. Since only D-brane states couple to the latter, massive $U(1)$s remain as global symmetries, exact in string perturbation theory. They are violated by non-perturbative D-brane instantons coupling to the a_k, see Chapter 13.

10.3.2 Beyond tori: D-branes on special lagrangian 3-cycles*

It is possible to generalize the above construction to general CY compactifications. Consider type IIA compactified on a CY geometry X_6; the closed string sector preserves 4d

$\mathcal{N} = 2$ supersymmetry. We would like to include D6-branes spanning 4d spacetime and wrapping 3-cycles Π_a which are volume-minimizing, so that the stability of the D6-branes is ensured. Such 3-cycles are mathematically described as *special lagrangian* 3-cycles, and are defined by the conditions

$$J\big|_{\Pi} = 0, \quad \mathrm{Im}\,(e^{-i\varphi}\,\Omega_3)\big|_{\Pi} = 0 \quad \text{for some fixed } \varphi, \tag{10.27}$$

where J and Ω_3 are the CY Kähler 2-form and holomorphic 3-form, and "$|_{\Pi}$" denotes the restriction to the 3-cycle. The volume of a special lagrangian 3-cycle is given by

$$\mathrm{Vol}(\Pi) = \int_{\Pi} \mathrm{Re}\,(e^{-i\varphi}\,\Omega_3). \tag{10.28}$$

The meaning of the conditions (10.27) is that the tangent spaces at different points of Π are related to each other by $SU(3)$ rotations. Hence special lagrangian 3-cycles are also known as supersymmetric 3-cycles, since a D6-brane wrapped on them preserves a 4d $\mathcal{N} = 1$ subalgebra of the underlying 4d $\mathcal{N} = 2$; the particular preserved subalgebra is determined by the angle φ for which the 3-cycle satisfies (10.27). Incidentally, this phase has already appeared in the decomposition of 4d $\mathcal{N} = 2$ multiplets into 4d $\mathcal{N} = 1$ in Section 2.4.1, and is also related to the BPS phase mentioned in Section 2.4.2. Two D6-branes preserve a common 4d $\mathcal{N} = 1$ supersymmetry if their phases φ are aligned. As mentioned, see also Section 10.6, globally supersymmetric models, however, require the introduction of O6-planes, which fix a preferred BPS phase, say $\varphi = 0$. In these, conditions (10.27) can be interpreted as the vanishing of the F- and D-term potentials from the 4d effective theory viewpoint, see Section 12.4 for details on the latter.

The conditions (10.27) agree with our toroidal or flat 10d discussion of supersymmetry; using complex coordinates z_i in the three 2-planes of $\mathbf{R}^6$, we have

$$J \simeq dz_1 d\bar{z}_1 + dz_2 d\bar{z}_2 + dz_3 d\bar{z}_3, \quad \Omega_3 \simeq dz_1\,dz_2\,dz_3. \tag{10.29}$$

With some choice of orientation a factorized 3-cycle Π can be parametrized as, e.g.,

$$z_1 = r_1\,e^{i\varphi_1}, \quad z_2 = r_2\,e^{i\varphi_2}, \quad z_3 = r_3\,e^{i\varphi_3}, \tag{10.30}$$

with fixed angles φ_i, and $r_i \in \mathbf{R}^+$. On the 3-cycle we have $dz_i = dr_i e^{-i\varphi_i}$ and $d\bar{z}_{\bar{\imath}} = dr_i e^{i\varphi_i}$; the condition $J|_{\Pi} = 0$ is automatic, while the second condition in (10.27) holds for $\varphi = \varphi_1 + \varphi_2 + \varphi_3$. Thus the condition for two D6-branes to preserve a common supersymmetry, $\varphi = \varphi'$, is that their relative angles $\theta_i = \varphi_i - \varphi_i'$ satisfy $\theta_1 + \theta_2 + \theta_3 = 0$, precisely reproducing (10.10). Other conditions (10.9) correspond to different choices of Ω, which flip the signs of some φ_i's, θ_i's, i.e. of different $\mathcal{N} = 1$ subalgebras of $\mathcal{N} = 4$.

Given a set of D6$_a$-brane stacks with multiplicities N_a wrapped on (not necessarily mutually supersymmetric) special lagrangian 3-cycles Π_a, the open string spectrum falls in two main sectors, which can be described quite explicitly:

- 6_a-6_a: Produces $U(N_a)$ vector multiplets of the 4d $\mathcal{N} = 1$ supersymmetry preserved by the D6$_a$-brane. In addition there may be massless modes arising from internal components of the gauge fields along Π_a, i.e. Wilson lines, whose number is $b_1(\Pi_a)$, the first

Betti number of Π_a (see Section B.1); they combine with additional geometric moduli of the 3-cycle (describing deformations consistent with the special lagrangian condition), and fermion partners, into $b_1(\Pi_a)$ chiral multiplets in the adjoint. For instance, factorized 3-cycles on $\mathbf{T}^6$ have $\mathbf{T}^3$ topology, with $b_1 = 3$, so there are three complex scalars describing three $\mathbf{T}^3$ Wilson lines and three positions of the D6-brane in the transverse space. The $\mathcal{N} = 1$ vector multiplet and these three adjoint chiral multiplets fill out a vector multiplet of the enhanced 4d $\mathcal{N} = 4$.

- 6_a-6_b+6_b-6_a: Produces I_{ab} chiral fermions in the representation $(\mathbf{N_a}, \overline{\mathbf{N_b}})$ (plus light scalars, with masses determined by the relative angles $(\theta_i)_{ab}$, and which become massless for supersymmetry preserving intersections). Here $I_{ab} = [\Pi_a] \cdot [\Pi_b]$ is the topological intersection number of the 3-cycles.

The chiral spectrum involves only purely topological data of the configuration, as expected. Hence, the discussion of RR tadpoles and anomaly cancellation can be borrowed directly from previous sections; it was in fact expressed with hindsight in terms of the homology classes $[\Pi_a]$ and a general internal space $\mathbf{X}_6$.

Intersecting D6-brane models on general CY spaces provide in principle a huge class of new models. However, it is difficult to characterize special lagrangian 3-cycles in CY spaces, since their odd number of dimensions prevents the use of the powerful tools of complex geometry. In practice, most intersecting brane models on CYs are obtained using toroidal orbifolds, as will be illustrated in Section 10.6.

10.4 Introducing O6-planes

As already mentioned, RR tadpole cancellation implies that models without O6-planes are necessarily non-supersymmetric. Heuristically, having zero total charge forces the implicit introduction of brane and antibrane charges. In more precise terms, a putative fully supersymmetric configuration of D6-branes would be, as a whole, a BPS state of type IIA on $\mathbf{X}_6$. For a BPS state the tension is proportional to the RR charge, and since the latter vanishes (due to RR tadpole cancellation), so must the former; hence the only supersymmetric configuration is just type IIA on $\mathbf{X}_6$, with no D6-branes at all.

The way out of this impasse is to introduce O6-planes, which have negative tension and negative RR charge, yet preserve the same supersymmetry as the D6-branes. We are thus led to the construction of orientifolds of type IIA on $\mathbf{X}_6$ with intersecting D6-branes. The construction is similar in spirit to the construction of type I theory in Section 4.4.3, in that one constructs an unoriented closed string sector, and introduces D-branes, i.e. open string sectors, to cancel its RR tadpoles.

10.4.1 O6-planes and D6-branes

The main properties of O6-planes and parallel D6-branes can be determined from T-duality to type I compactified on $\mathbf{T}^3$, by extending the construction in Section 5.3.4, and are briefly reviewed below. We are however interested in systems of intersecting D6-branes

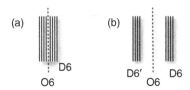

Figure 10.7 Configurations of D6-brane stacks parallel to an O6-plane. (a) The situation of D6-branes on top of the O6-plane, and (b) corresponds to the branes separated from it. Although the branes within a stack are coincident, they are shown slightly separated, for clarity.

in the presence of O6-planes. The properties relevant to the computation of their massless open string spectrum can be studied using configurations in flat 10d space. Such systems are easily described as configurations in flat 10d space invariant under $\Omega \mathcal{R}(-1)^{F_L}$, with $\mathcal{R} : (x^7, x^8, x^9) \rightarrow (-x^7, -x^8, -x^9)$. They contain D6-branes and their orientifold images, denoted D6'-branes, distributed in a symmetric fashion. The open string spectrum in the orientifold quotient theory is obtained by simply projecting onto invariant states. Note that in obtaining the latter one must take into account the orientation flip for open strings implied by Ω. We now study the massless open string spectrum for various O6/D6 systems:

- *Coincident* O6/D6: Consider a system of n D6-branes on top of an O6-plane, see Figure 10.7(a). The open string spectrum before the orientifold action is a $U(n)$ vector multiplet. The orientifold action on Chan–Paton indices is given by the $O^{\pm}$ (or SO/Sp) projections (3.116). We focus on the O^- case (SO projection), which corresponds to negatively charged O6-planes, with $Q_{O6} = -4$ as in (5.52). The Chan–Paton projection at the massless level is $\lambda = \lambda^T$, and leads to an $SO(n)$ vector multiplet with respect to the 16 supercharges unbroken by the O6/D6 configuration.
- *Parallel* O6/D6: Consider n coincident D6-branes, parallel to but separated from the O6-plane, and their n orientifold image D6'-branes, see Figure 10.7(b). The massless spectrum before the orientifold projection is a $U(n) \times U(n)'$ gauge group plus superpartners. The orientifold action identifies the degrees of freedom of both $U(n)$ factors, and only a linear combination survives, i.e. just one $U(n)$ vector multiplet. This result agrees with the spectrum on isolated D6-branes, as the massless open string sector is insensitive to the presence of distant objects like the O6-plane.

  Note that due to the orientation flip by Ω, open strings starting/ending on the D6-brane stack are identified with open strings ending/starting on the D6'-brane stack. Consequently, the group $U(n)$ is identified with its image $U(n)'$ such that the fundamental representation $\square$ of $U(n)$ is identified with the anti-fundamental $\overline{\square}'$ of $U(n)'$. This will be a useful fact in more involved upcoming configurations.

We now turn to intersecting D6-branes (and their images) in the presence of O6-planes. All the D6-branes and O6-planes span 4d Minkowski space, so that we have the by now familiar intersections of 3-planes in the six extra dimensions. We describe the gauge group

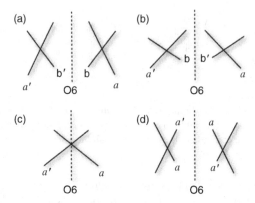

Figure 10.8 Intersecting D6-branes in the presence of an O6-plane. (a) The intersections of two stacks a and b, away from the O6-plane. (b) The intersection of a stack a with the image b' of another. (c) and (d) The intersection of the stacks a and its image a' on top of the O6-plane and away from it, respectively.

and chiral fermion spectrum at intersections, for various relative geometries of the intersection and the O6-plane:

- *ab intersections in O6/D6 system*: Consider two D6-brane stacks, labeled a and b, intersecting away from the O6-plane (and so with an $a'b'$ intersection of the image D6′-branes), see Figure 10.8(a). Before the orientifold projection, the gauge group is $U(N_a) \times U(N_b) \times U(N_a)' \times U(N_b)'$, and the ab and $a'b'$ intersections support 4d chiral fermions in the representation $(\square_a, \overline{\square}_b)$ and $(\square_{b'}, \overline{\square}_{a'})$. After the orientifold identification, we obtain a gauge group $U(N_a) \times U(N_b)$ and a 4d chiral fermion in the $(\square_a, \overline{\square}_b)_{(+1,-1)}$, with $U(1)$ charges indicated as subscripts.
- *ab′ intersections in O6/D6 system*: Consider the intersection of D6$_a$- with D6$_{b'}$-branes, namely the orientifold image of D6$_b$-branes, see Figure 10.8(b). Before the orientifold, the gauge group is $U(N_a) \times U(N_b)' \times U(N_a)' \times U(N_b)$, with 4d chiral fermions in the $(\square_a, \overline{\square}_{b'})_{(+1,-1)}$ and $(\square_b, \overline{\square}_{a'})_{(+1,-1)}$. After the orientifold, we have a gauge group $U(N_a) \times U(N_b)$ and a 4d chiral fermion in the $(\square_a, \square_b)_{(+1,+1)}$.
- *aa′ intersections on top of an O6*: Consider an intersection of D6$_a$-branes with their own image, on top of the O6-plane, see Figure 10.8(c). Before the orientifold, we have a gauge group $U(N_a) \times U(N_a)'$ and a 4d fermion in the $(\square_a, \overline{\square}_{a'})_{(+1,-1)}$. The orientifold reduces the gauge group to $U(N_a)$. The original 4d fermion transforms under it as $\square_{a,+1} \otimes \overline{\square}_{a',-1} = \square_{a,+1} \otimes \square_{a,+1} = \boxminus_{a,+2} + \boxtimes_{a,+2}$, and after the orientifold projection only the 4d fermions in the $\boxminus_{a,+2}$ component survive.
- *aa′ intersections away from O6*: Consider an intersection of D6$_a$- and D6$_{a'}$-branes, away from the O6-plane, see Figure 10.8(d). Before the orientifold we have a gauge group $U(N_a) \times U(N_a)'$ and 4d chiral fermions in the $2(\square_a, \overline{\square}_{a'})_{(+1,-1)}$, due to the two intersections. The orientifold reduces the gauge group to $U(N_a)$, and identifies both intersections, leaving one set of 4d fermions in the $\boxminus_{a,+2} + \boxtimes_{a,+2}$.

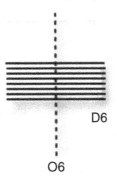

Figure 10.9 Configurations of D6-branes with some directions orthogonal to an O6-plane.

- *Orthogonal O6/D6:* There can also be configurations of D6-branes orthogonal to the O6-plane in some of the directions, so that the D6-brane stack is still mapped to itself under the orientifold action. An often appearing configuration, preserving 4d $\mathcal{N} = 2$ supersymmetry, is an O6-plane along, e.g., directions 0123456 and a D6-brane along 0123478, see Figure 10.9. A stack of n such D6-branes produces a 4d $\mathcal{N} = 2$ $USp(n)$ vector multiplet (hence n must be even) and a 4d $\mathcal{N} = 2$ hypermultiplet in the two-index symmetric (reducible) representation. The difference of gauge group for the cases of parallel and orthogonal O6/D6 system is related (by T-duality) to the different gauge groups for D9- and D5-branes in type I theory, see Section 6.1.3. The spectrum at intersections involving D6-branes parallel or orthogonal to the O6-planes can be obtained using arguments similar to the above, and we skip their detailed discussion.

10.4.2 Orientifold compactifications with intersecting D6-branes

We now put to use the information on the spectra of local intersections and discuss orientifolds of type IIA on a CY $\mathbf{X}_6$, subsequently particularizing to $\mathbf{T}^6$.

General construction

We consider type IIA on a CY $\mathbf{X}_6$, orientifolded by $\Omega\mathcal{R}(-)^{F_L}$, with $\mathcal{R}$ an antiholomorphic $\mathbf{Z}_2$ symmetry of $\mathbf{X}_6$, i.e. acting as $(z_1, z_2, z_3) \rightarrow (\bar{z}_1, \bar{z}_2, \bar{z}_3)$ on local complex coordinates. This is equivalent, up to relabeling, to $(x^7, x^8, x^9) \rightarrow (-x^7, -x^8, -x^9)$ and defines O6-planes that are locally identical to those in the previous section. In the language of Section 10.3.2, they are described globally by special lagrangian 3-cycles in $\mathbf{X}_6$, with $\varphi = 0$; the closed string sector preserves 4d $\mathcal{N} = 1$. We will not need the detailed O6-plane geometry, as most results depend only on the total homology class $[\Pi_{O6}]$ of the 3-cycles spanned by the O6-planes in the configuration.

The models also include N_a D6$_a$-branes wrapped on 3-cycles Π_a, and their image D6$_{a'}$-branes on 3-cycles denoted $\Pi_{a'}$. The D6-branes preserve the 4d $\mathcal{N} = 1$ supersymmetry of the model if they preserve a common supersymmetry with the O6-planes, i.e. if their BPS phase in (10.27) is $\varphi = 0$. In simpler terms, their (local) relative angles with the O6-planes must satisfy (10.10)

$$\theta_1 + \theta_2 + \theta_3 = 0. \tag{10.31}$$

The RR tadpole cancellation conditions include the contributions from D6-branes, image D6′-branes, and O6-planes (with -4 units of D6-brane charge). They read

$$\sum_a N_a \left[\Pi_a\right] + \sum_a N_a \left[\Pi_{a'}\right] - 4 \times \left[\Pi_{O6}\right] = 0. \tag{10.32}$$

The open string spectrum in orientifold models can be easily computed from the multiplicity of brane intersections, and how many sit on top of O6-planes. For instance, in models with no D6-brane coinciding with its image D6′-brane, the light spectrum in the different sectors is (omitting $U(1)$ charges from now on):

- $aa + a'a'$: Contains $U(N_a)$ gauge bosons (plus possible additional adjoint fields).
- $ab + ba + b'a' + a'b'$: Gives I_{ab} chiral fermions in the representation $(\mathbf{N_a}, \overline{\mathbf{N_b}})$, plus light (possibly massless) scalars.
- $ab' + b'a + ba' + a'b$: Contains $I_{ab'}$ chiral fermions in the representation $(\mathbf{N_a}, \mathbf{N_b})$, plus light (possibly massless) scalars.
- $aa' + a'a$: Contains $n_{\mathrm{sym},a}$ 4d chiral fermions in the representation $\square\square_a$ and $n_{\mathrm{asym},a}$ in the $\square\hspace{-0.3em}\square_a$, with

$$n_{\mathrm{sym},a} = \tfrac{1}{2}(I_{aa'} - I_{a,O6}), \quad n_{\mathrm{asym},a} = \tfrac{1}{2}(I_{aa'} + I_{a,O6}), \tag{10.33}$$

where $I_{a,O6} = [\Pi_a] \cdot [\Pi_{O6}]$ is the number of aa' intersections on top of O6-planes. In addition, there are light (possibly massless) scalars.

As expected, the new RR tadpole conditions in the presence of O6-planes guarantee the cancellation of 4d anomalies of the new chiral spectrum, in an easy generalization of the argument in Section 10.3.1. Also analogously, anomalous and non-anomalous $U(1)$s may acquire masses from their couplings to RR 2-form fields. The condition for a $U(1)$ to remain massless is an orientifold version of (10.25)

$$\sum_a N_a(Q_{ak} - Q_{a'k})c_a = 0 \quad \text{for all } k. \tag{10.34}$$

Concerning gravitational anomalies, in the orientifold case there are in general mixed gravitational triangle anomalies. They cancel via new Green–Schwarz contributions arising from both D6-brane and O6-plane worldvolume couplings.

Toroidal orientifold models

Orientifolds of $\mathbf{T}^6$ compactifications (or orbifolds thereof, see later), provide a very explicit realization of the above construction. As in Section 10.3.1, we focus on factorized tori $\mathbf{T}^6 = \mathbf{T}^2 \times \mathbf{T}^2 \times \mathbf{T}^2$, parametrized by $z_i = x^i + iy^i$. We take the orientifold action $\Omega \mathcal{R}(-1)^{F_L}$, with $\mathcal{R} : z_i \to \bar{z}_i$, i.e. $y^i \to -y^i$ leaving x^i invariant. There are two possible kinds of 2-tori

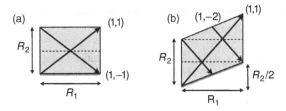

Figure 10.10 Cycles and their orientifold images in rectangular and tilted 2-tori.

compatible with this symmetry, as shown in Figure 10.10; it is possible to have rectangular tori, defined by the periodicities $z \sim z + iR_2$, $z \sim z + R_1$, and tilted tori, defined by $z \sim z + iR_2$, $z \sim z + R_1 + iR_2/2$. The orientifold action on (n, m)-cycles is different in both cases, with $(n, m) \rightarrow (n, -m)$ for rectangular tori, and $(n, m) \rightarrow (n, -m - n)$ for tilted tori. For simplicity we focus on geometries with three rectangular 2-tori, leaving the discussion of tilted geometries for some explicit examples in Section 10.5.1.

For a geometry with three rectangular 2-tori, there are O6-planes spanning the 3-cycle parametrized by x_i, and sitting at the locations $y^i = 0, \frac{1}{2}(R_2)_i$. There are thus eight O6-planes, all along the 3-cycle with wrapping numbers $(n_i, m_i) = (1, 0)$, so their total homology class is $[\Pi_{O6}] = 8[a_1][a_2][a_3]$. The models contain stacks of N_a D6$_a$-branes, which we take factorizable and characterized in terms of the wrapping numbers (n_a^i, m_a^i), and images with wrapping numbers $(n_a^i, -m_a^i)$. The RR tadpole conditions (10.32) read

$$\sum_a N_a n_a^1 n_a^2 n_a^3 = 16,$$

$$\sum_a N_a n_a^1 m_a^2 m_a^3 = 0, \text{ and permutations.} \tag{10.35}$$

As compared with the un-orientifolded case (10.18), the number of conditions is halved. This is because the orientifold projection removes half of the components of the RR fields. Equivalently, because branes and their orientifold images cancel each others' contributions to some of the RR tadpoles, which are thus satisfied automatically and not present in (10.35).

The spectrum of the orientifold theory is given by the general CY result above, with the specific intersection numbers computed using (10.12). Explicit examples of complete models can be found in later sections.

*K-theory charge cancellation**

In the presence of orientifold planes, there are additional discrete $\mathbf{Z}_2$-valued charges of non-BPS D-branes, classified by K-theory, as discussed in Section 6.5.3 for the type I non-BPS $\widehat{\text{D7}}$-brane. In compact models, global consistency requires the cancellation of these discrete charges, leading to additional conditions. In the above toroidal orientifolds, there are conditions requiring the cancellation of $\mathbf{Z}_2$-valued charges of D6-branes along $(1, 0) \times (1, 0) \times (0, 1)$, and permutations. Namely

$$\sum_a N_a m_a^1 n_a^2 n_a^3 \in 2\mathbf{Z}, \quad \text{and permutations of } 1, 2, 3. \tag{10.36}$$

These relate to the cancellation of $\mathbf{Z}_2$-valued $\widehat{D7}$-brane charges in type I by T-dualities turning the O6-planes into an O9-plane. In certain models containing USp gauge factors in the 4d spectrum, these conditions are related to cancellation of global gauge anomalies, which enforces the number of Weyl fermions in the fundamental representation to be even.

10.5 Non-supersymmetric particle physics models

We now exploit this class of string theory vacua to build models of particle physics. Namely intersecting brane models in orientifold compactifications, with the gauge group and massless spectrum reproducing those of the Standard Model (SM), or a simple extension thereof.

We start by constructing a large class of non-supersymmetric models with intersecting D6-branes and O6-planes. Even though a key motivation for O6-planes is to obtain supersymmetric models, they lead to interesting features in the non-supersymmetric context as well. For instance, a general argument shows that D-brane models without orientifold planes cannot lead to just the SM spectrum, rather necessarily include additional $SU(2)$ doublets, as follows. In non-orientifold models the $SU(2)$ factor is actually part of a $U(2)$ group, and the extra $U(1)$ charge allows to distinguish the **2** and the $\bar{\mathbf{2}}$. The RR tadpole cancellation conditions imply that the number of fundamentals and anti-fundamentals for each $U(N)$ factor must be equal, also for this $U(2)$. Since any non-orientifold model with SM group has the left-handed quarks in a representation $3(\mathbf{3}, \bar{\mathbf{2}})$ of $SU(3) \times SU(2)$, this contributes nine anti-fundamentals of $SU(2)$. The complete spectrum must necessarily contain nine fundamentals of $SU(2)$, three of which may be interpreted as left-handed leptons; the remaining six doublets are however exotic chiral fermions, beyond the spectrum of the SM.

The introduction of orientifold planes allows to avoid this issue in several ways, as we describe in this section. As a consequence, they allow the remarkable achievement of string compactifications with the chiral spectrum of just the SM.

10.5.1 The U(2) class

In these models one escapes the above argument by realizing that, in the presence of orientifold planes, there are two different kinds of possible bi-fundamental representations, namely $(\square, \bar{\square})$ and $(\square, \square)$. This allows an alternative construction of the SM chiral fermion spectrum, satisfying the RR tadpole constraints and without exotics, as follows. Realizing the three families of left-handed quarks as $(\mathbf{3}, \bar{\mathbf{2}}) + 2(\mathbf{3}, \mathbf{2})$, this contributes three net $\bar{\mathbf{2}}$s of $SU(2)$. The three **2**s of $SU(2)$ required by RR tadpole cancellation then correspond

Table 10.1 *Standard model fermion spectrum and $U(1)$ charges corresponding to the intersection numbers in (10.37)*

Intersection	Matter fields	$SU(3) \times SU(2)$	Q_a	Q_b	Q_c	Q_d	Y
(ab)	Q_L	$(\mathbf{3}, \mathbf{2})$	1	-1	0	0	$1/6$
(ab')	q_L	$2(\mathbf{3}, \mathbf{2})$	1	1	0	0	$1/6$
(ac)	U_R	$3(\bar{\mathbf{3}}, \mathbf{1})$	-1	0	1	0	$-2/3$
(ac')	D_R	$3(\bar{\mathbf{3}}, \mathbf{1})$	-1	0	-1	0	$1/3$
(bd')	L	$3(\mathbf{1}, \mathbf{2})$	0	-1	0	-1	$-1/2$
(cd)	E_R	$3(\mathbf{1}, \mathbf{1})$	0	0	-1	1	1
(cd')	N_R	$3(\mathbf{1}, \mathbf{1})$	0	0	1	1	0

simply to the three left-handed leptons. Note that this argument nicely connects the number of generations to the number of colors.

It is possible to propose a set of intersection numbers, such that any configuration of D6-branes wrapped on 3-cycles realizing them reproduces the chiral spectrum of the SM. Consider four stacks of D6-branes, denoted[6] a, b, c, d (and their images), giving rise to a gauge group $U(3)_a \times U(2)_b \times U(1)_c \times U(1)_d$. If the intersection numbers of the corresponding 3-cycles are given by

$$I_{ab} = 1 \quad I_{ab'} = 2 \quad I_{ac} = -3 \quad I_{ac'} = -3,$$
$$I_{bd} = 0 \quad I_{bd'} = -3 \quad I_{cd} = -3 \quad I_{cd'} = 3, \tag{10.37}$$

then the chiral spectrum of the model has the non-abelian quantum numbers of the SM chiral fermions (plus right-handed neutrinos). In order to reproduce exactly the SM spectrum, one also needs to require that the linear combination of $U(1)$'s

$$Q_Y = \frac{1}{6}Q_a - \frac{1}{2}Q_c + \frac{1}{2}Q_d, \tag{10.38}$$

which reproduces the hypercharge quantum numbers, remains as the only massless $U(1)$ in the model. The spectrum of massless fermions corresponding to the above intersection numbers is shown in Table 10.1. Such patterns of branes and intersections can be graphically displayed in diagrams of intersecting lines, sometimes loosely referred to as "quivers,"[7] see Figure 10.11(a) for the case at hand. The branes a, d have intersections supporting quarks and leptons, respectively, and are often dubbed the baryonic and leptonic branes. Similarly, the branes b, c are called left and right branes.

It is important to emphasize that at this level, we have not constructed any explicit model, but just specified what to look for; however, it is easy to provide explicit constructions

---

[6] Beware of the notation clash, also present in the literature, between this use of a, and its earlier use as an index labeling D6-brane stacks.

[7] A quiver is actually a graph in which gauge factors are denoted by dots, and bi-fundamentals as oriented arrows joining them. The intersecting line diagrams are similar, by replacing lines by dots and intersections by (possibly multiple) arrows.

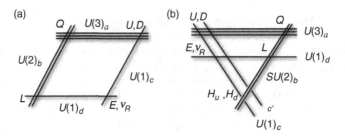

Figure 10.11 Schematic views of the intersecting brane models producing SM-like spectra, (a) based on $U(3)_a \times U(2)_b \times U(1)_c \times U(1)_d$ gauge group; (b) based on $U(3)_a \times SU(2)_b \times U(1)_c \times U(1)_d$. The latter allows for enhancement up to a Pati–Salam group $U(4)_{a,d} \times SU(2)_b \times SU(2)_c$ by letting branes a to coincide with d, and c with c'. Intersections are implicitly understood to be triplicated.

Table 10.2 *D6-brane wrapping numbers realizing the SM*
intersection numbers (10.37)

N	$\left(n_i^1, \tilde{m}_i^1\right)$	$\left(n_i^2, \tilde{m}_i^2\right)$	$\left(n_i^3, \tilde{m}_i^3\right)$
$N_a = 3$	$(1/\beta^1, 0)$	$\left(n_a^2, \epsilon\beta^2\right)$	$(1/\rho, 1/2)$
$N_b = 2$	$\left(n_b^1, -\epsilon\beta^1\right)$	$(1/\beta^2, 0)$	$(1, 3\rho/2)$
$N_c = 1$	$\left(n_c^1, 3\rho\epsilon\beta^1\right)$	$(1/\beta^2, 0)$	$(0, 1)$
$N_d = 1$	$(1/\beta^1, 0)$	$\left(n_d^2, -\beta^2\epsilon/\rho\right)$	$(1, 3\rho/2)$

with this precise structure, as we now illustrate with examples of orientifolds of $\mathbf{T}^6$. The construction of these models requires the use of both rectangular and tilted 2-tori, which we distinguish by discrete parameters $b_i = 0, \frac{1}{2}$. Recall that for rectangular 2-tori the orientifold image of the wrapping numbers $\left(n_a^i, m_a^i\right)$ is $\left(n_a^i, -m_a^i\right)$, while for tilted tori the image wrapping numbers are $\left(n_a^i, -n_a^i - m_a^i\right)$, see Figure 10.10. To unify their description, we introduce $\tilde{m}_a = m_a + bn_a$, so that branes and images have wrapping numbers $(n_a, \tilde{m}_a)$ and $(n_a, -\tilde{m}_a)$ respectively; note that in the $b = 1/2$ case the numbers $\tilde{m}$ are integer or half integer, when the corresponding number n is even or odd, respectively. It is straightforward to write down the RR tadpole conditions for models with rectangular or tilted geometries in the three 2-tori. The general class of models with factorizable 3-cycles realizing the intersection numbers (10.37) is shown in Table 10.2. These general solutions are parametrized by a sign $\epsilon = \pm 1$, the choice of tilting on the first two 2-tori via $\beta^i = 1 - b^i = 1, 1/2$, four integers $n_a^2, n_b^1, n_c^1, n_d^2$, and a parameter $\rho = 1, 1/3$. The third 2-torus is tilted for the whole class. The parameters are such that the RR tadpole conditions of type nmm (and permutations) are automatically satisfied. The remaining nnn tadpole condition is given by

$$\frac{3n_a^2}{\rho\beta^1} + \frac{2n_b^1}{\beta^2} + \frac{n_d^2}{\beta^1} + 2\frac{N_h}{\beta^1\beta^2} = 16, \qquad (10.39)$$

where we have allowed for N_h additional D6-branes along the O6-plane direction with wrapping numbers $(1/\beta^1, 0)(1/\beta^2, 0)(2, 0)$. The reader may check that these generically have no intersection with the SM branes and hence form a hidden sector of the theory. It is easy to find plenty of solutions for equations (10.39), with or without such hidden sector branes.

In order for the hypercharge $U(1)$ (10.38) to remain massless, those parameters have to verify the extra constraint (10.34) which yields

$$n_c^1 = \frac{\beta^2}{2\beta^1} \left(n_a^2 + 3\rho\, n_d^2\right). \tag{10.40}$$

The other three $U(1)$s in general mix with RR fields and get massive, although for particular parameter choices the anomaly-free $U(1)_{B-L}$ boson can remain massless.

Even though the models are non-supersymmetric, for each model one can find large regions in the complex structure moduli space in which there are no tachyons in the spectrum (recall that intersection angles, and thus scalar masses, depend on complex structure moduli through (10.11)). Alternatively, some of these tachyons may play the role of SM Higgs fields, for which there are candidate scalars in the bc and bc' sectors, if the b and c branes are close enough in the second torus.

Since the models are non-SUSY, if the string scale is large one must allow for a fine tuning so that the Higgs scalars remain light. Alternatively, the string scale could be just above the EW scale, recall Section 1.4.2. In such case, the three extra $U(1)$s beyond hypercharge could be produced at colliders like the LHC, as will be described in Section 16.4.2. One interesting property of these models is that the baryon number $U(1)$ is gauged, so the proton is stable even for low string scale. Lepton number is also gauged, and Majorana neutrino masses are forbidden, although Dirac masses are allowed. Nevertheless, there may be non-perturbative instanton effects violating these symmetries, as discussed in Chapter 13.

Even in non-tachyonic regions, these models are not completely stable. In non-supersymmetric models, tadpoles for NSNS fields are no longer related to the RR tadpoles, and cancellation of the latter does not imply cancellation of the former. Uncanceled NSNS tadpoles signal that the model sits at the slope of a non-trivial 4d scalar potential for complex structure moduli, arising from the non-zero total tension of the D6/O6 system, see Section 12.4 for its interpretation as an FI term. The ultimate stability of these models, however, depends on the possible introduction of extra ingredients, like field strength fluxes, see Chapters 14 and 15.

In any event, this type of non-SUSY models constitutes possibly the simplest class of string models with just the SM gauge group and fermion content (plus right-handed neutrinos).

The proposal to obtain the SM using realizations of the intersection numbers (10.37) is not restricted to the non-SUSY toroidal orientifold setup, or even to geometric compactifications. Indeed, these concepts can be put to work in more abstract worldsheet CFT "compactifications," to construct a large class of *supersymmetric* models realizing

Table 10.3 *D6-brane wrapping numbers realizing the SM*
intersection numbers (10.41)

	N	(n^1, m^1)	(n^2, m^2)	(n^3, m^3)
a	$N_a = 3$	$(1, 0)$	$(1/\rho, 3\rho)$	$(1/\rho, -3\rho)$
b	$N_b = 1$	$(0, 1)$	$(1, 0)$	$(0, -1)$
c	$N_c = 1$	$(0, 1)$	$(0, -1)$	$(1, 0)$
d	$N_d = 1$	$(1, 0)$	$(1/\rho, 3\rho)$	$(1/\rho, -3\rho)$

the above SM structure, as overviewed in Section 10.7.1. There are also SUSY orientifold models based, e.g., on the $\mathbf{Z}_2 \times \mathbf{Z}_2$ orbifold, with a similar realization of the SM but with Higgs doublets contributing to the $U(2)$ anomalies, see the example in Section 13.3.3.

10.5.2 The USp(2) class

A second possibility to avoid the problem of the extra $SU(2)$ doublets on page 320, is to exploit the fact that D6-branes in the presence of O6-planes may contain $USp(N)$ gauge factors (see Section 10.4.1); for these, all representations are real, and RR tadpole conditions do not impose any constraint on the matter content. It is possible to use D6-branes with group $USp(2) \equiv SU(2)$ to realize the electroweak gauge factor and circumvent the constraints on the number of doublets.

Indeed, it is possible to propose a set of intersection numbers, such that any configuration of D6-branes wrapped on 3-cycles realizing them reproduces the chiral spectrum of the SM in this way, see Figure 10.11(b) for its "quiver" depiction. Consider four stacks of D6-branes, labeled a, b, c, d, with gauge group $U(3)_a \times USp(2)_b \times U(1)_c \times U(1)_d$, and with intersection numbers

$$I_{ab} = 3 \quad I_{ab'} = 3 \quad I_{ac} = -3 \quad I_{ac'} = -3,$$
$$I_{db} = 3 \quad I_{db'} = 3 \quad I_{dc} = -3 \quad I_{dc'} = -3. \tag{10.41}$$

The $U(1)$ required to be massless in order to reproduce the SM hypercharge is

$$Q_Y = \frac{1}{6}Q_a - \frac{1}{2}Q_c - \frac{1}{2}Q_d. \tag{10.42}$$

As above, explicit realizations of D6-branes on 3-cycles with those intersection numbers (and with massless hypercharge) can be realized in toroidal orientifolds, in this case with rectangular two-tori. Consider the set of D6-branes in Table 10.3 (and their images), with $\rho = 1, 1/3$ being a discrete parameter. The stack b overlaps with their images to yield $USp(2) \equiv SU(2)$. The model must include additional branes to satisfy the RR tadpole conditions, whose detailed description we skip. As announced, the gauge group is $U(3) \times SU(2)_L \times U(1) \times U(1)$ and the spectrum contains three SM generations plus three right-handed neutrinos.

We conclude with some short remarks. The model admits a simple modification to a Pati–Salam model $U(4) \times USp(2)_L \times USp(2)_R$, by taking the branes a to coincide with d, and c with c'. Also note that by appropriate choice of 2-tori moduli, the above D6-brane configuration can preserve supersymmetry, see next section; however, explicit examples show that the additional branes required for tadpole cancellation are incompatible with this supersymmetry, so the complete purely toroidal models are non-supersymmetric, in agreement with the general arguments below. Finally, we again emphasize that the intersection numbers (10.41) are not restricted to toroidal orientifolds, and can be realized in other type IIA orientifold compactification setups, as in the next section and in Section 10.7.1.

10.6 Supersymmetric particle physics models in $T^6/Z_2 \times Z_2$ orientifolds

In order to obtain supersymmetric intersecting brane models, it is necessary to go beyond orientifolds of toroidal compactifications. This can be understood from the RR tadpole conditions (10.35) as follows. The last conditions imply that there are in general D6-brane contributions to RR charges for which no O6-plane contribution exists. Hence, any consistent tadpole-free model secretly contains brane–antibrane pairs associated to those charges, and is necessarily non-supersymmetric. The key to the construction of supersymmetric models is thus the introduction of O6-planes diverse enough to span all possible RR charges in the model.

This can be easily realized using orientifolds of toroidal orbifolds. The basic properties of orbifold constructions have been described in Chapter 8, when used for heterotic compactification. The computation of the closed string spectrum of type IIA orientifolds can be carried out similarly, so in this section we rather focus on the computation of the open string sector. We moreover restrict to one particularly simple orbifold space, $T^6/(Z_2 \times Z_2)$, which already appeared[8] in Section 8.2.6. Recall that the generators θ, ω have twists $v = \left(\frac{1}{2}, -\frac{1}{2}, 0\right)$ and $w = \left(0, \frac{1}{2}, -\frac{1}{2}\right)$, and so act on the T^6 complex coordinates z_i as

$$\theta : (z_1, z_2, z_3) \rightarrow (-z_1, -z_2, z_3),$$
$$\omega : (z_1, z_2, z_3) \rightarrow (z_1, -z_2, -z_3). \tag{10.43}$$

We consider the model with a factorizable underlying toroidal lattice, and for simplicity take the 2-tori to be rectangular, rather than tilted. We mod out this theory by $\Omega \mathcal{R}(-)^{F_L}$, with

$$\mathcal{R} : (z_1, z_2, z_3) \rightarrow (\bar{z}_1, \bar{z}_2, \bar{z}_3). \tag{10.44}$$

The model contains four kinds of O6-planes, associated to the actions of $\Omega\mathcal{R}(-)^{F_L}$, $\Omega\mathcal{R}\theta(-)^{F_L}$, $\Omega\mathcal{R}\omega(-)^{F_L}$, $\Omega\mathcal{R}\theta\omega(-)^{F_L}$, as shown in Figure 10.12.

In order to cancel the corresponding RR tadpoles, we introduce D6-branes wrapped on 3-cycles. The simplest 3-cycles in the orbifold space are obtained from factorized 3-cycles in the underlying toroidal geometry, characterized in terms of wrapping numbers $\left(n_a^i, m_a^i\right)$.

---

[8] More precisely, with the choice of discrete torsion leading to $h_{1,1} = 51$, $h_{2,1} = 3$.

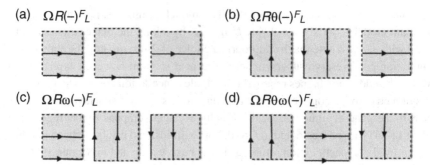

Figure 10.12 Structure of 3-cycles wrapped by the diverse O6-planes in the orientifold of $\mathbf{T}^6/(\mathbf{Z}_2 \times \mathbf{Z}_2)$. The arrows denote the orientation of the cycles.

Also, for simplicity we take each D6-brane stack to pass through some $\mathbf{Z}_2 \times \mathbf{Z}_2$ fixed point, so it coincides with its orbifold (but not their orientifold) images. The RR tadpole conditions have the familiar form

$$\sum_a N_a \, [\Pi_a] + \sum_a N_a \, [\Pi_{a'}] - 4[\Pi_{O6}] = 0, \tag{10.45}$$

where $[\Pi_{O6}]$ is the homology charge of the complete set of O6-planes. More explicitly, using (10.13) for rectangular tori one obtains

$$\sum_a N_a n_a^1 n_a^2 n_a^3 - 16 = 0, \qquad \sum_a N_a n_a^1 m_a^2 m_a^3 + 16 = 0,$$

$$\sum_a N_a m_a^1 n_a^2 m_a^3 + 16 = 0, \qquad \sum_a N_a m_a^1 m_a^2 n_a^3 + 16 = 0. \tag{10.46}$$

It can be shown that, since each stack of branes contains D6-branes and their orbifold images, a set of N D6-branes leads to a $U(N/2)$ gauge symmetry, so N must be even. With this information, the chiral spectrum in 4d $\mathcal{N}=1$ terms reads:

Sector	Representation
aa	$U(N_a/2)$ vector multiplet
	3 Adj. chiral multiplets
$ab + ba$	I_{ab} ($\square_a, \overline{\square}_b$) chiral multiplets
$ab' + b'a$	$I_{ab'}$ ($\square_a, \square_b$) chiral multiplets
$aa' + a'a$	$\frac{1}{2}(I_{aa'} - I_{a,O6})$ $\square\square_a$ ch. multiplets
	$\frac{1}{2}(I_{aa'} + I_{a,O6})$ $\boxminus_a$ ch. multiplets

where $I_{a,O6} = [\Pi_a] \cdot [\Pi_{O6}]$. There are also discrete K-theory constraints, which can be computed to be

$$\sum_a N_a m_a^1 m_a^2 m_a^3 \in 4\mathbf{Z}, \qquad \sum_a N_a n_a^1 n_a^2 m_a^3 \in 4\mathbf{Z}, \text{ and permutations.} \tag{10.47}$$

Table 10.4 *D-brane wrapping numbers giving rise to a 4d $\mathcal{N} = 1$ LR symmetric SUSY model, in the $\mathbf{T}^6/(\mathbf{Z}_2 \times \mathbf{Z}_2)$ orientifold*

N_α	$\left(n_\alpha^1, m_\alpha^1\right)$	$\left(n_\alpha^2, m_\alpha^2\right)$	$\left(n_\alpha^3, m_\alpha^3\right)$
$N_a = 6$	$(1, 0)$	$(3, 1)$	$(3, -1)$
$N_b = 2$	$(0, 1)$	$(1, 0)$	$(0, -1)$
$N_c = 2$	$(0, 1)$	$(0, -1)$	$(1, 0)$
$N_d = 2$	$(1, 0)$	$(3, 1)$	$(3, -1)$
$N_{h_1} = 2$	$(-2, 1)$	$(-3, 1)$	$(-4, 1)$
$N_{h_2} = 2$	$(-2, 1)$	$(-4, 1)$	$(-3, 1)$
$N_f = 40$	$(1, 0)$	$(1, 0)$	$(1, 0)$

These are actually $\mathbf{Z}_2$-valued constraints, since N_a are already even integers.

The condition that the system of branes preserves 4d $\mathcal{N} = 1$ supersymmetry is that the relative angles between each stack of D6-branes and the O6-planes satisfy (10.10), i.e. $\theta_1 + \theta_2 + \theta_3 = 0$. For fixed wrapping numbers (n^i, m^i), the condition translates into a constraint on the ratio of the two radii on each torus. Denoting $\chi_i = (R_2/R_1)_i$, with R_2, R_1 the vertical and horizontal radii, the constraint is

$$\arctan\left(\chi_1 \frac{m_1}{n_1}\right) + \arctan\left(\chi_2 \frac{m_2}{n_2}\right) + \arctan\left(\chi_3 \frac{m_3}{n_3}\right) = 0. \tag{10.48}$$

This defines a fairly general class of supersymmetric intersecting brane models on a toroidal orbifold. To show its model building potential, we provide an illustrative example realizing the intersection numbers (10.41) and leading to a 3-generation $SU(3) \times SU(2)_L \times SU(2)_R \times U(1)_{B-L}$ model (plus an almost hidden sector). The sets of branes are shown in Table 10.4, where h_1, h_2 are additional branes to satisfy RR tadpole cancellation, as the reader may easily check. A pictorial representation of this model is shown in Figure 12.5 in Section 12.6. The model belongs to the class in Section 10.5.2; in particular it admits an enhancement to a Pati–Salam unified theory, by letting stacks a and d coincide into an $SU(4)$ one (also denoted a in the following). The full gauge group includes an extra factor $USp(40)$, and the complete charged chiral spectrum is shown in Table 10.5, in Pati–Salam notation for the sake of clarity. After accounting for the $B \wedge F$ couplings there is only one massless $U(1)$, given by the combination $U(1)' = \frac{1}{3}U(1)_a + 2[U(1)_{h_1} - U(1)_{h_2}]$ (recall that $U(1)_{B-L}$ is contained inside $SU(4)$).

Note that, compared to typical orbifold heterotic vacua (e.g. see Appendix C) the spectrum is relatively simple. Moreover, it simplifies even further by allowing for the recombination of branes h_1, h_2' and f, corresponding to giving vevs to chiral fields at the (fh_1), (fh_2) and (h_1h_2') intersections along D- and F-flat directions. The resulting spectrum, again in Pati–Salam language, is shown in Table 10.6. There is no chiral matter arising from ah, ah', hh' or charged under $USp(40)$.

Table 10.5 *Chiral spectrum of the 4d $\mathcal{N} = 1$ 3-family Pati–Salam model of Table 10.4*

Sector	Matter	$SU(4) \times SU(2) \times SU(2) \times [USp(40)]$	Q_a	Q_{h_1}	Q_{h_2}	Q'
(ab)	F_L	$3(\mathbf{4, 2, 1})$	1	0	0	1/3
(ac)	F_R	$3(\mathbf{\bar{4}, 1, 2})$	−1	0	0	−1/3
(bc)	H	$(\mathbf{1, 2, 2})$	0	0	0	0
(ah_1)		$6(\mathbf{\bar{4}, 1, 1})$	−1	1	0	5/3
(ah'_2)		$6(\mathbf{4, 1, 1})$	1	0	1	−5/3
(bh_1)		$8(\mathbf{1, 2, 1})$	0	1	0	2
(bh_2)		$6(\mathbf{1, 2, 1})$	0	0	1	−2
(ch_1)		$6(\mathbf{1, 1, 2})$	0	1	0	2
(ch_2)		$8(\mathbf{1, 1, 2})$	0	0	1	−2
$(h_1 h'_1)$		$46(\mathbf{1, 1, 1})$	0	2	0	4
$(h_2 h'_2)$		$46(\mathbf{1, 1, 1})$	0	0	2	−4
$(h_1 h'_2)$		$196(\mathbf{1, 1, 1})$	0	1	1	0
$(f h_1)$		$(\mathbf{1, 1, 1; 40})$	0	−1	0	2
$(f h_2)$		$(\mathbf{1, 1, 1; 40})$	0	0	−1	−2

Table 10.6 *Spectrum for the D-brane model in Table 10.4 after D-brane recombination*

Sector	Matter	$SU(4) \times SU(2) \times SU(2) \times [USp(40)]$
(ab)	F_L	$3(\mathbf{4, 2, 1})$
(ac)	F_R	$3(\mathbf{\bar{4}, 1, 2})$
(bc)	H	$(\mathbf{1, 2, 2})$
(bh)		$2(\mathbf{1, 2, 1})$
(ch)		$2(\mathbf{1, 1, 2})$

This model is probably the simplest example of MSSM-like model from intersecting branes and is often used as a template for the computation of different quantities of physical interest. A useful T-dual type IIB D7-brane version is described in Section 11.4.2, and its Yukawa couplings, Kähler metrics, and gauge kinetic functions are described in Section 12.6. It is also used in flux compactifications in Section 14.3.2. The model is quite simple, but has the shortcoming (common to all toroidal based models) of containing massless adjoint chiral multiplets for all gauge factors. Even if they get TeV scale masses upon SUSY breaking, these fields would cause the gauge couplings to blow-up before unification, and are thus phenomenologically undesirable. There are, however, some variants in which the orbifold projection eliminates such adjoint fields (in the language of Section 10.3.2 these are rigid 3-cycles i.e. $b_1(\Pi) = 0$); we direct the reader to the Bibliography for more details. A second problem is the presence of exotic leptons with fractional charge $\pm 1/2$ from the *bh* and *ch* sectors, which may be phenomenologically problematic, see Section 17.1.3. This is a generic feature of $\mathbf{Z}_2 \times \mathbf{Z}_2$ orientifold models, but which is not necessarily present in models based on other orbifolds; indeed there exist explicit examples of 3-generation models without chiral exotics using, e.g., the $\mathbf{Z}'_6$ orientifold.

10.7 Generalizations and related constructions

There are further compactifications setups closely related to the intersecting D6-brane models described above. We briefly overview the construction of type II Gepner model orientifolds, and M-theory compactification on G_2 holonomy spaces.

10.7.1 RCFT orientifolds

The above strategies for model building in type II orientifolds with D-branes can be extended beyond the geometric realm, and explored for abstract CFT "compactifications," e.g. free fermion constructions or Gepner models, as done for the heterotic in Sections 8.5 and 8.6. As illustration, we overview the construction of a large class of semi-realistic models using orientifolds of type II Gepner model compactifications.

Boundary and crosscap states

To define orientifolds of type II "compactifications" where the internal space is described by an abstract CFT, one needs to generalize the concepts of D-branes and O-planes. This is done through the so-called boundary states $|B\rangle$, and crosscap states $|C\rangle$. They are linear combinations of states in the closed oriented string theory, which are, however, not normalizable, and so do not belong to the closed string Hilbert space. Their role is to describe the effect of the introduction of boundaries and crosscaps in the closed oriented string theory. For instance, for Dp-branes in bosonic string theory, the boundary states are defined by

$$
\begin{aligned}
\left(\partial_\sigma X^\mu_L - \partial_\sigma X^\mu_R \right) |B\rangle = 0, & \quad \text{i.e.} \quad \left(\alpha^\mu_n - \tilde{\alpha}^\mu_n \right) |B\rangle = 0, \quad \mu = 2, \ldots, p, \\
\left(\partial_t X^i_L - \partial_t X^i_R \right) |B\rangle = 0, & \quad \text{i.e.} \quad \left(\alpha^i_n + \tilde{\alpha}^i_n \right) |B\rangle = 0, \quad i = p+1, \ldots, 25.
\end{aligned}
\tag{10.49}
$$

Namely they impose the effects of Neumann or Dirichlet boundary conditions on worldsheet fields; there are similar conditions defining crosscap states. In general boundary and crosscap states describe the effect on the closed string sector of a reflection off a boundary or a crosscap, see Figure 10.13.

An orientifold model is obtained by introducing a crosscap state $|C\rangle$, and a set of boundary states $|B_a\rangle$, with Chan–Paton multiplicities N_a. It is useful to introduce a basis of (normalizable) closed string states $\{|m\rangle\}$, known as Ishibashi states, satisfying conditions similar to those of boundary and crosscap states. The latter can then be expanded as linear combinations of Ishibashi states as

$$
|B_a\rangle = \sum_m R_{am} |m\rangle, \quad |C\rangle = \sum_m U_m |m\rangle.
\tag{10.50}
$$

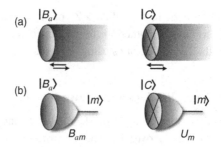

Figure 10.13 Interpretations of boundary and crosscap states as describing the reflection properties of closed string modes off boundaries and crosscaps (a) or as encoding disk and crosscap amplitudes with one closed string mode insertion (b).

Ishibashi states are the closed string states which can propagate off a boundary or a crosscap; the boundary and crosscap coefficients R_{am}, U_m describe disk and crosscap diagrams with one insertion of the Ishibashi state $|m\rangle$, i.e. the coupling of D-branes and O-planes to closed strings. The restrictions of boundary and crosscap states to the massless sector are analogous to the D-brane and O-plane homology classes in geometric models, so, e.g., the RR tadpole cancellation conditions are

$$\sum_a N_a \, R_{am} + U_m = 0 \quad \text{for massless RR states } m. \tag{10.51}$$

Also, the coefficient R_{a0}, where 0 denotes the CFT vacuum state, describes the D-brane coupling to the dilaton, hence the 4d gauge coupling constants.

Overlaps of boundary and crosscap states produce the Klein bottle, Moebius strip, and cylinder/annulus amplitudes in the tree closed string channel,

$$Z_{\mathcal{C}} = N_a \, N_b \, \langle B_a | e^{-t\ell H} | B_b \rangle \,, \quad Z_{\mathcal{M}} = N_a \, \langle B_a | e^{-t\ell H} | C \rangle + \text{h.c.}, \quad Z_{\mathcal{K}} = \langle C | e^{-t\ell H} | C \rangle,$$

where H is the closed string Hamiltonian, and $t\ell$ is the length of closed string propagation. Upon suitable modular transformations, they produce the open string loop amplitudes, from which one extracts the open string spectrum. Introducing the characters χ_i of Virasoro representations and their modular matrices S, T, as in Section E.5.1, these amplitudes have the structure

$$Z_{\mathcal{C}} = \frac{1}{2} N_a \, N_b \, A^i_{ab} \, \chi_i \left(\frac{t}{2} \right) \,, \quad Z_{\mathcal{M}} = \frac{1}{2} N_a \, M^i_a \, \hat{\chi}_i \left(\frac{1}{2} + \frac{t}{2} \right) \,, \quad Z_{\mathcal{K}} = \frac{1}{2} K^i \, \chi_i (2t).$$

Here $\hat{\chi}_i = T^{-1/2} \chi_i$, with $T^{1/2} = \exp i\pi (L_0 - c/24)$, where L_0 is a Virasoro algebra generator, see Section E.5.1. The multiplicities have a simple expression in terms of the boundary and crosscap coefficients

$$A^i_{ab} = N_a N_b \sum_m \frac{S^i_m R_{am} R^*_{bm}}{S_{i0}}, \qquad K^i = \sum_m \frac{S^i_m U_m U^*_m}{S_{i0}},$$

$$M^i_a = N_a \sum_m \frac{P^i_m R_{am} U^*_m}{S_{i0}}, \qquad \text{with } P \equiv T^{1/2} S T^2 S T^{1/2}. \tag{10.52}$$

Restricting to the open string massless level, the annulus multiplicities define the analogs of the intersection numbers for geometric compactifications.

Boundary/crosscap states must satisfy certain properties, known as *sewing constraints*, ensuring they can be consistently combined to generate amplitudes for worldsheets with arbitrary numbers of boundaries and crosscaps. A set of necessary conditions (possibly not sufficient, although extremely restrictive) is the requirement that the loop channel multiplicities must be integers, and that the annulus coefficients, as well as the combinations $\frac{1}{2}(Z_{ii} + K^i)$ and $\frac{1}{2}\left(A^i_{aa} + M^i_a\right)$ – the unoriented closed and open string partition function coefficients – must be non-negative integers. In practice, this can suffice to construct boundary and crosscap states in certain solvable CFT compactifications, as described next.

*Construction of RCFT orientifolds**

The above ideas can be applied to the construction of orientifolds of type II compactifications on Gepner models. The procedure is carried out very efficiently regarding Gepner models as rational conformal field theories (RCFTs) defined by simple current extensions, see Sections E.5.2 and E.5.3, to which we refer for notation; then Gepner orientifolds are a particular application of the formalism to construct RCFT orientifolds explained in the following.

The starting point is an oriented closed string theory with the Virasoro algebra extended by a set of simple currents, generating an abelian group $\mathcal{H}$. For Gepner models these are the alignment currents and spectral flow currents described in Section E.5.3. The modular invariant partition function is (E.40)

$$Z(\tau, \overline{\tau}) = \sum_{i,j} Z_{ij} \chi_i \overline{\chi}_j. \tag{10.53}$$

The integer multiplicities Z_{ij} count the number of simple currents satisfying (E.41).

We need to specify the behavior of these currents upon reflection off boundary and crosscap states. We restrict to the simplest possibility that the extension currents are preserved, e.g.

$$(J_n + (-1)^{h_J} \tilde{J}_n,)\,|B\rangle = 0, \quad (J_n + (-1)^{h_J+n} \tilde{J}_n,)\,|C\rangle = 0, \tag{10.54}$$

where J_n, $\tilde{J}_n$ are modes of left- and right-moving currents. These are the analogs of conditions (10.49); indeed when applied to the spacetime momentum current $J = i\partial X^\mu$ of the bosonic theory, they reproduce the translational invariant Neumann boundary conditions in (10.49). Since the left and right spectral flow currents underly 4d $\mathcal{N} = 2$ spacetime supersymmetry, the preserved current implies 4d $\mathcal{N} = 1$ supersymmetry of the resulting orientifold model. Although we do not consider them, it is possible to construct models with supersymmetric closed string sector but non-supersymmetric branes (or orientifolds), by using conditions preserving the Virasoro algebra, but not its extension.

The structure of simple current RCFTs naturally suggests an ansatz for Ishibashi states, and boundary and crosscap coefficients, which can be shown to satisfy the above mentioned integrality properties. We offer some flavour of its ingredients, leaving the detailed description for the references.

Ishibashi states are closed string states which can propagate between crosscap and/or boundary states, which exchange left and right sectors. Hence, they are states contributing to the left–right symmetric part $i = j$ of the torus partition function (10.53). Recalling (E.41), they are labeled by pairs $\boldsymbol{m} \equiv (m, J)$, such that

$$m = Jm$$

$$Q_L(m) + X(L, J) = 0 \bmod 1, \quad \forall L \in \mathcal{H}. \tag{10.55}$$

Boundaries are labeled by an index $\boldsymbol{a} \equiv [a, \psi_a]$, where a correspond to ($\mathcal{H}$-orbits of) primaries, and ψ_a defines an additional label required in cases with fixed points (it is the group theoretical character of a representation of the central stabilizer $\mathcal{C}_a$ mentioned below). The boundary coefficients $R_{\boldsymbol{am}} = R_{[a,\psi_a](m,J)}$ in (10.50) are

$$R_{\boldsymbol{am}} = \sqrt{\frac{|\mathcal{H}|}{|\mathcal{C}_a||\mathcal{S}_a|}} \, \psi_a^*(J) \, S_{am}^J. \tag{10.56}$$

This formula includes some extra technical ingredients, which we define for completeness. S_{am}^J is the fixed point resolution matrix, which implements the modular transformation S on the set of primaries fixed by J; it is unitary and obeys

$$S_{Li,j}^J = F_i(L, J) \, e^{2\pi i \, Q_L(j)} \, S_{ij}^J, \tag{10.57}$$

where $Q_L(j)$ is the monodromy charge (E.36). The phases $F_i(L, J)$ are called simple current twists. The *stabilizer* $\mathcal{S}_a$ is the subgroup of $\mathcal{H}$ leaving a invariant, and the *central stabilizer* $\mathcal{C}_a$ is the subgroup of $\mathcal{S}_a$ whose elements J satisfy $F_i(L, J) \, e^{2\pi i \, X(L,J)} = 1$ for all $L \in \mathcal{S}_a$.

Crosscaps are introduced in terms of a new ingredient, the *Klein bottle current* K; this is a simple current (possibly beyond $\mathcal{H}$), charge neutral under all order-2 currents in $\mathcal{H}$, i.e. $Q_J(K) = 0$ for all $J \in \mathcal{H}$ with $J^2 = 0$. This defines the analog of the geometric orientifold action, and must be supplemented by a choice of additional signs (analogous to the choice of O-plane charge signs). The choice of signs is encoded in a choice of phases $\beta(L)$ for all $L \in \mathcal{H}$, with $\beta(0) = e^{\pi i \, h_K}$ and satisfying

$$\beta(L)\beta(L') = \beta(LL')e^{2\pi i \, X(L,L')}, \quad \forall L, L' \in \mathcal{H}. \tag{10.58}$$

They can be used to define $\sigma(L) = \beta(L)e^{\pi i \, (h_{LK} - h_K)}$, which can be shown to be signs. The crosscap coefficients $U_{\boldsymbol{m}} = U_{(m,J)}$ in (10.50) are

$$U_{\boldsymbol{m}} = \frac{1}{\sqrt{|\mathcal{H}|}} \sum_{L \in \mathcal{H}} \sigma(L) \, P_{LK,m} \, \delta_{J,0}, \tag{10.59}$$

where we recall that $P = T^{1/2}ST^2ST^{1/2}$. The boundary coefficients for the orientifold image $\boldsymbol{a}'$ of the D-brane $\boldsymbol{a}$ are

$$R^*_{am} = g^{\Omega,m}_{JJ'} R_{a'm} \quad \text{with } g^{\Omega,m}_{JJ'} = \frac{S_{m0}}{S_{mK}} \beta(J) \delta_{J',J^*}. \qquad (10.60)$$

The boundary and crosscap coefficients (10.56), (10.59) can be used to compute the amplitudes (10.52) and to extract the open string spectrum of the models.

Model building

Systematic computer searches for Gepner orientifold models with MSSM-like structure have been performed by Schellekens and collaborators. They actually considered the mirror case of type IIB Gepner orientifolds, which would rather correspond to, e.g., magnetized D7/D3-brane models, see the next chapter. It is, however, more intuitive to discuss the models in the language of intersecting branes, so their inclusion at this point is timely.

We concentrate in the particular case of the so-called *Madrid quivers*, i.e. models with four intersecting brane stacks a, b, c, d, and hypercharge embedding (10.38), as in Section 10.5, see Figure 10.11. They are taken to realize the following six possible classes of gauge groups,

$$
\begin{aligned}
\text{Type 0} \quad & U(3) \times USp(2) \times U(1) \times U(1), \\
\text{Type 1} \quad & U(3) \times U(2) \times U(1) \times U(1), \\
\text{Type 2} \quad & U(3) \times USp(2) \times O(2) \times U(1), \\
\text{Type 3} \quad & U(3) \times U(2) \times O(2) \times U(1), \\
\text{Type 4} \quad & U(3) \times USp(2) \times USp(2) \times U(1), \\
\text{Type 5} \quad & U(3) \times U(2) \times USp(2) \times U(1).
\end{aligned}
\qquad (10.61)
$$

More additional hidden branes are in general required to cancel RR tadpoles. Checking the latter conditions becomes increasingly complicated as the number of extra branes increases so in the search there is a computational bias in favor of models with a reduced number of stacks. Note that, once anomalous $U(1)$s become massive, the gauge group is actually that of the SM (types 0–3) or its left–right (LR) symmetric extension (types 4 and 5). As in the toroidal orientifolds, in types 0–3 one must ensure that the hypercharge gauge boson does not get massive through $B \wedge F$ couplings. In these types of models there is a non-anomalous gauged $U(1)_{B-L}$ which may or may not become massive by that mechanism. On the other hand, in types 4 and 5 the $U(1)_{B-L}$ must be forced to be massless in order to have it contribute to hypercharge upon breaking of the LR symmetry.

The chiral particle spectrum contains exactly three MSSM generations (including three right-handed neutrinos). In addition, it contains a number of Higgs multiplets $H_u + H_d$, and additional non-chiral combinations of states; these can include extra quarks and leptons, singlets, adjoints, and two-index symmetric and antisymmetric representations of the SM factors, as well as hidden sector matter fields, not coupling directly to the SM fields, or exotics with SM and hidden quantum numbers.

The search uses the basic 168 Gepner models with A and E invariants and considers all modular invariant partition functions (MIPF) defined by simple current extensions. There are 5403 such MIPFs yielding 873 different combinations of Hodge numbers. For each of these MIPFs, one considers all possible consistent orientifolds quotients, whose total number is 49 304. Within each such orientifold, the search scans for sets of boundary states (i.e. branes) including the SM a, b, c, d branes (with massless hypercharge or $B - L$ gauge bosons) plus additional boundary states needed to cancel the (typically few dozens of) RR tadpole conditions. For a given choice of visible sector, the search stops once it produces one hidden sector canceling RR tadpoles and leading to no chiral exotics charged under the SM. The scan yields around 116 000 models of type 4, 50 000 of type 2 and 10 500 of type 0, and smaller numbers for the remaining possibilities, with a total of order 200 000. Around 4% of type 0 models have a massive $U(1)_{B-L}$ (but a massless hypercharge boson). This is an enormous set of MSSM-like string vacua, although none of them has yet been subject to detailed phenomenological study. An important limitation in this respect is that the models correspond to particular points in the moduli space of CY orientifolds, so their moduli dependent effective action, relevant for issues like moduli fixing or SUSY-breaking, is difficult to obtain. Also, Yukawa couplings are in principle computable, although they are not yet available. Still there are interesting statistical results for, e.g., gauge couplings (see Section 16.1.3), and number of generations, Higgs fields, and vector-like extra matter (see Section 17.2).

10.7.2 M-theory on G_2 manifolds*

Supersymmetric intersecting D6-brane models can be related to purely geometrical compactifications of M-theory on seven-dimensional spaces of G_2 holonomy; thus M-theory on G_2 manifolds provide a definition of intersecting brane models beyond the perturbative regime. Moreover, there exist G_2 manifolds not admitting reduction to type IIA models, so this setup provides a new playground for model building.

Supersymmetry and G_2 holonomy manifolds

In order to lift intersecting brane models to M-theory, we describe the lift of their basic ingredients. Recall from Section 6.3.3 that N D6-branes lift to a purely geometric background in M-theory, described by an N-center Taub-NUT space (6.30)

$$ds^2 = \frac{V(\vec{x})}{4} d\vec{x}^2 + \frac{V(\vec{x})^{-1}}{4} \left(dx^{10} + \vec{\omega} \cdot d\vec{x}\right)^2,$$

$$V(\vec{x}) = 1 + \sum_{a=1}^{N} \frac{1}{|\vec{x} - \vec{x}_a|}, \qquad \vec{\nabla} \times \vec{\omega} = \vec{\nabla} V(\vec{x}). \tag{10.62}$$

The purely geometric nature of the M-theory lift follows because all supergravity fields sourced by the D6-brane originate from the 11d metric. This also holds for O6-planes, whose lift is a purely geometrical background, known as Atiyah–Hitchin space. For large

$|\vec{x}|$, its geometry is approximately (10.62) with $V = 1 - 4/|\vec{x}|$, modded out by $(\vec{x}, x^{10}) \to (-\vec{x}, -x^{10})$. The formal value $N = -4$ reproduces the O6-plane RR charge in the covering space. At short distances, there are exponentially suppressed corrections to this approximation, which violate the $U(1)$ shift isometry in x^{10}.

Since both D6-branes and O6-planes lift to purely geometric backgrounds in M-theory, 4d type IIA orientifolds with O6-planes and D6-branes lift to purely geometric compactifications of M-theory on a seven-dimensional manifold. Arguing as in Section 7.2.1, 4d $\mathcal{N} = 1$ supersymmetric models must correspond to compactifications on spaces with special holonomy. Indeed, the holonomy group must be G_2, the subgroup of $SO(7)$ with one preserved spinor, as follows from the decomposition

$$SO(7) \to G_2,$$
$$\mathbf{8} \to \mathbf{7} + \mathbf{1}. \tag{10.63}$$

Hence, the lifts of supersymmetric intersecting D6-brane models correspond to compactifications of M-theory on 7d spaces of G_2 holonomy (G_2 manifold, for short).

An alternative definition of G_2 is as the subgroup of $SO(7)$ preserving the following 3-form in $\mathbf{R}^7$

$$\varphi = dx^{127} + dx^{136} + dx^{145} + dx^{246} - dx^{235} - dx^{347} - dx^{567}, \tag{10.64}$$

where $dx^{123} = dx^1 \wedge dx^2 \wedge dx^3$, etc. Hence, G_2 manifolds can be defined by the condition of having a covariantly constant 3-form φ, known as the "associative form"; equivalently, a nowhere vanishing 3-form satisfying $d\varphi = d * \varphi = 0$, with local expression as in (10.64). The preserved spinor ξ and 3-form φ in a G_2 manifold are related, in analogy with (7.9) for CY spaces, by $\varphi_{mnp} = \xi \, \Gamma_m \Gamma_n \Gamma_p \, \xi$, with Γ_m the 7d Dirac matrices. A further property of G_2 manifolds $\mathbf{X}_7$ is that they are simply connected, i.e. $b_1(\mathbf{X}_7) = 0$, so the independent Betti numbers are $b_2(\mathbf{X}_7)$ and $b_3(\mathbf{X}_7)$; these arise naturally in the computation of the massless spectrum, recall Section 7.2.4.

Manifolds of G_2 holonomy are notoriously difficult to construct. Since they have odd dimension, the powerful tools of complex geometry are inapplicable. In addition, there is no analog of Yau's theorem translating the existence of a special holonomy metric into simple topological conditions. Their mathematical study thus requires the explicit construction of metrics, successful so far only in a few non-compact examples. The only known explicit compact G_2 manifolds are constructed as orbifolds; a popular example is the Joyce manifold, a $\mathbf{T}^7/\mathbf{Z}_2^3$ orbifold obtained by quotienting a square $\mathbf{T}^7$, with unit lengths, by the $\mathbf{Z}_2^3$ actions

$$\alpha : (x^1, \ldots, x^7) \quad \to \quad (-x^1, -x^2, -x^3, -x^4, x^5, x^6, x^7),$$
$$\beta : (x^1, \ldots, x^7) \to \left(-x^1, \tfrac{1}{2} - x^2, x^3, x^4, -x^5, -x^6, x^7\right),$$
$$\gamma : (x^1, \ldots, x^7) \to \left(\tfrac{1}{2} - x^1, x^2, \tfrac{1}{2} - x^3, x^4, -x^5, x^6, -x^7\right), \tag{10.65}$$

which clearly preserve (10.64). Each individual generator has 16 fixed $\mathbf{T}^3$'s; these fixed sets are disjoint, so e.g. β and γ swap different α-fixed $\mathbf{T}^3$'s, etc. Hence in the quotient there are $3 \times 16/4 = 12$ independent fixed $\mathbf{T}^3$'s, with local A_1 singularities (i.e. locally $\mathbf{C}^2/\mathbf{Z}_2$, recall Section 6.3.3). It is possible to blow up the singularities to obtain a smooth G_2 manifold with $b_2 = 12$, $b_3 = 43$, although M-theory is well-defined even in the singular orbifold space. In fact, as will soon become clear, 4d physics of M-theory compactifications is particularly interesting for singular G_2 spaces.

M-theory on smooth G_2 manifolds

Let us consider first compactifications of M-theory on G_2 manifolds in the supergravity approximation. This is suitable for compactifications on smooth G_2 manifolds in the large radius limit. The 4d $\mathcal{N} = 1$ massless spectrum can be obtained from KK reduction, in analogy with the CY compactifications in Sections 7.3.1 and 10.1.1. Focusing on the bosonic sector, the massless 11d fields are the metric $G_{\hat{M}\hat{N}}$ and the 3-form $C_{\hat{M}\hat{N}\hat{P}}$. The number of massless modes of the 3-form are given by the Betti numbers of $\mathbf{X}_7$; on the other hand, the 3-form φ allows to establish a one-to-one correspondence between zero modes g_{mn} of the G_2 holonomy metric and harmonic 3-forms ω_{mpq}, via the contraction

$$\omega_{mpq} = \varphi_{n[pq}g^n_{m]}. \tag{10.66}$$

This gives $b_3(\mathbf{X}_7)$ real scalar metric moduli. The KK bosonic zero modes are:

M-theory		Gravity	b_2 Vector	b_3 Chiral
G	$\rightarrow$	$g_{\mu\nu}$		ω_{mnp}
C	$\rightarrow$		$C_{mn\,\mu}$	C_{mnp}

Namely, we get the 4d $\mathcal{N} = 1$ gravity multiplet, coupled to $U(1)^{b_2}$ vector multiplets and b_3 neutral chiral multiplets. The spectrum of M-theory on smooth G_2 manifolds is 4d $\mathcal{N} = 1$, but completely uninteresting from the viewpoint of particle physics model building; there are no non-abelian gauge symmetries, and all chiral multiplets are neutral. This is not surprising, however, as it agrees with Witten's proof, mentioned in Section 1.4, that compactifications of 11d supergravity on smooth 7-manifolds cannot lead to 4d chiral matter. Indeed, the realization of non-abelian gauge symmetry and chiral matter in M-theory requires compactifications on singular geometries, as easily shown from the M-theory lift of D6-brane configurations.

Codimension-4 singularities and gauge symmetry

Recall from Section 6.3.3 that M-theory on an ADE singularity $\mathbf{C}^2/\Gamma$, with $\Gamma \subset SU(2)$, produces an ADE enhanced gauge symmetry, localized on the fixed set of the orbifold. For the A and D cases this is related to the lift of overlapping D6-branes (coincident with an O6-plane for the D case), whereas the E cases have no perturbative type IIA realization.

ADE singularities have four dimensions transverse to the singular locus and are said to be codimension-4.

This realization of non-abelian gauge symmetries also holds in compactifications to 4d. Consider sets of N_a D6-branes on special lagrangian 3-cycles Π_a in a CY (orientifold) compactification. The M-theory lift corresponds to a G_2 manifold, given essentially by an S^1 fibration with degenerations leading to $\mathbf{C}^2/\mathbf{Z}_{N_a}$ singularities over the 3-cycles Π_a on the base. The local singularities lead to enhanced non-abelian gauge symmetries, which upon KK reduction on the 3-cycles lead to 4d gauge symmetries.

Denoting J, Ω_3 the Kähler and holomorphic 3-forms of the underlying CY, the 3-form on the M-theory G_2 space $\mathbf{X}_7$ is locally given by

$$\varphi = \mathrm{Re}\,\Omega_3 + J \wedge dx^{10}. \tag{10.67}$$

The special lagrangian conditions (10.27) on the CY 3-cycles translate into the statement that the singular loci Π of codimension-4 singularities in $\mathbf{X}_7$ satisfy $\varphi|_\Pi = 0$; such loci are called associative 3-cycles, and provide the analog of supersymmetric cycles in CY manifolds. Although a general G_2 manifold is not necessarily related to a type IIA model, the main lessons extend trivially when phrased in this general language. In a G_2 manifold, an ADE singularity along an associative 3-cycle Π leads to an enhanced ADE gauge symmetry on Π. Subsequent KK reduction on Π leads to 4d ADE vector multiplets and $b_1(\Pi)$ adjoint chiral multiplets.

An interesting subtlety is that A_N singularities in M-theory lead to $SU(N)$ rather than $U(N)$ gauge symmetries. This is the strong coupling avatar of the disappearance of $U(1)$s due to $B \wedge F$ couplings in type IIA models. Although the type IIA $U(1)$s remain as perturbative global symmetries, they are violated by non-perturbative brane instantons, see Chapter 13, and can be regarded as directly absent in the strong coupling M-theory setup. Accordingly, our upcoming discussions of M-theory lifts of IIA configurations are a bit cavalier in treating $U(1)$ factors.

For future reference in Section 12.1.3, we describe a simple class of singular G_2 manifolds with non-abelian gauge symmetries (but non-chiral matter content). Consider $\mathbf{X}_7 = (\mathbf{X}_6 \times S^1)/\mathbf{Z}_2$, where $\mathbf{X}_6$ is a CY, and $\mathbf{Z}_2$ acts as $x^{10} \rightarrow -x^{10}$ on S^1, and as an antiholomorphic involution $\mathcal{R}$ on $\mathbf{X}_6$, i.e. $\mathcal{R}(J) = -J$, $\mathcal{R}(\Omega_3) = \overline{\Omega}_3$. This is a G_2 manifold, with (10.67) as its preserved 3-form. The geometry is an S^1 fibration over a base $\mathbf{X}_6/\mathcal{R}$. The fixed points of the $\mathbf{Z}_2$ action lead to $SO(4)$ gauge symmetries localized at the fixed points of $\mathcal{R}$ on the base. By shrinking the S^1, the model corresponds to type IIA on $\mathbf{X}_6$ modded out by $\Omega\mathcal{R}(-)^{F_L}$, with $(-)^{F_L}$ associated to the flip of x^{10} in M-theory. The 3-cycles fixed under $\mathcal{R}$ are wrapped by O6-planes, whose RR tadpole is canceled locally by the introduction of 4 overlapping D6-branes, reproducing the $SO(4)$ gauge factors.

Codimension-7 singularities and chiral fermions

Codimension-4 singularities produce gauge symmetries, but not 4d charged chiral fermions. The latter arise from a different kind of singularity in G_2 spaces, whose structure is suggested from the lift of intersecting D6-brane configurations.

Consider first a local model of two stacks of N and M D6-branes in flat 10d space, intersecting over a 4d space, so they yield a 4d bi-fundamental chiral multiplet. In the transverse 6d the D6-branes span the 3-planes $x^5 = x^7 = x^9 = 0$ and $x^5 = \tan\theta_1 x^4$, $x^7 = \tan\theta_2 x^6$, $x^9 = \tan\theta_3 x^8$, with angles satisfying (10.10). The M-theory lift is a G_2 space with A_{N-1} and A_{M-1} singularities over 3-planes, intersecting at a worse singularity at the origin on the base; since the $\mathbf{S}^1$ fiber also shrinks at this point, the singularity is codimension-7. Chiral matter is thus localized at codimension-7 singularities in G_2 manifolds; this cleverly circumvents the smoothness assumption in the no-go result for chirality mentioned in Section 1.4.

The G_2 holonomy metric in the above example can be approximately described as two intersecting Taub-NUT geometries as follows. Consider the metric (10.62) for $N = N_a + N_b$, with base coordinates $\vec{x} = (x^5, x^7, x^9)$ and center positions $\vec{x}_a = 0$ for $a = 1, \ldots, N + M$; this is the M-theory lift of $N + M$ D6-branes at $\vec{x} = 0$. In order to describe the lift of intersecting D6-branes, we fiber non-trivially the $(N + M)$-center Taub-NUT geometry over the three extra coordinates (x^4, x^6, x^8), by letting the center coordinates vary over them, as

$$\vec{x}_a = 0 \text{ for } a = 1, \ldots, N,$$
$$\vec{x}_a = (\tan\theta_1 x^4, \tan\theta_2 x^6, \tan\theta_3 x^8) \text{ for } a = N + 1, \ldots, N + M.$$

This describes A_{N-1} and A_{M-1} singularities over the loci $x^5 = x^7 = x^9 = 0$ and $x^5 = \tan\theta_1 x^4$, $x^7 = \tan\theta_2 x^6$, $x^9 = \tan\theta_3 x^8$, respectively. They intersect at the origin, at which all centers coincide and enhance the singularity to a local A_{N+M}. The geometry can be described as an "unfolding" of singularities $A_{N+M} \to A_{N-1} \times A_{M-1}$. This unfolding is qualitatively similar to the unfolding in F-theory in Section 11.5.2, although the dimensions of the singular loci and their intersections are different in both cases.

The unfolding picture is particularly useful to read off the gauge representation of the localized chiral matter. The above system can be regarded as a configuration of coincident $N + M$ D6-branes, with $U(N + M)$ symmetry, deformed by the relative rotation breaking the group to $U(N) \times U(M)$, see Figure 10.14; the 4d bi-fundamental chiral matter at the intersection arises from states in the adjoint of $U(N + M)$ which are not in the adjoint of the actual $U(N) \times U(M)$ gauge group,

$$U(N + M) \quad \to \quad U(N) \times U(M),$$
$$\mathbf{Adj} \quad \to \quad (\mathbf{Adj}, 1) + (1, \mathbf{Adj}) + [\, (\mathbf{N}, \overline{\mathbf{M}}) + \text{c.c.}\,]. \tag{10.68}$$

Similarly, in G_2 compactifications the 4d chiral matter content can be read out from the group-theoretical decomposition of the symmetries involved in the unfolding. In the M-theory unfolding of singularities (10.68), the massless chiral multiplet arises from M2-branes on additional collapsed 2-cycles localized at the codimension-7 singularity. These can be constructed in the intersecting Taub-NUT geometry as fibrations of the $\mathbf{S}^1$ over segments joining the two singular loci on the base.

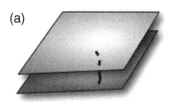 

Figure 10.14 Unfolding of enhanced symmetry loci in M-theory is analogous to tilting parallel D6-branes in type IIA theory.

This picture can be generalized to codimension-7 singularities corresponding to unfoldings $G \to \otimes_r G_r$, of any ADE singularity, with associated symmetry G, into several ADE singularities, with symmetries G_r. The chiral matter representations arising at the codimension-7 singularities can be read off from the decomposition of the adjoint of G under the subgroup $\otimes_r G_r$. For instance, unfolding a local E_6 singularity into a D_5 singularity corresponds to

$$
\begin{aligned}
E_6 &\to && SO(10) \times U(1), \\
\mathbf{78} &\to && \mathbf{45} + \mathbf{1} + [\, \mathbf{16} + \text{c.c.} \,].
\end{aligned}
\tag{10.69}
$$

The intersection supports a 4d chiral multiplet in the spinor representation $\mathbf{16}$ of $SO(10)$. Since it involves a local E_6 enhancement, this local configuration is genuinely M-theoretical and admits no perturbative type IIA realization. This is in agreement with the fact that spinor representations cannot arise from perturbative D-brane constructions. Somewhat more trivially, other unfoldings can provide the matter multiplets for $SU(5)$ GUT model building

$$
\begin{aligned}
SU(6) &\to && SU(5) \times U(1), \\
\mathbf{35} &\to && \mathbf{24} + \mathbf{1} + [\, \mathbf{5} + \text{c.c.} \,]. \\
SO(10) &\to && SU(5) \times U(1), \\
\mathbf{45} &\to && \mathbf{24} + \mathbf{1} + [\, \mathbf{10} + \text{c.c.} \,].
\end{aligned}
\tag{10.70}
$$

for the $\mathbf{10}$, $\bar{\mathbf{5}}$, or their conjugate representations. These unfoldings admit a type IIA interpretation in terms of intersecting D6-branes, with O6-planes for $SO(10)$.

These local rules allow in principle to envisage the possibility to construct G_2 manifolds leading to particle physics models, by engineering singularities over loci with suitable intersections. However, the difficulties in the construction of compact G_2 manifolds short-circuit most attempts at explicit model building. On the other hand, general results for the effective action of M-theory on G_2 manifolds in Section 12.1.3, together with the above intuitions, can be exploited in the construction of M-theory inspired phenomenological models.

11

Type IIB orientifolds

In this chapter we construct orientifolds of type IIB 4d compactifications. There are several constructions in this class, including toroidal orientifolds, systems of D3-branes at singularities, and magnetized D-branes, providing the mirrors of type IIA models in Chapter 10. We also introduce F-theory and its compactifications. These setups provide a rich arena for particle physics model building in string theory.

11.1 Generalities of type IIB orientifold actions

Type IIB orientifolds are obtained by considering IIB theory on a CY $\mathbf{X}_6$ and modding out by $\Omega \mathcal{R}$, where $\mathcal{R}$ is a geometric symmetry acting holomorphically on the complex coordinates on $\mathbf{X}_6$. For instance, the simplest orientifold action is just Ω, with trivial $\mathcal{R}$, thus acting holomorphically $z_i \to z_i$ in a trivial way. These models involve 10d spacetime filling orientifold planes (O9-planes) and open string sectors (D9-branes), and so correspond to type I compactifications on $\mathbf{X}_6$, which are thus included as particular type IIB orientifold compactifications. Additional constructions are obtained for non-trivial actions $\mathcal{R}$, as follows:

- O7/D7 *models*: Consider the orientifold action $\Omega R_i (-1)^{F_L}$, where R_i acts as $z_i \to -z_i$ leaving other complex coordinates invariant; the factor $(-1)^{F_L}$, with F_L being the left-moving fermion number, is necessary for the orientifold action to square to the identity operator. The quotient introduces O7-planes at the fixed points of R_i, namely transverse to the coordinate z_i, and wrapped on a 4-cycle parametrized by the remaining two complex coordinates. Their RR charges must be canceled by D7-branes along similar 4-cycles, usually denoted D7$_i$-branes. These models are closely related to the F-theory constructions in Section 11.5.
- O5/D5 *models*: Consider the orientifold action $\Omega R_i R_j$, with $i \neq j$. The action $R_i R_j$ flips the coordinates z_i, z_j, leaving invariant the remaining complex coordinate z_k, with $i \neq j \neq k \neq i$. The model contains O5-planes transverse to the directions z_i, z_j, and wrapped on a 2-cycle parametrized by z_k. Their RR charge is canceled by D5-branes, denoted D5$_k$-branes, transverse to z_i, z_j and spanning z_k.

- O3/D3 *models*: Consider the orientifold action $\Omega R_1 R_2 R_3 (-1)^{F_L}$, where the geometric action flips all complex coordinates. The models contain O3-planes, sitting at points in the internal space $\mathbf{X}_6$, and whose RR charge is canceled by D3-branes.

In models with O3- or O7-planes, the geometric part of the orientifold action $\mathcal{R}$ acts as $J \to J$, $\Omega_3 \to -\Omega_3$, whereas in models with O5- or O9-planes $\mathcal{R}$ acts as $J \to J$, $\Omega_3 \to \Omega_3$. Certain compactifications can include O-planes and D-branes of different kinds, as long as they preserve a common 4d $\mathcal{N} = 1$ supersymmetry in the compactification. This is the case, e.g., for D7- and D3-branes, since there are spinors satisfying conditions (6.5) for $p = 7, 3$ simultaneously, and for D9- and D5-branes, with preserved spinors satisfying (6.5) for $p = 9, 5$. Type IIB orientifolds with O3- and/or O7-planes (and D3- and/or D7-branes) are referred to as D3/D7-models (also O3/O7-models); similarly for D9/D5-models (also known as O9/O5-models).

In toroidal compactifications (or orbifolds thereof), different type IIB orientifolds are related by T-dualities, as follows from Section 5.3.4; for instance, they relate D9/D5-models and D7/D3-models, as further discussed in Section 11.2.3. In more general CY spaces it is not possible to perform such T-dualities, so D9/D5- and D7/D3-models are inequivalent. Still, the realization of type IIA/IIB mirror symmetry as T-duality, in Section 10.1.2, persists even in the presence of O-planes and D-branes. Hence, any orientifold of type IIB on a CY $\mathbf{X}_6$ can be mapped into an orientifold of type IIA on the mirror CY $\mathbf{Y}_6$. A simple realization of this relation will relate the magnetized brane models of Section 11.4 to the intersecting brane models of Chapter 10.

Closed string spectrum

Before entering a more explicit study of type IIB D-brane models, we conclude with the discussion of the closed string spectrum in type IIB CY orientifold compactifications. This is obtained by considering the closed string spectrum of type IIB on $\mathbf{X}_6$, recall Section 10.1.1, and truncating to states invariant under the orientifold action. For instance, the 4d $\mathcal{N} = 2$ gravity multiplet projects down to the $\mathcal{N} = 1$ gravity multiplet, and the Kähler structure hypermultiplets reduce to $h_{1,1}$ chiral multiplets. To these one has to add the complex dilaton S chiral multiplet. Finally, the $h_{2,1}$ 4d $\mathcal{N} = 2$ vector multiplets truncate to $h_{2,1}^+$ $\mathcal{N} = 1$ vector multiplets and $h_{2,1}^-$ chiral multiplets, where $h_{2,1}^{\pm}$ denote the numbers of (2, 1)-forms even or odd under the geometrical action $\mathcal{R}$ in the orientifold operation. This is as expected from mirror symmetry on the IIA result in Section 10.1.3. Note this is consistent with the IIA result in Section 10.1.3, with the mirror symmetry exchange $h_{1,1}^{\pm} \leftrightarrow h_{2,1}^{\pm}$. The microscopic description of these multiplets in terms of 10d fields is described in Section 12.1.2.

11.2 Type IIB toroidal orientifolds

Toroidal orbifolds provide a particularly simple class of compactifications displaying many of the properties of CY compatifications, while keeping an exact description from the worldsheet viewpoint. Thus it is natural to construct orientifolds of type IIB compactifications on toroidal orbifolds, known as type IIB toroidal orientifolds. The closed string sector can be computed using orbifold techniques analogous to those in Section 8.2, and implementing the orientifold projection as described above; the results will be briefly provided in our examples below. Our main interest, however, lies in the D-brane sector, which we discuss in several illustrative examples.

11.2.1 Ω-orientifold of type IIB on T^6/Z_3

As a case study, we consider the Ω-orientifold of type IIB on T^6/Z_3 – this orbifold already appeared in Section 8.2 in the context of heterotic compactification. Equivalently, this can be regarded as type I on T^6/Z_3. We work this example in some detail, as it provides the basis for many extensions discussed later on.

The starting point is type IIB on the toroidal orbifold T^6/Z_3, recall Figure 8.4. The parent torus is based on the $SU(3)^3$ root lattice, i.e. the product of three 2-tori parametrized by complex coordinates z_k, $k = 1, 2, 3$, with identifications $z_k \simeq z_k + 1$ and $z_k \simeq z_k + e^{2\pi i/3}$, where here we take radii $R_i = 1$ for simplicity of presentation. The orbifold generator θ acts with the twist vector $v = \frac{1}{3}(1, 1, -2)$, i.e.

$$\theta : (z_1, z_2, z_3) \rightarrow (e^{2\pi i v_1} z_1, e^{2\pi i v_2} z_2, e^{2\pi i v_3} z_3). \tag{11.1}$$

Before the orientifold projection, the 4d $\mathcal{N} = 2$ model has the following massless spectrum: the untwisted sector contains the gravity multiplet, and 9 θ-invariant Kähler moduli hypermultiplets, while all complex structure moduli (vector multiplets) are projected out; in the twisted sector, there are 27 fixed points, and each contains one Kähler modulus hypermultiplet. Geometrically, the latter are blowing-up modes replacing the singular points by compact 4-cycles. The Hodge numbers corresponding to the model are $(h_{1,1}, h_{2,1}) = (36, 0)$, recall Table 8.2. Modding out by Ω, the geometric action is trivial and we have $h_{1,1}^- = h_{2,1}^\pm = 0$, so the closed string sector truncates to the 4d $\mathcal{N} = 1$ gravity multiplet, the dilaton chiral multiplet and 36 Kähler moduli chiral multiplets. In particular there are no RR $U(1)$ gauge bosons.

The orientifold quotient by Ω introduces one O9-plane filling 4d Minkowski spacetime times T^6/Z_3, which generates a tadpole canceled by the introduction of D9-branes. In the following we describe the orbifold and orientifold quotients on this D-brane system, in a rather general language useful for later examples. For this purpose we momentarily keep the number N of D9-branes arbitrary, although the familiar type I RR tadpole cancellation constraint (recall Section 4.4.3) will eventually impose $N = 32$.

Algebraic constraints and general spectrum

The system of N D9-branes wrapping T^6, has open string states with structure $\lambda \mathcal{O}|0\rangle$, where λ is the Chan–Paton factor and $\mathcal{O}$ contains the oscillators. For massless states, the latter are efficiently described in the bosonized formulation of Section 4.2.6 in terms of an $SO(8)$ weight vector r:

$$
\begin{array}{ccc}
 & \text{NS} & \text{R} \\
V: & r = (0, 0, 0, \pm) & r = \pm\left(\tfrac{1}{2}, \tfrac{1}{2}, \tfrac{1}{2}, -\tfrac{1}{2}\right), \\[2mm]
\Phi_i: & r = (\underbrace{\pm, 0, 0, 0}_{i}) & r = \pm\left(\underbrace{-\tfrac{1}{2}, \tfrac{1}{2}, \tfrac{1}{2}, \tfrac{1}{2}}_{i}\right) \quad i = 1, 2, 3,
\end{array}
\tag{11.2}
$$

where the underbrace indicates that the different entry is at the ith position, and the last entry encodes the 4d Lorentz quantum numbers. These fields correspond to a 4d $\mathcal{N} = 1$ vector multiplet V, and three chiral multiplets Φ_i, $i = 1, 2, 3$, in the adjoint representation, the D9-brane Wilson line moduli along the three 2-tori.

The open string sector suffers the actions of Ω and the orbifold generator θ. They generate the orientifold group

$$
G_{\text{orient}} = \{1, \theta, \theta^2, \Omega, \Omega\theta, \Omega\theta^2\}.
\tag{11.3}
$$

It is convenient to first impose the orbifold projection, and then mod out by Ω. Open strings do not admit twisted boundary conditions, so the effect of the orbifold is just to truncate onto θ-invariant states, exactly as for untwisted sectors. Using an enlarged $v = (v_1, v_2, v_3, 0)$, the θ action on the operator part $\mathcal{O}$ of states (11.2) is

$$
\theta : |r\rangle \to e^{2\pi i\, r \cdot v} |r\rangle.
\tag{11.4}
$$

This corresponds to $V \to V$, $\Phi_i \to e^{2\pi i\, v_i} \Phi_i$, consistently with the geometric action (11.1). In addition, the action of θ on the Chan–Paton factor is implemented through an $N \times N$ unitary matrix $\gamma_{\theta,9}$, satisfying $\gamma_{\theta,9}^3 = \mathbf{1}_N$. Without loss of generality it can be diagonalized and described as

$$
\gamma_{\theta,9} = \text{diag}\,(\mathbf{1}_{n_0}, \alpha\, \mathbf{1}_{n_1}, \alpha^2\, \mathbf{1}_{n_2}),
\tag{11.5}
$$

where $\alpha = e^{2\pi i/3}$ and $n_0 + n_1 + n_2 = N$. Namely n_a denotes the number of Chan–Paton indices (i.e. D9-branes) with θ-eigenvalue $e^{2\pi i\, a/3}$. The Chan–Paton matrix λ is in the adjoint representation of $U(N)$, so θ acts by conjugation

$$
\lambda \to \gamma_{\theta,9}\lambda\gamma_{\theta,9}^{-1}.
\tag{11.6}
$$

Note that, since θ generates the orbifold group, we have $\gamma_{\theta^k,9} = (\gamma_{\theta,9})^k$. The states surviving the combined geometric and Chan–Paton orbifold projection are

Multiplet	Condition	Fields	
Vector	$\lambda = \gamma_{\theta,9}\,\lambda\,\gamma_{\theta,9}^{-1}$	$U(n_0) \times U(n_1) \times U(n_2)$,	(11.7)
Chiral	$\lambda = e^{2\pi i/3}\,\gamma_{\theta,9}\,\lambda\,\gamma_{\theta,9}^{-1}$	$3\,[\,(\square_0, \overline{\square}_1, \mathbf{1}) + (\mathbf{1}, \square_1, \overline{\square}_2) + (\overline{\square}_0, \mathbf{1}, \square_2)\,]$.	

As in Chapter 10, we omit $U(1)$ charges, since fundamental/anti-fundamental representations of $SU(N)$ always carry charge ± 1 under the corresponding $U(1)$. Note that the gauge group arises from open strings with both ends on the same kind of D9-branes (i.e. with same Chan–Paton θ-eigenvalue), while chiral matter arises from open strings stretching between D9-branes of different kinds; the analogy with intersecting brane models is a consequence of mirror symmetry to type IIA models.

We now impose invariance under Ω. Its action on the operator part $\mathcal{O}$ of massless states is a -1 factor, as in 10d type I in Section 4.4.3. The action on Chan–Paton factors is defined by a matrix $\gamma_{\Omega,9}$, acting as

$$\lambda \rightarrow \gamma_{\Omega,9}\,\lambda^T\,\gamma_{\Omega,9}^{-1}, \tag{11.8}$$

where the transposition implements the fact that worldsheet orientation reversal exchanges ab and ba open strings. As in Section 3.4.3, applying this twice and requiring $\Omega^2 = 1$ leads to the constraint (3.115)

$$\gamma_{\Omega,9} = \pm\gamma_{\Omega,9}^T. \tag{11.9}$$

The negative and positive signs correspond to the Sp and SO projections, respectively; as in 10d type I, the cancellation of the RR 10-form tadpole is possible only for the latter, hence $\gamma_{\Omega,9} = \gamma_{\Omega,9}^T$. One can analogously derive mutual algebraic consistency conditions of the orientifold and orbifold actions, $\gamma_{\Omega,9}$ and $\gamma_{\theta,9}$. In particular, one finds

$$\gamma_{\Omega,9}\,\gamma_{\theta^k,9}\,\gamma_{\Omega,9} = \gamma_{\theta^{-k},9}. \tag{11.10}$$

In other words, the orientifold action exchanges sets of D9-branes with conjugate Chan–Paton phases in $\gamma_{\theta,9}$. Hence an Ω invariant configuration of D9-branes must have $n_1 = n_2$. In the basis (11.5), the matrix $\gamma_{\Omega,9}$ takes the form

$$\gamma_{\Omega,9} = \begin{pmatrix} \mathbf{1}_{n_0} & & \\ & & \mathbf{1}_{n_1} \\ & \mathbf{1}_{n_1} & \end{pmatrix}, \tag{11.11}$$

where absent entries are zero. States surviving the orientifold projection must have Chan–Paton Ω-transformation compensating the -1 factor on $\mathcal{O}$, as follows:

Multiplet	Condition	Fields
Vector	$\lambda = -\gamma_{\Omega,9}\,\lambda^T\,\gamma_{\Omega,9}^{-1}$	$SO(n_0) \times U(n_1)$
Chiral	$\lambda = -\gamma_{\Omega,9}\,\lambda^T\,\gamma_{\Omega,9}^{-1}$	$3\left[\,(\square_0, \overline{\square}_1) + \left(\mathbf{1}, \boxminus_1\right)\,\right].$

$$\tag{11.12}$$

Intuitively, the D9-branes leading to $U(n_0)$ are mapped to themselves under the orientifold action, yielding $SO(n_0)$. The original $U(n_1)$ and $U(n_2)$ factors are swapped, and lead to one invariant linear combination. For chiral multiplets, the fundamental representations of $U(n_a)$ are exchanged with anti-fundamental representations of $U(n_{-a})$, with a understood mod 3. Hence, each $(\square_0, \overline{\square}_1)$ is exchanged with one $(\overline{\square}_0, \square_2)$, leading to an invariant combination, while the $(\square_1, \overline{\square}_2)$ are mapped to themselves and project down to two-index antisymmetric representations.

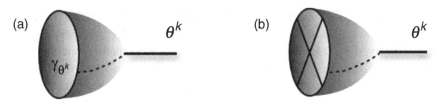

Figure 11.1 Twisted tadpoles for disk and crosscap diagrams. The θ^k-twisted closed string insertion introduces a branch cut, denoted by the dashed line, across which worldsheet fields suffer a θ^k action, in particular a γ_{θ^k} factor for boundary Chan–Paton indices.

The model contains non-abelian gauge interactions, and replicated charged chiral matter. The three sets of chiral multiplets are related to the three complex planes in the 6d space; this will be exploited to build 3-generation models in Section 11.3.3.

Cancellation of RR tadpoles

As in Section 10.3.1 for type IIA models, the consistency of type IIB orientifolds also requires the cancellation of RR tadpoles. In orientifolds of toroidal orbifolds, tadpoles for untwisted RR fields are known as untwisted tadpoles; for instance, in the present $\mathbf{Z}_3$ model, the orientifold action Ω in (11.3) produces a RR 10-form tadpole, which as in 10d type I theory in Section 4.4.3, fixes the total number of D9-branes to be

$$N = n_0 + 2n_1 = 32. \tag{11.13}$$

In addition, there are twisted tadpoles, i.e. associated to RR forms in the twisted sector, arising from disk and crosscap diagrams with one θ^k-twisted sector closed string insertion, see Figure 11.1. These can be computed from factorization of the one-loop Klein bottle, Moebius strip, and cylinder amplitudes, see Section D.2 for details, leading to the result

$$\text{tr } \gamma_{\theta,9} = \text{tr } \gamma_{\theta^2,9} = -4 \quad \Longrightarrow \quad n_0 - 2n_1 = -4. \tag{11.14}$$

It will be shortly interpreted more physically in terms of 4d anomaly cancellation. The tadpole conditions (11.13), (11.14) are solved by $n_0 = 8$, $n_1 = 12$, so the massless spectrum (11.12) becomes

$$
\begin{aligned}
&\text{Vector mult.} &&SO(8) \times U(12), \\
&\text{Chiral mult.} &&3\,[\,(\mathbf{8}, \overline{\mathbf{12}})_{-1} + (\mathbf{1}, \mathbf{66})_2\,],
\end{aligned}
\tag{11.15}
$$

where we have listed the $U(1)$ charges as subscripts. The model is not particularly appealing from the phenomenological viewpoint, but provides a good starting point for more interesting upcoming constructions.

Cancellation of cubic non-abelian anomalies

As in earlier orientifold models, the close relationship between RR tadpole cancellation and anomaly cancellation also holds in type IIB orientifolds. In the above $\mathbf{Z}_3$ example the

RR tadpole conditions can in fact be recovered from anomaly cancellation arguments, as follows. First note that away from the orbifold fixed points, the local dynamics is that of a 10d type I theory, so cancellation of 10d anomalies locally in the bulk requires the presence of 32 D9-branes, namely (11.13). Secondly, cancellation of non-abelian $SU(n_1)^3$ anomalies for the general 4d open string spectrum (11.12) is precisely the condition (11.14). This example nicely illustrates the general lesson for type II orientifolds that RR tadpole cancellation guarantees cancellation of 4d anomalies. The former may however be stronger than the latter, e.g. as some RR tadpole conditions (like the condition $N = 32$) may not relate to 4d anomalies, but rather to cancellation of anomalies in higher dimensions.

In addition to cubic non-abelian anomalies, there are also mixed $U(1)$ gauge and gravitational anomalies; their cancellation involves a 4d Green–Schwarz mechanism, studied in the next section for general IIB toroidal orientifolds.

Other toroidal orientifolds

The above techniques can be used to construct other toroidal orientifolds. Although chiral, they do not have direct phenomenological interest, and we skip their detailed discussion, simply quoting some examples in Table 11.1. An interesting novelty is that the orientifold group for even order orbifolds contains actions like $\Omega R_1 R_2$ in addition to Ω, requiring the introduction of D5-branes in addition to the D9-branes, leading to 9-9, 5-5, and 9-5 open string sectors. For even order $\mathbf{Z}_N$ orbifolds, cancellation of RR tadpoles can in general be achieved only for $(\gamma_\theta)^N = -\mathbf{1}$, usually referred to as "without vector structure." For future reference, the orbifold Chan–Paton action is

$$\gamma_{\theta,9} = \text{diag}\,(\alpha^{1/2}\mathbf{1}_{n_1},\, \alpha^{3/2}\mathbf{1}_{n_2},\, \ldots,\, \alpha^{(2N-1)/2}\mathbf{1}_{n_N})\,, \quad \alpha^{1/2} = e^{\pi i/N}, \tag{11.16}$$

and similarly for D5-branes. The orientifold action exchanges D-branes with opposite phases in γ_θ, hence $n_a = n_{-a+1}$, with a understood modulo N. Recalling from Section 6.1.3 the opposite orientifold projection for D9- and D5-branes, the matrices have the structure (with some obvious reordering of entries)

$$\gamma_{\Omega,9} = \text{diag}\,(\mathbf{1}_{n_1}, \ldots, \mathbf{1}_{n_{N/2}}) \otimes \begin{pmatrix} 0 & 1 \\ 1 & 0 \end{pmatrix};\quad \gamma_{\Omega,5} = i\,\text{diag}\,(\mathbf{1}_{m_1}, \ldots, \mathbf{1}_{m_{N/2}}) \otimes \begin{pmatrix} 0 & 1 \\ -1 & 0 \end{pmatrix}.$$

We note that some orbifolds do not lead to consistent 4d $\mathcal{N} = 1$ orientifolds. Computations generalizing those in section D.2 show that orbifolds containing the twist $\frac{1}{4}(1, 1, -2)$ produce twisted RR tadpoles which cannot be canceled with supersymmetric sets of D-branes.

The $\mathbf{Z}_2 \times \mathbf{Z}_2$ toroidal orbifold produces a model, not listed in Table 11.1, containing D9-branes and D5$_i$-branes wrapped on the ith $\mathbf{T}^2$ for $i = 1, 2, 3$. In the present setup it leads to non-chiral spectra, e.g. with a gauge group $USp(16)^4$ when all D5-branes are located at orbifold (and orientifold) fixed points. Despite its present seeming little phenomenological interest, the addition of D-brane worldvolume magnetic fluxes will induce the appearance

Table 11.1 *Gauge group and charged chiral multiplets in some* $\mathbf{Z}_N$ *and* $\mathbf{Z}_N \times \mathbf{Z}_M$ *4d*
$\mathcal{N} = 1$ *type IIB orientifolds. Only models with at most one set of 5-branes are shown, and*
they are taken to sit all at the fixed point at the origin

Point group Gauge group	(99)/(55) matter	(95) matter
$\mathbf{Z}_3$ $U(12) \times SO(8)$	$3\,(\mathbf{12}, \mathbf{8}_V) + 3\,(\overline{\mathbf{66}}, \mathbf{1})$	–
$\mathbf{Z}_3 \times \mathbf{Z}_3$ $U(4)^3 \times SO(8)$	$(\mathbf{4}, \mathbf{1}, \mathbf{1}, \mathbf{8}_V) + (\overline{\mathbf{4}}, \overline{\mathbf{4}}, \mathbf{1}, \mathbf{1}) + (\mathbf{6}, \mathbf{1}, \mathbf{1}, \mathbf{1})$	–
$\mathbf{Z}_7$ $U(4)^3 \times SO(8)$	$(\mathbf{4}, \mathbf{1}, \mathbf{1}, \mathbf{8}_V) + (\overline{\mathbf{4}}, \overline{\mathbf{4}}, \mathbf{1}, \mathbf{1}) + (\mathbf{6}, \mathbf{1}, \mathbf{1}, \mathbf{1})$ $+ (\overline{\mathbf{4}}, \mathbf{4}, \mathbf{1}, \mathbf{1}) + (\mathbf{1}, \overline{\mathbf{4}}, \mathbf{4}, \mathbf{1}) + (\mathbf{4}, \mathbf{1}, \overline{\mathbf{4}}, \mathbf{1})$	–
$\mathbf{Z}_6$ $(U(6)^2 \times U(4))^2$	$2\,(\mathbf{15}, \mathbf{1}, \mathbf{1}) + 2\,(\mathbf{1}, \overline{\mathbf{15}}, \mathbf{1})$ $+ 2\,(\overline{\mathbf{6}}, \mathbf{1}, \mathbf{4}) + 2\,(\mathbf{1}, \mathbf{6}, \overline{\mathbf{4}})$ $+ (\overline{\mathbf{6}}, \mathbf{1}, \overline{\mathbf{4}}) + (\mathbf{1}, \mathbf{6}, \mathbf{4}) + (\mathbf{6}, \overline{\mathbf{6}}, \mathbf{1})$	$(\mathbf{6}, \mathbf{1}, \mathbf{1}; \mathbf{6}, \mathbf{1}, \mathbf{1}) + (\mathbf{1}, \overline{\mathbf{6}}, \mathbf{1}; \mathbf{1}, \overline{\mathbf{6}}, \mathbf{1}) +$ $(\mathbf{1}, \mathbf{6}, \mathbf{1}; \mathbf{1}, \mathbf{1}, \overline{\mathbf{4}}) + (\mathbf{1}, \mathbf{1}, \overline{\mathbf{4}}; \mathbf{1}, \mathbf{6}, \mathbf{1}) +$ $(\overline{\mathbf{6}}, \mathbf{1}, \mathbf{1}; \mathbf{1}, \mathbf{1}, \mathbf{4}) + (\mathbf{1}, \mathbf{1}, \mathbf{4}; \overline{\mathbf{6}}, \mathbf{1}, \mathbf{1})$
$\mathbf{Z}'_6$ $(U(4)^2 \times U(8))^2$	$(\overline{\mathbf{4}}, \mathbf{1}, \mathbf{8}) + (\mathbf{1}, \mathbf{4}, \overline{\mathbf{8}}) + (\mathbf{6}, \mathbf{1}, \mathbf{1}) +$ $(\mathbf{4}, \mathbf{1}, \mathbf{8}) + (\mathbf{1}, \overline{\mathbf{4}}, \overline{\mathbf{8}}) + (\overline{\mathbf{4}}, \mathbf{4}, \mathbf{1}) + (\mathbf{1}, \mathbf{1}, \mathbf{28})$ $+ (\mathbf{1}, \mathbf{1}, \overline{\mathbf{28}}) + (\mathbf{4}, \mathbf{4}, \mathbf{1}) + (\overline{\mathbf{4}}, \overline{\mathbf{4}}, \mathbf{1})$	$(\overline{\mathbf{4}}, \mathbf{1}, \mathbf{1}; \overline{\mathbf{4}}, \mathbf{1}, \mathbf{1}) + (\mathbf{1}, \mathbf{4}, \mathbf{1}; \mathbf{1}, \mathbf{4}, \mathbf{1}) +$ $(\mathbf{1}, \overline{\mathbf{4}}, \mathbf{1}; \mathbf{1}, \mathbf{1}, \mathbf{8}) + (\mathbf{1}, \mathbf{1}, \mathbf{8}; \mathbf{1}, \overline{\mathbf{4}}, \mathbf{1}) +$ $(\mathbf{4}, \mathbf{1}, \mathbf{1}; \mathbf{1}, \mathbf{1}, \overline{\mathbf{8}}) + (\mathbf{1}, \mathbf{1}, \overline{\mathbf{8}}; \mathbf{4}, \mathbf{1}, \mathbf{1})$
$\mathbf{Z}_3 \times \mathbf{Z}_6$ $(U(2)^6 \times U(4))^2$	$(\mathbf{2}, \mathbf{2}, \mathbf{1}^5) + (\mathbf{1}^2, \mathbf{2}, \mathbf{2}, \mathbf{1}^3) + (\mathbf{1}^4, \mathbf{2}, \mathbf{2}, \mathbf{1}) +$ $(\mathbf{1}^4, \mathbf{2}, \mathbf{1}, \mathbf{4}) + (\mathbf{1}^5, \mathbf{2}, \overline{\mathbf{4}}) + (\mathbf{1}, \mathbf{2}, \mathbf{1}^2, \mathbf{2}, \mathbf{1}^2)$ $+ (\mathbf{1}^3, \mathbf{2}, \mathbf{1}, \mathbf{2}, \mathbf{1}) + (\mathbf{2}, \mathbf{1}^5, \mathbf{4}) + (\mathbf{2}, \mathbf{1}^4, \mathbf{2}, \mathbf{1})$ $+ (\mathbf{1}^2, \mathbf{2}, \mathbf{1}^3, \overline{\mathbf{4}}) + (\mathbf{1}, \mathbf{2}, \mathbf{1}^4, \overline{\mathbf{4}}) +$ $(\mathbf{1}^2, \mathbf{2}, \mathbf{1}, \mathbf{2}, \mathbf{1}^2) + (\mathbf{1}^3, \mathbf{2}, \mathbf{1}^2, \mathbf{4}) + 4\,(\mathbf{1}^7)$	$(\mathbf{2}, \mathbf{1}^6; \mathbf{1}, \mathbf{2}, \mathbf{1}^5) + (\mathbf{1}^2, \mathbf{2}, \mathbf{1}^4; \mathbf{1}^3, \mathbf{2}, \mathbf{1}^3) +$ $(\mathbf{1}^4, \mathbf{2}, \mathbf{1}^2; \mathbf{1}^5, \mathbf{2}, \mathbf{1}) + (\mathbf{1}^5, \mathbf{2}, \mathbf{1}; \mathbf{1}^6, \overline{\mathbf{4}})$ $+ (\mathbf{1}^4, \mathbf{2}, \mathbf{1}^2; \mathbf{1}^6, \mathbf{4})$ + same with groups reversed
$\mathbf{Z}_{12}$ $(U(3)^4 \times U(2)^2)^2$	$(\overline{\mathbf{3}}, \mathbf{1}, \overline{\mathbf{3}}, \mathbf{1}, \mathbf{1}, \mathbf{1}) + (\mathbf{3}, \mathbf{1}, \mathbf{1}, \mathbf{1}, \mathbf{2}, \mathbf{1}) +$ $2(\mathbf{1}, \mathbf{3}, \mathbf{1}, \mathbf{1}, \mathbf{2}, \mathbf{1}) + 2(\mathbf{3}, \mathbf{1}, \mathbf{1}, \mathbf{1}, \mathbf{1}, \mathbf{2}) +$ $2(\mathbf{1}, \mathbf{1}, \mathbf{1}, \mathbf{3}, \mathbf{2}, \mathbf{1}) + 2(\mathbf{1}, \mathbf{1}, \mathbf{3}, \mathbf{1}, \mathbf{1}, \mathbf{2}) +$ $(\mathbf{3}, \mathbf{1}, \mathbf{1}, \mathbf{1}, \mathbf{1}, \mathbf{1})$	$(\overline{\mathbf{3}}, \mathbf{1}^5; \mathbf{1}, \overline{\mathbf{3}}, \mathbf{1}^4) + (\mathbf{1}, \mathbf{3}, \mathbf{1}^4; \mathbf{1}^5, \mathbf{2}) +$ $(\mathbf{3}, \mathbf{1}^5; \mathbf{1}^4, \mathbf{2}, \mathbf{1}) + (\mathbf{1}^2, \mathbf{3}, \mathbf{1}^3; \mathbf{1}^4, \mathbf{2}, \mathbf{1}) +$ $(\mathbf{1}^2, \overline{\mathbf{3}}, \mathbf{1}^3; \mathbf{1}^3, \overline{\mathbf{3}}, \mathbf{1}^2) + (\mathbf{1}^3, \mathbf{3}, \mathbf{1}^2; \mathbf{1}^5, \mathbf{2})$ + same with groups reversed

of chirality, allowing the explicit construction of interesting particle physics models, see Section 11.4.2.

Incidentally, several toroidal orientifolds contain closed string RR $U(1)$ gauge bosons, arising from $h^+_{2,1}$ as described in Section 11.1, see also Section 12.3.3. This is the case for instance for the $\mathbf{Z}'_6$ and $\mathbf{Z}_{12}$ models, with three and one RR $U(1)$ gauge bosons in the twisted sectors, respectively. Possible phenomenological implications of such $U(1)$s will be discussed in Section 16.4.3.

11.2.2 Cancellation of U(1) anomalies

As explained in the above $\mathbf{Z}_3$ example, in general type II orientifolds cubic non-abelian triangle anomalies are automatically zero as a consequence of the RR tadpole cancellation conditions. On the other hand, mixed $U(1)$ gauge and gravitational anomalies cancel by a 4d Green–Schwarz mechanism, mediated by exchange of RR fields. The basic mechanism

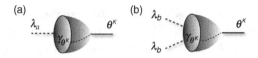

Figure 11.2 Couplings of D-brane worldvolume gauge bosons and θ^k-twisted RR fields mediating the anomaly cancellation mechanism.

is very analogous to that studied in type IIA orientifolds in Section 10.3.1, as expected from mirror symmetry. The relevant Green–Schwarz diagrams, for the particular case of mixed $U(1)$ non-abelian anomalies, are in fact as in Figure 10.6. We turn to the microscopic description of these couplings.

Green–Schwarz mechanism in type IIB toroidal orientifolds

We focus on $\mathbf{Z}_N$ orbifolds, with D9- (and possibly D5-) branes; generalization to other models is straightforward. The Green–Schwarz couplings of D-brane gauge bosons to the RR fields arise from the worldsheet disk diagrams in Figure 11.2. These can be computed using CFT techniques, but their basic structure can be guessed as follows. The closed string fields involved are RR forms associated to the non-trivial cycles in the compactification, which in toroidal orbifolds arise in twisted sectors. From the insertions of closed and open external states in the diagram, the coupling of a $U(1)_a$ gauge boson and a θ^k-twisted 2-form B_k has the structure

$$\mathrm{tr}\left(\gamma_\theta^k \lambda_a\right) B_k \wedge \mathrm{tr}\, F_{U(1)_a}, \tag{11.17}$$

where λ_a is the Chan–Paton wave function for the $U(1)_a$ gauge boson string state and γ_θ^k implements the transformation of Chan-Paton indices in surrounding the insertion of the θ^k-twisted vertex operator; for untwisted RR fields such couplings vanish, since $\mathrm{tr}\, \lambda_a \equiv 0$ from cancellation between D-branes and their images.

Similarly the RR 2-form B_k couples to two non-abelian gauge bosons via

$$\mathrm{tr}\left(\gamma_\theta^k \lambda_b^2\right) \partial^{[\mu} B_k^{\nu\rho]} \left(\omega_3^b\right)_{\mu\nu\rho} \quad \Longrightarrow \quad \mathrm{tr}\left(\gamma_\theta^k \lambda_b^2\right) a_k\, \mathrm{tr}\,(F_b \wedge F_b), \tag{11.18}$$

where ω_3^b is the Chern–Simons 3-form (4.82) for the non-abelian gauge factor G_b. In the second expression we have integrated by parts, dualized the 2-form B_k into a scalar a_k, and used $d\omega_3 = \mathrm{tr}\, F^2$. For mixed gravitational anomalies, there is a similar coupling, with the gravitational Chern–Simons ω_3^{grav} (with $d\omega_3^{\mathrm{grav}} = \mathrm{tr}\, R^2$), and λ_b replaced by the identity matrix to reproduce the universal coupling of gravity.

The full Green–Schwarz amplitude contributing to the mixed $U(1)_a \times G_b^2$ anomaly receives contributions from all twisted RR forms, and has the structure

$$A_{ab} = \frac{-i}{2N} \sum_k C_k^{\alpha\beta}\, \mathrm{tr}\left(\gamma_{\theta^k,\alpha}\, \lambda_a\right)\, \mathrm{tr}\left(\gamma_{\theta^k,\beta}\, \lambda_b^2\right)^*, \tag{11.19}$$

where the factor of $-i/2$ is introduced for later convenience, and the indices α, β indicate the e.g. D9/D5-brane sector containing the gauge factors a, b, respectively. The coefficients $C_k^{\alpha\beta}$ depend on these sectors and are given by

$$C_k^{\alpha\beta} = \prod_{i=1}^{3} (2 \sin \pi k v_i) \qquad \alpha = \beta, \text{ i.e. both in D9s or same D5}_i\text{s},$$

$$C_k^{\alpha\beta} = 2 \sin \pi k v_i \qquad \alpha \neq \beta. \tag{11.20}$$

In the second line, i refers to the unique complex plane with DD or NN boundary conditions for open strings in the ab sector. These crucial numerical factors arise from the computation of the cylinder diagram, see section D.2 for analogous factors.

We can make (11.19) more explicit as follows. For odd order orbifolds, in the Chan–Paton matrix generalizing (11.5) the $\gamma_{\theta k}$-eigenvalue for the ath set of D-branes is $e^{2\pi i \, ak/N}$, with a mod N. Also, the ath gauge boson has Chan–Paton wave function $\lambda_a = \text{diag} (\mathbf{1}_a, -\mathbf{1}_{-a})$, with subindices denoting the location of the non-zero entries, at the D-brane and its image, and the relative minus sign arising from the orientifold exchange $\square_a \leftrightarrow \overline{\square}_{-a}$. For even order orbifolds, the ath set of D-branes has $\gamma_{\theta k}$-eigenvalue $e^{2\pi i \, k(2a-1)/2N}$, and $\lambda_a = \text{diag} (\mathbf{1}_a, -\mathbf{1}_{-a+1})$. We thus have

Odd : $\text{tr} \left(\gamma_\theta^k \lambda_a \right) = 2i \, n_a \sin \frac{2\pi ka}{N}$, **Even :** $\text{tr} \left(\gamma_\theta^k \lambda_a \right) = 2i \, n_a \, e^{-\frac{2\pi i k}{N}} \sin \frac{2\pi ka}{N}$,

$\text{tr} \left(\gamma_\theta^k \lambda_b^2 \right) = \cos \frac{2\pi kb}{N}$, $\text{tr} \left(\gamma_\theta^k \lambda_b^2 \right) = e^{-\frac{2\pi i k}{N}} \cos \frac{2\pi kb}{N}$.

$$\tag{11.21}$$

The factor of n_a in the first line arises from the $U(1)$ normalization. In both cases the Green–Schwarz contribution (11.19) for mixed non-abelian anomalies reads

$$A_{ab} = \frac{1}{N} \sum_k C_k^{\alpha\beta} n_a \sin \frac{2\pi ka}{N} \cos \frac{2\pi kb}{N}. \tag{11.22}$$

For $U(1)_a$-$U(1)_b^2$ anomalies there is a similar result, up to a factor of $2n_b$ from the $U(1)_b$ normalization. For mixed gravitational anomalies, one must include a normalization factor $\frac{3}{4} = \frac{24}{32}$, and sum over all D-branes and images

$$A_a = \frac{3}{4} \frac{1}{N} \sum_b \sum_k C_k^{\alpha\beta} n_a \sin \frac{2\pi ka}{N} 2n_b \cos \frac{2\pi kb}{N}. \tag{11.23}$$

These amplitudes suitably cancel the triangle contributions to mixed anomalies. Their underlying general structure is further clarified in a related setup in Section 11.3.2, but we now illustrate their surprising numerics with an example.

Examples

Let us verify the Green–Schwarz anomaly cancellation in the $\mathbf{Z}_3$ orientifold of Section 11.2.1. The reader may check that the massless spectrum (11.15) has the following

$U(1)$ mixed and cubic triangle anomalies with respect to $SU(12)$, $SO(8)$

$$A_{U(1)}(SU(12), SO(8), U(1), \text{grav}) = (18, -36, 432, 108). \tag{11.24}$$

Using $C_1^{\alpha\beta} = -C_2^{\alpha\beta} = -3\sqrt{3}$, the Green–Schwarz contributions give

$A_{U(1)}(SU(12), SO(8), U(1), \text{grav})$

$$= \tfrac{1}{3} 2 \left(-3\sqrt{3}\right) \left(12 \tfrac{\sqrt{3}}{2}\right) \left(-\frac{1}{2}, 1, 24 \left(-\frac{1}{2}\right), \frac{3}{4}(-4)\right) = (18, -36, 432, 108), \tag{11.25}$$

where the factor 2 comes from the sum over $k = 1, 2$. These contributions exactly cancel those from the triangle diagrams.

Further discussions of anomaly cancellation, also involving factors in different D-brane sectors, will come up in Section 11.3.2. The modifications that anomaly cancellation implies for gauge kinetic functions are discussed in Section 12.3.2. The couplings of (anomalous or not) $U(1)$ gauge bosons to RR 2-forms, the corresponding masses and FI terms are discussed in Section 12.4. In coming examples of particle physics models it will be necessary to ensure that the hypercharge generator does not have this kind of coupling.

11.2.3 Additional model building ingredients

There are additional ingredients which can enrich the model building possibilities of type IIB toroidal orientifolds. These include general Wilson lines and arbitrary D-brane positions (which are related by T-duality); also the introduction of antibranes, which relaxes the constraints of supersymmetry.

T-duality and D3/D7-brane models

In addition to models with D9- and/or D5-branes, there exist type IIB orientifold models whose untwisted RR tadpole conditions require the introduction of D3- and/or D7$_i$-branes. As mentioned in Section 11.1, they correspond to orientifold actions $\Omega R_i (-1)^{F_L}$, or $\Omega R_1 R_2 R_3 (-1)^{F_L}$, with $R_i : z_i \to -z_i$. For toroidal orientifolds, these D3/D7-models are related by T-duality to D9/D5-models. Indeed, T-duality along the two directions in $(\mathbf{T}^2)_i$ changes the orientifold operation by a factor of $R_i(-1)^{F_L}$, and maps D-branes wrapped on $(\mathbf{T}^2)_i$ into unwrapped D-branes, and vice versa. For instance, an $\Omega R_1 (-1)^{F_L}$ orientifold, with D7$_1$-branes, can be T-dualized along the two directions of $(\mathbf{T}^2)_1$ into an Ω orientifold with D9-branes. Similarly, T-duality along all $\mathbf{T}^6$ directions maps it to an $\Omega R_2 R_3$ orientifold with D5$_1$-branes. Note that such T-dualities can also relate models with D9-branes to models with D5-branes, and vice versa. These T-dualities relate toroidal models with same orbifold groups, but different orientifold action and D-brane content.

Despite this relation, it is important to understand the direct construction of D3/D7-models, since the equivalence of D9/D5- and D3/D7-models holds only for toroidal orientifolds, but not for more general CYs. Also, even within toroidal orientifolds, some features may be more manifest in some particular picture; for instance, the upcoming interpretation of Wilson lines as D-brane positions in the T-dual.

For this purpose we briefly review the direct construction of the $\mathbf{Z}_3$ orientifold with D3-branes, i.e. with orientifold action $\Omega R_1 R_2 R_3 (-1)^{F_L}$. There are four orientifold fixed points on each two-torus, leading to 64 O3-planes, with charge $-\frac{1}{2}$ according to (5.52), and so requiring 32 D3-branes for (untwisted) RR tadpole cancellation. The O3-plane at the origin in all three complex planes is at an orbifold fixed point, and generates a twisted RR tadpole, to be canceled by D3-branes sitting at the origin, with the condition

$$\mathrm{tr}\, \gamma_{\theta,3} = \mathrm{tr}\, \gamma_{\theta^2,3} = -4, \tag{11.26}$$

where $\gamma_{\theta,3}$ is the orbifold Chan–Paton matrix, which has the general structure (11.5). There are no further points fixed simultaneously under the orbifold and orientifold action, and so no further RR twisted tadpoles to be canceled with D3-branes.

For simplicity we choose all 32 D3-branes at the origin. The computation of the open string spectrum just requires imposing the orbifold and orientifold projection on the $\mathcal{N} = 4$ spectrum in the parent toroidal model. The orbifold projection leads to the $U(n_0) \times U(n_1) \times U(n_2)$ theory in (11.7), and the subsequent orientifold projection enforces $n_1 = n_2$ and produces the $SO(n_0) \times U(n_1)$ theory in (11.12). The integers n_0, n_1 are constrained by the tadpole conditions, solved by

$$\gamma_{\theta,3} = \mathrm{diag}\,(\mathbf{1}_8, \alpha\,\mathbf{1}_{12}, \alpha^2 \mathbf{1}_{12}), \tag{11.27}$$

with $\alpha = e^{2\pi i/3}$. The resulting spectrum is the $SO(8) \times U(12)$ theory (11.15). Clearly these computations as well as other physical results are identical to those for the D9-brane model in Section 11.2.1, as expected from T-duality.

Wilson lines

Recall from Section 5.3.3 that Wilson lines on wrapped D-branes are backgrounds for the internal components of their worldvolume gauge fields. In toroidal orientifolds, they can be described in a language analogous to that in Section 8.3, as embedding the space group translations (i.e. the shifts in the 6d lattice defining the underlying toroidal geometry) in the gauge group. For each toroidal direction i, the space group shift is embedded on the ath wrapped D-brane stack through a matrix $\gamma_{W_i,a}$ acting on its Chan–Paton indices. Wilson lines along toroidal directions related by orbifold rotations are actually not independent; in general, except for 2-tori rotated only by order 2 twists, there is one choice of Wilson line per each 2-torus (for factorized $\mathbf{T}^6$).

The consistency conditions for $\gamma_{W_i,a}$ follow from the requirement that the gauge action is a representation of the space group. Consider for example the $\mathbf{Z}_3$ model with D9-branes, and introduce one non-trivial Wilson line $\gamma_{W,9}$ along the first 2-torus, and no Wilson lines along the other two. In the first 2-torus there are three fixed points, associated to the space group elements $\theta, \theta + e_1, \theta + e_1 + e_2$, where e_1, e_2 generate the $\mathbf{T}^2$ shifts. The three fixed points are associated with the Chan–Paton actions $\gamma_{\theta,9}, \gamma_{\theta,9}\gamma_{W,9}$ and $\gamma_{\theta,9}\gamma_{W,9}^2$. These local Chan–Paton matrices around each fixed point must obey algebraic constraints similar to those of $\gamma_{\theta,9}$, namely

$$(\gamma_{\theta,9} \, \gamma_{W,9})^3 = \left(\gamma_{\theta,9} \, \gamma_{W,9}^2 \right)^3 = \mathbf{1}_3,$$

$$\gamma_{\Omega,9} \left(\gamma_{\theta^k,9} \, \gamma_{W,9}^n \right) \gamma_{\Omega,9}^{-1} = \left(\gamma_{\theta^k,9} \, \gamma_{W,9}^n \right)^{-1}, \ \text{etc.} \tag{11.28}$$

They must also satisfy the twisted RR tadpole cancellation constraints, at each point fixed under the orbifold and orientifold actions, namely

$$\mathrm{tr}\, \gamma_{\theta,9} = \mathrm{tr}\, \gamma_{\theta^2,9} = -4,$$

$$\mathrm{tr}\, \gamma_{\theta,9} \, \gamma_{W,9} = \mathrm{tr}\, \gamma_{\theta^2,9} \, \gamma_{W,9}^2 = -4,$$

$$\mathrm{tr}\, \gamma_{\theta,9} \, \gamma_{W,9}^2 = \mathrm{tr}\, \gamma_{\theta^2,9} \, \gamma_{W,9} = -4. \tag{11.29}$$

A simple set of solutions is obtained for $\gamma_{W,9}$ commuting with $\gamma_{\theta,9}$, for instance

$$\gamma_{\theta,9} = (\mathbf{1}_8; \ \alpha \, \mathbf{1}_{12}; \ \alpha^2 \, \mathbf{1}_{12}), \tag{11.30}$$

$$\gamma_{W,9} = (\mathbf{1}_{8-2k}, \ \alpha \, \mathbf{1}_k, \ \alpha^2 \, \mathbf{1}_k; \ \mathbf{1}_{12-2k}, \ \alpha \, \mathbf{1}_k, \ \alpha^2 \, \mathbf{1}_k; \ \mathbf{1}_{12-2k}, \ \alpha \, \mathbf{1}_k, \ \alpha^2 \, \mathbf{1}_k),$$

with $\alpha = e^{2\pi i/3}$, and k an integer $0 \le k \le 3$. In this basis, $\gamma_{\Omega,9}$ is a symmetric matrix exchanging conjugate orbifold and Wilson lines phases.

Recalling Section 5.3.3, Wilson lines shift the momentum along their toroidal direction, and give additional contributions to the 4d mass of states. The surviving massless spectrum can be obtained from the original one with no Wilson line, by imposing a Wilson line projection condition[1] on the Chan–Paton wave function

$$\lambda = \gamma_{W,9} \, \lambda \, \gamma_{W,9}^{-1}. \tag{11.31}$$

Massless states are thus invariant under orientifold, orbifold and Wilson line actions.

For illustration consider the above example (11.30). Imposing first the Wilson line projection, the massless spectrum splits into the three eigenspaces of $\gamma_{W,9}$ with eigenvalues $1, \alpha, \alpha^2$; these are three 4d $\mathcal{N} = 4$ decoupled sectors with groups $U(32 - 6k)$, $U(3k)$, and $U(3k)$. The $U(32 - 6k)$ sector suffers the orbifold and orientifold projections, leading to a structure (11.12)

$$SO(8 - 2k) \times U(12 - 2k)$$
$$3\,[\,(\square_0, \overline{\square}_1) + (\mathbf{1}, \boxminus_1)\,]. \tag{11.32}$$

The two $U(3k)$ sectors have conjugate $\gamma_{W,9}$ eigenvalues, and so are exchanged by the orientifold action. We can focus on a single one, and impose just the orbifold projection, which produces a structure (11.7), namely

$$U(k)_2 \times U(k)_3 \times U(k)_4$$
$$3\,[\,(\square_2, \overline{\square}_3, \mathbf{1}) + (\mathbf{1}, \square_3, \overline{\square}_4) + (\overline{\square}_2, \mathbf{1}, \square_4)\,]. \tag{11.33}$$

---

[1] Equivalently, imposing simultaneous invariance under all local Chan–Paton actions $\gamma_{\theta,9} \gamma_{W,9}^n$.

The complete spectrum is the combination of (11.32), (11.33), and nicely illustrates the efficient use of Wilson lines to introduce decoupled gauge sectors; they will shortly receive a T-dual interpretation as separated D-brane stacks.

The above example is a particular case of a more general Wilson line, depending on continuous parameters, and in general not commuting with $\gamma_{\theta,9}$. Consider extracting from (11.30) k sets of 3 D9-branes, each with θ action diag $(1, \alpha, \alpha^2)$ and Wilson line $W =$ diag (α, α, α), and their orientifold images. The Wilson line piece for each such set can be generalized to

$$
W = \begin{pmatrix} w & a & a \\ a & w & a \\ a & a & w \end{pmatrix} \quad \text{with } w^3 - 2a^2w + 2a^3 = 1, \tag{11.34}
$$

where the constraint arises from the first condition in (11.28) restricted to the set of three D9-branes; the Wilson line W depends on one independent complex parameter. Taking the same W for the k sets for simplicity, the full matrices are

$$
\gamma_{\theta,9} = \text{diag} \left(\mathbf{1}_{8-2k},\ \alpha\,\mathbf{1}_{12-2k},\ \alpha^2\,\mathbf{1}_{12-2k};\ (1, \alpha, \alpha^2) \otimes \mathbf{1}_k;\ (1, \alpha^2, \alpha) \otimes \mathbf{1}_k \right),
$$
$$
\gamma_{W,9} = \text{diag} \left(\mathbf{1}_{8-2k},\ \alpha\,\mathbf{1}_{12-2k},\ \alpha^2\,\mathbf{1}_{12-2k};\ W \otimes \mathbf{1}_k;\ W^* \otimes \mathbf{1}_k \right). \tag{11.35}
$$

For $w = \alpha$, $a = 0$ we recover the Wilson line (11.30), up to reordering. For general Wilson line, the spectrum is given by the $SO(8-2k) \times U(12-2k)$ sector (11.32), and a decoupled $\mathcal{N} = 4$ $U(k)$ gauge theory. The continuous Wilson line will shortly receive a very intuitive interpretation in terms of T-dual D-brane positions.

D-brane positions

In previous examples with D-branes not filling all 10d dimensions, we have restricted to configurations with all D-branes coincident and at an orbifold fixed point. We now consider more general distributions of D-branes in the transverse dimensions. As announced, this is T-dual to the introduction of Wilson lines discussed above. In order to preserve the orbifold and orientifold symmetry, the D-branes must either sit at fixed points, or distribute in stacks mapped to each other by these symmetries. In the former case, the spectrum on the D-branes is obtained by imposing the relevant projections; in the latter case, we must impose only the projections of actions mapping each stack to itself, since other actions simply introduce orbifold and/or orientifold images of the D-brane system.

As concrete example, consider the above $\Omega R_1 R_2 R_3(-1)^{F_L}$ orientifold of $\mathbf{T}^6/\mathbf{Z}_3$, with the 32 D3-branes now taken at more general positions. On each complex plane there are three orbifold fixed points, denoted by "•," "×," "∘," recall Figure 8.4. The point "•" at $z = 0$ is also fixed under the orientifold action, while the points "×" and "∘" are exchanged by the orientifold action. Consider locating all D3-branes at $z_2 = z_3 = 0$, but distributed in the first torus with $32 - 6k$ D3-branes at point "•," $3k$ at point "∘" and their $3k$ orientifold images at point "×." The twisted RR tadpoles for D3-branes at $z_1 = 0$, i.e. the point "•," and their solution, are

$$\text{tr } \gamma_{\theta,3,\bullet} = \text{tr } \gamma_{\theta^2,3,\bullet} = -4 \;\Rightarrow\; \gamma_{\theta,3,\bullet} = \text{diag} \left(\mathbf{1}_{8-2k}, \alpha \, \mathbf{1}_{12-2k}, \alpha^2 \, \mathbf{1}_{12-2k} \right). \quad (11.36)$$

The orientifold Chan–Paton matrix exchanges entries with conjugate θ eigenvalues, and the massless spectrum is given by the $SO(8 - 2k) \times U(12 - 2k)$ sector (11.32). The $3k$ D3-branes at point "$\circ$" (and the images at "$\times$") are fixed under the orbifold, but not under the orientifold. The twisted RR tadpole cancellation conditions do not involve contributions from O3-planes, and give

$$\text{tr } \gamma_{\theta,3,\circ} = \text{tr } \gamma_{\theta^2,3,\circ} = 0 \;\Rightarrow\; \gamma_{\theta,3,\circ} = \text{diag} \left(\mathbf{1}_k, \alpha \mathbf{1}_k, \alpha^2 \mathbf{1}_k \right), \quad (11.37)$$

(and similarly for the image point "$\times$"). The solution follows from $1 + \alpha + \alpha^2 = 0$. The spectrum is obtained by imposing just the orbifold projection (and no orientifold projection), and gives the $U(k)^3$ theory (11.33). The complete model is clearly T-dual to the above D9-brane model with the Wilson line (11.30).

The above D3-brane distribution is a particular case of the following more general choice. Keep $32 - 6k$ D3-branes at the origin with the Chan–Paton matrix (11.36), and locate k D3-branes at generic points $p_1, \ldots, p_k$ in $\mathbf{T}^6$, along with their orbifold and orientifold images. Their spectrum is that of k separated D3-branes in flat space, i.e. $\mathcal{N} = 4$ $U(1)^k$ vector multiplets, possibly enhanced to $U(k)$ for coincident branes. This is T-dual to the non-commuting Wilson line (11.35).

The picture of D-brane positions illuminates the different gauge enhancements for coincident D3-branes at or away from orbifold and orientifold fixed points. For instance, consider three D3-branes forming a $\mathbf{Z}_3$ invariant set, brought from the bulk into an orbifold (but not orientifold) fixed point (like point "$\times$" above). Initially they produce a 4d $\mathcal{N} = 4$ $U(1)$ sector, but this is enhanced at the fixed point: the three D3-branes are related by cyclic permutation, so when coincident their Chan–Paton matrix can be diagonalized as $\gamma_\theta = \text{diag}(1, \alpha, \alpha^2)$, which produces a $U(1)^3$ theory of the kind (11.33). Namely, there are chiral multiplets Φ^i_{12}, Φ^i_{23}, Φ^i_{31}, for $i = 1, 2, 3$, where Φ^i_{ab} has charges $(+1, -1)$ under $U(1)_a \times U(1)_b$.

Conversely, a set of D3-branes at an orbifold point and with a Chan–Paton matrix including all possible phases, can move off to the bulk as an orbifold-invariant system of D3-branes. The transition $U(1)^3 \to U(1)$ is described as a Higgsing in the 4d worldvolume field theory, in terms of flat directions parametrized by

$$\left\langle \Phi^i_{12} \right\rangle = \left\langle \Phi^i_{23} \right\rangle = \left\langle \Phi^i_{31} \right\rangle \equiv z^i. \quad (11.38)$$

As suggested by the notation, these vevs correspond to the D3-brane locations in the ith transverse coordinate (with the orbifold point taken as the origin). For generic vevs, the worldvolume theory is broken down to a $\mathcal{N} = 4$ $U(1)$ theory, reproducing the microscopic D3-brane result. Such descriptions as effective field theory Higgsing exist for other motions of D3-branes away from loci of enhanced gauge symmetry.

The decoupling of gauge sectors by use of locality in the transverse dimensions is at the heart of many properties of D-brane models, and will be further exploited in the construction of local models in Section 11.3.

A 3-generation $SU(5)$ GUT toy model

The freedom to choose general D-brane locations can be used to improve the particle physics model building prospects of toroidal type IIB orientifolds. As a simple example, we construct a (not fully realistic, yet interesting) 3-family $SU(5)$ GUT based on the $\mathbf{T}^6/\mathbf{Z}_3$ orientifold with D3-branes. The model admits equivalent, but less intuitive, T-dual constructions using e.g. D9-branes with Wilson lines.

We take the $\Omega R_1 R_2 R_3 (-1)^{F_L}$ orientifold of $\mathbf{T}^6/\mathbf{Z}_3$, and locate 11 D3-branes at the origin with

$$\gamma_{\theta,3} = \mathrm{diag}\,(\mathbf{1}_1, \alpha \mathbf{1}_5, \alpha^2 \mathbf{1}_5), \qquad (11.39)$$

which satisfies $\mathrm{tr}\,\gamma_{\theta,3} = -4$. There are no other twisted RR tadpoles to cancel, so the remaining 21 D3-branes need not (although can) be at orbifold fixed points. We take, e.g., 12 of them in two six-plets invariant under the orbifold and orientifold actions, and locate the nine remaining at O3-planes away from orbifold points, in a configuration whose details we skip for simplicity. The 11 D3-branes at the origin are fixed under the orbifold and orientifold actions, and produce a spectrum (11.12) with $n_0 = 1$, $n_1 = 5$. The gauge factor $SO(1)$ is actually trivial, so we have

$$\begin{array}{ll} \text{Vector mult.} & U(5), \\ \text{Chiral mult.} & 3\,(\overline{\mathbf{5}}_{-1} + \mathbf{10}_{+2}). \end{array} \qquad (11.40)$$

The remaining D3-branes produce decoupled (and uninteresting) sectors. The spectrum (11.40) is chiral but anomaly-free, with mixed $U(1)$ anomalies canceled by Green–Schwarz couplings, which render the $U(1)$ massive. The low-energy spectrum thus reduces to a 3-family $SU(5)$ GUT-like theory; the model is however not realistic, since there are neither adjoint fields for GUT gauge symmetry breaking, nor $\mathbf{5} + \overline{\mathbf{5}}$ multiplets to play as electroweak Higgs doublets. Moreover, it cannot generate a large top mass, since the up-type Yuwaka coupling $\mathbf{10} \cdot \mathbf{10} \cdot \mathbf{5}$ is forbidden by the $U(1)$ symmetry in $U(5)$, which remains as a perturbatively exact global symmetry (while the down-type $\mathbf{10} \cdot \overline{\mathbf{5}} \cdot \overline{\mathbf{5}}$ is allowed, and indeed present). This is a general problem of $SU(5)$ GUTs in perturbative type II orientifolds, although can be overcome using non-perturbative instanton effects (Section 13.3.3) or in F-theory models (Section 11.5). Despite these issues, this example provides a simple toy model illustrating the additional possibilities opened up by the new ingredients of Wilson lines/general D-brane locations.

Models with antibranes

An additional possibility to relax and generalize the construction rules of type IIB orientifolds is to allow for the introduction of anti-D-branes, using the tools introduced in Section 6.5. Models with antibranes break supersymmetry in the open string sector, although the closed string sector is still supersymmetric at leading order. This suggests an interesting source of supersymmetry breaking, with potential phenomenological applications, e.g. using hidden sectors of anti-D-branes.

An important question in non-supersymmetric models is stability. The open strings stretching between branes and antibranes have a non-supersymmetric GSO projection, and can lead to tachyonic modes at small inter-brane distances. However, branes and antibranes can be stuck at orbifold or orientifold fixed points, and may not be free to approach each other and annihilate; or, even in the case of coincident branes and antibranes, the projections can remove the open string tachyons. Thus brane–antibrane systems in toroidal orientifolds have improved stability properties, as compared with flat space ones.

Even when annihilation of branes and antibranes is not possible, the models have interesting non-supersymmetric dynamics, with possible applications for moduli stabilization. In supersymmetric models, cancellation of RR charges implies the cancellation of tadpoles for NSNS fields, as expected for a BPS configuration; in non-supersymmetric models including anti-D-branes, NSNS tadpoles are in general non-vanishing, and can introduce non-trivial potentials for the 4d moduli. This scalar potential may compete or combine with other sources of moduli stabilization present in the model, and indeed anti-D-branes play an important role in moduli stabilization and the construction of de Sitter string vacua in Section 15.3.1.

11.3 D-branes at singularities

The interplay of locality and D-brane physics motivates the construction of local models of D-branes at CY singularities, which allow for a bottom-up approach to the embedding of the SM into string theory.

11.3.1 Local models and bottom-up

In all our previous discussions we have constructed phenomenological string models as global compactifications. This is clearly necessary in order to obtain 4d gravity; moreover, in some setups like heterotic compactifications, the massless 4d spectrum of gauge fields and charged matter depends on global properties of the compactification CY space. The situation is different in compactifications with lower-dimensional D-branes, e.g. D3- or D7-brane models, since the worldvolume gauge theory is determined by the local geometry around them. This is already manifest in the D3-brane models in Section 11.2.3. In fact, one can take the compactification volume to infinity, corresponding to the decoupling of gravity, producing a non-compact local model with the same D3-brane physics. Also, in general CY models with GUT-like theories arising from wrapped branes, e.g. 7-branes on 4-cycles, one may even argue that the (admittedly mild) hierarchy of gauge and gravitational scales $M_{\mathrm{GUT}} \ll M_p$ hints to a decoupling of the D-branes from the bulk; in such models, the unification scale is the wrapped cycle compactification scale, while M_p is controlled by the whole internal volume, so $M_{\mathrm{GUT}} \ll M_p$ implies that the wrapped cycle is small compared with the total 6d volume, and can hence be well-described in a local model.

The decoupling property endows models of D-branes with a modular structure, widely exploited in model building. Different sets of D-branes may be located in different regions

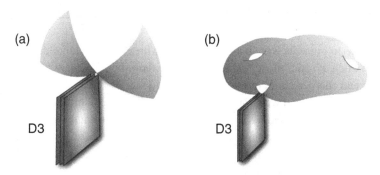

Figure 11.3 Pictorial representation of the bottom-up approach to the embedding of the Standard Model in string theory. In (a) the SM is realized in a local system of D-branes in a non-compact CY space, e.g. a D3/D7-brane system at a singularity. In (b) this local configuration is embedded in a fully fledged global compactification. Many interesting phenomenological issues depend essentially only on the local structure and are quite insensitive to the details of the compactification. The global model in general contains additional structures (like other branes, antibranes, fluxes) not shown in the figure.

of the internal CY space, with many of their properties depending only on the local geometry around them, insensitive to the global structure of the compactification. Such properties can thus be studied in much simpler local models of D-branes in non-compact CY spaces. The study of local models can lead to a simple construction of interesting building blocks, realizing diverse physically relevant gauge sectors: embedding of the Standard Model, supersymmetry breaking, inflation, and so on. Such local D-brane systems must be regarded as part of an eventual full-fledged global compactification, but in a first approximation they can be taken as decoupled from each other; clearly, the interactions among such sectors plays a very important role in the dynamics of the compact model in which they are simultaneously embedded.

This is the *bottom-up* approach to string theory phenomenological model building, illustrated in Figure 11.3. One first constructs a simple D-brane building block realizing a particular gauge sector of physical interest, for instance a SM sector. In a second step one subsequently embeds this local sector in a consistent global compactification, which defines a complete string theory vacuum. The key point is that many phenomenologically interesting questions are independent of the global compactification, and can already be studied at the level of the first step, thereby allowing for a simpler determination of robust and model-independent features.

In this section we apply this bottom-up strategy to the construction of local models of type IIB D3/D7-brane systems, leading to gauge sectors very close to the MSSM (or simple extensions thereof). These could be subsequently embedded in global models, like toroidal orientifolds or otherwise. Another realization of this idea is described in Section 11.5 using 7-branes on 4-cycles. More generally, the intuition of localization of physics subsectors on D-brane worldvolumes is far-reaching, and permeates many discussions of D-brane model building.

11.3.2 D3/D7-branes at orbifold singularities

We would like to consider systems of D3-branes whose worldvolume theory potentially allows for particle physics model building. Namely they should contain non-abelian gauge interactions and charged chiral fermions in replicated families. We will later focus on models with three families, and with gauge group and chiral matter content as close as possible to the SM, but already the requirement of getting a chiral spectrum is very constraining. As just argued, the massless spectrum on the D3-branes is determined by the local geometry around their location. If they sit at a smooth point in the internal space, the local geometry is just flat space and the D3-brane worldvolume theory is $\mathcal{N} = 4$ supersymmetric, and so automatically non-chiral; this holds even if the global compactification preserves only $\mathcal{N} = 1$ supersymmetry. To overcome this problem, we consider D3-branes at singular points of the compactification space. These are geometrically singular points, yet well-behaved in string theory, as in the prototypical example of orbifold singularities, already appeared earlier in the book. As already shown in Section 11.2.3, systems of D3-branes at orbifold fixed points lead to chiral gauge sectors.

Construction of the models: spectrum and superpotential

We mainly focus on 4d $\mathcal{N} = 1$ D-brane systems at abelian orbifold singularities. The local orbifold geometries are $\mathbf{C}^3 / \Gamma$, with Γ a $\mathbf{Z}_N$ or $\mathbf{Z}_N \times \mathbf{Z}_M$ abelian subgroup of $SU(3)$. Notice that the local model need not belong to a global toroidal orbifold, so we are not restricted to crystallographic actions, and can allow for a larger set of orbifold groups. Also the singularities need not sit on top of orientifold planes in the global model, so we consider local models with no orientifold projection.

In addition to the D3-branes, the local model may include other D-branes, partially wrapped on some of the non-compact direction in the CY. In 4d $\mathcal{N} = 1$ models we may include D7$_i$-branes transverse to the complex direction z_i, and filling 4d Minkowski space. These D7-branes pass through the singularity and lead to additional 4d charged chiral multiplets from D3-D7 open strings, which are localized at the D3-brane position, and thus are independent of the global compactification. On the other hand, the D7-D7 open strings lead to fields propagating on higher-dimensional subspaces, and are in principle sensitive to the global properties of the compact model. Thus the spectrum of massless 77 states may not be reliably computed just in terms of the local geometry. Still, some properties of the 77 sector are relevant to the local model, e.g. the D7-brane gauge groups behave as global symmetries of the local model, which can be either gauged or mildly broken by compactification effects in the full global model.

We thus consider supersymmetric systems of D3- and D7$_i$-branes at abelian orbifolds $\mathbf{C}^3 / \Gamma$, and focus on $\Gamma = \mathbf{Z}_N$, the generalization to $\mathbf{Z}_N \times \mathbf{Z}_M$ being straightforward. The generator θ has twist $v = (v_1, v_2, v_3) = \frac{1}{N}(a_1, a_2, a_3)$, with $a_i \in \mathbf{Z}$ and $\sum_i a_i \in 2\mathbf{Z}$, and D3-brane Chan–Paton action

$$\gamma_{\theta,3} = \mathrm{diag}\left(\mathbf{1}_{n_0}, \, e^{2\pi i \frac{1}{N}} \mathbf{1}_{n_1}, e^{2\pi i \frac{2}{N}} \mathbf{1}_{n_2}, \ldots, \, e^{2\pi i \frac{N-1}{N}} \mathbf{1}_{n_{N-1}}\right). \tag{11.41}$$

Namely, there are n_a D3-branes with Chan–Paton θ-phase $e^{2\pi i a/N}$. If present, the action on, e.g., D7$_3$-branes (i.e. spanning the 4-cycle $z_3 = 0$) is

$$\gamma_{\theta,7_3} = \text{diag}\left(\mathbf{1}_{u_0}, e^{2\pi i/N}\,\mathbf{1}_{u_1}, \ldots, e^{2\pi i(N-1)/N}\,\mathbf{1}_{u_{N-1}}\right) \qquad \text{for } a_3 = \text{even},$$

$$\gamma_{\theta,7_3} = \text{diag}\left(e^{\pi i\frac{1}{N}}\,\mathbf{1}_{u_0}, e^{2\pi i\frac{3}{N}}\,\mathbf{1}_{u_1}, \ldots, e^{2\pi i\frac{2N-1}{N}}\,\mathbf{1}_{u_{N-1}}\right) \qquad \text{for } a_3 = \text{odd}. \quad (11.42)$$

The difference between the even and odd a_3 cases will be clear later on. There are similar matrices for D7$_1$- and D7$_2$-branes, if present in the model.

Let us turn to the computation of the spectrum and superpotential of these D3/D7-brane systems. In the D3-D3 open string sector, the parent theory of D3-branes in flat space is a 4d $\mathcal{N} = 4$ $U(n)$ gauge theory, i.e. one $\mathcal{N} = 1$ vector multiplet V and three adjoint chiral multiplets Φ^i, $i = 1, 2, 3$, with superpotential (2.61)

$$W_{33} = \epsilon_{ijk}\,\text{tr}\,\Phi^i\,\Phi^j\,\Phi^k. \qquad (11.43)$$

The orbifold projection for vector and chiral multiplets is, in analogy with (11.7),

$$V \qquad \lambda = \gamma_{\theta,3}\,\lambda\,\gamma_{\theta,3}^{-1},$$

$$\Phi^i \qquad \lambda = e^{-2\pi i a_i/N}\,\gamma_{\theta,3}\,\lambda\,\gamma_{\theta,3}^{-1}. \qquad (11.44)$$

The resulting spectrum in the 33 sector is

$$\text{Vector mult.} \qquad \prod_{a=0}^{N-1} U(n_a), \qquad (11.45)$$

$$\text{Ch. mult. } \Phi^i_{a,a+a_i} \qquad \sum_{a=0}^{N-1}\left[\,(\mathbf{n}_a, \bar{\mathbf{n}}_{a+a_1}) + (\mathbf{n}_a, \bar{\mathbf{n}}_{a+a_2}) + (\mathbf{n}_a, \bar{\mathbf{n}}_{a+a_3})\,\right],$$

with subindices defined mod N. The superpotential is just the truncation of (11.43),

$$W_{33} = \sum_a \text{tr}\left(\Phi^1_{a,a+a_1}\,\Phi^2_{a+a_1,a+a_1+a_2}\,\Phi^3_{a+a_1+a_2,a}\right). \qquad (11.46)$$

Replication of chiral multiplets requires several complex planes to have identical twist entry a_r (mod N). Hence triplication of families is possible only for the $\mathbf{Z}_3$ orbifold $\frac{1}{3}(1, 1, -2)$, which will play an important role in model building, see Section 11.3.3. For the moment, however, we proceed with the general discussion.

Consider now the D3-D7$_i$ and D7$_i$-D3 open strings, which are similar to the D9/D5-brane system in Section 6.1.3. In the parent theory the massless sector is a 4d $\mathcal{N} = 2$ bi-fundamental hypermultiplet, i.e. two $\mathcal{N} = 1$ chiral multiplets Φ^{37_i}, $\Phi^{7_i 3}$ in conjugate bi-fundamentals. Recalling Section 2.4.1, they couple to 33 states through a superpotential

$$W_{37_i} = \Phi^i\,\Phi^{37_i}\,\Phi^{7_i 3}, \qquad (11.47)$$

with implicit contraction of gauge indices. For example, for D7$_3$-branes, the θ-projection is

$$\lambda_{37_3} = e^{-i\pi a_3/N}\,\gamma_{\theta,3}\,\lambda\,\gamma_{\theta,7_3}^{-1} \quad, \quad \lambda_{7_3 3} = e^{-i\pi a_3/N}\,\gamma_{\theta,7_3}\,\lambda\,\gamma_{\theta,3}^{-1}. \qquad (11.48)$$

For odd a_3 the phase is not an Nth root of unity, but this compensates against the relative phase between (11.41), (11.42). The spectrum from the 37+73 sector reads

$$a_3 = \text{even} \rightarrow \sum_{a=0}^{N-1} \left[\left(\mathbf{n}_a, \bar{\mathbf{u}}_{a+\frac{1}{2}a_3} \right) + \left(\mathbf{u}_a, \bar{\mathbf{n}}_{a+\frac{1}{2}a_3} \right) \right],$$

$$a_3 = \text{odd} \rightarrow \sum_{a=0}^{N-1} \left[\left(\mathbf{n}_a, \bar{\mathbf{u}}_{a+\frac{1}{2}(a_3-1)} \right) + \left(\mathbf{u}_a, \bar{\mathbf{n}}_{a+\frac{1}{2}(a_3+1)} \right) \right]. \tag{11.49}$$

There is a similar structure for other D7$_i$-branes, resulting in fields denoted by $\Phi^{37_i}_{a,a+a_i/2}$, $\Phi^{7_i3}_{a,a+a_i/2}$, with 37$_i$-7$_i$3-33 superpotential from the truncation of (11.47):

$$W_{37} = \Phi^i_{a,a+a_i} \, \Phi^{37_i}_{a,a+a_i/2} \, \Phi^{7_i3}_{a+a_i/2,a+a_i}. \tag{11.50}$$

In situations with the same orbifold twist in several complex planes, the local geometry has enhanced isometries. This allows to include more general D7-branes, for instance D7-branes wrapping the 4-plane $c_1z_1 + c_2z_2 + c_3z_3 = 0$ in $\mathbf{C}^3/\mathbf{Z}_3$. Their spectrum and interactions are easily obtained by simple generalization.

RR tadpoles and anomalies

We can now compute the non-abelian and mixed gauge anomalies of the above spectra, and establish the constraints for consistent anomaly-free theories. As usual in type II orientifolds, cancellation of non-abelian anomalies is closely related to RR tadpole cancellation. Mixed $U(1)$ anomalies are canceled by a 4d Green–Schwarz mechanism, essentially as in Section 11.2.2.

Consider a general D3/D7$_i$ system, with even a_i for simplicity of notation, and denote u^i_a the number of entries with phase $e^{2\pi i\, a/N}$ in $\gamma_{\theta,7_i}$. The $SU(n_a)$ non-abelian anomaly cancellation conditions are

$$SU(n_a)^3 \quad \rightarrow \quad \sum_{i=1}^{3} \left(n_{a+a_i} - n_{a-a_i} \right) + \sum_{i=1}^{3} \left(u^i_{a+\frac{1}{2}a_i} - u^i_{a-\frac{1}{2}a_i} \right) = 0. \tag{11.51}$$

To relate these conditions to the twisted RR tadpole cancellation, note that

$$n_a = \frac{1}{N} \sum_{k=1}^{N} e^{-2\pi i \frac{ka}{N}} \, \text{tr}\, \gamma_{\theta^k,3} \quad ; \quad u^i_a = \frac{1}{N} \sum_{k=1}^{N} e^{-2\pi i \frac{ka}{N}} \, \text{tr}\, \gamma_{\theta^k,7_i}, \tag{11.52}$$

and substitute in (11.51) to obtain

$$\frac{2i}{N} \sum_{k=1}^{N} e^{-2\pi i \frac{ka}{N}} \left[\sum_{i=1}^{3} \sin\left(2\pi \frac{ka_i}{N} \right) \text{tr}\, \gamma_{\theta^k,3} + \sum_{i=1}^{3} \sin\left(\pi \frac{ka_i}{2N} \right) \text{tr}\, \gamma_{\theta^k,7_i} \right] = 0. \tag{11.53}$$

Using the identity

$$\sum_{i=1}^{3} \sin\left(2\pi \frac{ka_i}{N} \right) = 4 \prod_{i=1}^{3} \sin\left(\pi \frac{ka_i}{N} \right), \tag{11.54}$$

the Fourier-transformed anomaly cancellation condition is recast as

$$\left[\prod_{i=1}^{3} 2\sin\left(\pi\frac{ka_i}{N}\right)\right] \operatorname{tr}\gamma_{\theta^k,3} + \sum_{i=1}^{3} 2\sin\left(\pi\frac{ka_i}{N}\right)\operatorname{tr}\gamma_{\theta^k,7_i} = 0. \qquad (11.55)$$

These are precisely the twisted RR tadpole cancellation conditions for D3- and D7$_r$-brane disks, as can be computed in analogy with Section D.2.

Note that in certain cases some conditions are trivially satisfied. For instance, in models with only D3-branes, twists θ^k with fixed planes have zero θ^k-twisted anomaly/tadpole, due to vanishing sine factors. This nicely dovetails the geometric interpretation of tadpoles: θ^k-twisted RR form fields are supported on the non-compact fixed plane, so its flux-lines can escape to infinity, and no charge cancellation condition must be imposed.

The mixed $U(1)$ anomalies are canceled by a Green–Schwarz mechanism mediated by closed string twisted RR fields, essentially identical to that in Section 11.2.2. The only difference is the absence of orientifold projection, hence

$$\operatorname{tr}\left(\gamma_{\theta^k,3}\lambda_a\right) = n_a\, e^{2\pi i\frac{ka}{N}}; \quad \operatorname{tr}\left(\gamma_{\theta^k,3}\lambda_b^2\right) = e^{2\pi i\frac{kb}{N}}, \qquad (11.56)$$

and similarly for D7-branes. For both $U(1)_a$ and $SU(n_b)$ on D3-branes, the mixed anomaly becomes, after use of (11.51)

$$A_{ab} = \frac{1}{2}n_a \sum_{i=1}^{3}\left(\delta_{b,a+a_i} - \delta_{b,a-a_i}\right). \qquad (11.57)$$

Using $\delta_{ab} = \frac{1}{N}\sum_k e^{2\pi i k(a-b)}$ and (11.54), this can be rewritten as

$$A_{ab} = \frac{-i}{2N}\sum_{k=1}^{N-1}\left\{\left[\prod_{i=1}^{3} 2\sin\left(\pi\frac{ka_i}{N}\right)\right]n_a\exp\left(2i\pi\frac{ka}{N}\right)\exp\left(-2i\pi\frac{kb}{N}\right)\right\}, \qquad (11.58)$$

which using (11.56) has the precise structure to be canceled by the Green–Schwarz formula (11.19) with coefficients C_k as in (11.20). Cancellation of mixed anomalies between groups on D3- and D7-branes, and cubic $U(1)$ anomalies, works similarly. Note that although some of these are 4d global symmetries in the non-compact model, they can eventually become gauged in a full global compactification, so their cancellation mechanism must be at work even in the non-compact model. Finally, mixed $U(1)_a$-gravitational anomalies cancel automatically, since they are proportional to the number of fundamentals $\square_a$ (carrying $U(1)_a$ charge $+1$) minus anti-fundamentals $\overline{\square}_a$ (carrying charge -1), which cancels by the RR tadpole conditions (just like for cubic non-abelian anomalies).

An interesting fact is that for D3-branes at singularities, there is always a non-anomalous $U(1)$ linear combination

$$Q_{\text{diag.}} = \sum_{a=0}^{N-1}\frac{Q_a}{n_a}. \qquad (11.59)$$

This "diagonal" $U(1)$ couples to the RR untwisted 2-form, which is projected out in global models with O3-planes, so generically remains massless. It will play a prominent role in the particle physics models of Section 11.3.3. In orbifolds with fixed complex planes, there exist other non-anomalous $U(1)$ linear combinations. Although non-anomalous, in general they couple to the corresponding twisted RR 2-forms, and hence can become massive, in analogy to the discussion on page 312. This is however somewhat model-dependent, since they may remain massless if the corresponding local RR 2-form field does not extend globally to the full compact model; this is a version of a mechanism exploited in Section 11.5.3 to keep massless a phenomenologically interesting $U(1)$ gauge boson.

11.3.3 Particle physics models from D3-branes at $\mathbf{C}^3/\mathbf{Z}_3$

The above systems of D3/D7-branes at singularities can be used to build explicit particle physics models. In order to recover family triplication, we focus on the $\mathbf{C}^3/\mathbf{Z}_3$ geometry, with generator twist $v = \frac{1}{3}(1, 1, -2)$, and Chan–Paton matrices

$$\gamma_{\theta,3} = \text{diag}\left(\mathbf{1}_{n_0}, \alpha\mathbf{1}_{n_2}, \alpha^2\mathbf{1}_{n_3}\right); \quad \gamma_{\theta,7_1} = -\text{diag}\left(\mathbf{1}_{u_0^1}, \alpha\mathbf{1}_{u_2^1}, \alpha^2\mathbf{1}_{u_3^1}\right),$$

$$\gamma_{\theta,7_3} = \text{diag}\left(\mathbf{1}_{u_0^3}, \alpha\mathbf{1}_{u_1^3}, \alpha^2\mathbf{1}_{u_2^3}\right); \quad \gamma_{\theta,7_2} = -\text{diag}\left(\mathbf{1}_{u_0^2}, \alpha\mathbf{1}_{u_2^2}, \alpha^2\mathbf{1}_{u_3^2}\right), \quad (11.60)$$

with $\alpha = e^{2\pi i/3}$. The notation for the D7$_i$-brane entries differs slightly from the previous section, to make the symmetry between the three complex planes more manifest. The full 4d $\mathcal{N} = 1$ spectrum is given by

$$
\begin{array}{ll}
\mathbf{33} & U(n_0) \times U(n_1) \times U(n_2) \\
& 3\left[\,(\mathbf{n}_0, \overline{\mathbf{n}}_1) + (\mathbf{n}_1, \overline{\mathbf{n}}_2) + (\mathbf{n}_2, \overline{\mathbf{n}}_0)\,\right] \\
\mathbf{37}_i, \mathbf{7}_i\mathbf{3} & \left(\mathbf{n}_0, \overline{\mathbf{u}}_1^i\right) + \left(\mathbf{n}_1, \overline{\mathbf{u}}_2^i\right) + \left(\mathbf{n}_2, \overline{\mathbf{u}}_0^i\right) \\
& + \left(\mathbf{u}_0^i, \overline{\mathbf{n}}_1\right) + \left(\mathbf{u}_1^i, \overline{\mathbf{n}}_2\right) + \left(\mathbf{u}_2^i, \overline{\mathbf{n}}_0\right).
\end{array}
\quad (11.61)
$$

The complete superpotential is

$$W = \sum_{a=0}^{2} \sum_{i,j,k=1}^{3} \epsilon_{ijk} \text{tr}\left(\Phi^i_{a,a+1}\Phi^j_{a+1,a+2}\Phi^k_{a+2,a}\right)$$

$$+ \sum_{a=0}^{2} \sum_{i=1}^{3} \text{tr}\left(\Phi^i_{a,a+1}\Phi^{37_i}_{a+1,a+2}\Phi^{7_i3}_{a+2,a}\right). \quad (11.62)$$

The twisted tadpole/anomaly cancellation conditions read

$$\text{tr}\,\gamma_{\theta,7_3} - \text{tr}\,\gamma_{\theta,7_1} - \text{tr}\,\gamma_{\theta,7_2} + 3\text{tr}\,\gamma_{\theta,3} = 0. \quad (11.63)$$

The signs from the sine prefactors cancel those in (11.60), so both are consistently ignored in what follows.

Table 11.2 *Spectrum of the* $SU(3) \times SU(2) \times U(1)$ *model, with non-abelian and* $U(1)^9$ *quantum numbers. The first three* $U(1)$*s come from the D3-brane sector. The remaining six arise in the* $D7_r$*-brane sectors, and are written in two columns. The last column gives the charges under the hypercharge generator* (11.65)

Matter fields	Q_3	Q_2	Q_1	$Q_{u_1^i}$	$Q_{u_2^i}$	Y
33 sector						
$3\,(\mathbf{3},\mathbf{2})$	1	−1	0	0	0	1/6
$3\,(\mathbf{\bar{3}},\mathbf{1})$	−1	0	1	0	0	−2/3
$3\,(\mathbf{1},\mathbf{2})$	0	1	−1	0	0	1/2
37$_r$ sector						
$(\mathbf{3},\mathbf{1})$	1	0	0	−1	0	−1/3
$(\mathbf{\bar{3}},\mathbf{1};\mathbf{2}')$	−1	0	0	0	1	1/3
$(\mathbf{1},\mathbf{2};\mathbf{2}')$	0	1	0	0	−1	−1/2
$(\mathbf{1},\mathbf{1};\mathbf{1}')$	0	0	−1	1	0	1

Standard Model and branes at a C^3/Z_3 singularity

We now use this general framework to construct a local D3/D7-brane model with spectrum close to the MSSM. To produce the SM gauge group, we take six D3-branes with

$$\gamma_{\theta,3} = \text{diag}\,(\mathbf{1}_3, \alpha\mathbf{1}_2, \alpha^2\mathbf{1}_1). \tag{11.64}$$

The gauge group is $U(3) \times U(2) \times U(1)$, with two anomalous $U(1)$s and the diagonal anomaly-free linear combination (11.59) given by

$$Q_Y = \frac{1}{3}Q_{(3)} + \frac{1}{2}Q_{(2)} + Q_{(1)}, \tag{11.65}$$

with subindices denoting the rank of the corresponding $U(n)$ factor.

The simplest way to satisfy the tadpole conditions (11.63) is to introduce only one set of D7-branes, e.g. D7$_3$-branes, with Chan–Paton embedding $u_0^3 = 0$, $u_1^3 = 3$, $u_2^3 = 6$. This leads to a large D7$_3$-brane flavour group $U(3) \times U(6)$, which may be eventually broken by compactification effects. An alternative to obtain a smaller group on the D7-branes is to use all three kinds of D7-branes, e.g. $u_0^i = 0$, $u_1^i = 1$, $u_1^i = 2$, for $i = 1, 2, 3$. Each kind of D7-brane then carries a $U(1) \times U(2)$ flavour group.

The spectrum for this latter model is given in Table 11.2. The last column shows the charges under the anomaly-free combination (11.65). Very remarkably, it gives the correct hypercharge assignments for SM matter fields; this elegant appearance of hypercharge is a very appealing feature of this model. The model contains some additional fields beyond the minimal SM content; most notably, extra sets of triplets and doublets, with vector-like quantum numbers under the SM group. In the absence of some global projection or obstruction, there are additional states from the $\mathbf{7_r7_r}$ sectors transforming like $(\mathbf{1}, \mathbf{2}')$ under the

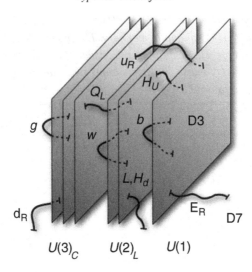

Figure 11.4 D-brane configuration of a SM $\mathbf{Z}_3$ orbifold model. Six D3-branes are located on a $\mathbf{Z}_3$ orbifold singularity and the symmetry is broken to $U(3) \times U(2) \times U(1)$. For the sake of visualization the D3-branes are depicted at different locations, although they are in fact coincident. Open strings with both ends on the same set of D3-branes lead to gauge bosons; those stretching between different sets produce left-handed quarks, right-handed U-quarks, and one set of Higgs fields. Leptons and right-handed D-quarks correspond to open strings between D3- and D7-branes (whose worldvolumes span the whole figure).

flavour groups and coupling to the extra triplets. Vevs for those states render the extra color triplets massive. On the other hand, the appearance of extra doublets is in fact expected from the IIB version of the argument at the introduction of Section 10.5. Still, it is remarkable that such a simple configuration produces a spectrum so close to the SM. A pictorial representation of this type of model is given in Figure 11.4.

Left–Right Symmetric Models and the C^3/Z_3 Singularity

We can also use these systems to engineer 3-family models with a left–right (LR) symmetric gauge group. The resulting models have a quite economical massless spectrum and are probably the simplest semi-realistic local models from branes at singularities. We take seven D3-branes with

$$\gamma_{\theta,3} = \mathrm{diag}\,(\mathbf{1}_3, \alpha\mathbf{1}_2, \alpha^2\mathbf{1}_2),\tag{11.66}$$

so the D3-brane gauge group is $U(3) \times U(2)_L \times U(2)_R$. The non-anomalous $U(1)$ combination (11.59) is

$$Q_{B-L} = -2\left(\frac{1}{3}\,Q_3 + \frac{1}{2}\,Q_L + \frac{1}{2}\,Q_R\right),\tag{11.67}$$

Table 11.3 *Spectrum of* $SU(3) \times SU(2)_L \times SU(2)_R \times U(1)_{B-L}$ *model,*
with their $U(1)^9$ *quantum numbers. The first three* $U(1)$*s arise from the*
D3-brane sector. The remaining six come from the $D7_r$*-brane sectors, and*
are written in two columns. The last column gives the $(B - L)$ *charges*
under (11.67)

Matter fields	Q_3	Q_L	Q_R	$Q_{U_1^i}$	$Q_{U_2^i}$	$B - L$
33 sector						
$3(\mathbf{3}, \mathbf{2}, \mathbf{1})$	1	−1	0	0	0	1/3
$3(\mathbf{\bar{3}}, \mathbf{1}, \mathbf{2})$	−1	0	1	0	0	−1/3
$3(\mathbf{1}, \mathbf{2}, \mathbf{2})$	0	1	−1	0	0	0
37$_i$ sector						
$(\mathbf{3}, \mathbf{1}, \mathbf{1})$	1	0	0	−1	0	−2/3
$(\mathbf{\bar{3}}, \mathbf{1}, \mathbf{1})$	−1	0	0	0	1	2/3
$(\mathbf{1}, \mathbf{2}, \mathbf{1})$	0	1	0	0	−1	−1
$(\mathbf{1}, \mathbf{1}, \mathbf{2})$	0	0	−1	1	0	1

and will elegantly play the role of $B - L$. To cancel RR tadpoles, we introduce for instance
D7$_i$-branes, $i = 1, 2, 3$ with the symmetric choice $u_0^i = 0$, $u_1^i = u_2^i = 1$. Each set of
D7$_i$-branes leads to a $U(1)^2$ symmetry. The spectrum for this model, with the relevant
$U(1)$ quantum numbers is given in Table 11.3.

The spectrum is quite close to a minimal LR model. There are extra color triplets in the
37$_i$ sectors, which may again get massive by giving vevs to singlets in the 7$_i$7$_i$ sectors,
to which they couple. The final spectrum is that of a LR model with three generations of
both fermions and Higgs multiplets. This model has the interesting property that gauge
couplings unify at an intermediate scale of order 10^{12} GeV (see Section 16.1.2), if the
left–right symmetry is spontaneously broken at a scale of order 1 TeV; an obvious signature for this kind of model would be the discovery at the LHC of right-handed W
bosons and a Z' associated to $U(1)_{B-L}$. The superpotential (11.46) contains quark Yukawa
couplings, giving masses to only two quark families at the renormalizable level. Lepton
Yukawa couplings are forbidden by the anomalous $U(1)$s in the model (see the charges in
Table 11.2), although they could be induced by non-perturbative D-brane instanton effects
(see Chapter 13).

As mentioned in the introduction, the local models just described must be regarded
as part of a full global compactification. There exist global embeddings of the above
SM and LR models, in which the cancellation of D3- and D7-brane charges is achieved
either by introducing antibranes (leading to non-supersymmetric models) or by orientifold
planes or generalizations (leading to supersymmetric models). We direct the reader to the
Bibliography for further details on these compact models.

11.3.4 Other systems of D3-branes at singularities*

We now briefly overview D3-branes at more general singularities, with an eye on their application to particle physics model building.

Non-abelian orbifolds and the Δ_{27} theory

A natural generalization is to consider orbifold singularities with non-abelian orbifold group $\Gamma \subset SU(3)$. The computation of the spectrum is just a natural group-theoretical generalization of the abelian case.

The $\mathbf{C}^3/\Gamma$ singularity is obtained by modding $\mathbf{C}^3$ by an action of Γ specified by a three-dimensional representation $\mathcal{R}^{(3)}$ of Γ, i.e. a set of 3×3 matrices $\mathcal{R}^{(3)}(g)$ implementing the action of $g \in \Gamma$ on $\mathbf{C}^3$. The group Γ can also act on the Chan–Paton matrices of the D3-branes, which can thus be classified in terms of their transformation properties under Γ. Denote $\{\mathcal{R}_a\}$, $a = 1, \ldots, N$ the irreducible representations of Γ, with N the number of conjugacy classes of Γ. The action on D3-brane Chan–Paton indices is given by a representation $\mathcal{R}^{\mathrm{CP}}$, decomposing as

$$\mathcal{R}^{\mathrm{CP}} = \sum_{a=1}^{N} n_a \mathcal{R}_a. \tag{11.68}$$

The field theory on D3-branes at $\mathbf{C}^3/\Gamma$ is obtained by projecting the 4d $\mathcal{N} = 4$ theory of D3-branes in flat space with the combined geometric plus Chan–Paton action of Γ. The invariance conditions under $g \in \Gamma$ are

$$\text{Vector mult. :}\ \lambda_V = \mathcal{R}^{CP}(g)\,\lambda_V\,\mathcal{R}^{CP}(g)^{-1} \qquad \rightarrow \qquad \prod_{a=1}^{N} U(n_a) \tag{11.69}$$

$$\text{Chiral mult. :}\ \lambda_{\Phi^i} = \mathcal{R}^{(3)}(g)_{ij}\,\mathcal{R}^{CP}(g)\,\lambda_{\Phi^j}\,\mathcal{R}^{CP}(g)^{-1} \qquad \rightarrow \qquad \oplus_{a,b}\,\mathbf{a}^3_{ab}\,(\square_a, \overline{\square}_b)$$

where λ_V, λ_{Φ^i} are the Chan–Paton matrices for the vector and chiral multiplets, respectively. The so-called adjacency matrix $\mathbf{a}^3_{ab}$ is defined by the decomposition

$$\mathcal{R}^{(3)} \otimes \mathcal{R}_a = \sum_{b=1}^{N} \mathbf{a}^3_{ab}\,\mathcal{R}_b, \tag{11.70}$$

and determines the multiplicities of the corresponding bi-fundamental representations. The superpotential is obtained by substituting the surviving chiral multiplets Φ^i_{ab} in the $\mathcal{N} = 4$ superpotential, but we will not need it explicitly.

The system is subject to twisted RR tadpole cancellation. The simplest choice satisfying this constraint is the regular representation for the Chan–Paton action $n_a = \dim \mathcal{R}_a$. Other choices of D3-brane Chan–Paton actions may be rendered consistent upon addition of suitable D7-branes, which we do not discuss. As usual, cancellation of RR tadpoles guarantees cancellation of 4d chiral anomalies, involving a Green–Schwarz mechanism for mixed $U(1)$ anomalies.

As illustrative example within this class, we consider the Δ_{27} orbifold group, which provides the only non-abelian orbifold with triplication of some bi-fundamental multiplet. Its action on $\mathbf{C}^3$ is given by a $\mathbf{Z}_3 \times \mathbf{Z}_3$ abelian action, with generator twists $\frac{1}{3}(1, 0, -1)$ and $\frac{1}{3}(0, 1, -1)$, and a further generator acting as a permutation $(z_1, z_2, z_3) \rightarrow (z_3, z_1, z_2)$. The D3-brane gauge group and matter content is

$$\prod_{a=1}^{9} U(n_a) \times U(n_{10}) \times U(n_{11})$$

$$\sum_{a=1}^{9} (\mathbf{n}_a, \overline{\mathbf{n}}_{10}) + 3 \,(\mathbf{n}_{10}, \overline{\mathbf{n}}_{11}) + \sum_{a=1}^{9} (\mathbf{n}_{11}, \overline{\mathbf{n}}_a). \tag{11.71}$$

For D3-branes in a single copy of the regular representation we have $n_{10} = n_{11} = 3$, $n_1 = \cdots = n_9 = 1$, leading to a non-anomalous spectrum. The gauge group is $U(3) \times U(3)' \times U(1)^9$, and there are chiral multiplets Q in the $(\mathbf{3}, \mathbf{3}'; 0)$, L_a in the $(\mathbf{1}, \mathbf{3}'; (-1)_a)$, and $\overline{Q}_a$ in the $(\overline{\mathbf{3}}, \mathbf{1}'; (+1)_a)$, where the last entry denotes the $U(1)_a$ charges for $a = 1, \ldots, 9$. Turning on closed string moduli vevs (corresponding to partial geometrical blow-up of the singularity) introduces FI terms, as described in Section 12.4, which can trigger a Higgsing down to $SU(3) \times SU(2)$ times a number of $U(1)$ factors, yielding three SM generations, plus a few exotics.

Non-orbifold toric singularities and dimer diagrams

There is a broad class of 4d $\mathcal{N} = 1$ chiral D3-brane gauge theories that can be obtained from the so-called dimer diagrams. They arise from D3-branes at toric singularities, i.e. non-compact CY singularities defined using toric geometry, introduced in Section 7.2.3.

The gauge group, matter content and superpotential of these gauge theories can be encoded in a simple diagram, known as brane tiling or dimer diagram. This is a tiling of $\mathbf{T}^2$ (or a periodic tiling of $\mathbf{R}^2$) defined by a graph with black and white nodes, with no edges connecting nodes of the same color, and with an even number of edges bounding each face. The dictionary is as follows:

- Each face F_a in the dimer diagram corresponds to a $U(n_a)$ gauge factor in the field theory. The choices of ranks are constrained by RR tadpole cancellation, or equivalently by anomaly cancellation, but we keep them arbitrary for convenience.
- Each edge E_{ab} separating faces F_a, F_b corresponds to a chiral multiplet Φ_{ab} in the bi-fundamental $(\square_a, \overline{\square}_b)$. The ordering $a \rightarrow b$ is determined by the prescription that, e.g., edges should be crossed anticlockwise/clockwise around black/white nodes.
- Each node at the convergence of edges $E_{a_1 a_2}, E_{a_2 a_3}, \ldots, E_{a_k a_1}$, ordered with the above prescription, corresponds to a term $W = \pm \mathrm{tr} \,(\Phi_{a_1 a_2} \Phi_{a_2 a_3} \cdots \Phi_{a_k a_1})$ in the superpotential. The sign choice is determined by the color of the node.

For illustration, consider the two dimer diagrams in Figure 11.5. The gauge theory for Figure 11.5(a) is known as conifold theory, for reasons explained below. Adapting notation to the literature, its gauge group, matter content and superpotential are

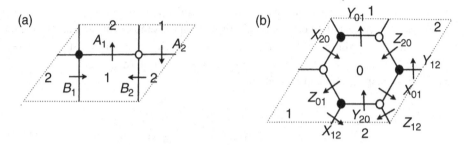

Figure 11.5 Two examples of dimer diagrams. They describe the worldvolume gauge theory on D3-branes at the conifold singularity (a) and at the $\mathbf{C}^3/\mathbf{Z}_3$ orbifold (b).

$$
\begin{aligned}
\text{Gauge} \qquad & U(n_1) \times U(n_2) \\
\text{Chiral} \qquad & A_1, A_2 : \ (\Box_1, \overline{\Box}_2) \\
& B_1, B_2 : \ (\overline{\Box}_1, \Box_2)
\end{aligned}
$$

$$
W = \mathrm{tr}\,(\, A_1 B_1 A_2 B_2 - A_1 B_2 A_2 B_1 \,). \tag{11.72}
$$

The theory for Figure 11.5(b), also in adapted notation, is

$$
\text{Gauge} \qquad U(n_0) \times U(n_1) \times U(n_2)
$$

$$
\text{Chiral } X_{a,a+1}, Y_{a,a+1}, Z_{a,a+1} : \ (\Box_a, \overline{\Box}_{a+1}) \qquad a = 0, 1, 2 \ \text{mod} \ 3
$$

$$
W = \mathrm{tr}\,(\, X_{a,a+1} Y_{a+1,a+2} Z_{a+2,a} - X_{a,a+1} Z_{a+1,a+2} Y_{a+2,a} \,). \tag{11.73}
$$

It reproduces the theory of D3-branes at $\mathbf{C}^3/\mathbf{Z}_3$; it illustrates the general fact that abelian orbifolds are particular cases of toric singularities.

Gauge theories described by dimer diagrams arise on the worldvolume theory on D3-branes at toric CY singularities. There are several techniques to read off the CY geometry corresponding to a given dimer gauge theory, based on the following idea. The theory with all ranks n_a equal to n is always tadpole-free and describes n sets of branes which can bind together and move off the singularity as n bulk D3-branes. As described in Section 11.2.3 for orbifolds, this is a flat direction of the field theory. Hence for $n = 1$ the flat directions of the gauge theory parametrize the D3-brane position in the transverse CY space. Turning this around, the CY geometry can be constructed as the set of gauge invariant operators (i.e. D-flat directions), modulo F-term relations $\partial W/\partial \Phi \equiv 0$, of the $n_a = 1$ theory. The dimer graph can be efficiently used for this analysis, since gauge invariant operators map to closed paths crossing edges in the dimer, and F-term equivalence is equivalence of paths under deformation across nodes. We skip the general discussion and present a simplified procedure sufficiently practical for the simplest singularities.

Toric manifolds have been introduced in Section 7.2.3, as ambient spaces of *compact* CYs, defined by a set of homogeneous holomorphic equations. In the present setup, however, we are interested in non-compact CY spaces, which can be directly constructed as toric manifolds as follows. Consider complex coordinates $(z_1, \ldots, z_{r+3}) \in \mathbf{C}^{r+3}$, and (after removing a subset of points) mod out by the $(\mathbf{C}^*)^r$ actions, given by $z_i \to \lambda^{q_{in}} z_i$, for $n = 1, \ldots, r$, with $q_{in} \in \mathbf{Z}$. The resulting toric space $\mathbf{C}^{n+3}/(\mathbf{C}^*)^3$ is CY if $\sum_i q_{in} = 0$ for all n.

A popular example is the conifold, defined by coordinates $(a_1, a_2, b_1, b_2) \in \mathbf{C}^4$, quotiented by a $\mathbf{C}^*$ under which they have charges $(1, 1, -1, -1)$. The quotient space can be parametrized by the $\mathbf{C}^*$-invariants $x = a_1 b_1$, $y = a_2 b_2$, $z = a_1 b_2$, $w = a_2 b_1$. These are not independent, but satisfy

$$xy = zw. \tag{11.74}$$

Hence the conifold is often equivalently described as a subspace of $\mathbf{C}^4$ parametrized by (x, y, z, w) defined by (11.74).

It is easy to show that the theory (11.72) corresponds to D3-branes at a conifold singularity. Consider the theory with $n_1 = n_2 = 1$, for which $W \equiv 0$, and construct the gauge invariant operators $x = A_1 B_1$, $y = A_2 B_2$, $z = A_1 B_2$, $w = A_2 B_1$. Their vevs obey the constraint (11.74) and thus parametrize a conifold.

Non-compact toric CY spaces include abelian orbifold singularities as particular cases. For instance, the $\mathbf{C}^3/\mathbf{Z}_3$ singularity can be described using coordinates $(z_1, z_2, z_3, w) \in \mathbf{C}^4$, and quotienting by a $\mathbf{C}^*$ under which they have charges $(1, 1, 1, -3)$. Using the $\mathbf{C}^*$ action to gauge away w, i.e. to fix it to $w = 1$, we are left with $(z_1, z_2, z_3) \in \mathbf{C}^3$ modded out by a left-over $\mathbf{Z}_3$ discrete subgroup, precisely acting as $z_i \to e^{2\pi i/3} z_i$. The $\mathbf{C}^3/\mathbf{Z}_3$ singularity can be also described in terms of the $\mathbf{C}^*$-invariants $x = z_1^3 w$, $y = z_2^3 w$, $z = z_3^3 w$, $t = z_1 z_2 z_3 w$, satisfying

$$xyz = t^3. \tag{11.75}$$

We can show that the theory (11.73) describes D3-branes at $\mathbf{C}^3/\mathbf{Z}_3$ in this language. For the theory with $n_0 = n_1 = n_2 = 1$, consider the gauge invariant operators

$$x = X_{12} X_{23} X_{31} , \quad y = Y_{12} Y_{23} Y_{31} , \quad z = Z_{12} Z_{23} Z_{31} , \quad t = X_{12} Y_{23} Z_{32}. \tag{11.76}$$

Using F-term relations, they satisfy (11.75), so their vevs parametrize $\mathbf{C}^3/\mathbf{Z}_3$.

We conclude with an illustrative example of the model building possibilities of non-orbifold toric singularities. Consider the dimer in Figure 11.6, corresponding to

Gauge $\qquad\qquad U(n_0) \times U(n_1) \times U(n_2) \times U(n_3)$

Chiral $\quad 3(\square_1, \bar\square_2) + 2(\square_2, \bar\square_3) + (\square_2, \bar\square_0) + 2(\square_0, \bar\square_1) + (\square_3, \bar\square_1) + (\square_3, \bar\square_0)$

$$W = \mathrm{tr}\, (\, \Phi_{30} X_{01} Y_{12} Z_{23} - \Phi_{30} Z_{01} Y_{12} X_{23} + X_{23} Y_{31} Z_{12} - Z_{23} Y_{31} X_{12} +$$

$$+ X_{12} Y_{20} Z_{01} - Z_{12} Y_{20} X_{01} \,). \tag{11.77}$$

It corresponds to D3-branes at a singularity mathematically described as a complex cone over the del Pezzo complex surface dP_1 (see Section 11.5.3 for information on del Pezzo

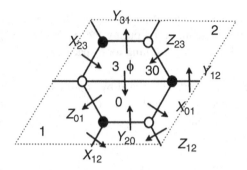

Figure 11.6 Dimer diagram for the dP_1 theory. Upon removal of the edge corresponding to Φ_{30}, it becomes the $\mathbf{C}^3/\mathbf{Z}_3$ theory.

surfaces, in a different context), and is hence termed the dP_1 theory. The cone over dP_1 is one of the few toric singularities leading to triplication of some chiral multiplet. This can be exploited to build a particle physics model, by choosing $n_1 = 3$, $n_2 = 2$, $n_3 = n_0 = 1$. Cancellation of tadpoles/anomalies can be achieved by adding suitable D7-branes, whose discussion we skip. Suffice it to say that the result is an $SU(3) \times SU(2) \times U(1)_Y$ model with three quark-lepton families with correct gauge quantum number assignments. As usual, the model contains some additional $U(1)$ factors, and a few exotics.

The dP_1 singularity, and therefore the above gauge theory, are closely related to $\mathbf{C}^3/\mathbf{Z}_3$. Geometrically, the dP_1 singularity can be partially blown-up to $\mathbf{C}^3/\mathbf{Z}_3$, and this manifests at the level of the D3-brane gauge theory as follows. As will be explained in Section 12.4, the blowing-up is parametrized by a twisted sector modulus, coupling to the D3-branes as a worldvolume FI term, which triggers vevs for suitable charged multiplets. In this case, blowing-up the dP_1 singularity to $\mathbf{C}^3/\mathbf{Z}_3$ forces Φ_{30} to acquire a vev, breaking the gauge factors 0 and 3 to the diagonal combination. This Higgsing is easily shown to turn the dP_1 theory (11.77) into the $\mathbf{C}^3/\mathbf{Z}_3$ theory (11.73); in the dimer, this amounts to removing the Φ_{30} edge in the dP_1 dimer, turning it into the $\mathbf{C}^3/\mathbf{Z}_3$ one. The additional parameter provided by the vev enters the superpotential, so the Higgsed dP_1 theory provides a deformed version of the $\mathbf{C}^3/\mathbf{Z}_3$ theories in Section 11.3.3. The slightly reduced global symmetry $SU(3) \rightarrow SU(2) \times U(1)$ allows for more flexible pattern of Yukawa couplings.

11.4 Magnetized D-brane models

In the previous sections we have considered type IIB models in which the D-brane worldvolume gauge field background is trivial.[2] In this section we consider an interesting alternative, magnetized brane models, in which there are non-trivial worldvolume gauge field strength fluxes. This leads to additional flexibility in model building. Moreover, in the

---

[2] Actually, as explained in Section 11.4.3, models with D-branes at orbifolds can be regarded as secretly including a non-trivial gauge background at the orbifold singularities.

context of toroidal models (and certain orbifolds thereof) it provides the mirror realization of the toroidal intersecting brane models of Section 10.4.2. Magnetized brane models also admit a natural generalization to CY geometries, which is further discussed in Section 11.4.3, and which introduces some basic ingredients for the F-theory models in Section 11.5.

11.4.1 Magnetization of D-branes on R^2 and T^2

We start considering open strings between D-branes magnetized on a single 2-plane in flat space, taken first to be non-compact $\mathbf{R}^2$, and latter compactified to $\mathbf{T}^2$.

D-branes with magnetization on R^2

Consider two D-branes labeled a, b, spanning a common $\mathbf{R}^2$ parametrized by x^4, x^5, and carrying constant magnetic fields F_a, F_b for their worldvolume $U(1)$ gauge fields. For open strings stretched between them, the worldsheet action in conformal gauge, including the coupling (3.14) to background gauge fields, is

$$S = \frac{1}{4\pi\alpha'} \int_\Sigma d^2\xi \, \partial_a X^m \partial_a X^m + \int dt \, (A_a)_m \partial_t X^m \big|_{\sigma=0} - \int dt \, (A_b)_m \partial_t X^m \big|_{\sigma=\ell}.$$

For constant magnetic fields, we set $A_m = \frac{1}{2} F_{mn} X^n$. Boundary conditions follow from requiring the vanishing of boundary terms upon variation of this action, i.e.

$$\left(\partial_\sigma X^m + 2\pi\alpha' F_{mn} \partial_t X^n \right) \Big|_{\sigma=0,\ell} = 0, \tag{11.78}$$

where we implicitly mean $F_{mn} = (F_a)_{mn}$ at $\sigma = 0$ and $F_{mn} = -(F_b)_{mn}$ at $\sigma = \ell$. We focus on the directions 4, 5, and for simplicity consider $(F_a)_{45} = 0$. Parametrizing $2\pi\alpha'(F_b)_{45} = -\tan\theta$ in terms of an angle θ, we have

$$\partial_\sigma X^4 \big|_{\sigma=0} = 0, \quad \left(\cos\theta \, \partial_\sigma X^4 + \sin\theta \, \partial_t X^5 \right) \big|_{\sigma=\ell} = 0,$$

$$\partial_\sigma X^5 \big|_{\sigma=0} = 0, \quad \left(-\sin\theta \, \partial_t X^4 + \cos\theta \, \partial_\sigma X^5 \right) \big|_{\sigma=\ell} = 0. \tag{11.79}$$

The structure is analogous to (10.3) for D-branes intersecting at an angle θ, up to the exchange $\partial_t X^5 \leftrightarrow \partial_\sigma X^5$. This corresponds to T-duality along x^5, whose action in this non-compact setup is meaningful for the ab sector (since it describes localized modes). Hence the mode expansion obeying (11.79), and its quantization are isomorphic to those corresponding to a system of D-branes at angles, and we may just borrow the results. By simple generalization, an ab open string stretched between two D-branes with magnetic fields F_a, F_b on a two-plane leads to the same spectrum as an open string stretching between two D-branes with relative angle

$$\theta_{ab} = \arctan(2\pi\alpha' F_a) - \arctan(2\pi\alpha' F_b). \tag{11.80}$$

These ideas generalize to magnetization in several complex planes. For instance, compactifications with D9-branes carrying magnetic fields in all three complex planes lead to 4d charged chiral fermions, see Sections 11.4.2 and 12.5.4. Incidentally, this mechanism to generate chirality is essentially field-theoretical, and identical to that for heterotic CY models, Section 7.3. Chiral 4d fermions also arise in configurations with 7-branes intersecting at angles in two complex planes and magnetized in the common complex direction, as discussed at the end of Section 11.4.2, and in Section 11.5 in the context of F-theory models.

D-branes with magnetization on T^2

For magnetized directions along a compact $\mathbf{T}^2$, rather than $\mathbf{R}^2$, there are some additional features. We describe them in the simple situation of a rectangular 2-torus, with vanishing NSNS 2-form. First, D-branes may wrap the $\mathbf{T}^2$ a multiple number of times, which we denote by m. Second, as described in Section B.3, the total magnetic flux on $\mathbf{T}^2$ obeys a Dirac quantization condition (B.26), namely

$$m \int_{\mathbf{T}^2} F = 2\pi\, n. \tag{11.81}$$

In our present setup, the factor of m arises from the normalization of the $U(1)$ factor in the $U(m)$ gauge theory of the multi-wrapped D-branes. The two topological integers (n, m) characterizing a magnetized D-brane wrapped on a $\mathbf{T}^2$ are T-dual to the wrapping numbers (n, m) of a D-brane on a 1-cycle on the dual $\mathbf{T}^2$. As already advanced, the magnetic field F maps to an angle θ by $2\pi\alpha' F = \tan\theta$. This also relates conditions (11.81) and (10.11), up to exchange of directions of the IIA $\mathbf{T}^2$.

The T-dual picture is useful in making some properties manifest. For instance, recalling Section 10.3.1, the integers (n, m) must be coprime, a property not obvious from the IIB perspective. A second interesting property is that magnetized D-branes with $(n, m) = (1, 0)$, formally corresponding to zero wrapping number yet non-trivial magnetic flux quanta, actually make sense and correspond to lower-dimensional branes localized on $\mathbf{T}^2$. This useful notation allows to treat D-branes of different dimension on similar footing, and will be implicitly employed in our examples.

Another remarkable property of magnetized branes in compactifications is the replication in the open string spectrum, just like for intersecting branes. This replication can be understood in purely field theoretic terms, an approach valid in more general upcoming setups, like CY compactifications and F-theory – and analogous to heterotic CY compactifications. Consider two stacks of N_a and N_b branes wrapped m_a and m_b times on $\mathbf{T}^2$, and with n_a, n_b magnetic flux quanta. Focus on the large volume regime where the gauge field strengths are diluted and can be regarded as small perturbations around the flux-less configuration. The gauge groups in the latter are $U(N_a m_a)$ and $U(N_b m_b)$, subsequently broken down to $U(N_a) \times U(N_b)$ by the magnetic flux background, via the branching

$$U(N_a m_a) \times U(N_b m_b) \to U(N_a)^{m_a} \times U(N_b)^{m_b} \to U(N_a) \times U(N_b), \tag{11.82}$$

where the intermediate step is for convenience and the last step corresponds to breaking to the diagonal subgroup. Open ab strings lead to fermions transforming in the bifundamental $(\square_a, \overline{\square}_b)$ of the original $U(N_a m_a) \times U(N_b m_b)$ group, which under (11.82) splits as

$$(\square_a, \overline{\square}_b) \rightarrow (\underline{\square_a, \ldots}; \underline{\overline{\square}_b, \ldots}) \rightarrow m_a m_b (\square_a, \overline{\square}_b). \tag{11.83}$$

Recalling Section 7.2.4, each higher-dimensional chiral fermion yields upon $\mathbf{T}^2$ compactification a number of lower-dimensional chiral fermions, with multiplicity given by the index of the suitable Dirac operator on $\mathbf{T}^2$. The index for a fermion with charges $(+1_a, -1_b)$ under $U(1)_a \times U(1)_b$ can be extracted from (7.47), and reads

$$\text{ind } \not{D}_{ab} = \frac{1}{2\pi} \int_{\mathbf{T}^2} (F_a - F_b) = \frac{n_a}{m_a} - \frac{n_b}{m_b}, \tag{11.84}$$

where we have used the quantization (11.81). Including the factor in (11.83), the multiplicity of fermions in the $(\square_a, \overline{\square}_b)$ in the lower-dimensional theory is

$$I_{ab} = n_a m_b - m_a n_b, \tag{11.85}$$

in agreement with the T-dual $\mathbf{T}^2$ intersection number (B.19). Hence, family replication for magnetized branes arises from the multiplicity of fermion zero modes of the Dirac operator for fields coupled to the gauge background. For toroidal models, one can actually solve the Dirac equation and compute the internal profile of these fermion zero modes, see Section 12.5.4.

The above field theory argument is topological and thus applies also in regimes where the gauge field strength is not particularly diluted. For instance (11.85) is valid for delta-function supported magnetic fields, i.e. branes with $(n, m) = (0, 1)$.

11.4.2 Magnetized D-branes in toroidal models

We now consider magnetized D9-branes on toroidal models and their chiral 4d spectra. We start with toroidal type IIB models, and subsequently introduce orientifold planes. For simplicity we take a factorized $\mathbf{T}^6 = \mathbf{T}^2 \times \mathbf{T}^2 \times \mathbf{T}^2$ geometry, and factorized magnetic fluxes, which suffice for our purposes.

Magnetized D9-branes on T^6

Consider sets of N_a magnetized D9-branes, labeled D9$_a$-branes, with multi-wrapping m_a^i and magnetic flux quanta n_a^i on the ith 2-torus $(\mathbf{T}^2)_i$. Equivalently, there is a worldvolume $U(1)_a$ magnetic field $F_a^i = (F_a)_{x^i y^i}$, with quantization condition

$$m_a^i \int_{\mathbf{T}_i^2} F_a^i = 2\pi n_a^i. \tag{11.86}$$

With this information, we can already compute the chiral part of the 4d spectrum of the model, given by

$$aa \text{ sector} \quad \prod_a U(N_a),$$
$$ab + ba \text{ sector} \sum_{a,b} I_{ab} (N_a, \overline{N}_b), \tag{11.87}$$

where I_{ab} is the index of the Dirac operator for a fermion charged under the gauge background $F_a - F_b$, given by the product of the index (11.85) for the three $(\mathbf{T}^2)_i$,

$$I_{ab} = \prod_{i=1}^{3} \left(n_a^i m_b^i - m_a^i n_b^i \right). \tag{11.88}$$

The RR tadpoles can be read off from the D-brane Chern–Simons couplings (6.16)

$$\int_{D9_a} C_{10} ; \quad \int_{D9_a} C_8 \wedge \text{tr } F_a ; \quad \int_{D9_a} C_6 \wedge \text{tr } F_a^2 ; \quad \int_{D9_a} C_4 \wedge \text{tr } F_a^3. \tag{11.89}$$

Reducing to 4d by performing the integrals, the RR tadpole conditions are

$$\int_{\mathbf{T}^6} C_{10} \quad \rightarrow \quad \sum_a N_a m_a^1 m_a^2 m_a^3 = 0,$$

$$\int_{(\mathbf{T}^4)_{2,3}} C_8 \quad \rightarrow \quad \sum_a N_a n_a^1 m_a^2 m_a^3 = 0 \quad \text{and permutations of } 1, 2, 3,$$

$$\int_{(\mathbf{T}^2)_3} C_6 \quad \rightarrow \quad \sum_a N_a n_a^1 n_a^2 m_a^3 = 0 \quad \text{and permutations of } 1, 2, 3,$$

$$C_4 \quad \rightarrow \quad \sum_a N_a n_a^1 n_a^2 n_a^3 = 0. \tag{11.90}$$

This amounts to cancellation of D9-brane charge, as well as the induced D7-, D5-, and D3-brane charges. The conditions agree with (10.18) as expected from mirror symmetry/T-duality. Such relations are even more manifest using the language of Appendix B, and introducing the homology classes $[\mathbf{0}]_i$ and $[\mathbf{T}^2]_i$ for the class of the point and of the two-torus for each $(\mathbf{T}^2)_i$. The $D9_a$-brane has an associated homology class

$$[\mathbf{Q}_a] = \prod_{i=1}^{3} \left(m_a^i [\mathbf{T}^2]_i + n_a^i [\mathbf{0}]_i \right), \tag{11.91}$$

playing the same role as $[\Pi_a]$ in Chapter 10. For instance, (11.90) becomes

$$\sum_a N_a [\mathbf{Q}_a] = 0. \tag{11.92}$$

Also, the multiplicities (11.88) can be written $I_{ab} = [\mathbf{Q}_a] \cdot [\mathbf{Q}_b]$, in terms of an antisymmetric bilinear product $[\mathbf{0}]_i \cdot [\mathbf{T}^2]_i = -[\mathbf{T}^2]_i \cdot [\mathbf{0}]_i = 1$, which we dub "intersection number" with some abuse of language.

The T-duality between magnetized and intersecting D-branes allows to directly translate many results from Chapter 10. But just to illustrate some direct derivation in the IIB picture, we derive the Green–Schwarz couplings involved in mixed $U(1)$ anomaly cancellation. The relevant D9-brane worldvolume couplings are

$$\int_{D9_a} C_0 \operatorname{tr} F_a^5 ; \qquad \int_{D9_a} C_2 \operatorname{tr} F_a^4 ; \qquad \int_{D9_a} C_4 \operatorname{tr} F_a^3 ,$$
$$\int_{D9_a} C_6 \operatorname{tr} F_a^2 ; \qquad \int_{D9_a} C_8 F_a ; \qquad \int_{D9_a} C_{10} . \tag{11.93}$$

Upon reduction to 4d, the 10d RR forms lead to 2-forms and their 4d dual axions,

$$C_2 ; \qquad\qquad\qquad B_0 = \int_{\mathbf{T}^6} C_6 ,$$
$$C_2^i = \int_{(\mathbf{T}^2)_i} C_4 ; \qquad B_0^i = \int_{(\mathbf{T}^2)_j \times (\mathbf{T}^2)_k} C_4 \quad i \neq j \neq k \neq i ,$$
$$B_2^i = \int_{(\mathbf{T}^2)_j \times (\mathbf{T}^2)_k} C_6 ; \qquad C_0^i = \int_{(\mathbf{T}^2)_i} C_2 ,$$
$$B_2 = \int_{\mathbf{T}^6} C_8 ; \qquad\qquad C_0 . \tag{11.94}$$

Performing the 4d reduction, we obtain the 4d couplings

$$N_a\, n_a^1 n_a^2 n_a^3 \int_{\mathbf{M}_4} C_2 \wedge F_a ; \qquad m_b^1 m_b^2 m_b^3 \int_{\mathbf{M}_4} B_0 \operatorname{tr}(F_b \wedge F_b) ,$$
$$N_a\, m_a^i n_a^j n_a^k \int_{\mathbf{M}_4} C_2^i \wedge F_a ; \qquad m_b^j m_b^k n_b^i \int_{\mathbf{M}_4} B_0^i \operatorname{tr}(F_b \wedge F_b) ,$$
$$N_a\, m_a^j m_a^k n_a^i \int_{\mathbf{M}_4} B_2^i \wedge F_a ; \qquad m_b^i n_b^j n_b^k \int_{\mathbf{M}_4} C_0^i \operatorname{tr}(F_b \wedge F_b) ,$$
$$N_a\, m_a^1 m_a^2 m_a^3 \int_{\mathbf{M}_4} B_2 \wedge F_a ; \qquad n_b^1 n_b^2 n_b^3 \int_{\mathbf{M}_4} C_0 \operatorname{tr}(F_b \wedge F_b) , \tag{11.95}$$

where the N_a prefactors arise from the $U(1)_a$ normalization. The Green–Schwarz amplitude, recall Figure 10.6, has the precise form to cancel the triangle contribution

$$-N_a\, n_a^1 n_a^2 n_a^3\, m_b^1 m_b^2 m_b^3 + N_a \sum_i m_a^i n_a^j n_a^k\, m_b^j m_b^k m_b^i - N_a \sum_i m_a^i m_a^j n_a^k\, m_b^k n_b^i n_b^j$$
$$+ N_a\, m_a^1 m_a^2 n_a^3\, n_b^1 n_b^2 n_b^3 = N_a \prod_i \left(m_a^i n_b^i - n_a^i m_b^i \right) = -N_a I_{ab} . \tag{11.96}$$

The above relative signs arise from signs in the 4d Hodge duality among fields.

As a further illustration of the use of T-duality/mirror symmetry to translate properties of IIA intersecting brane systems to IIB magnetized brane language, we consider the supersymmetry condition for a system of branes. The condition that two magnetized D-branes preserve a common supersymmetry is given by (10.9), where the angles are defined in terms of the magnetic fields via (11.80). The discussion of light scalar dynamics and possible tachyon condensation in the ab open string sector can similarly be borrowed from Section 10.2.2.

The supersymmetry condition $\sum_i \theta_i = 0$ can be recast in geometric terms, by noticing that it implies the identity $\sum_i \tan \theta_i = \prod_i \tan \theta_i$, and using $\tan \theta_i = 2\pi \alpha F_i$ to obtain

$$F_a \wedge J \wedge J = (2\pi\alpha')^2 \, F_a \wedge F_a \wedge F_a, \tag{11.97}$$

where J is the Kähler form on $\mathbf{T}^6$. It is interesting to particularize the above supersymmetry condition to the large volume limit, in which the field strengths are dilute and the right hand side (i.e. α' corrections) can be neglected. Then the above condition, together with $F_{(2,0)} = F_{(0,2)} = 0$ (automatic for factorized fluxes), reproduces the supersymmetry conditions (7.49) of heterotic models, as expected from heterotic/type I duality.

Magnetized D-branes in toroidal orientifolds

Let us discuss the introduction of orientifolds in toroidal models, focusing on models with O3- and/or O7-planes. We start with $\mathbf{T}^6$ quotiented by $\Omega R_1 R_2 R_3 (-1)^{F_L}$, containing O3-planes, and later consider further orbifolds, which introduce also O7-planes. An important subtlety is that the supersymmetry condition in models with O3- and/or O7-planes, in terms of the usual angles $\theta_a^i = \arctan\left(2\pi\alpha' F_a^i\right)$, is

$$\theta_a^1 + \theta_a^2 + \theta_a^3 = \frac{3\pi}{2}. \tag{11.98}$$

We thus consider type IIB on a factorized[3] $\mathbf{T}^6$ modded out by $\Omega R_1 R_2 R_3 (-1)^{F_L}$. We introduce magnetized D9-branes, namely stacks of N_a D9$_a$-brane with topological numbers $\left(n_a^i, m_a^i\right)$ and their orientifold images, denoted D9$_{a'}$-branes, with numbers $\left(n_a^i, -m_a^i\right)$. The relative sign in the wrapping multiplicity arises from D9-brane charge being odd under the O3-plane action. Using the RR charge vector (11.91), the analogous $[\mathbf{Q}_{a'}]$ for the D9$_{a'}$-branes, and $[\mathbf{Q}_{O3}] = [\mathbf{0}]_1 \times [\mathbf{0}]_2 \times [\mathbf{0}]_3$ for the O3-planes, the RR tadpole cancellation conditions read

$$\sum_a N_a [\mathbf{Q}_a] + \sum_a N_a [\mathbf{Q}_{a'}] - 32 \, [\mathbf{Q}_{O3}] = 0, \tag{11.99}$$

whose explicit expression, using (11.91), reproduces (10.35), as expected from T-duality. This corresponds to cancellation of induced D7- and D3-brane charge, as they are sources for the RR fields C_8, C_4 surviving the orientifold action. There are also discrete K-theory $\widehat{D5}$-brane charge cancellation conditions, reproducing (10.36).

---

[3] The orientifold allows for a non-trivial discrete NSNS 2-form background on the $\mathbf{T}^2$s, which we ignore for simplicity. It is T-dual to the tilting of 2-tori in intersecting brane models.

The spectrum follows from the Dirac index in the relevant sectors, and can be read off from the T-dual intersecting brane model in Section 10.4.2. It is given by

$$
\begin{aligned}
&aa + a'a' \text{ sector} && \prod_a U(N_a), \\
&ab + ba + b'a' + a'b' \text{ sector} && \sum_{a,b} I_{ab} \, (\mathbf{N}_a, \overline{\mathbf{N}}_b), \\
&ab' + b'a + ba' + a'b \text{ sector} && \sum_{a,b} I_{ab'} \, (\mathbf{N}_a, \mathbf{N}_b), \\
&aa' + a'a \text{ sector} && n_{\text{sym},a} \, \boxplus_a + n_{\text{asym},a} \, \boxminus_a,
\end{aligned}
\tag{11.100}
$$

where $I_{ab} = [\mathbf{Q}_a] \cdot [\mathbf{Q}_b]$ and $I_{ab'} = [\mathbf{Q}_a] \cdot [\mathbf{Q}'_b]$. Also $n_{\text{sym},a} = \frac{1}{2}(I_{aa'} - I_{a,\text{O3}})$, and $n_{\text{asym},a} = \frac{1}{2}(I_{aa'} + I_{a,\text{O3}})$, with $I_{a,\text{O3}} = [\mathbf{Q}_a] \cdot [\mathbf{Q}_{\text{O3}}]$.

Magnetized D7-brane models on $\mathbf{T}^6/(\mathbf{Z}_2 \times \mathbf{Z}_2)$ orientifolds

As a concrete realization of the above ideas, we consider a model of magnetized D-branes in a $\mathbf{T}^6/(\mathbf{Z}_2 \times \mathbf{Z}_2)$ orientifold, mirror/T-dual to that in Section 10.6. Incidentally the gauge sectors of the model are localized on intersecting D7-branes.

Consider type IIB on the $\mathbf{T}^6/(\mathbf{Z}_2 \times \mathbf{Z}_2)$ orbifold,[4] defined by the actions θ, ω in (10.43), and mod out by $\Omega R_1 R_2 R_3 (-1)^{F_L}$. The model contains 64 O3-planes, and four $\text{O}7_i$-planes for $i = 1, 2, 3$, with overall RR charges -32 times the classes

$$
\begin{aligned}
&[\mathbf{Q}_{\text{O3}}] = [\mathbf{0}]_1 \times [\mathbf{0}]_2 \times [\mathbf{0}]_3; && [\mathbf{Q}_{\text{O7}_1}] = -[\mathbf{0}]_1 \times [\mathbf{T}^2]_2 \times [\mathbf{T}^2]_3, \\
&[\mathbf{Q}_{\text{O7}_2}] = -[\mathbf{T}^2]_1 \times [\mathbf{0}]_2 \times [\mathbf{T}^2]_3; && [\mathbf{Q}_{\text{O7}_3}] = -[\mathbf{T}^2]_1 \times [\mathbf{T}^2]_2 \times [\mathbf{0}]_3.
\end{aligned}
$$

The signs are related to the defining orbifold action, as read off from the T-dual intersecting brane model. In order to cancel the tadpoles we introduce stacks of N_a magnetized $\text{D}9_a$-branes with topological numbers $\left(n_a^i, m_a^i\right)$, and their orientifold images $\text{D}9_{a'}$-branes with $\left(n_a^i, -m_a^i\right)$, which are T-dual to the intersecting D6-branes in Section 10.6. The RR tadpole conditions read

$$
\sum_a N_a [\mathbf{Q}_a] + \sum_a N_a [\mathbf{Q}_{a'}] - 32 [\mathbf{Q}_{\text{O}p}] = 0,
\tag{11.101}
$$

with $[\mathbf{Q}_{\text{O}p}] = [\mathbf{Q}_{\text{O3}}] + [\mathbf{Q}_{\text{O7}_1}] + [\mathbf{Q}_{\text{O7}_2}] + [\mathbf{Q}_{\text{O7}_3}]$. Using (11.91), these formulae reproduce precisely the T-dual expression (10.46).

The spectrum and other properties of the model can be directly computed in the IIB magnetized brane picture, by imposing orbifold quotients as in Sections 11.2.1 and 11.3.2. However, it is far simpler to just translate the results from the T-dual construction in Section 10.6, from which we borrow the examples to avoid redundancy. In particular, the brane system in Table 10.4, translated into the present type IIB picture, corresponds to a visible MSSM sector embedded in a configuration of D7-branes with magnetic fluxes. The branes

---

[4] Our choice of discrete torsion corresponds to the orbifold with Hodge numbers $(h_{11}, h_{21}) = (3, 51)$, which corresponds to the mirror/T-dual of the model in Section 10.6, as mentioned in Section 8.2.6.

h_1, h_2 correspond to an additional sector of D9-branes, with orientifold image $\overline{\text{D9}}$-branes, which remarkably preserve 4d $\mathcal{N} = 1$ supersymmetry thanks to their worldwolume magnetization. Upon supersymmetric D9-$\overline{\text{D9}}$ recombination, the D9-brane charges disappear and they become a hidden sector of supersymmetric D7-branes, decoupled from the visible sector. This interesting construction provides a prototype of a more general class of models, further discussed in the next section.

11.4.3 Magnetized D7-branes beyond tori

The above example can be regarded as belonging to the general class of magnetized D7-brane models in CY orientifolds. Namely, we can consider a CY orientifold with O7-planes, whose RR tadpoles are canceled by stacks of D7-branes wrapped on holomorphic 4-cycles S_a, and carrying supersymmetry preserving gauge backgrounds. The latter condition can be described in analogy with (11.97), as

$$F^{(2,0)} = F^{(0,2)} = 0, \quad F^{(1,1)} \wedge J = 0, \tag{11.102}$$

where the super-index denotes the Hodge type. The magnetic flux is said to be "holomorphic" and "primitive." These conditions can be interpreted as the vanishing of the F- and D-term scalar potentials in the 4d $\mathcal{N} = 1$ effective theory. Incidentally, they are equivalent to the self-duality $F = *_S F$ of the gauge background on the wrapped 4-cycle.

In analogy with (10.28), the tension of such a supersymmetric magnetized D7-brane is given by the integral of a suitable form, namely

$$\int_{S_a} \text{Re} \left(e^{-i\varphi_a} e^{J + i(2\pi\alpha' F + B)} \right). \tag{11.103}$$

In addition to the gauge multiplets, the D7-branes may contain adjoint chiral multiplets, arising from geometric deformations of the wrapped 4-cycle, i.e. zero modes of the 8d scalar ϕ^i parametrizing the transverse position. These can be turned into worldvolume $(2, 0)$-forms by contracting with the holomorphic 3-form Ω_3, as $\alpha_{ij} \sim \Omega_{ijkl}\phi^i$, hence their number is $h_{2,0}(S)$. However, non-trivial worldvolume gauge fields may lift some of these moduli, due to the constraints (11.102).

Concerning charged matter, two sets a, b of D7-branes generically intersect over 2-cycles Σ_{ab}, supporting chiral 6d fermions, which couple to the gauge background restricted to Σ_{ab}, and thus produce chiral 4d fermions according to the index of the Dirac operator. We refrain from a detailed discussion of these constructions in the type IIB orientifold setup, as their structure is closely related to the more general setup of F-theory models, studied in Section 11.5.

We conclude by remarking that D3-branes at singularities can be regarded as a particular realization of this construction, albeit away from the large volume regime. Indeed, upon blowing up the singular point, i.e. turning on twisted sector moduli vevs as will be described in Section 12.4, the D3-branes actually correspond to D7-branes wrapped on the

4-cycle replacing the singular point, and carrying worldvolume gauge backgrounds. The different gauge factors arise from differently magnetized D7-branes, while their magnetized intersections produce the chiral matter. For instance the blown-up geometries for the $\mathbf{C}^3/\mathbf{Z}_3$, dP_1 and $\mathbf{C}^3/\Delta_{27}$ are smooth local CYs containing a 4-cycle corresponding to the delPezzo complex surface $dP_0 \equiv \mathbf{P}_2$, dP_1 and dP_8 respectively. Despite the intuitive appeal of this geometric picture, its validity to study systems of D3-branes at singularities is limited to quantities not receiving α' corrections. On the other hand, magnetized 7-branes on blown-up 4-cycles in the large volume regime can be used to construct interesting local models. This is explored in Section 11.5 in the more general setup of F-theory.

*11.4.4 Back to type IIA: magnetized D8-branes**

The new intuitions about magnetized brane motivate us to revisit type IIA orientifolds for a moment. Naively, D6-branes wrapping special lagrangian 3-cycles as in Chapter 10 exhaust all the possibilities for 4d spacetime filling BPS D-branes in type IIA orientifolds, since D4- and D8-branes should wrap on 1- or 5-cycles, which are actually homologically trivial in a general CY space. In fact this is not quite true for D8-branes, because a non-trivial worldvolume magnetic flux F induces a non-trivial D6-brane charge through the coupling (6.16)

$$\int_{\mathbf{M}_4 \times \Pi_5} F \wedge C_7. \qquad (11.104)$$

The induced D6-brane charge corresponds, in terminology of Section B.1, to the 3-homology class $[\Pi_3]$, Poincaré dual to the cohomology class $[F]$ on Π_5. Such BPS D8-branes are known as coisotropic D8-branes in the mathematical literature. Their existence is in fact required by mirror symmetry, which in toroidal models, for example, relates them to magnetized D9/D5-branes with off-diagonal (also called oblique) magnetic fluxes, i.e. $F_{x^i x^j}$, $F_{x^i y^j}$, $F_{y^i y^j}$, for $i \neq j$.

We now proceed to the construction of such coisotropic D8-branes on $\mathbf{T}^6$ as a simple example. Although it admits non-trivial 1- and 5-cycles, and is thus a very special case, it is a useful starting point for more representative examples, like orbifolds. For simplicity we take a factorized $\mathbf{T}^2 \times \mathbf{T}^2 \times \mathbf{T}^2$ torus (or a $\mathbf{Z}_2 \times \mathbf{Z}_2$ orbifold thereof). We consider a D8-brane wrapped on the 1-cycle $\left(n_a^1, m_a^1\right)$ in $(\mathbf{T}^2)_1$, and spanning $(\mathbf{T}^2)_2 \times (\mathbf{T}^2)_3$ with constant quantized magnetic flux, given by the integer cohomology class

$$
\begin{aligned}
\frac{1}{2\pi} F &= n_{xx}\, dx_2\, dx_3 + n_{xy}\, dx_2\, dy_3 + n_{yx}\, dy_2\, dx_3 \\
&\quad + n_{yy}\, dy_2\, dy_3 + \tilde{n}_2\, dx_2\, dy_2 + \tilde{n}_3\, dx_3\, dy_3, \\
[\Pi_3] &= \left(n^1 [a_1] + m^1 [b_1]\right) \otimes \left(n_{xx}\, [b_2][b_3] + n_{xy}\, [b_2][a_3]\right. \\
&\quad \left. + n_{yx}\, [a_2][b_3] + n_{yy}\, [a_2][a_3] + \tilde{n}_2\, [a_3][b_3] + \tilde{n}_3\, [a_2][b_2]\right),
\end{aligned}
\qquad (11.105)
$$

with wedge products understood. In compactifications with two such D8-branes, denoted a, b, wrapped on different 1-cycles in the first torus and carrying different fluxes F_a, F_b in the remaining two, the ab open strings produce 4d chiral fermions with a multiplicity

$$I_{ab} = \left(n_1^a m_1^b - m_1^a n_1^b \right) \frac{1}{8\pi^2} \int_{(T^2)_2 \times (T^2)_3} (F_b - F_a)^2 \tag{11.106}$$

$$= \left(n_1^b m_1^a - m_1^b n_1^a \right) \left[\left(n_{xx}^b - n_{xx}^a \right) \left(n_{yy}^b - n_{yy}^a \right) - \left(n_{xy}^b - n_{xy}^a \right) \left(n_{yx}^b - n_{yx}^a \right) \right].$$

Chiral fermions also appear if the two D8-brane 1-cycles C_a, C_b lie on different T^2's, with multiplicity proportional to $\int_{C_a \times C_b \times T^2} (F_b - F_a)^2$; finally, chiral fermions also appear at the intersection of D6$_b$- and D8$_a$-branes with multiplicity proportional to $\left(n_1^a m_1^b - m_1^a n_1^b \right)$ $\int_{\Pi_{3,b} \cap \Pi_{5,a}} (F_b - F_a)$. In general, the origin of chirality and fermion replication is a combination of the intersecting and magnetized brane mechanisms. The fermion multiplicities in general may be directly computed as intersection numbers of the corresponding (actual or induced) D6-brane charges, i.e.

$$I_{ab} = \left[\Pi_3^{D8_a} \right] \cdot \left[\Pi_3^{D8_b} \right] \quad \text{or} \quad I_{ab} = \left[\Pi_3^{D6_a} \right] \cdot \left[\Pi_3^{D8_b} \right]. \tag{11.107}$$

The possibility to introduce magnetized D8-branes can be exploited to build new particle physics models in type IIA orientifolds, or improve certain features of existing ones. For example, consider the $\mathbf{Z}_2 \times \mathbf{Z}_2$ orientifold with MSSM-like spectrum in Section 10.6, and replace the D6-branes a and d in Table 10.4 by D8-branes with wrapping numbers $\left(n_a^1, m_a^1 \right) = (1, 0)$ and magnetic fluxes $(n_{xx}, n_{xy}, n_{yx}, n_{yy}) = (1, 3, -3, -10)$, leaving the D6-branes b and c unchanged. The reader can check that the SM spectrum remains the same, and one can easily find extra D6/D8-branes to cancel RR tadpoles. One novel feature, however, is that the D6-brane charge induced on a D8-brane by the flux does not correspond to a factorizable 3-cycle of the kind considered in Chapter 10; at most it can be regarded as a superposition of two such cycles, since $[\Pi_3]$ in (11.105) can be recast as

$$[\Pi_3] = [a_1] \left([b_2][b_3] + 3 [b_2][a_3] - 3 [a_2][b_3] - 10[a_2] \otimes [a_3] \right)$$

$$= [a_1] (-3[a_2] + [b_2]) (3[a_3] + [b_3]) - [a_1][a_2][a_3]. \tag{11.108}$$

Another new feature of models with magnetized D8-branes is that the flux generates superpotential couplings involving the Kähler moduli T_i, which may be useful for moduli fixing. These couplings arise from the F-term supersymmetry condition for these D8-branes, which can be shown to be

$$(F + J_c)^2 |_{\Pi_5} = 0, \tag{11.109}$$

where $J_c = B_2 + iJ$, as defined later in (12.9). This condition plays the role of the first condition in (10.27). Explicitly, in the above toroidal example (11.105), we can expand $J_c = i \sum_i T_i dx^i dy^i$ and easily obtain

$$F^2 = (n_{xy}n_{yx} - n_{xx}n_{yy} + \tilde{n}_2\tilde{n}_3)\, dx_2 \wedge dy_2 \wedge dx_3 \wedge dy_3,$$

$$(J_c)^2|_{\Pi_5} = -T_2T_3 dx_2 \wedge dy_2 \wedge dx_3 \wedge dy_3,$$

$$F \wedge J_c|_{\Pi_5} = -i(\tilde{n}_2 T_3 + \tilde{n}_3 T_2)dx_2 \wedge dy_2 \wedge dx_3 \wedge dy_3.$$

The reader may check that the F-term condition (11.109) then yields

$$(T_2 + i\tilde{n}_2)(T_3 + i\tilde{n}_3) = n_{xy}n_{yx} - n_{xx}n_{yy}, \tag{11.110}$$

which (redefining the Im T_i) may be interpreted as coming from the superpotential

$$W_1 = \Phi_1 (T_2 T_3 - f_1); \quad f_1 = n_{xy}n_{yx} - n_{xx}n_{yy}, \tag{11.111}$$

where Φ_1 is the open string modulus in $(\mathbf{T}^2)_1$, encoding the D8-brane position and Wilson line. The SUSY condition $F_{\Phi_1} = 0$ would seemingly fix a combination of the Kähler moduli T_2, T_3. However, toroidal or $\mathbf{Z}_2 \times \mathbf{Z}_2$ models also contain open string cubic superpotentials $\Phi_1\Phi_2\Phi_3$, so the full superpotential at best fixes combinations of Kähler and open string moduli. In any event the introduction of magnetized D8-branes produces new moduli superpotentials, potentially useful in moduli fixing (see Chapter 14 for other class of moduli dependent superpotentials induced by closed string rather than open string fluxes). We direct the reader to the references for details on explicit models with partial fixed moduli.

11.5 F-theory model building

In this section we study particle physics model building in the context of F-theory. F-theory models are a non-perturbative generalization of type IIB D7-brane models, allowing for richer group theoretical structures and with additional flexibility in model building. In particular, using F-theory 7-branes it is possible to obtain spinor representations **16** of $SO(10)$, and thus construct $SO(10)$ GUT models. It also allows the construction of $SU(5)$ models in which the $\mathbf{10} \cdot \mathbf{10} \cdot \mathbf{5}$ Yukawa coupling is directly present, with an $\mathcal{O}(1)$ coefficient; this is in contrast with perturbative type IIB models, where this coupling is forbidden in perturbation theory, as already mentioned in the 3-family $SU(5)$ GUT toy model in Section 11.2.3. F-theory nicely combines some virtues of heterotic vacua (like $SO(10)$ spinorials and top Yukawas) with those of type IIB compactifications (e.g. modular bottom-up structure, and flux moduli stabilization as in Chapters 14 and 15). The price to pay is that, being non-perturbative, perturbative string theory techniques do not apply, and this limits the computation of non-chiral features of the spectrum, and of the 4d low-energy effective action.

11.5.1 Basics of F-theory

F-theory can most simply be defined as a particular limit of certain compactifications of M-theory. Recall from Section 6.3.4 the duality relation between M-theory on a $\mathbf{T}^2$ and 10d type IIB string theory. Denoting the M-theory $\mathbf{T}^2$ radii by R_1, R_2, and their relative angle by θ, the moduli of $\mathbf{T}^2$ correspond to the area A and complex structure τ

$$A = R_1 R_2 \sin\theta, \quad \tau = \frac{R_2}{R_1}\, e^{i\theta}. \tag{11.112}$$

Taking the limit of vanishing area $A \to 0$ while keeping τ fixed, the 9d theory grows a new dimension that lifts the theory to 10d type IIB theory, with complex coupling constant $\tau = C_0 + i e^{-\phi}$. The type IIB $SL(2, \mathbf{Z})$ S-duality turns into a geometric symmetry (modular transformations) of the compactification $\mathbf{T}^2$ in M-theory; it acts on the $\mathbf{T}^2$ basis 1-cycles $[a]$, $[b]$, and on τ as

$$\begin{pmatrix} [a] \\ [b] \end{pmatrix} \to \begin{pmatrix} a & b \\ c & d \end{pmatrix} \begin{pmatrix} [a] \\ [b] \end{pmatrix}, \quad \tau \to \frac{a\tau + b}{c\tau + d}, \tag{11.113}$$

with $a, b, c, d \in \mathbf{Z}$ and $ad - bc = 1$. F-theory on $\mathbf{T}^2$ is defined as the zero area limit of M-theory on $\mathbf{T}^2$, and corresponds to 10d type IIB theory, with constant τ.

The idea now is to consider fibering this duality over a base space. Consider a compactification of M-theory on a so-called elliptically fibered K3 space. This is just a two-complex dimensional CY space,[5] described as a base $\mathbf{P}_1$ (i.e. a two-sphere $\mathbf{S}^2$) parametrized by a complex coordinate z, and a $\mathbf{T}^2$ erected on top of each point z (known as *elliptic fiber*[6]). The space is not a simple product, however, as the complex structure of the $\mathbf{T}^2$ fiber in general varies over the base, and is given by a (holomorphic) function $\tau(z)$. F-theory on K3 is defined as a limit of this M-theory K3 compactification, in which the area of the $\mathbf{T}^2$ fiber is sent to zero, keeping the function $\tau(z)$ fixed. In this limit, the 7d theory grows a new dimension and lifts to an 8d compactification of type IIB theory on $\mathbf{P}_1$, with a complex coupling $\tau(z)$ which varies over the compactification space; this last feature is crucial to make the compactification preserve supersymmetry, even though $\mathbf{P}_1$ is not a CY space.

A crucial property of $\mathbf{T}^2$ fibrations is that the fiber can become singular over some locus on the base. In the above example, there are points in the base where $\tau(z)$ diverges to $\tau \to i\infty$. This corresponds to a geometry in which one of the $\mathbf{T}^2$ 1-cycles collapses to zero size (so that one radius is infinitely larger than the other), so the $\mathbf{T}^2$ pinches off. In general, at each degeneration point there is a (p, q) 1-cycle of the $\mathbf{T}^2$ fiber that shrinks to zero size. These special points in $\mathbf{P}_1$ have an interesting interpretation in the type IIB picture; they correspond to (p, q) 7-branes sitting at a point in $\mathbf{P}_1$ and filling the non-compact 8d Minkowski space. Recall from Section 6.3.1 that (p, q) 7-branes are non-perturbative objects in type IIB theory, which are related by $SL(2, \mathbf{Z})$ S-duality transformations to

---

[5] K3 is the only compact manifold with $SU(2)$ holonomy. Its non-zero Hodge numbers are $h_{1,1} = 20$, $h_{0,0} = h_{2,0} = h_{0,2} = h_{2,2} = 1$. A simple realization is as an orbifold $\mathbf{T}^4/\mathbf{Z}_2$, with four untwisted (1,1)-forms, and the 16 remaining from the twisted sector at the 16 orbifold fixed points.

[6] "Elliptic curve" is a mathematical name for a $\mathbf{T}^2$ (with one marked point, i.e. a choice of origin).

the standard D7-branes – which are $(1, 0)$ 7-branes in this language. Since D7-branes are magnetically charged with respect to the RR scalar C_0 of type IIB theory, when one moves around a D7-brane in a small circle on $\mathbf{P}_1$ there is a monodromy $C_0 \to C_0 + 1$, i.e. an $SL(2, \mathbf{Z})$ transformation $\tau \to \tau + 1$; this monodromy is nicely encoded in the F-theory K3 geometry, because around a $(1, 0)$ degeneration the basic $\mathbf{T}^2$ 1-cycles $[a] = (1, 0)$, $[b] = (0, 1)$ suffer a monodromy $[a] \to [a]$, $[b] \to [b] + [a]$, namely $a = b = d = 1$, $c = 0$ in (11.113), hence $\tau \to \tau + 1$. The $SL(2, \mathbf{Z})$ monodromy around a general (p, q) 7-brane is

$$M_{p,q} = \begin{pmatrix} 1 - pq & p^2 \\ -q^2 & 1 + pq \end{pmatrix}. \tag{11.114}$$

These monodromies in general prevent the existence of a weak coupling limit with small string coupling everywhere on $\mathbf{P}_1$. F-theory is an inherently non-perturbative description of type IIB vacua with varying coupling constant, and (p, q) 7-branes.

In order to characterize the location and nature of these degenerations in more detail, one introduces a mathematical description of the elliptic fibration. A $\mathbf{T}^2$ can be described algebraically by an equation in two complex coordinates x, y

$$y^2 = x^3 + f x + g, \tag{11.115}$$

where f, g are complex parameters related to the complex structure τ in a way not explicitly needed here; this is the so-called Weierstrass description[7] of a $\mathbf{T}^2$. In order to describe a geometry where the $\mathbf{T}^2$ fiber changes over a base $\mathbf{P}_1$, one simply promotes the constants f, g to functions of the base coordinate z

$$y^2 = x^3 + f_8(z) x + g_{12}(z), \tag{11.116}$$

where f_8, g_{12} must be polynomials of degree at most 8, 12 for the full geometry to satisfy the CY condition. These functions specify the $\mathbf{T}^2$ fibration completely and in particular the location of the degenerate fibers, i.e. the type IIB 7-branes. They are located at points z given by the roots of the so-called discriminant function

$$\Delta_{24}(z) = 4 g_{12}(z)^2 + 27 f_8(z)^3. \tag{11.117}$$

The equation $\Delta_{24}(z) = 0$ has 24 roots (including solutions at infinity) so there are 24 (p, q) 7-branes in the corresponding type IIB picture. Generically the 24 roots are located at different points in $\mathbf{P}_1$, in which case the total space is smooth, even if the fiber degenerates. There is an abelian gauge group $U(1)^{20}$, which can be interpreted in the M-theory picture as arising from the 11d 3-form with two indices along 2-forms on the K3 (since $b_2 = h_{1,1} + h_{2,0} + h_{0,2} = 20$). Equivalently, the local geometry around each degenerate fiber

---

[7] Intuitively, y describes a double cover of the x-plane (i.e. for every value of x there are two values of y), with branch points introducing branch cuts connecting the two sheets of the double cover. There are four branch points corresponding to the roots of the polynomial in x, and to infinity. The x-plane plus the point at infinity describes a 2-sphere, so the equation describes two 2-spheres joined by two tubes corresponding to two branch cuts joining the four branch points pairwise. The resulting geometry is a $\mathbf{T}^2$, with the branch point at infinity as preferred point, i.e. an elliptic curve.

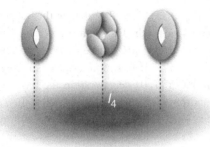

Figure 11.7 Degeneration of the $\mathbf{T}^2$ fiber in F-theory. The $\mathbf{T}^2$ fiber is generically smooth, but degenerates at a point in the base, e.g. by pinching into four 2-spheres. The latter display the structure of the (extended) Dynkin diagram of an A_3 algebra. Correspondingly, the singular fiber is of type I_4 in Table 11.4 and leads to an $SU(4)$ gauge symmetry.

a supports a harmonic $(1, 1)$-form ω_2^a, so there are zero modes of the 11d 3-form with the structure

$$C_3 = \omega_2^a \wedge A_1^a. \tag{11.118}$$

The massless 1-forms A_1^a are $U(1)$ gauge bosons on the worldvolume of the ath 7-brane. Note, however, that since $b_2 = 20$, the 24 2-forms ω_2^a are not all independent. In general $U(1)$s cannot be unambiguously assigned to a single 7-brane, although our slightly cavalier treatment of $U(1)$s will suffice for the models of interest.

When different degenerations coincide, i.e. when $\Delta_{24}(z)$ has some multiple root, the geometry has a singularity and the theory develops an enhanced non-abelian gauge symmetry. In the type IIB picture this arises from coincident 7-branes (generalizing the enhanced $SU(n)$ symmetry from n coincident D7-branes). In the M-theory picture, this arises from collapsed 2-cycles on which wrapped M2-branes produce massless gauge bosons enhancing the gauge symmetry, as studied in Section 6.3.3. Indeed, the (extended) Dynkin diagram of the enhanced gauge symmetry becomes visible in the geometry of the 2-cycles arising from a multiply pinched $\mathbf{T}^2$ fiber. This is shown in Figure 11.7 for a degeneration associated to n coincident D7-branes, namely an order-n pinching of the $(1, 0)$-cycle. The general situation is characterized mathematically in the so-called Kodaira classification of possible degenerate fibers and their ADE enhanced gauge symmetries, shown in Table 11.4.

Orientifold limit and F-theory/type IIB connection[*]

As a simple example, let us construct the F-theory description of type IIB on $\mathbf{T}^2$ modded out by the orientifold $\Omega \mathcal{R}(-1)^{F_L}$, with $\mathcal{R} : z \to -z$. In the quotient space, the model contains four O7-planes with charge -4 (equivalently -8 in the toroidal covering space), and 16 D7-branes (32 in the covering space, i.e. 16 D7-branes and their orientifold images). We consider the configuration with four D7-branes on top of each O7-plane, canceling the RR charge locally, and leading to an $SO(8)^4$ gauge symmetry. Due to local RR charge

Table 11.4 *Kodaira classification of singular fibers in elliptic fibrations, with their mathematical names in the first column. The last column describes the ADE enhanced gauge symmetry associated to a singularity characterized by the order of vanishing of the functions f, g, Δ at the relevant point on the base, shown in the three middle columns*

Name	$\mathrm{ord}(f)$	$\mathrm{ord}(g)$	$\mathrm{ord}(\Delta)$	Singularity-type
smooth	≥ 0	≥ 0	0	none
I_n	0	0	n	A_{n-1}
II	≥ 1	1	2	none
III	1	≥ 2	3	A_1
IV	≥ 2	2	4	A_2
I_n^*	2	≥ 3	$n+6$	D_{n+4}
I_n^*	≥ 2	3	$n+6$	D_{n+4}
IV^*	≥ 3	4	8	E_6
III^*	3	≥ 5	9	E_7
II^*	≥ 4	5	10	E_8

cancellation, the RR scalar has trivial monodromies and τ is constant over the base space. Configurations of this kind have a tunable string coupling which can be used to define a weak coupling limit; such limits do not exist in other generic F-theory models.

The F-theory geometry corresponds to a $\mathbf{T}^4/\mathbf{Z}_2$ orbifold limit of K3. This can indeed be regarded as a $\mathbf{T}^2$ fibration, with constant complex structure, over a base with geometry $\mathbf{T}^2/\mathcal{R} \simeq \mathbf{P}^1$. The four fixed points on the base are loci around which there is a non-trivial $SL(2, \mathbf{Z})$ transformation (11.113) with $a = d = -1$, $c = d = 0$, which leaves τ invariant but flips the sign of the NSNS and RR 2-forms, which from (6.28) transform as an $SL(2, \mathbf{Z})$ doublet; this is precisely the $\Omega\mathcal{R}(-)^{F_L}$ effect on closed string states. In order to understand the gauge symmetry of this F-theory model, we introduce the Weierstrass form of the elliptic fibration, given by

$$f_8 = \alpha\,\phi_4(z)^2, \quad g_{12} = \phi_4(z)^3 \quad \text{with } \phi_4(z) = \prod_{i=1}^{4}(z - z_i), \tag{11.119}$$

where α, z_i are complex parameters. The discriminant is given by

$$\Delta_{24}(z) = (4\alpha^3 + 27) \prod_{i=1}^{4}(z - z_i)^6. \tag{11.120}$$

At the four points at $z = z_i$ the fiber is degenerate, with orders of vanishing $\mathrm{ord}(f, g, \Delta) = (2, 3, 6)$; this corresponds to an I_0^* Kodaira fiber, leading to an enhanced $SO(8)$ gauge factor at each point, in agreement with the gauge group on the D7/O7 system.

It is possible to deform the above model by moving the D7-branes slightly off the O7-plane. This is described by slightly deforming f_8 and g_{12} in the elliptic fibration, and can be used to determine the individual F-theory lifts of the D7-branes and the O7-plane. As described above, the D7-branes are described as $(1, 0)$ 7-branes, i.e. loci where the $(1, 0)$ 1-cycle in the $\mathbf{T}^2$ fiber pinches off, inducing a monodromy $\tau \to \tau + 1$. Interestingly, an O7-plane does not lift to a single object, but rather to a pair of 7-branes with (p, q) labels $(3, -1)$ and $(1, -1)$. Their combined monodromy $M_{3,-1}M_{1,-1}$ acts, using (11.114), as $\tau \to \tau - 4$, in agreement with the RR charge of a type IIB O7-plane. The $SO(8)^4$ configuration, known as orientifold limit of F-theory on K3, has the 24 degenerate fibers distributed in four sets, each containing four D7-branes plus two 7-branes reproducing the O7-plane. In this regime the independent $U(1)$'s in F-theory can be regarded as supported on D7-branes, with no independent $U(1)$ gauge bosons for the 7-branes corresponding to the O7-planes.

11.5.2 F-theory on CY fourfolds

We would like to consider 4d compactifications of F-theory on an (elliptically fibered) 8d space $\mathbf{X}_8$. This can be defined in terms of M-theory compactified to 3d on $\mathbf{X}_8$, in the limit of vanishing fiber area, in which a new dimension grows to render the theory 4d. Considerations familiar from Section 7.2.1 show that 4d $\mathcal{N} = 1$ SUSY requires $\mathbf{X}_8$ to have $SU(4)$ holonomy, i.e. must be a CY fourfold, recall Section 7.2.2. The base of the elliptic fibration is a three complex dimensional space denoted $\mathbf{B}_6$. In type IIB language, we are dealing with a compactification on $\mathbf{B}_6$ to 4d, with a non-trivial holomorphic background for the complex string coupling $\tau(z_1, z_2, z_3)$, where the z_i denote complex coordinates on $\mathbf{B}_6$; the compactification contains (p, q) 7-branes, whose location is defined by the vanishing of the discriminant, namely

$$\Delta(z_1, z_2, z_3) = 0. \tag{11.121}$$

Since this is one complex equation in three complex variables, it defines a 4-cycle. Namely, the 7-branes in the configuration span 4d Minkowski space and wrap the discriminant 4-cycle in the space $\mathbf{B}_6$. In general, the discriminant may factorize as a product of polynomials $\Delta(z_1, z_2, z_3) = \prod_a p_a(z_1, z_2, z_3)$, corresponding to a configuration with several 7-branes wrapped on different 4-cycles S_a, defined by the equation $p_a(z_1, z_2, z_3) = 0$, e.g. see Figure 11.9. On the 4-cycle S_a the Kodaira type of the fiber degeneration defines the gauge group G_a carried by the corresponding coincident 7-branes, denoted 7_a-branes.

The system is a non-perturbative generalization of configurations already appeared in Section 11.4.3, namely type IIB orientifolds with D7-branes on 4-cycles. We can now borrow intuitions from these configurations to understand further details on these F-theory models. For instance, two 4-cycles S_a and S_b will in general intersect along a 2-cycle C_{ab} (or several, depending on their number of intersections; we will ignore this possibility for simplicity). We expect the presence of 6d chiral matter localized on these complex curves (which are hence referred to as "matter curves"). The appearance and quantum numbers of

(a) 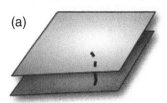   (b)

Figure 11.8 Unfolding of enhanced symmetry loci in F-theory is analogous to tilting parallel D-branes in type IIB theory.

this 6d matter can be derived by realizing that the fiber over C_{ab} is more singular than the generic fibers over S_a or S_b. In other words, the 7-branes wrapped on S_a and S_b coincide over the locus C_{ab}, and lead to an enhanced structure of the singularity. This leads to additional collapsed 2-cycles in the F-theory geometry, and therefore to additional massless states localized at the intersection C_{ab}. Since the branes are not exactly coincident, these states are not vector multiplets, but rather 6d matter multiplets, charged under the gauge groups G_a, G_b carried by the 7_a- and 7_b-branes.

In general, the 7-branes involved are not perturbative D7-branes, so these massless states do not arise from perturbative open strings. The computation of the massless spectrum is however possible using the following "unfolding trick," see also Section 10.7.2. Since the states are localized at the intersection C_{ab}, it is enough to carry out the argument locally, in terms of 7-branes in flat space. We first consider the case of intersecting type IIB D7-branes and proceed by analogy. Consider two stacks of N_a and N_b D7-branes in flat 10d space intersecting at angles over a 6d space (corresponding to 4d Minkowski space plus two dimensions along C_{ab}). The configuration can be regarded as a deformation, by the relative rotations, of a system of coincident D7-branes. The latter would lead to an enhanced gauge symmetry $U(N_a + N_b)$, which is broken to $U(N_a) \times U(N_b)$ by the rotation. The 6d bi-fundamental chiral matter at the intersection arises from states in the adjoint of the enhanced group at the intersection $U(N_a + N_b)$, which are not in the adjoint of $U(N_a) \times U(N_b)$. Namely they arise from the decomposition

$$U(N_a + N_b) \quad \rightarrow \quad U(N_a) \times U(N_b)$$
$$\mathbf{Adj} \quad \rightarrow \quad (\mathbf{Adj}_a, \mathbf{1}) + (\mathbf{1}, \mathbf{Adj}_b) + [\,(\mathbf{N}_a, \overline{\mathbf{N}}_b) + \text{c.c.}\,]. \quad (11.122)$$

Conversely, bi-fundamental chiral matter can be engineered by "unfolding" a set of coincident branes with $U(N_a + N_b)$ symmetry into two sets carrying $U(N_a) \times U(N_b)$ symmetry, see Figure 11.8.

The idea generalizes automatically to F-theory 7-branes. Consider two sets of 7-branes carrying gauge groups G_a, G_b intersecting over C_{ab}, where the symmetry locally enhances to a larger group G_{ab}. The 6d matter at the intersection transforms in the representations arising in the decomposition of the adjoint of G_{ab} under $G_a \times G_b$. For instance, take two sets of 7-branes carrying $SO(10)$ and $U(1)$ symmetries, with an enhanced E_6 symmetry over the intersection locus. The decomposition

$$E_6 \quad \rightarrow \quad SO(10) \times U(1)$$
$$\mathbf{78} \quad \rightarrow \quad \mathbf{45}_0 + \mathbf{1}_0 + [\,\mathbf{16}_{-3} + \text{c.c.}\,] \qquad (11.123)$$

implies that the intersection supports 6d matter in the $\mathbf{16}$ of $SO(10)$. This is already a remarkable feature of F-theory 7-branes, since spinor representations cannot arise from perturbative D-branes. The crucial ingredient is the existence of points of enhanced exceptional symmetry at the intersection. Somewhat more trivially (as they can be realized with perturbative D-branes), other unfoldings can provide the matter multiplets for $SU(5)$ GUT model building, i.e. $\mathbf{10}$, $\overline{\mathbf{5}}$, or their conjugates

$$SU(6) \quad \rightarrow \quad SU(5) \times U(1)$$
$$\mathbf{35} \quad \rightarrow \quad \mathbf{24}_0 + \mathbf{1}_0 + [\,\mathbf{5}_1 + \text{c.c.}\,],$$
$$SO(10) \quad \rightarrow \quad SU(5) \times U(1)$$
$$\mathbf{45} \quad \rightarrow \quad \mathbf{24}_0 + \mathbf{1}_0 + [\,\mathbf{10}_4 + \text{c.c.}\,]. \qquad (11.124)$$

The construction of F-theory GUT models is described in Section 11.5.3.

The matter localized on the intersection curves is 6d, so it is not chiral if directly reduced to 4d. In order to obtain chiral 4d matter we need to introduce a non-trivial background for the 7-brane worldvolume gauge fields, namely we need to magnetize the 7-branes (as for D7-brane modes in Section 11.4.3). Since 7-brane $U(1)$ gauge bosons arise from the (dual M-theory) 3-form as in (11.118), worldvolume magnetic fields correspond to components of the field strength $G_4 = dC_3$ with the structure

$$G_4 = \sum_a \omega_2^a \wedge F_2^a. \qquad (11.125)$$

Compactifications with field strength fluxes for p-form fields will be studied in Chapter 14, in particular in addressing the problem of moduli stabilization; we postpone a more detailed discussion of such configurations, and just we treat the G_4 flux simply as a worldvolume flux on the different 7-branes. In the presence of magnetic fields, the 6d charged matter on C_{ab} produces 4d chiral fermions with multiplicity given by the index of the relevant Dirac operator, as usual in magnetized brane models.

Note that sets of three or more 7-branes can intersect simultaneously at points in the base space $\mathbf{B}_6$, see Figure 11.9. For instance, already in type IIB models D7$_1$- , D7$_2$-, and D7$_3$-branes (wrapped on 4-cycles defined locally by $z_i = 0$, $i = 1, 2, 3$ respectively) intersect at $z_1 = z_2 = z_3 = 0$. At this point there are no additional massless degrees of freedom; rather, these are points of triple intersection of the matter curves C_{12}, C_{23}, C_{31}, and support cubic Yukawa couplings of fields $7_1 7_2 - 7_2 7_3 - 7_3 7_1$, as in Figure 12.2 in Section 12.5.2. This can be determined for type IIB D7-branes using open string techniques, which are however not available in F-theory; but it can also be derived from unfolding, and generalized readily, as follows. The key observation is that the Yukawa coupling arises at a point of rank-2

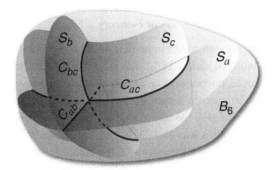

Figure 11.9 General structure of 7-branes and their intersections on the base of an F-theory compactification to 4d. The 7-branes wrap 4-cycles S_a, S_b, etc., pairs of 7_a- and 7_b-branes intersect over 2-cycles C_{ab}, supporting matter multiplets, and triple intersection points produce Yukawa couplings.

enhancement of the gauge symmetry. Namely the configuration of three kinds of D7-branes intersecting pairwise along curves, and with a triple intersection point, can be obtained by deforming overlapping D7-branes by two relative rotations. The matter fields entering the Yukawa coupling can indeed be obtained from the decomposition

$$U(N_1 + N_2 + N_3) \;\rightarrow\; U(N_1) \times U(N_2) \times U(N_3) \tag{11.126}$$
$$\mathbf{Adj} \;\rightarrow\; \mathbf{Adj}_1 + \mathbf{Adj}_2 + \mathbf{Adj}_3 + [\,(\mathbf{N}_1, \overline{\mathbf{N}}_2) + (\mathbf{N}_2, \overline{\mathbf{N}}_3) + (\mathbf{N}_3, \overline{\mathbf{N}}_1)$$
$$+ \text{ c.c.}\,],$$

and the Yukawa coupling is $(\mathbf{N}_1, \overline{\mathbf{N}}_2) \cdot (\mathbf{N}_2, \overline{\mathbf{N}}_3) \cdot (\mathbf{N}_3, \overline{\mathbf{N}}_1)$; the latter can also be understood as inherited from the 8d SYM action of the theory of overlapping 7-branes before unfolding, see Section 12.5.4. Analogously, couplings $\mathbf{10} \cdot \overline{\mathbf{5}} \cdot \overline{\mathbf{5}}$ in an $SU(5)$ theory arise at points of enhanced $SO(12)$ symmetry, unfolding as

$$SO(12) \;\rightarrow\; SU(5) \times U(1) \times U(1) \tag{11.127}$$
$$\mathbf{66} \;\rightarrow\; \mathbf{24}_{0,0} + \mathbf{1}_{0,0} + \mathbf{1}_{0,0} + [\,\mathbf{10}_{4,0} + \overline{\mathbf{5}}_{-2,2} + \overline{\mathbf{5}}_{-2,-2} + \text{c.c.}\,].$$

Similarly, the up-type quark Yukawa coupling $\mathbf{10} \cdot \mathbf{10} \cdot \mathbf{5}$, which is forbidden in D-brane models at the perturbative level, can arise in F-theory from a point of enhanced exceptional E_6 symmetry, as follows from the unfolding pattern

$$E_6 \;\rightarrow\; SU(5) \times U(1) \times U(1) \tag{11.128}$$
$$\mathbf{78} \;\rightarrow\; \mathbf{24}_{0,0} + \mathbf{1}_{0,0} + \mathbf{1}_{0,0} + [\,\mathbf{10}_{-1,-3} + \mathbf{10}_{4,0} + \mathbf{5}_{-3,3} + \text{c.c.}\,] + \mathbf{1}_{5,3}$$
$$+ \; \mathbf{1}_{-5,-3}.$$

As already mentioned, the possibility of having up-type quark Yukawa couplings is one of the main motivations for the construction of $SU(5)$ F-theory GUTs.

11.5.3 Local F-theory GUTs

We now put the above ingredients to work and describe F-theory models of particle physics. The motivation is to build models which, due to the non-perturbative character of the framework, have improved phenomenological properties as compared with their perturbative type IIB cousins; it is hence natural to try and construct F-theory models with an underlying $SU(5)$ GUT, realizing the above mechanism to produce order one $\mathbf{10} \cdot \mathbf{10} \cdot \mathbf{5}$ Yukawa couplings. Although this was possible already in the heterotic compactifications in Chapters 7 and 8, the realization in F-theory is better suited for further analysis of moduli stabilization and supersymmetry breaking, described in later chapters.

The construction of compact CY fourfolds and the computation of the effective 4d physics is a notoriously difficult subject. In our present setup, however, we are mainly interested in the study of an $SU(5)$ gauge sector localized on a set of 7-branes wrapped on a 4-cycle. Many properties of this gauge sector, like its chiral spectrum, will depend only on the local configuration around the 4-cycle, and are quite independent of the details of the compactification. Hence we may adopt a bottom-up approach similar to that in Section 11.3; namely one first constructs a local configuration leading to models of phenomenological interest, and subsequently describes possible embeddings into a full-fledged compactification.

The basic strategy is to first construct a simple non-compact base threefold $\mathbf{B}_6$, containing compact 4-cycles, and engineer the particle physics model by introducing sets of 7-branes, specifying their wrapped 4-cycles, gauge symmetries, intersection pattern, and worldvolume fluxes. Local threefold geometries with compact 4-cycles correspond to blow-ups of singularities, where the singular point is replaced by one or several compact 4-cycles; hence, models of 7-branes wrapping compact holomorphic 4-cycles in local geometries can be regarded as a large volume version of models of branes at singularities, making the application of the bottom up approach natural in both setups. It is possible to construct the blowup of, e.g., abelian orbifold singularities, but this in general leads to geometries with several compact 4-cycles. Since we are interested in a sector with a single gauge factor, a simpler strategy is to search for a local base geometry containing a single compact 4-cycle, denoted S, wrapped by the $SU(5)$ 7-branes. Local threefolds with a single compact 4-cycle are actually very restricted, and exist only when the 4-cycle is a so-called del Pezzo (complex) surface. The simplest example is the blow-up of the $\mathbf{C}^3/\mathbf{Z}_3$ singularity, studied in Section 11.3.3, where the 4-cycle is topologically a complex projective space $\mathbf{P}_2$, which corresponds to a dP$_0$ surface. Similarly, the blow-up of the $\mathbf{C}^3/\Delta_{27}$ singularity in Section 11.3.4 contains a compact dP$_8$ surface.

Given their prominent role in local F-theory models, we describe some geometrical properties of del Pezzo surfaces. The 0th del Pezzo surface, denoted dP$_0$, is the complex projective plane $\mathbf{P}_2$, defined as $\mathbf{C}^3 - \{(0, 0, 0)\}$ modded out by the equivalence relation $(z_1, z_2, z_3) \sim (\lambda z_1, \lambda z_2, \lambda z_3)$, with $\lambda \in \mathbf{C} - \{0\}$; roughly speaking, it is $\mathbf{C}^2$ with the point at infinity added. It contains a non-trivial homology 2-cycle, whose class we denote by H, obtained by considering any hyperplane defined by a linear equation, e.g. $z_1 = 0$ (plus the point at infinity). The nth del Pezzo surface is denoted dP_n, and is obtained by blowing up $\mathbf{P}_2$ at n generic points.[8] Namely, each of the n points is replaced by a 2-cycle, with $\mathbf{P}_1$ topology, whose homology classes are denoted by E_r, $r = 1, \ldots, n$. In order to get a well-defined local geometry with a unique compact 4-cycle, there is a constraint on the possible values of $n = 0, \ldots, 8$, so there are nine different del Pezzo surfaces.

The cycles H, E_r form a basis of 2-cycles in dP_n, with intersection numbers

$$H \cdot H = 1, \quad H \cdot E_r = 0, \quad E_r \cdot E_s = -\delta_{rs}. \tag{11.129}$$

Namely, two hyperplanes intersect at a point, a generic hyperplane does not intersect the blown-up cycles, and different blown-up cycles are located in different regions and do not intersect. As in Section 7.2.2, we may use the complex nature of dP_n to define refined homology, in which terms in the above classes are $(1, 1)$. The only non-zero Hodge numbers are $h_{0,0} = h_{2,2} = 1$, $h_{1,1} = n + 1$.

In particular $h^{2,0} = 0$, which recalling Section 11.4.3 implies that the holomorphic 4-cycle S has no geometric moduli. Hence 7-branes on dP_n have no massless chiral multiplets in the adjoint representation, i.e. no GUT-Higgs to break the $SU(5)$ GUT symmetry. Moreover, the additional property $\Pi_1(dP_n) = 0$ further implies that it is not possible to implement such breaking by discrete Wilson lines, as exploited in Sections 7.3.3 and 7.4 in the heterotic setup. Fortunately, F-theory models with 7-branes wrapped on dP_n allow for an alternative symmetry breaking mechanism in terms of a hypercharge magnetic flux, as described later.

In the local configuration there are $SU(5)$ 7-branes wrapped on the compact 4-cycle $S = dP_n$. There are also other 7-branes wrapped on non-compact 4-cycles of the local model, denoted S_a, which intersect S on curves C_a which support 6d matter multiplets; the kind of matter multiplet is determined by the enhanced symmetry at the intersection, as explained in Section 11.5.2. The non-compact 7-branes generically carry non-trivial gauge backgrounds for their worldvolume symmetries, which we will take to be generically $U(1)$. These should be regarded as global symmetries from the 4d viewpoint – which may be gauged in the eventual full compactifications, albeit in a way dependent on the global properties of the compact space. For the local model, the only relevant data about these extra 7-branes are the intersection curves C_a with S, the kind of localized 6d matter, and the magnetic field quanta to which the 6d matter couples

$$n_a = \frac{1}{2\pi} \int_{C_a} F_a. \tag{11.130}$$

This information can be sketched in a diagram representing S and the matter curves C_a as in Figure 11.10, with the understanding that each real dimension in the diagram corresponds to a complex dimension in the local geometry.

Using these ideas, it is easy to construct local configurations of 7-branes leading to a 3-family $SU(5)$ GUT. Consider an example based on dP_n, $n \geq 5$, with several extra 7-branes,

---

[8] In addition, the space $\mathbf{F}_0 = \mathbf{P}_1 \times \mathbf{P}_1$ is also usually included in the del Pezzo family, as it also leads to local models with a single 4-cycle.

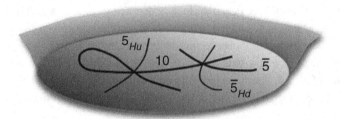

Figure 11.10 The picture shows the local del Pezzo surface and the intersection curves supporting the matter and Higgs multiplets of a $SU(5)$ GUT model, as well as the intersection points leading to Yukawa couplings.

labeled by an index a, with the following intersections, local symmetry enhancements, and worldvolume fluxes:

Field	Curve class C_a	Enhancement	n_a
$\mathbf{10}_m$	$2H - E_1 - E_5$	$SO(10)$	3
$\mathbf{\bar{5}}_m$	H	$SU(6)$	-3
$\mathbf{5}_{H_u}$	$H - E_1 - E_3$	$SU(6)$	1
$\mathbf{\bar{5}}_{H_d}$	$H - E_2 - E_4$	$SU(6)$	-1

where the sub-index m denotes "matter" fields. A depiction of the relevant curves is sketched in Figure 11.10. It is an $SU(5)$ model with three families of $\mathbf{10} + \mathbf{\bar{5}}$, and a vector-like pair of Higgses in $\mathbf{5}, \mathbf{\bar{5}}$. The particular choices of curves are motivated to allow for a simple modification to be discussed shortly. Also, the two Higgs multiplets are localized on different curves, for reasons to be explained at the end of this chapter. Finally, the curves in Figure 11.10 are chosen to intersect at points in a way that yields the desired Yukawa couplings.

As explained above, D7-branes on del Pezzo 4-cycles neither have massless adjoints nor discrete Wilson lines to break the GUT gauge symmetry. There is however a possibility to achieve the breaking of $SU(5)$, by turning on a worldvolume magnetic field for the hypercharge generator in $SU(5)$,

$$Y \simeq \text{diag}\,(2, 2, 2, -3, -3), \tag{11.131}$$

supported on the homologically non-trivial 2-cycles of dP_n. The magnetized branes indeed experience a symmetry breaking of $SU(5)$ to the SM group. In the process, the $SU(5)$ gauginos can potentially produce exotic chiral matter charged under the SM group; their quantum numbers are given by the branching

$$\begin{aligned} SU(5) &\rightarrow & SU(3) \times SU(2)_L \times U(1)_Y \\ \mathbf{24} &\rightarrow & (\mathbf{8, 1})_0 + (\mathbf{1, 3})_0 + (\mathbf{1, 1})_0 + [\,(\mathbf{3, 2})_5 + \text{c.c.}\,], \end{aligned} \tag{11.132}$$

where the hypercharge normalization has been chosen to avoid fractional charges later on. Since the above fields have (hyper)charge 5, the quantization conditions (B.26) for the hypercharge flux over the basis of 2-cycles H, E_r in S are

$$5 \int_H \frac{F_Y}{2\pi} = n_{Y,H} \ , \ 5 \int_{E_r} \frac{F_Y}{2\pi} = n_{Y,E_r}, \tag{11.133}$$

with integer flux quanta $n_{Y,H}$, $n_{Y,E_r} \in \mathbf{Z}$. Equivalently, the homology dual is

$$\left[\frac{F_Y}{2\pi} \right] = \frac{1}{5} \left(n_{Y,H} H + \sum_r n_{Y,E_r} E_r \right). \tag{11.134}$$

The multiplicity of the above $(\mathbf{3}, \mathbf{2})_5$ exotics is given by the index of the relevant Dirac operator coupled to the hypercharge flux. For del Pezzo surfaces, there exist certain "vanishing theorems," which ensure that this index vanishes if the class $5 \left[\frac{F_Y}{2\pi} \right]$ has self-intersection -2, e.g. $5 \left[\frac{F_Y}{2\pi} \right] = E_r - E_s$, with $r \neq s$, as exploited later.

As is familiar from previous brane constructions, achieving the actual SM group at low energies requires that the hypercharge $U(1)_Y$ remains massless. Indeed, a worldvolume coupling on the $SU(5)$ 7-branes can produce a $B \wedge F$ term for $U(1)_Y$,

$$\int_{8d} C_4 \wedge F_Y^2 \rightarrow \int_{4d} B_2 \wedge F_Y, \tag{11.135}$$

with

$$B_2 = \int_S C_4 \wedge F_Y = \int_{[F_Y]} C_4, \tag{11.136}$$

where $[F_Y]$ is momentarily shorthand for $5 \left[\frac{F_Y}{2\pi} \right]$. This potential problem is identical to that in heterotic compactification with $U(1)$ bundles, in Section 7.4.3, and would spoil the interest of the present F-theory models; there is however an interesting way out, already noted at the end of Section 11.3.2 in a different context. The idea uses the global compactification geometry to eliminate the very existence of the closed string zero mode B_2. From (11.136), this is achieved if $[F_Y]$ is homologically trivial on the full base geometry $\mathbf{B}_6$, even if non-trivial on the 4-cycle S, that is $[F_Y] \in H_2(S)$ but $[F_Y] \notin H_2(\mathbf{B}_6)$; in other words, it is the boundary of a 3-chain Σ_3 in $\mathbf{B}_6$, but not lying in S, see Figure 11.11. In our above ansatz, $[F_Y] \simeq E_r - E_s$, this corresponds to requiring that $[E_r] = [E_s]$ in $\mathbf{B}_6$, despite these classes being different in S. Note that the triviality of $[F_Y]$ in $\mathbf{B}_6$ is a constraint on the global compactification, not determined purely in terms of the local geometry. Explicit examples of global CY four-fold geometries with this property can be constructed by different techniques; their description is beyond our present scope, so in what follows we will simply assume this mechanism to be at work in the local models under discussion.

At the intersection curves C_a of S with other 4-cycles S_a, there is 6d matter charged under both the $SU(5)$ (hence coupling to the hypercharge flux F_Y) and the $U(1)_a$ factors

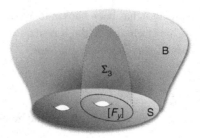

Figure 11.11 Sketch of a a 4-cycle S in a global geometry $\mathbf{B}_6$, such that there is a homology 2-cycle class $[F_Y]$, non-trivial in S but trivial in $\mathbf{B}_6$, due to the 3-chain Σ_3.

on the non-compact 7_a-branes. The multiplicity of 4d chiral fermions is given by the index of the Dirac operator coupled to the gauge backgrounds of the two intersecting 7-branes. For a matter field localized on C_a, in a representation $\mathbf{R}_i$ of the SM group, with hypercharge $y_{i,a}$ and $U(1)_a$ charge q_a, the index is

$$\#_{\mathbf{R}_i} = y_{i,a} \int_{C_a} \frac{F_Y}{2\pi} + q_a \int_{C_a} \frac{F_a}{2\pi} = y_{i,a} \frac{n_{Y,a}}{5} + q_a \frac{n_a}{5} \ , \quad \text{no sum in } a. \qquad (11.137)$$

where $n_{Y,a} = 5 \left[\frac{F_Y}{2\pi} \right] \cdot [C_a]$ and n_a are integer hypercharge and $U(1)_a$ fluxes over C_a. Since the hypercharge of these states is in general not a multiple of 5, the $U(1)_a$ magnetic flux quanta are $n_a/5$, i.e. fractional to yield an integer multiplicity.

Since different fields in a given $SU(5)$ multiplet have different hypercharge, they couple differently to the hypercharge flux, and can have in general different number of zero modes. This leads to a nice mechanism to produce light $SU(2)$ doublet Higgses, while keeping their triplet GUT partners heavy, by requiring that only the former have zero modes. On the other hand, since quarks and leptons actually form full $SU(5)$ multiplets, they should be most naturally located on curves over which the integrated hypercharge flux vanishes, to avoid splitting of GUT multiplets in the matter sector. An example of this idea is provided in the example below.

As usual, the non-trivial gauge backgrounds must satisfy certain conditions to preserve supersymmetry. In F-theory they can be understood from the supersymmetry conditions on the 11d G_4 flux in M-theory, to be discussed in Section 14.1.2, i.e. using (14.35) with the ansatz (11.125). The result, however, is identical to (11.102) for magnetized D7-branes in general geometric compactifications

$$F^{(2,0)} = F^{(0,2)} = 0, \quad F^{(1,1)} \wedge J = 0, \qquad (11.138)$$

where the super-index denotes the Hodge type, and J is the Kähler form on S. The first condition is automatically satisfied, as del Pezzo surfaces do not support $(2, 0)$- or $(0, 2)$-forms, while the second is non-trivial but can be satisfied for our choices of F_Y (and of the other gauge backgrounds) in suitable regions in moduli space.

We conclude with a concrete model, based on the earlier example. We take $5\left[\frac{F_Y}{2\pi}\right] = E_3 - E_4$, with self-intersection -2, and the following data:

Matter	Curve class C_a	Enhancement	$n_{Y,a}$	q_a	n_a	4d field	SM field
$\mathbf{10}_m$	$2H - E_1 - E_5$	$SO(10)$	0	1	15	$3 \times \mathbf{10}$	Q_L, U, E
$\overline{\mathbf{5}}_m$	H	$SU(6)$	0	1	15	$3 \times \overline{\mathbf{5}}$	D, L
$\mathbf{5}$	$H - E_1 - E_3$	$SU(6)$	1	1	2	$(\mathbf{1, 2})_3$	H_u
$\mathbf{5}$	$H - E_2 - E_4$	$SU(6)$	-1	1	-2	$(\mathbf{1, 2})_{-3}$	H_d

where in the last two columns we have indicated the surviving massless 4d fields from (11.137). Note that the vanishing hypercharge flux through the $\mathbf{10}$ and $\overline{\mathbf{5}}$ curves (for which $\left[\frac{F_Y}{2\pi}\right] \cdot [C_a] = 0$) produces 4d matter in full $SU(5)$ multiplets, while the non-trivial flux through the Higgs curves produces a zero mode for the $SU(2)$ doublet but not for the triplets, according to (11.137), for the decomposition

$$
\begin{aligned}
SU(5) &\rightarrow SU(3) \times SU(2)_L \times U(1)_Y \\
\mathbf{5} &\rightarrow (\mathbf{3, 1})_{-2} + (\mathbf{1, 2})_3.
\end{aligned}
\tag{11.139}
$$

Notice that even though the Higgs triplets are heavy, they can still lead to too fast proton decay through dimension five operators. This could indeed happen if the two Higgs doublets would arise as a non-chiral set localized on a single curve. In such case, the two massive triplets T, T' lead to couplings, e.g.

$$
W = Q_L L T + T' Q_L Q_L + m T T' \rightarrow \frac{1}{m} L Q_L Q_L Q_L.
\tag{11.140}
$$

This is automatically avoided in the above models, where the two Higgs multiplets arise from different curves, so that the triplets T, T' do not have mutual couplings.

Some further properties of these $SU(5)$-based F-theory GUTS are described in Sections 12.7 and 16.1.2. As mentioned previously, it would be possible to use similar techniques to build $SO(10)$ GUT models in F-theory; however, due to the larger size of the group, its breaking by $U(1)$ fluxes produces exotic matter multiplets on S, all of which cannot be avoided simultaneously.

In analogy with Section 11.3, these local models must be regarded as part of a full global compactification. There exist global embeddings satisfying most of the above mentioned phenomenological requirements, and for which we refer the reader to the Bibliography.

12

Type II compactifications: effective action

We turn now to the structure of the effective action of type II orientifold compactifications. We mostly focus on the type IIA orientifolds of Chapter 10, and the type IIB models with D3/D7-branes (both in orbifolds, and magnetized) of Chapter 11, with brief discussions of D9/D5-brane models. We also often stick to one particular picture, relying on T-duality/mirror symmetry for the derivation of results in other constructions. We are mostly interested in 4d $\mathcal{N} = 1$ compactifications, and hence focus on the Kähler potential K, the gauge kinetic functions f_a, and the superpotential W. Although the effective action shares some features with that of heterotic string compactifications, there is a substantial number of novel features. In particular, we describe the implications of anomalous $U(1)$s and their corresponding FI terms, and carry out detailed microscopic descriptions of Yukawa couplings in intersecting brane models, magnetized brane models, and local F-theory local models.

12.1 The closed string moduli in type II orientifolds

We start with a description of the closed string moduli in type II orientifolds, in particular of the connection between the supergravity moduli fields and the geometric data; this is slightly tricky in the way it involves the p-form field backgrounds.

As discussed in Section 10.1, compactification of type II string theories on a CY manifold $\mathbf{X}_6$ gives rise to 4d $\mathcal{N} = 2$ supersymmetry. This can be further reduced to 4d $\mathcal{N} = 1$ by subsequent orientifold actions, whose geometric $\mathbf{Z}_2$ action $\mathcal{R}$ on the CY $\mathbf{X}_6$ we recall from Sections 10.1.3 and 11.1

$$\text{Type IIA with O6-planes}: \quad \mathcal{R}J = -J, \quad \mathcal{R}\Omega_3 = \overline{\Omega}_3, \tag{12.1}$$

$$\text{Type IIB with O3/O7-planes}: \quad \mathcal{R}J = J, \quad \mathcal{R}\Omega_3 = -\Omega_3, \tag{12.2}$$

$$\text{Type IIB with O5/O9-planes}: \quad \mathcal{R}J = J, \quad \mathcal{R}\Omega_3 = \Omega_3, \tag{12.3}$$

where J is the CY Kähler form and Ω_3 is the holomorphic 3-form. As we show next, the orientifold operation projects out a subset of the original $\mathcal{N} = 2$ fields and reduces the closed string spectrum.

12.1.1 Type IIA orientifolds with O6 planes

Recalling Section 10.1.1, compactification of type IIA string theory on a CY manifold $\mathbf{X}_6$ produces a 4d $\mathcal{N} = 2$ theory with Kähler moduli in $h_{1,1}$ vector multiplets, and the complex structure moduli and dilaton in $h_{2,1} + 1$ hypermultiplets. Let us recall their microscopic origin in terms of the harmonic forms on $\mathbf{X}_6$, focusing on the bosonic components for simplicity. We introduce a basis $\{\omega_P\}_{P=1,\ldots,h_{1,1}}$ of (1, 1)-forms, and a basis $\{\alpha_K, \beta^K\}_{K=0,\ldots,h_{2,1}}$ of 3-forms, satisfying

$$\int_{\mathbf{X}_6} \alpha_K \wedge \beta^L = \delta_K^L. \tag{12.4}$$

From Section 7.2.4, the zero modes in the KK reduction of the 10d p-form fields correspond to the expansion along harmonic forms on $\mathbf{X}_6$. In the present case,

$$
\begin{aligned}
C_1 &= C_1(x), \\
B_2 &= b_2(x) + b^P \,\omega_P, && P = 1, \ldots, h_{1,1}, \\
C_3 &= A_1^P(x) \wedge \omega_P + C^K(x)\,\alpha_K - \tilde{C}_K(x)\,\beta^K, && K = 0, 1, \ldots, h_{2,1}. \tag{12.5}
\end{aligned}
$$

Here b^P, C^K, $\tilde{C}_K$ are 4d scalars, while $C_1(x)$, $A_1^P(x)$ are 4d $U(1)$ gauge bosons, and b_2 is a 4d 2-form, which can be dualized into a scalar a, as in Section 9.1.1. There are additional massless scalars from the CY geometric moduli, given by $h_{1,1}$ Kähler moduli real scalars v^P, and $h_{2,1}$ complex structure complex scalars z^K. The Kähler moduli v^P combine with b^P, A_1^P to form $h_{1,1}$ vector multiplets (A_1^P, v^P, b^P). The complex structure moduli z^K combine with C^K, $\tilde{C}_K$, $K = 1, \ldots, h_{2,1}$, to form $h_{2,1}$ hypermultiplets $(z^K, C^K, \tilde{C}_K)$. There is finally one hypermultiplet $(a, \phi, C^0, \tilde{C}^0)$, and the gravity multiplet $(g_{\mu\nu}, C_1)$.

Let us consider now the $\Omega\mathcal{R}(-1)^{F_L}$ orientifold projection, where $\mathcal{R}$ acts on J and Ω_3 as in (12.1). The intrinsic action of $\Omega(-1)^{F_L}$ on the original type IIA 10d fields can be easily obtained from the Ω action in type I, Section 4.4.2, upon formal application of three T-dualities. The NSNS 2-form B_2 is intrinsically odd (and the metric G and dilaton ϕ are even), while the RR 1-form C_1 is intrinsically odd and the RR 3-form C_3 is even; this follows since T-duality in three transverse directions maps C_1 into the Ω-odd type IIB RR 4-form, and C_3 into an Ω-even RR 6-form (as it relates to the Ω-even RR 2-form in type I by 10d duality in the sense of Section B.3).

Harmonic forms are classified according to their behavior under the orientifold action. We thus define a basis of odd (even) (1, 1)-forms ω_a (ω_α), with $a = 1, \ldots, h_{1,1}^-$, $\alpha = 1, \ldots, h_{1,1}^+$, and their dual (2, 2)-forms $\tilde{\omega}_a$ ($\tilde{\omega}_\alpha$) by

$$\int \omega_a \wedge \tilde{\omega}^b = \delta_a^b; \quad \int \omega_\alpha \wedge \tilde{\omega}^\beta = \delta_\alpha^\beta. \tag{12.6}$$

The harmonic 3-forms are similarly split into even and odd under the $\mathbf{Z}_2$ geometric action $\mathcal{R}$. In type IIA orientifolds it is possible to choose the basis $\{\alpha_K, \beta^K\}$ such that α_K are even and β^K are odd. As described above, both J and B_2 are odd under the orientifold action, hence their zero modes arise from expansions into odd harmonic $(1, 1)$ forms

$$J = v^a(x)\,\omega_a, \quad B_2 = b^a(x)\,\omega_a, \quad a = 1, \ldots, h^-_{1,1}. \tag{12.7}$$

These correspond to $h^-_{1,1}$ (complexified) Kähler moduli, as described later. Similarly one can show that the number of scalars from the $h_{2,1}$ complex structure moduli is reduced by half, leaving $h_{2,1}$ real scalars. Finally, C_3 is even under the orientifold action and can be expanded as

$$C_3 = A^\alpha_1(x) \wedge \omega_\alpha + C^K(x)\,\alpha_K, \tag{12.8}$$

where C^K are $h_{2,1} + 1$ real scalars. They combine with the surviving z^K real scalars and the dilaton to form $h_{2,1} + 1$ chiral multiplets. The A^α_1 give rise to $h^+_{1,1}$ vector bosons which are contained in vector multiplets. The closed string massless spectrum thus contains, in addition to the gravity multiplet, $h^-_{1,1}$ Kähler moduli T_a, $h_{2,1}$ complex structure moduli U_K and the complex dilaton $U_0 = S$. As shown in the Bibliography, they are defined in terms of the microscopic string parameters through the expansion in harmonic forms of suitably complexified Kähler 2-form and holomorphic 3-form

$$J_c = B_2 + iJ = i\sum_{a=1}^{h^-_{1,1}} T_a\,\omega_a,$$

$$\Omega_c = C_3 + i\,\mathrm{Re}\,(C\Omega_3) = iS\,\alpha_0 - i\sum_{K=1}^{h_{2,1}} U_K\,\alpha_K. \tag{12.9}$$

Here C is a normalization factor defined by

$$C = e^{-\phi_4}e^{K_{cs}/2}; \quad K_{cs} = -\log\left(-\frac{i}{8}\int_{\mathbf{X}_6}\Omega_3 \wedge \overline{\Omega}_3\right), \tag{12.10}$$

where K_{cs} is the complex structure moduli Kähler potential (9.29), and ϕ_4 is the 4d dilaton defined by

$$e^{\phi_4} = \frac{e^\phi}{\sqrt{\mathrm{Vol}(\mathbf{X}_6)}}. \tag{12.11}$$

As in the heterotic case in Section 9.1.1, the general form of the closed string 4d effective action can be derived from the KK reduction of the 10d type IIA action. We skip this computation and simply quote the resulting Kähler potential

$$\kappa^2_4 K_{\mathrm{IIA}} = -\log\left[\frac{4}{3}\int J \wedge J \wedge J\right] - 2\log\left[2\int \mathrm{Re}(C\Omega_3) \wedge *_{6d}\mathrm{Re}(C\Omega_3)\right]. \tag{12.12}$$

The first term depends on the Kähler moduli T_a and, as in the heterotic case in (9.29), is the logarithm of the volume of $\mathbf{X}_6$. The second term is a function of the complex dilaton

S and complex structure moduli. The explicit form of this Kähler potential for toroidal orientifolds is described in Section 12.1.4.

12.1.2 Type IIB orientifolds with O3/O7-planes

A similar analysis can be carried out to compute the closed string moduli and their 4d effective action for type IIB orientifolds, as we review in this section for models with O3- and/or O7-planes. The case of O9/O5-plane models is discussed in the toroidal setup at the end of Section 12.1.4.

For type IIB with O3/O7-planes, the orientifold action is $\Omega \mathcal{R}(-1)^{F_L}$, with $\mathcal{R}$ acting as in (12.2). The intrinsic action of $\Omega(-1)^{F_L}$ on the 10d fields can be easily obtained from the Ω action in type I, Section 4.4.2, upon formal application of six T-dualities. The NSNS 2-form B_2 is intrinsically odd (and the metric G and dilaton ϕ are even), while the RR 0-form C_0 is intrinsically even (as it maps to the RR 6-form, whose Hodge dual, in the sense of Section B.3, is the Ω-even RR 2-form in type I), the RR 2-form is odd (as it T-dualizes into the Ω-odd RR 4-form), and the RR 4-form is even (as it T-dualizes into the Ω-even RR 2-form).

Again the harmonic forms split as even or odd under $\mathcal{R}$, and we introduce odd and even (1, 1)-forms ω_a, ω_α, (2, 2)-forms $\tilde{\omega}_a$, $\tilde{\omega}_\alpha$, and 3-forms α_k, $\alpha_{\tilde{k}}$. The dilaton combines with the even C_0 field into the complex dilaton S. The odd fields B_2 and C_2, and the even C_4 have the expansions

$$
\begin{aligned}
B_2 &= b^a(x)\,\omega_a, \\
C_2 &= c^a(x)\,\omega_a, \quad a = 1, \ldots, h^-_{1,1}, \\
C_4 &= A^{\tilde{k}}_1 \wedge \alpha_{\tilde{k}} + C_\alpha(x)\,\tilde{\omega}^\alpha, \quad \tilde{k} = 1, \ldots, h^+_{2,1}, \ \alpha = 1, \ldots, h^+_{1,1},
\end{aligned}
\tag{12.13}
$$

where b^a, c^a and C_α are 4d scalars and $A^{\tilde{k}}_1$ are $U(1)$ gauge bosons. There are additional massless scalars from geometric CY moduli. Since $\mathcal{R}J = J$, there are $h^+_{1,1}$ surviving real Kähler moduli v^α; on the other hand, $\mathcal{R}\Omega_3 = -\Omega_3$, and there are $h^-_{2,1}$ complex structure moduli. The massless closed string sector contains the gravity multiplet, plus $h^+_{1,1}$ chiral multiplets T_α containing the scalars (v^α, C_α), $h^-_{2,1}$ chiral multiplets U_A containing the complex structure moduli, $h^-_{1,1}$ chiral multiplets G^a containing the scalars (b^a, c^a), $h^+_{2,1}$ vector multiplets, and the complex dilaton chiral multiplet S. As shown in the Bibliography, the relationships between the supergravity variables and the geometric data are

$$
\begin{aligned}
T_\alpha &= -i \int \mathcal{J}_c \wedge \omega_\alpha, \quad \mathcal{J}_c = C_4 + \frac{i}{2}e^{-\phi} J \wedge J + (C_2 - iSB_2) \wedge B_2, \\
U_A &= i \int \Omega_3 \wedge \alpha_A, \\
S &= e^{-\phi} + iC_0, \\
G^a &= c^a - iS\,b^a.
\end{aligned}
\tag{12.14}
$$

The 4d effective action may be described in terms of the Kähler potential

$$\kappa_4^2 K_{\text{IIB}} = -\log\left(-i \int \Omega_3 \wedge \overline{\Omega}_3\right) - \log\left(S + S^*\right) - 2\log\left(e^{-\frac{3}{2}\phi} \int J \wedge J \wedge J\right).$$
(12.15)

The Kähler potential is in general an implicit function of the moduli, with no explicit expression in terms of T_α, G_a and S; there is, however, an explicit expression for the large volume Kähler potential for the overall modulus T, leading to the familiar result

$$\kappa_4^2 K_{\text{IIB}} = -\log\left(-i \int \Omega_3 \wedge \overline{\Omega}_3\right) - \log(S + S^*) - 3\log(T + T^*).$$
(12.16)

This has the same no-scale structure already encountered in (9.17) and (9.30) for heterotic compactifications. The above Kähler potential is at leading order in α'; there are in general higher order perturbative corrections in α', as well as non-perturbative α' corrections (worldsheet instantons), similar to those mentioned in Section 9.1.3. Already before orientifolding, there is a computable leading correction (in an expansion in inverse powers of the volume) to the Kähler potential for the 4d $\mathcal{N} = 2$ Kähler moduli of type IIB on $\mathbf{X}_6$; it arises from the KK reduction of a computable R^4 correction in the 10d action. Type IIB orientifolds inherit this correction, which modifies the 4d $\mathcal{N} = 1$ Kähler potential (12.15) by an additional piece inside the log, i.e.

$$-2\log\left[e^{-\frac{3}{2}\phi}\left(\int J \wedge J \wedge J + \frac{\xi}{2}\right)\right], \quad \xi = -\frac{\zeta(3)}{16\pi^3}\chi(\mathbf{X}_6),$$
(12.17)

where $\chi(\mathbf{X}_6) = 2(h_{1,1} - h_{2,1})$ is the Euler number of $\mathbf{X}_6$, and ζ is the Riemann zeta function with $\zeta(3) = \sum_{k=1}^{\infty} 1/k^3 \simeq 1.2$; this perturbative correction will be relevant for moduli stabilization in the framework of Section 15.3.2.

In type II CY orientifold compactifications, Kähler and complex structure moduli have imaginary parts given by integrals of RR potentials. The 10d gauge invariance of the latter implies a shift symmetry of the 4d RR scalars, exact in string perturbation theory. This shift symmetry, together with holomorphy, prevents the appearance of these fields in the superpotential, hence they are exact moduli in string perturbation theory. Non-perturbative D-brane instanton effects, however, can produce superpotential contributions for these moduli, as described in Chapter 13. Also, compactifications including additional backgrounds, like p-form fields strength fluxes, have ingredients violating these symmetries and thus produce non-trivial superpotentials for moduli, as described in Chapter 14.

12.1.3 Moduli in M-theory compactified on G_2 manifolds*

As we reviewed in Section 10.7.2, one can obtain 4d $\mathcal{N} = 1$ supersymmetric theories by compactifying M-theory on a 7d manifold $\mathbf{X}_7$ of G_2 holonomy. In this section we review the structure of the 4d effective action obtained from KK reduction of 11d $\mathcal{N} = 1$ supergravity, leaving its computation for the references. Recall that the non-abelian gauge group and chiral content of such compactifications arise from singularities in the compact manifold, and hence cannot be obtained in this approximation; the analysis thus restricts to the moduli, and the 4d $U(1)$ gauge bosons.

Introduce the basis $\{\phi_i\}_{i=1,\ldots,b_3}$ and $\{\omega_\alpha\}_{\alpha=1,\ldots,b_2}$ of harmonic 3- and 2-forms on X_7, respectively. The M-theory 3-form C_3 produces massless scalars c^i and gauge bosons A^α from the KK expansion

$$C_3 = c^i(x)\,\phi_i + A^\alpha \wedge \omega_\alpha, \; i = 1,\ldots,b_3(X_7), \; \alpha = 1,\ldots,b_2(X_7). \tag{12.18}$$

The metric moduli are encoded in the associative 3-form φ introduced in Section 10.7.2, which can be expanded as

$$\varphi = \sum_i s^i(x)\phi_i. \tag{12.19}$$

The scalars c^i, s^i combine into the complex moduli

$$S^i = c^i + is^i, \tag{12.20}$$

which belong to $b^3(X_7)$ 4d $\mathcal{N}=1$ chiral multiplets. The dimensional reduction of the 11d action can be shown to produce for them the following Kähler potential

$$\kappa_4^2 K = -3\log\left(\frac{1}{\kappa_{11}^2}\frac{1}{7}\int_{X_7} \varphi \wedge *\varphi\right), \tag{12.21}$$

where the term in brackets is proportional to the volume of X_7. Concerning the abelian gauge bosons A_α, their gauge kinetic function is given by

$$f_{\alpha\beta} = \frac{i}{2\kappa_{11}^2}S^i \int_X \phi_i \wedge \omega_\alpha \wedge \omega_\beta. \tag{12.22}$$

An interesting template for comparison are the G_2 holonomy manifolds introduced in Section 10.7.2, of the form

$$X_7 = (X_6 \times S^1)/Z_2, \tag{12.23}$$

where X_6 is a CY, and Z_2 acts as $x^{10} \to -x^{10}$ on S^1, and as an antiholomorphic involution $\mathcal{R}$ on X_6, i.e. $\mathcal{R}(J) = -J$, $\mathcal{R}(\Omega_3) = \overline{\Omega}_3$; they correspond to the M-theory lift of the $\Omega\mathcal{R}(-)^{F_L}$-orientifold of type IIA on X_6, with the fixed points of $\mathcal{R}$ in X_6 corresponding to the O6-planes (with overlapping D6-branes). This connection with type IIA orientifolds allows to compare the effective action from the above KK reduction of the 11d supergravity with the results in Section 12.1.1. Indeed, recall the structure (10.67) of the associative 3-form φ,

$$\varphi = J \wedge dx^{10} + \text{Re}\left(\sqrt{8}\,C\,\Omega_3\right), \tag{12.24}$$

where this more precise formula includes a factor C, defined in (12.10); we also normalize $\int_{S_1} dx^{10} = 2\pi R$, and set $\kappa_{10}^2 = \kappa_{11}^2/2\pi R = 1$. Then plugging φ into (12.21) one recovers the type IIA orientifold Kähler potential (12.12).

Concerning the gauge bosons A^α, they fall in two classes. First, those associated to harmonic forms supported at degenerations of the S^1 fiber correspond in the type IIA orientifold to $U(1)$'s on D6-brane worldvolumes. Second, gauge bosons arising from 11d 3-form components $C_{\mu i \bar{j}}$ along the $h^+_{1,1}$ harmonic 2-forms even under the orientifold action, correspond in the type IIA orientifold to RR $U(1)$ gauge bosons, further discussed in Section 12.1.1.

12.1.4 The closed string moduli space in toroidal orientifolds

In this section we flesh out the results of Sections 12.1.1, and 12.1.2 in the simplest case of toroidal orientifolds. This allows to provide explicit expressions for several somewhat obscurely defined quantities, and to display their geometric interpretation.

We first fix our notation for the geometric moduli on the tori. For simplicity we consider factorized $\mathbf{T}^6 = (\mathbf{T}^2)_1 \times (\mathbf{T}^2)_2 \times (\mathbf{T}^2)_3$, with coordinates (x^i, y^i) of periodicity 1 on each $(\mathbf{T}^2)_i$, $i = 1, 2, 3$. A basis of 3-forms is

$$\alpha_0 = dx^1 \wedge dx^2 \wedge dx^3; \quad \beta_0 = dy^1 \wedge dy^2 \wedge dy^3,$$

$$\alpha_1 = dx^1 \wedge dy^2 \wedge dy^3; \quad \beta_1 = dy^1 \wedge dx^2 \wedge dx^3,$$

$$\alpha_2 = dy^1 \wedge dx^2 \wedge dy^3; \quad \beta_2 = dx^1 \wedge dy^2 \wedge dx^3,$$

$$\alpha_3 = dy^1 \wedge dy^2 \wedge dx^3; \quad \beta_3 = dx^1 \wedge dx^2 \wedge dy^3, \tag{12.25}$$

normalized to satisfy (12.4). A basis of 2- and 4-forms is given by

$$\omega_i = -dx^i \wedge dy^i; \quad \tilde{\omega}_i = dx^j \wedge dy^j \wedge dx^k \wedge dy^k \quad \text{for } i \neq j \neq k \neq i, \tag{12.26}$$

satisfying $\int_{\mathbf{T}^6} \omega_i \wedge \tilde{\omega}_j = \delta_{ij}$. The geometry of each $(\mathbf{T}^2)_j$ is defined by two basis vectors $\{\vec{e}_{jx}, \vec{e}_{jy}\}$ along x^j, y^j, of radii R^j_x, R^j_y, with area denoted $(2\pi)^2 A_j$. Here we measure geometric lengths in units of $\alpha'^{1/2}$, so all geometric quantities are adimensional, with correct dimensions restored by rescaling $R^j_{x,y} \to \alpha'^{-1/2} R^j_{x,y}$. In terms of these data, the $(\mathbf{T}^2)_j$ complex structure parameter is

$$\mathcal{U}_j = \frac{1}{\vec{e}^2_{jx}} (A_j + i \, \vec{e}_{jx} \cdot \vec{e}_{jy}), \tag{12.27}$$

as in (9.81) with $G_{11} = \vec{e}_y \cdot \vec{e}_y$, $G_{22} = \vec{e}_x \cdot \vec{e}_x$, $G_{12} = \vec{e}_y \cdot \vec{e}_x$; we then have $dz_j = dx^j + i \mathcal{U}_j dy^j$, and the Kähler form and holomorphic 3-forms read

$$J = \sum_{i=1}^{3} A_i \omega_i,$$

$$\Omega_3 = (dx^1 + i \mathcal{U}_1 dy^1) \wedge (dx^2 + i \mathcal{U}_2 dy^2) \wedge (dx^3 + i \mathcal{U}_3 dy^3), \tag{12.28}$$

with the latter clearly admitting an expansion in the 3-form basis (12.25).

Consider first the type IIA orientifold $\mathbf{T}^6/[\Omega(-1)^{F_L}\mathcal{R}_A]$, with $\mathcal{R}_A$: $(x^i, y^i) = (x^i, -y^i)$, and so $\mathcal{R}_A(J) = -J$ and $\mathcal{R}_A(\Omega_3) = \bar{\Omega}_3$. The model contains O6-planes along the x^i directions, e.g. for rectangular tori $\mathcal{U}_j = R_y^j/R_x^j$ and $A_j = R_x^j R_y^j$. We focus on the seven "diagonal" moduli of this IIA orientifold, namely the dilaton S, the three Kähler moduli T_i, and three complex structure moduli U_i, for the three 2-tori. As described above, the moduli can be concisely described in terms of the complexified forms (12.9). In this case the normalization factor (12.10) is $C = e^{-\phi}$, and

$$T_i = \alpha'^{-1} R_x^i\, R_y^i + i \int_{\mathbf{T}^6} B_2 \wedge \tilde{\omega}_i,$$

$$S = e^{-\phi}\alpha'^{-3/2} R_x^1\, R_x^2\, R_x^3 + i \int_{\mathbf{T}^6} C_3 \wedge \beta_0,$$

$$U_i = e^{-\phi}\alpha'^{-3/2} R_x^i\, R_y^j\, R_y^k - i \int_{\mathbf{T}^6} C_3 \wedge \beta_i, \quad i \neq j \neq k \neq i, \qquad (12.29)$$

where we have made now explicit the length dimensions of $R_{x,y}^i$. With these definitions the Kähler potential (12.12) for the moduli reads

$$\kappa_4^2 K_{II} = -\log(S + S^*) - \sum_{i=1}^{3} \log\left(U_i + U_i^*\right) - \sum_{i=1}^{3} \log\left(T_i + T_i^*\right). \qquad (12.30)$$

Consider now the type IIB orientifold $\mathbf{T}^6/[\Omega(-1)^{F_L}\mathcal{R}_B]$, with $\mathcal{R}_B$: $(x^i, y^i) = (-x^i, -y^i)$ and so $\mathcal{R}_B(J) = J$ and $\mathcal{R}_B(\Omega_3) = -\Omega_3$. The model contains O3-planes, but the results below are valid for models with O7-planes as well. Using the above basis, we have $h_{1,1}^- = h_{2,2}^- = h_{2,1}^+ = 0$ and so there are neither G^a moduli nor RR gauge bosons. We again focus on the seven diagonal closed string moduli, i.e. the dilaton, and the Kähler and complex structure moduli of the three 2-tori; they are also denoted S, T_i, and U_i, even though their relation with the 10d microscopic parameters differs from the case of type IIA or heterotic models. In (12.14), the last term in $\mathcal{J}_c$ drops since B_2 and C_2 are projected out. The expressions for moduli in type IIB orientifolds with O3/O7-planes are

$$T_i = e^{-\phi}\alpha'^{-2} R_x^j\, R_y^j\, R_x^k\, R_y^k - i \int_{\mathbf{T}^6} C_4 \wedge \omega_i, \quad i \neq j \neq k \neq i,$$

$$S = e^{-\phi} + iC_0,$$

$$U_i = \mathcal{U}_i. \qquad (12.31)$$

Incidentally, the 4d complex structure fields are directly the 2-tori complex structures. The complete moduli Kähler potential is again given by (12.30). The equality of both IIA and IIB expressions is in fact expected from mirror symmetry, which as described in Section 10.1.1 exchanges Kähler and complex structure moduli,

$$T_i \longleftrightarrow U_i, \quad S \longleftrightarrow S, \qquad (12.32)$$

or, more simply, from T-duality along x^1, x^2, and x^3, which recalling (3.89) acts as

$$R_x^i \to \frac{\alpha'}{R_x^i}, \qquad e^{-\phi} \to e^{-\phi} \alpha'^{-3/2} R_x^1 R_x^2 R_x^3. \qquad (12.33)$$

Toroidal type IIB O3/O7 models can be related to type IIB O9/O5 models (in particular, type I on $\mathbf{T}^6$) by a T-duality in all $\mathbf{T}^6$ directions. Applying to the moduli in (12.31) the above T-duality rules, along with the action (5.6) on RR fields, the moduli for IIB orientifolds with O9/O5-planes can be shown to be

$$T_i = e^{-\phi} \alpha'^{-1} R_x^i \, R_y^i - i \int_{T^6} C_2^i \wedge \tilde{\omega}_i,$$

$$S = e^{-\phi} \alpha'^{-3} R_x^1 \, R_x^2 \, R_x^3 \, R_y^1 \, R_y^2 \, R_y^3 + i \, a,$$

$$U_i = \frac{1}{\mathcal{U}_i} \to \mathcal{U}_i, \qquad (12.34)$$

where a is the 4d dual of the RR 2-form C_2, and we have relabeled x^i and y^i to set $U_i = \mathcal{U}_i$. Using the T-duality, the moduli Kähler potential for type IIB with O9/O5-planes has also the form (12.30); this is a reflection of the heterotic/type I duality of Section 6.3.5.

Note that although toroidal heterotic, type IIA and IIB models have the same Kähler potential for the diagonal moduli, the expression of the latter in terms of microscopic string quantities differs in the different constructions. For instance, Re T_i are volumes of 2-tori in heterotic and type IIB O9/O5 models, while it relates to volumes of transverse 4-tori in type IIB O3/O7 models; also, in type IIB models the dilaton dependence is encoded in S, T_i, while in type IIA it appears in U_i.

In the above discussion we have not included the open string sector. Actually, mixing between both sectors is important, and in fact the Kähler potential for closed moduli, or rather the very definition of the 4d supergravity fields, is modified in the presence of D-branes, as we describe in next section.

12.2 Kähler metrics of matter fields in toroidal orientifolds

As in the heterotic case, the Kähler metrics of the charged matter fields in type II orientifolds depend on geometric moduli, but in the orientifold case they will depend also on the complex dilaton. They are difficult to compute in general CY orientifolds, but are explicitly computable in toroidal orientifolds, as we review now for type IIB O3/O7-models with D3-branes and magnetized D7$_i$-branes of Section 11.4. The results for type IIB toroidal orientifold with O9/O5-planes, or type IIA with O6-planes, can be obtained using T-dualities as described above.

For concreteness we focus on the toroidal orientifold models introduced in Section 11.4.2. We consider type IIB on $\mathbf{T}^6/\mathbf{Z}_2 \times \mathbf{Z}_2$ modded out by $\Omega R_1 R_2 R_3 (-1)^{F_L}$, with R_i : $(x^i, y^i) \to (-x^i, -y^i)$. The models contain D3-branes and magnetized D7$_i$-branes, which we describe in the notation of magnetized D9-branes, i.e. using the wrapping numbers m_a^i

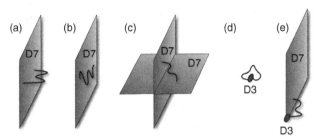

Figure 12.1 Different open string chiral fields in O3/O7 orientifold models.

and magnetic flux quanta n_a^i; D7$_i$-branes are recovered for $m^i = 0$, while D3-branes correspond to $m^1 = m^2 = m^3 = 0$. For convenience we recall the definition (11.80) of the "angles" associated to each D-brane

$$\theta_a^i = \arctan(2\pi\alpha' F_a^i) = \arctan\left(\frac{\alpha' n_a^i}{m_a^i A_i}\right),\qquad(12.35)$$

where $(2\pi)^2 A_i$ is the area of the $(\mathbf{T}^2)_i$.

Charged matter fields arise from different open string sectors, depicted in Figure 12.1, and falling in two main classes. We denote by "untwisted" the states from open strings with both endpoints on the same stack of branes, and "twisted" those from open strings between different stacks. On a stack of D7$_i$-branes, there is a vector multiplet of a gauge factor G_i, and three massless chiral multiplets $C_j^{7_i}$, $j = 1, 2, 3$, in the adjoint of G_i; these are moduli of the D7$_i$-brane, with $C_i^{7_i}$ associated to the brane position in the transverse $(\mathbf{T}^2)_i$, see Figure 12.1(a), and $C_j^{7_i}$, $i \neq j$, associated to the Wilson lines along $(\mathbf{T}^2)_j$, see Figure 12.1(b). In twisted sectors D7$_i$-D7$_j$ there are massless chiral multiplets, denoted $C^{7_i 7_j}$, transforming as bi-fundamentals of $G_i \times G_j$, Figure 12.1(c). On a stack of D3-branes, there is a vector multiplet and three "untwisted" chiral multiplets C_j^3 describing the brane position, Figure 12.1(d). For D3-branes coinciding with some D7$_i$-branes, there are extra "twisted" matter fields, denoted C^{37_i}, Figure 12.1(e).

The metric of these matter fields can be obtained from an explicit computation of string scattering amplitudes; these computations are beyond our scope and we simply quote the results, starting with the fields $C_j^{7_i}$ on a D7$_i$-brane. These are in the adjoint in magnetized toroidal orientifolds, but can correspond to bi-fundamental or two-index tensor representations in toroidal $\mathbf{Z}_N$ orbifolds, such as those constructed in Section 11.2 in the T-dual D9/D5-brane description. For the D7$_i$-brane position moduli $C_i^{7_i}$, the Kähler metric is

$$K_{C_i^{7_i}} = \frac{e^{\phi_4}}{\alpha'^2 (U_i + U_i^*)} \frac{\sqrt{\alpha' A_i}}{\sqrt{A_j A_k}} \prod_{l \neq i} \left| m_i^l A_l + i\alpha' n_i^l \right|, \quad i \neq j \neq k \neq i,\qquad(12.36)$$

with 4d dilaton $e^{\phi_4} = e^{\phi} \alpha'^{3/2}/\sqrt{A_1 A_2 A_3}$. The result for Wilson line moduli $C_j^{7_i}$, $i \neq j$ is

$$K_{C_j^{7_i}} = \frac{e^{\phi_4}}{\left(U_j + U_j^*\right)} \frac{\sqrt{\alpha' A_j}}{\sqrt{A_i A_k}} \left| \frac{m_i^k A_k + i\alpha' n_i^k}{m_i^j A_j + i\alpha' n_i^j} \right|, \quad i \neq j \neq k \neq i. \tag{12.37}$$

The results (12.36) and (12.37) can also be obtained from KK dimensional reduction in type I, followed by T-duality transformations on $(\mathbf{T}^2)_i$, turning the D9-branes into D7$_i$-branes.

For D7$_i$-branes with no worldvolume magnetic flux, we have $(n^i, m^i) = (1, 0)$ and $(n^j, m^j) = (0, 1)$ for $j \neq i$. The above metrics simplify and using (12.31) read

$$K_{C_i^{7_i}} = \frac{1}{(U_i + U_i^*)(S + S^*)}, \quad K_{C_j^{7_i}} = \frac{1}{(U_j + U_j^*)(T_k + T_k^*)}, \quad i \neq j \neq k. \tag{12.38}$$

Twisted fields between D7-branes, e.g. D7$_1$-D7$_2$ states, are the most interesting from the model-building point of view, since these bi-fundamentals contain quarks and leptons in specific constructions leading to models of particle physics. Their Kähler metric can be obtained from string scattering amplitudes, which have actually been computed in the literature in the mirror/T-dual type IIA intersecting D6-brane models. The result is

$$K_{C^{7_1 7_2}} = e^{\phi_4} \prod_{j=1}^{3} \left(U_j + U_j^*\right)^{-\nu_j} \sqrt{\frac{\Gamma(1 - \nu_j)}{\Gamma(\nu_j)}}, \tag{12.39}$$

in terms of the Euler Γ-function, and where $\nu_j = (\theta_1^j - \theta_2^j)/\pi$, with the convention $0 \leq \nu_j < 1$ and $\nu_1 + \nu_2 + \nu_3 = 2$. The Kähler metric depends on Kähler moduli through the ν_j, and there is also an additional dependence inside e^{ϕ_4}. For D7$_1$-, D7$_2$-branes with no worldvolume fluxes, the result simplifies to

$$K_{C^{7_1 7_2}} = \frac{1}{\left(U_1 + U_1^*\right)^{1/2} \left(U_2 + U_2^*\right)^{1/2} \left(T_3 + T_3^*\right)^{1/2} (S + S^*)^{1/2}}. \tag{12.40}$$

Analogous results apply to D7$_2$-D7$_3$ and D7$_1$-D7$_3$ intersections.

D3-branes do not carry worldvolume fluxes, so the Kähler metrics for the matter fields C_j^3 are analogous to the untwisted sector of a heterotic orbifold, i.e.

$$K_{C_j^3} = \frac{1}{\left(U_j + U_j^*\right)\left(T_j + T_j^*\right)}. \tag{12.41}$$

Finally, for D3-branes embedded into, e.g., a D7$_3$-brane, the Kähler metric for $C^{3 7_3}$ is

$$K_{C^{3 7_3}} = \frac{1}{\left(U_1 + U_1^*\right)^{\theta_1/\pi} \left(U_2 + U_2^*\right)^{\theta_2/\pi} \left(T_1 + T_1^*\right)^{1/2} \left(T_2 + T_2^*\right)^{1/2}}, \tag{12.42}$$

where θ_1, θ_2 are related through (12.35) with the fluxes in the first and second tori wrapped by the D7$_3$-brane. Analogous results hold for D3-D7$_1$ and D3-D7$_2$ twisted fields. Note that for $\theta_1 = \theta_2 = \pi/2$ the above result agrees with (12.40) upon T-duality along the second and third tori. Note also that all the above metrics are diagonal, i.e. there are no mixed kinetic terms for the matter fields.

To illustrate the structure of the above matter metrics, let us consider models with no worldvolume flux, and take $U_i = 1$ for simplicity. The above metrics and closed string Kähler potential are encoded in a complete Kähler potential

$$
\kappa_4^2 K = -\log\left(S + S^* - \sum_{i=1}^{3} \left|C_i^{7_i}\right|^2\right) - \sum_{i=1}^{3} \log\left(T_i + T_i^* - \left|C_i^3\right|^2 - \sum_{j,k=1}^{3} D_{ijk}\left|C_j^{7_k}\right|^2\right)
$$
$$
+ \frac{1}{2}\sum_{i,j,k=1}^{3} D_{ijk} \frac{\left|C^{7_j 7_k}\right|^2}{(S+S^*)^{1/2}\left(T_i+T_i^*\right)^{1/2}}
$$
$$
+ \frac{1}{2}\sum_{i,j,k=1}^{3} D_{ijk} \frac{\left|C^{37_i}\right|^2}{\left(T_j+T_j^*\right)^{1/2}\left(T_k+T_k^*\right)^{1/2}}, \tag{12.43}
$$

with $D_{ijk} = 1$ for $i \neq j \neq k \neq i$, and $D_{ijk} = 0$ otherwise. The lowest-order term in an expansion in matter fields reproduces the T and S dependence of the above metrics in the flux-less case; however, expression (12.43) is valid to all orders in the untwisted matter fields $C_i^{7_j}$, C_i^3. The structure inside the logs can be obtained from T-duality to the result for toroidal type I models, for which the untwisted matter dependence can be computed from KK reduction of the 10d theory; it amounts to a shift of the closed string moduli, as in (9.9) in heterotic models. For instance, T-duality along $\mathbf{T}^6$ turns the D9- into D3-branes and reproduces the first line in (12.43).

It is interesting to compare the above metrics for the flux-less case with those in heterotic toroidal orbifolds in Section 9.3.1. In the latter case the leading order T_i dependence of matter fields metrics in (9.44) is $\prod_{i=1}^{3}\left(T_i + T_i^*\right)^{n_i}$. The modular weights n_i are 0 or -1 for untwisted fields, and fractional numbers for twisted sector fields. In contrast, in type IIB orientifolds the analogs of the modular weights are always -1, 0, or $-1/2$; the only "twisted" modular weight is $1/2$, arising in mixed open string sectors, for which worldsheet oscillators have moddings shifted by $1/2$.

A second difference concerns the dependence on the complex dilaton S. In heterotic orbifolds the Kähler metrics do not depend perturbatively on S, while in IIB orientifolds this dependence appears in the Kähler metric of the D7-brane field $C_i^{7_i}$, and of the D7-brane intersection fields $C^{7_j 7_k}$. This will lead to relevant consequences in the structure of SUSY-breaking soft terms in flux compactifications in Section 15.5.3. From string duality, we would expect $SO(32)$ heterotic models to yield the additional S dependence only in the presence of non-perturbative NS5-branes, but inclusion of these non-perturbative objects would invalidate the use of the worldsheet tools required to treat the orbifold model.

On the other hand we already encountered such S-dependence for matter metrics in the heterotic M-theory effective action, see (9.41).

The computation of Kähler metrics for matter fields for non-toroidal type II orientifolds remains a challenge. However, we will see in Section 15.5.3 that, for certain classes of models with matter fields on a localized set of D7-branes, one may use scaling arguments to obtain the Kähler moduli dependence of the matter metrics. Similar scaling arguments also apply to local F-theory GUT models.

We conclude by mentioning that all these formulae equally apply to the mirror type IIA orientifolds with intersecting D6-branes, obtained via T-duality along the three x^i coordinates. This amounts to an exchange $T_i \leftrightarrow U_i$ in the above formulae.

12.3 The gauge kinetic function

We move on to the description of the gauge kinetic functions for type II orientifold models. We focus on D-brane worldvolume gauge fields, and consider models of intersecting/magnetized D-branes, and of D-branes at singularities. We finally also describe the gauge kinetic function for $U(1)$s arising in the RR closed string sector.

12.3.1 Gauge couplings for intersecting/magnetized branes

In perturbative heterotic CY compactifications the gauge kinetic function is universal and essentially given by the complex dilaton S. In contrast, in type II orientifolds, gauge bosons are localized on Dp-branes and the gauge kinetic function depends on p and the volume of the wrapped cycle Π_{p-3}. This translates into a dependence on the dilaton S and complex structure moduli U_i in type IIA models, and on S and Kähler moduli T_i in type IIB models. The gauge kinetic function may be computed from KK dimensional reduction of the D-brane DBI+CS action (6.6), (6.15), whose relevant structure for present purposes is

$$S_{\mathrm{D}p,\mathrm{DBI}} = -\mu_p \int_{\mathrm{D}p} e^{-\phi} \sqrt{\det(G + B - 2\pi\alpha'F)}, \quad \mu_p = \frac{(\alpha')^{-(p+1)/2}}{(2\pi)^p},$$

$$S_{\mathrm{D}p,\mathrm{CS}} = \mu_p \int_{\mathrm{D}p} \sum_q C_q \mathrm{tr}\, e^{2\pi\alpha'F - B} + \cdots, \tag{12.44}$$

where G, B are the metric and NSNS 2-form, implicitly pulled back onto the brane worldvolume, F the worldvolume gauge field strength, $C^{(n)}$ are RR forms and, exterior products of forms are understood. Expanding S_{DBI} to second order in F, the inverse of the gauge coupling constant is

$$\frac{1}{g_{\mathrm{D}p}^2} = e^{-\phi} \frac{(\alpha')^{(3-p)/2}}{(2\pi)^{p-2}} \mathrm{Vol}(\Pi_{p-3}). \tag{12.45}$$

It relates to the $(p + 1)$-dimensional coupling (6.14) by a factor of $\mathrm{Vol}(\Pi_{p-3})$, the volume of the wrapped cycle (absent for $p = 3$). The CS action provides for the extra axion coupling rendering the gauge kinetic function holomorphic in the moduli, e.g. through the term

$$\int_{\Pi_{p-3}} C_{p-3} \int_{4d} \text{tr}\,(F \wedge F). \tag{12.46}$$

Such couplings already appeared in Sections 10.3.1 and 11.4.2, when discussing $U(1)$ anomaly cancellation in orientifold compactifications.

Let us note that the above coupling constants and the corresponding gauge kinetic functions for D-brane groups derived below use the gauge group generator normalization $\text{Tr}\,(T_a T_b) = \delta_{ab}$. To compare with the SM gauge couplings, usually expressed in the convention $\text{Tr}\,(T_a T_b) = \frac{1}{2}\delta_{ab}$, one has to divide the gauge kinetic functions by a factor of 2.

Gauge couplings for intersecting D6-brane models

Let us consider first the case of D6-branes in type IIA orientifolds. In the general setup of Section 10.3.2, the above expressions give the following gauge kinetic function for the ath stack of D6-branes

$$f_a^{\text{D6}} = \frac{(\alpha')^{-3/2}}{(2\pi)^4} \left[e^{-\phi} \int_{\Pi_a} \text{Re}(e^{-i\varphi_a}\Omega_3) + i \int_{\Pi_a} C_3 \right], \tag{12.47}$$

where D6-branes wrap special lagrangian 3-cycles Π_a of BPS phase φ_a defined in (10.27), and whose volume is given by (10.28), i.e. $\int_{\Pi_a} \text{Re}\,(e^{-i\varphi_a}\Omega_3)$.

A particular case allowing simple explicit computations is provided by D6$_a$-branes with wrapping numbers (n_a^i, m_a^i) on a factorized $\mathbf{T}^6$. The CS term leads to

$$\text{Im}\,\left(f_a^{\text{D6}}\right) = \frac{1}{2\pi} n_a^1 n_a^2 n_a^3 C^0 - \frac{1}{2\pi} \sum_i n_a^i m_a^j m_a^k C^i, \quad i \neq j \neq k \neq i, \tag{12.48}$$

where the 4d RR scalars C^0, C^i are defined in terms of the basis (12.25) as

$$C^K = (4\pi^2\alpha')^{-3/2} \int_{\mathbf{T}^6} C_3 \wedge \beta^K, \quad K = 0, 1, 2, 3. \tag{12.49}$$

On the other hand, the inverse coupling constant from (12.45) reads

$$\frac{1}{g_a^2} = e^{-\phi} \frac{(\alpha')^{-3/2}}{(2\pi)^4} \prod_{i=1}^3 \left[\left(n_a^i\, 2\pi R_x^i\right)^2 + \left(m_a^i\, 2\pi R_y^i\right)^2 \right]^{\frac{1}{2}} \tag{12.50}$$

where the last factor gives the wrapped volume, just the product of the lengths of the three 1-cycles in each 2-torus, taken rectangular for simplicity; in the case of branes sitting on top of an orientifold plane the volume is halved, and so is the inverse coupling. The above expressions are valid in general, even in the absence of supersymmetry. In supersymmetric configurations (12.50) and (12.48) combine into a holomorphic function. This is shown by simplifying the non-linear structure of (12.50) using the supersymmetry condition (10.10) $\sum_{i=1}^{3} \theta_a^i = 0$, which implies $\sum_i \tan\left(\theta_a^i\right) = \prod_i \tan\left(\theta_a^i\right)$; then, plugging the trigonometric identity

$$\prod_{i=1}^{3} \left(1 + \tan^2 \theta_a^i\right)^{\frac{1}{2}} = 1 - \sum_{i \neq j} \tan\theta_a^i \tan\theta_a^j, \tag{12.51}$$

in (12.50), the resulting linear expression combines with (12.48) into the holomorphic coupling

$$2\pi f_a^{D6} = n_a^1 n_a^2 n_a^3 S - \sum_{i \neq j \neq k \neq i} n_a^i m_a^j m_a^k U_i, \tag{12.52}$$

with S, U_i as in (12.29). This result also follows from (12.47) with Ω_3 as in (12.28).

Gauge couplings for D7- and D3-brane models

Consider now type IIB orientifolds with D3- and magnetized D7-branes. In the general setup of Section 11.4.3, D7-branes on holomorphic 4-cycles Π_a, carrying worldvolume gauge fields F_a, have a gauge kinetic function

$$f_a^{D7} = \frac{(\alpha')^{-2}}{(2\pi)^5} \left[e^{-\phi} \int_{\Pi_a} \mathrm{Re}\left(e^{-i\varphi_a} e^{J + i2\pi\alpha' \mathcal{F}_a}\right) + i \int_{\Pi_a} \sum_k C_{2k} e^{2\pi\alpha' \mathcal{F}_a} \right], \tag{12.53}$$

where $2\pi\alpha'\mathcal{F} = 2\pi\alpha' F + B$, and we have used (11.103), allowing for a general BPS phase φ_a in analogy with (12.47). Upon expanding the exponential, the first term produces the wrapped volume, while the subsequent describe contributions from the worldvolume gauge flux, i.e. from lower-dimensional induced branes.

A particular case allowing simple explicit computations is D7$_i$-branes on a factorized $\mathbf{T}^6$, with wrapping numbers m^j and magnetic fluxes n^j, for which one finds

$$\frac{1}{g_i^2} = \frac{e^{-\phi}}{(2\pi)^5 \alpha'^2} \prod_{j \neq i} \left| m_i^j A^j + i\alpha' n_i^j \right|. \tag{12.54}$$

In the supersymmetric case, the holomorphic gauge kinetic function is the mirror of (12.52) with $(n^i, m^i) = (1, 0)$, now with S and T^i defined as in (12.31).

$$2\pi f_i^{D7} = n_i^j n_i^k S - m_i^j m_i^k T_i, \quad i \neq j \neq k \neq i. \tag{12.55}$$

For D3-branes we have the simple result

$$2\pi f_{D3} = S, \tag{12.56}$$

which can also be recovered from the above by using $(n^j, m^j) = (1, 0)$ for $i = 1, 2, 3$.

In F-theory unification models with 7-branes wrapping a local 4-cycle S, as in Section 11.5.3, the gauge kinetic function in the large volume limit is also $2\pi f \propto T_S$, with T_S the local Kähler modulus whose real part is the volume of S. In an $SU(5)$ F-theory GUT there are, however, additional corrections from the hypercharge flux F_Y required to break the symmetry down to the SM; those corrections appear from the expansion of (12.53) at second order in the flux, and are different for the three unbroken gauge factors $SU(3)$, $SU(2)$ and $U(1)_Y$, giving rise to sub-leading corrections to the $SU(5)$ unification of coupling constants. We direct the reader to the Bibliography for further details.

Gauge couplings for D9- and D5-brane models

The results for D9- and D5-branes can be obtained analogously. In particular, for toroidal orientifolds they can be obtained by application of T-duality, as in Section 12.1.4. For instance, in toroidal orientifolds with D9/D5-branes (with no magnetization), we have

$$2\pi f_{D9} = S, \quad 2\pi f_{D5_k} = T_k. \tag{12.57}$$

A general feature of type IIA and IIB orientifolds is that the value of gauge couplings is given by the volume wrapped by the corresponding brane. In a model with the different MSSM gauge factors localized on different branes, gauge couplings are in general not unified. One possibility to implement unification, as seemingly suggested by experimental data, is to simply tune these volumes; in this approach, the threshold corrections (which, just as for heterotic orbifolds in Section 9.6.1, are computable in toroidal models or their quotients) are less relevant since such corrections are minute compared to tree level modifications of the gauge couplings. An aesthetically more pleasing possibility is to construct orientifolds with a $SU(5)$ GUT symmetry, in which this unification is automatic; however, they have the drawback that perturbatively exact $U(1)$ symmetries forbid the $\mathbf{10} \cdot \mathbf{10} \cdot \mathbf{5}$ Yukawa coupling, so that either non-perturbative D-brane instanton effects (Chapter 13) or non-perturbative F-theory constructions (as in Section 11.5) must be invoked. We will come back to the question of gauge coupling unification in the different string constructions in Section 16.1.

12.3.2 Gauge couplings for D-branes at singularities

The above formulae are essentially geometric in nature, and need modifications for systems of D-branes in orbifolds, such as those in Sections 11.2 and 11.3. Their gauge kinetic functions contain, on top of the expected contributions described above (divided by the order of the orbifold for branes fixed under it), new contributions from twisted sector moduli, implied by the anomaly cancellation mechanism, e.g. in Section 11.2.2. Indeed, the complexification of the axion coupling (11.18) results in an additional term in the gauge kinetic function for the ath set of D-branes, given by

$$\Delta f_a \simeq \mathrm{Tr} \left(\gamma_\theta^k \lambda_a^2 \right) \Phi_k, \tag{12.58}$$

with Φ_k the θ^k-twisted sector moduli, whose imaginary parts are the RR axions.

For instance, consider a model with D9-branes in a type IIB toroidal $\mathbf{Z}_N$ orientifold, with odd N; using the D9-brane gauge kinetic function (12.57), and adding (12.58) with the Chan–Paton traces (11.21), we get

$$2\pi f_{D9,a} = \frac{1}{N} S + \frac{1}{N} \sum_k \sum_P n_a \cos\left(\frac{2\pi\,ka}{N}\right) \frac{\Phi_k^{(P)}}{c_k}, \tag{12.59}$$

where c_k^2 gives the number of θ_k fixed points, labeled by P. In applications the factor $1/N$ is often absorbed by redefining the moduli.

12.3.3 Kinetic function of RR U(1)s

As described in Section 12.1, in generic type II orientifolds (albeit more rarely in toroidal orbifolds) there are in general massless $U(1)$s from the RR sector. Since there are no massless states charged under them, they cannot be identified with any known physical $U(1)$, like hypercharge; however, there are possible phenomenological consequences of these gauge bosons, as briefly discussed in Section 16.4.3, so we quickly overview their gauge kinetic functions.

In the type IIA case there are $h_{1,1}^+$ such gauge bosons. Performing a dimensional reduction from the 10d theory their gauge kinetic function can be shown to be

$$f_{\alpha\beta}(T_c) = -D_{\alpha\beta c} T_c, \tag{12.60}$$

where $\alpha, \beta = 1, \ldots, h_{1,1}^+$, the T^c are the Kähler moduli labeled by $c = 1, \ldots, h_{1,1}^-$, and $D_{\alpha\beta c}$ are the CY triple intersection numbers (9.37). In the case of type IIB orientifolds with O7/O3-planes, there are $h_{2,1}^+$ RR $U(1)$s with gauge kinetic function

$$f_{\alpha,\beta}(U_{\tilde{k}}) = \mathcal{F}_{\alpha\beta}(U_{\tilde{k}}), \tag{12.61}$$

where now $\alpha, \beta = 1, \ldots, h_{2,1}^+$ and $\tilde{k} = 1, \ldots, h_{2,1}^+$. Here $\mathcal{F}_{\alpha\beta}$ is a holomorphic function of the complex structure moduli $U^{\tilde{k}}$, given by the second derivative of the underlying $\mathcal{N} = 2$ prepotential (see Section 2.4.1) – which has already appeared as $\mathcal{G}(U)$ in the heterotic context in Section 9.1.3. In the large complex structure regime, in the sense of Section 10.1.2, (12.61) is a linear function of $U_{\tilde{k}}$, in agreement with the expected mirror symmetry with (12.60). Note that the moduli dependence of the gauge kinetic functions are reversed compared to D-brane worldvolume gauge bosons; in type IIA the gauge kinetic functions for D6-brane gauge groups depend on complex structure moduli, whereas for RR $U(1)$s they depend on the Kähler moduli, and conversely for type IIB models.

12.4 U(1)'s and FI terms

As discussed, e.g., in Section 10.3.1, in type II orientifold compactifications some $U(1)$ gauge bosons become massive due to $B \wedge F$ couplings to RR closed string 2-forms. Such couplings play a crucial role in the 4d Green–Schwarz cancellation of mixed $U(1)$

anomalies, but can also exist for non-anomalous $U(1)$s. As shown in Section 9.5.2, super-symmetry in 4d $\mathcal{N} = 1$ models relates such couplings to field dependent FI terms. In type IIA orientifolds they are controlled by the dilaton S and complex structure fields U_n, whereas in type IIB orientifolds they are controlled by S and the Kähler moduli T_i. Recall that $U(1)_a$ FI parameters ξ^a modify the D-term scalar potential in the 4d effective action as in (2.29), i.e.

$$V_D = \sum_a \frac{1}{2\,g_{U(1)_a}^2} \left(\sum_n q_a^n |\phi_n|^2 + \xi_a \right)^2, \tag{12.62}$$

where the sum in a runs over all $U(1)$s, and the sum in n over all charged scalars. We have also reabsorbed here a factor $g_{U(1)_a}$ in the gauge field.

The FI contribution to the scalar potential can be understood as a contribution to the vacuum energy arising from D-brane configurations slightly deformed away from the supersymmetric vacuum conditions, as we show next. For concreteness we focus on type IIA intersecting D6-brane models, the analysis for mirror type IIB models being identical. The vacuum energy arises from the tensions of the D6-branes (including images, implicit in what follows) and O6-planes as

$$V = \mu_6\, e^{-\phi} \left[\sum_a N_a \, \mathrm{Vol}(\Pi_a) - 4\, N_{O6} \mathrm{Vol}(\Pi_{O6}) \right], \tag{12.63}$$

where $\mu_6 = (2\pi)^{-6} \alpha'^{-7/2}$, and N_a and N_{O6} are the numbers of D6-branes on Π_a and O6-planes on Π_{O6}. For a supersymmetric configuration, the D6/O6 tensions equal their RR charge, so RR tadpole cancellation implies that the above total energy vanishes; namely, in the language of Section 10.3.2, if Π_a and Π_{O6} are special lagrangian 3-cycles with equal phase $\varphi = 0$ in (10.27), the volumes (10.28) are such that (12.63) vanishes. Let us now consider a small deviation from this configuration, by changing complex structure moduli $\Omega_3 \to \Omega_3'$, with $\Omega_3'|_{\Pi_a} = (1 + i \tan \delta_a)\Omega_3|_{\Pi_a}$, so δ_a is a small non-zero BPS phase, i.e. $\mathrm{Im}\left(e^{-i\delta_a} \Omega_3' \right)|_{\Pi_a} = 0$ (and $\delta = 0$ for Π_{O6}). The vacuum energy is

$$V = \mu_6\, e^{-\phi} \left(\sum_a N_a \int_{[\Pi_a]} |\Omega_3'| - 4\, N_{O6} \int_{[\Pi_{O6}]} \mathrm{Re}\, \Omega_3' \right)$$

$$= \mu_6 e^{-\phi} \sum_a N_a \left(\int_{[\Pi_a]} |\Omega_3'| - \int_{[\Pi_a]} \mathrm{Re}\, \Omega_3 \right) = \mu_6 e^{-\phi} \sum_a N_a \int_{[\Pi_a]} |\Omega_3'| \, (1 - \cos \delta_a)$$

$$= \sum_a \frac{1 - \cos \delta_a}{g_{U(1)_a}^2 (2\pi \alpha')^2} \approx \sum_a \frac{1}{2\, g_{U(1)_a}^2} \left(\frac{\delta_a}{2\pi \alpha'} \right)^2. \tag{12.64}$$

In the second equality we have used $\mathrm{Re}\, \Omega_3'|_{\Pi_{O6}} = \mathrm{Re}\, \Omega_3|_{\Pi_{O6}}$, and the fact that in the SUSY case $\sum_a \int_{\Pi_a} \mathrm{Re}\, \Omega_3 - 4 N_{O6} \int_{\Pi_{O6}} \mathrm{Re}\, \Omega_3 = 0$, from the RR tadpole condition

$\sum_a N_a[\Pi_a] - 4N_{O6} = 0$; in the third, we used that $\mathrm{Im}\,\Omega|_{\Pi_a} = 0$ and hence $\mathrm{Re}\,\Omega_3|_{\Pi_a} = |\Omega_3||_{\Pi_a} = \cos\delta_a\,|\Omega_3'|\,|_{\Pi_a}$. The final FI term is

$$\xi_a = \frac{\delta_a}{2\pi\alpha'}, \tag{12.65}$$

controlled by the string scale and the angles (determined by complex structure moduli).

As a more explicit example, consider the type IIA $\mathbf{T}^6/(\mathbf{Z}_2\times\mathbf{Z}_2)$ orientifold. The volumes wrapped by the $D6_a$-brane (and its image a') and the whole set of O6-planes are

$$\mathrm{Vol}(\Pi_a) = (2\pi)^3\left(\prod_{i=1}^3 n_a^i\,R_x^i - \sum_{i=1}^3 n_a^i\,m_a^j\,m_a^k\,R_x^i\,R_y^j\,R_y^k\right),\quad i\neq j\neq k\neq i,$$

$$4N_{O6}\mathrm{Vol}(\Pi_{O6}) = (2\pi)^3\left(4\prod_{i=1}^3 R_x^i + 4\sum_{i=1}^3 R_x^i\,R_y^j\,R_z^k\right), \tag{12.66}$$

to be replaced in (12.63) to get the vacuum energy. It is easy to show that in supersymmetric models it cancels as a consequence of the RR tadpole cancellation conditions (10.46). The above explicit formula allows for a more explicit derivation of the expression for the FI term, e.g. for $D6_a$-branes with slightly non-supersymmetric angles $(0, \theta_a, -\theta_a + \delta_a)$ with $\delta_a \ll 1$:

$$V = e^{-\phi}\mu_6\sum_a N_a\,\mathrm{Vol}(\Pi_a)\,(1 - \cos\delta_a) \approx \sum_a \frac{1}{2\,g_{U(1)_a}^2}\left(\frac{\delta_a}{2\pi\alpha'}\right)^2, \tag{12.67}$$

We can also derive the FI term from the effective action. In this $\mathbf{Z}_2\times\mathbf{Z}_2$ orientifold, invariance under the shift (10.26) requires the Kähler potential for the dilaton S and complex structure fields U_i to include the $U(1)_a$ vector superfields as follows:

$$\kappa_4^2 K(U, U^*) = -\log\left(S + S^* - \sum_a Q_a^0 V_a\right) - \sum_{i=1}^3 \log\left(U_i + U_i^* - \sum_a Q_a^i\,V_a\right), \tag{12.68}$$

with $Q_a^0 = m_a^1 m_a^2 m_a^3$, $Q_a^i = -m_a^i n_a^j n_a^k$, $i \neq j \neq k \neq i$. Expanding to linear order in V_a, we get the following FI term

$$\frac{\xi_a}{g_{U(1)_a}^2} = \frac{e^{-\phi}}{2\pi\alpha'^{\frac{5}{2}}}\left(N_a\,m_a^1\,m_a^2\,m_a^3\,R_y^1\,R_y^2\,R_y^3 - \sum_i N_a\,m_a^i\,n_a^j\,n_a^k\,R_y^i\,R_x^j\,R_x^k\right), i\neq j\neq k,$$

where we have used $\kappa_4^2 = (\pi\alpha'^4 e^{2\phi})/\left(R_x^1\,R_x^2\,R_x^3\,R_y^1\,R_y^2\,R_y^3\right)$, easily obtained from (4.47). The FI term ξ_a is thus proportional to $\prod_i \tan\theta_a^i - \sum_i \tan\theta_a^i$, which vanishes for a SUSY configuration; on the other hand, for slightly non-SUSY angles, e.g. $(0, \theta_a, -\theta_a + \delta_a)$, the reader may check that $\xi_a = \sin\delta_a/(2\pi\alpha') \simeq \delta_a/(2\pi\alpha')$ as in (12.64). The mass matrix of the $U(1)$ bosons may be obtained expanding (12.68) to second order on the vector multiplets. One obtains

$$M_{ab}^2 = \sum_{K=0}^{3} \frac{Q_a^K Q_b^K}{\kappa_4^2 \left(U_K + U_K^*\right)^2},$$

(12.69)

with $U_0 = S$. Recalling the above expression for κ_4^2, one easily checks that (at least for isotropic tori) the non-vanishing masses are of order of the string scale; in particular for square tori with $R_{x,y}^1 = R_{x,y}^2 = R_{x,y}^3$ one obtains $M_{ab}^2 = \sum_K Q_a^K Q_b^K /(4\pi\alpha')$.

The mixed term in the D-term scalar potential (12.62) produces masses for the charged scalars, which actually reproduce the microscopic string theory mass formula (10.6) for light scalars at D6-brane intersections. Take a bi-fundamental scalar with charges $(q_a, q_b) = (+1, -1)$ under two D6-branes with SUSY misalignments δ_a, δ_b, with $\delta = \theta_1 + \theta_2 + \theta_3$. The mass from (12.62), using (12.65), is

$$m_{ab}^2 = q_a\xi_a + q_b\xi_b \simeq \frac{\delta_a - \delta_b}{2\pi\alpha'} = \frac{\theta_{ab}^1 + \theta_{ab}^2 + \theta_{ab}^3}{2\pi\alpha'},$$

(12.70)

reproducing the result in Section 10.2.2.

Note that the structure of FI terms in perturbative type II orientifolds is somewhat different from that in Section 9.5 for perturbative heterotic CY compactifications (without $U(1)$ gauge backgrounds) or toroidal orbifold compactifications. In these heterotic models the FI terms arise at one-loop, whereas in type II orientifolds they appear at tree level; in fact FI terms in type II orientifolds do not get further quantum corrections, as follows from the topological nature of the $B \wedge F$ couplings to which they relate by 4d $\mathcal{N} = 1$ supersymmetry. A further difference is that in the heterotic case the FI terms are controlled by the dilaton, so they cannot vanish in an interacting theory, and trigger unavoidably a Higgs mechanism by some charged scalars. In contrast, in type II orientifolds they are controlled by closed string geometric moduli (Kähler moduli in IIB and complex structure moduli in IIA models), and are proportional to the deviation from the supersymmetric configuration; hence, they can be consistently set to zero by simply choosing supersymmetric D-brane configurations, or turned on continuously if the symmetry breaking is desirable. Incidentally, this gauge symmetry breaking is the D-brane recombination described in Section 10.2.2.

We conclude by mentioning the structure of FI terms in type IIB models with D-branes in orbifolds, as in Sections 11.2 and 11.3. As described there, the $B \wedge F$ couplings involve Kähler moduli Φ_k in orbifold twisted sectors, hence the FI terms are controlled by their real parts Re Φ_k. Specifically, to leading order in the blowing-up modes one has kinetic terms

$$K_{\Phi_k} = \left(\Phi_k + \Phi_k^* + d_k V_{U(1)}\right)^2,$$

(12.71)

where $d_k \propto \text{tr}\,(\gamma_k \lambda_{U(1)})$. Then the FI term $\xi \propto d_k \text{Re}\,(\Phi_k)$, and so blowing-up the singularity turns on an FI term on the D-brane system. In the blown-down orbifold limit Re $\Phi_k = 0$ the FI terms are vanishing, so again, unlike the heterotic case, the FI terms may be safely put to zero and no gauge symmetry breaking is triggered. The corresponding $U(1)$, however, gets a mass of order the string scale.

12.5 Superpotentials and Yukawa couplings in type II orientifolds
12.5.1 Generalities

The computation of charged matter superpotentials in type II orientifolds is in general a difficult task, but a phenomenologically relevant one, as they lead to Yukawa couplings for SM fields in models of particle physics. The superpotential, and thus the Yukawa couplings, are functions of closed string moduli and may be in principle computed from disk diagrams; in practice, their computation has been carried out in models described by a free worldsheet CFT, namely toroidal models and orbifolds and orientifolds thereof, as described later on, although some ingredients have wider applicability.

Before focusing on these particular classes, we review some general features of perturbative superpotentials in type II orientifolds. An important property is that (in the absence of fluxes) the holomorphic superpotential for type IIA CY orientifolds depends only on Kähler moduli, while for type IIB CY orientifolds it depends only on complex structure moduli. This property follows from considerations of the microscopic worldsheet computation, and will be clearer in our derivation of Yukawa couplings in Sections 12.5.3 and 12.5.4. Note that this is consistent with the mirror symmetry exchanging both constructions and both kinds of moduli. Hence at the level of Yukawa couplings for massless chiral multiplets C^I we have

$$W_{\mathrm{IIA}} = h_{IJK}^{\mathrm{IIA}}(T_i)\, C^I C^J C^K + \cdots, \quad W_{\mathrm{IIB}} = h_{IJK}^{\mathrm{IIB}}(U_i)\, C^I C^J C^K + \cdots. \quad (12.72)$$

Despite this "decoupling" of moduli at the level of the holomorphic superpotential, the physical Yukawa couplings are not holomorphic functions of the moduli, due to additional factors from rescaling the participating fields to get canonical kinetic terms. Recalling (2.109), physical Yukawa couplings are thus

$$Y_{IJK} = (K_I K_J K_K)^{-1/2} e^{G/2} h_{IJK}, \quad (12.73)$$

where K_I is the Kähler metric of the matter field C^I and G is the Kähler function (2.47); the presence of these factors is crucial to disentangle the holomorphic and physical Yukawa couplings in explicit computations, as we describe below.

A final important property of perturbative Yukawa couplings in type II orientifolds is the selection rules imposed by the $U(1)$ symmetries arising from the $U(N)$ gauge groups on each D-brane stack. The perturbative superpotential is invariant under these symmetries, as follows from a simple analysis of the Chan–Paton indices in worldsheet diagrammatics. This is true even for those $U(1)$s which are massive due to $B \wedge F$ couplings, discussed in Section 12.4, which remain as perturbatively exact global symmetries. They are generically present, and impose selection rules beyond the mere invariance under the actual 4d gauge group; as already mentioned in Section 11.5, for example, they lead to important restrictions on Yukawa couplings. These selection rules can be overcome by considering non-perturbative setups, like the F-theory models in Section 11.5, or including non-perturbative instanton effects in the type II orientifold setups, as in Chapter 13.

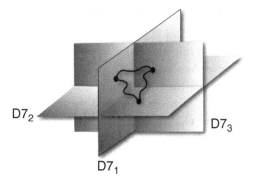

Figure 12.2 Schematic Yukawa coupling among three chiral fields $C^{7_i 7_j}$ living at D7-brane intersections. Each dimension represents one complex coordinate z_i, $i = 1, 2, 3$.

12.5.2 Yukawas in type IIB toroidal orientifolds and D7/D3-branes at singularities

The Yukawa couplings and their moduli dependence can be explicitly computed in certain setups, like the type IIB toroidal orientifolds in Section 11.2. Their couplings in fact have a simple structure and admit intuitive interpretations, as we now illustrate for D3/D7-brane models. For concreteness we focus on $\mathbf{Z}_2 \times \mathbf{Z}_2$ orientifolds (other cases admitting a similar analysis), in the absence of worldvolume magnetic fluxes. The model contains D7$_i$-branes wrapping the $\mathbf{T}^4$ transverse to the ith 2-torus. They have adjoint matter fields from position moduli $C_i^{7_i}$ and Wilson line moduli $C_i^{7_j}$ $i \neq j$, and bi-fundamental fields at intersections $C^{7_i 7_j}$, $i \neq j$. There are Yukawa couplings among these fields of the form (ignoring gauge indices for simplicity)

$$W_7 = C^{7_1 7_2} C^{7_2 7_3} C^{7_3 7_1} + \sum_{i=1}^{3} \left[D_{ijk} \, C_j^{7_i} C^{7_i 7_k} C^{7_i 7_k} + C_1^{7_i} C_2^{7_i} C_3^{7_i} \right], \tag{12.74}$$

with $D_{ijk} = 1$ for $i \neq j \neq k \neq i$, and zero otherwise. The last term in this expression is analogous to the untwisted Yukawa in heterotic orbifolds, as expected from T-duality to type I, and S-duality to heterotic. The second term shows that continuous Wilson lines $\langle C_j^{7_i} \rangle$ break the gauge symmetry and give masses to some of the $C^{7_i 7_k}$ fields living at intersections. Finally, the first Yukawa coupling corresponds to one among three intersecting D7-branes, see Figure 12.2.

There are additional Yukawa couplings in the presence of D3-branes, involving the D3-brane position moduli C_i^3, and the D3-D7$_i$ fields C^{37_i}. Concretely,

$$W_{37} = C_1^3 C_2^3 C_3^3 + \sum_i C_i^3 C^{37_i} C^{37_i} + \sum_i C_i^{7_i} C^{37_i} C^{37_i} + \sum_{i \neq j \neq k} D_{ijk} \, C^{7_j 7_k} C^{37_i} C^{37_k}. \tag{12.75}$$

The first coupling is T-dual of the third term in W_7 above, and has already appeared in (11.43). Other terms may be understood geometrically, e.g. the second term, already appeared in (11.47), describes the mass for D3-D7$_i$ fields C^{37i} when the D3-brane is moved away through a vev $\langle C_i^3 \rangle$. All the above superpotential couplings are of order one and satisfy the H-momentum conservation rule introduced for heterotic models in Section 9.3.2, now arising from worldsheet correlators on the disk.

In models of particle physics based on D7-brane configurations, the SM quarks and leptons often reside at D7$_i$-D7$_j$ intersections, and the relevant Yukawa couplings arise from the first term in W_7. Using (12.40), (12.30) the supergravity formula (12.73) gives for the physical Yukawa coupling for three intersecting D7-branes,

$$Y_{\text{inter}} = \frac{(S + S^*)^{1/4}}{\left[\left(T_1 + T_1^*\right) \left(T_2 + T_2^*\right) \left(T_3 + T_3^*\right) \right]^{1/4}}, \qquad (12.76)$$

for the flux-less case. Incidentally, this coupling is a geometric mean of the gauge couplings of the three intersecting D7-branes. For the other two terms in (12.74), the physical Yukawa couplings are $(T_i + T_i^*)^{-1/2}$, and hence equal the corresponding gauge coupling constants; this is expected, as these terms are related to gauge interactions by an enhanced $\mathcal{N} = 2$ SUSY preserved by the fields involved.

For other toroidal orientifolds, the Yukawa coupling superpotentials are directly inherited from the above, by simply truncating the fields to those surviving the orbifold projection, as described in Section 11.3.2. In particular, this truncation applied to (12.75) leads to the superpotential (11.46), (11.50) for local systems of D3/D7-branes at $\mathbf{C}^3/\mathbf{Z}_N$ abelian orbifold singularities, since it involves only fields localized at the singularity. There are also generalizations of these expressions for D3-brane systems at toric singularities, with some terms in the D3-brane superpotentials in (11.72) for the conifold and (11.77) for the dP$_1$ theory. We will not need to delve further into their description.

12.5.3 Type IIA orientifolds: Yukawas from disk worldsheet instantons

In the context of type IIA orientifolds, Yukawa couplings between fields living at D6-brane intersections arise from worldsheet instantons, in a way somewhat analogous to the Yukawa couplings in heterotic orbifolds described in Section 9.3.2. These are string worldsheets wrapped on a holomorphic 2d surface with disk topology and with boundaries on the intersecting D6-branes; hence, they pass through the different intersection points, each of which introduces an open string vertex operator changing the boundary conditions. For instance, Figure 12.3 shows the quark Yukawa coupling $H Q_L q_R$ in a configuration of three D6-branes on 3-cycles Π_a, Π_b, and Π_c, corresponding to the baryonic, left, and right stacks in the particle physics model in Section 10.6.

The requirement that the 2d surface on $\mathbf{X}_6$ spanned by the worldsheet should be holomorphic ensures that these instantons contribute to the holomorphic superpotential; their area is given by the integral of the Kähler form on $\mathbf{X}_6$. Denoting A_{ijk} the area (in string units) of a disk $\mathcal{D}_{ijk}$ joining fields at three intersections i, j, k, the corresponding Yukawa couplings is roughly of the form

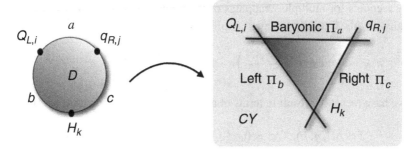

Figure 12.3 Worldsheet instantons involving three different boundary conditions, from the 2d worldsheet and spacetime points of view.

$$Y_{ijk} = h_{\text{qu.}} \sum_{\mathcal{D}_{ijk}} e^{-A_{ijk}}, \tag{12.77}$$

where the sum runs over possible holomorphic embeddings of disks joining the intersections. The coefficient $h_{\text{qu.}}$ stands for the contribution from worldsheet quantum corrections, i.e. fluctuations around the minimal area semiclassical solution; just as for heterotic orbifolds in Section 9.3.2, such contributions factorize from the semiclassical sum. Although the expression (12.77) is real, the inclusion of couplings of the string worldsheet to the background NSNS 2-form and to D6-brane Wilson lines makes them complex, and in fact holomorphic on the Kähler and D6-brane moduli, as shown later for toroidal models.

Let us describe the explicit computation of the worldsheet instanton sum for cubic superpotential couplings in factorized toroidal geometries, the results being valid for orbifold and orientifolds thereof. The 3-cycles wrapped by the D6-branes, denoted a, b, c, correspond to straight lines in each $\mathbf{T}^2$, and the holomorphic disks among three D6-branes project to triangles defined by three lines on each $\mathbf{T}^2$. Due to the factorized Kähler form (12.28), the area of the disk is the sum of the triangle areas. For a single $\mathbf{T}^2$ the triangle area defined by three ab, bc, ca intersection points can be computed to be

$$A_{ijk}(l) = \frac{(2\pi)^2 A}{2} |I_{ab} I_{bc} I_{ca}| \left(\delta(i, j, k) + l\right)^2, \tag{12.78}$$

where A is the $\mathbf{T}^2$ area, I_{ab}, I_{bc}, I_{ca} are the intersection numbers on a single $\mathbf{T}^2$, and i, j, k label the different intersection points of each kind, i.e.

$$i = 0, 1, \ldots, |I_{ab}| - 1, \quad j = 0, 1, \ldots, |I_{ca}| - 1, \quad k = 0, 1, \ldots, |I_{bc}| - 1. \tag{12.79}$$

Here we take the integers I_{ab}, I_{bc}, I_{ca} to be coprime, which ensures that there exists a triangle connecting every triplet (i, j, k). One also defines

$$\delta(i, j, k) = \frac{i}{I_{ab}} + \frac{j}{I_{ca}} + \frac{k}{I_{bc}} + \tilde{\epsilon}, \quad \tilde{\epsilon} = \frac{I_{ab}\epsilon_c + I_{ca}\epsilon_b + I_{bc}\epsilon_a}{I_{ab} I_{bc} I_{ca}}. \tag{12.80}$$

Here $\tilde{\epsilon}$ parametrizes the relative position of the branes, with ϵ_a, ϵ_b, ϵ_c giving the D6-brane transverse distances to the origin, in units of $A/\text{Vol}(\Pi)$, e.g. see Figure 12.5. Finally, the integer l accounts for multiple wrappings of the triangle around the $\mathbf{T}^2$. The contribution of triangles on this $\mathbf{T}^2$ to the Yukawa coupling is the sum

$$h_{ijk} \sim \sum_{l \in \mathbf{Z}} \exp\left(-\frac{A_{ijk}(l)}{2\pi\alpha'}\right) \sim \vartheta\begin{bmatrix}\delta\\\phi\end{bmatrix} = \sum_{l \in \mathbf{Z}} q^{\frac{1}{2}(\delta+l)^2} e^{2\pi i(\delta+l)\phi}, \qquad (12.81)$$

where we have recast the result in terms of a modular theta function (A.6), with

$$\delta = \delta(i,j,k), \quad \phi = 0, \quad q = \exp\left(-2\pi\frac{A}{\alpha'}|I_{ab}I_{bc}I_{ca}|\right). \qquad (12.82)$$

As mentioned above, there are in general phase contributions to the above Yukawa couplings. The string worldsheet couples to a NSNS 2-form field background via (3.10); this can be accounted for by complexifying the $\mathbf{T}^2$ area as $iA \to iA + B$, so the modular parameter becomes

$$q = \exp\left(2\pi i\frac{B+iA}{\alpha'}|I_{ab}I_{bc}I_{ca}|\right) \equiv \exp(2\pi i J |I_{ab}I_{bc}I_{ca}|), \qquad (12.83)$$

where J denotes the integral of the complexified Kähler 2-form over the 2-torus. Furthermore, the worldsheet boundary couples also to D6-brane Wilson line backgrounds around the compact directions via (3.13). Since in general we consider models with no gauge symmetry breaking by Wilson lines, we restrict to gauge backgrounds along the diagonal $U(1)$ factors, and denote $\exp(2\pi i\beta_a)$, $\exp(2\pi i\beta_b)$ and $\exp(2\pi i\beta_c)$ the Wilson lines phases for each D6-brane on its wrapped 1-cycle on $\mathbf{T}^2$, respectively. The worldsheet sweeping out the triangle picks up a total phase depending on the relative length fractions x_a, x_b, x_c, of its sides, namely

$$e^{2\pi i x_a\beta_a} \cdot e^{2\pi i x_b\beta_b} \cdot e^{2\pi i x_c\beta_c} = e^{2\pi i(I_{bc}\beta_a + I_{ca}\beta_b + I_{ab}\beta_c)(\delta+l)}. \qquad (12.84)$$

The Yukawa couplings including this contribution from the Wilson lines can be recast as the theta function with characteristics

$$h_{ijk} \sim \vartheta\begin{bmatrix}\delta\\\phi\end{bmatrix}(J|I_{ab}I_{bc}I_{ca}|), \qquad (12.85)$$

where now

$$\delta = \delta(i,j,k), \quad \phi = I_{ab}\beta_c + I_{ca}\beta_b + I_{bc}\beta_a. \qquad (12.86)$$

Note that this is in general a complex function, which is a relevant property to achieve a non-trivial CKM CP-violating phase in semi-realistic models.

Since the complete $\mathbf{T}^6$ model is factorizable, the total value of the Yukawa is obtained as the product of the three $\mathbf{T}^2$ contributions, namely

$$Y_{ijk} = h_{\text{qu.}} \prod_{r=1}^{3} \vartheta\begin{bmatrix}\delta^r\\\phi^r\end{bmatrix}(J^r |I_{ab}^r I_{bc}^r I_{ca}^r|), \qquad (12.87)$$

where the parameters δ^r, ϕ^r, and J^r, are given by (12.80), (12.86), and (12.83) for each $(\mathbf{T}^2)_r$. Remarkably, this semiclassical piece contains all the flavour-dependent information on the Yukawas.

The quantum part can be computed to be given by

$$h_{\text{qu.}} = e^{\phi/2} \prod_{r=1}^{3} \left[\frac{\Gamma\left(1 - \nu_{ab}^r\right) \Gamma\left(1 - \nu_{ca}^r\right) \Gamma\left(\nu_{ab}^r + \nu_{ca}^r\right)}{\Gamma\left(\nu_{ab}^r\right) \Gamma\left(\nu_{ca}^r\right) \Gamma\left(1 - \nu_{ab}{}^r - \nu_{ca}^r\right)} \right]^{1/4}, \tag{12.88}$$

in terms of Euler's Γ function with $\nu_{ab}^r = \left(\theta_a^r - \theta_b^r\right)/\pi$. The fact that the flavour dependence is restricted to a subsector of moduli (Kähler), while the remaining moduli (complex structure) are flavour-blind, is physically relevant in addressing the flavour problem in supersymmetry breaking, see Section 15.6.2.

12.5.4 Type IIB orientifolds: Yukawas from overlap integrals

We now describe the computation of Yukawa couplings in type IIB orientifolds with magnetized branes. We focus on toroidal models, which are actually mirrors of the previous toroidal type IIA D6-brane models; it is however useful to derive the Yukawa couplings directly in type IIB language, since the main ideas generalize to other geometries, as we exploit in Section 12.7.

We consider magnetized D9-brane models in the large volume regime, in which 4d chiral matter arises from zero modes in the KK compactification of 10d fields, and the cubic Yukawa couplings arise from cubic interactions already present in the 10d spacetime action. The Yukawa coupling coefficients are given by overlap integrals of these internal zero modes, in analogy with heterotic Yukawa couplings in Section 9.1.3 – which is not surprising due to the relation between heterotic and type I compactifications. An important property about these zero modes is their localization in the internal space: from the string theory viewpoint, open strings with endpoints on differently magnetized D-branes have boundary conditions freezing the center of mass position; from the spacetime viewpoint, such strings behave as charged particles in a magnetic field, with groundstates corresponding to lowest Landau levels. Their profiles are Gaussian and lead to exponentially suppressed overlaps, in agreement with mirror symmetry to the type IIA exponential Yukawa couplings (12.77).

To make the discussion more explicit, the 10d lagrangian on D9-branes with $U(N)$ gauge group reduces at low energies (i.e. large radius regime) to 10d super-Yang–Mills (SYM)

$$L = -\frac{1}{4}\text{Tr}\left(F^{MN}F_{MN}\right) + \frac{i}{2}\text{Tr}\left(\bar{\Psi}\Gamma^M D_M \Psi\right). \tag{12.89}$$

The worldvolume magnetic fields F_a, F_b, etc., break the gauge group upon compactification as $U(n) \to U(n_a) \times U(n_b) \times \cdots$

$$F = \begin{pmatrix} F_a & & \\ & F_b & \\ & & \ddots \end{pmatrix}, \quad \Rightarrow \quad A = \begin{pmatrix} A_a & & \\ & A_b & \\ & & \ddots \end{pmatrix}.$$

The 10d $U(N)$ gaugino field leads to massless 4d gauginos and bi-fundamental chiral fermions, from diagonal and off-diagonal blocks, respectively. For the case of two factors, we have

$$\Psi = \left(\begin{array}{c|c} A & B \\ \hline C & D \end{array} \right) = \left(\begin{array}{c|c} \text{4d } U(n_a) & \text{4d bif.}(\mathbf{n}_a, \bar{\mathbf{n}}_b) \\ \text{gaugino} & \text{chiral fermion} \\ \hline & \text{4d } U(n_b) \\ \text{CPT conj.} & \text{gaugino} \end{array} \right).$$

In the compactification to 4d, the 10d fields are expanded in harmonics as

$$\Psi(x^\mu, y^m) = \sum_k \chi_{(k)}(x^\mu) \otimes \psi_{(k)}(y^m), \tag{12.90}$$

$$A_n(x^\mu, y^m) = \sum_k \varphi_{(k)}(x^\mu) \otimes \phi_{(k),n}(y^m),$$

where x^μ and y^m are 4d and internal coordinates, respectively. The internal components of the off-diagonal blocks $\psi_{(k)}^{ab}$, $\phi_{(k),m}^{ab}$ are eigenfunctions of the internal kinetic operators $\tilde{\slashed{D}}_{6d}$, $\tilde{\Delta}_{6d}$, and zero modes are solutions of

$$\tilde{\slashed{D}}_{6d} \, \psi_{(0)}^{ab} = 0, \quad \tilde{\Delta}_{6d} \, \phi_{(0),m}^{ab} = 0, \tag{12.91}$$

where tildes indicate the coupling to the gauge background $F_a - F_b$. We focus on SUSY models, where the boson and fermion 6d wave functions are identical.

The 4d Yukawa couplings between these massless fields arise from KK reduction of the cubic coupling $A \cdot \Psi \cdot \Psi$ from the 10d fermion lagrangian in (12.89). As illustrated in Figure 12.4, the Yukawa coupling coefficients are the overlap integrals

$$Y_{ijk} = \frac{g}{2} \int_{\mathbf{X}_6} \psi_i^{\alpha\dagger} \, \Gamma^m \, \psi_j^\beta \, \phi_{k\,m}^\gamma \, f_{\alpha\beta\gamma}, \tag{12.92}$$

where g is the 10d gauge coupling, α, β, γ are $U(n)$ gauge indices and $f_{\alpha\beta\gamma}$ are $U(n)$ structure constants; also ψ, ϕ are fermionic and bosonic zero modes respectively, and i, j, k label the different zero modes in a given charge sector, i.e. the families in semi-realistic models. The Yukawa couplings are thus obtained as overlap integrals of the three zero mode wave functions in $\mathbf{X}_6$.

In the following we consider such wave functions and their overlap integrals explicitly for factorized $\mathbf{T}^6$ compactifications, the results being useful for orientifolds and orbifolds

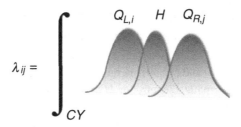

$$\lambda_{ij} = \int_{CY}$$

Figure 12.4 Pictorial representation of the computation of Yukawa couplings as triple overlap integrals of zero modes.

thereof, see Section 11.4. Consider first the simple case of a $U(1)$ gauge group in a single $\mathbf{T}^2$ with quantized constant flux given by

$$F_{z\bar{z}} = \frac{i\pi}{\operatorname{Im}\tau}M, \quad M \in \mathbf{Z}, \tag{12.93}$$

where $\tau = i\,\mathcal{U}$ in terms of the $\mathbf{T}^2$ complex structure parameter in (12.27), and z is a $\mathbf{T}^2$ complex coordinate. Consider a 2d two-component spinor $\psi = (\psi_+, \psi_-)$ on $\mathbf{T}^2$, with charge q under the $U(1)$. The corresponding zero mode equations are

$$D\psi_+ = D^{\dagger}\psi_- = 0, \tag{12.94}$$

with D the 2d gauge covariant derivative. To describe their solutions, we introduce the functions

$$\psi^{j,N}(\tau, \nu) = \mathcal{N}_j\, e^{i\pi N\nu \operatorname{Im}\nu/\operatorname{Im}\tau}\, \vartheta\begin{bmatrix} \frac{j}{N} \\ 0 \end{bmatrix}(N\nu, N\tau), \tag{12.95}$$

where $\mathcal{N}_j$ are normalization factors $\mathcal{N}_j = (2\operatorname{Im}\tau|M|)^{1/4}A^{-1/2}$, with A the $\mathbf{T}^2$ area; also ϑ is the Jacobi theta-function with characteristics

$$\vartheta\begin{bmatrix} a \\ b \end{bmatrix}(\nu, \tau) = \sum_{l\in\mathbf{Z}} e^{\pi i(a+l)^2\tau}\, e^{2\pi i(a+l)(\nu+b)}. \tag{12.96}$$

In terms of (12.95), we can write down the solutions to (12.94) as

$$\psi_+^j \equiv \psi^{j,qM}(\tau, z+\zeta) \quad \left(\psi_+^j\right)^* = \psi^{-j,-qM}(\bar{\tau}, \bar{z}+\bar{\zeta}),$$

$$\psi_-^j \equiv \psi^{j,qM}(\bar{\tau}, \bar{z}+\bar{\zeta}) \quad \left(\psi_-^j\right)^* = \psi^{-j,-qM}(\tau, z+\zeta), \tag{12.97}$$

where $j = 0, \ldots, |M| - 1$, the star denotes complex conjugation, and $\zeta = \zeta_1 + \tau\zeta_2$ is a complexified Wilson line. One can explicitly check that

$$D\psi_+^j = 0 \quad (q = +1), \quad D^{\dagger}\left(\psi_+^j\right)^* = 0 \quad (q = -1), \tag{12.98}$$

$$D^{\dagger}\psi_-^j = 0 \quad (q = +1), \quad D\left(\psi_-^j\right)^* = 0 \quad (q = -1). \tag{12.99}$$

We can loosely interpret ψ_+^j as the zero mode wave functions corresponding to massless 4d left-handed fermions, while $(\psi_+^j)^*$ correspond to their antiparticles. Right-handed fermions are then associated to ψ_-^j and their antiparticles to $(\psi_-^j)^*$. Notice the important fact that the solutions (12.98) and (12.99) are mutually exclusive, in the sense that the theta function defining ψ_+^j is normalizable only for $M > 0$, whereas ψ_-^j is only well-defined for $M < 0$; hence, the non-trivial flux $M \neq 0$ automatically selects a preferred chirality of the 2d (and hence the 4d) spinors. Moreover, we obtain several replicas of such chiral fermions, from the $|M|$ independent solutions (lowest Landau levels) of the Dirac equation. This is the microscopic origin of chirality and replication in the type IIB language, reproducing the index theorem result in Section 11.4.1.

Instead of a single $U(1)$, we are rather interested on a product of non-abelian factors, resembling if possible the SM structure. In order to have non-trivial Yukawa couplings, we introduce three magnetized stacks, in terms of a flux of the form

$$
F_{z\bar{z}} = \frac{\pi i}{\text{Im}\,\tau} \begin{pmatrix} \frac{n_a}{m_a}\mathbf{1}_{n_a} & & \\ & \frac{n_b}{m_b}\mathbf{1}_{n_b} & \\ & & \frac{n_c}{m_c}\mathbf{1}_{n_c} \end{pmatrix}, \tag{12.100}
$$

where we take $m_\alpha \in \mathbf{N}^+$ and n_α is an integer multiple of m_α, $\alpha = a, b, c$. The initial gauge group $U(m)$ is broken to $U(m_a) \times U(m_b) \times U(m_c)$, We are now interested in the zero modes corresponding to chiral bi-fundamental fields with respect to the three gauge groups involved. One finds wave functions similar to those of (12.97) upon the replacements $N \to \tilde{I}_{\alpha\beta} = n_\alpha/m_\alpha - n_\beta/m_\beta = I_{\alpha\beta}/m_\alpha m_\beta$; the $\tilde{I}_{\alpha\beta}$ are analogs of the intersection numbers in the intersecting D6-brane type IIA dual, since they give the number of zero mode solutions. Generalizing to $\mathbf{T}^2 \times \mathbf{T}^2 \times \mathbf{T}^2$ the zero mode wave functions are products of the three $\mathbf{T}^2$ wave functions,

$$
\psi^{j^r, \tilde{I}_{ab}^r}(\tau_r, z^r + \zeta_{ab}^r) = \mathcal{N}_r \exp\left[i\pi \, \tilde{I}_{ab}^r \left(z^r + \zeta_{ab}^r \right) \frac{\text{Im}\,\left(z^r + \zeta_{ab}^r \right)}{\text{Im}\,\tau_r} \right]
$$

$$
\times \vartheta \begin{bmatrix} \frac{j^r}{\tilde{I}_{ab}^r} \\ 0 \end{bmatrix} \left(\tilde{I}_{ab}^r \left(z^r + \zeta_{ab}^r \right), \tilde{I}_{ab}^r \tau_r \right), \tag{12.101}
$$

where $j^r = 0, \ldots, |\tilde{I}_{ab}^r| - 1$ label the number of zero modes on $(\mathbf{T}^2)_r$, $r = 1, 2, 3$ and the Wilson line parameters $\zeta_{\alpha\beta}^r$ on $(\mathbf{T}^2)_r$ are defined as

$$
\zeta_{\alpha\beta} = \frac{n_\alpha m_\beta \zeta_a - n_\beta m_\alpha \zeta_\beta}{I_{\alpha\beta}}. \tag{12.102}
$$

The corresponding Laplace equation for the scalar fields leads to analogous wave functions. In a factorized torus $\mathbf{T}^2 \times \mathbf{T}^2 \times \mathbf{T}^2$ the Yukawa coupling is the product of three integrals of the form

$$
y_{ijk} = \int_{\mathbf{T}^2} dz\,d\bar{z}\; \psi^{i, \tilde{I}_{ab}}(z + \zeta_{ab}) \cdot \psi^{j, \tilde{I}_{ca}}(z + \zeta_{ca}) \cdot \left(\psi^{k, \tilde{I}_{cb}}(z + \zeta_{cb}) \right)^*, \tag{12.103}
$$

where i, j, k label the flavours. Despite the seemingly complicated structure of the zero mode wave functions, their triple overlaps can be computed explicitly. The final result on $\mathbf{T}^6$ is

$$Y_{ijk} = e^{\phi/2} \prod_{r=1}^{3} \left[\frac{2\mathrm{Im}\,\tau_r}{(A_r/\alpha')^2} \right]^{\frac{1}{4}} \left| \frac{\tilde{I}_1^r \tilde{I}_2^r}{\tilde{I}_1^r + \tilde{I}_2^r} \right|^{\frac{1}{4}} e^{\frac{H^r}{2}} \vartheta \begin{bmatrix} \delta_{ijk}^r \\ 0 \end{bmatrix} \left(\tilde{\zeta}^r, \tau_r \left| I_{ab}^r I_{bc}^r I_{ca}^r \right| \right),$$

$$(12.104)$$

Here $|\tilde{I}_1^r|$ and $|\tilde{I}_2^r|$ are the two smallest numbers among $|\tilde{I}_{ab}^r|$, $|\tilde{I}_{bc}^r|$, and $|\tilde{I}_{ca}^r|$, ϕ is the 10d dilaton, and

$$\delta_{ijk}^r = \frac{i^r}{I_{ab}^r} + \frac{j^r}{I_{ca}^r} + \frac{k^r}{I_{bc}^r}, \qquad (12.105)$$

$$H^r = 2\pi i \left| I_{ab}^r I_{bc}^r I_{ca}^r \right|^{-1} \frac{\tilde{\zeta}^r \mathrm{Im}\,\tilde{\zeta}^r}{\mathrm{Im}\,\tau_r}, \qquad (12.106)$$

$$\tilde{\zeta}^r = I_{ab}^r \tilde{\zeta}_c^r + I_{bc}^r \tilde{\zeta}_a^r + I_{ca}^r \tilde{\zeta}_b^r, \qquad (12.107)$$

with $\tilde{\zeta}_\alpha = n_\alpha \zeta_\alpha / m_\alpha$. This is already the physical Yukawa coupling, i.e. with fields having canonical kinetic terms. Comparing with the supergravity formula (12.73) one is led to identify the holomorphic piece with the holomorphic superpotential Yukawa coupling

$$h_{ijk}(\tau_r, \tilde{\zeta}^r) = \prod_{r=1}^{3} \vartheta \begin{bmatrix} \delta_{ijk}^r \\ 0 \end{bmatrix} \left(\tilde{\zeta}^r, \tau_r \left| I_{ab}^r I_{bc}^r I_{ca}^r \right| \right). \qquad (12.108)$$

Recalling (12.72), this is indeed a holomorphic function of the complex structure and the open string moduli (complexified Wilson lines). In particular, since it is independent of Kähler moduli, this holomorphic superpotential is remarkably valid even away from large volume. The remaining factors in (12.104) can be shown to correspond to the non-holomorphic pieces in (12.73), and do depend on Kähler moduli, hence the above expression only holds in the large volume regime.

This result should be equivalent by T-duality/mirror symmetry to the results found for intersecting D6-branes models in Section 12.5.3. In particular, writing $z_r = x^r + \tau_r y^r$, T-duality along the y^r-directions turns the intersecting D6-branes models into type I compactifications with magnetized D9-branes. The intersection angles $\theta_{\alpha\beta}^r$ map into $\alpha' \tilde{I}_{\alpha\beta}^r / A$ and the type IIB large volume regime corresponds to the type IIA regime of small angles. One can indeed show that both results for Yukawas agree after the T-duality replacements

$$\tau \longleftrightarrow J, \quad \tilde{\zeta} = \tilde{\zeta}_1 + \tau \tilde{\zeta}_2 \longleftrightarrow \nu = \phi + J\epsilon, \qquad (12.109)$$

in each $\mathbf{T}^2$. Namely, exchanging complex structure and Kähler moduli, and type IIA and IIB open string moduli. The matching of both computations is a remarkable test of T-duality/mirror symmetry. Both computations are technically and conceptually quite distinct, since in type IIB it involves compactification of 10d SYM *field theory*, whereas in type IIA the result arises from *stringy* effects, i.e worldsheet instantons. However, both describe the same 4d physical quantities, and nicely agree.

12.6 Effective action of an MSSM-like example

To illustrate the above general computations in an explicit model, it is worthwhile to describe the effective action of an explicit toroidal orientifold yielding an MSSM-like model of particle physics. The model is a generalization with k families of the $\mathbf{T}^6/(\mathbf{Z}_2 \times \mathbf{Z}_2)$ orientifold with MSSM-like spectrum in Section 10.6, interpreted in type IIB framework as magnetized D7$_i$-brane models as in Section 11.4.2. The relevant data for the MSSM sector are:

Branes	N_a	(n_a^1, m_a^1)	(n_a^2, m_a^2)	(n_a^3, m_a^3)	$(\theta_a^1, \theta_a^2, \theta_a^3)$
D7$_1$	6	(1,0)	$(k, 1)$	$(k, -1)$	$\left(\dfrac{\pi}{2}, \pi\delta_2, \pi - \pi\delta_3\right)$
D7$_2$	2	(0,1)	$(1, 0)$	$(0, -1)$	$\left(0, \dfrac{\pi}{2}, \pi\right)$
D7$_3$	2	(0,1)	$(0, -1)$	$(1, 0)$	$\left(0, \pi, \dfrac{\pi}{2}\right)$

$$\text{(12.110)}$$

with θ_a^i defined by (12.35) and $\pi\delta_i = \arctan(\alpha' k / A_i)$. This sector is 4d $\mathcal{N} = 1$ supersymmetric, with condition (11.98), for a suitable choice of Kähler moduli. In particular, the SUSY condition for D7$_1$-branes requires $A_2 = A_3$. Also, recalling from Section 11.4.2 that the model includes additional hidden branes, the SUSY condition $\sum_i \tan\theta_a^i = \prod_i \tan\theta_a^i$ requires $12A_1 + 8A_2 + 6A_3 = A_1 A_2 A_3 / \alpha'^2$. Let us discuss now some relevant ingredients of the effective action.

Gauge kinetic functions

For simplicity, we describe the effective action for the model in the Pati–Salam case, where some brane stacks coincide and the gauge group enhances to $SU(3 + 1) \times SU(2)_L \times SU(2)_R \times U(1)'$. The gauge kinetic functions from (12.55) are

$$2\pi f_{3+1} = T_1 + k^2 S; \quad 2\pi f_L = \frac{1}{2} T_2; \quad 2\pi f_R = \frac{1}{2} T_3, \tag{12.111}$$

where the factors $1/2$ in $f_{L,R}$ arise from the fact that the D7$_2$- and D7$_3$-branes sit on top of orientifold planes. Possible gauge coupling unification within this model is discussed in Section 16.1.2.

Matter fields kinetic terms

The SM fields are localized at intersections of the $D7_i$ branes. The Higgs field lives at the intersection of D7$_2$- and D7$_3$-branes, which carry no worldvolume magnetic fluxes, so from (12.40) we have

$$K_H = \frac{1}{(s\, t_1\, u_2\, u_3)^{1/2}}, \tag{12.112}$$

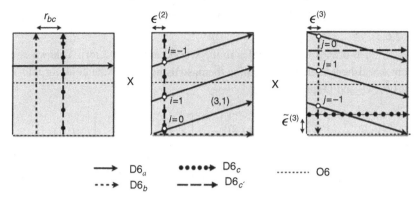

$$\longrightarrow \quad D6_a \qquad \bullet\bullet\bullet\bullet\bullet\bullet \quad D6_c \qquad \cdots\cdots\cdots \quad O6$$
$$\cdots\cdots\rightarrow \quad D6_b \qquad \text{---} \rightarrow \quad D6_{c'}$$

Figure 12.5 Brane configuration corresponding to the MSSM-like model described in the text, for the choice $k = 3$. For simplicity, we have not depicted the stacks d, d', a'.

where we define $s = (S + S^*)$, $t_i = (T_i + T_i^*)$, $u_i = (U_i + U_i^*)$. Left-handed fermions Q_L, L, are localized at D7$_1$-D7$_2$ intersections, and, after some algebra, (12.39) yields

$$K_{Q_L,L} = \frac{1}{s^{1/2}(t_2 t_3)^{1/4}} \frac{\Gamma\left(\frac{1}{2} - \delta\right)}{\Gamma(1 - \delta)}, \tag{12.113}$$

with the same result for the metric $K_{Q_R,R}$ of right-handed SM multiplets from D7$_1$-D7$_3$ intersections. Here we have not included the dependence on the complex structure fields. In the large volume regime, fluxes are dilute and we can expand these expressions for small δ. One finds for $K_{Q_L,L}$

$$K_{Q_L,L} = K_{Q_R,R} \xrightarrow{\delta \to 0} \frac{\pi}{(st)^{1/2}} \left(1 + \frac{k \log 4}{\pi} \sqrt{\frac{s}{t}}\right), \tag{12.114}$$

where we have set $t_2 = t_3 = t$ because $A_2 = A_3$. Kähler metrics analogous to these will be relevant in the computation of SUSY-breaking soft terms in Section 15.5.3.

Yukawa couplings

The structure of Yukawa couplings is more intuitive in the language of the mirror type IIA orientifolds with intersecting D6-branes. In order to illustrate the flavour structure in a realistic case, we focus on the 3-family model $k = 3$, shown in Figure 12.5 in the purely toroidal (not orbifold) case. We denote a, b, c the D6-branes which, together with their orientifold images, give rise to the $SU(4)$, $USp(2)_L$ and $USp(2)_R$ gauge factors in Pati–Salam language. Recall that the $SU(4)_a$ may be broken to $SU(3)_a \times U(1)_d$ by splitting the stack a; in the toroidal case one can also separate the brane c from its image to break $SU(2)_R \to U(1)_c$. The relevant gauge group then becomes the SM one, enlarged by a gauged $U(1)_{B-L}$. The D6-brane locations are characterized by the parameters ϵ, and r_{bc} denotes the b-c distance in $(\mathbf{T}^2)_1$. The latter is the vev of a modulus X coupling to the Higgs pairs as $X H_u H_d$ (since they arise in the bc open string sector, and get a mass from stretching across r_{bc}), so it plays a role similar to the NMSSM singlet mentioned in Section 2.6.3.

The a, b, c D6-branes form triangles in the second and third 2-tori. Yukawa couplings arise from worldsheet instantons spanning them, so using (12.87) the holomorphic Yukawa couplings for U- and D-quarks are

$$
Y_{ij}^U \sim \vartheta \left[\begin{array}{c} \frac{i}{3} + \epsilon^{(2)} \\ \phi^{(2)} \end{array} \right] \left(3J^{(2)}\right) \times \vartheta \left[\begin{array}{c} \frac{j}{3} + \epsilon^{(3)} + \tilde{\epsilon}^{(3)} \\ \phi^{(3)} + \tilde{\phi}^{(3)} \end{array} \right] \left(3J^{(3)}\right),
$$

$$
Y_{ij*}^D \sim \vartheta \left[\begin{array}{c} \frac{i}{3} + \epsilon^{(2)} \\ \phi^{(2)} \end{array} \right] \left(3J^{(2)}\right) \times \vartheta \left[\begin{array}{c} \frac{j*}{3} + \epsilon^{(3)} - \tilde{\epsilon}^{(3)} \\ \phi^{(3)} - \tilde{\phi}^{(3)} \end{array} \right] \left(3J^{(3)}\right),
$$

(12.115)

where the geometric meaning of $\epsilon^{(2)}$, $\epsilon^{(3)}$, and $\tilde{\epsilon}^{(3)}$ is indicated in Figure 12.5, and $\phi^{(2)}$, $\phi^{(3)}$, $\tilde{\phi}^{(3)}$ are Wilson line variables whose full definition is unnecessary for our purposes. The structure of the mass matrices can be recast as

$$
Y^U \sim A \cdot \begin{pmatrix} 1 & 1 & 1 \\ 1 & 1 & 1 \\ 1 & 1 & 1 \end{pmatrix} \cdot B, \quad Y^D \sim A \cdot \begin{pmatrix} 1 & 1 & 1 \\ 1 & 1 & 1 \\ 1 & 1 & 1 \end{pmatrix} \cdot \tilde{B},
$$

(12.116)

where A, B, $\tilde{B}$ are diagonal matrices with entries bilinear in ϑ-functions. This texture leads to one massive and two massless eigenstates, an encouraging result reproducing qualitatively the observed structure of fermion masses, with the third quark-lepton family much heavier than the first two. We may expect that further corrections from other sources (e.g. instantons, see next chapter) may give rise to smaller but non-negligible contributions to the rest of the masses.

In order to obtain the physical Yukawa couplings, these terms in the holomorphic superpotential must be multiplied by the non-holomorphic factors as in (12.73). This can be done using the above expressions for the Kähler metrics. Very remarkably, since different families have equal intersection angles, this amounts to an overall factor, with no flavour dependence.

The result for Yukawa couplings in the type IIB intersecting D7-brane picture can be obtained by replacing the Kähler parameter J^r, the brane positions ϵ^r, $\tilde{\epsilon}^r$, and Wilson line parameters ϕ^r by the complex structure U^r and the Wilson line parameters ζ_1^r, ζ_2^r.

This simple model nicely illustrates the possibilities of realistic D-brane model building in string theory, in setups explicit enough to allow the computation of their 4d effective action. It is remarkably easy to obtain non-trivial appealing features, like a realistic gauge group and a correct chiral spectrum of three quark-lepton families, along with reasonable leading order Yukawa coupling structure. On the other hand, the model is not fully realistic, e.g. since it contains unwanted massless adjoint chiral multiplets from D-brane moduli. Some of these features can be improved upon supersymmetry breaking and moduli fixing, described in later chapters.

12.7 Yukawa couplings in local F-theory models*

The techniques to compute Yukawa couplings in the type IIB setting in principle generalize to more complicated curved geometries and non-constant gauge fluxes. In practice, an obvious difficulty is that the computation of the zero mode wave functions requires a knowledge of the global compact geometry. The problem is however more tractable within the context of a bottom-up approach as described in Section 11.3.1. The idea is that in models with localized SM brane sector, it suffices to have a local description of the geometry around the branes to compute the relevant SM physical parameters. This strategy applies both to local configurations of D7-branes wrapping intersecting 4-cycles inside CYs and to the local F-theory models in Section 11.5.3, on which we now focus.

As discussed in Section 11.5.2, 7-branes wrapped on a localized 4-cycle S carry a worldvolume GUT gauge group G_S, e.g. $SU(5)$. The quark and lepton multiplets of the SM reside at matter curves $\Sigma_a \subset S$, corresponding to the intersection with additional 7_a-branes. Yukawa couplings come from the triple overlaps of these matter curves involving quarks, leptons and Higgs fields, see Figure 11.10 for an $SU(5)$ example. As already mentioned, one of the advantages of F-theory GUT unification is the natural presence of $\mathbf{10} \cdot \mathbf{10} \cdot \mathbf{5}_H$ Yukawa couplings, which are perturbatively absent in $SU(5)$ type IIB orientifolds. A point worth mentioning in this respect is that, as shown in the figure, the $\mathbf{10}$ matter curves in F-theory must self-intersect (or pinch) for such a Yukawa coupling to be viable; indeed, if the two intersecting $\mathbf{10}$ matter curves were distinct, the texture of the resulting U-quark Yukawas would have zero entries in the diagonal, leading to a phenomenologically unreasonable pattern of one zero eigenvalue and two other comparable eigenvalues. It turns out that the pinched structure is fortunately quite common in F-theory, arising from monodromies of the singularity structure relating naively different matter curves.

The localization of the Yukawa around a point is actually not exact, since the different massless fields have a non-zero spread, associated to their internal zero mode profiles. Actually, the Yukawa couplings again arise as overlap integrals

$$Y_{ij} = \int_S \psi_i \, \psi_j \, \phi_H. \tag{12.117}$$

Since the Yukawa coupling is localized, its computation through (12.117) requires the zero mode wave functions only near the intersection point. The latter can be characterized using the unfolding introduced in Section 11.5.2. Consider an unfolding $G \to G_S \times U(1)$ describing a matter curve; from the 7-brane worldvolume perspective, this is described as a deformation of an 8d field theory with gauge group G, by a linearly changing vev. The detailed structure is beyond our scope, and for present purposes it suffices to describe the local dynamics in terms of a 4d $\mathcal{N} = 1$ gauge multiplet (A_μ, η), and three chiral multiplets $(A_{\bar{1}}, \psi_{\bar{1}})$, $(A_{\bar{2}}, \psi_{\bar{2}})$ and $(\varphi_{12}, \chi_{12})$, all in the adjoint of a gauge group G; the latter correspond to components of the 8d gauge fields along the two complex coordinates z_1, z_2 of S, and to the 7-brane position in the transverse direction. The unfolding $G \to G_S \times U(1)$ is implemented through a varying vev for the position moduli, along the generator Q of the $U(1) \subset G$,

$$\langle \phi \rangle = M_*^2 z_1 Q. \tag{12.118}$$

Here M_* is the fundamental F-theory scale (the analog of the string scale in IIB orientifolds). At $z_1 \neq 0$ the symmetry is broken to $G_S \times U(1)$, while at $z_1 = 0$ it is enhanced to G; hence the locus $z_1 = 0$ defines the matter curve Σ, supporting fields in representations of $G_S \times U(1)$ given by decomposing the adjoint of G.

The local structure of the zero mode wave function for these matter fields can be obtained from the local Dirac equations, descending from the original 8d theory. Locally we may take a flat Kähler form $dz_1 d\bar{z}_1 + dz_2 d\bar{z}_2$, and the equations of motion can be shown to be

$$\sqrt{2}\,\partial_1 \eta - M_*^2 z_1 q\, \psi_{\bar{2}} = 0; \qquad \bar{\partial}_{\bar{1}} \psi_{\bar{2}} - \sqrt{2}\, M_*^2 \bar{z}_1 q\, \eta = 0, \tag{12.119}$$

$$\partial_1 \psi_{\bar{1}} - M_*^2 \bar{z}_1 q\, \chi = 0; \qquad \bar{\partial}_{\bar{1}} \chi - M_*^2 z_1 q\, \psi_{\bar{1}} = 0, \tag{12.120}$$

where q is the $U(1)$ charge of the matter field, $\chi \equiv \chi_{12}$, and the pieces linear in z_1 arise from the linear vev (12.118). From the above equations one can see that there are no localized solutions for η and $\psi_{\bar{2}}$, and indeed it is consistent to set $\eta = \psi_{\bar{2}} = 0$; on the other hand, the coupled system for χ and $\psi_{\bar{1}}$ has solutions (taking $q = 1$)

$$\chi = f(z_2)\, e^{-M_*^2 |z_1|^2}; \qquad \psi_{\bar{1}} = -f(z_2)\, e^{-M_*^2 |z_1|^2}, \tag{12.121}$$

where $f(z_2)$ is an arbitrary holomorphic function of the coordinate along Σ. Note that there is a single independent zero mode with two (non-independent) components χ and $\psi_{\bar{1}}$. The zero mode is peaked around $z_1 = 0$, with width $\simeq 1/M_*^2$, and corresponds to the 6d matter field localized on Σ. The replication of 4d chiral fermions arises upon compactification on Σ with magnetic fluxes, which can produce multiple zero modes on Σ; at the local level this is captured by the general holomorphic function $f(z_2)$, as will be clear shortly.

We are now interested in evaluating the Yukawa coupling of three chiral fields coming from three intersecting matter curves Σ_a, Σ_b, Σ_c, which may be locally described by $z_1 = 0$, $z_2 = 0$ and $z_1 + z_2 = 0$ in the surface S. The Yukawa couplings are obtained as overlap integrals over S of the three wave functions involved, i.e.

$$Y = \int_S \left(\psi_{\bar{1}}^a \psi_{\bar{2}}^b - \psi_{\bar{2}}^a \psi_{\bar{1}}^b \right) \chi^c \, dz_1 d\bar{z}_1 dz_2 d\bar{z}_2 + \text{cyclic permutations in } abc, \tag{12.122}$$

where a, b, c denote the three matter curves. Given the matter field localization, the above integral is essentially supported at $z_1 = z_2 = 0$. In Yukawa couplings for semi-realistic models we can, without loss of generality, associate quarks and leptons to the curves a, b and the Higgs to the matter curve c. The existence of three quark and lepton families corresponds to the existence of three independent zero mode wave functions on Σ_a, Σ_b; near $z_1 = z_2 = 0$, their local behavior corresponds to a basis of holomorphic functions $f^a(z_2) = (1, z_2, z_2^2)$ on Σ_a, and $g^b(z_1) = (1, z_1, z_1^2)$ on Σ_b, while the single Higgs zero mode on Σ_c behaves as a constant. Given the exponential dumping of the wave functions, the integral to be performed is not sensitive to the global geometry of the curves and one can extend the integral on z_1, z_2 over $\mathbf{C}^2$. Ignoring momentarily the effect of magnetic fluxes, we can just use the zero mode wave functions in (12.121) for Σ_a, and its analogs for Σ_b, Σ_c. Since the integration measure is invariant under $z_i \to e^{i\theta} z_i$, the integral vanishes except for $f(z_2) = 1$, $g(z_1) = 1$, in which case $Y = \pi^2$; thus only one flavour of quark and leptons can get a mass at this level, which is a reasonable starting point to explain fermion mass textures. This result is quite similar to those already described in toroidal type IIA and IIB models, e.g. the example in Section 12.6.

As in the toroidal case, one expects that further ingredients may generate corrections and induce smaller Yukawa couplings for the rest of the generations. One might have expected to induce such corrections from the inclusion of magnetic fluxes, necessarily present to get 4d chirality; however, this is not the case, and magnetic fluxes modify the zero mode profiles, but not the Yukawa couplings. A heuristic explanation based on perturbative type IIB intuitions is as follows. Consider carrying out the above computation, with non-trivial magnetic fluxes, for perturbative type IIB D7-brane models, to produce their holomorphic Yukawa superpotential. From (12.72) the result cannot depend on Kähler moduli, and so must be given by the infinite volume result, in which the fluxes are infinitely diluted and can be taken to zero. A more precise argument (valid also in the non-perturbative F-theory setup) is that there is always a gauge choice in which the background gauge field A_m is purely holomorphic, and so, e.g., the last equation in (12.120) is unchanged upon covariantizing derivatives; such unchanged equations (which can be interpreted as F-term equations in the 8d SYM theory) turn out to be the only relevant ones to compute the holomorphic Yukawa coupling. Nevertheless, there are further generically present ingredients which can induce the desired corrections. In particular, D-brane instantons, or the addition of IASD $(1, 2)$ fluxes discussed in Section 15.4.1, have been shown to induce corrections to these Yukawa couplings. We refer to the Bibliography section for more details.

13

String instantons and effective field theory

Most 4d string vacua constructions in previous chapters have been carried out in the realm of perturbation theory (around the corresponding weak coupling limit). However, there are several instances in which non-perturbative effects play an important role. In this chapter we introduce brane instantons in 4d string vacua, analyze their impact on the low-energy effective action, and discuss some applications for particle physics model building.

13.1 Instantons in field theory and string theory

Non-perturbative effects are a crucial ingredient in quantum field theory and string theory. An important class of non-perturbative effects can be described in terms of instantons, i.e. semiclassical configurations providing saddle points in the euclidean path integral of the spacetime fields of the theory. In this section we review instantons in field theory and their generalization to brane instantons in string theory.

13.1.1 Instantons in gauge field theory

Many ingredients of instanton effects in string theory are already present in the more famil-iar setup of field theory. We briefly review gauge theory instantons, restricting to those aspects relevant for the brane instantons in the following sections.

Generalities

Instantons are classical solutions to the euclidean equations of motion in field theory. The prototypical example is given by instantons in 4d gauge theories, which are gauge field configurations obeying the self-duality condition in euclidean space

$$F_{\mu\nu} = \frac{1}{2}\epsilon_{\mu\nu\rho\sigma} F_{\rho\sigma} \quad \text{i.e. } F = *_{4d} F. \tag{13.1}$$

An explicit solution for $SU(2)$ gauge groups is given, in the gauge $\partial_\mu A_\mu^a = 0$, by

$$A_\mu^a = 2\bar{\eta}_{\mu\nu}^a \frac{(x - x_0)_\nu}{(x - x_0)^2} \frac{\rho^2}{(x - x_0)^2 + \rho^2}. \tag{13.2}$$

Here x_0 and ρ are the position and size of the instanton, and correspond to bosonic zero modes in the language introduced later; the $\bar{\eta}^a_{\mu\nu}$ are the so-called 't Hooft symbols, which realize the ath $SU(2)$ generator as a generator of the $SO(4) \simeq SU(2)^2$ rotation group. The above expression implies an alignment of the internal $SU(2)$ direction of the gauge field and the 4d spacetime direction, such that as one moves around the $\mathbf{S}^3$ at infinity in 4d, the gauge field winds around the $SU(2)$.

In general, an instanton configuration can be characterized as providing a map from $\mathbf{S}^3$ at infinity in $\mathbf{R}^4$ to the gauge group G, and is therefore classified by the homotopy group $\Pi_3(G)$, as reviewed in Section B.4. For any simple group, $\Pi_3(G) = \mathbf{Z}$, which implies that instanton configurations are labeled by an integer winding number, or topological charge, referred to as the instanton number

$$k = \frac{1}{8\pi^2} \int_{4d} \text{Tr} \, (F \wedge F) . \tag{13.3}$$

Negative values are realized by anti-instantons, which satisfy an anti-self-duality relation. Using the self-duality (13.1), the classical action for an instanton configuration is

$$S_{\text{cl.}} = \frac{8\pi^2}{g_{\text{YM}}^2} |k| . \tag{13.4}$$

As is familiar from quantum mechanics, a classical solution to the euclidean equations of motion describes the saddle point approximation to a tunneling process in the minkowskian theory, with strength $e^{-S_{\text{cl.}}} \simeq e^{-k/g_{\text{YM}}^2}$, clearly a non-perturbative contribution to the path integral of the theory. If the gauge theory has a non-trivial θ angle, as in (1.29), the instanton amplitude is, schematically

$$e^{-S_{\text{cl.}}} = \exp\left[-k \left(\frac{8\pi^2}{g_{\text{YM}}^2} + i\theta \right) \right] . \tag{13.5}$$

Hence instanton configurations break the perturbative continuous shift symmetry of the θ parameter to a periodicity of 2π.

Gauge instantons and fermion zero modes

There is an interesting effect when introducing fermions, related to the chiral anomaly. Consider for concreteness a theory with a Weyl fermion in a representation of G (assumed free of gauge anomalies), and with charge $+1$ under an additional $U(1)$ global symmetry e.g. fermion number. The mixed $U(1) - G^2$ anomaly implies that under a global $U(1)$ transformation

$$\delta_\epsilon S_{4d} \simeq \epsilon \int_{4d} \text{tr} \, (F \wedge F) \simeq \epsilon k , \tag{13.6}$$

i.e. an instanton number k background violates $U(1)$ charge conservation by k units.

This can be understood microscopically by considering the semiclassical expansion of the theory around the instanton background. In a saddle point approximation to the path integral, we need to integrate over fluctuations around the saddle configuration, and in particular over fermion field configurations obeying the Dirac equation coupled to the instanton background. By explicit evaluation, or using the index theorem (7.47), it can be shown the fermion field has k zero modes $\not{D}\psi_0^i = 0$, $i = 1, \ldots, k$, which therefore do not appear in the action at the quadratic level. Since these are Grassmann variables, satisfying $\int d\theta = 0$, $\int \theta d\theta = 1$ as in (2.9), the path integral of the instanton process automatically vanishes unless it includes one external leg for each fermion field with a zero mode. For each fermion field integration we schematically have

$$\int [\mathcal{D}\psi] e^{-S} \psi(x_1) \cdots \psi(x_k) \rightarrow \int \left[\mathcal{D}\psi_0^1\right] \cdots \left[\mathcal{D}\psi_0^k\right] e^{-S_{\mathrm{cl.}}} \psi_0^1 \cdots \psi_0^k \neq 0, \quad (13.7)$$

with possible restrictions on k arising from the Grassmann nature of the fermion zero modes, and from the necessary contraction of gauge indices (ignored above for simplicity).

Since instantons are localized configurations in 4d spacetime, the effect of the instanton process to the amplitude can be described as an effective so-called 't Hooft vertex

$$e^{-S_{\mathrm{cl.}}} [\psi(x)]^k \equiv e^{-S_{\mathrm{cl.}}} \underbrace{\psi(x) \cdots \psi(x)}_{k}. \quad (13.8)$$

The insertion of Grassmann variables in the integrand, required to yield non-vanishing amplitudes, is referred to as saturation of fermion zero modes.

The instanton location, as x_0 in (13.2), is actually a bosonic zero mode of the configuration, arising from the fluctuations of the gauge field around the instanton background. These zero modes are Goldstone modes of the translational symmetries broken by the instanton. In the path integral we integrate over them and this restores the 4d Poincaré invariance of the theory. In addition, we need to sum over processes including multiple instanton vertices. Since they are indistinguishable, the resulting sum can be exponentiated, resulting in a correction to the 4d effective action. Considering a situation with several fermion species $\psi_{(a)}$, $a = 1, \ldots, n$, which for simplicity we take to have the same $U(1)$ charge and representation under G, the instanton correction to the 4d effective action has the structure

$$\Delta S_{4d} \simeq A \int d^4x \, e^{-S_{\mathrm{cl.}}} [\psi_{(1)}(x)]^k \cdots [\psi_{(n)}(x)]^k, \quad (13.9)$$

where the factor A indicates further contributions from the integration over additional bosonic zero modes (like the instanton size, or its orientation in the gauge group), as well as the integration over non-zero modes for the different fields.

The generalization to different charges and representations, and to several $U(1)$ symmetries, is straightforward. In particular, in a theory with vector-like fermions, there may be couplings among their putative fermion zero modes, actually lifting them at the level of the interacting theory. Saturation of such fermion modes (often still denoted "zero modes" by abuse of language) in their Grassmann integrals can proceed via these interaction terms, so they do not require the introduction of external legs. Thus the number of external legs in instantons corrections like (13.9) actually reproduces the net number of fermions, in agreement with the anomaly argument.

We will be interested in the structure of instantons in 4d $\mathcal{N}=1$ supersymmetric theories, to which we turn now. Supersymmetric theories include fermions, making the study of instanton fermion zero modes particularly relevant. Instantons break some or all the supersymmetries of the theory, and therefore have goldstino fermion zero modes, which are crucial in order to determine the kind of 4d superspace interaction term that is induced by the instanton. Instantons preserving half the supersymmetries are denoted BPS, with the relation between instanton charge (13.3) and classical action (13.4) playing the role of the BPS bound (2.63). BPS instantons break two supersymmetries and have two goldstinos θ^α, which are exact fermion zero modes of the configuration. In the absence of other fermion zero modes (or if they are lifted by interactions as described above), the instanton induces a correction to the 4d superpotential

$$\Delta S_{4d} \simeq \int d^4x \, d^2\theta \, e^{-\tau} \, \mathcal{O}(\Phi_1, \ldots, \Phi_n), \tag{13.10}$$

where τ is the chiral multiplet whose scalar component is the gauge kinetic function, with value $\tau = 8\pi^2/g_{\text{YM}}^2 + i\theta$, and $\mathcal{O}$ is a possible operator involving charged chiral multiplets, in a supersymmetric generalization of (13.9). Note that the integration over the fermion zero modes θ restores the SUSY invariance of the interaction.

Let us emphasize the important property that only instantons with exactly two fermion zero modes (just those required to saturate the $d^2\theta$ integration) can contribute to the superpotential. BPS instantons with additional fermion zero modes induce 4d F-term corrections involving higher-derivative interactions, or multi-fermion terms (containing more than two fermions). These are therefore less relevant operators, and ignored in the following. Finally, instantons breaking all supersymmetries have (at least) four fermion zero modes, the goldstinos θ^α, $\bar{\theta}_{\dot\alpha}$. Integration over them leads to a D-term interaction in the 4d action. These could be potentially interesting corrections to Kähler potentials, but as they originate from non-supersymmetric configurations, they are in general not amenable to precise computation.

Instanton superpotentials in 4d $\mathcal{N}=1$ SQCD

There are several examples of instanton effects in 4d $\mathcal{N}=1$ supersymmetric gauge field theories, in particular the SQCD theories introduced in Section 2.5.1. These have $SU(N_c)$ gauge symmetry and N_f flavours, i.e. massless chiral multiplets Q_i, $\tilde{Q}_{\bar\jmath}$ in the fundamental and anti-fundamental representations respectively. For instance, $SU(N_c)$ instantons in $N_f = N_c - 1$ SQCD generate a non-perturbative superpotential

$$W = \frac{\Lambda^{2N_c+1}}{\det(Q_i \tilde{Q}_{\bar\jmath})}, \tag{13.11}$$

where $\Lambda^{2N_c+1} \equiv e^{-\tau} \mu^{2N_c+1}$ is the dynamical (or confinement) scale of the theory, with μ being a scale required on dimensional grounds. The above expression is a particular case of the general $N_f < N_c$ Affleck–Dine–Seiberg superpotential (2.65). The general expression can be derived from (13.11) by adding large flavour mass terms $m_i \gg \Lambda$, and integrating them out. In particular this allows one to recover the $\mathcal{N} = 1$ SYM superpotential (2.66), whose structure can be written

$$W_{\mathrm{SYM}} = N_c\, e^{-\frac{\tau_{\mathrm{eff}}}{N_c}}\, \mu^3, \tag{13.12}$$

in terms of the SYM gauge kinetic function $\tau_{\mathrm{eff}} = \left(8\pi^2/g_{\mathrm{YM}}^2 + i\theta\right) + \cdots$, where the dots denote additional corrections, most prominently the one-loop contribution. Again μ is a scale required by dimensional analysis. Since the exponent in (13.12) is N_c times smaller than the instanton action, the superpotential is sometimes said to originate from "fractional instantons." Incidentally these will receive a natural interpretation in the string theory implementations in Section 13.2.2.

13.1.2 Worldsheet and brane instantons in string theory

String theory also contains non-perturbative effects from euclidean instanton configurations, not surprisingly since string theory contains field theory (coupled to gravity) as its low-energy limit. Indeed in 4d string theory vacua there are euclidean configurations, localized in 4d Minkowski space, and characterized by a topological charge, hence also called "instantons."

The most obvious possibility is configurations of euclidean closed string worldsheets completely wrapped on 2-cycles C of the internal compactification geometry, and localized at a point in 4d Minkowski space. These are the worldsheet instantons already mentioned in Sections 9.1.3 and 10.1.2, and whose contribution to the superpotential in heterotic orbifolds was described in Section 9.3.2. In theories with open strings, there are analogous open worldsheet instantons, whose contribution to the superpotential was computed in Section 12.5.3. Worldsheet instantons are BPS when C is a holomorphic 2-cycle, their topological charge is the 2-homology class $[C]$, and their strength is

$$e^{-t} = \exp\left[-\frac{1}{2\pi\alpha'}\int_C (J + iB_2)\right], \tag{13.13}$$

where J is the CY Kähler form, and B_2 is the 2-form coupling to the fundamental string. The factor of i in the coupling to the 2-form field, as compared with (3.10) arises in going to the euclidean theory. The strength of worldsheet instantons is thus controlled by the 2d worldsheet area, so they are non-perturbative in α', but perturbative (in fact, tree level) in g_s. Note that in heterotic compactifications the shift symmetries of some of the 4d scalars from the 10d 2-form are violated already at tree level in g_s. We are more interested in genuinely non-perturbative instantons in string theory. Clearly these should be attributed to objects whose tension is an inverse power of g_s, i.e. branes in string theory. Hence 4d string compactifications have brane instanton effects arising from possible branes wrapped on different cycles of the internal CY, and localized in 4d Minkowski space.

For instance, in 4d heterotic compactifications there are instantons from NS5-branes completely wrapped on the internal space $\mathbf{X}_6$, and localized in 4d. Recalling from

Section 6.2 the NS5-brane tension $T_{NS5} \sim 1/g_s^2$, we find that the strength of NS5-brane instantons is schematically

$$\exp\left[-\left(\frac{V_{\mathbf{X}_6}}{g_s^2} + i \int_{\mathbf{X}_6} B_6\right)\right] = e^{-S}, \tag{13.14}$$

with modulo numerical factors in the exponent. Here B_6 is the 10d dual to the 2-form B_2, so its integral is the universal heterotic axion a, with the factor of i due to the euclidean signature. Also, $V_{\mathbf{X}_6}$ is the volume of $\mathbf{X}_6$, and S is the 4d complex dilaton in (9.9). The amplitude (13.14) is non-perturbative in g_s, and in fact, noting the gauge kinetic functions (9.21) for the 4d gauge groups, it behaves like a gauge instanton. Indeed, as noted in Section 7.5, the NS5-branes can continuously turn into gauge instantons in heterotic theory.

There are similar NS5-brane instantons in type IIA orientifolds and type IIB O3/O7 models. However, NS5-branes lack a proper microscopic description in string theory, and are not very tractable objects, so we will not discuss them further. We rather focus on D-brane instantons in type II orientifolds, describing them via open strings. This allows a precise determination of their zero modes by quantizing the open strings with one or both endpoints on the D-brane instanton, see Section 13.2. We also focus on 4d $\mathcal{N} = 1$ models, and concentrate on BPS instantons, although certain properties depend just on topological arguments, and extend to non-BPS instantons, or non-supersymmetric compactifications. Also for concreteness we use the language of geometric CY compactifications, but many properties extend to other setups in close analogy.

Consider a 4d $\mathcal{N} = 1$ type II orientifold compactification with 4d spacetime filling D-branes (denoted "gauge branes" in the following). The conditions for a wrapped euclidean brane to define a BPS instanton are the same kind of supersymmetry conditions as for gauge branes, e.g. recall Sections 10.3.2 and 11.4.3. This follows because gauge and instanton branes have four ND directions along 4d Minkowski space, and preserve supersymmetry if their internal cycles are mutually supersymmetric as well. Thus in type IIA intersecting brane models, with D6-branes wrapped on 3-cycles, BPS instantons arise from euclidean D2-branes wrapped on (not necessarily the same) special lagrangian 3-cycles (10.27) with suitable phase. In type IIB orientifold models with D3-branes at points and D7-branes on 4-cycles, BPS instantons arise from D(-1) branes at points in the CY, and from euclidean D3-branes wrapped on (not necessarily the same) holomorphic 4-cycles (possibly with non-trivial supersymmetric gauge backgrounds). In type IIB models with D5- and D9-branes, BPS instantons arise from euclidean D1-branes on holomorphic 2-cycles, and euclidean D5-branes wrapped on the whole CY (possibly with non-trivial supersymmetric gauge backgrounds). For simplicity, we use "cycle" (with quotes) to mean all the data specifying its internal structure on the CY, namely the wrapped cycle plus any possible gauge background on its worldvolume. Note that mirror symmetry exchanges type IIA and IIB orientifold models, and acts consistently on the relevant instantons. Therefore we often restrict our discussion to the type IIA setup, and merely translate the corresponding results to the mirror type IIB. This is particularly useful since in type IIA all BPS instantons are given by euclidean D2-branes.

Let us consider a Dp-brane instanton wrapped on a $(p + 1)$-cycle W_{p+1}, with trivial NSNS 2-form and worldvolume gauge background for simplicity. From the D-brane action (6.6), (6.15), its strength is

$$\exp\left[-\left(\frac{\mu_p}{g_s}\int_{W_{p+1}} d^{p+1}x \sqrt{\det(G)} + i\,\mu_p \int_{W_{p+1}} C_{p+1}\right)\right], \qquad (13.15)$$

with G being the worldvolume metric induced from spacetime. The factor of i in the CS piece arises from the euclidean signature. The above expression clearly defines a non-perturbative effect scaling as e^{-1/g_s}, as expected. The instanton action is controlled by the wrapped volume, and its phase is given by the 4d scalar $a_{(p+1)}$ arising from the $(p + 1)$-form. The latter coupling breaks the perturbatively exact shift symmetry of the RR scalar to a discrete subgroup,

$$a_{(p+1)} \to a_{(p+1)} + 2\pi, \qquad (13.16)$$

in analogy with gauge instantons. Hence D-brane instantons are the perfect candidates to induce certain perturbatively forbidden couplings, as will be exploited later on. Note that for Dp-branes carrying worldvolume magnetic fluxes, there are induced lower-dimensional D-branes, which modify the classical action of the configuration and its couplings to 4d RR scalars.

For BPS instantons the exponent in (13.15) is a complexified modulus, as in Section 12.3.1. This is a complex structure modulus in type IIA models, measuring the 3-cycle (complexified) volume wrapped by the D2-brane, and correspondingly a Kähler modulus in type IIB models, controlling the (complexified) volume of the wrapped "cycle" (i.e. including gauge field contributions).

General D-brane instantons include gauge field theory instantons as a particular case. Indeed, as mentioned around (6.17) in Section 6.1.2, Dp-branes on top of D$(p+4)$-branes behave as gauge instantons in the D$(p + 4)$-brane gauge theory. Hence in type II orientifolds, gauge instantons in a sector of D$(p + 4)$-branes wrapped on a "cycle" W_{p+1} are described as euclidean Dp-branes on W_{p+1}. In fact, their strength (13.15) agrees with the gauge instanton strength (13.5) upon using the gauge kinetic functions in Section 12.3.1.

String theory models however include novel non-perturbative instanton effects, from D-branes wrapped on "cycles" different from those of gauge branes. They do not admit the interpretation of gauge instantons, and are often termed "non-gauge," "stringy" or "exotic," instantons. A most relevant property for our purposes is that their strengths are not controlled by gauge coupling constants, and therefore need not be extremely suppressed, and may lead to interesting effects in realistic particle physics models, as explored in Section 13.3.

13.1.3 Field theory operators from D-brane instantons

We have seen that D-brane instantons violate the shift symmetries of the RR scalars to which they couple. In type II orientifolds with gauge D-branes, some of these RR axion shifts are intertwined with D-brane $U(1)$ gauge transformations, as in (10.26). D-brane instantons coupling to such RR scalars can violate, in a proper sense, these $U(1)$ symmetries, and induce perturbatively forbidden couplings, as we now describe in the type IIA picture. Note that the following arguments are topological and hold for general CYs and also for non-BPS instantons and non-supersymmetric models.

Consider a type IIA compactification with D6$_a$-branes wrapped on 3-cycles Π_a, momentarily ignoring orientifold projections. As in Section 10.3.1, we introduce a basis $\{[\alpha_k]\}$ of 3-cycles, and its dual basis $\{[\beta^k]\}$, i.e. $[\alpha_k] \cdot [\beta^l] = \delta_k^l$, and define

$$Q_{ak} = [\Pi_a] \cdot [\alpha_k]. \tag{13.17}$$

The $B \wedge F$ coupling in (10.23) implies a shift (10.26) of the RR scalar $a_k = \int_{\beta^k} C_3$ under D6-brane $U(1)_a$ gauge transformations, i.e.

$$A_\mu^a \rightarrow A_\mu^a + \partial_\mu \Lambda_a, \quad a_k \rightarrow a_k + \sum_a N_a\, Q_{ak}\, \Lambda_a. \tag{13.18}$$

Recall that the corresponding $U(1)$ gauge bosons are actually massive, and that the $U(1)$ symmetries may be anomalous or not. They remain as perturbatively exact global symmetries, but broken, in a suitable sense, by D2-brane instantons as follows.

These models contain euclidean D2-brane instantons wrapped on 3-cycles, thus coupling to the RR scalars and naively breaking their shift symmetry. To see this, consider a D2-brane instanton on a 3-cycle Π_{inst} of volume V_3, with an expansion

$$[\Pi_{\text{inst}}] = \sum_k q_{\text{inst}}^k\, [\alpha_k], \quad \text{with } q_{\text{inst}}^k = [\Pi_{\text{inst}}] \cdot [\beta^k]. \tag{13.19}$$

Using (13.15), the naive instanton amplitude is roughly

$$e^{-S_{\text{cl.}}} = \exp\left[-\left(\frac{V_3}{g_s} + i \int_{\Pi_{\text{inst}}} C_3\right)\right] = \exp\left(-\frac{V_3}{g_s} - i \sum_k q_{\text{inst}}^k a_k\right). \tag{13.20}$$

This amplitude picks up a phase under the RR scalar shift (13.18),

$$e^{-S_{\text{cl.}}} \rightarrow e^{-S_{\text{cl.}}} \exp\left(-i \sum_{a,k} N_a\, Q_{ak}\, q_{\text{inst}}^k\, \Lambda_a\right) = e^{-S_{\text{cl.}}} \exp\left(-i \sum_a N_a\, I_{a,\text{inst}}\, \Lambda_a\right), \tag{13.21}$$

where $I_{a,\text{inst}} = [\Pi_a] \cdot [\Pi_{\text{inst}}]$ is the intersection number of the gauge and instanton 3-cycles. This would seem to clash with the RR scalar shift being actually part of the $U(1)$ gauge symmetry (13.18).

The puzzle is solved by the fact that the instanton in general has fermionic zero modes charged under the $U(1)$ gauge symmetries, which lead to additional charged matter field insertions. Namely, open strings stretching between the D2-brane and the D6$_a$-brane lead to $I_{a,\text{inst}}$ (net) fermion zero modes in the fundamental representation of $U(N_a)$. Let us

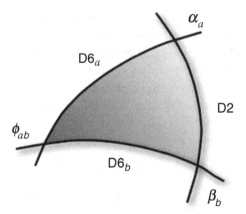

Figure 13.1 Worldsheet disk amplitude inducing a cubic coupling in the euclidean D2-brane instanton action. The cubic coupling involves the 4d charged chiral multiplets at the intersection ab of the D6-branes, and instanton fermion zero modes at the intersections of the D2-brane with the D6-branes a and b.

denote them $\gamma^{i_a}_{a,p_a}$ with $p_a = 1, \ldots, N_a$ a $U(N_a)$ gauge index, and $i_a = 1, \ldots, |I_{a,\text{inst}}|$ labeling intersections. The instanton amplitude contains an integration over such fermion zero modes, i.e.

$$\int \prod_{a,p_a,i_a} \left[d\gamma^{i_a}_{a,p_a} \right] e^{-S_{\text{cl.}} + S_{\text{z.m.}}}, \tag{13.22}$$

where $S_{\text{z.m.}}$ contains possible interaction terms of the instanton zero modes, see later. It is easy to check that the phase rotation (13.21) is actually canceled by the $U(1)_a$ phase rotations of the charged fermion modes in the instanton measure.

Assuming some knowledge about possible interaction terms of the fermion zero modes γ, we can provide a more explicit expression for the resulting instanton amplitude, and understand this cancellation at the level of the 4d effective action. As discussed for gauge theory instantons, there are two basic ways to saturate these additional fermion zero modes in the path integral: through interactions lifting the zero modes, or through external fermionic leg insertions. Consider the former, in a simplified description. Split the D6$_a$-branes in two sets, labeled $a\pm$ according to the sign of $I_{a,\text{inst.}}$. Hence we have fermion zero modes α^i_{a-} in the $\overline{\Box}_{a-}$, and β^j_{a+} in the $\Box_{a+}$. Here we assume them to be in equal numbers to achieve pairing by interactions. The systems typically contain interaction terms of the form

$$S_{\text{z.m.}} = c_{ij,k}\, \alpha^i_{a-}\, \Phi^{(k)}_{ab}\, \beta^j_{b+}, \tag{13.23}$$

where $\Phi^{(k)}_{ab}$ are 4d chiral multiplets in the $(\Box_a, \overline{\Box}_b)$ at D6$_a$-D6$_b$ brane intersections (labeled by $k = 1, \ldots, I_{ab}$). These terms arise via open string disk worldsheet instantons, see Figure 13.1, analogous to those producing Yukawa couplings in Section 12.5.3. These interactions can be used to saturate the fermion zero mode integrations, so the instanton amplitude has the structure

$$e^{-S_{\text{cl.}}} \det(\Phi_{ab}), \tag{13.24}$$

where $\det(\Phi_{ab})$ denotes a polynomial in the fields $\Phi_{ab}^{(k)}$ of degree $\sum_{a-} N_{a-} |I_{a-,\text{inst}}| = \sum_{b+} N_{b+} |I_{b+,\text{inst}}|$. It indeed reduces to a determinant when there is a single field at the ab intersection. Its role is analogous to the operator $\mathcal{O}$ in (13.10). It is now straightforward to show that the phase rotation (13.21) is canceled by the charge of $\det(\Phi_{ab})$. The mechanism generalizes to situations with a more involved structure of the couplings (13.23), see later for examples. Note that in what concerns the charged matter content of the theory, the instanton generates couplings which are forbidden by the $U(1)$ symmetries to all orders in perturbation theory, and which are possible non-perturbatively due to the non-trivial transformation (13.21). In this sense, the intersection number $I_{a,\text{inst}}$ of the D2-brane instanton with the D6$_a$-branes determines the amount of violation of the $U(1)_a$ gauge symmetry. The above discussion generalizes easily to models with orientifold action, by simply replacing $I_{a,\text{inst}} \to I_{a,\text{inst}} - I_{a',\text{inst}}$ in the relevant formulae.

In cases where there are unpaired charged fermion zero modes, which cannot be saturated with interaction terms, insertion of external legs is required, as in (13.9). The end result is an expression similar to (13.24), with $\det(\Phi)$ replaced by an operator in the chiral multiplets $\overline{D\Phi}_{ab}$, whose lowest component is a multi-fermion term saturating all unpaired fermion zero modes. Clearly, the resulting effective vertex reproduces the net amount of charge violation.

The charge violation by D-brane instantons reproduces as a particular case the violation of anomalous $U(1)$ charges by gauge instantons of Section 13.1.1, as follows. Consider N_b D6$_b$-branes on a 3-cycle Π_b, leading to a $U(N_b)$ gauge factor, and N_a D6$_a$-branes on Π_a producing a $U(N_a)$ factor, on which abelian $U(1)_a$ part we focus. There is a $U(1)_a - SU(N_b)^2$ mixed anomaly (10.21) proportional to $N_a I_{ab}$ (no sum over a), so from (13.6) a $k = 1$ $SU(N_b)$ gauge theory instanton violates $U(1)_a$ charge by $N_a I_{ab}$ units. In the string theory setup, the gauge instanton is played by a euclidean D2-brane on Π_b, i.e. on top of the gauge D6$_b$-branes, and its $U(1)_a$ violation is $[\Pi_a] \cdot [\Pi_{\text{inst}}] = I_{ab}$. Note, however, that the $U(1)$ charge violation is more general in the string theory setup, since "stringy" brane instantons produce vertices violating even non-anomalous $U(1)$s.

In applications, we are in general interested in $U(1)$-violating instantons leading to operators with the lowest possible dimension. In particular, we are interested in non-perturbative superpotentials (in general involving the charged matter fields), rather than other higher-dimensional F-term operators. As for field theory instantons, the conditions for a BPS instanton to contribute to the superpotential relate to its structure of (neutral) fermion zero modes, as we describe in the next section.

13.2 Fermion zero modes for D-brane instantons

As for instantons in gauge field theory, the structure of fermion zero modes on D-brane instantons determines the superspace structure of the non-perturbative correction. As noted above, we are eventually interested in contributions to the non-perturbative superpotential, which arise from brane instantons with two exact fermion zero modes. We first discuss the computation of fermion zero modes, and subsequently present several instances in which all zero modes but two are eliminated, either via orientifold projections, or via saturation with interactions.

13.2.1 Computation of D-brane instanton zero modes

For concreteness we focus on 4d $\mathcal{N} = 1$ type IIA orientifolds with intersecting D6-branes, and on a BPS instanton from a D2-brane on a special lagrangian 3-cycle Π_{inst}. We consider the two cases of Π_{inst} being invariant under the orientifold action ("invariant" instanton), or mapped to an image on Π'_{inst} ("non-invariant" instanton).

The computation of D2-brane instanton zero modes is very analogous to the computation of the massless spectrum on a putative D6-brane wrapped on Π_{inst}. Hence we may borrow results from Chapter 10, with a few minor modifications: First, since D2-branes do not stretch in the 4d Minkowski space, it is not possible to use the light-cone gauge for the corresponding open strings. Consequently there are additional physical open string modes, e.g. arising from oscillators in the two additional 4d dimensions. Second, the orientifold projection on certain 2-2 and 2-2' open strings has an extra sign as compared with the analogous 6-6 and 6-6' sectors, due to the change between NN and DD boundary conditions in the 4d directions. This is similar to the relative sign between orientifold projections in 99 and 55 sectors in Section 6.1.3. Note that the 4d space-filling D6-brane on Π_{inst} is an auxiliary tool, and it need not be one of the gauge D6-branes present in the model.

For illustration, and to emphasize some important properties, we first turn to the discussion of some relevant sectors.

Neutral fermion zero modes

Let us start with the sector of zero modes neutral under the 4d D6-brane gauge groups. This arises from open strings with both endpoints on the D2-branes (or stretching between the D2- and the D2'-branes, if different).

- *Universal sector: non-invariant instanton*: There is a universal sector of 2-2 zero modes, present for any BPS D-brane instanton, independently of the properties of the wrapped cycle. For an instanton not invariant under the orientifold action, and with multiplicity k, there is a worldvolume $U(k)$ gauge symmetry. Notice that, although there are no gauge bosons in $0 + 0$ dimensions, the gauge group is still well-defined, since it acts on charged states (open strings ending on the instanton brane). In this universal sector there are four real bosonic zero modes in the adjoint representation, whose eigenvalues describe the instanton positions. The "center of mass" degrees of freedom are four translational Goldstone bosons x^μ. In addition there are four fermion zero modes, in the adjoint, whose "center of mass" degrees of freedom are goldstinos of the accidental 4d $\mathcal{N} = 2$ supersymmetry in this sector. They split in two sets, according to their chirality under the $SO(4)$ 4d euclidean rotational symmetry. Two of them, denoted θ^α, are the actual goldstinos of the 4d $\mathcal{N} = 1$ supersymmetry. The remaining two, denoted $\bar{\tau}_{\dot\alpha}$, are present even in the minimal case $k = 1$. Hence such $U(1)$-type instantons cannot contribute to superpotential terms, unless additional effects lift the extra zero modes, see Section 13.2.2.

- *Universal sector: invariant instanton*: For an instanton mapped to itself under the orientifold action, the projection truncates the above universal 2-2 zero mode content as follows. The worldvolume group is projected down to orthogonal or symplectic groups, which we use to denote the instantons as being of $O(k)$- or $USp(k)$-type. For $O(k)$-type instantons, the four bosonic zero modes and two fermion zero modes (those associated to θ^α) are projected down to the two-index symmetric representation, while the other two fermion zero modes (related to $\bar{\tau}_{\dot\alpha}$) project onto the antisymmetric representation. Hence for $O(1)$-type instantons, the universal sector contains just two fermion zero modes, the goldstinos θ^α, so they are optimal candidates to generate non-perturbative superpotentials (if no additional zero modes arise from other non-universal sectors).

  For $USp(k)$-type instantons, the orientifold projects the four bosonic zero modes and two fermion zero modes to the two-index antisymmetric representation, and the other two fermion zero modes to the symmetric representation. Hence, even for the minimal case of $USp(2)$-type instantons, there are at least two fermion zero modes in the triplet representation, in addition to the two 4d $\mathcal{N} = 1$ goldstinos. Hence these instantons cannot contribute to the superpotential, unless additional effects lift the extra zero modes, see Section 13.2.2.

- *Deformation modes*: In addition to the above universal modes, there may be bosonic and/or fermionic zero modes related to geometric moduli of the cycle wrapped by the D-brane instanton (or their fermionic analogs). In our applications we are mostly interested in instantons contributing to the superpotential, hence with only the two universal θ^α fermion zero modes. We will therefore focus on (or assume) instantons with no additional deformation zero modes. Hence our present discussion is sketchy, leaving the detailed description for the references.

  Deformation modes can arise in the 2-2 sector for invariant instantons, and in the 2-2 and 2-2′ sectors (and their images) for non-invariant instantons. In the 2-2 sector, the zero modes before the orientifold are (the dimensional reduction of) $b_1(\Pi_{\text{inst}})$ adjoint chiral multiplets, in analogy with the result for D6-brane geometric moduli. In the 2-2′ sector of non-invariant instantons, the zero modes before the orientifold projection are given by (the dimensional reduction of) $[\Pi_{\text{inst}}] \cdot [\Pi'_{\text{inst}}]$ chiral multiplets in the bi-fundamental representation. The orientifold projection of all these multiplets is easily determined from the analogous projection for D6-branes, using the above-mentioned sign differences.

Charged fermion zero modes

Consider now sectors of open strings stretching between the D2-brane (or its image) and the 4d gauge D6-branes in the model (or its images). The zero mode multiplicities are again closely related to the spectrum in a putative D6-brane on Π_{inst}. Note, however, that replacing a D2-D6 to a D6-D6 system, changes the 4d Minkowski direction boundary conditions from ND to NN. These additional twists change the mass formula by shifting the zero point energy by 1/2, rendering all bosonic modes in the D2-D6 sector "massive," while relating the chiral multiplets in the D6-D6 intersections to purely fermionic zero modes at D2-D6 intersections. Still their multiplicities are preserved, and for any $D6_a$-brane, the (net) number of instanton fermion zero modes in the fundamental representation $\Box_a$ is

$$\text{Invariant instanton} \quad [\Pi_a] \cdot [\Pi_{\text{inst}}],$$
$$\text{Non-invariant instanton} \ [\Pi_a] \cdot [\Pi_{\text{inst}}] + [\Pi_a] \cdot \left[\Pi'_{\text{inst}}\right], \qquad (13.25)$$

as anticipated in the discussion in Section 13.1.3. In the above description we have considered that a runs through D6-branes and their orientifold images. In the latter case, the fundamental should be flipped to the corresponding anti-fundamental, according to the general rule $\Box_a \equiv \overline{\Box}_{a'}$ in Section 10.4.1. In applications, we assume the above charged fermion zero modes to have couplings to 4d charged chiral multiplets, allowing for their saturation.

13.2.2 Non-perturbative superpotentials from D-brane instantons

Instantons contributing to the superpotential must have the two 4d $\mathcal{N} = 1$ goldstinos θ^α as the only exact fermion zero modes. Any additional fermion zero modes must either be absent, or appear in interactions allowing for their saturation. In this section we describe several mechanisms to eliminate the pervasive universal modes $\overline{\tau}_{\dot\alpha}$, leading (in the absence of non-universal exact fermion zero modes) to instantons contributing to the superpotential. If the instanton has zero net intersections with the 4d gauge branes, these superpotentials can involve only closed string moduli, and be of the form

$$W_{\text{n.p}} = A(z) \, e^{-S_{\text{cl.}}}, \quad \text{no charged fermion zero modes,} \qquad (13.26)$$

where $S_{\text{cl.}}$ is essentially the wrapped "cycle" volume modulus, a Kähler modulus in type IIB and a complex structure modulus in type IIA models. The prefactor $A(z)$, arising from the integral over non-zero modes, is a holomorphic function of the moduli z, which are complex structure moduli in type IIB and Kähler moduli in type IIA models.

As described in Section 13.1.3, if the instanton has fermion zero modes α, β charged under the 4d gauge branes, and coupled with 4d chiral multiplets via interactions (13.23), the superpotential has instead the structure (13.24), schematically

$$W_{\text{n.p}} = A(z) \, e^{-S_{\text{cl.}}} \det \Phi_{ab}, \quad \text{with charged fermion zero modes.} \qquad (13.27)$$

Note that this structure implies a violation of the perturbative decoupling in (12.72).

Both kinds of superpotentials have interesting model building applications, which we discuss in Section 13.3 for (13.27), and Section 15.3.1 for (13.26).

There are several mechanisms lifting the universal $\overline{\tau}_{\dot\alpha}$ fermion zero modes, and allowing the instantons to generate superpotentials of the above kinds (in the absence of extra non-universal fermion zero modes). In particular:

- $O(1)$-*type instantons*: As already mentioned in Section 13.2.1, instantons with world-volume $O(1)$ symmetry have their $\bar{\tau}_{\dot\alpha}$ fermion zero modes removed by the orientifold projection.

- *Gauge instantons*: D-brane instantons wrapped on the same cycle as gauge D-branes correspond to gauge instantons, and can generate superpotentials as in Section 13.1.1. This is possible because their $\bar{\tau}_{\dot\alpha}$ fermion zero modes have interactions with modes from open strings stretching between the instanton and gauge D-branes, allowing for their saturation. Interestingly, this lifting mechanism works also for $U(1)$ gauge D-branes. In this case, the D-brane instanton does not admit a gauge theory interpretation, as there are no instantons in $U(1)$ gauge theories.

- *Closed string fluxes*: As will be discussed in Section 14.3.3, the presence of field strength fluxes for closed string fields can induce interactions lifting fermion zero modes, in particular the universal $\bar{\tau}_{\dot\alpha}$.

*Instantons in F-theory and the arithmetic genus**

The use of $O(1)$-type instantons is perhaps the simplest way to ensure the appearance of non-perturbative superpotentials in type II orientifolds. Interestingly, for type IIB models the conditions for a D3-brane instanton to generate a superpotential admit a generalization to F-theory, as follows. Recall from Section 11.5 the relation between type IIB/F-theory and M-theory. Namely M-theory on a CY fourfold $\mathbf{X}_8$ elliptically fibered over a base three-fold $\mathbf{B}_6$, in the limit of vanishing $\mathbf{T}^2$ fiber size, is equivalent to F-theory on $\mathbf{X}_8$ i.e. type IIB on $\mathbf{B}_6$ with (p, q) 7-branes wrapped on the 4-cycles over which the (p, q) 1-cycle of the $\mathbf{T}^2$ fiber degenerates. In this duality, instantons in type IIB/F-theory from D3-branes wrapped on a 4-cycle C_{inst} of $\mathbf{B}_6$ map to M-theory instantons from M5-branes wrapped on a 6-cycle D_{inst}, given by the $\mathbf{T}^2$ fibration over C_{inst}. The condition that the D3-brane instanton generates a superpotential can be recast as geometric conditions for D_{inst}. The KK reduction of the M5-brane action shows, via the index theorem, that the net number of neutral fermion zero modes is twice the arithmetic genus $\chi(D_{\text{inst}})$, defined below. Thus a necessary condition for the instanton to generate a non-perturbative superpotential is

$$\chi(D_{\text{inst}}) = 1, \quad \text{with } \chi(D_{\text{inst}}) = \sum_{n=0}^{3} (-1)^n \, h_{n,0}(D_{\text{inst}}). \tag{13.28}$$

This only counts the net number of fermion zero modes, with the individual $h_{n,0}$ providing their actual multiplicities. Thus the sufficient condition is

$$h_{0,0}(D_{\text{inst}}) = 1, \quad h_{n,0}(D_{\text{inst}}) = 0 \quad \text{for } n = 1, 2, 3. \tag{13.29}$$

We stress that these conditions apply only to instantons carrying no worldvolume fluxes (as these modify the counting of fermion zero modes) and in the absence of additional effects which may lift some of the fermion zero modes, e.g. bulk p-form field strength fluxes, see Section 14.3.3.

It is interesting to check that these conditions for superpotential contributions reproduce the need of $O(1)$-type instantons for F-theory vacua related to perturbative type IIB orientifolds. Indeed, consider a type IIB model with a $U(1)$-type D3-brane instanton, not intersecting any 7-brane in the model. In the M-theory picture, the $\mathbf{T}^2$ fiber does not degenerate at any point in C_{inst}, hence the M5-brane 6-cycle is topologically $D_{\text{inst}} = C_{\text{inst}} \times \mathbf{T}^2$. Since the arithmetic genus is multiplicative and $\chi(\mathbf{T}^2) = 0$, we have $\chi(D_{\text{inst}}) = 0$, which does not fulfill the necessary condition (13.28). This agrees with the result in Section 13.2.1 that $U(1)$-type instantons have extra zero modes and cannot contribute to the superpotential. Consider now a D3-brane instanton such that C_{inst} intersects only 7-branes of same (p, q) kind. Then D_{inst} is a $\mathbf{T}^2$ fibration whose only degenerations over C_{inst} are pinchings of the (p, q) 1-cycle. Since the orthogonal 1-cycle does not pinch, D_{inst} contains an $\mathbf{S}^1$ factor and again $\chi(D_{\text{inst}}) = 0$. This agrees with the type IIB statement that intersections of the D3-brane with D7-branes cannot lift the $\overline{\tau}$ modes. Clearly, condition (13.28) requires that two independent 1-cycles in the $\mathbf{T}^2$ fiber degenerate over C_{inst}. This is in particular achieved by type IIB D3-brane instantons of $O(1)$-type, as follows. The corresponding 4-cycle intersects some O7-plane, which, recalling Section 11.5.1, lifts to F-theory as two 7-branes with different (p, q) labels. Namely, D_{inst} is a $\mathbf{T}^2$ fibration over C_{inst} with degenerations along the two independent 1-cycles, thus making (13.28) possible. In terms of the type IIB D3-brane, the $\overline{\tau}$ fermion zero modes are removed by the monodromy around the (p, q) degenerations, i.e. the O7-plane projection. The actual appearance of a superpotential requires the additional conditions (13.29), which are related to the absence of additional (deformation) fermion zero modes.

It is worthwhile to mention that the F-theory realization of instantons provides a physical interpretation of the fractional instantons generating the non-perturbative superpotential (13.12) of $\mathcal{N} = 1$ pure SYM. Consider for concreteness type IIB with a set of N_c D7-branes wrapped on a rigid 4-cycle C, i.e. $h_{1,0} = h_{2,0} = 0$, thus leading to a 4d $\mathcal{N} = 1$ pure $SU(N_c)$ gauge sector (times an irrelevant $U(1)$ factor). The non-perturbative superpotential from D3-brane instantons in this system can be efficiently discussed in terms of F-theory, equivalently as M5-brane instantons in the dual M-theory picture. Recall from Section 11.5.1 that in F-theory the $SU(N_c)$ gauge group arises from an I_{N_c} degeneration, in which the $\mathbf{T}^2$ fiber splits into N_c 2-cycles of $\mathbf{S}^2$ topology, which we denote C_i, $i = 1, \ldots, N_c$, with the homological relation $\sum_i [C_i] = [\mathbf{T}^2]$. The $\mathcal{N} = 1$ SYM superpotential is generated by euclidean M5-brane instantons wrapped on the 6-cycles $D_i = C_i \times C$, which satisfy (13.28) and also (13.29). The fractional nature of the cycles $[C_i]$, whose volume modulus is morally N times smaller compared with the $[\mathbf{T}^2]$ modulus, explains the fractional exponent in (13.12).

Despite the above nice matches, note that general F-theory compactifications, and hence their instantons, generically do not admit a perturbative type IIB interpretation.

13.3 Phenomenological applications

The generation of non-perturbative superpotentials from D-brane instantons has several possible phenomenological applications. They differ from gauge theory instantons in two important respects. Unlike the latter, which are normally very much suppressed, the exponential factor $\exp(-S_{\text{cl.}})$ for non-gauge D-brane instantons is not necessarily too small, since $S_{\text{cl.}}$ is in general unrelated to the SM gauge couplings. Also, gauge theory instantons only violate anomalous $U(1)$ symmetries (like $B + L$ or PQ-like symmetries in the SM), whereas D-brane instantons in string theory can violate anomalous and anomaly free $U(1)$ symmetries (like $B - L$).

We consider several different applications in turn below.

13.3.1 Neutrino masses

As mentioned in Section 1.2.6, the small masses for the observed neutrinos may be explained by the existence of right-handed neutrinos with large Majorana masses. Most SM-like models obtained from type II orientifolds have three right-handed neutrino fields ν_R^i, $i = 1, 2, 3$, in their massless spectrum. For concreteness let us concentrate on type IIA intersecting D6-brane models with four stacks of branes, a, b, c, d, and images, with gauge symmetry $SU(3) \times SU(2) \times U(1)_a \times U(1)_b \times U(1)_c \times U(1)_d$, as in Section 10.5.1. Right-handed neutrinos arise at the intersection of the c and d' branes and have charges $(1, 1)$ under $U(1)_c \times U(1)_d$. Thus a direct Majorana mass for these right-handed neutrinos is forbidden by these perturbatively exact $U(1)$ symmetries. However, instantons may generate a superpotential coupling of the form

$$e^{-S_{\text{cl.}}} \, \nu_R \nu_R, \tag{13.30}$$

where ν_R now denotes the chiral multiplet. This seemingly violates $U(1)_c \times U(1)_d$, and hence $B - L$, in two units. In fact, as described in Section 13.1.3, the exponential factor transforms as in (13.21), and can compensate for this non-invariance if

$$I_{\text{inst},c} = 2; \quad I_{\text{inst},d'} = -2, \tag{13.31}$$

for an orientifold invariant instanton. There are hence four charged fermionic zero modes $\alpha_i, \gamma_i, i = 1, 2$ corresponding respectively to those intersections. Disk amplitudes like that in Figure 13.1 generate cubic couplings of the form

$$S_{\text{z.m.}} \propto d_a^{ij} \, (\alpha_i \nu^a \gamma_j), \quad a = 1, 2, 3. \tag{13.32}$$

Integration over these charged zero modes produces a factor

$$\int d^2\alpha \, d^2\gamma \, e^{-d_a^{ij} \, (\alpha_i \nu^a \gamma_j)} \propto -\nu_a \nu_b \int d^2\alpha \, d^2\gamma \, \alpha_i \alpha_j \gamma_k \gamma_l \, d_a^{ik} d_b^{jl} = \nu_a \nu_b \left(\epsilon_{ij} \epsilon_{kl} d_a^{ik} d_b^{jl} \right),$$

yielding a right-handed Majorana neutrino mass term of the form

$$\nu_a \nu_b \, M_s \left(\epsilon_{ij} \epsilon_{kl} d_a^{ik} d_b^{jl} \right) e^{-S_{\text{cl.}}}. \tag{13.33}$$

The right-handed Majorana masses obtained are of order

$$M_R \simeq M_s \, d^2 \exp\left(-\frac{V_3}{g_s} \right), \tag{13.34}$$

and are not necessarily much suppressed, since V_3 is unrelated to the volumes wrapped by the gauge D6-branes and hence to the SM (inverse) gauge coupling constants. In particular, for $M_s \simeq 10^{16}$ GeV, a suppression of a few orders of magnitude produces a mass M_R in the preferred see-saw range.

It is worth emphasizing that this generation of non-perturbative neutrino masses requires a gauged $U(1)_{B-L}$ with gauge boson made massive by $B \wedge F$ couplings. Also, since this is an anomaly-free symmetry (in the SM with right-handed neutrinos), the effect can be realized only by stringy, rather than gauge, instantons.

Instanton effects may also give rise to the dimension-5 Weinberg operator (1.24) of Section 1.2.6, through a superpotential

$$W = \frac{\lambda}{M}(LH_uLH_u). \tag{13.35}$$

Once the Higgs field gets a vev v this operator gives rise directly to Majorana masses for the left-handed neutrinos of order $\simeq \lambda v^2/M$. In this case an $O(1)$-type instanton should have intersection numbers

$$I_{\text{inst},c} = -2, \quad I_{\text{inst},d'} = 2. \tag{13.36}$$

These intersection numbers are precisely opposite to those required for ν_R Majorana masses. There are four fermionic zero modes $\alpha_i, \gamma_i, i = 1, 2$, corresponding to the intersections of the instanton with the branes c, d'. These zero modes can have couplings involving the L and H_u chiral multiplets

$$S_{\text{z.m.}} \propto c_a^{ij} [\alpha_i(L^aH_u)\gamma_j]. \tag{13.37}$$

Integration over these fermion zero modes produces a Weinberg operator with a coefficient $\sim c^2 e^{-S_{\text{cl.}}}/M_s$. In a given compactification instantons generating Weinberg operators may coexist with instantons generating ν_R masses. The latter seesaw contribution is expected to dominate over the Weinberg operator contribution, which is suppressed by both the $1/M_s$ and exponential factors. On the other hand, for moderate exponential suppression, it may compete with the seesaw contribution.

The flavour structure, which is relevant for neutrino masses and mixings, is encoded in the coefficients d_a^{ij} and c_a^{ij}. These are model dependent and may be computable in particular models. On general grounds, this flavour structure of the instanton contributions is totally uncorrelated to that of Yukawa couplings, so it is expected to produce typically large mixings in the neutrino sector, as observed experimentally. In order to extract more specific results for the flavour structure of the Majorana neutrino masses one needs more information about the d_a^{ij} and c_a^{ij} coefficients. A particular situation where they are partially fixed by symmetry are $USp(2)$ instantons (assuming that their extra zero modes are lifted), for which i, j are $USp(2)$ indices and the symmetry requires $d_a^{ij} = d_a\epsilon^{ij}$. They can be used

Table 13.1 *Zero mode multiplicities required to generate lepton/baryon number violating superpotential operators from O(1)-type instanton*

4d Operator	$I_{\text{inst},a}$	$I_{\text{inst},a'}$	$I_{\text{inst},b}$	$I_{\text{inst},c}$	$I_{\text{inst},c'}$	$I_{\text{inst},d'}$	$I_{\text{inst},d}$
$\nu_R\nu_R$	0	0	0	2	0	−2	0
LH_uLH_u	0	0	0	−2	0	2	0
LH_u	0	0	0	−1	0	1	0
QDL	0	0	−1	0	1	1	0
UDD	−1	0	0	1	2	0	0
LLE	0	0	−1	0	1	1	0
$QQQL$	1	0	−2	0	0	1	0
$UUDE$	−1	0	0	2	2	−1	0

to illustrate the generic hierarchical structure of masses. Considering several contributing instantons, labeled by r, the right-handed neutrino mass matrix reads

$$M_R^{ab} \simeq M_s \sum_r d_a^{(r)} d_b^{(r)} e^{-S_{\text{cl.}}^{(r)}}. \tag{13.38}$$

This produces a structure of the form

$$M_R \simeq M_s \sum_r e^{-S_{\text{cl.}}^{(r)}} \left(d_1^{(r)}, d_2^{(r)}, d_3^{(r)} \right) \begin{pmatrix} 1 & 1 & 1 \\ 1 & 1 & 1 \\ 1 & 1 & 1 \end{pmatrix} \begin{pmatrix} d_1^{(r)} \\ d_2^{(r)} \\ d_3^{(r)} \end{pmatrix}. \tag{13.39}$$

For three or more contributing instantons the rank of the matrix is three and all neutrinos get massive, with generically hierarchical masses due to the exponential factors. A similar hierarchical structure could also appear if the left-handed neutrino masses came directly from a Weinberg operator as above.

13.3.2 Other B/L-violating superpotential terms

Other baryon/lepton number violating superpotentials may be generated by instantons, including R-parity violating dimension-4 operators (2.70), as well as dimension five proton decay operators as in Section 2.6.2. A list of such operators is given in Table 13.1, which also shows the required instanton intersection numbers in SM-like brane configurations with four D-brane stacks (assuming $SU(2)_L$ to be realized as $USp(2)$ for simplicity).

For instance, the LH_u bilinear operator is essentially the square root of the Weinberg operator, and may be induced by an $O(1)$-type instanton with intersections

$$I_{\text{inst},c} = -1; \quad I_{\text{inst},d'} = 1, \tag{13.40}$$

and suitable zero mode couplings, with coefficient $\sim M_s \exp(-S_{\text{inst}})$.

It is similarly possible to induce dimension-4 operators. For example, the QDL operator may be generated by an $O(1)$-type instanton with intersection numbers

$$I_{\text{inst},b} = -1; \quad I_{\text{inst},c'} = 1; \quad I_{\text{inst},d'} = 1 \tag{13.41}$$

and zero mode couplings

$$S_{\text{z.m.}} \propto c_{ab} \left[\alpha \left(D^a Q_j^b \right) \gamma^j \right] + c'_a \left(\beta L_j^a \gamma_j \right). \tag{13.42}$$

Here α, β, γ are zero modes from $(\text{inst} - c')$, $(\text{inst} - d')$, and $(b - \text{inst})$ intersections, j is an $SU(2)_L$ index, and a, b are flavour indices. Analogous trilinear or quartic disk amplitudes involving two charged zero modes should exist to generate the remaining operators in Table 13.1.

All R-parity violating couplings should not be simultaneously present in the low-energy field theory since they would induce too fast proton decay. Thus models in which R-parity is respected are specially attractive. On the other hand, lepton number violation is required to have neutrino Majorana masses. These phenomenological considerations point to the interesting possibility of having instantons inducing neutrino masses but no R-parity violating operators. A relevant remark in this respect is that if only $USp(2)$ instantons are present, R-parity conservation is automatic. This follows because all charged zero modes necessarily come in $USp(2)$ doublets, so the induced 4d operators must involve an even number of 4d charged fields. On the other hand, such instantons can generate neutrino masses, if the extra universal zero mode triplets are lifted by some mechanism.

13.3.3 Yukawa couplings

In perturbative type II orientifolds some of the required SM Yukawa couplings may be forbidden by $U(1)$ symmetries. An already familiar example is the U-type Yukawa coupling $\mathbf{10} \cdot \mathbf{10} \cdot \mathbf{5}_H$ in D-brane realizations of $SU(5)$ GUTs, which are forbidden by the $U(1)$ factor in $U(5)$. In flipped $SU(5)$ the opposite situation occurs, since now the $\mathbf{10} \cdot \mathbf{10} \cdot \mathbf{5}_H$ couplings contain D-quark Yukawas instead. The $U(1)$ symmetries forbidding the U-quark Yukawa couplings are in general anomalous, and hence massive, and so D-brane instantons may induce the forbidden couplings. It is easy to figure out what is the structure of instanton fermion zero modes required in order to get these Yukawas. Consider a stack 1 of five D6-branes with gauge group $U(5)_1$ intersecting a single D6-brane 2 with gauge group $U(1)_2$. The minimal structure of D6-brane intersections required to get a $SU(5)$ GUT is as in Table 13.2.

The subindices show the $U(1)_1 \times U(1)_2$ charges, and primes denote orientifold image D6-branes. The $\mathbf{10}$'s and $\bar{\mathbf{5}}$'s arise from the $11'$ and $1'2'$ intersections, respectively, whereas the Higgs fields reside at $12'$ intersections. As mentioned, the D-type Yukawas

$$\mathbf{10}_{(2,0)} \cdot \bar{\mathbf{5}}_{(-1,1)} \cdot \bar{\mathbf{5}}^H_{(-1,-1)} \tag{13.43}$$

are allowed by the $U(1)$ symmetries, whereas the U-type coupling

$$\mathbf{10}_{(2,0)} \cdot \mathbf{10}_{(2,0)} \cdot \mathbf{5}^H_{(1,1)} \tag{13.44}$$

Table 13.2 *Configuration of intersecting*
D6-branes realizing an SU(5) GUT

Intersection	I_{ab}	$U(5)_1 \times U(1)_2$
$11'$	3	$\mathbf{10}_{(2,0)}$
$1'2'$	3	$\bar{\mathbf{5}}_{(-1,1)}$
$22'$	3	$\mathbf{1}_{(0,-2)}$
$12'$	1	$\mathbf{5}^H_{(1,1)} + \bar{\mathbf{5}}^H_{(-1,-1)}$

is forbidden. The latter may however be generated by an $O(1)$-type instanton with intersection numbers $I_{\text{inst},1} = I_{2,\text{inst}} = 1$, yielding fermion zero modes α^i, $i = 1, \ldots, 5$ (transforming in a $\bar{5}$) and γ, and with couplings to the GUT multiplets

$$S_{\text{z.m.}} \simeq d_a \alpha^i \mathbf{10}^a_{ij} \alpha^j + d' \alpha^i \mathbf{5}^H_i \gamma, \tag{13.45}$$

where $a = 1, 2, 3$ is a family index. Integrating over the zero modes yields a U-type Yukawa superpotential

$$W_Y = e^{-S_{\text{cl.}}} \epsilon_{ijklm} d_a d_b d' \left(\mathbf{10}^a_{ij} \mathbf{10}^b_{kl} \mathbf{5}^H_m \right). \tag{13.46}$$

The Yukawa coupling flavour structure is encoded in the product $d_a d_b$, and hence a single instanton gives rise to a rank-1 Yukawa matrix. However if several different instantons contribute with a slight hierarchy in their prefactors one can obtain a rank-3 matrix with hierarchical eigenvalues, as in (13.39). The exponential prefactors, although they need not be very small, suggest that such contributions may not suffice to reproduce the $\mathcal{O}(1)$ top Yukawa coupling. As mentioned in Section 11.5, this is one of the motivations to realize GUT-based models in the non-perturbative setting of F-theory.

A similar analysis yields the conditions to generate the forbidden Yukawa couplings in flipped $SU(5)$ models. The use of instantons in this scheme is perhaps more attractive, since the Yukawa coupling in question is of D-type, so the exponential instanton factor could possibly explain the top/bottom quark mass hierarchy, as the former arises at tree level, while the latter arises non-perturbatively.

Instanton effects may also play an important role in modifying the perturbative structure of Yukawa couplings in non-unified SM-like models. This can be illustrated with a new MSSM-like type IIA $\mathbf{Z}_2 \times \mathbf{Z}_2$ toroidal orientifold model, with D6-branes on the 3-cycles in Table 13.3. The model is dubbed "local" in the sense that it does not satisfy the RR tadpole conditions, although this may be easily solved by adding suitable extra D6-branes, which we do not display for simplicity. The gauge group is $SU(3) \times SU(2) \times U(1)_a \times U(1)_b \times U(1)_c \times U(1)_d$, and its massless spectrum is shown in Table 13.4. It corresponds to the MSSM with the hypercharge defined as $Y = \frac{1}{6}Q_a - \frac{1}{2}Q_c - \frac{1}{2}Q_d$, with a gauged $U(1)_{B-L}$. The anomalous $U(1)_b$ becomes massive in the usual way. The local model is supersymmetric for $\mathbf{T}^2$ complex structure moduli satisfying $\text{Re}\, U_1 = \text{Re}\, U_2 = \text{Re}\, U_3/2$.

Table 13.3 *Wrapping numbers of a MSSM-like local intersecting D6-brane model*

MSSM-2	$\left(n_i^1, m_i^1\right)$	$\left(n_i^2, m_i^2\right)$	$\left(n_i^3, \tilde{m}_i^3\right)$
$N_a = 6 + 2$	$(1, 0)$	$(3, 1)$	$(3, -1/2)$
$N_b = 4$	$(1, 1)$	$(1, 0)$	$(1, -1/2)$
$N_c = 2$	$(0, 1)$	$(0, -1)$	$(2, 0)$

Table 13.4 *Spectrum and U(1) charges of the MSSM-like model*

Intersection	Matter fields		Q_a	Q_b	Q_c	Q_d	Q_Y
ab	Q_L	$(\mathbf{3}, \mathbf{2})$	1	-1	0	0	$1/6$
ab'	q_L	$2(\mathbf{3}, \mathbf{2})$	1	1	0	0	$1/6$
ac	U_R	$3(\bar{\mathbf{3}}, \mathbf{1})$	-1	0	1	0	$-2/3$
ac'	D_R	$3(\bar{\mathbf{3}}, \mathbf{1})$	-1	0	-1	0	$1/3$
bd	L	$(\mathbf{1}, \mathbf{2})$	0	-1	0	1	$-1/2$
bd'	l	$2(\mathbf{1}, \mathbf{2})$	0	1	0	1	$-1/2$
cd	ν_R	$3(\mathbf{1}, \mathbf{1})$	0	0	1	-1	0
cd'	E_R	$3(\mathbf{1}, \mathbf{1})$	0	0	-1	-1	1
bc	H_d	$(\mathbf{1}, \mathbf{2})$	0	-1	1	0	$-1/2$
bc'	H_u	$(\mathbf{1}, \mathbf{2})$	0	-1	-1	0	$1/2$

The perturbatively allowed couplings are

$$q_{L(1,0,0)}\, H_{u(-1,-1,0)}\, U_{R(0,1,0)} \quad q_{L(1,0,0)}\, H_{d(-1,1,0)}\, D_{R(0,-1,0)},$$
$$l_{(1,0,1)}\, H_{d(-1,1,0)}\, E_{R(0,-1,-1)} \quad l_{(1,0,1)}\, H_{u(-1,-1,0)}\, \nu_{R(0,1,-1)}, \qquad (13.47)$$

where the subscripts denote the charges under $U(1)_b \times U(1)_c \times U(1)_d$. Thus the $U(1)$ symmetries allow for perturbative Yukawa matrices of the form

$$m_{U,D,L,N} \simeq \begin{pmatrix} Y_{11} & Y_{12} & Y_{13} \\ Y_{21} & Y_{22} & Y_{23} \\ 0 & 0 & 0 \end{pmatrix}. \qquad (13.48)$$

There are also couplings which are perturbatively forbidden by $U(1)_b$, namely

$$Q_{L(-1,0,0)}\, H_{u(-1,-1,0)}\, U_{R(0,1,0)} \quad Q_{L(-1,0,0)}\, H_{d(-1,1,0)}\, D_{R(0,-1,0)},$$
$$L_{(-1,0,1)}\, H_{d(-1,1,0)}\, E_{R(0,-1,-1)} \quad L_{(-1,0,1)}\, H_{u(-1,-1,0)}\, \nu_{R(0,1,-1)}, \qquad (13.49)$$

and provide the missing entries in (13.48). These can be non-perturbatively generated by instantons with fermion zero modes α^i, $i = 1, 2$ in an $SU(2)_L$ doublet from $I_{\text{inst},b} = -1$, two singlet zero modes γ^+, γ^- from $I_{\text{inst},c} = I_{\text{inst},c'} = 1$, and zero mode couplings of the form, e.g. $\alpha_i Q_i^a U_b \gamma^-$ and $\alpha_i H_u^i \gamma^+$. This generates a U-quark Yukawa coupling with structure

$$\int d^2\alpha d^2\gamma e^{-S_{\text{cl.}}} \, d_b \epsilon_{ij} \epsilon_{kl} \alpha^i (Q^j U_b) \gamma^- \alpha^k H_u^l \gamma^+, \tag{13.50}$$

where $b = 1, 2, 3$ is a flavour index. And analogously for the remaining perturbatively forbidden Yukawa couplings in (13.49). Incidentally, note that in this example instantons cannot generate ν_R Majorana masses since $U(1)_{B-L}$ is not made massive by Stückelberg couplings.

13.3.4 μ-term

In the context of the MSSM, there is one phenomenologically relevant mass bilinear superpotential term, the μ-term in (2.69), i.e.

$$\mu \, H_u H_d. \tag{13.51}$$

In some MSSM-like D-brane models such a term may be forbidden by $U(1)$ symmetries, as for instance in the $\mathbf{Z}_2 \times \mathbf{Z}_2$ example in the previous section (in which the μ-term violates $U(1)_b$). It can, however, be generated by an instanton with intersection numbers $I_{\text{inst},b} = -1$, $I_{\text{inst},c} = I_{\text{inst},c'} = 1$ (incidentally, the same as in the above example) and cubic couplings

$$\alpha_i H_u^i \gamma^+, \quad \alpha_j H_d^j \gamma^-. \tag{13.52}$$

Upon integration over fermion zero modes one obtains $\mu \simeq \exp(-S_{\text{cl}})M_s$, thus offering an explanation for the smallness of the μ-term even for large M_s. One potential problem is that the same instanton generating the μ-term may also contribute to some entry for a Yukawa matrix, leading to conflict due to the two very different required couplings. This issue is, however, very model dependent.

This is an attractive approach to the μ-problem, already mentioned in Section 2.6.3. The μ-term is forbidden perturbatively by an anomalous $U(1)$ symmetry, and is generated by instantons with a hierarchical small value. The mechanism does however not explain why the SUSY-breaking soft terms turn out to have values of the same order of magnitude as μ. An alternative mechanism to generate the μ-term, addressing this last point, will be described in Chapter 15, based on closed string fluxes.

13.3.5 Instanton induced Polonyi terms

Instantons may generate superpotential terms not only for SM matter fields but also for fields in a hidden sector involved in the SUSY-breaking of the theory. A simple example of such sector is the Polonyi model in Section 2.3, described by a superpotential linear in a single field z. Such sectors can be engineered using two stacks of single D6-branes carrying gauge group $U(1)_1 \times U(1)_2$ intersecting at one point, which supports a SM singlet z, with

$U(1)$ charges $(1, -1)$. The linear superpotential is forbidden by these $U(1)$ symmetries, but may be generated by a D-brane instanton with intersection numbers $I_{\text{inst},1} = I_{b,\text{inst}} = 1$, and a cubic coupling $\alpha z \gamma$. This would lead to

$$W_P = M_s^2 \exp(-S_{\text{cl.}})z. \tag{13.53}$$

Hence D-brane instantons may play an interesting role in contributing to the dynamics of SUSY breaking, see also Section 15.7. This is yet another example of how string theory instanton effects can generate a variety of couplings of phenomenological relevance.

14

Flux compatifications and moduli stabilization

In all string compactifications studied in previous chapters, the massless spectrum contains a large number of moduli, including the dilaton and the CY geometric moduli. These are problematic, as they couple to matter particles and easily lead to deviations from universality of gravitational interactions (fifth forces), which have not been observed experimentally. Therefore, any serious attempt to reproduce realistic string models of particle physics and/or cosmology, must address the issue of moduli stabilization; namely, the generation of a scalar potential fixing the vevs of the moduli fields and giving them large enough masses to overcome such phenomenological problems. In this chapter we review a very general and systematic mechanism to fix large numbers of (or even all) moduli. It is based on a generalization of the simplest compactification ansatz, allowing for non-trivial backgrounds for additional 10d fields. Most prominently, the compactifications include non-trivial fluxes for the field strength tensor of the diverse p-form fields in the 10d theory. Application of string dualities to these motivates additional possible backgrounds, termed geometric and non-geometric fluxes. Compactifications including field strength fluxes, or their generalizations, are thus known as "flux compactifications." This chapter focuses on their definition and impact on moduli stabilization, while their role in SUSY breaking is further explored in Chapter 15.

14.1 Type IIB with 3-form fluxes

A prototypical class of flux compactifications is obtained from type IIB (orientifolds) on CY geometries with non-trivial fluxes for the NSNS and RR 3-form field strengths (and their F-theory generalization, which we briefly touch upon). These compactifications admit a simple 10d supergravity description, similar to a standard CY compactification, but with a non-trivial warp factor (and a 5-form field strength) from the backreaction of the fluxes. In the large volume regime, the fluxes are dilute and the moduli stabilization can be studied in terms of the 4d effective theory, by the simple addition of a superpotential to the effective action of the flux-less CY compactification. This description provides a template for more general discussions of moduli stabilization in later sections.

14.1.1 Ten-dimensional supergravity solution

We consider type IIB compactification on 6d spaces $\mathbf{X}_6$ with CY topology, but in the presence of non-trivial NSNS and RR 3-form field strength fluxes H_3, F_3. Since fluxes gravitate, their backreaction implies that the actual 10d spacetime is not just the product of 4d Minkowski space times a CY. The full solution is however simple and can be spelled out quite explicitly, as we now describe.

The type IIB supergravity action (4.45), written in the Einstein frame, becomes

$$
S_{\text{IIB}} = \frac{1}{2\kappa_{10}^2} \int d^{10}x \sqrt{-g} \left(R - \frac{\partial_M \bar{\tau} \partial^M \tau}{2(\operatorname{Im}\tau)^2} - \frac{1}{2}|F_1|^2 - \frac{|G_3|^2}{2\operatorname{Im}\tau} - \frac{1}{4}|\tilde{F}_5|^2 \right)
$$
$$
+ \frac{1}{2\kappa_{10}^2} \int C_4 \wedge \frac{G_3 \wedge \bar{G}_3}{4i \operatorname{Im}\tau} + S_{\text{local}}.
\tag{14.1}
$$

Here,

$$
\tau = C_0 + ie^{-\phi}, \quad G_3 = F_3 - \tau H_3, \quad \text{with } F_3 = dC_2 \text{ and } H_3 = dB_2,
\tag{14.2}
$$

and

$$
\tilde{F}_5 = F_5 - \frac{1}{2}C_2 \wedge H_3 + \frac{1}{2}B_2 \wedge F_3, \quad \text{with } *_{10d}\,\tilde{F}_5 = \tilde{F}_5.
\tag{14.3}
$$

The term S_{local} includes local sources of the 10d supergravity fields present in the compactification, like D3-branes or O3-planes, see below.

We consider compactifications with $F_1 = 0$, and fluxes F_3, H_3 with no sources,

$$
dF_3 = 0, \quad dH_3 = 0.
\tag{14.4}
$$

They thus determine cohomology classes in $\mathbf{X}_6$; in fact, minimization of the kinetic energy stored in the flux implies that F_3, H_3, are given by the harmonic representatives (with respect to the underlying CY metric) of these classes. The fluxes are determined by their integrals over 3-cycles γ in $\mathbf{X}_6$, which from the arguments in Section B.26, must obey a Dirac quantization condition (B.26), i.e.

$$
\frac{1}{(2\pi)^2\alpha'} \int_\gamma F_3 = m_\gamma \in \mathbf{Z}, \qquad \frac{1}{(2\pi)^2\alpha'} \int_\gamma H_3 = n_\gamma \in \mathbf{Z},
\tag{14.5}
$$

where the factors $(2\pi\alpha')^{-1}$ arise from the elementary (F1- and D1-brane) charges under B_2, C_2. For odd quanta, there is a potential subtlety in flux quantization in orientifold models, so we assume even quanta in order to avoid it.

The internal 3-form fluxes gravitate, and their effect on the metric follows from the 6d piece of their kinetic term

$$
\mathcal{L}_G = -\frac{1}{24\kappa_{10}^2} \int_{\mathbf{X}_6} d^6y\, g^{\frac{1}{2}} \frac{(G_3)_{mnp}\,(\bar{G}_3)^{mnp}}{\operatorname{Im}\tau} = -\frac{1}{8\kappa_{10}^2} \int_{\mathbf{X}_6} d^6y\, \frac{G_3 \wedge *_6 \bar{G}_3}{\operatorname{Im}\tau}.
\tag{14.6}
$$

In addition the simultaneous presence of NSNS and RR 3-form flux provides a source for the RR 4-form, as follows from the Chern–Simons interaction

$$\mathcal{L}_{C_4} = \frac{1}{2\kappa_{10}^2} \int C_4 \wedge \frac{G_3 \wedge \bar{G}_3}{4i \operatorname{Im} \tau}. \tag{14.7}$$

This manifests as a contribution to the Bianchi identity for the RR 5-form,

$$d\tilde{F}_5 = d * \tilde{F}_5 = H_3 \wedge F_3 + 2\kappa_{10}^2 \mu_3 \rho_3^{\text{local}}, \tag{14.8}$$

where ρ_3^{local} is the localized source contribution e.g. from D3-branes and O3-planes.

Since fluxes carry tension and charge, they behave in many respects as a distribution of D3-brane charge; this suggests an ansatz for the supergravity solution produced by the flux background, similar to the Dp-brane solution (6.25) for $p = 3$. We assume 4d Poincaré invariance, constant dilaton, and maintain independent metric and 5-form backgrounds, postponing their actual relation, and propose

$$ds_{10}^2 = e^{2A(y)} \eta_{\mu\nu} dx^\mu dx^\nu + e^{-2A(y)} \tilde{g}_{mn} dy^m dy^n,$$

$$\tilde{F}_{(5)} = (1 + *_{10d}) \left[d\alpha \wedge dx^0 \wedge dx^1 \wedge dx^2 \wedge dx^3 \right], \tag{14.9}$$

where $\tilde{g}$ is the underlying CY metric, which is also implicitly used to raise/lower indices in what follows. The warp factor $e^{2A(y)}$ and the function $\alpha(y)$ are determined by Laplace equations sourced by fluxes and gradients (and localized sources, which we ignore for the moment), i.e.

$$\tilde{\nabla}^2 e^{4A} = e^{2A} \frac{(G_3)_{mnp} (\bar{G}_3)^{mnp}}{12 \operatorname{Im} \tau} + e^{-6A} \left[\partial_m \alpha \, \partial^m \alpha + \partial_m e^{4A} \, \partial^m e^{4A} \right],$$

$$\tilde{\nabla}^2 \alpha = i e^{2A} \frac{(G_3)_{mnp} (*_6 \bar{G}_3)^{mnp}}{12 \operatorname{Im} \tau} + 2 e^{-6A} \partial_m \alpha \, \partial^m e^{4A}, \tag{14.10}$$

with $\tilde{\nabla}$ computed with the underlying CY metric. From equations (14.6) and (14.7), the tension and 4-form charge induced by the fluxes are equal if they satisfy the imaginary self-duality (ISD) condition

$$*_{6d} G_3 = i G_3. \tag{14.11}$$

For this class of fluxes we can consistently set

$$\alpha = e^{4A}, \tag{14.12}$$

so both equations in (14.10) are equivalent, and the supergravity solution becomes analogous to that of BPS D3-branes; the ISD condition is indeed similar to a BPS condition, although its interplay with supersymmetry is trickier, as discussed later.

The ISD condition on fluxes is not just a convenient choice. From (14.10) we have

$$\tilde{\nabla}^2 (e^{4A} - \alpha) = \frac{e^{2A}}{6 \operatorname{Im} \tau} \left| i G_3 - *_6 G_3 \right|^2 + e^{-6A} \left| \partial (e^{4A} - \alpha) \right|^2. \tag{14.13}$$

On a compact manifold the left-hand side integrates to zero, and the right-hand side is positive definite, so all terms must vanish; hence consistency of the 4d Poincaré invariant ansatz requires ISD fluxes.

Since fluxes source the RR 4-form, its RR tadpole cancellation condition, i.e. the integrability of the Bianchi identity (14.8), becomes

$$\frac{1}{2} N_{\text{flux}} + N_{D3} - \frac{1}{4} N_{O3} = 0. \tag{14.14}$$

Here and in what follows, N_{D3} denotes the number of D3-branes in the quotient space (equivalently, the number of D3-branes in the parent space, not counting orientifold images). The term N_{D3} also includes contributions from induced D3-brane charge on magnetized D7-branes; similarly N_{O3} includes, besides O3-planes, other negative D3-brane charge sources, e.g. see (14.38) in the upcoming F-theory generalizations of the IIB setup. Finally N_{flux} is the contribution from the fluxes,

$$N_{\text{flux}} = \frac{1}{(2\pi)^4 (\alpha')^2} \int_{\mathbf{X}_6} H_3 \wedge F_3, \tag{14.15}$$

determined by the flux quanta as follows. Introduce a basis of 3-cycles $\{\alpha_K, \beta^K\}$, $K = 0, \ldots, h_{21}$, with $[\alpha_K] \cdot [\beta^L] = \delta_K^L$, and denote m^K, m_K, and n^K, n_K the corresponding F_3 and H_3 flux quanta (14.5). Then

$$N_{\text{flux}} = \frac{1}{(2\pi)^4 (\alpha')^2} \sum_K \left(\int_{\alpha_K} H_3 \int_{\beta^K} F_3 - \int_{\beta^K} H_3 \int_{\alpha_K} F_3 \right) = \sum_K (n^K m_K - n_K m^K).$$

For ISD fluxes, this 3-form flux tadpole contribution N_{flux} is positive, as follows. Displaying the ISD condition as

$$*_6 H_3 / g_s = -(F_3 - C_0 H_{(3)}), \tag{14.16}$$

we have

$$\int_{\mathbf{X}_6} H_3 \wedge F_3 = -\frac{1}{g_s} \int_{\mathbf{X}_6} H_3 \wedge *_6 H_3 \sim \frac{1}{g_s} \int_{\mathbf{X}_6} \tilde{g}^{\frac{1}{2}} |H_3|^2 > 0. \tag{14.17}$$

Hence the RR tadpole cancellation requires the presence of negative D3-brane charges, e.g. O3-planes; since these extra sources are BPS, their effects on the metric and 5-form Laplace equations (14.10) are still consistent with (14.12).

The above system has a natural generalization to F-theory, which thus includes also non-trivial holomorphic background for the 10d complex coupling τ. These are best described in terms of M-theory on a CY fourfold $\mathbf{X}_8$, with non-trivial flux G_4 for the field strength 4-form, as described later on.

The ISD condition (14.11) is actually not a condition on the fluxes, which are fixed by the integer quanta (14.5), but rather a condition fixing the CY moduli which enter the ISD condition through the 6d Hodge operation. The analysis of moduli fixing is most efficiently carried out in the 4d effective field theory description, worked out in the next section. Certain features are however manifest at this level, like the presence of a universal unfixed modulus, corresponding to the overall size R of $\mathbf{X}_6$; this is a particular case of the upcoming general pattern that Kähler moduli are not fixed by 3-form fluxes.

14.1.2 Effective field theory description

From the above discussion, type IIB compactifications with 3-form fluxes lead to (possibly partial) stabilization of the dilaton and complex structure moduli. The mass scale of the lifted moduli is of order the scale of local flux densities, and (modulo extra factors from possible local effects) typically scales with the (unfixed) overall size R as

$$m_{\text{moduli}} \simeq \frac{\alpha'}{R^3}. \qquad (14.18)$$

In the large volume regime, this is parametrically smaller than the KK scale $\simeq 1/R$, and the moduli stabilization dynamics can be studied purely in terms of the 4d effective theory. For large R, the flux densities and their effect on the Laplace equations (flux-laplace) dilute away, and the metric ansatz (14.9) is close to a CY compactification. Hence, the 4d effective theory between the KK scale and the flux scale is the $\mathcal{N}=1$ effective action in Chapter 12; at scales (14.18), the effective action must include new terms to account for the fluxes present in the background.

The flux superpotential

The most relevant such term corresponds to a superpotential; other corrections, e.g. to the moduli Kähler potential, are suppressed in the large volume regime and ignored in the following. The 4d $\mathcal{N}=1$ superpotential induced by the fluxes can be heuristically computed using the following argument, which describes the introduction of fluxes in terms of BPS domain walls.

Consider a type IIB orientifold compactification on a CY $\mathbf{X}_6$, and introduce a D5-brane wrapped on a special lagrangian 3-cycle $[\Pi]$, spanning the coordinates x^0, x^1, x^2 in 4d spacetime, and located at $x^3 = 0$. This describes a BPS domain wall separating the two regions $x^3 < 0$ and $x^3 > 0$; the two regions differ in their F_3 flux, as we now show. The D5-brane sources F_3 as

$$dF_3 = \delta_3(\Pi) \wedge \delta_0(x^3)\,dx^3, \qquad (14.19)$$

where $\delta_0(x^3)$ is a standard Dirac delta function, and $\delta(\Pi)$ is the Poincaré dual of Π, i.e. a 3-form on the CY, localized on Π and with indices in the three directions transverse to Π. Now (14.19) implies a jump in the F_3 flux across the 3-cycle $\tilde{\Pi}$ dual to Π (in the sense that $[\Pi] \cdot [\tilde{\Pi}] = 1$), as follows. Consider a path $I = \{x^3 \in (-\epsilon, \epsilon)\}$ crossing the domain wall in 4d, and compute the difference in flux across $\tilde{\Pi}$

$$\int_{\tilde{\Pi}, x^3=\epsilon} F_3 - \int_{\tilde{\Pi}, x^3=-\epsilon} F_3 = \int_{\tilde{\Pi} \times I} dF_3 = \int_I dx^3 \delta(x^3) \int_{\tilde{\Pi}} \delta_3(\Pi) = [\Pi] \cdot [\tilde{\Pi}] = 1,$$

where we have used Stokes theorem in the first equality and (14.19) in the second.

The D5-brane domain wall has a tension given by the volume of the wrapped 3-cycle Π. For a BPS domain wall in a 4d $\mathcal{N} = 1$ theory, its tension is associated to the difference of the superpotentials between the two vacua. Assuming vanishing flux for $x^3 < 0$, the superpotential for the CY compactification with F_3 flux at $x^3 > 0$ is essentially

$$W_{F_3} = \int_\Pi \Omega_3 = \int_{\mathbf{X}_6} F_3 \wedge \Omega_3, \qquad (14.20)$$

where the localization of $[F_3]$ on $[\tilde{\Pi}]$ is used to extend the integral to the whole $\mathbf{X}_6$.

The superpotential for general NSNS and RR fluxes can be obtained analogously by using domain walls with NS5/D5-brane charge; equivalently, by covariantizing the above expression with respect to the $SL(2, \mathbf{Z})$ S-duality of type IIB theory, under which the two 3-form fluxes form a doublet. The result is known as the Gukov–Vafa–Witten superpotential

$$W_{\mathrm{GVW}} = \int_{\mathbf{X}_6} G_3 \wedge \Omega_3 = \int_{\mathbf{X}_6} (F_3 - i\,S\,H_3) \wedge \Omega_3, \qquad (14.21)$$

where we have replaced the complex coupling τ by the 4d dilaton notation S. This result may also be obtained by an explicit KK reduction of the original type IIB 10d theory. The superpotential depends explicitly on S, and implicitly on complex structure moduli U through Ω_3; the resulting scalar potential generically stabilizes all these moduli.

Supersymmetry conditions and no-scale structure

The above 4d effective action description allows to study the structure of the scalar potential, its minima, and their supersymmetry properties. For simplicity, we consider the case with only the overall Kähler modulus T, and take $\kappa_4^2 = 1$.

Recall the moduli Kähler potential (12.16) in the large volume regime

$$K = -3 \log(T + T^*) - \log (S + S^*) - \log \left(-i \int_{\mathbf{X}_6} \Omega_3 \wedge \overline{\Omega}_3 \right). \qquad (14.22)$$

The scalar potential (2.50) is

$$V = e^K \left(g^{a\bar{b}} D_a W\, D_{\bar{b}}\overline{W} - 3|W|^2 \right), \qquad (14.23)$$

where we sum over all moduli. The structure of (14.22) produces a cancellation between the overall Kähler modulus T contribution, and the second term above; the remaining scalar potential is

$$V = e^K \left(g^{i\bar{j}} D_i W\, D_{\bar{j}}\overline{W} \right), \qquad (14.24)$$

where we sum over the dilaton and complex structure moduli only. It is positive definite, and vanishes on configurations $D_i W = 0$, for which it leads to Minkowski vacua. These vacua are supersymmetric only if they satisfy the additional condition $D_T W = 0$. Even without supersymmetry, these vacua have zero 4d cosmological constant, due to the no-scale supergravity structure, already mentioned in Sections 9.1.1 and 12.1.2. This structure holds in the classical and large volume approximation, and is modified by quantum and α' corrections, which are in fact relevant for the Kähler moduli stabilization in Section 15.3. Despite corrections, the leading no-scale structure plays an important role in supersymmetry breaking and its (gravity) mediation to the Standard Model, as studied in Sections 15.4 and 15.5.

Let us be more explicit on the conditions for a Minkowski vacuum, and for supersymmetry. From the dilaton Kähler potential (14.22) and the superpotential (14.21), the dilaton multiplet auxiliary field is

$$D_S W = -\frac{1}{S + S^*} \int_{\mathbf{X}_6} \overline{G}_3 \wedge \Omega_3. \tag{14.25}$$

For the complex structure moduli, we need the relation

$$\partial_{U_i} \Omega_3 = -K_{U_i} \Omega_3 + \chi_i, \tag{14.26}$$

where $K_{U_i} = \partial_{U_i} K$, and the χ_i form a complete basis of $(2, 1)$ forms. We obtain

$$D_{U_i} W = \int_{\mathbf{X}_6} G_3 \wedge \chi_i. \tag{14.27}$$

Therefore, denoting $G_3|_{(p,q)}$ the (p, q) component of the complex tensor G_3, the conditions $D_S W = D_{U_i} W = 0$ to have a (supersymmetric or not) 4d Minkowski vacuum read

$$G_3|_{(3,0)} = 0, \quad G_3|_{(1,2)} = 0. \tag{14.28}$$

These are equivalent to the ISD condition (14.11), as follows. Since the 6d Hodge operation $*_{6d}$ on complex tensors introduces a factor $\pm i$ for each holomorphic or antiholomorphic index, respectively, the imaginary (anti)self-dual parts of G_3 decompose as

$$\begin{aligned} G_3|_{\mathrm{ISD}} &= G_3|_{(2,1),\mathrm{P}} + G_3|_{(1,2),\mathrm{NP}} + G_3|_{(0,3)}, \\ G_3|_{\mathrm{IASD}} &= G_3|_{(3,0)} + G_3|_{(2,1),\mathrm{NP}} + G_3|_{(1,2),\mathrm{P}}, \end{aligned} \tag{14.29}$$

where "P" denotes the so-called "primitive" term, obeying $G_3 \wedge J = 0$, and NP stands for "non-primitive"; in a CY compactification there are no harmonic 5-forms, so $G_3 \wedge J \equiv 0$, and there are no NP components. The conditions (14.28) therefore imply the vanishing of the IASD 3-form flux, i.e. we recover that 4d Minkowski vacua require ISD fluxes.

The ISD condition allows for $(2, 1)$ and $(0, 3)$ flux components. The condition for 4d $\mathcal{N} = 1$ supersymmetry requires vanishing of

$$D_T W = \frac{-3}{T + T^*} \int_{\mathbf{X}_6} G_3 \wedge \Omega_3. \tag{14.30}$$

Namely

$$G_3|_{(0,3)} = 0, \tag{14.31}$$

and hence supersymmetric Minkowski vacua are obtained for $(2, 1)$ fluxes.

When considering the introduction of fluxes on spaces with non-trivial homology 1-cycles, like $\mathbf{T}^6$ orientifolds, the dual 5-form can support a non-zero $G_3 \wedge J$; in these cases, supersymmetry requires the additional condition of its vanishing

$$G_3 \wedge J = 0, \tag{14.32}$$

This arises from D-term constraints related to the superpotential by the underlying enhanced supersymmetry in such compactifications.

F-theory generalization*

The above description of the effect of fluxes in the low-energy effective action generalizes to F-theory. Recalling Section 11.5, F-theory compactifications are defined in terms of M-theory on a CY fourfold $\mathbf{X}_8$, in this case with 4-form flux G_4 for the 3-form gauge potential. Using arguments similar to the above with an M5-brane domain wall, the superpotential is

$$W = \int_{\mathbf{X}_8} G_4 \wedge \Omega_4. \tag{14.33}$$

This reduces to the type IIB expression (14.21) by using the relation between F-theory and IIB quantities away from degenerations of the $\mathbf{T}^2$ fiber

$$G_4 = \frac{i\, G_3\, d\overline{w}}{S + \overline{S}} + \text{h.c.}, \quad \Omega_4 = \Omega_3 \wedge dw. \tag{14.34}$$

Here $dw = dx + i S dy$, where x, y are coordinates on the $\mathbf{T}^2$ fiber, with complex structure $\tau = iS$, and so $\int_{\mathbf{T}^2} i\, dw\, d\overline{w}/(S + \overline{S}) = 1$. The F-theory perspective is useful in showing that type IIB 3-form fluxes can stabilize 7-brane position moduli, since they correspond to complex structure moduli of the CY fourfold $\mathbf{X}_8$; this will be recovered in Section 15.4.1 from a perturbative type IIB perspective.

The SUSY conditions on G_4 can be analyzed in analogy with the type IIB case. They require G_4 to be (2, 2) and primitive

$$G_4 = G_4\big|_{(2,2)}, \quad G_4 \wedge J_{\mathbf{X}_8} = 0, \quad \text{with } J_{\mathbf{X}_8} = J_{\mathbf{B}_6} + dw\, d\overline{w}. \tag{14.35}$$

It is easy to use (14.34) to recover the SUSY conditions on the type IIB G_3.

As described in Section 11.5.1, degenerations of the $\mathbf{T}^2$ fiber correspond to type IIB 7-branes. Such degenerations support harmonic (1, 1)-forms ω_a, localized on holomorphic 4-cycles S_a on the base $\mathbf{B}_6$, corresponding to the 4-cycle wrapped by the type IIB 7_a-branes. From (11.118), components of G_4 flux along ω descend to 7-brane worldvolume magnetic fields F in the latter, via

$$G_4 = \sum_a \omega_a \wedge F_a. \tag{14.36}$$

Applying the F-theory G_4 SUSY conditions (14.35) to such components (noting $\omega_a \wedge dw\, d\overline{w} = 0$), one easily reproduces the SUSY conditions (11.102) for general magnetized 7-branes.

It is worthwhile to mention the generalization of the RR tadpole cancellation condition (14.14) to F-theory; in M-theory on $\mathbf{X}_8$ there are different sources contributing to the 3-form tadpole, sketchily

$$\sum_i N_i \int_{M2_i} C_3 + \int_{\mathbf{M}_3 \times \mathbf{X}_8} G_4 \wedge G_4 \wedge C_3 + \int_{\mathbf{M}_3 \times \mathbf{X}_8} Y_8(R) \wedge C_3$$

$$\Longrightarrow \sum_i N_i + \int_{\mathbf{X}_8} G_4 \wedge G_4 + \frac{\chi(\mathbf{X}_8)}{24} = 0. \tag{14.37}$$

The first term is the contribution from possible M2-branes at points in $\mathbf{X}_8$, and the second is the 4-form flux contribution from the Chern–Simons coupling in the 11d supergravity action (4.51); the third is a quantum correction in M-theory, with Y_8 a polynomial quartic in the curvature 2-form whose only relevant property for us is $\int_{\mathbf{X}_8} Y_8 = \chi(\mathbf{X}_8)/24$, where $\chi(\mathbf{X}_8) = \sum_{i=0}^8 b_i(\mathbf{X}_8)$ is the fourfold Euler characteristic. Translating to type IIB/F-theory language, these correspond to sources of RR 4-form tadpole. Using the G_4 components (14.34), (14.36), and the property $\int_{\mathbf{T}^2} \omega_a \wedge \omega_b = \delta_{ab} \delta(S_a)$, the schematic tadpole condition is

$$\sum_i N_i + \frac{1}{8\pi^2} \sum_{7_a} \int_{S_a} F_2^a \wedge F_2^a + \frac{1}{2} \int_{\mathbf{B}_6} F_3 \wedge H_3 + \frac{\chi(\mathbf{X}_8)}{24} = 0. \tag{14.38}$$

This generalizes (14.14) as follows. The first and second contributions correspond to N_{D3} and describe actual D3-branes, or induced ones on magnetized 7-branes, while the third is the N_{flux} contribution (14.15) from NSNS and RR 3-form fluxes; the last term generalizes the O3-plane contribution $\frac{1}{2} N_{O3}$, including also the gravitational Cher–Simons couplings on 7-branes, $\int_{7_a} C_4 \text{tr}(R^2)$, recall (6.16).

14.1.3 Warped throats*

The warp factor in (14.9) is an interesting new ingredient beyond the standard KK compactification. In particular it describes a gravitational redshift with dependence on the internal coordinates, and implies that the 4d length scales change as one moves in the internal space. This can be probed by a D-brane sector localized in $\mathbf{X}_6$, e.g. D3-branes. The effect is generically negligible in the large volume limit, since uniformly distributed fluxes become dilute, and so does the warp factor; however, configurations with partially localized fluxes can lead to deep gravitational wells in the internal geometry, known as throats. They can be used to realize the Randall–Sundrum scenario of Section 1.4.3 in string theory, and generate spectacular hierarchies in the 4d physics. We briefly detour to discuss the prototype system of this kind, the deformed conifold with fluxes, and to mention generalizations including rich gauge sectors at the bottom of the throat. Other model building applications of warped throats appear in Sections 15.3.1 and 16.6.2.

The deformed conifold with fluxes

Recall from Section 11.3.4 the description of the conifold as the non-compact singular CY space (11.74), which, changing variables, we write as $w_1^2 + w_2^2 + w_3^2 + w_4^2 = 0$. This singular space admits a smoothing, known as complex deformation, which replaces the singular point by an $\mathbf{S}^3$ with size controlled by a complex structure modulus (hence the name). The deformed conifold is described by the modified defining equation

$$w_1^2 + w_2^2 + w_3^2 + w_4^2 = z. \tag{14.39}$$

Here the complex parameter z is the modulus controlling the $\mathbf{S}^3$ size; e.g. for z real and positive, the equation for the $\mathbf{S}^3$ arises by restricting the w_i to real values. The conifold describes the local geometry of a CY near an $\mathbf{S}^3$, which we refer to as A-cycle. Its dual B-cycle is non-compact in the local model, but is eventually compact on a real CY compactification; this can be mimicked in the local model by introducing a cutoff in the radial direction of the conifold space.

On this local CY we introduce the following 3-form flux quanta on these cycles

$$\frac{1}{2\pi\alpha'} \int_A F_3 = 2\pi M, \qquad \frac{1}{2\pi\alpha'} \int_B H_3 = -2\pi K, \tag{14.40}$$

and so $N_{\text{flux}} = MK$. As we show shortly, the hierarchy is generated for flux choices satisfying $K \gg Mg_s$ (since the dilaton field is not localized in the conifold geometry, we treat it as an external parameter, possibly fixed by other ingredients in the compact model).

The explicit solution for the resulting 10d supergravity background is known, but many of the relevant physical properties of the model can be understood using the simpler 4d effective theory description. The superpotential (14.21) becomes

$$W = \int_{\mathcal{M}} G_3 \wedge \Omega_3 = (2\pi)^2\alpha' \left(M \int_B \Omega_3 - i KS \int_A \Omega_3 \right). \tag{14.41}$$

The integrals of Ω_3 for the conifold geometry are known to be

$$z = \int_A \Omega_3, \qquad \int_B \Omega_3 = \frac{z}{2\pi i} \log z + \text{holomorphic}, \tag{14.42}$$

so the explicit superpotential is

$$W = (2\pi)^2\alpha' \left(M \frac{z}{2\pi i} \log z - i KS z \right). \tag{14.43}$$

Setting $C_0 = 0$, $S = 1/g_s$, the leading terms in $D_z W = 0$ for $K \gg Mg_s$ are

$$\frac{M}{2\pi i} \log z - i\frac{K}{g_s} = 0, \tag{14.44}$$

leading to stabilization of the $\mathbf{S}^3$ modulus at an exponentially small value

$$z \sim \exp\left(-\frac{2\pi K}{Mg_s} \right). \tag{14.45}$$

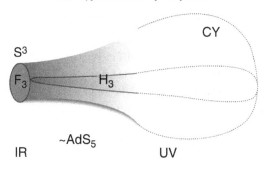

Figure 14.1 Warped throat from the deformed conifold with 3-form fluxes.

The exponentially small $\mathbf{S}^3$ size allows for a useful approximate description of the throat. The deformed conifold can be approximated by a singular conifold, with a cone structure

$$ds_{6d}^2 = \tilde{g}_{mn}\, dy^m\, dy^n = dr^2 + r^2\, ds_{T_{1,1}}^2, \tag{14.46}$$

with $T_{1,1} = \mathbf{S}^2 \times \mathbf{S}^3$ being the 5d base of the cone, whose details are not particularly relevant here. Fluxes are localized around the $\mathbf{S}^3$ and their effect on the supergravity solution can be approximately described as a stack of N_{flux} D3-branes at $r = 0$. The solution is analogous to that of D3-branes in flat space (6.33), with $Z = e^{-4A}$, by simply replacing the transverse $\mathbf{R}^6$ metric by (14.46).

$$ds^2 = e^{2A(r)}\, \eta_{\mu\nu}\, dx^\mu\, dx^\nu + e^{-2A(r)}\left(dr^2 + r^2\, ds_{T_{1,1}}^2\right), \tag{14.47}$$

with the warp factor harmonic function

$$e^{-4A(r)} = 1 + \frac{R^4}{r^4}, \quad \text{with } R^4 = 4\pi g_s N_{\text{flux}}\, \alpha'^2. \tag{14.48}$$

At very large radius $r \gg R$ the solution becomes the conifold metric, which glues to the rest of the compact CY space. On the other hand, for $r \ll R$ we can neglect the constant term, and obtain

$$ds^2 = \frac{r^2}{R^2}\, \eta_{\mu\nu}\, dx^\mu dx^\nu + \frac{R^2}{r^2}\, dr^2 + R^2\, ds_{T_{1,1}}^2. \tag{14.49}$$

The 4d Minkowski directions combine with the throat radial direction to produce an AdS$_5$ geometry; the angular coordinates parametrize a constant size $T_{1,1}$ space. This approximation holds at locations r large compared with the $\mathbf{S}^3$ size $\tilde{r} = z^{1/3}$, at which the fluxes cut off the AdS throat, with a value for the warp factor

$$e^{A_{\min}} \simeq \tilde{r} \simeq \exp\left(-\frac{2\pi K}{3M g_s}\right), \tag{14.50}$$

providing an exponential hierarchy of energy scales of physical systems localized at the bottom of the throat, due to their large redshift.

The system provides a string theory realization of the Randall–Sundrum (RS) warped dimensions in Section 1.4.3, see Figure 14.1. The throat (14.49) corresponds, upon a change of coordinates, to the AdS$_5$ geometry in the RS construction (with the compact $T_{1,1}$ space being irrelevant for this discussion). The cutoff at $\tilde{r}$ corresponds to the infrared (IR) brane in the RS construction. Finally, the cutoff at very large radius where the throat joins the remaining CY geometry plays the role of the ultraviolet (UV or Planck) brane in the RS construction. Note that the negative tension of the latter is actually realized in the string theory setup, since RR tadpole cancellation requires the rest of the CY space to produce a negative 4-form charge $-N_{\text{flux}}$, and, by the BPS condition, a corresponding negative tension.

The present discussion fits well the AdS/CFT holographic relation described in Section 6.4, as suggested by the analogy between (14.49) and (6.35). The approximate AdS$_5$ region is holographically dual to an approximately conformal 4d field theory, with the radial direction corresponding to the field theory energy scale. The large radius region is the ultraviolet of the field theory, which can in fact be described quite explicitly: it is given by the theory (11.72) of D3-branes at a conifold geometry in Section 11.3.4, with gauge group $SU(N) \times SU(N+M)$ and $N = N_{\text{flux}} = KM$. The approximately AdS$_5$ bulk is dual to the renormalization group flow of this field theory, which for large K is almost conformal, but still non-trivial. The end of the throat at $\tilde{r}$ is dual to the infrared of the field theory, described effectively by a pure $SU(M)$ SYM, which confines at a scale computed to be $\Lambda = \exp\left(-2\pi K/3Mg_{\text{YM}}^2\right)$. The nice agreement with (14.50) implies that the exponential hierarchy generated by the warp factor is holographically dual to the exponential hierarchy generated by dimensional transmutation.

The SM on a warped throat

There are diverse model building applications of warped throats. Following the proposal in Section 1.4.3, one interesting possibility is to locate the SM degrees of freedom at the IR end of the throat, to generate the electroweak/Planck hierarchy. String theory provides an arena for such explicit model building, by localizing the SM on a set of D-branes. In order to achieve chirality, the latter must belong to one of the type IIB classes of constructions studied in Chapter 11. A simple possibility is to consider a throat geometry with a $\mathbf{C}^3/\mathbf{Z}_3$ orbifold singularity at its IR end, at which sets of D3- (and possibly D7-) branes realize a chiral gauge theory, as in Section 11.3.3. Although the conifold does not have a $\mathbf{Z}_3$ symmetry to generate a $\mathbf{Z}_3$ singularity (of the appropriate kind) by quotienting, it is easy to construct spaces with the required properties using the toric singularities in Section 11.3.4.

Consider the toric singularity defined as a quotient of $\mathbf{C}^6$ by the $\mathbf{C}^*$ actions

	z_1	z_2	z_3	z_4	z_5	z_6
$(\mathbf{C}^*)_1$	1	1	-1	-1	0	0
$(\mathbf{C}^*)_2$	0	1	0	-2	1	0
$(\mathbf{C}^*)_3$	0	1	1	1	0	-3

Let us describe the resulting geometry in terms useful for our discussion. Using $(\mathbf{C}^*)_3$ to set $z_6 = 1$, the remaining $(\mathbf{C}^*)^2$ quotient of $\mathbf{C}^5$ is a singular space, known as the "suspended pinch point" (SPP) singularity; it can be described via $(\mathbf{C}^*)^2$ invariant coordinates $x = z_1 z_4 z_5^2$, $y = z_2^2 z_3 z_4$, $v = z_1 z_3$, $w = z_2 z_4 z_5$, subject to $xy = vw^2$. Actually, since z_6 has

$(\mathbf{C}^*)_3$ charge -3, there is a left-over discrete $\mathbf{Z}_3$ identification acting on the remaining coordinates as $z_i \to e^{2\pi i/3} z_i$ for $i = 2, 3, 4$. Hence, the actual geometry is the quotient of the SPP geometry by the $\mathbf{Z}_3$ action

$$x \to e^{2\pi i/3} x, \quad y \to e^{2\pi i/3} y, \quad v \to e^{2\pi i/3} v, \quad w \to e^{4\pi i/3} w. \tag{14.51}$$

This space admits a complex deformation, as follows. In the SPP covering space, the coordinates z_1, z_2, z_3, z_4 quotiented by $(\mathbf{C}^*)_1$ describe a conifold singularity, admitting a complex deformation as described in (14.39). This induces a deformation of the SPP singularity to a smooth space, described by

$$xy = vw^2 + \epsilon w, \tag{14.52}$$

where ϵ denotes the parameter controlling the 3-cycle size. The deformed equation is still invariant under the $\mathbf{Z}_3$ action (14.51), so its quotient describes a complex deformation of the space $\mathrm{SPP}/\mathbf{Z}_3$; the resulting geometry contains a non-trivial 3-cycle with $\mathbf{S}^3/\mathbf{Z}_3$ topology, and has a $\mathbf{C}^3/\mathbf{Z}_3$ singularity arising from the $\mathbf{Z}_3$ fixed point $x = y = v = w = 0$ in the covering SPP space.

Turning on fluxes on the 3-cycle and its dual leads to a warped throat in complete analogy with the deformed conifold case, but with a $\mathbf{C}^3/\mathbf{Z}_3$ singularity at its tip. Locating stacks of D3-branes at the latter allows the construction of chiral gauge sectors at the IR end of the throat. The simplest possibility is to use D3-branes (and no D7-branes) to realize the $SU(3)^3$ trinification model of Section 1.2.4; including D7-branes on non-compact 4-cycles (partially stretching along the radial direction on the throat) allows to realize the models in Section 11.3.3. This provides an explicit realization of a (supersymmetric) Randall–Sundrum construction with semi-realistic spectrum in string theory. In analogy with the conifold case, it is possible to describe in detail the gauge theory holographically dual to the throat, which corresponds to an approximately conformal theory with partial confinement at an IR scale, below which the SM fields arise as emergent (or composite) degrees of freedom.

14.2 Fluxes in type II toroidal orientifolds

The above general type IIB flux compactifications can be studied very explicitly in the framework of toroidal models (and quotients thereof); these also allow for an analogous discussion of type IIA flux compactifications. The expectation that type IIA/B flux compactifications are related by T-duality/mirror symmetry will suggest the existence of larger classes of fluxes, whose discussion we also start in this section. Our notation and conventions for this section follow Chapter 12.

14.2.1 Type IIB orientifolds and fluxes

Let us first consider the type IIB orientifold $\mathbf{T}^6/\Omega\mathcal{R}(-1)^{F_L}$, with $\mathcal{R}$ acting as $(x^i, y^i) \to (-x^i, -y^i)$, so there are 64 O3-planes. We focus on the diagonal moduli, i.e. the dilaton S, the three Kähler moduli T_i and the three complex structure moduli U_i, which appear e.g. in the $\mathbf{Z}_2 \times \mathbf{Z}_2$ models in Sections 10.6 and 11.4.2.

We can introduce arbitrary NSNS and RR 3-form fluxes, since all components survive the orientifold action. We fix a normalization absorbing the $(4\pi^2\alpha')$ factor in (14.5), so that fluxes are integer linear combinations of the 3-forms in (12.25),

$$F_3 = -m\,\alpha_0 - e_0\beta_0 + \sum_{i=1}^{3}(e_i\alpha_i - q_i\beta_i),$$

$$H_3 = h_0\beta_0 - \sum_{i=1}^{3}a_i\alpha_i + \bar{h}_0\alpha_0 - \sum_{i=1}^{3}\bar{a}_i\beta_i, \tag{14.53}$$

with integer m, e_0, e_i, q_i, h_0, a_i, $\bar{h}_0$, $\bar{a}_i$. Replacing into the superpotential (14.21), and using the definitions (12.31), (12.28) of the moduli and Ω_3, one obtains

$$W_{\text{flux}} = e_0 + i\sum_{i=1}^{3}e_i\,U_i - q_1\,U_2U_3 - q_2\,U_1U_3 - q_3\,U_1U_2 + i\,m\,U_1U_2U_3$$

$$+ S\left[i\,h_0 - \sum_{i=1}^{3}a_i\,U_i + i\bar{a}_1\,U_2U_3 + i\bar{a}_2\,U_1U_3 + i\bar{a}_3\,U_1U_2 - \bar{h}_0\,U_1U_2U_3\right].$$

$$\tag{14.54}$$

As announced, the superpotential depends on the dilaton and complex structure moduli, but not on the Kähler moduli. The reader may check that the RR tadpole condition (14.14) becomes

$$N_{\text{D3}} + \tfrac{1}{2}\left[mh_0 - e_0\bar{h}_0 + \sum_i(q_ia_i + e_i\bar{a}_i)\right] = 16. \tag{14.55}$$

Section 14.2.4 presents explicit examples of flux compactifications in this orientifold.

14.2.2 Type IIA orientifolds and fluxes

Let us now consider the introduction of field strength fluxes in the type IIA mirror of the above toroidal orientifold. The model without flux is obtained by applying T-duality along the directions x^i, and corresponds to a type IIA orientifold $\mathbf{T}^6/\Omega\mathcal{R}(-1)^{F_L}$, with $\mathcal{R} : (x^i, y^i) \to (x^i, -y^i)$. We focus again on the seven diagonal moduli, and on NSNS and RR field strength fluxes, introduced in close analogy with the previous section. We postpone more general possibilities to later sections.

The NSNS 3-form flux H_3 is intrinsically odd under the orientifold action, see Section 12.1, and expands in odd 3-forms in the basis (12.25),

$$H_3 = \sum_{I=0}^{3}h_I\beta_I. \tag{14.56}$$

The RR field strength F_4 is intrinsically even, while F_2 and F_6 are odd under the orientifold projection; in addition there is a 0-form flux F_0, the Romans parameter mentioned in Section 4.2.5, and which is even under the orientifold action. Restricting to factorized fluxes, the expansions in the basis (12.26) are

$$F_0 = -m; \qquad F_6 = e_0 \, d\text{vol}_6; \qquad F_2 = \sum_{i=1}^{3} q_i \omega_i; \qquad F_4 = \sum_{i=1}^{3} e_i \widetilde{\omega}_i, \qquad (14.57)$$

with $d\text{vol}_6$ the volume form (B.11) on $\mathbf{X}_6$. We normalize our basis p-forms so that coefficients in (14.56), (14.57) are integer.

A novelty of this type IIA compactification is that the NSNS and RR fluxes have odd and even degree, respectively, so the corresponding flux superpotentials depend on both complex structure and Kähler moduli, respectively. Indeed, the superpotential, from the type IIA version of the argument in Section 14.1.2, reads

$$W_{\text{RR-flux}} = \int_{\mathbf{X}_6} e^{J_c} \wedge F_{RR}, \qquad W_{\text{NSNS-flux}} = \int_{\mathbf{X}_6} \Omega_c \wedge H_3, \qquad (14.58)$$

where $F_{RR} = F_0 + F_2 + F_4$ is a formal sum of RR fluxes and J_c, Ω_c are defined in (12.9). These expressions may also be obtained by KK reduction of the type IIA 10d theory, see the Bibliography. The expected dependence on all moduli becomes manifest upon replacement of the flux expansions (14.56), (14.57) in (14.58), i.e.

$$W_{\text{RR-flux}} = e_0 + i \sum_{i=1}^{3} e_i T_i - q_1 T_2 T_3 - q_2 T_1 T_3 - q_3 T_1 T_2 + i \, m \, T_1 T_2 T_3,$$

$$W_{\text{NSNS-flux}} = i h_0 S - i \sum_{i=1}^{3} h_i U_i. \qquad (14.59)$$

The RR tadpole conditions, including D6$_a$-branes on factorized 3-cycles, are

$$\sum_a N_a n_a^1 n_a^2 n_a^3 + \frac{1}{2} m h_0 = 16, \qquad \sum_a N_a n_a^1 m_a^2 m_a^3 + \frac{1}{2} m h_1 = 0,$$

$$\sum_a N_a m_a^1 n_a^2 m_a^3 + \frac{1}{2} m h_2 = 0, \qquad \sum_a N_a m_a^1 m_a^2 n_a^3 + \frac{1}{2} m h_3 = 0. \qquad (14.60)$$

Comparing the type IIA and IIB results, there is a seeming contradiction with mirror symmetry. For instance, the superpotentials (14.54) and (14.59) do not agree upon the exchange $T_i \leftrightarrow U_i$; also, type IIA fluxes contribute to the four RR tadpole conditions, while type IIB ones contribute to the D3-brane tadpole, but not to the D7-brane one. The explanation is that NSNS 3-form fluxes turn into more general fluxes upon T-duality/mirror symmetry. The complete set of fluxes required for consistency with T-duality/mirror symmetry is described in Section 14.4.

14.2.3 Geometric fluxes

The motivation to introduce CY geometries as compactification spaces in string theory arose in Chapter 7 as a natural (supersymmetry preserving) ansatz for compactifications, in which only the 10d metric acquires a non-trivial background. In flux compactifications there are additional backgrounds turned on, so it is natural to consider more general possibilities for compactification spaces. The classification of 10d backgrounds preserving 4d Poincaré or AdS supersymmetry has been carried out in terms of quite advanced mathematical tools, in terms of the so-called "$SU(3)$ structures" and "generalized complex geometry." In practice, many interesting examples correspond to "twisted tori," which can be described in terms of an underlying toroidal compactification with additional twists of the periodic directions. They introduce non-trivial components of the curvature 2-form, and are hence known as geometric fluxes. Metric fluxes are not invariant under the orientifold action of type IIB O3-planes, hence we focus on geometric fluxes in the type IIA toroidal orientifolds with O6-planes of the previous section.

Consider a 6d internal space with coordinates x^M, $M = 1, \ldots, 6$, and introduce the (co)tangent 1-forms $\eta^M(x)$; in $\mathbf{T}^6$ compactifications, $\eta^M = dx^M$, and $d\eta^M \equiv 0$. A twisted $\mathbf{T}^6$ is defined by considering the 1-forms to be linear in the x^M, so that

$$d\eta^P = -\frac{1}{2}\omega^P_{MN}\eta^M \wedge \eta^N, \tag{14.61}$$

where ω^P_{MN} are constant coefficients, antisymmetric in the lower indices. They admit several physical interpretations. For instance, ω^P_{MN} describes a twisting of the direction η^P over the directions M, N, hence the name twisted torus. It also can be regarded as a quantized magnetic flux along M, N for the $U(1)$ gauge boson arising from the metric in the KK compactification along x^P, recall Sections 1.4.1 and 3.2.3, hence the term "geometric flux."

The ω^P_{MN} also admit an algebraic interpretation as follows. Each 1-form η^M has a dual tangent vector, or momentum operator Z^M

$$\eta^M = A^M_N(x)dx^N \quad \rightarrow \quad Z_M = (A^{-1})^N_M \frac{\partial}{\partial x^N}. \tag{14.62}$$

The metric fluxes ω^P_{MN} are the structure constants of the Lie algebra they generate,

$$[Z_M, Z_N] = \omega^P_{MN} Z_P. \tag{14.63}$$

The Jacobi identity of the algebra implies that geometric fluxes must satisfy

$$\omega^P_{[MN}\omega^S_{R]P} = 0. \tag{14.64}$$

Equivalently, this follows from requiring $d^2\eta^P = 0$; also, requiring $d\mathrm{vol}_6 = \eta^1 \ldots \eta^6$ not to be exact implies $\sum_P \omega^P_{PN} = 0$. A last useful result from (14.61) is that, for a 2-form $X = \frac{1}{2}X_{MN}\eta^M \wedge \eta^N$ with coefficients independent of the x^N, we have

$$(dX)_{LMN} = \omega^P_{[LM}X_{N]P}, \tag{14.65}$$

denoted $dX = \omega \cdot X$ for short; similar formulae can be obtained for higher forms.

As already mentioned, geometric fluxes are not invariant under the O3-plane orientifold action of type IIB. We thus focus on type IIA orientifolds with O6-planes, defined by the geometric action on the twisted torus coordinates $\eta^i \to \eta^i$, $\eta^a \to -\eta^a$, $i = 1, 2, 3$, $a = 4, 5, 6$. Hence the only allowed geometric fluxes are of type ω^i_{ab}, ω^i_{jk}, ω^a_{ib}. As in Section 14.2.2, we focus on fluxes along factorized directions. To make the constraints (14.64) explicit, we introduce the convenient notation

$$
\begin{pmatrix} a_1 \\ a_2 \\ a_3 \end{pmatrix} = \begin{pmatrix} \omega^1_{56} \\ \omega^2_{64} \\ \omega^3_{45} \end{pmatrix} ; \quad
\begin{pmatrix} b_{11} & b_{12} & b_{13} \\ b_{21} & b_{22} & b_{23} \\ b_{31} & b_{32} & b_{33} \end{pmatrix} = \begin{pmatrix} -\omega^1_{23} & \omega^4_{53} & \omega^4_{26} \\ \omega^5_{34} & -\omega^2_{31} & \omega^5_{61} \\ \omega^6_{42} & \omega^6_{15} & -\omega^3_{12} \end{pmatrix} , \tag{14.66}
$$

in terms of which the Jacobi identities imply the twelve constraints

$$
b_{ij}a_j + b_{jj}a_i = 0; \quad i \neq j,
$$
$$
b_{ik}b_{kj} + b_{kk}b_{ij} = 0; \quad i \neq j \neq k. \tag{14.67}
$$

Some obvious solutions to these constraints are, e.g., $b_{ij} = 0$, $a_i \neq 0$; or $a_i = 0$, $b_{ij} = b_i \delta_{ij}$; or $a_i = a$, $b_{ij} = b$, $i \neq j$, $b_{ii} = -b$.

An interesting further interpretation for geometric fluxes is as T-duals of NSNS 3-form fluxes. Consider an example of type IIB on a flat $\mathbf{T}^6$ with NSNS flux

$$
ds^2 = (dx^1)^2 + \cdots + (dx^6)^2, \tag{14.68}
$$
$$
H_3 = -a_1 \, dx^1 \wedge dx^5 \wedge dx^6 - a_2 \, dx^4 \wedge dx^2 \wedge dx^6 - a_3 \, dx^4 \wedge dx^5 \wedge dx^3.
$$

In a particular gauge the NSNS 2-form components are $B_{15} \sim a_1 x^6$, $B_{26} \sim a_2 x^4$, $B_{34} \sim a_3 x^5$. Upon T-duality along x^1, x^2, x^3, recalling (5.6), they turn into off-diagonal components of the T-dual metric g'_{15}, g'_{26}, g'_{34}. The type IIA T-dual has vanishing NSNS flux, and a twisted torus metric

$$
ds'^2 = (dx^1 + a_1 x^6 dx^5)^2 + (dx^2 + a_2 x^4 dx^6)^2 + (dx^3 + a_3 x^5 dx^4)^2
$$
$$
+ (dx^4)^2 + (dx^5)^2 + (dx^6)^2. \tag{14.69}
$$

The twisted torus tangent 1-forms are indeed

$$
\begin{aligned}
\eta^1 &= dx^1 + a_1 x^6 dx^5; & \eta^4 &= dx^4, \\
\eta^2 &= dx^2 + a_2 x^4 dx^6; & \eta^5 &= dx^5, \\
\eta^3 &= dx^3 + a_3 x^5 dx^4; & \eta^6 &= dx^6.
\end{aligned} \tag{14.70}
$$

From (14.61), the resulting geometric fluxes are $\omega^1_{56} = a_1$, $\omega^2_{64} = a_2$, $\omega^3_{45} = a_3$.

A final interpretation of geometric fluxes is in terms of twisting the theory by a symmetry along an internal $\mathbf{S}^1$, a so-called Scherk–Schwarz compactification. Consider the $\mathbf{T}^2$ associated to (η^1, η^5) in the above example. Equation (14.70) implies that in moving along the $\mathbf{S}^1$ along η^6, i.e. $x^6 \to x^6 + 1$, the $\mathbf{T}^2$ undergoes a transformation $\eta^1 \to \eta^1 + a_1 \eta^5$, $\eta^5 \to \eta^5$. This is a geometric $SL(2, \mathbf{Z})$ symmetry of $\mathbf{T}^2$ for integer a_1, i.e. for quantized geometric flux; twisted torus compactification can thus be described as a modified KK $\mathbf{T}^6$ compactification, with additional symmetry twists along internal $\mathbf{S}^1$ directions. This viewpoint is also useful in generalization to non-geometric fluxes in Section 14.4; it also shows that geometric fluxes must be quantized for consistency of the twisted torus structure.

The algebra structure (14.63) underlies the description of the 4d effective action of compactifications with geometric fluxes in terms of gauged supergravity, but for simplicity we present a more pedestrian approach, based simply on the introduction of a generalized flux superpotential. This can be derived by a generalized KK compactification on the twisted torus, or using T-duality relations. The result is a generalized NSNS flux superpotential

$$W_{\text{NSNS}} = \int_{\mathbf{X}_6} \Omega_c \wedge (H_3 + d J_c), \tag{14.71}$$

with Ω_c, J_c defined as in (12.9) with the formal replacement $dx^m \to \eta^m$. The geometric fluxes appear in $d J_c$, which is computed using (14.65). The full superpotential, including the above W_{NSNS} and $W_{\text{RR-flux}}$ in (14.59), interestingly contains terms coupling U_i and T_i moduli,

$$W = e_0 + i h_0 S + \sum_{i=1}^{3} \left[\left(i e_i - a_i S - b_{ii} U_i - \sum_{j \neq i} b_{ij} U_j \right) T_i - i h_i U_i \right]$$
$$- q_1, T_2 T_3 - q_2 T_1 T_3 - q_3 T_1 T_2 + i m T_1 T_2 T_3. \tag{14.72}$$

Geometric fluxes contribute to the RR tadpole conditions, as can be derived from the equations of motion for the RR 7-form C_7. The relevant piece of the action is

$$\int_{\mathbf{M}_4 \times \mathbf{X}_6} [C_7 \wedge (m H_3 + d F_2)] + \sum_a N_a \int_{\mathbf{M}_4 \times \Pi_a} C_7. \tag{14.73}$$

The first term arises from the kinetic energy $\int G_2 \wedge *_{10d} G_2$, with $G_2 = m B_2 + F_2$ and $*_{10d} F_2 = d C_7 + \cdots$, with dots denoting possible Chern–Simons couplings, while the second describes the coupling to D6-branes (and O6-planes). The contribution from geometric fluxes arises because $d F_2 \neq 0$ on the twisted torus, e.g. $(d F_2)_{456} = (a_1 q_1 + a_2 q_2 + a_3 q_3)$. Consistency of the equations of motion leads, in our example, to the RR tadpole cancellation conditions

$$\sum_a N_a n_a^1 n_a^2 n_a^3 + \frac{1}{2}(h_0 m + a_1 q_1 + a_2 q_2 + a_3 q_3) = 16,$$

$$\sum_a N_a n_a^1 m_a^2 m_a^3 + \frac{1}{2}(m h_1 - q_1 b_{11} - q_2 b_{21} - q_3 b_{31}) = 0,$$

$$\sum_a N_a m_a^1 n_a^2 m_a^3 + \frac{1}{2}(m h_2 - q_1 b_{12} - q_2 b_{22} - q_3 b_{32}) = 0, \tag{14.74}$$

$$\sum_a N_a m_a^1 m_a^2 n_a^3 + \frac{1}{2}(m h_3 - q_1 b_{13} - q_2 b_{23} - q_3 b_{33}) = 0,$$

where we have included one O6-plane and D6-branes on factorized 3-cycles in $\mathbf{T}^6$.

14.2.4 Examples of moduli fixing

We now present some examples of moduli stabilization with the above flux superpotentials, with an illustrative rather than exhaustive purpose. We concentrate on the seven generic "bulk," i.e. untwisted, moduli S, U_i, T_i of the $\mathbf{Z}_2 \times \mathbf{Z}_2$ orientifold for both its simplicity and its usefulness to build semi-realistic models. The inclusion of interesting D-brane sectors is postponed to Section 14.3.

Assuming the familiar Kähler potential (12.30), the 4d scalar potential is

$$V = e^K \left[\sum_{\Phi=S,T_i,U_i} (\Phi + \Phi^*)^2 |D_\Phi W|^2 - 3|W|^2 \right], \tag{14.75}$$

where we have set $\kappa_4^2 = 1$. Examples below include 4d $\mathcal{N} = 1$ supersymmetric Minkowski vacua, Minkowski no-scale vacua, and AdS vacua.

Example of $\mathcal{N} = 1$ type IIB flux vacuum

The type IIB flux superpotential only includes the S and U_i moduli. The equations for supersymmetric Minkowski vacua are

$$D_T W \equiv W = 0, \quad D_S W = \partial_S W = 0, \quad D_{U_i} W = \partial_{U_i} W = 0. \tag{14.76}$$

Since the number of equations exceeds that of parameters by one, supersymmetric Minkowski vacua are non-generic and require one extra constraint, as follows. Using (14.54) and denoting $s = \mathrm{Re}\, S$, $t_i = \mathrm{Re}\, T_i$ and $u_i = \mathrm{Re}\, U_i$, the real part of the above vacuum equations give four homogeneous relations

$$a_1 u_1 + a_2 u_2 + a_3 u_3 = 0, \quad a_i s + q_j u_k + q_k u_j = 0, \quad i \neq j \neq k. \tag{14.77}$$

In order to have s, $u_i \neq 0$ the determinant of this system must vanish, leading to the constraint

$$(a_1 q_1 + a_2 q_2 - a_3 q_3)^2 = 4 a_1 q_1 a_2 q_2. \tag{14.78}$$

A simple solution is $a_2 = a_3 = -a_1/2$, $q_2 = q_3 = -q_1/2$. Choosing vanishing values for the rest of the fluxes, the induced superpotential has the form

$$W = -a_1 S U_1 - q_1 U_2 U_3 + \frac{1}{2}(U_2 + U_3)(a_1 S + q_1 U_1). \tag{14.79}$$

Minimization yields a SUSY Minkowski minimum at $U_1 = U_2 = U_3 = a_1 S/q_1$. The flux contribution to the RR 4-form tadpole is $(3 a_1 q_1/2)$, which must be positive for consistency. Altogether the three complex structure moduli are fixed in terms of S, while the Kähler moduli T_i, absent from the superpotential, remain undetermined.

Example of $\mathcal{N} = 0$ type IIB no-scale flux vacuum

We now consider Minkowski vacua with no-scale SUSY breaking, requiring $D_S W = D_{U_i} = 0$ but generically $W \neq 0$, so that $D_{T_i} W \neq 0$ and SUSY is broken. Such minima can be shown to exist only if $m \neq 0$, $h_0 \neq 0$, and $\gamma_1 = \gamma_2 = \gamma_3 = 0$, where

$$\gamma_i = me_i + q_j q_k; \quad i \neq j \neq k. \tag{14.80}$$

Consider the superpotential of the form

$$W = e_0 + ih_0 S + i \sum_i e_i U_i - q_1 U_2 U_3 - q_2 U_1 U_3 - q_3 U_1 U_2 + im\, U_1 U_2 U_3. \tag{14.81}$$

Minimization fixes the axions $\operatorname{Im} U_i = -q_i/m$, $\operatorname{Im} S = (e_0 m^2 - q_1 q_2 q_3)/h_0 m^2$, whereas for the real parts one gets $h_0 s = m u_1 u_2 u_3$. Hence one must have $h_0 m > 0$ and again the flux contribution to RR tadpoles is positive. The superpotential at the minimum is $W_0 = 2ih_0 s$, and the gravitino mass is $m_{3/2}^2 = h_0 m/32 t_1 t_2 t_3$, displaying a no-scale behavior. Only four axions and one combination of the real part of the moduli are fixed. This set of fluxes is exploited in Section 14.3.2 to trigger supersymmetry breaking in the semi-realistic particle physics model of Section 11.4.2.

Fixing all bulk moduli in AdS in type IIA

Type IIB examples like the above necessarily lead only to partial moduli stabilization, since Kähler moduli do not appear in the superpotential; one must then resort to further effects, like non-perturbative superpotentials as in Section 15.3.1, to achieve full moduli stabilization. Type IIA flux compactifications overcome this difficulty and lead to full moduli stabilization already at the perturbative level. As an example, we construct an AdS$_4$ $\mathcal{N} = 1$ SUSY model, with no geometric fluxes – introduction of geometric fluxes or generalization to non-supersymmetric AdS examples are easy and not discussed explicitly. The superpotential is given by the two terms in (14.59), and we assume $m \neq 0$. The vacuum structure is determined by the combinations γ_i in (14.80), required to satisfy $\gamma_i < 0$ in order to produce SUSY vacua, i.e. $D_{T_i} W = 0$, $D_S W = 0$, and $D_{U_i} W = 0$. Minimization fixes Kähler axions at $\operatorname{Im} T_i = -q_i/m$, whereas for the other axions

$$h_0 \operatorname{Im} S - \sum_i h_i \operatorname{Im} U_i = \frac{1}{m^2} \left[e_0 m^2 - q_1 q_2 q_3 + \sum_i q_i \gamma_i \right]. \tag{14.82}$$

The real parts are fixed at

$$t_1 = \sqrt{\frac{5|\gamma_2 \gamma_3|}{3m^2 |\gamma_1|}}; \quad t_2 = \frac{\gamma_1 t_1}{\gamma_2}; \quad t_3 = \frac{\gamma_1 t_1}{\gamma_3}; \quad s = -\frac{2\gamma_1 t_1}{3mh_0}; \quad u_i = \frac{2\gamma_1 t_1}{3mh_i}. \tag{14.83}$$

Hence there is stabilization of T_i, s and u_i, and a single linear combination of the axions $\operatorname{Im} S$ and $\operatorname{Im} U_i$. The fact that some axions remain unfixed is in fact welcome to allow for the introduction of D6-branes leading to chiral gauge theories, as we will discuss in Section 14.3.

Note that $s > 0$, $u_k > 0$ requires $mh_0 > 0$, $mh_k < 0$, hence the flux contribution to the RR tadpoles is again positive (in D6-brane units). In this example, only m, h_0 and h_k are restricted by RR tadpole conditions (14.60), while e_0, e_i and q_i are unconstrained. This allows for minima at arbitrarily large volume and small dilaton, as follows. Introducing overall factors c_2, c_4, which rescale the RR 2- and 4-form fluxes q_i, e_i respectively, the moduli vevs and the 4d and 10d dilaton for large fluxes scale with c_4 and c_2 respectively as

$$t \simeq s^{1/3} \simeq u_k^{1/3} \simeq \gamma^{1/2} \simeq \mathcal{O}(c_4^{1/2}), \ \mathcal{O}(c_2)$$
$$e^{\phi_4} \simeq \mathcal{O}(c_4^{-3/2}), \ \mathcal{O}(c_2^{-3}); \quad e^{\phi} \simeq \mathcal{O}(c_4^{-3/4}), \ \mathcal{O}(c_2^{-3/2}). \tag{14.84}$$

Thus the configuration is at arbitrarily large volume and weak coupling for large enough c_4 and/or c_2, a limit in which the 4d vacuum energy rapidly decreases.

Toroidal orbifold/orientifold compactifications with different fluxes produce explicit examples of AdS minima with full *bulk* moduli stabilization. Full moduli stabilization would also require the fixing of the twisted sector moduli in orbifolds. There are actually explicit models (e.g. based on a $\mathbf{Z}_3 \times \mathbf{Z}_3$ orientifold) with stabilization of twisted moduli, although the examples analyzed do not lead to Minkowski or de Sitter minima (with all moduli fixed). As discussed in Section 15.3.1, further ingredients may provide mechanisms to uplift fully stabilized AdS vacua to de Sitter vacua. However, the realization of these possibilities in the oversimplified setup of toroidal compactifications is questionable. Even if possible, the very symmetric underlying toroidal geometry is unlikely to produce SUSY breaking scales hierarchically below the string scale. Nevertheless, these models prove useful to show the explicit realization of full moduli fixing by fluxes.

14.3 D-branes and fluxes

In order to build compactifications leading to realistic particle physics models, we need to combine the above fluxes with the D-branes producing the SM fields, as illustrated in an example in this section. The simultaneous presence of both ingredients leads in general to interesting interplays. For instance, we have already seen that both D-branes and fluxes contribute to RR tadpoles. Also, as described next, there are compatibility conditions restricting possible D-brane wrappings in terms of the fluxes present, or vice versa. Finally, the generation of soft terms induced from non-supersymmetric fluxes on a set of supersymmetric D-branes is studied in Sections 15.4.1 and 15.5.3.

14.3.1 Freed–Witten consistency conditions

The combined introduction of fluxes and D-branes requires certain mutual compatibility (so-called Freed–Witten) conditions. We describe them for toroidal type IIA orientifolds with D6-branes and NSNS and geometric fluxes, although the derivation holds in more general compactifications, and in the dual type IIB orientifolds.

The consistency conditions can be derived from considerations of the 4d effective action as follows. Recall from Section 10.3.1 that in type IIA (orientifold) compactifications,

RR scalars in complex structure moduli shift under D6-brane worldvolume $U(1)$ gauge transformations according to (10.26). Since these RR scalars appear linearly in the superpotential W_{NSNS} in (14.71), the latter is in general not gauge invariant. The condition for it to remain gauge invariant is

$$\int_{\Pi_a} (H_3 + \omega J_c) = 0, \tag{14.85}$$

where we have used $dJ_c = \omega \cdot J_c$, according to (14.65).

This constraint, derived macroscopically from consistency of the 4d effective action, can also be obtained from a microscopic viewpoint, as we now show, without geometric fluxes for simplicity. Recall from Section 6.1.2 that the D-brane worldvolume field strength F_2 and the NSNS 2-form B_2 necessarily combine into the tensor (6.12), i.e. $2\pi\alpha'\mathcal{F}_2 = 2\pi\alpha'F_2 + B_2$. Since $dF_2 = 0$, the worldvolume Bianchi identity for $\mathcal{F}_2$ in the presence of NSNS 3-form flux is

$$d\mathcal{F}_2 = \frac{1}{2\pi\alpha'}H_3\big|_{\Pi_a}, \tag{14.86}$$

and integration over the wrapped 3-cycle Π_a yields the Freed–Witten constraint. Condition (14.85) is a generalization in the presence of geometric fluxes. The constraint is sometimes called cancellation of Freed–Witten anomalies, since it was originally derived from the study of worldsheet anomalies for open strings in the presence of NSNS 3-form flux.

In toroidal models, expanding in a basis of 3-cycles, the Freed–Witten condition in the case without geometric fluxes implies that any D6$_a$-brane should obey

$$\sum_I c_{aI}\, h_I = 0, \quad \text{with } c_{a0} = m_a^1 m_a^2 m_a^3,$$

$$c_{ai} = m_a^i n_a^j n_a^k, \text{ for } i \neq j \neq k. \tag{14.87}$$

This is in general a strong constraint on the possible D6-branes in flux compactifications. Physically this constraint guarantees that the RR scalar linear combinations becoming massive by mixing with $U(1)_a$ gauge bosons are orthogonal to those getting mass from the fluxes. In particular, in models with chiral D-brane sectors, the Freed–Witten constraints ensure that the RR scalars involved in the Green–Schwarz anomaly cancellation are not made massive by the fluxes. A final interesting implication of the Freed–Witten conditions is that mutual compatibility restricts the kinds of fluxes that may be coupled to a given D-brane sector, and vice versa. This may have implications for the space of realistic string vacua, as we discuss in Section 17.2.2.

14.3.2 An MSSM-like example with fluxes

To illustrate the previous points one can construct an explicit string compactification which combines flux moduli stabilization and sectors of D-branes yielding MSSM-like particle physics models. The $\mathcal{N} = 0$ type IIB no-scale flux vacua in Section 14.2.4 can be combined with the MSSM-like $\mathbf{T}^6/(\mathbf{Z}_2 \times \mathbf{Z}_2)$ orientifold discussed in Section 11.4.2, which is mirror to that in Section 10.6. The model has a D-brane content as in Table 10.4, where the integers $\left(n_i^a, m_i^a\right)$ correspond to the worldvolume magnetic flux and wrapping numbers, respectively, so the MSSM sector is realized on D7-branes. The tadpole cancellation (14.60) (extended to include the $O7_i$-plane contributions) can be satisfied by choosing $N_f = 8$ (instead of $N_f = 40$), and closed string fluxes $h_0 = m = 8$, with the rest of the fluxes vanishing. From the results in Section 14.2.4, one finds partial moduli stabilization, with the real parts of S and U_i related by $h_0 s = m u_1 u_2 u_3$, and their imaginary parts fixed. This simple example illustrates partial moduli fixing in a MSSM-like model.

This simple model has Freed–Witten anomalies due to the non-trivial H_3 flux on the D9-branes $h_{1,2}$ and their images, since these branes have $h_0 m_1 m_2 m_3 \neq 0$. This can be overcome by recombining the brane h_1 with the image h_2' of h_2, leading to a consistent recombined brane $h_1 + h_2'$.

14.3.3 Fluxes and D-brane instantons*

Fluxes may also have non-trivial effects on the D-brane instantons introduced in Chapter 13. There are Freed–Witten constraints restricting the possible D-brane instantons in a given flux compactification; also, background fluxes may lift extra fermion zero modes of D-brane instantons, allowing more instantons to contribute to the superpotential. Hence, fluxes have competing effects on the non-perturbative superpotential, removing some instanton contributions and bringing down new ones.

Freed–Witten condition on D-brane instantons

In Section 14.3.1 we provided a microscopic derivation of the Freed–Witten constraint on possible D-brane wrappings in a flux background. This derivation and hence constraints like (14.85) apply not only to gauge D-branes filling the 4d non-compact directions, but also to D-brane instantons. Namely, the presence of fluxes imposes additional topological conditions on the allowed D-brane instantons. The conditions have an interesting 4d effective theory interpretation also in this case; they ensure that the combinations of RR scalars to which the D-brane instanton couple are orthogonal to those made massive by the flux superpotential, as follows.

Let us focus on D2-brane instantons in type IIA compactification with NSNS 3-form flux H_3 (and no geometric fluxes). The argument bears some similarity with that in Section 14.3.1, from which we borrow the required notation, and is similarly valid beyond the toroidal setup. The flux superpotential W_{NSNS} (14.58) in type IIA compactifications depends linearly on the moduli U_I, and in general gives masses to the linear combinations of RR scalars $\sum_I h_I \text{Im} \, U_I$. A D2-brane instanton wrapped on a 3-cycle Π_a also couples to certain linear combinations of the RR scalars, since its non-perturbative contribution to the 4d action has the structure

$$S_{4d}^{(a)} = \int d^4 x \, e^{-\sum_I c_I^a U_I} \ldots , \tag{14.88}$$

where dots denote additional pieces irrelevant for our discussion. The Freed–Witten condition (14.87) for D2-brane instantons precisely ensures that allowed instantons couple to combinations of RR scalars not coupled to the fluxes.

Although important, this condition is in practice not too restrictive in certain classes of examples, most notably in the type IIB compactifications with 3-form fluxes of Section 14.1. There, non-perturbative effects arise from D(-1)-brane instantons, located at points in $\mathbf{X}_6$ and insensitive to total H_3 flux integrals, or from D3-brane instantons wrapped on 4-cycles; generic holomorphic 4-cycles in CY spaces do not have non-trivial 3-cycles, so integrals of H_3 on the D3-brane volume vanish, and the Freed–Witten conditions are satisfied. Incidentally, the pattern that 3-form fluxes and D3-brane instantons naturally couple to different moduli, arises in Section 15.3.1 in a proposed framework for full moduli stabilization.

Lifting of fermion zero modes

In the interplay of fluxes and D-brane instantons, a second interesting effect is the possible lifting of instanton fermion zero modes by the fluxes, as mentioned in Section 13.2.2. This arises because the fermionic completion of the D-brane worldvolume action contains couplings with the structure

$$\Theta \Gamma^{MNP} \Theta \, T_{MNP}. \tag{14.89}$$

Here Θ describes the open string Ramond groundstate, transforming as a 10d spinor whose decomposition describes the different worldvolume fermions; also, T^{MNP} is a 10d 3-index antisymmetric tensor defined in terms of field strength (or possible geometric) fluxes. These couplings can be obtained from a computation analogous to that of gaugino masses in Section 15.4.1, and is beyond our scope. For illustrative purposes, we quote the structure of the tensor T_{MNP} for D(-1)- and D3-brane instantons in type IIB O3-plane models with NSNS and RR 3-form fluxes, with $C_0 = 0$. Denoting with hatted and unhatted indices the directions along or transverse to the D-brane, the relevant tensor components can be computed to be

$$\mathrm{D}(-1): -i\, F_{mnp} - \frac{1}{g_s} H_{mnp} = -i\, (G_3)_{mnp},$$

$$\mathrm{D}3: -\frac{i}{2}\, \epsilon_{\hat{m}\hat{n}\hat{r}\hat{s}}\, F^{\hat{r}\hat{s}}_{\;p} + \frac{1}{g_s} H_{\hat{m}\hat{n}p}. \tag{14.90}$$

This lifting of fermion zero modes allows more instantons to contribute to the 4d superpotential, enriching the non-perturbative effects in flux compactifications. To illustrate this possibility, consider one D(-1)-brane instanton, at a fixed point of a $\mathbf{Z}_2 \times \mathbf{Z}_2$ orbifold, not fixed by orientifold actions. Its fermion zero modes are θ^α and $\overline{\tau}_{\dot\alpha}$, in the notation of Section 13.2.1, so it does not contribute to the superpotential. However, turning on the (orbifold invariant) 3-form flux $G_3 = G_{(3,0)} dz_1 dz_2 dz_3 + G_{(0,3)} d\overline{z}_1 d\overline{z}_2 d\overline{z}_3$ induces fermion zero mode interactions of the form

$$S_{\mathrm{z.m.}} \simeq G_{(3,0)}\, \theta\theta + G_{(0,3)}\, \overline{\tau}\overline{\tau}. \tag{14.91}$$

Hence ISD $(0, 3)$ flux components lift the $\bar{\tau}$ universal fermion zero modes, and allow the instanton to contribute to the superpotential of the 4d $\mathcal{N} = 1$ effective theory (spontaneously broken by the non-SUSY flux).

This fermion zero mode lifting effect is relevant in the proposed generic character of the scenario in Section 15.3.1 of full Kähler moduli stabilization via non-perturbative superpotentials.

14.4 Mirror symmetry, T-duality, and non-geometric fluxes*

The type IIA and type IIB toroidal orientifolds discussed in previous sections are related by mirror symmetry, namely T-duality along three directions. Recalling (5.6), a T-duality along x^M transforms RR fluxes as

$$F_{MN_1\cdots N_p} \longleftrightarrow F_{N_1\cdots N_p}. \tag{14.92}$$

Correspondingly, the RR flux superpotential pieces in (14.59) and (14.54) indeed match after the replacement $U_i \leftrightarrow T_i$ (i.e. a triple T-duality). On the other hand, as already mentioned, the NSNS flux terms in (14.59) and (14.54) do not match, reflecting that NSNS 3-form fluxes are not preserved under T-duality, and map to new kinds of fluxes. This was already partially described in the introduction of geometric fluxes in Section 14.2.3, although the superpotentials (14.72) and (14.54) are still not consistent with T-duality/mirror symmetry. In this section we describe the additional NS fluxes required for a T-duality invariant formulation.

Systematic application of T-duality suggests the introduction of two new types of NS fluxes, described by tensors Q_P^{MN} and R^{MNP}, according to the chain

$$-H_{MNP} \xleftrightarrow{\text{T}_M} \omega_{NP}^M \xleftrightarrow{\text{T}_N} Q_P^{MN} \xleftrightarrow{\text{T}_P} -R^{MNP}. \tag{14.93}$$

We have introduced some extra signs in order to agree with earlier conventions, e.g. in Section 14.2.3 we showed that $H_{156} = -a_1$ turns into $\omega_{56}^1 = a_1$ upon T-duality along x^1. The new tensors are completely antisymmetric in the upper indices.

The new fluxes Q and R do not admit a geometric interpretation and are known as non-geometric fluxes. For the fluxes Q, this may be understood as follows. Recall from Section 14.2.3 that, e.g., a geometric flux ω_{56}^1 produces a reparametrization of the 2-torus $(\mathbf{T}^2)_{15}$ as one moves around the direction 6 by $x^6 \to x^6 + 1$, i.e. an $SL(2, \mathbf{Z})_U$ modular transformation on the complex structure of $(\mathbf{T}^2)_{15}$. Consequently, its T-dual flux Q_1^{56} produces an $SL(2, \mathbf{Z})_T$ modular transformation on the *Kähler* parameter of the 2-torus $(\mathbf{T}^2)_{15}$ as $x^6 \to x^6 + 1$. The latter $SL(2, \mathbf{Z})_T$ is identical to that in Section 9.6 for heterotic theories, and can act non-trivially on the $(\mathbf{T}^2)_{15}$ area e.g. $A \to 1/A$, so there is no globally defined geometric interpretation for the internal space with Q fluxes. In other words, Q fluxes lead to locally well-defined 10d supergravity backgrounds, but transition functions glueing different patches involve T-dualities (which are a good symmetry of string theory but not of geometry), and so are globally non-geometric. The fluxes R are even more strongly non-geometric and do not admit even a local geometric interpretation.

Table 14.1 *Non-geometric and NSNS 3-form fluxes in O3-plane models, and their T-duals*

IIB/O3

$$\begin{pmatrix} Q_4^{23} & Q_5^{31} & Q_6^{12} \end{pmatrix} \qquad \begin{pmatrix} Q_1^{56} & Q_2^{64} & Q_3^{45} \end{pmatrix} \qquad \begin{pmatrix} H_{423} & H_{153} & H_{126} \end{pmatrix}$$

$$\begin{pmatrix} -Q_1^{23} & Q_5^{34} & Q_6^{42} \\ Q_4^{53} & -Q_2^{31} & Q_6^{15} \\ Q_4^{26} & Q_5^{61} & -Q_3^{12} \end{pmatrix} \quad \begin{pmatrix} -Q_4^{56} & Q_2^{61} & Q_3^{15} \\ Q_1^{26} & -Q_5^{64} & Q_3^{42} \\ Q_1^{53} & Q_2^{34} & -Q_6^{45} \end{pmatrix} \quad \begin{matrix} \begin{pmatrix} H_{156} & H_{426} & H_{453} \end{pmatrix} \\ H_{123} \\ H_{456} \end{matrix}$$

IIA/O6

$$-\begin{pmatrix} H_{423} & H_{153} & H_{126} \end{pmatrix} \quad -\begin{pmatrix} R^{156} & R^{426} & R^{453} \end{pmatrix} \quad -\begin{pmatrix} Q_4^{23} & Q_5^{31} & Q_6^{12} \end{pmatrix}$$

$$\begin{pmatrix} -\omega_{23}^1 & \omega_{53}^4 & \omega_{26}^4 \\ \omega_{34}^5 & -\omega_{31}^2 & \omega_{61}^5 \\ \omega_{42}^6 & \omega_{15}^6 & -\omega_{12}^3 \end{pmatrix} \quad \begin{pmatrix} -Q_4^{56} & Q_1^{26} & Q_1^{53} \\ Q_2^{61} & -Q_5^{64} & Q_2^{34} \\ Q_3^{15} & Q_3^{42} & -Q_6^{45} \end{pmatrix} \quad \begin{matrix} \begin{pmatrix} -\omega_{56}^1 & \omega_{64}^2 & \omega_{45}^3 \end{pmatrix} \\ R^{123} \\ H_{456} \end{matrix}$$

IIB/O9

$$\begin{pmatrix} \omega_{23}^4 & \omega_{31}^5 & \omega_{12}^6 \end{pmatrix} \quad \begin{pmatrix} \omega_{56}^1 & \omega_{64}^2 & \omega_{45}^3 \end{pmatrix} \quad \begin{pmatrix} R^{423} & R^{153} & R^{126} \end{pmatrix}$$

$$\begin{pmatrix} -\omega_{23}^1 & \omega_{34}^5 & \omega_{42}^6 \\ \omega_{53}^4 & -\omega_{31}^2 & \omega_{15}^6 \\ \omega_{26}^4 & \omega_{61}^5 & -\omega_{12}^3 \end{pmatrix} \quad \begin{pmatrix} -\omega_{56}^4 & \omega_{61}^2 & \omega_{15}^3 \\ \omega_{26}^1 & -\omega_{64}^5 & \omega_{42}^3 \\ \omega_{53}^1 & \omega_{34}^2 & -\omega_{45}^6 \end{pmatrix} \quad \begin{matrix} \begin{pmatrix} R^{156} & R^{426} & R^{453} \end{pmatrix} \\ R^{123} \\ R^{456} \end{matrix}$$

Flux quanta

$$-\begin{pmatrix} h_1 & h_2 & h_3 \end{pmatrix} \quad -\begin{pmatrix} \bar{h}_1 & \bar{h}_2 & \bar{h}_3 \end{pmatrix} \quad -\begin{pmatrix} \bar{a}_1 & \bar{a}_2 & \bar{a}_3 \end{pmatrix}$$

$$\begin{pmatrix} b_{11} & b_{12} & b_{13} \\ b_{21} & b_{22} & b_{23} \\ b_{31} & b_{32} & b_{33} \end{pmatrix} \quad \begin{pmatrix} \bar{b}_{11} & \bar{b}_{12} & \bar{b}_{13} \\ \bar{b}_{21} & \bar{b}_{22} & \bar{b}_{23} \\ \bar{b}_{31} & \bar{b}_{32} & \bar{b}_{33} \end{pmatrix} \quad \begin{matrix} -\begin{pmatrix} a_1 & a_2 & a_3 \end{pmatrix} \\ \bar{h}_0 \\ h_0 \end{matrix}$$

The Q's are intrinsically odd under the orientifold involution, while the R's are even (and recall that H_{MNP} is odd and ω_{NP}^M is even). Hence, in type IIB with O3-planes there are neither geometric fluxes nor non-geometric fluxes of type R. The only allowed fluxes are therefore non-geometric fluxes Q_P^{MN}, and NSNS 3-form fluxes H_{MNP}. Using T-duality along the directions 1,2,3, the fluxes present in the mirror type IIA model with O6-planes include components of all NSNS field strength, geometric, and non-geometric fluxes Q and R. Further T-dualities (along 4,5,6) relate the models to type I theory (i.e. type IIB with O9-planes), for which the orientifold involution is the identity and only intrinsically even fluxes, denoted ω and R, are possible. The different fluxes and their T-duality relations are shown in Table 14.1. Notice that the indices are ordered cyclically according to their corresponding sub-torus.

In analogy with field strength and geometric fluxes, non-geometric fluxes contribute both to the 4d superpotential and the RR tadpoles. We describe their computation for type IIB O3-plane models, with results for other models following straightforwardly from T-duality. It is useful to define, in analogy with (14.65), the contraction of a p-form X with Q to obtain a $(p-1)$-form QX with components

$$(QX)_{LM_1\cdots M_{p-2}} = \frac{1}{2} Q_{[L}^{AB} X_{M_1\cdots M_{p-2}]AB}. \tag{14.94}$$

The type IIB superpotential, completed to be compatible with T-duality, reads

$$W = \int_{\mathbf{T}^6} (F_3 - i S H_3 + \mathcal{Q}\mathcal{J}_c) \wedge \Omega_3, \tag{14.95}$$

with $\mathcal{J}_c = i \sum_{i=1}^{3} T_i \tilde{\omega}_i$, recall Section 12.1, and $\mathcal{Q}\mathcal{J}_c$ is a 3-form according to (14.94). In terms of flux quanta, the RR, NSNS 3-form, and $\mathcal{Q}$-induced terms are

$$W = e_0 + i \sum_{i=1}^{3} e_i U_i - q_1 U_2 U_3 - q_2 U_1 U_3 - q_3 U_1 U_2 + im\, U_1 U_2 U_3$$

$$+ S\left[ih_0 - \sum_{i=1}^{3} a_i U_i + i\bar{a}_1 U_2 U_3 + i\bar{a}_2 U_1 U_3 + i\bar{a}_3 U_1 U_2 - \bar{h}_0 U_1 U_2 U_3 \right]$$

$$+ \sum_{i=1}^{3} T_i\left[-ih_i - \sum_{j=1}^{3} U_j b_{ji} + i\bar{b}_{1i} U_2 U_3 + i\bar{b}_{2i} U_1 U_3 + i\bar{b}_{3i} U_1 U_2 + \bar{h}_i U_1 U_2 U_3 \right].$$

$$\tag{14.96}$$

In this type IIB orientifold the RR 4-form tadpole condition is (14.55). From T-duality we also expect the non-geometric fluxes to contribute to RR 8-form C_8 tadpoles (which also receive familiar contributions from D7/O7s if present). A natural guess for this coupling, furthermore motivated by T-duality, is

$$- \int_{\mathbf{M}_4 \times \mathbf{T}^6} C_8 \wedge \mathcal{Q} F_3, \tag{14.97}$$

where the contraction $\mathcal{Q} F_3$ is a 2-form according to (14.94). Explicit computation yields three independent tadpole conditions,

$$-N_{\mathrm{D}7_i} + \frac{1}{2}\left[mh_i - e_0\bar{h}_i - \sum_j (q_j b_{ji} + e_j \bar{b}_{ji}) \right] = 0, \tag{14.98}$$

related to the three possible $\mathrm{D}7_i$-branes, which we have included for illustration. Analogous results are obtained in other T-dual setups.

As with geometric fluxes in Section 14.2.3, non-geometric fluxes are subject to certain consistency conditions, generalizing (14.64). They can be derived as the Jacobi identities of an algebra extending (14.63) by the inclusion of new generators, which can be regarded as associated to translations in the T-dual coordinates. We direct the reader to the references for details.

The idea to use string dualities to derive the existence of new fluxes can be extended to non-perturbative dualities. In toroidal compactifications this leads to the introduction of the so-called "S-dual fluxes," which lie beyond our scope. The extension of flux compactifications to include non-geometric (or these more general) fluxes gives rise to several new terms in the superpotential, allowing for new patterns in the moduli potential and its minima. Refraining from a systematic discussion, we just mention the existence of explicit flux configurations leading to de Sitter or Minkowski minima with full moduli stabilization. More generally, a lesson from the very existence of these new flux degrees of freedom is the suggestion that generic string vacua with moduli stabilized may have a non-geometric character. A precise statement in this direction, however, requires a not yet achieved understanding of these generalized fluxes beyond toroidal setups.

14.5 Fluxes in other string constructions*

It is possible to consider the introduction of field strength (or more general) fluxes in other string or M-theory compactifications; these are often related to the fluxes in type II orientifold models studied in previous sections. For illustration, we consider heterotic string flux compactifications on a (not necessarily CY) 6d space, with 3-form fluxes. This class is related by heterotic/type I duality to type IIB compactifications with O9-planes, with RR 3-form and geometric fluxes. Familiar domain wall arguments, or direct KK compactification, lead to the superpotential

$$W_{\text{het}} = \int_{\mathbf{X}_6} \Omega_3 \wedge (H_3 + d J_c). \tag{14.99}$$

For $\mathbf{X}_6$ a twisted torus with geometric fluxes ω, we have $d J_c = \omega J_c$, and we recover the structure of the dual type IIB models. In this case, we may apply heterotic/type I duality to borrow explicit expressions from type IIB models with O9-planes. For instance, we expand the heterotic 3-form flux as

$$H_3 = -e_0 \alpha_0 + m \beta_0 - \sum_{i=1}^{3} (q_i \alpha_i + e_i \beta_i), \tag{14.100}$$

and use the notation in Table 14.1 for the geometric fluxes. Recalling the definition of moduli (12.34), $U_i = \mathcal{U}_i$, $J_c = i \sum_j T_j \omega_j$, the explicit superpotential is

$$W_{\text{het}} = m + i \sum_{i=1}^{3} q_i U_i + e_1 U_2 U_3 + e_2 U_1 U_3 + e_3 U_1 U_2 - i e_0 U_1 U_2 U_3$$

$$+ \sum_{i=1}^{3} T_i \left[-i \bar{h}_i + \sum_{j=1}^{3} \bar{b}_{ji} U_j + i b_{1i} U_2 U_3 + i b_{2i} U_1 U_3 + i b_{3i} U_1 U_2 - h_i U_1 U_2 U_3 \right]. \tag{14.101}$$

The structure of (14.99) is incidentally similar to the type IIB superpotential (14.21), with $H_3 + d J_c$ playing the role of G_3. This is in particular manifest by restricting (14.101) to $T_i = T$, so it becomes identical to (14.54) up to a replacing $T \to S$ and relabeling flux quanta. By arguments analogous to those in Section 14.1.2, compactifications with ISD $H_3 + d J_c$ lead to heterotic vacua with 4d Minkowski no-scale structure, with the complex dilaton S as unfixed modulus.

15

Moduli stabilization and supersymmetry breaking in string theory

We continue the study of moduli stabilization, now focusing on its interplay with SUSY breaking in string compactifications. We describe different mechanisms proposed for heterotic and type II orientifold compactifications, and elaborate on type IIB models, which allow for very explicit computations. Fluxes not only help in fixing moduli but generically induce SUSY breaking soft terms in the effective action. Some scenarios of SUSY breaking in string construction boil down to SUSY breaking in the dilaton or geometric moduli sector, and can be described in the 4d effective supergravity action in a quite model independent formalism. This setup allows the computation of soft terms and sparticle spectrum of string compactifications in terms of a few parameters, which may be tested at the LHC if low-energy SUSY is realized in Nature.

15.1 SUSY and SUSY breaking in string compactifications

In previous chapters we have described the construction of string compactifications, mainly focusing on 4d $\mathcal{N} = 1$ supersymmetric models; indeed, these have been far more studied than directly non-supersymmetric compatifications, for several reasons. SUSY constitutes a theoretically most compelling ingredient proposed for particle physics at the TeV scale, providing a natural solution to the hierarchy problem, as mentioned in Sections 1.3.3 and 2.6. This requires mechanisms breaking supersymmetry in string compactifications, at scales parametrically smaller than the fundamental string scale M_s. Directly non-supersymmetric models, i.e. breaking supersymmetry close to the fundamental scale, must either have a low fundamental scale, i.e. with a (possibly effective) string scale just above the weak scale, or else override naturalness as a misleading criterion and admit the existence of a fine-tuning on, e.g., statistical or anthropic grounds. From the theoretical perspective, 4d $\mathcal{N} = 1$ supersymmetric constructions are also advantageous, as they are under better control; for instance, non-supersymmetric configurations often contain tachyons and are unstable, as explicitly discussed in Section 10.2.2 for intersecting D6-branes, while supersymmetric models are automatically tachyon-free. Even overlooking this point, supersymmetric compactifications allow to carry further the computation of physically relevant features of the 4d models; for instance, the massless spectrum, in which bosons and fermions combine into supermultiplets, or the structure of the 4d low-energy effective field theory, which enjoys powerful non-renormalization properties for its holomorphic couplings.

The realization of TeV scale SUSY particle physics models in 4d $\mathcal{N} = 1$ string compactifications requires SUSY breaking at a scale much lower than the fundamental scales (like the string, Planck, or gauge coupling unification scales). This cannot arise from quantum corrections in g_s, i.e. loop corrections, due to the non-renormalization of the superpotential and other holomorphic quantities. Therefore the diverse ingredients playing a role in SUSY breaking fall in two classes. The first corresponds to modifications of the compactification, like the introduction of p-form or geometric fluxes in Chapter 14; these are usually described in terms of a (generalized) flux superpotential in the 4d $\mathcal{N} = 1$ effective action, potentially breaking SUSY spontaneously. The second corresponds to the inclusion of non-perturbative effects arising from brane instantons, which can correspond to strong gauge dynamics or stringy instantons, as described in Chapter 13. In some constructions several of these mechanisms are invoked simultaneously to achieve SUSY breaking.

An important issue is that SUSY breaking is a property of the (possibly local) minimum of the scalar potential, so it can be established and studied only once full moduli stabilization has been achieved. This relates the a priori independent problems of SUSY breaking and moduli stabilization. The latter was partially addressed in Chapter 14, in terms of flux compactifications, and we now continue its study in several proposals for full moduli stabilization, focusing on their interplay with SUSY breaking. The models fall into three main classes: (a) models with a common origin for SUSY breaking and moduli stabilization, so both are controlled by the same scale; (b) models with moduli stabilized at a high scale, so they decouple leaving a rigid 4d $\mathcal{N} = 1$ SUSY theory, which subsequently develops dynamical SUSY breaking at lower energies; (c) mixed situations, in which some of the moduli are fixed at a high scale, whereas the remaining moduli are stabilized by effects which break SUSY simultaneously. An example of case (a) is provided by heterotic models with gaugino condensation, Section 15.2. Case (b) can be envisaged in type II compactifications, since the multiple fluxes may potentially fix all moduli in a supersymmetric way; such setups are implicit in the studies of gauge mediated SUSY breaking in string theory, Section 15.7. In practice, the simplest full stabilization (type IIB) models realize case (c), with the dilaton and complex structure moduli fixed at a high scale, leaving an effective theory for the Kähler moduli, addressing simultaneously their stabilization and SUSY breaking, Section 15.3.

There are still many poorly understood aspects of SUSY breaking and moduli stabilization in string theory; yet there has been substantial progress in uncovering possible ingredients for full moduli stabilization, and in combining them to achieve SUSY breaking at hierarchically small scales, with the possibility of tuning the 4d cosmological constant. Although the explicit construction of complete realistic models realizing all required ingredients has not been achieved yet, general classes of models allow the study of many qualitative features, and the quantitative parametrization of phenomenologically relevant particle physics properties.

15.2 SUSY breaking and moduli fixing in heterotic models

The main proposal for SUSY breaking in heterotic compactifications is gaugino condensation in a hidden sector. The mechanism is most simply illustrated in the CY compactification with standard embedding of Section 7.3.2, although it applies to more general heterotic compactifications. Moreover, gaugino condensation or similar non-perturbative superpotentials apply in other string constructions, including type II orientifold compactifications, see Section 15.3.1.

15.2.1 Gaugino condensation in heterotic compactifications

Heterotic CY compactifications with standard embedding have an $E_6 \times E_8$ gauge group, with chiral matter charged under the E_6 (providing the visible particle physics sector) and decoupled from the E_8 hidden sector. The latter is a 4d $\mathcal{N} = 1$ pure SYM sector which, as described in Sections 2.5.1 and 13.1.1, develops a non-perturbative superpotential for the dilaton multiplet and a gaugino condensate

$$W(S) = \mu^3 \exp\left(\frac{3S}{2\beta_{E_8}}\right), \quad \langle \lambda\lambda \rangle \simeq \Lambda^3 = \mu^3 \exp\left(\frac{3}{2g_{E_8}^2 \beta_{E_8}}\right), \tag{15.1}$$

where $\mu \simeq \alpha'^{-1/2}$, S is as defined in Section 9.1.1, and g_{E_8} is the gauge coupling constant of the condensing E_8 group, whose one-loop β-function is $\beta_{E_8} = -90/(16\pi^2)$. A similar general structure holds in other $(0, 2)$ compactifications, with other groups yielding smaller, more realistic, condensing scale.

The resulting effects on the heterotic model can be systematically studied using 4d $\mathcal{N} = 1$ supergravity, with the above non-perturbative superpotential. Restricting for simplicity to the dynamics of the overall Kähler modulus T and the dilaton S, the Kähler potential is (9.17),

$$K = -\log(S + S^*) - 3\log(T + T^*), \tag{15.2}$$

where from now on we set $\kappa_4 = 1$ for simplicity. The resulting scalar potential is

$$V = e^K (K_{SS^*})^{-1} |D_S W|^2 = \frac{1}{(S + S^*)(T + T^*)^3} \left|\left(1 - \frac{3}{2}\frac{S + S^*}{\beta_{E_8}}\right)\mu^3 e^{\frac{3S}{2\beta_{E_8}}}\right|^2 .$$

This potential has a *runaway* behavior on both T and S, driving the theory towards a non-interacting ($S \to \infty$) decompactified ($T \to \infty$) limit, as sketched for S in Figure 15.1(a); this is a rather generic feature of simple gaugino condensation in heterotic compactifications, and motivates the introduction of additional ingredients to produce physical minima at finite moduli vevs.

A possible additional ingredient is to introduce 3-form fluxes in the compactification, as in Section 14.5. In the absence of metric fluxes the flux superpotential is independent of

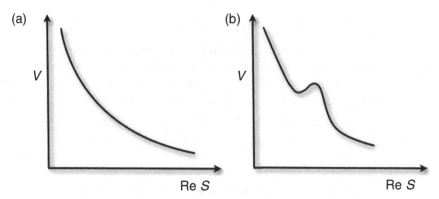

Figure 15.1 (a) Runaway behavior of a single gaugino condensate; (b) for double gaugino condensate a local minimum at finite ReS may develop.

S and T, hence for the present purposes can be regarded as a constant contribution c. The scalar potential becomes

$$V = \frac{(S+S^*)}{(T+T^*)^3} \left| \frac{c}{(S+S^*)} + \left(\frac{1}{S+S^*} - \frac{3}{2b_{E_8}} \right) \mu^3 e^{\frac{3S}{2b_{E_8}}} \right|^2. \tag{15.3}$$

Upon minimization, we have $D_S W = 0$ and the S vev is fixed, with $V \equiv 0$ and hence unfixed T vev. Since $D_T W = -3W/(T+T^*) \neq 0$, SUSY is broken along this classical flat direction, yet with vanishing 4d vacuum energy (i.e. cosmological constant); this realizes a no-scale structure, as the SUSY breaking scale slides with the T vev. Despite its seeming appeal, this structure does not survive higher corrections in α' or g_s to the above classical supergravity approximation; e.g. in the strong coupling regime, this manifests in the heterotic M-theory corrections in Section 9.2, which do indeed spoil the no-scale structure. A further issue is that the above potential generically describes high-scale SUSY breaking, since pure 3-form fluxes yield values of c which cannot be made parametrically small in string units. This might be improved by generalizing the flux compactification to allow for geometric fluxes (i.e. non-CY compactifications) as in Section 14.2.3, but then the Kähler moduli dependence of the corresponding flux superpotential spoils the no-scale structure.

A second possible ingredient, in more general non-standard embedding models, is the presence of several hidden gauge factors G_n with gaugino condensation, i.e.

$$W(S) = \sum_n d_n e^{3S/2b_n}, \tag{15.4}$$

where b_n are the G_n one-loop β-functions. A purely phenomenological tuning of the superpotential parameters shows that, e.g., for two gauge factors there exist local minima for S, as sketched in Figure 15.1(b), for which the value of Re S is furthermore consistent with the unified gauge coupling as extrapolated from the MSSM values. This can already be seen in a simple toy model with two gaugino condensates and, e.g., gauge group $SU(N)_1 \times SU(M)_2$, with $M > N$ and no matter transforming under the condensing groups. The gaugino condensation superpotential reads

$$W(S) = \mu_1^3 e^{-S/2N} + \mu_2^3 e^{-S/2M} e^{i\delta}, \tag{15.5}$$

where $\mu_{1,2}$ are of order the string scale and we have allowed for a relative phase δ. The model produces a minimum $\partial W / \partial S = 0$, which fixes the value of Re S at

$$\text{Re } S = \frac{2}{1 - N/M} \log \left(\frac{M \mu_1^3}{N \mu_2^3} \right), \tag{15.6}$$

with Im S such that the condensates have opposite sign, irrespective of the value of δ. One can easily find values for N, M and the scales $\mu_{1,2}$ such that Re $S = 1/g^2 \simeq 2$, as required for standard gauge coupling unification. This mechanism for fixing the dilaton is sometimes called the *racetrack* scenario; the actual construction of concrete heterotic models with suitable structure of visible and hidden gauge factors is non-trivial and beyond our scope.

15.2.2 Modular invariant gaugino condensation

In the simple scheme in the previous section, the Kähler modulus T remains undetermined, due to the T-independence of the superpotential. However the Kähler moduli generically enter the superpotential when considering the one-loop threshold corrections to the gauge kinetic functions. These can be explicitly computed in orbifold compactifications, and enjoy interesting modular properties under T-duality symmetries, as discussed in Section 9.6. We now study their impact on moduli stabilization upon gaugino condensation.

We focus on orbifolds with factorized underlying $\mathbf{T}^6 = \mathbf{T}^2 \times \mathbf{T}^2 \times \mathbf{T}^2$, and on the $SL(2, \mathbf{Z})$ T-duality symmetry on the T_i moduli, and for simplicity ignore the dependence on complex structure moduli. The gaugino condensation superpotentials (15.1), (15.4), are clearly invariant under the $SL(2, \mathbf{Z})_{T_i}$, in contrast with the result (9.90) that superpotentials must transform with modular weight -1. This puzzle is solved by the one-loop corrections to the effective action and their T-duality transformations, studied in Section 9.6.1; indeed, recall the one-loop corrected gauge kinetic functions (9.106)

$$f_a^{\text{tree}} + f_a^{\text{one-loop}} = k_a S - \frac{1}{16\pi^2} \sum_{i=1}^{3} (b_a^{i \, \prime} - k_a \delta_{\text{GS}}^i) \log \eta(T_i)^4. \tag{15.7}$$

For a single E_8 gaugino condensate, with gauge kinetic function f_{E_8} and Kac–Moody level $k = 1$, the superpotential is

$$W_{E_8} = \mu^3 e^{\frac{3 f_{E_8}}{2 \beta_{E_8}}} = \mu^3 e^{\frac{3S}{2 \beta_{E_8}}} \prod_{i=1}^{3} [\eta(T_i)]^{- \frac{6 \left(b_{E_8}^{i \, \prime} - \delta_{\text{GS}}^i \right)}{16\pi^2 \beta_{E_8}}}. \tag{15.8}$$

From (9.105), we have $\beta_{E_8} = 3b_8^{i\,\prime}/(16\pi^2)$, and so the $SL(2, \mathbf{Z})_{T_i}$ transformation of the piece depending on δ_{GS}^i is precisely canceled by the transformation (9.101) of the dilaton S, reproducing the duality transformation (9.90) for the superpotential

$$W_{E_8} \longrightarrow W_{E_8}(ic_i T_i + d_i)^{-1}. \tag{15.9}$$

In order to describe the implications for moduli stabilization, we consider for simplicity cases with $\delta_{GS}^i = 0$ (e.g. the $\mathbf{Z}_2 \times \mathbf{Z}_2$ orbifold) and concentrate on the dependence on the overall Kähler modulus $T_1 = T_2 = T_3 \equiv T$. The above single E_8 gaugino condensate superpotential, generalized to several gauge factors (allowed to have charged matter), motivates the following structure for the superpotential:

$$W_{\text{cond}} = \frac{\Omega(S)}{\eta(T)^6} = \frac{\sum_n d_n e^{\frac{3S}{2\beta_n}}}{\eta(T)^6}, \tag{15.10}$$

where now the coefficients d_n depend on the matter content of the gauge factor G_n, and $\Omega(S)$ is a shorthand notation. The supergravity scalar potential is

$$V = \frac{|\eta(T)|^{-12}}{(S + S^*)(T + T^*)^3} \left\{ |(S + S^*)\Omega_S - \Omega|^2 + \left[\frac{3(T + T^*)^2}{4\pi^2} |\hat{G}_2|^2 - 3 \right] |\Omega|^2 \right\},$$

where $\hat{G}_2 = G_2 - 2\pi/(T + T^*)$, with $G_2 = -4\pi \eta(T)^{-1} \partial\eta(T)/\partial T$ being the holomorphic Eisenstein function. Using (9.100), the reader may easily check that this potential is modular invariant under $SL(2, \mathbf{Z})$ T-duality transformations.

This *duality completed* multiple gaugino condensation scalar potential has an extremum with respect to the complex dilaton S, defined by

$$(S + S^*)\Omega_S - \Omega = 0. \tag{15.11}$$

After closer numerical scrutiny it can be shown to correspond to a minimum which fixes S. The analysis can proceed further for the simplest case of two condensates. Suitable choices of the underlying parameters can again lead to minima with reasonable values of Re $S \simeq 2$, i.e. $\alpha_G \simeq 1/24$, the unification coupling constant. For this two-condensate model, extremization with respect to T also leads to a minimum fixing Re $T \simeq 1.2$. Thus in this simplified example, the S and T moduli are simultaneously fixed, and there is SUSY breaking along T, namely $F_S = 0$, $F_T \neq 0$.

The stabilization at finite values of T is very generic, and follows from T-duality invariance, since $\eta(T) \to 0$ when Re $T \to \infty$; alternatively, T-duality invariance effectively renders the Kähler moduli space finite, thus preventing runaway. The detailed physics around the minimum may however lie beyond perturbative analysis, due to the potentially large one-loop correction to the gauge kinetic function in (15.7), which may even cancel the tree level contribution, leading to a strong coupling regime.

Results for the above simplified model are expected to generalize to more realistic situations, e.g. involving all untwisted moduli; it is also reasonable to expect analogous dynamics in other heterotic compactifications (e.g. on CY spaces). The main lesson is that multiple condensates, once one-loop corrections to gauge kinetic functions are included, leads naturally to moduli stabilization and SUSY breaking. However, the mechanism has several potential drawbacks, as for instance it typically leads to only moderately large Kähler moduli vevs, rendering the effective field theory approximation questionable due to α' corrections; stabilization in a reliable regime thus requires a careful choice of the condensate parameters, whose microscopic realization in a realistic compactification is challenging. In addition, as in examples in Section 2.5.1, gauge sectors with charged matter may lead to superpotentials involving the charged matter multiplets, and not just the moduli.

We conclude this section with an important issue not addressed in the above discussion. Racetrack potentials typically have minima at negative value of the potential, i.e. 4d cosmological constant, and thus describe 4d anti-de-Sitter (AdS$_4$) vacua. In order to agree with the observation of a tiny but positive cosmological dark energy, the scenario must include further sources of potential energy, for which no widely accepted proposals exist in the heterotic setup. As discussed in the next section, the situation is in better shape in type II orientifold models.

15.3 SUSY breaking and moduli fixing in type II orientifolds

Type II orientifolds include new ingredients as compared to the heterotic case, like the rich set of p-form fluxes, and the generic presence of localized D-brane sources. Also, the better computational control allows to address new issues, like full moduli stabilization with tuning of the 4d vacuum energy to a small positive value.

15.3.1 The KKLT setting

As described in Section 14.1, type IIB with NSNS and RR 3-form fluxes generically leads to stabilization of the dilaton and complex structure moduli, but leaves Kähler moduli unfixed. Moreover, shift symmetries in the RR scalars prevent Kähler moduli from appearing in the superpotential in perturbation theory. These symmetries are broken by non-perturbative D-brane instanton effects (recall Chapter 13), which can produce non-perturbative superpotentials fixing the Kähler moduli. This setup, known as the Kachru–Kallosh–Linde–Trivedi (KKLT) scenario, suggests a generic picture for full moduli stabilization in type IIB orientifold compactifications (or their F-theory generalizations). It moreover suggests a possible mechanism for tuning the vacuum energy, even to small positive values.

Full moduli stabilization

For our purposes it suffices to focus on the dynamics of the overall T-modulus, as in Section 14.1.2. In the large volume regime the Kähler potential is (12.16), i.e.

$$K = -\log(S + S^*) - 3\log(T + T^*) - \log\left(-i\int_{\mathbf{X}_6} \Omega_3 \wedge \overline{\Omega}_3\right), \qquad (15.12)$$

where recall that $(\operatorname{Re} T)^{3/2}$ gives the overall compactification volume. The 3-form fluxes H_3, F_3 induce a superpotential (14.21) and, due to the no-scale structure of (15.12), the scalar potential is positive definite and given by (14.24), i.e.

$$V = e^K \sum_{i,j} g^{i\bar{j}} D_i W D_{\bar{j}} \overline{W}, \tag{15.13}$$

where the sum runs over the dilaton and complex structure moduli. As described in Section 14.1.2, minima correspond to configurations with imaginary self-dual (ISD) G_3, generically containing a supersymmetric $(2, 1)$ piece and a non-supersymmetric $(0, 3)$ piece. The latter induces a non-zero value W_0 for the superpotential at the minimum. Assuming that flux stabilization of the dilaton and complex structure moduli occurs at a parametrically high scale, compared to Kähler moduli stabilization, we proceed to the discussion of the latter regarding W_0 as a constant.

In analogy with the heterotic analysis in the previous section, non-perturbative effects are expected to contribute to Kähler moduli stabilization. In the present type IIB models with D3/D7-branes, there are non-perturbative superpotentials from D3-brane instantons wrapped on 4-cycles, hence yielding T-dependent superpotentials; they also include gauge theory non-perturbative effects, like gaugino condensate superpotentials, as particular cases when the D3-brane instantons wrap the same 4-cycles as some background D7-brane stack. In any event, the superpotential including non-perturbative contributions, has the structure

$$W = W_0 + c e^{-2\pi a T}, \tag{15.14}$$

where a is a (positive) model dependent quantity, and c is in general a holomorphic function of the complex structure fields, assumed already fixed at a higher scale, hence is regarded as a constant in what follows. The scalar potential for T is minimized for $D_T W = 0$, and setting $\operatorname{Im} T = 0$, $\operatorname{Re} T = \sigma$ for simplicity, yields

$$W_0 = -c e^{-2\pi a \sigma} \left(1 + \frac{4\pi a}{3} \sigma \right). \tag{15.15}$$

For sufficiently small W_0, the σ vev is fixed at moderately large values. Since $D_T W = 0$, SUSY remains unbroken, in an AdS$_4$ vacuum with negative cosmological constant given by the potential energy at the minimum

$$V_{\min} = -3 e^K |W|^2 = -\frac{2\pi^2 a^2 c^2 e^{-4\pi a \sigma}}{3\sigma}. \tag{15.16}$$

The qualitative dependence of the potential on σ is depicted in Figure 15.2(a).

A natural question is whether an exponentially small W_0 is reasonable and/or possible. Such very small values require delicate cancellations among contributions from many different flux degrees of freedom; although a priori unnatural, this discrete tuning may be justified on general grounds by the large number of possible flux vacua for a given CY orientifold compactification, so that some of them realize accidentally small values, see Section 17.3. Incidentally, a small W_0 is also required in the setup of Section 15.5.3, where

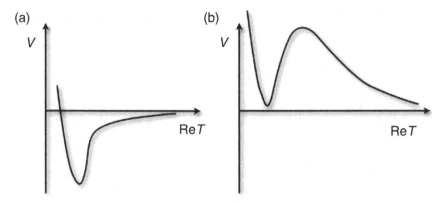

Figure 15.2 (a) Qualitative form of the T-dependent potential in compactification with 3-form fluxes and non-perturbative effects; there is a supersymmetric AdS$_4$ minimum. (b) After uplifting, the minimum is a non-SUSY dS$_4$ with small cosmological constant.

it sets the order parameter of gravity mediated SUSY breaking, so that $W_0/M_p^3 \leq 10^{-15}$ for TeV low-energy SUSY.

Uplifting to de-Sitter vacua

The flexibility of type IIB orientifold compactifications suggests an improvement of the above models, allowing to envisage a structure with small and positive potential energy at the minimum, thus leading to 4d de-Sitter (dS$_4$) vacua. The mechanism, usually referred to as *uplifting*, amounts to include additional sources of tension in the compactification, implicitly from the beginning of the construction. The simplest possibility is to include anti-D3-branes ($\overline{\text{D3}}$-branes) sitting at points in the compact space, and satisfying a RR tadpole constraint generalizing (14.14)

$$N_{D3} - N_{\overline{D3}} + N_{\text{flux}} - \frac{1}{2}N_{O3} = 0, \qquad (15.17)$$

or a similar generalization of the tadpole condition (14.38) for F-theory models.

The $\overline{\text{D3}}$-branes break supersymmetry explicitly, and give a positive contribution to the scalar potential from their tension, which in the Einstein frame is of the form

$$\delta V = \frac{D}{(T + T^*)^3}, \qquad (15.18)$$

where D is proportional to the number of anti-D3-branes, and may contain model-dependent suppression factors. As will become clear in Section 15.4.1, the interactions with the NSNS and RR 3-form fluxes dynamically fix the $\overline{\text{D3}}$-brane location on the compact space $\mathbf{X}_6$, such that they naturally fall to the bottom of possibly present warped throats; this leads to exponential suppression warp factors $e^{4A_{\min}}$ in D, allowing δV to balance the exponentially small T-dependent superpotential contributions. The full potential has now the form

$$V = \frac{\pi a c e^{-2\pi a\sigma}}{\sigma^2} \left(\frac{2\pi a c\sigma e^{-2\pi a\sigma}}{3} + W_0 + c e^{-2\pi a\sigma} \right) + \frac{D}{8\sigma^3}. \qquad (15.19)$$

It is reasonable to expect that the value of D may be (discretely) fine-tuned by varying the fluxes controlling the warping; thus the vacuum energy can be tuned to a small and positive value, exponentially suppressed with respect to fundamental scales of the microscopic theory. The structure of the potential is sketched in Figure 15.2(b). The dS$_4$ vacuum is meta-stable, with an obvious tunneling channel to the $\sigma \to \infty$ decompactification regime, and other less obvious channels of flux-$\overline{D3}$-brane annihilation; the lifetime of such meta-stable vacua has however been shown to be much larger than the age of the universe, for the range of parameters of interest.

The above approach should be regarded as a *road map* for full moduli stabilization in a dS$_4$ vacuum, since the construction of truly explicit models realizing all ingredients simultaneously is beyond present computational techniques; for instance, an important bottleneck is the computation of complete non-perturbative superpotentials in models with several Kähler moduli. In addition, the assumed separation of scales of complex structure and Kähler moduli stabilization can be problematic in general, and mixing can potentially lead to tachyonic directions. Concerning the uplifting, its realization in terms of $\overline{D3}$-branes is conceptually problematic due to the explicit breaking of SUSY. This may be overcome in other realizations, like e.g. non-supersymmetric gauge flux backgrounds on D7-branes, interpreted as non-vanishing D-terms as in Section 12.4, and leading to uplifting potentials of the form (15.18); or by simple flux compactifications with no SUSY breaking objects, but with meta-stable vacua in their flux-induced scalar potentials (15.13), yielding a vacuum energy with a $1/(T + T^*)^3$ dependence, due to the factor e^K. In any event, complete explicit models are notoriously hard to realize.

A qualification to the importance of explicit constructions is however the following. This approach to full moduli stabilization in a dS$_4$ vacuum requires a delicate tuning of fluxes, to achieve small values of W_0 and D. However, for each such vacuum there are further effects, from α' and quantum corrections, which in general modify significantly the vacuum energy, yet are beyond present computational abilities. The discussion in the above setup, and other related constructions, is however meaningful, in that it establishes the possibility to engineer, with fairly explicit and realistic ingredients, potentials with minima deep enough to survive the inclusion of such corrections. Although the properties of these minima (e.g. the 4d cosmological constant) suffer important modifications, the large number of models, e.g. of flux choices for a given CY orientifold, raises the reasonable expectation that a subset of them realizes the proposal, even after inclusion of these corrections. Further ideas based on the large multiplicity of flux vacua are discussed in Section 17.3.

There are finally several open questions in going from the toy models of moduli fixing to more realistic compactifications including gauge D-brane sectors, realizing particle physics models. For instance, stabilization of *all* Kähler moduli through non-perturbative effects solely may not be completely viable in the presence of realistic SM-like D-brane sectors. This is because necessarily some D-brane instantons intersect the gauge D-branes, and lead to superpotentials involving the (SUSY) SM charged fields, rather than just the Kähler moduli, recall Section 13.1.3. Hence realistic compactifications possibly require further ingredients in the moduli dynamics.

15.3.2 Large volume generalization

One of the drawbacks of the above scheme for moduli fixing is the requirement of very small values of W_0. Although achievable by suitably tuned fluxes, this raises the question of moduli stabilization for generic values of W_0. The smallness of W_0 is necessary to achieve Kähler moduli stabilization in the large radius regime, although in practice the volumes cannot be made arbitrarily large, and often stabilize barely above the string length. For instance, taking $W_0/M_p^3 \simeq 10^{-15}$ (e.g. to obtain TeV scale SUSY breaking in gravity mediation as in Section 15.5.3), and assuming $c \simeq a \simeq 1$, eq. (15.15) yields $\sigma \simeq 6$. Since $\sigma \simeq g_s^{-1} R^4$, in the perturbative regime of small g_s we have $R \simeq \mathcal{O}(1)$ in string units.

This argument suggests that the supergravity approximation may need the inclusion of α' corrections to describe the moduli stabilization for generic W_0. Indeed, it has been shown that in type IIB orientifold compactifications the leading correction to the Kähler potential (12.17) can compete with non-perturbative corrections to the superpotential, and lead to generically very large volume moduli stabilization.

We consider type IIB compactifications with 3-form fluxes, assumed to fix the dilaton and complex structure moduli at a parametrically large scale, below which the effective dynamics reduces to the Kähler moduli. The main ingredients of the "large volume moduli stabilization" scenario can be illustrated in a toy model with two Kähler moduli, T_b and T_s, controlling the overall volume and a 4-cycle size, eventually shown to stabilize at very large ("big") and moderately large ("small") sizes, respectively; the mechanism also requires $\chi(\mathbf{X}_6) < 0$. An explicit example is provided by the CY $\mathbf{CP}^4_{(1,1,1,6,9)}$[18], which has $h_{1,1} = 2$, $h_{2,1} = 272$. In such models, the Kähler potential including the leading α' correction (12.17), written in terms of the real parts $\tau_b = \mathrm{Re}\, T_b$, $\tau_s = \mathrm{Re}\, T_s$, reads

$$K = -2 \log\left[\frac{1}{9\sqrt{2}}(\tau_b^{3/2} - \tau_s^{3/2}) + \frac{\xi}{2g_s^{3/2}} \right], \qquad (15.20)$$

with $\xi > 0$. The superpotential includes a non-zero constant contribution W_0 related to the 3-form fluxes, and we moreover assume a non-perturbative contribution from D3-brane instantons on the "small" 4-cycle, namely $W = W_0 + A_s e^{-a_s T_s}$. The scalar potential for the volume $\mathcal{V} \simeq \tau_b^{3/2}$ and τ_s has the structure

$$V \simeq \frac{\lambda a_s^2 |A_s|^2 \sqrt{\tau_s} e^{-2a_s \tau_s}}{\mathcal{V}} - \frac{\mu a_s W_0 A_s \tau_s e^{-a_s \tau_s}}{\mathcal{V}^2} + \frac{\nu \xi |W_0|^2}{g_s^{3/2} \mathcal{V}^3}, \qquad (15.21)$$

where λ, μ, ν are $\mathcal{O}(1)$ constants. The first two terms are analogous to those in (15.19), with the relative minus sign due to a phase factor arising from the value of the Im T_s axion at the minimum.

The structure of the potential is as follows. For $\tau_b \to \infty$ with $a_s \tau_s = \log \mathcal{V}$, the second term dominates and the potential approaches zero from below; for smaller $\mathcal{V}$, τ_s, the other terms dominate and are positive. Thus the potential must have a local AdS$_4$ minimum at intermediate values, which explicit minimization fixes at

$$\mathcal{V} \propto e^{a_s \tau_s} \gg 1, \quad \tau_s \propto \frac{\xi^{2/3}}{g_s}. \tag{15.22}$$

For moderate a_s, the overall volume $\mathcal{V}$ is much bigger than the size of the small 4-cycle governed by τ_s. The hierarchy arises from competition of the non-perturbative super-potential e^{-T_s} and the perturbative correction in negative powers of τ_b. As announced, the minimum exists for generic values of W_0, with no specific tuning. In contrast to Section 15.3.1, the AdS$_4$ minimum is not supersymmetric, with SUSY broken by F_{T_b} and F_{T_s}; however, for $|W_0|^2 \ll 1$, the third term in (15.21), i.e. the α' correction, becomes negligibly small, and we recover the scenario of Section 15.3.1.

As in the previous section, any realistic model should contain additional ingredients, e.g. $\overline{\text{D3}}$-branes, to uplift the minimum to a dS$_4$ vacuum with small and positive cosmological constant. The important difference is that, in the present setup, the SUSY breaking effects from these extra sources are sub-dominant compared to the Kähler moduli auxiliary field vevs; hence the SUSY breaking can be modeled as modulus dominated, see Section 15.5, with order parameter the gravitino mass

$$\frac{m_{3/2}}{M_p} \simeq \frac{|W_0|/M_p^3}{(\text{Re } T_b)^{3/2}}. \tag{15.23}$$

In the large volume scheme the overall volume may be arbitrarily large, and is in fact decoupled from W_0. Thus the hierarchy of scales can be generated either by fine-tuning $W_0 \ll M_p^3$, or through a naturally achieved large volume Re T_b.

This impacts on the value of the string scale, since $M_s^2 \simeq M_p^2/\mathcal{V}$. An attractive possibility is an intermediate string scale $M_s \simeq 10^{11}$ GeV with $\mathcal{V} \simeq 10^{16}$, requiring a generic $W_0/M_p^3 \simeq 1$, and yielding $m_{3/2} \simeq 1$ TeV. A high string scale, e.g. $M_s \simeq 10^{16}$ GeV, is possible for $\mathcal{V} \simeq 10^3$, but requires a fine-tuned $W_0/M_p^3 \simeq 10^{-13}$. The extreme alternative of TeV string scale would be problematic (for $W_0 \simeq 1$), because the too light moduli could give rise to unobserved deviations from Newton's law.

The large volume scenario should again be regarded as a road map towards full moduli stabilization in type IIB compactifications. In this respect, it is interesting that this moduli stabilization pattern generalizes to cases with additional Kähler moduli, for the so-called "swiss cheese" CYs. These are spaces with one Kähler modulus T_b controlling the (eventually very large) overall volume, and a number of additional moduli $T_{s,i}$ whose (eventually moderately large) vevs describe the size of 4-cycles decreasing the CY volume, see Figure 15.3. However, fully explicit models realizing all required properties

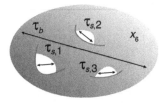

Figure 15.3 Pictorial view of a "swiss cheese" CY, where the modulus τ_b defines the overall size, while the modulus τ_s describes the size of "hole" 4-cycles.

(in particular, the non-perturbative contributions) are lacking. Still the flexibility of the construction allows to envisage including sectors of D3- or D7-branes realizing particle physics models; given the large overall size of $\mathbf{X}_6$, the setup is well-suited to the bottom-up strategy introduced in Section 11.3.1. An appealing scenario is to consider a supersymmetric particle physics D-brane sector, with SUSY breaking due to fluxes, as we explore in coming sections.

15.4 Soft terms from fluxes in type IIB orientifolds

Up to now we have studied the stabilization and SUSY breaking effects of fluxes in the moduli sector. We now wish to explore their effects on the D-brane open string sectors potentially associated to SM fields. This can be explored by using the D-brane DBI + CS action coupled to the non-SUSY flux (supergravity) background, which shows the appearance of additional terms in the D-brane worldvolume effective action. These are typically SUSY breaking soft terms, as well as certain SUSY contributions to the superpotential. Interestingly, these terms may often be understood as arising from spontaneous SUSY breaking of a low-energy supergravity action, as in the gravity mediation scenario discussed in Section 2.6.5.

15.4.1 Microscopic description from the D-brane action

The effects of (possibly SUSY breaking) fluxes on a set of D-branes can be computed, from a microscopic viewpoint, by coupling the D-brane to the 10d supergravity background of the fluxes using the DBI and CS actions (6.6), (6.15). This is particularly suitable in situations where the D-branes are localized on a small region of the compactification space $\mathbf{X}_6$, so that a local non-compact approximation to the latter can be used.

For concreteness we focus on the well-understood setup of type IIB compactifications with NSNS and RR 3-form fluxes, and systems of D3- or D7-branes, see Figure 15.4. We consider the following ansatz for the 10d supergravity flux background

$$ds^2 = Z_1(x^m)^{-1/2}\eta_{\mu\nu}dx^\mu dx^\nu + Z_2(x^m)^{1/2}\tilde{g}_{mn}dx^m dx^n,$$
$$F_5 = (1 + *_{10d})dC_4 \quad \text{with } C_4 = \chi(x^m)dx^0 dx^1 dx^2 dx^3,$$

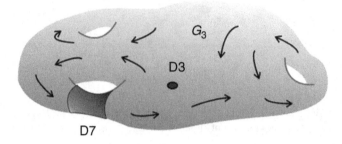

Figure 15.4 A type IIB CY (orientifold) compactification with D3- and D7-branes in the presence of NSNS and RR 3-form fluxes G_3.

$$\tau = \tau(x^m),$$
$$G_3 = \frac{1}{3!} G_{lmn}(x^m) dx^l dx^m dx^n, \tag{15.24}$$

where $\tilde{g}_{mn}$ is the underlying CY metric. For ISD 3-form flux, the equations of motion enforce $Z_1 = Z_2$, $\chi_4 = Z_1^2$, so the ansatz reduces to the supergravity background (14.9) with (14.12), but we keep the above more general form. We are interested in the background near the D-brane location; for instance, for a D3-brane stack, the background can be expanded to the relevant order around its location $x^m = 0$ as

$$Z_1^{-1/2} = 1 + \frac{1}{2} K_{mn} x^m x^n + \cdots, \qquad Z_2^{1/2} = 1 + \cdots,$$
$$\tau = \tau_0 + \frac{1}{2} \tau_{mn} x^m x^n, \qquad \tilde{g}_{mn} = \delta_{mn} + \cdots,$$
$$\chi = \text{const.} + \frac{1}{2} \chi_{mn} x^m x^n + \cdots, \qquad G_{lmn}(x^m) = G_{lmn} + \cdots, \tag{15.25}$$
$$F_5 = \frac{1}{2} (\chi_{mn} + \chi_{nm}) x^m dx^n dx^0 dx^1 dx^2 dx^3 + \cdots.$$

The local constant 3-form flux G_3 has 20 independent components, and transforms in a reducible $\overline{\mathbf{10}} + \mathbf{10}$ representation under the local $SO(6)$ rotation group, corresponding to the imaginary self-dual (ISD) and imaginary anti self-dual (IASD) pieces. It is convenient to split them under the CY $SU(3)$ subgroup as $\mathbf{10} = \mathbf{6} + \mathbf{3} + \mathbf{1}$, leading to Table 15.1. There, we have defined

$$S_{ij} = \frac{1}{2} (\epsilon_{ikl} G_{j\bar{k}\bar{l}} + \epsilon_{jkl} G_{i\bar{k}\bar{l}}), \quad A_{\bar{i}j} = \frac{1}{2} (\epsilon_{\bar{i}\bar{k}\bar{l}} G_{kl\bar{j}} - \epsilon_{\bar{j}\bar{k}\bar{l}} G_{kl\bar{i}}), \tag{15.26}$$

and similarly for $S_{\bar{i}\bar{j}}$, $A_{i\bar{j}}$; also, the subindices P, NP indicate whether or not the flux satisfies the primitivity condition $G_3 \wedge J = 0$. NP fluxes are absent in compact CY spaces, since there are no harmonic 5-forms, so we ignore them in the following.

Table 15.1 *Different components of the complexified 3-form flux G_3*

	ISD			IASD	
$SU(3)$ rep.	Form	Tensor	$SU(3)$ rep.	Form	Tensor
$\bar{1}$	$(0,3)$	$G_{\bar{1}\bar{2}\bar{3}}$	1	$(3,0)$	G_{123}
$\bar{6}$	$(2,1)_P$	$S_{\bar{i}\bar{j}}$	6	$(1,2)_P$	S_{ij}
$\bar{3}$	$(1,2)_{NP}$	A_{ij}	3	$(2,1)_{NP}$	$A_{\bar{i}\bar{j}}$

Soft terms on D3-branes

Let us now describe the coupling of D3-branes to the above flux background. The world-volume theory is, in 4d $\mathcal{N}=1$ language, a $U(N)$ SYM theory with three adjoint chiral multiplets Φ_i, $i=1,2,3$. The non-abelian version of the DBI + CS bosonic action (6.6), (6.15), known as the Myers action, reads

$$S_{DBI} = -\mu_3 \int d^4x \, e^{-\phi} \text{Tr} \left(\sqrt{-\det(P[\tilde{E}] - 2\pi\alpha' F_{\mu\nu}) \det(Q)} \right),$$

$$S_{CS} = \mu_3 \int \text{Tr} \left(P[e^{2\pi i\alpha' i_\phi i_\phi} \sum_q C_q e^{-B}] e^{2\pi\alpha' F} \right),$$

with

$$\tilde{E}_{\mu\nu} = E_{\mu\nu} + E_{\mu m}(Q^{-1} - \delta)^{mn} E_{n\nu}, \qquad (15.27)$$

$$E_{MN} = G_{MN} + B_{MN}, \qquad Q^m{}_n = \delta^m{}_n + 2\pi i\alpha'[\phi^m, \phi^P] E_{pn},$$

where we have ignored the curvature CS coupling. Here $P[M]$ denotes the pullback of the 10d background onto the D3-brane worldvolume, as in (6.8), and $i_\phi C_p$ denotes contraction of a leg of the p-form, transverse to the D3-brane, with the associated worldvolume scalar. The supersymmetric completion of the above action contains also fermionic terms, skipped for simplicity.

One can now replace the ansatz (15.24) with the expansions (15.25) in the action. In doing so, the coordinates x^m in (15.25) are regarded as D3-brane worldvolume scalars via $x^m = 2\pi\alpha'\phi_m$. The expansion of the resulting D3-brane action is the flux-less flat space one with additional terms, which in general correspond to SUSY breaking soft terms on the D3-brane worldvolume theory; for example, plugging the 4-form $C_4 = \frac{1}{2}\chi_{mn}x^m x^n + \cdots$ in the CS action produces soft scalar masses, upon the identification $x^m \sim \phi_m$. We skip the full computation and simply quote the resulting soft term lagrangian, given by

$$\mathcal{L} = -\left[2K_{i\bar{j}} - \chi_{i\bar{j}} + g_s(\text{Im}\,\tau)_{i\bar{j}}\right]\Phi^i\Phi^{\bar{j}} - \frac{1}{2}\left[2K_{ij} - \chi_{ij} + g_s(\text{Im}\,\tau)_{ij}\right]\Phi^i\Phi^j + \text{h.c.}$$

$$+ g_s\sqrt{2\pi}\left(\frac{1}{3}G_{123}\epsilon_{ijk}\Phi^i\Phi^j\Phi^k + \frac{1}{2}\epsilon_{\bar{i}\bar{j}\bar{l}}S_{lk}\Phi^{\bar{i}}\Phi^{\bar{j}}\Phi^k + \text{h.c.}\right)$$

$$+ \frac{g_s^{1/2}}{2\sqrt{2}}\left(G_{123}\lambda\lambda + \frac{1}{2}S_{ij}\Psi^i\Psi^j + \text{h.c.}\right), \tag{15.28}$$

where $g_s = e^\phi$, and we have normalized the kinetic terms and used complex notation for the indices. Also, λ, Ψ^i are the gaugino and the three fermionic partners of the complex scalars Φ^i, all in the adjoint representation, and gauge traces are implicit. As announced, (15.28) contains the different soft terms introduced in Section 2.6.3.

For pure flux background with no additional sources, the 10d supergravity equations of motion relate the different background parameters, allowing to compute the soft terms lagrangian in terms of the local 3-form flux density; the only piece remaining essentially unconstrained is the second scalar mass term in (15.28), which corresponds to a *B-term* in the notation of Section 2.6.5. In the notation therein, one gets soft terms of the form

$$m_{i\bar{j}}^2 = \frac{g_s}{6}\left[|G_{123}|^2 + \sum_{ij}|S_{ij}|^2 - \text{Re}\left(G_{123}G_{\bar{1}\bar{2}\bar{3}} + \frac{1}{4}S_{lk}S_{\bar{l}\bar{k}}\right)\right],$$

$$A'_{ijk} = -h_{ijk}\frac{g_s^{1/2}}{\sqrt{2}}G_{123}, \quad M_a = \frac{g_s^{1/2}}{\sqrt{2}}G_{123}, \quad \mu_{ij} = -\frac{g_s^{1/2}}{2\sqrt{2}}S_{ij}, \tag{15.29}$$

where h_{ijk} are the coupling constants of the $\mathcal{N} = 4$ superpotential in this complex basis, $W_{\mathcal{N}=4} = h_{ijk}\Phi^i\Phi^j\Phi^k/3!$. Interestingly, all soft terms vanish for purely ISD flux, e.g. there are no terms involving the ISD non-SUSY (0, 3) component $G_{\bar{1}\bar{2}\bar{3}}$ alone. The underlying reason is a cancellation between the DBI and CS contributions for ISD fluxes; this is in fact expected since ISD fluxes, even the non-SUSY (0, 3) component, satisfy a BPS-like relation, so that their contribution to the energy density ("tension") is related to their contribution to the 4-form RR tadpole ("charge"), and they have no interaction with the D3-branes in the system. As we show in Section 15.5.3, this is consistent with the no-scale structure in the effective action.

On the other hand, for $\overline{\text{D3}}$-branes, the sign flip in the CS piece eliminates the cancellations, and ISD fluxes do induce soft terms. The physical interpretation is that $\overline{\text{D3}}$-branes fall into regions of maximum flux density, corresponding to potential wells for them, and scalar masses at the minimum describe the energetic cost in climbing this potential; incidentally, this explains our statement, in Section 15.3.1, that $\overline{\text{D3}}$-branes naturally stabilize at the bottom of warped throats. If only (0, 3) fluxes are present the above soft terms have the relationships

$$A'_{ijk} = -h_{ijk}M_a; \quad m_{i\bar{j}}^2 = \frac{1}{3}\delta_{i\bar{j}}|M_a|^2. \tag{15.30}$$

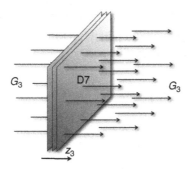

Figure 15.5 The 4-cycle wrapped by a D7-brane is pierced by the 3-form fluxes, which induce modifications of the 8d worldvolume action.

These SUSY breaking soft terms can be realized in semi-realistic particle physics models obtained from stacks of $\overline{D3}$-branes on local singularities e.g. $\mathbf{C}^3/\mathbf{Z}_N$, along the lines of Section 11.3. Conversely, there are non-trivial terms on D3-branes from IASD fluxes. For instance, the pattern (15.30) arises for D3-branes in the presence of IASD $(3,0)$ fluxes. We will see in Section 15.5.3 that these correspond to the relationships in supergravity with SUSY breaking induced by the dilaton. Also, $(1,2)$ fluxes S_{ij}, although IASD, can be checked to induce a SUSY superpotential mass μ_{ij} for the chiral multiplets.

We thus see that ISD fluxes do not induce interesting soft terms on D3-branes. Although they induce soft terms on $\overline{D3}$-branes, the latter models are explicitly non-supersymmetric (i.e. even before the introduction of fluxes) and bring about additional difficulties. As we now discuss, a more appealing setup is provided by D7-branes, for which ISD fluxes do induce SUSY breaking soft terms.

Soft terms on D7-branes

The D7-brane action coupled to the flux background describes an 8d theory with flux induced terms, which subsequently must be compactified on the wrapped 4-cycle Σ_4 to yield the 4d soft term lagrangian, see Figure 15.5. To avoid complications due to this KK compactification, we consider the simplified situation of backgrounds constant over the wrapped 4-cycle, and moreover focus on the simplest local geometry $\mathbf{T}^4 \times \mathbf{C}$. We thus consider an expansion like (15.25) only in the two transverse directions, so the relevant local symmetry is just $SO(4) \times SO(2)$.

In contrast to the D3-brane case, ISD fluxes lead to non-trivial soft terms on D7-branes. For example, expansion of the DBI action produces a term $B_{mn}B^{mn}$, with indices along the D7-brane worldvolume. Writing $B_{mn} = H_{mnk}x^k + \cdots$ with x^k in the transverse directions, produces mass terms $H_{mnk}H^{mn}{}_l\phi^k\phi^l$ for the transverse scalars fields; there is no similar coupling in the CS piece, and thus no cancellation. The fact that ISD fluxes stabilize D7-brane position moduli is consistent with the F-theory picture, in which the latter are complex structure moduli of the CY fourfold, and thus are generically stabilized by fluxes.

We skip the full computation of the soft term lagrangian from the DBI + CS action, and just quote the result, restricting to purely ISD fluxes for simplicity. For a D7-brane transverse to the z_3 complex plane, one obtains

$$m^2_{\Phi^3} = \frac{g_s}{18}\left(|G_{\bar{1}\bar{2}\bar{3}}|^2 + \frac{1}{4}|S_{\bar{3}\bar{3}}|^2\right); \quad M_a = \frac{g_s^{1/2}}{3\sqrt{2}}\,(G_{\bar{1}\bar{2}\bar{3}})^*; \quad A'_{ijk} = -h_{ijk}\frac{g_s^{1/2}}{3\sqrt{2}}(G_{\bar{1}\bar{2}\bar{3}})^*,$$

$$\mu_{33} = -\frac{g_s^{1/2}}{6\sqrt{2}}(S_{\bar{3}\bar{3}})^*; \quad B'_{33} = -\frac{g_s}{18}(G_{\bar{1}\bar{2}\bar{3}})^*(S_{\bar{3}\bar{3}})^* = 2M_a\mu_{33}, \tag{15.31}$$

where M_a is the gaugino mass, Φ^3 is D7-brane position modulus, and μ_{33}, B_{33} are μ- and B-terms for this field. There are two additional adjoint chiral multiplets, corresponding to Wilson lines along $\mathbf{T}^4$, which remain massless at this level; note however that they are generically absent in more realistic wrapped 4-cycle topologies. Note that the different soft terms obey relations

$$A'_{ijk} = -h_{ijk}M_a, \quad m^2_{\Phi^{1,2}} = 0, m^2_{\Phi^3} = |M_a|^2, \quad B'_{33} = 2M_a\mu_{33}. \tag{15.32}$$

This soft term pattern is of phenomenological interest in compactifications realizing particle physics models, as discussed later in Section 15.6. In particular, note that ISD fluxes on D7-branes induce both μ- and B-terms for the Φ_3 superfield; they are in general of the same order of magnitude as the SUSY breaking soft terms, so this provides in principle an elegant solution to the μ-problem of the MSSM discussed in Section 2.6.3. The appearance of non-trivial soft terms on D7-branes by ISD fluxes solving the supergravity equations of motion provides a nice picture of SUSY breaking mediation in flux compactification, and motivates the construction of string models in which SM fields reside at D7-branes.

The above formalism to compute flux induced soft terms is valid for models of D3- and/or D7-branes in orbifolds, as in Sections 11.2 and 11.3; one just truncates the above soft term lagrangian onto invariants of the corresponding $\mathbf{Z}_N$ symmetry. However, in models of intersecting magnetized D7-branes, quarks and leptons arise from fields at D7-brane intersections; the effect of fluxes on such fields is difficult to obtain using the above techniques, since the DBI + CS action does not describe the "twisted" open strings corresponding to intersecting branes. This case will be addressed in Section 15.5 using a more general and model independent approach, using the supergravity 4d effective action.

15.4.2 Macroscopic description in the 4d effective action

The presence of ISD and IASD fluxes has an interesting interpretation in the 4d $\mathcal{N}=1$ supergravity effective theory. As suggested by the analysis of SUSY conditions in Section 14.1.2, 3-form fluxes correspond to non-vanishing vevs for the auxiliary fields of the dilaton S, Kähler moduli T_i, and complex structure moduli.

For simplicity we focus on the dilaton S and overall Kähler modulus T. Recall the supergravity auxiliary field for a chiral field (2.51), i.e.

$$\overline{F}^{\bar{i}} = -e^{K/2} K^{\bar{i}j} D_j W. \tag{15.33}$$

Using the Kähler potential (14.22), and the derivatives (14.25), (14.30), we have

$$\overline{F}^S = (S + S^*)^{1/2}(T + T^*)^{-3/2} \int_{\mathbf{X}_6} \overline{G}_{(3)} \wedge \Omega,$$

$$\overline{F}^T = (S + S^*)^{-1/2}(T + T^*)^{-1/2} \int_{\mathbf{X}_6} G_{(3)} \wedge \Omega. \tag{15.34}$$

Thus $(3, 0)$ and $(0, 3)$ flux components correspond to vevs for F^S and F^T, respectively. In the next section we show that these vevs act as spurions for SUSY breaking, and reproduce the earlier D3- and D7-brane soft term patterns; moreover this formalism allows the computation of soft terms in more involved D7-brane configurations, including intersections and magnetization.

Before entering details, it is illustrative to discuss the scale of soft terms; this is fixed by the flux density G_3, which scales roughly as $G_3 \simeq f\alpha'/R^3$, with f related to the flux quanta. The soft term scale on D3- or $\overline{\text{D3}}$-branes is

$$m_{\text{soft}} = \frac{g_s^{1/2}}{\sqrt{2}} G_3 = \frac{f g_s^{1/2}}{\sqrt{2}} \frac{\alpha'}{R^3} = \frac{f M_s^2}{M_p}, \tag{15.35}$$

where we have used $M_p \simeq (\alpha')^{-2} R^3$. Comparing with (2.92), this is consistent with $\langle F \rangle \simeq M_s^2$, i.e. SUSY breaking of order the string scale. In particular, for an intermediate string scale $M_s \simeq 10^{10}$ GeV, soft terms fall in the ballpark of 10^2–10^3 GeV. Realization of this scale pattern, e.g. in the large volume scenario of Section 15.3.2, relates the electroweak hierarchy to the large overall compactification volume. Note, however, that flux SUSY breaking does not necessarily imply the intermediate string scale, since in general fluxes may be inhomogeneously distributed, and the smallness of soft terms could rely on local suppression factors, like warping. Alternatively, this smallness could come from the cancellation of many contributions to a flux-induced constant superpotential W_0, as in Sections 15.3.1 and 17.3.

15.5 General parametrization of moduli/dilaton induced SUSY breaking

Gravity mediation is one of the most elegant schemes for low-energy SUSY breaking in a way consistent with phenomenological constraints. String compactifications are particularly well-suited for that proposal, as they generically have the dilaton, Kähler and complex structure moduli S, T_i, U_n, with Planck scale suppressed couplings to ordinary (SM) matter. It is natural to assume that some of these fields may play an important role in mediating SUSY breaking to the SM. This is in fact explicitly realized in type IIB compactifications with SUSY breaking by 3-form fluxes, as discussed in the previous section.

The consequences of general SUSY breaking in the moduli sector can be explored in a fairly model independent formalism. This leads to the prediction of SUSY breaking soft term patterns in large classes of models, characterized by the form of their 4d $\mathcal{N} = 1$ supergravity action, i.e. the gauge kinetic functions and Kähler metrics for matter fields. The basic assumption in this approach is that the bulk of SUSY breaking can be encoded in vevs, of unspecified origin, for auxiliary fields F_S, F_{T^i} for the dilaton and/or the Kähler moduli. The strategy is analogous to the successful parametrization of EW symmetry breaking in particle physics in terms of a vev (of somewhat unspecified origin) for the Higgs field.

We start with the description of the general formalism, common to all compactifications, and subsequently spell out the results for the different setups, like heterotic compactifications, D3-brane models, or intersecting D7-brane models.

15.5.1 General formalism

We consider compactifications with a MSSM sector, with general gauge kinetic functions f_a, and (diagonal) Kähler metrics K_α for the chiral matter fields C^α, i.e.

$$
\begin{aligned}
f_a &= f_a(S, T_i, U_n), \\
K &= K(S, S^*; T_i, T_i^*; U_n, U_n^*) + K_\alpha(S, S^*; T_i, T_i^*; U_n, U_n^*)|C^\alpha|^2 \\
&\quad + \left[\frac{1}{2} Z_{\alpha\beta}(S, S^*; T_i, T_i^*; U_n, U_n^*) C^\alpha C^\beta + \text{h.c.} \right].
\end{aligned} \tag{15.36}
$$

(We maintain the convention $\kappa_4^2 = 1$ in this section.) Diagonal Kähler metrics to leading order are quite generic in string compactifications, hence this restriction is well-motivated. Note that the term in Z, if present for the MSSM Higgs doublet bilinear $H_u H_d$, may give rise to induced μ- and B-terms, as mentioned on page 56.

In addition there is a total superpotential W, including the standard Yukawa couplings, as well as possibly constant and moduli dependent pieces,

$$
W = W_{\text{mod.}} + \frac{1}{2} \mu_{\alpha\beta} C^\alpha C^\beta + \frac{1}{6} Y_{\alpha\beta\gamma} C^\alpha C^\beta C^\gamma + \cdots, \tag{15.37}
$$

where $W_{\text{mod.}}$, μ and $Y_{\alpha\beta\gamma}$ in general depend on the moduli. We in principle allow for explicit μ-term type of couplings, e.g. induced by fluxes, instantons, or otherwise.

In the following we focus on heterotic and type IIB orientifold compactifications, in the large volume regime, and on the dilaton S and overall Kähler modulus T, which are universally present; a different variant is also considered at the end of Section 15.5.3. Type IIB results can be translated to type IIA language via mirror symmetry. In both heterotic and type II models, the moduli Kähler potential is

$$
K = -\log(S + S^*) - 3\log(T + T^*). \tag{15.38}
$$

Under our assumption that SUSY breaking is dominated by F^S and F^T vevs, the 4d $\mathcal{N} = 1$ supergravity scalar potential leads to the vacuum energy

$$V_0 = K_{S\bar{S}}|F^S|^2 + K_{T\bar{T}}|F^T|^2 - 3e^K, \tag{15.39}$$

where $\overline{F}^S = e^{K/2} K_{S\bar{S}}^{-1} D_S W$ and $\overline{F}^T = e^{K/2} K_{T\bar{T}}^{-1} D_T W$. It is convenient to parametrize

$$F^S = \sqrt{3} m_{3/2} (K_{S,\bar{S}})^{-1/2} \sin\theta e^{-i\gamma_S},$$
$$F^T = \sqrt{3} m_{3/2} (K_{T,\bar{T}})^{-1/2} \cos\theta e^{-i\gamma_T}, \tag{15.40}$$

where we defined the gravitino mass $m_{3/2} = \exp(G/2)$, with $G = K + \log|W|^2$; this parametrization is automatically consistent with the observed approximate value $V_0 = 0$. The *goldstino angle* θ measures the relative magnitudes of S- and T-induced SUSY breaking. There are arbitrary phases γ_S, γ_T, potentially relevant for CP violation in the theory, but which we ignore in the following for simplicity.

Regarding the auxiliary field vevs as spurions for SUSY breaking, the 4d $\mathcal{N} = 1$ supergravity lagrangian produces the following soft terms

$$\mathcal{L}_{\text{soft}} = \frac{1}{2}(M_a \widehat{\lambda}^a \widehat{\lambda}^a + \text{h.c.}) - m_\alpha^2 \widehat{C}^{*\bar{\alpha}} \widehat{C}^\alpha$$
$$- \left(\frac{1}{6} A_{\alpha\beta\gamma} \widehat{Y}_{\alpha\beta\gamma} \widehat{C}^\alpha \widehat{C}^\beta \widehat{C}^\gamma + B\widehat{\mu} \widehat{H}_u \widehat{H}_d + \text{h.c.} \right), \tag{15.41}$$

with $M_a, m_\alpha, A_{\alpha\beta\gamma}$ and B defined by (2.104)–(2.107), and $\widehat{C}_\alpha, \widehat{Y}_{\alpha\beta\gamma}, \widehat{\mu}$ defined by (2.108), (2.109), (2.110). From now on we omit hats, and turn to the application of these general formulae to the different classes of string embeddings of the SM.

15.5.2 Heterotic compactifications

Recall from Chapter 9 the leading order gauge kinetic functions and charged matter Kähler metrics

$$f_a = k_a S; \quad K^\alpha = (T + T^*)^{n_\alpha}, \tag{15.42}$$

where k_a are the gauge Kac–Moody levels, and n_α are the modular weights. In large volume smooth CY compactifications $n_\alpha = -1$, while for toroidal orbifolds other (integer) values are possible. The soft terms computed from the above formulae are

$$M_a = \sqrt{3} m_{3/2} \sin\theta,$$
$$m_\alpha^2 = m_{3/2}^2 (1 + n_\alpha \cos^2\theta),$$
$$A_{\alpha\beta\gamma} = -\sqrt{3} m_{3/2} (\sin\theta + \cos\theta \omega_{\alpha\beta\gamma}), \tag{15.43}$$

where

$$\omega_{\alpha\beta\gamma} = \frac{1}{\sqrt{3}} \left[3 + n_\alpha + n_\beta + n_\gamma - (T + T^*) \frac{\partial_T Y_{\alpha\beta\gamma}}{Y_{\alpha\beta\gamma}} \right]. \tag{15.44}$$

Incidentally, the last term in ω vanishes in cases with constant Yukawa couplings, e.g. for untwisted sector fields or $\mathbf{Z}_2$ twisted sector fields in toroidal orbifolds; it is also negligible for Yukawas among fields at the same fixed point, as they are approximately constant in the large T limit.

Concerning the B-term for the Higgs fields H_u, H_d, one obtains the results

(a) $\quad B_\mu = -m_{3/2}\left[1 + \sqrt{3}\sin\theta + \cos\theta(3 + n_{H_u} + n_{H_d})\right],$ (15.45)

(b) $\quad B_Z = 2m_{3/2}(1 + \cos\theta),$ (15.46)

with case (a) when there is an explicit μ-term in the superpotential, and case (b) if it is induced by the presence of a non-vanishing Z-term in the Higgs field Kähler metrics. The latter possibility arises, e.g., in certain toroidal orbifolds of even order, as explained in Section 9.3.1, page 275. An actual realization of this mechanism would require the construction of a realistic model where the fields C_3^T, C_3^U in (9.48) play the role of Higgs fields. No such models have been constructed hitherto, so the idea remains a tantalizing suggestion rather than a working mechanism.

From (15.43) one observes that a general feature of modulus mediated SUSY breaking in heterotic models is that scalar masses are always smaller than gaugino masses. This pattern survives upon renormalization group running down to the weak scale, as described later in Section 15.6.1. An interesting feature is that scalars may actually become tachyonic for $\cos^2\theta > 1/|n_\alpha|$, thus constraining the value of the goldstino angle in terms of the modular weights present in the model. We now turn to further soft term patterns in two interesting limits.

Dilaton domination

For $\sin\theta = 1$ the auxiliary field of the dilaton F^S dominates SUSY breaking, and leads to the remarkably simple and universal soft terms

$$m_\alpha = m_{3/2}, \quad M_a = \pm\sqrt{3}m_{3/2}, \quad A_{\alpha\beta\gamma} = -M_a,$$ (15.47)

where the gaugino mass sign relates to the choice of phase γ_S. The model independence of soft terms in dilaton domination follows from the universal couplings of the dilaton in all heterotic compactifications; it is worth emphasizing that this implies universality of soft scalar masses, a good starting point to avoid FCNC in the low-energy theory. In practice the main challenge is to actually realize heterotic compactifications in which the dilaton dominates SUSY breaking.

T-modulus domination

For $\sin\theta = 0$, SUSY breaking is dominated by the auxiliary field of the overall Kähler modulus T. From the above expressions, tachyons are avoided only for $n_\alpha = -1$. In that case we also have $\omega_{\alpha\beta\gamma} = 0$, and the soft terms are

$$m_\alpha = M_a = A_{\alpha\beta\gamma} = 0,$$ (15.48)

namely, a no-scale structure with vanishing soft terms. Concerning the B-terms, $\cos\theta = \pm 1$ produces $B_\mu = -m_{3/2}(1 \pm 1)$ and $B_Z = 2m_{3/2}(1 \pm 1)$. So, for $\cos\theta = -1$, we also have $B_\mu = B_Z = 0$, while $\cos\theta = 1$ yields $B_\mu = -2m_{3/2}$ and $B_Z = 4m_{3/2}$.

Given the vanishing of the above soft terms at classical level, one-loop corrections to the gauge kinetic functions and the Kähler potential become relevant. Those corrections are rather model dependent, so it is difficult to draw general lessons. Still some partial results can be obtained in abelian toroidal orbifolds, in which such corrections are computable, recall (9.102), (9.106). The one-loop gauge kinetic function (9.106) can indeed be used to obtain one-loop corrected gaugino mass, with the result

$$M_a = \frac{k_a Y}{\mathrm{Re}\, f_a} \sqrt{3} m_{3/2} \left[\sin\theta + \frac{B_a'(T+T^*)\hat{G}_2(T, T^*)}{16\sqrt{3}\pi^3 k_a Y} \left(1 - \frac{\delta_{GS}}{24\pi^2 Y}\right)^{-1/2} \cos\theta \right],$$

with

$$Y = S + S^* - \frac{\delta_{GS}}{8\pi^2} \log(T + T^*), \qquad B_a' = b_a - \delta_{GS}, \qquad (15.49)$$

where δ_{GS} is the constant introduced in Section 9.6.1. Including the one-loop S-T mixing term in the Kähler potential, the scalar masses read

$$m_\alpha^2 = m_{3/2}^2 \left[1 - \left(1 - \frac{\delta_{GS}}{24\pi^2 Y}\right)^{-1} \cos^2\theta \right]. \qquad (15.50)$$

In the $\sin\theta \to 0$ limit we have

$$m_\alpha^2(\sin\theta \to 0) \simeq m_{3/2} \left(\frac{-\delta_{GS}}{24\pi^2 Y}\right) \simeq -m_{3/2}^2 \delta_{GS} \times 10^{-3}. \qquad (15.51)$$

In all known orbifold examples δ_{GS} is a negative (integer) number, and so tachyons are avoided; in such examples, and for not too large $\mathrm{Re}\, T$, scalar masses turn out to be larger than gaugino masses, in contrast with the dilaton domination picture. Another novel feature of T-modulus dominance is that soft terms are suppressed compared to the gravitino mass, a potentially interesting mechanism to overcome the cosmological gravitino problem mentioned in Section 16.6.3. Despite these interesting properties, the main drawbacks of T-modulus domination are its model dependence, and the lack of analytic control over one-loop corrections and other possible competing contributions, like anomaly mediation contributions.

15.5.3 Type IIB orientifolds

Let us now apply the general formalism of Section 15.5.1 to type IIB orientifold compactifications. We focus on configurations where the MSSM sector is localized on D3- and/or D7-branes, since the soft term structure in D9/D5-brane systems is similar in the simplest examples. Also, we do not consider type IIA orientifolds, whose results are related by mirror symmetry.[1]

Soft terms on D3-branes

Recalling (12.56), (12.41), D3-branes on a compact CY with overall Kähler modulus T have gauge kinetic functions $f_a = S$ (absorbing a 2π factor in the definition of the moduli from now on) and a chiral multiplet Kähler metric $K_{33} = 1/(T + T^*)$. These are identical to the heterotic case with modular weight $n_\alpha = -1$; since the Kähler potential is also given by (15.38), the computation of soft terms is identical to the heterotic case, and yields

$$M_a = \sqrt{3}m_{3/2}\sin\theta,$$
$$m_\alpha^2 = m_{3/2}^2 \sin^2\theta,$$
$$A_{\alpha\beta\gamma} = -\sqrt{3}m_{3/2}. \sin\theta. \qquad (15.52)$$

The result is slightly simpler because all D3-D3 fields have "modular weight" -1, and their Yukawa couplings are moduli independent, and so $\omega_{\alpha\beta\gamma} = 0$. The above soft terms apply also to models of D3-branes at singularities, studied in Section 11.3, and therefore to the semi-realistic particle physics models therein. As in the heterotic case, there are two interesting limits:

- For $\sin\theta = 1$ there is dilaton dominance producing the soft terms (15.47). These can be checked to agree with the soft terms on D3-branes from IASD $(3, 0)$ 3-form fluxes in Section 15.4.1; this dovetails the identification of such flux component with a non-vanishing vev for F_S, recall Section 15.4.2. Since pure IASD fluxes do not in principle provide minima of the flux potential, it is difficult to achieve pure dilaton dominated soft terms in type IIB models with D3-branes.
- For $\sin\theta = 0$ we have modulus domination, and we recover a no-scale result (15.48), with vanishing soft terms to leading order. Again, one-loop and other sub-leading effects may now become relevant, e.g. the α'-correction (12.17), which spoils the no-scale structure and thus produces non-vanishing contributions to soft terms. Other corrections may appear from the contributions of twisted moduli to gauge kinetic functions of D3-branes at singularities, recall Section 12.3.2. In addition, one-loop anomaly mediation terms may also be non-negligible. As in the heterotic case, soft terms in T-modulus domination are in general very model dependent.

Soft terms on D7-branes

Consider now soft terms for open string fields on stacks of coincident D7-branes, with vanishing worldvolume fluxes, wrapped on a 4-cycle which for simplicity we take to be $\mathbf{T}^4$. The gauge kinetic function is $f_a = T$ and the Kähler metrics for Wilson line moduli

---

[1] Note that mirror symmetry maps, e.g., type IIB T-modulus dominance into type IIA U-modulus dominance. However, this is the interesting case from the type IIA perspective as well, since its gauge kinetic functions do not depend on Kähler moduli, so T-modulus dominance produces vanishing gaugino masses.

C_{WL} and D7-brane position moduli $C_{\text{Pos.}}$ (denoted C_j^{7i} and C_i^{7i} in Section 12.2) may be obtained from (12.38),

$$K_{\text{WL}} = \frac{1}{T + T^*}; \quad K_{\text{Pos.}} = \frac{1}{S + S^*}. \tag{15.53}$$

Recalling (12.74), there is also a Yukawa coupling $C_{\text{WL}}C_{\text{WL}}C_{\text{Pos.}}$. We consider the most interesting case of T-modulus dominance, since it occurs for SUSY breaking $(0, 3)$ 3-form fluxes, which are ISD and thus provide minima of the flux superpotential. The soft terms obtained from (2.104), (2.105), (2.106), and (2.107) read

$$M = m_{3/2}, \quad m_{\text{WL}}^2 = 0, \quad m_{\text{Pos.}}^2 = M^2, \quad A = -M, \quad B = -2M, \tag{15.54}$$

where the B-term value assumes an explicit μ-term. In contrast with the D3-brane case, T-modulus dominance produces non-vanishing soft terms for most fields. As expected, the above soft terms agree with the microscopic computation in Section 15.4.1 of soft terms induced by ISD $(0, 3)$ fluxes (15.32) (modulo a sign convention for B).

Wilson line moduli remain massless (due to their underlying nature of 8d gauge boson components), but they are typically absent for more generic choices of the wrapped 4-cycle. If present, the $C_{\text{WL}}C_{\text{WL}}C_{\text{Pos.}}$ Yukawa couplings motivates the model building possibility of letting $C_{\text{Pos.}}$ and C_{WL} play the role of Higgs fields and quark/lepton multiplets respectively; the resulting soft term pattern of massless squarks and sleptons and massive gauginos and Higgs multiplets is sometimes called *gaugino domination*.

Consider now a stack of coincident D7-branes, with non-trivial but identical worldvolume magnetic flux F. This modifies the gauge kinetic function and the Kähler metric for the D7-brane position modulus. Using (12.55) and (12.36) to get an estimate, we find

$$\text{Re } f_a = \text{Re } T(1 + |F|^2), \quad K_{\text{Pos.}} = \frac{1}{S + S^*}(1 + |F|^2), \tag{15.55}$$

where one has $F \simeq n(\text{Re } S/\text{Re } T)^{1/2}$, with n proportional to the flux quanta; the Kähler metrics for Wilson line moduli C_{WL} are not corrected by worldvolume fluxes. For large volume (diluted fluxes) these corrections are suppressed, and the soft terms for T-modulus dominance read

$$m_{\text{WL}}^2 = 0, \quad m_{\text{Pos.}}^2 = M^2(1 - 2\rho), \quad A = -M(1 - \rho), \tag{15.56}$$

where $\rho \propto 1/(T + T^*)$. Hence, the presence of worldvolume fluxes does not modify the general qualitative structure of the soft terms.

Soft terms for matter at D7-brane intersections

Some of the most interesting MSSM-like string vacua are based on configurations of intersecting (and magnetized) D7-branes, e.g. the example in Sections 11.4.2 and 12.6, and the SM matter fields are localized at D7-brane intersections; the S, T dependent part of their Kähler metric, in the absence of magnetic fluxes, is (12.40):

$$K_{C^{7_i 7_j}} = \frac{1}{(S + S^*)^{1/2}(T + T^*)^{1/2}}. \tag{15.57}$$

Table 15.2 *Structure of soft terms for different kinds of fields in intersecting D7-brane configurations*

Coupling	M	m_L^2	m_R^2	m_H^2	A	B
$(C_{WL}\text{-}C_{WL}\text{-}C_{Pos.})$	M	0	0	$\|M\|^2$	$-M$	$-2M$
$(C_{7_i7_j}\text{-}C_{7_j7_i}\text{-}C_{WL})$	M	$\frac{\|M\|^2}{2}$	$\frac{\|M\|^2}{2}$	0	$-M$	0
$(C_{7_i7_j}\text{-}C_{7_j7_k}\text{-}C_{7_k7_i})$	M	$\frac{\|M\|^2}{2}$	$\frac{\|M\|^2}{2}$	$\frac{\|M\|^2}{2}$	$-3/2M$	$-M$

There are several possible Yukawa coupling structures, as summarized in Section 12.5.2. For the Yukawa coupling $C_{7_i7_j} C_{7_j7_k} C_{7_k7_i}$ involving three intersecting D7-branes, modulus dominance yields

$$M = m_{3/2}, \quad m^2_{C^{7_i7_j}} = \frac{|M|^2}{2}, \quad A = -\frac{3}{2}M, \quad B = -M, \tag{15.58}$$

where again an explicit μ-term is assumed. The most relevant point is that all sparticles get masses. For the Yukawa coupling $C_{7_i7_j} C_{7_j7_i} C_{WL}$, the soft terms are

$$M = m_{3/2}, \quad m^2_{C^{7_i7_j}} = \frac{|M|^2}{2}, \quad m^2_{WL} = 0, \quad A = -M, \quad B = 0. \tag{15.59}$$

A summary of the three types of modulus dominance soft terms for different D7-brane configurations is given in Table 15.2. For illustration, consider the toroidal $\mathbf{Z}_2 \times \mathbf{Z}_2$ MSSM-like model in Sections 11.4.2 and 12.6. All SM matter fields localize at D7-brane intersections, and only the $SU(3+1)$ D7-branes have worldvolume magnetic flux. In the dilute flux limit the soft terms from modulus dominance are given by the third line in Table 15.2.

Soft terms on local D7-brane configurations

In the above discussion, SUSY breaking on D7-branes is characterized in terms of a single overall Kähler modulus T, in the large volume regime. This may not be an appropriate description for D7-branes wrapping 4-cycles whose volume is independent from (and may be parametrically smaller than) the overall CY size; for instance, in the large volume scenario of Section 15.3.2, or the F-theory models in Section 11.5.

These situations are better modeled in terms of two Kähler moduli, T_s, T_b, controlling the wrapped 4-cycle and overall CY sizes, respectively, as in the simplest setting in the large volume scenario. The soft terms on the D7-brane may be computed from the above general formalism, and turn out to be mainly determined by the auxiliary field of T_s, as we now show. We consider a general topology for the wrapped 4-cycle, since the present models are realized in non-toroidal setups. Although the matter field Kähler metrics are not known, their dependence on the moduli can be determined by scaling arguments in the regime $\tau_b \gg \tau_s \gg 1$ (where recall that $\tau_b = \mathrm{Re}\, T_b$, $\tau_s = \mathrm{Re}\, T_s$). Specifically, one proposes the Kähler metric for matter fields at intersections to follow the ansatz

$$K_\alpha = \frac{\tau_s^{(1-\xi_\alpha)}}{\tau_b}, \tag{15.60}$$

where ξ_α plays the role of modular weights, whose leading behavior we compute in the following. Recall that the physical Yukawa couplings $\widehat{Y}_{\alpha\beta\gamma}$ are related to the holomorphic $Y^{(0)}_{\alpha\beta\gamma}$ by (12.73), i.e.

$$\widehat{Y}_{\alpha\beta\gamma} = e^{K/2} \frac{Y^{(0)}_{\alpha\beta\gamma}}{(K_\alpha K_\beta K_\gamma)^{1/2}}. \tag{15.61}$$

Also, from (12.72), in type IIB models $Y^{(0)}$ is independent of the Kähler moduli. Then using (15.20), the physical Yukawa coupling scales as

$$\widehat{Y}_{\alpha\beta\gamma} \simeq \tau_s^{(\xi_\alpha+\xi_\beta+\xi_\gamma-3)/2}. \tag{15.62}$$

The dependence on the large modulus τ_b drops at leading order in τ_s/τ_b, as expected for a model whose physics is essentially localized on the 4-cycle. This justifies *a posteriori* the τ_b dependence of the ansatz (15.60). The τ_s dependence can be analyzed by recalling that Yukawa coupling arises from local zero mode overlap integrals

$$\widehat{Y}_{\alpha\beta\gamma} \simeq \int \Psi_\alpha \Psi_\beta \Psi_\gamma, \tag{15.63}$$

with zero mode wave functions normalized as

$$\int |\Psi_\alpha|^2 = \int |\Psi_\beta|^2 = \int |\Psi_\gamma|^2 = 1. \tag{15.64}$$

For the case of three fields localized on 2d subspaces at D7-brane intersections, this normalization implies a scaling $\Psi \simeq \tau_s^{-1/4}$ for the zero modes, and hence $\widehat{Y} \simeq \tau^{-3/4}$ for the physical Yukawa. Comparing with (15.62) yields $\xi_\alpha = 1/2$ for the three fields involved. Hence for fields at D7-brane intersections

$$K_\alpha = \frac{\tau_s^{1/2}}{\tau_b}. \tag{15.65}$$

Incidentally, note that setting $\tau_b \simeq \tau_s$ reproduces the metrics of fields at intersections in toroidal models; the result (15.65) is, however, general in the limit $\tau_s \ll \tau_b$.

Using the gauge kinetic function $f_a = T_s$ and the Kähler metric (15.65), we can compute the soft terms induced by F_{T_b} and F_{T_s}. The result is actually independent of F_{T_b}, as expected from locality, and is again given by Table 15.2, now with $M = F_{T_s}/(T_s + T_s^*)$; the modulus dominance regime thus corresponds to T_s-modulus dominance. Note that the above derivations and results rely only on the geometric properties of 7-branes and their intersections, and hence apply also to the local F-theory models in Section 11.5, in which 7-branes wrap local del Pezzo surfaces.

15.6 Modulus/dilaton dominated SUSY breaking spectra and the LHC

The assumption of dilaton and/or modulus domination in SUSY breaking may be quite predictive in specific cases. For instance, in dilaton domination in heterotic and type IIB D3-brane models, or in T-modulus domination in type IIB intersecting D7-brane models. In these cases, essentially all soft terms are determined by two parameters, a universal gaugino mass M and the Higgs μ-term. In other cases, like T-modulus domination in heterotic or type IIB D3-brane models, soft terms vanish at leading order and sub-leading contributions, like one-loop corrections, become relevant. Since the latter are very model dependent, for illustrative purposes we focus on cases with non-vanishing soft terms at leading order.

15.6.1 Computation of sparticle spectrum

The soft terms computed in previous sections should be taken as boundary conditions at a large string/GUT scale, for subsequent renormalization group (RG) running to low energies, to be experimentally tested in the coming years.

There are a number of constraints that the resulting low-energy parameters must satisfy. For instance, the RG evolution of Higgs doublet masses should trigger electroweak (EW) symmetry breaking at the right scale, as discussed in Section 2.6.3. This extra condition reduces the two free parameters to just one. The condition (2.83) of correct $SU(2)_L \times U(1)_Y$ breaking may be rewritten as

$$\mu^2 = \frac{-m_{H_u}^2 \tan^2 \beta + m_{H_d}^2}{\tan^2 \beta - 1} - \frac{1}{2} M_Z^2, \qquad (15.66)$$

with $\sin 2\beta = 2|B\mu|/(\mu_u^2 + \mu_d^2)$, $\tan \beta = v_u/v_d$, all taken at the EW scale. It is convenient to trade μ for $\tan \beta$ and use M, $\tan \beta$ (and the sign of μ) as the free parameters before imposing EW symmetry breaking. This low-energy condition yields complicated relationships among the input soft parameters. In practice the renormalization group equations are solved with standard numerical codes, to run soft terms from the string to the weak scale, and to compute the low-energy SUSY spectrum and the Higgs potential.

The resulting low-energy spectrum must in addition pass a number of experimental constraints, including: present mass bounds for SUSY particles and the lightest Higgs boson; constraints from rare processes like $b \to s\gamma$ and $B_s^0 \to \mu^+\mu^-$; limits from measurements of the muon anomalous magnetic moment a_μ. If in addition one requires neutralinos to provide the correct cosmological dark matter density, there are extra constraints from, e.g., the WMAP data. In the situation with a single independent parameter, this long list of experimental constraints leads to a predictive scenario, where an eventual measurement of a single sparticle mass in principle determines the whole SUSY spectrum.

As an illustration, consider the case of modulus dominance in models with quarks and leptons on type IIB intersecting D7-branes. The two Yukawa coupling structures, leading to soft terms (15.59), (15.58), may be analyzed jointly by formally taking the Higgs modular weight ξ_H as a free parameter, with $\xi_H = 1, 1/2$ reproducing the two limits. Then the soft terms for the different MSSM fields are

$$m_{L,E,Q,U,D}^2 = |M|^2/2, \quad m_{H_u,H_d}^2 = (1 - \xi_H)|M|^2,$$
$$A_{U,D,L} = -M(2 - \xi_H), \quad B = -2M(1 - \xi_H). \tag{15.67}$$

These high-scale boundary conditions should be run down to low energies. For illustration, we take the high scale to be the gauge coupling unification scale $M_{GUT} \simeq 2 \times 10^{16}\,$GeV, and present a sample of the resulting SUSY spectra, for $M = 400\,$GeV and $\mu < 0$; it is shown in Figure 15.6 as a function of $\tan \beta$, and for both cases with $\xi_H = 1/2, 1$. In this figure g stands for the gluino, b is the lightest b-squark, and t the lightest top squark. The squarks of the first two families have masses similar to that of the b squark for small $\tan \beta$, and charged sleptons have masses of the order of the sneutrino ν. Also $\chi^\pm$, χ^0 are the charginos and the lightest neutralino, and h^0, A^0 are the lightest scalar and pseudoscalar Higgses respectively. Both figures yield quite similar results, approximately compatible with the ratios

$$M_{\tilde{g}} : m_{\tilde{q}} : m_{\tilde{l}} : M_{\chi^\pm} : M_{\chi^0} = 1 : 0.84 : 0.34 : 0.4 : 0.18, \tag{15.68}$$

for the first two generations of squarks and sleptons. Note that, for not too large $\tan \beta$, the lightest neutralino is the lightest supersymmetric particle (LSP); for large $\tan \beta$ the lightest stau becomes lighter, because its Yukawa is proportional to $1/\cos \beta$, and the running to low energies decreases the squared masses for large Yukawas. Preventing it from becoming tachyonic implies the bounds $\tan \beta \leq 45\,$GeV for $\xi_H = 1/2$, and $\tan \beta \leq 55$ for $\xi_H = 1$.

It is also interesting to see the present experimental and theoretical constraints on the parameters; these are shown in the $(M, \tan \beta)$-plane in Figure 15.7, for $\xi_H = 0.6$ and $\xi_H = 1$. Dark grey regions are excluded by experimental bounds, as follows: the area below the thin dashed line is ruled out by the lower bound on the lightest Higgs mass; the region below the thin dotted line is excluded by the lower bounds on the stau and chargino masses; the area below the thick dashed line is excluded by $b \to s\gamma$; the region below the double-dot/dashed line is excluded by $B_s^0 \to \mu^+\mu^-$; finally, the thin dot-dashed lines correspond to the lower and upper constraint on the muon anomalous magnetic moment a_μ. The area to the right of the solid line has the stau as the LSP, and is depicted in light grey when experimental constraints are fulfilled. In the remaining white area the neutralino is the LSP.

The thick dot-dashed lines correspond to parameters consistent with the boundary conditions for the B-parameter, which are theoretically more uncertain, since they rely on the assumption of having an explicit μ-term; this more model dependent constraint thus restricts M and $\tan \beta$ to those lines. For those models, the neutralino is mostly bino, and if stable produces a thermal relic density easily exceeding the WMAP constraints on dark matter. This is known to happen for most of the parameter space of the CMSSM, as already remarked in Section 2.6.1. In our case the correct neutralino abundance is reproduced only for neutralino masses close to the stau mass, to enhance coannihilation and so

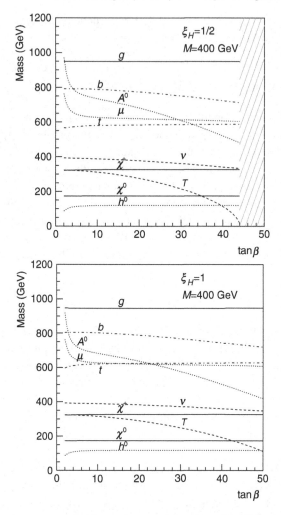

Figure 15.6 Low-energy supersymmetric spectrum as a function of $\tan \beta$ for $\xi_H = 1/2$, (left) and $\xi_H = 1$ (right) with $M = 400\,\text{GeV}$ and $\mu < 0$. The dashed area for large $\tan \beta$ is excluded by the occurrence of tachyons in the slepton sector.

deplete the neutralino density. The surviving regions correspond to very narrow bands in the vicinity of the area with stau LSP, favoring a very narrow range of (typically large) values for $\tan \beta$. Using the boundary condition for the B-term one may obtain appropriate relic neutralino abundance for $\xi_H \simeq 0.6$, in which case the thick dot-dashed curve lies in the vicinity of the coannihilation region (thus justifying the choice of showing the $\xi_H = 0.6$ plot in Figure 15.7, rather than $\xi_H = 1/2$). This strongly resembles the result for configurations with all particles at 7-brane intersections, i.e. $\xi_H = 0.5$, with small deviations possibly due to sub-leading corrections from, e.g., worldvolume fluxes. Alternatively, the whole white area would be consistent with standard cosmology if there is R-parity violation and the LSP is unstable, in which case dark matter should come from another source.

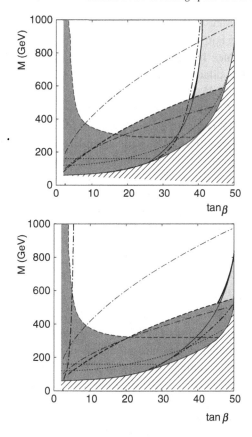

Figure 15.7 Effect of various experimental constraints on the $(M, \tan\beta)$-plane for Higgs modular weights $\xi_H = 0.6$ and $\xi_H = 1$.

In the R-parity preserving case, observation (or bounds from the non-observation) of missing energy signals from squarks and gluinos at the LHC will allow to test this intersecting 7-brane scheme for certain ranges of M. For instance, a luminosity of 1 fb^{-1} at 14 TeV allows to explore the region $M \lesssim 650$ GeV, enough to start probing regions of consistent radiative EW symmetry breaking fulfilling all experimental constraints and with viable neutralino dark matter; in these areas, gluino and squark masses are $m_{\tilde{g}} \lesssim 1.5$ TeV and $m_{\tilde{q}} \lesssim 1.3$ TeV, while neutralino masses are $m_{\tilde{\chi}^0} \lesssim 300$ GeV. With a luminosity of 10 fb^{-1} at 14 TeV, the LHC would be able to explore the whole region of the parameter space with $M \lesssim 900$ GeV; this corresponds to gluinos and squarks with $m_{\tilde{g}} \lesssim 2$ TeV and $m_{\tilde{q}} \lesssim 1.8$ TeV, and neutralinos with $m_{\tilde{\chi}^0} \lesssim 400$ GeV.

A similar analysis may be performed for the dilaton domination boundary conditions (15.47). The running from a high scale $\simeq 10^{16}$ GeV down to the weak scale yields a structure for the spectrum with

$$M_{\tilde{g}} : m_{\tilde{Q}} : m_{\tilde{L}} = 1 : 0.93 : 0.32, \tag{15.69}$$

which is qualitatively similar to (but still distinct from) that of modulus domination with intersecting D7-branes. In both cases gluinos are slightly heavier than squarks, and sleptons are a factor $\simeq 3$ times lighter. In the case with a B_Z term, radiative electroweak symmetry breaking does not work in dilaton domination, but it can be implemented in the alternative case of a B_μ term, as in (15.45).

These two examples of modulus dominance in intersecting 7-branes and dilaton domination, show how, if SUSY is found, we might be able to obtain important information on the underlying string compactification. General natural assumptions about the origin of the SM spectrum and the origin of SUSY breaking in string theory lead to testable predictions. A final point to remark is that, in both scenarios, the minima with correct electroweak symmetry breaking are not absolute minima; rather there are other lower minima, typically breaking charge and color symmetries. This is however not problematic, since the decay lifetime of the meta-stable vacua has been shown to be longer than the age of the universe.

15.6.2 Soft terms, FCNC, and complex phases

There are other possible signatures of SUSY breaking soft terms in low-energy physics, like the already mentioned $b \to s\gamma$, $B_s^0 \to \mu^+\mu^-$ and a_μ. In general SUSY models, there are also potentially more dangerous FCNC contributions to kaon physics, if soft terms (scalar masses and A-parameters) are not sufficiently universal, i.e. flavour independent. As we have seen, soft-terms coming from dilaton or T-modulus dominance are, at leading order in α' and loop effects, universal. This does not contradict the existence of rich flavour patterns in the Yukawa sector, since SUSY breaking and Yukawas are controlled by different moduli sectors; for instance, in type IIB models the holomorphic Yukawas depend on complex structure moduli, which play no role in the SUSY breaking dynamics in the scenario of Kähler modulus domination. So there is a built-in decoupling between the Yukawa structure and modulus dominated SUSY breaking.

A possible exception to soft term universality may occur in heterotic models with different modular weights for the different families, which would result in different soft masses; however in all constructions to date the SM fields have all the same overall modular weight, either $n = -1$ or $n = -2$.

As noted in Section 2.6.3, another generic problem of low-energy SUSY models is the possible generation of large contributions to CP violating processes, like the electric dipole moment of the neutron. Indeed, if soft terms like gaugino masses, A and B parameters, and μ, are generated with arbitrary phases, their contributions to the neutron electric dipole moment are generically too large, contradicting experimental bounds. In the dilaton and/or modulus SUSY breaking scenario, such phases arise from the phases γ_S, γ_{T_i} of auxiliary fields (15.40), which would thus be severely constrained. There is however a natural mechanism to overcome these potential problems; if SUSY breaking is due to a single auxiliary field vev, e.g. F_T, its phase is unphysical and can be rotated away from all soft terms. Thus absence of large CP violation effects may be an indication that, in this scheme, SUSY breaking is essentially sourced by a single field.

15.7 Other mediation mechanisms in string theory

Gravity and moduli appear as natural mediators of SUSY breaking in string theory. Still it is natural to ask whether other mechanisms mentioned in Section 2.6.5 are also possible. With a supergravity effective action, it is clear that one-loop anomaly mediation SUSY breaking contributions will be generically present; indeed, we already mentioned that in some cases in which moduli SUSY breaking contributions are suppressed the anomaly mediation terms cannot be neglected. However, a situation with full anomaly mediation dominance requires *sequestering* the SM sector away from the SUSY breaking sector, in particular preventing moduli from participating in the SUSY breaking transmission. Although localization of the SM sector on warped throats could in principle help in its sequestering from bulk CY dynamics, fully realistic implementations of the sequestering idea within string theory have proven difficult.

Another popular idea for SUSY breaking and transmission is gauge mediation. A general feature of this scenario is that the required ingredients are purely field theoretical, and therefore insensitive to UV physics. Still, it is reasonable to consider string theory embeddings of models of gauge mediated SUSY breaking; this requires full moduli stabilization at a high energy, leaving a 4d $\mathcal{N} = 1$ rigid supersymmetric theory at lower energies, in which dynamical SUSY breaking (and its mediation to the SM) can subsequently occur. Due to this prior requirement of full SUSY moduli stabilization with $\mathcal{N} = 1$, the potential construction of explicit models is notoriously hard already from the start; in addition, given the lack of a canonical or minimal model of gauge mediation, most attempts to describe gauge mediation in string theory in practice focus on the realization of gauge sectors with dynamical SUSY breaking. This has been most successfully addressed in the context of type II models with D-branes, given their flexibility to engineer non-trivial gauge sectors, and the geometric implementation of decoupling gravitational and moduli dynamics using non-compact models.

Despite the UV insensitivity of the whole scenario, there are several instances in which string theory ingredients or concepts are useful for dynamical SUSY breaking. The first is the rich non-perturbative dynamics in string theory, which as described in Section 13.1 includes D-brane instanton effects, in addition to those in gauge theories. These can induce contributions to the superpotential relevant for dynamical SUSY breaking, with naturally hierarchically small coefficients due to the e^{-1/g_s} suppression. The second is the use of holography in the context of the gauge/gravity correspondence, Sections 6.4 and 14.1.3; in particular, the construction of gravitational duals of gauge theories with dynamical SUSY breaking, in terms of warped throats with a SUSY breaking source at its infrared end.

Meta-stable field theory vacua in SQCD and string theory*

Gauge field theories with spontaneous SUSY breaking at their global minimum exist, but are usually rather complicated, and often non-calculable due to the strong dynamics. There are, however, very simple models with non-SUSY local minima, which are meta-stable but long-lived against decay to an absolute SUSY minimum; they can be used as SUSY breaking sectors for all practical purposes.

Remarkably enough, such situation can be realized in the SQCD theories introduced in Section 2.5.1, in the so-called Intriligator–Seiberg–Shih (ISS) model. It is based on 4d $\mathcal{N} = 1$ $SU(N_c)$ SQCD, with N_f massive flavours with masses much smaller than the scale Λ of strong gauge dynamics. In the free magnetic range $N_c + 1 \leq N_f < \frac{3}{2}N_c$, the Seiberg dual theory is weakly coupled, and describes most efficiently the exact infrared dynamics. The dual is a $SU(N_f - N_c)$ theory, with N_f flavours q, $\tilde{q}$, transforming in the $(\square; \overline{\square}, \mathbf{1})$, $(\overline{\square}; \mathbf{1}, \square)$ under the gauge and $SU(N_f)_L \times SU(N_f)_R$ global chiral symmetry, and gauge singlets Φ in the $(\mathbf{1}; \square, \overline{\square})$, with superpotential (assuming equal flavour masses)

$$W = h\, q\, \Phi\, \tilde{q} + h\mu^2 \, \text{tr} \, \Phi, \tag{15.70}$$

with μ essentially the original (small) flavour mass, and h an $\mathcal{O}(1)$ parameter.

The F-term condition $\partial W / \partial \Phi = \tilde{q}q - \mathbf{1}_{N_f} = 0$ cannot be satisfied, since $\tilde{q}q$ has rank at most N_c, strictly smaller than N_f, and there is SUSY breaking. There is a classical flat direction of non-SUSY vacua along Φ, which is lifted by one-loop effects, leading to a local non-SUSY minimum at $\langle \Phi \rangle = 0$. Inclusion of $SU(N_f - N_c)$ non-perturbative effects shows the existence of N_c SUSY vacua in the large $\langle \Phi \rangle$ region, as expected from the original $SU(N_c)$ theory. The minimum at the origin is however meta-stable and long-lived, leading to SUSY breaking at a scale controlled by the dimensionful parameter μ, which in this simple model must be tuned to parametrically low values.

Given its simplicity, it is natural to consider the realization of this SUSY breaking model in string theory. There are fairly explicit such implementations based on local models of D3-branes at CY singularities of Section 11.3.4, in particular quotients of the conifold theory (11.72). The realization of the SQCD theory is easily achieved by, e.g., engineering D3-brane theories with gauge group $SU(N_c) \times SU(N_f)$, with bi-fundamentals $(\square, \overline{\square})$, $(\overline{\square}, \square)$, and taking the $SU(N_f)$ coupling very small to effectively reduce it to a global symmetry. Although there are no perturbative mass terms for these flavours, they can be generated by non-perturbative instantons from D-branes wrapped on the collapsed cycles in the singular geometry. The instanton fermion zero modes and their interactions can be easily described in terms of dimer diagrams. In suitable examples, they can produce the desired flavour masses in analogy with the mechanism in Section 13.1.3. This non-perturbative origin in string theory thus provides a natural explanation for the smallness of these mass terms in the ISS model.

The meta-stable non-SUSY vacuum is difficult to establish using the weak coupling description of the D-brane system, due to the role of strong gauge dynamics implicit in the Seiberg duality. However systems of D3-branes at CY singularities, at strong coupling, admit a holographic dual description in terms of warped throats; in particular, models based on quotients of the conifold inherit the structure of the deformed conifold warped throat in Section 14.1.3. In this holographic description, the non-SUSY vacua are convincingly related to systems of antibranes at the bottom of the throat, in an interesting link with the uplifting mechanism used in Section 15.3.1. The meta-stability against decay to a SUSY configuration with no antibranes is associated to possible flux/brane annihilation processes, allowed because they carry opposite charges, as manifest in the tadpole conditions (15.17).

This example thus nicely illustrates the two uses of string theory ingredients in the discussion of dynamical SUSY breaking.

16

Further phenomenological properties.
Strings and cosmology

In previous chapters we have considered the main properties of the 4d effective action
of string compactifications, regarding low-energy particle spectra, anomaly cancellation,
Yukawa couplings, supersymmetry breaking soft terms, and moduli fixing. In this chapter
we discuss in more detail several other general phenomenological features of string theory
models of particle physics. These include the possible range of the string scale, gauge
coupling unification, axions, baryon and lepton number violation, Z'-bosons, and possible
signatures of low-scale string models. We also introduce some aspects of string cosmology
and inflation.

16.1 Scales and unification in string theory

In string compactifications there are two most important scales, the string scale $M_s = \alpha'^{-1/2}$, which controls the mass of string excitations, and the Planck scale $M_p = 1.2 \times 10^{19}$ GeV, which controls the strength of gravitational interactions; within each kind of
string compactifications, they are related by the gauge coupling and/or internal volume
factors, but these relations differ for the various classes of compactifications. In addition
there are other scales, associated to the geometry of the internal manifold, which control
the mass of KK replicas. In this section we describe the structure of scales in different
types of compactifications, and overview the status of gauge coupling unification and its
scale; a brief summary is provided at the end of Section 16.1.3.

16.1.1 Heterotic compactifications: scales and unification

Perturbative heterotic models

In Section 9.1.2 we showed that in perturbative heterotic compactifications the tree level
relation between the Planck and string scales is

$$M_s^2 = \frac{k}{16} \alpha M_p^2, \tag{16.1}$$

where α is the fine structure constant of the gauge interaction and k the corresponding
Kac–Moody level, see Section 9.4. In realistic particle physics models, the values for SM

or GUT couplings fix $\alpha \simeq 1/24$, and hence there is little flexibility for the value of the string scale, which must lie slightly below the Planck scale.

In a compactification with the SM group below the string scale, the corresponding gauge couplings (at the string scale) are related by

$$k_3\alpha_3 = k_2\alpha_2 = k_1\alpha_1 = \frac{16M_s^2}{M_p^2}, \tag{16.2}$$

where, as in section 9.4, k_2, k_3 are the $SU(2)$ and $SU(3)$ Kac–Moody levels, and k_1 is the $U(1)_Y$ normalization factor. In the usual compactifications, $k_2 = k_3 = 1$, whereas k_1 is model dependent, with the GUT-like value $k_1 = 5/3$ arising in models with an underlying GUT group (e.g broken by discrete Wilson lines). Therefore, in the simplest perturbative heterotic compactifications there is unification of gauge coupling constants, even in the absence of an explicit 4d GUT symmetry.

Assuming further that the model contains just the MSSM spectrum below the string scale, it is possible to run down the coupling constants to, e.g., the electroweak (EW) scale $\simeq M_Z$, by using the renormalization group equations (RGEs). For $k_2 = k_3 = 1$ the one-loop relationships are

$$\sin^2\theta_W(M_Z) = \frac{1}{1+k_1}\left\{1 + \frac{k_1\alpha(M_Z)}{2\pi}\left(b_2 - \frac{1}{k_1}b_1\right)\log\frac{M_G}{M_Z}\right\}, \tag{16.3}$$

$$\frac{1}{\alpha_s(M_Z)} = \frac{1}{1+k_1}\left\{\frac{1}{\alpha(M_Z)} - \frac{1}{2\pi}\left[b_1 + b_2 - (1+k_1)b_3\right]\log\frac{M_G}{M_Z}\right\},$$

where $M_G = 5.27 \times g_s \times 10^{17}$ GeV is the effective string scale appropriate for the one-loop RGEs, as in Section 9.6.1. As discussed in Section 2.6.2 the RGE running for $k_1 = 5/3$ yields unification around the scale 2×10^{16} GeV, with $\sin^2\theta_W(M_Z) = 0.23$ and $\alpha_3(M_Z) = 0.12$, in good agreement with experimental data. Thus the string unification scale M_G is about a factor 20 higher than the standard GUT scale, and gives $\sin^2\theta_W = 0.22$ and $\alpha_3 = 0.19$ at the weak scale, in disagreement with data; this is the gauge unification problem of perturbative heterotic compactifications.

This may not be a severe problem, since the heterotic unification picture is still valid *grosso modo*, and there may be further corrections potentially improving the quantitative aspects. For instance, the hypercharge normalization factor k_1 may differ from the canonical $SU(5)$ GUT value. Indeed, a slightly smaller value $k_1 \simeq 1.4$ substantially improves the agreement with the low-energy couplings, as depicted in Figure 16.1(a). Such small values for k_1 can actually be realized, e.g. in examples of heterotic compactifications in the fermionic construction, although (due to extra matter beyond the MSSM) there is no explicit example solving the unification problem based on this idea.

A second source of modifications is that most of the realistic heterotic compactifications contain additional matter fields beyond the MSSM, which can modify the gauge coupling running. Even for heterotic models with no such additional fields at low energies, there are generically extra matter fields around the unification/string scale, and massive KK

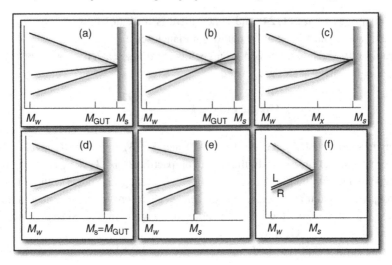

Figure 16.1 Different possibilities for gauge coupling unification in string models. (a) A slightly modified k_1 allows to postpone the unification scale to M_S; (b) large threshold corrections at the string scale; (c) extra matter fields at some intermediate scale M_x focus the couplings to unification at M_S; (d) in type I or Hořava–Witten models it is possible to have $M_S = M_{GUT}$; (e) in intersecting D6-brane models (or their IIB mirror) the wrapped volumes can be tuned to reproduce the low-energy gauge couplings; (f) in models of D3-branes at singularities with a left–right symmetric gauge sector, couplings unify at an intermediate scale which may be identified with M_S.

and winding states, which can give important threshold contributions to gauge coupling unification. At the string scale the couplings are actually given by

$$\frac{16\pi^2}{g_a^2(M_G)} = k_a\,(16\pi^2)\,\mathrm{Re}\,S + \Delta_{\mathrm{Thr}}^a\left(T_n, T_n^*, U_m, U_m^*\right), \qquad (16.4)$$

where Δ_{Thr}^a includes moduli dependent threshold corrections. Such corrections can be computed in the orbifold case, as discussed in Section 9.6.1; a general feature in this setup is that at large compactification radius R, threshold corrections scale as

$$\Delta^a \simeq \frac{\left(b_a' - k_i \delta_{GS}\right)}{12}\,R^2, \qquad (16.5)$$

where b_a' are the model-dependent coefficients (9.105) for the overall modulus. It is possible to show that for $k_1 = 5/3$, moderate values $R \simeq 2 - 4$, and appropriate choices of the coefficients b_a' (compatible with reasonable patterns of modular weights for SM fields), one can achieve quantitative agreement with the low-energy gauge couplings, as depicted in Figure 16.1(b). There is, however, no known explicit heterotic model where the required modular weights are actually realized.

Yet another possibility, illustrated in Figure 16.1(c), is the presence of additional matter fields at an intermediate scale M_x, focusing the gauge coupling running towards unification. Again, the realization of this scenario in explicit models is possible but challenging.

Strong coupling and dual type I and Hořava–Witten models

Remarkably, there is a quite elegant solution to the perturbative heterotic unification problem, by moving onto the strongly coupled regime. For compactifications of the heterotic $SO(32)$ theory, this is achieved by using the dual type I string compactifications. In this set up, gravitational and gauge interactions arise from closed and open string sectors, respectively, so their strengths have different dependence on the microscopic parameters of the compactification. Indeed, the 10d type I action in (4.109) contains the terms

$$S_{10} = \int \frac{d^{10}x}{(2\pi)^7} (-G)^{\frac{1}{2}} \left(\frac{1}{g_s^2 \alpha'^4} R - \frac{1}{4\sqrt{2} g_s \alpha'^3} \operatorname{Tr} F_{\mu\nu} F^{\mu\nu} + \cdots \right), \tag{16.6}$$

where $g_s = e^\phi$ and we have used $2\kappa_{10}^2 = (2\pi)^7 \alpha'^4$ and $\kappa_{10}^2/g_{10}^2 = \alpha'/(2\sqrt{2})$. Note the different dilaton dependence of the gravitational and gauge terms. In the heterotic case both terms have the same dilaton dependence, and cancel out in the coupling strength ratios, leading to (16.1); in type I instead, upon compactification to 4d (in which one picks up a 6d volume factor), the relation is

$$M_s^2 = \frac{\alpha_G \, g_s}{4\sqrt{2}} M_p^2. \tag{16.7}$$

Setting α_G to the GUT value, the string scale M_s can still be lowered to coincide with $M_G \sim 10^{16}$ by choosing an appropriately small g_s, see Figure 16.1(d). This is a quite elegant and economical option for unification in type I models (equivalently, in the strongly coupled limit of the $SO(32)$ heterotic). Incidentally, one may try to lower the string scale all the way down to the weak scale; however, doing this while maintaining $\alpha_G \simeq 1/24$ fixed requires going to compactification volumes smaller than the string length scale, so the 10d supergravity effective action is no longer a good approximation. In toroidal compactifications, though, it is possible to apply T-duality to obtain a model with large compactification volume and D3-branes rather than D9-branes, in which case the theory remains under control. Unification in D3-brane models is described in the next section.

An analogous idea may be implemented in $E_8 \times E_8$ heterotic string compactification, whose strong coupling regime is described in terms of the Hořava–Witten theory, see Section 7.5, namely M-theory on $\mathbf{X}_6 \times \mathbf{S}^1/\mathbf{Z}_2$. As described in Section 9.2, the gravitational constant G_N and gauge fine-structure constant α_G are given in terms of the CY volume V_6 and the length of the interval ρ by

$$G_N = \frac{\kappa_{11}^2}{16\pi^2 V_6 \, \rho}, \quad \alpha_G = \frac{\left(4\pi \kappa_{11}^2 \right)^{2/3}}{2V_6}, \tag{16.8}$$

with κ_{11} the 11d gravitational constant. The unification mass can be identified as the CY compactification scale $M_G = V_6^{-1/6} = (8\pi^2)^{-1/2} \alpha_G^{1/2} (G_N \rho)^{-1/3}$; plugging the values of

G_N and $\alpha_G = \alpha_{GUT}$, the value $M_G = M_{GUT} = 10^{16}\,\mathrm{GeV}$ can be reproduced by setting $1/\rho \propto 10^{14}\,\mathrm{GeV}$. This also fixes the fundamental M-theory scale at $\kappa_{11}^{-2/9} \simeq 2M_{GUT}$. The resulting hierarchy of scales is as follows: at energies below $10^{14}\,\mathrm{GeV}$ there is an effective theory of 4d gravitational and gauge interactions, with the latter assumed to be the MSSM; above this energy regime, a new dimension opens up, and there is an effective description in terms of gravity on a 5d space, with the MSSM still localized on its 4d boundaries; thus gauge couplings have 4d running, and unify at the cutoff M_{GUT}, defined by the CY compactification scale, slightly above which lies the fundamental 11d scale.

The structure of gravitational and gauge interactions and their strengths in Hořava–Witten theory is a particular realization of the brane-world scenario of Section 1.4.2. This is more systematically exploited in the next section.

16.1.2 Type II orientifold and F-theory scales and unification

In type II string compactifications (and orientifolds thereof), the strengths of gravitational and gauge interactions are independent, since they arise from the bulk closed string sector or the D-brane open string sectors, respectively. Recalling (6.14) and performing a KK compactification, the 4d gauge coupling constant on a Dp-brane wrapped on a $(p-3)$-cycle on the CY is

$$\frac{1}{g_{Dp}^2} = \frac{\alpha'^{(3-p)/2}}{g_s (2\pi)^{p-2}} V_{p-3}, \tag{16.9}$$

with V_{p-3} being the wrapped volume (and $V_0 = 1$ for D3-branes). On the other hand, dimensionally reducing (4.45), (4.47), the 4d Planck mass is

$$M_p^2 = \frac{8 V_6}{(2\pi)^6 g_s^2 \alpha'^4}. \tag{16.10}$$

The type I case corresponds to D9-branes, i.e. $p = 9$. For $p < 9$ the wrapped volumes V_{p-3} are in general independent from the overall volume V_6, and so the ratio between the Planck and string scales depends on the kind of D-branes in which the SM is localized. We thus consider different possibilities in turn.

The SM on D3-branes

This situation is realized by D3-branes at singularities, recall Section 11.3. From Section 12.3, the tree-level gauge kinetic function has the form $f = S + \sum_i r_k \Phi_k$, where Re Φ_k are the blowing-up modes; so, in the configuration strictly at a blown-down singularity, all gauge couplings unify with $\alpha_{D3} = g_s/2$, and

$$M_s^4 = \alpha_{D3} \frac{(2\pi)^3 M_p}{\sqrt{2V_6}}. \tag{16.11}$$

The string scale M_s can be made arbitrarily small by taking the compact volume V_6 large enough. Some interesting choices are discussed below.

Although non-abelian gauge couplings unify, the hypercharge $U(1)_Y$ does not in general have the canonical $SU(5)$ normalization, $k_1 = 5/3$. For a model of D3-branes at a general $\mathbf{Z}_N$ orbifold singularity, with group $U(3) \times U(2) \times U(1)^{N-2}$, the only generically non-anomalous and massless $U(1)$ linear combination is (11.59), namely

$$Q_{\text{diag}} = -\left(\frac{1}{3}Q_3 + \frac{1}{2}Q_2 + \sum_{s=1}^{N-2} Q_1^{(s)}\right). \qquad (16.12)$$

This should be identified with hypercharge, and so

$$k_1 = 5/3 + 2(N-2), \qquad (16.13)$$

in the normalization tr $T_a^2 = \frac{1}{2}$ for $U(n)$ generators. This leads to a weak angle

$$\sin^2\theta_W = \frac{1}{k_1 + 1} = \frac{3}{6N - 4} \qquad (16.14)$$

at the string scale. The canonical $SU(5)$ result $\sin^2\theta_W = 3/8$ is reproduced only for a $\mathbf{Z}_2$ singularity, however, which produces a non-chiral spectrum and is thus unrealistic; for the $\mathbf{Z}_3$ and other singularities of interest, the hypercharge normalization is not the canonical $SU(5)$ one. However, this is not necessarily a fatal drawback for unification, as discussed in an explicit example later on.

Concerning the string scale, there are at least three interesting possible values:

- $M_s = M_{\text{GUT}}$, namely the gauge coupling unification scale. In this case, unification of the MSSM gauge couplings would signal the existence of the fundamental string scale at unification, as in Figure 16.1(d). This choice of string scale, however, reproduces gauge coupling unification only for canonical hypercharge normalization. This is not automatically achieved, except if the D3-branes realize an actual $SU(5)$ GUT symmetry; however, such constructions lead to additional difficulties, like the perturbative absence of top Yukawa couplings, or the explicit violation of R-parity.
- $M_s \simeq 1$ TeV, i.e. string scale close to the EW scale. This is compatible with the observed 4d Planck scale by taking a large enough compact volume V_6. Assuming for simplicity a single typical length scale in all directions in $\mathbf{X}_6$ (*isotropic* compactification), the required compactification scale is ~ 10 MeV. However, there are no light KK replicas of SM gauge or charged fields, since they are localized on the D3-branes, and do not propagate on the internal space (which is felt only by closed string fields like gravitons and their KK replicas). This is the simplest string theory implementation of the large extra dimensions introduced in Section 1.4.2, some of whose phenomenological implications are discussed in Section 16.5. The low fundamental scale reduces the energy range of gauge coupling running and prevents their unification. The low-energy couplings should at best be reproduced by suitable (non-unified) couplings at the scale M_s, see Figure 16.1(e); for D3-branes at singularities, this can be achieved (albeit not naturally) by tuning vevs for the blowing-up moduli appearing in the gauge kinetic function.

- $M_s \simeq \sqrt{M_W M_p} \simeq 10^{11}$ GeV, i.e. string scale at a geometric mean intermediate scale. This scale appears naturally, e.g. in the large volume compactifications of Section 15.3.2, in which SUSY breaking by fluxes with $M_s \sim 10^{11}$ GeV results in soft terms of order $m_{\text{soft}} \simeq M_s^2/M_p \simeq M_W$, around the EW scale. Such intermediate scale is moreover interesting for additional phenomenological reasons. For instance, a pseudoscalar in the closed string sector may play the role of QCD axion with a decay constant $F_a \simeq 10^{11}$ GeV, consistent with astrophysical and cosmological bounds in Section 1.3.2, see also Section 16.2. Also, in models with right-handed neutrinos, a Majorana mass of order 10^{11} GeV reproduces, as in Section 1.2.6, seesaw neutrino masses $m_\nu \simeq M_W^2/M_s$ in the correct range.

An intermediate scale is not compatible with standard MSSM gauge coupling unification. A possibility to reproduce the low-energy couplings is to tune the gauge coupling at M_s, using the extra moduli fields in the gauge kinetic functions, as mentioned above. A possibly more appealing alternative is to have gauge coupling unification at $M_s \simeq 10^{11}$ GeV in models including extra fields beyond the MSSM. One explicit realization is the left–right symmetric model from D3-branes at a $\mathbf{Z}_3$ singularity in Section 11.3.3, with a spectrum as in Table 11.3. The boundary conditions for the $SU(3) \times SU(2)_L \times SU(2)_R \times U(1)_{B-L}$ gauge couplings at the string scale are $g_3^2 = g_L^2 = g_R^2 = (32/3)g_{B-L}^2$; the normalization of the hypercharge generator $Y = -T_R^3 + \frac{1}{2}Q_{B-L}$ is $k_1 = k_R + \frac{1}{4}k_{B-L} = 11/3$, and hence $\sin^2 \theta_W = 3/14 = 0.214$ at tree level. To run down to the EW scale, we assume the LR model to be valid from M_s down to a scale M_R, with $SU(3) \times SU(2)_L \times SU(2)_R \times U(1)_{B-L}$ one-loop β-function coefficients B_3, B_L, B_R and B_{B-L}; at the scale M_R, the LR model is assumed to break spontaneously to the MSSM (with three EW Higgs sets), with one-loop β-function coefficients b_i. The complete one-loop running yields

$$\sin^2 \theta_W(M_Z) = \frac{1}{1+k_1} \left\{ 1 + k_1 \frac{\alpha_e(M_Z)}{2\pi} \left[\left(B_L - \frac{1}{k_1} B_1' \right) \log \frac{M_s}{M_R} \right. \right.$$
$$\left. \left. + \left(b_2 - \frac{1}{k_1} b_1 \right) \log \frac{M_R}{M_Z} \right] \right\}, \tag{16.15}$$

$$\frac{1}{\alpha_e(M_Z)} - \frac{1+k_1}{\alpha_3(M_Z)} = \frac{1}{2\pi} \left\{ \left[b_1 + b_2 - (1+k_1) b_3 \right] \log \frac{M_R}{M_Z} \right.$$
$$\left. + \left[B_1' + B_L - (1+k_1) B_3 \right] \log \frac{M_s}{M_R} \right\}, \tag{16.16}$$

where we define

$$B_1' = B_R + \tfrac{1}{4} B_{(B-L)}, \quad k_1 = k_R + \tfrac{1}{4} k_{B-L}. \tag{16.17}$$

The spectrum of the LR model gives

$$B_3 = -3, \quad B_L = +3, \quad B_R = +3, \quad B_{B-L} = +16, \tag{16.18}$$

and using the MSSM b_i values (2.71), and $k_1 = 11/3$, we obtain

$$\sin^2 \theta_W(M_Z) = \frac{3}{14} \left[1 + \frac{\alpha_e(M_Z)}{2\pi} \left(4 \log \frac{M_s}{M_R} - \frac{22}{3} \log \frac{M_R}{M_Z} \right) \right],$$

$$\frac{1}{\alpha_e(M_Z)} - \frac{14}{3\alpha_3(M_Z)} = \frac{1}{2\pi} \left(26 \log \frac{M_R}{M_Z} + 24 \log \frac{M_s}{M_R} \right). \tag{16.19}$$

Using $\alpha_e(M_Z)^{-1} = 127.9$ as input, the choice $M_s \simeq 10^{12}$ GeV leads, for $M_R \simeq 1$ TeV, to values $\alpha_3 = 0.12$, $\sin^2 \theta_W = 0.23$, in reasonable agreement with data. Thus the model shows gauge coupling unification, albeit in a non-standard fashion, see Figure 16.1(f). Since the LR gauge symmetry is broken in the TeV range, it could serve as a low-energy signature for an intermediate fundamental scale.

The string model as presented above should be completed with extra branes, in order to introduce additional scalars, with quantum numbers of right-handed sneutrinos, to trigger the LR symmetry breaking; these extra multiplets do not spoil the one-loop running of gauge couplings and their unification. More serious potential difficulties of this model are possible FCNCs from the presence of three sets of SM-like Higgs multiplets. Also, a generic issue in LR models with relatively low M_R scale is that neutrino masses tend to be not sufficiently small.

The SM on D7-branes

Let us consider the status of gauge coupling unification in semi-realistic models with the SM sector on 7-branes, such as the magnetized D7-brane models in Section 11.4.2, or the F-theory 7-brane models in Section 11.5.3. The former are mirror to type IIA intersecting D6-brane models, which are thus not discussed explicitly.

An illustrative magnetized brane example is the MSSM-like model in Table 10.4, constructed as a type IIB orientifold as described in Section 11.4.2. As mentioned there and in Section 12.6, the MSSM sector is actually realized on three sets of D7$_i$-branes, $i = 1, 2, 3$, transverse to the ith 2-torus; the relevant data for the MSSM sector are given in (12.110). Using Pati–Salam notation for simplicity, the $SU(3+1) \times SU(2)_L \times SU(2)_R$ gauge kinetic functions are (12.111), i.e.

$$2\pi f_{3+1} = T_1 + k^2 S; \quad 2\pi f_L = \frac{1}{2} T_2; \quad 2\pi f_R = \frac{1}{2} T_3. \tag{16.20}$$

As shown in Section 12.6, the SUSY conditions require $A_2 = A_3$, and $12A_1 + 8A_2 + 6A_3 = A_1 A_2 A_3/(\alpha'^2)$. Hence, the $SU(2)_L$ and $SU(2)_R$ gauge couplings are equal at the string scale, even though the gauge factors are not unified into a larger gauge group. One can also tune the areas A_1, A_2 to achieve further unification with the $SU(3+1)$ coupling; for instance, with $A_2 = A_3 = 6.3\alpha'$, the SUSY condition implies $A_1 = 3.18\alpha'$, and there is approximate unification of the coupling constants, i.e. Re $f_{3+1} \simeq$ Re $f_L =$ Re f_R. Thus, although the SUSY condition does not enforce unification with the $SU(3+1)$ gauge coupling, it can accommodate it. However, one can check that, in this example, unification with $\alpha_G \simeq 1/25$ forces the string coupling to be large, $g_s \simeq 3$. Alternatively, the couplings may not unify and the string scale may be lowered as in Figure 16.1(e).

The unification of couplings is clearly more natural in 7-brane models with an underlying GUT group, like the $SU(5)$ in the F-theory models in Section 11.5.3. Considering for simplicity a perturbative type IIB D7-brane realization, the unified coupling constant is given by

$$\frac{1}{\alpha_G} = \frac{1}{8\pi^4 g_s} \left(\frac{V_4}{\alpha'^2} \right), \tag{16.21}$$

where V_4 is the volume of the wrapped 4-cycle. To display the structure of scales, we parametrize the total compact volume as $V_{X_6} = V_4 R^2$; then the string and Planck scales are related by

$$M_s^2 = \frac{M_P}{R} (g_s \alpha_G)^{1/2} \pi, \tag{16.22}$$

and the string scale may be lowered by taking larger transverse size R, or smaller string coupling.

In this class of 7-brane GUT models, the GUT symmetry is broken down to the SM by additional gauge backgrounds (i.e. the hypercharge flux in the $SU(5)$ case), whose natural scale is $V_4^{1/4}$. The GUT unification scale is therefore $M_G \simeq V_4^{-1/4}$, which for suitable choices of microscopic parameters can give $M_G \simeq 2 \times 10^{16}$ GeV, thus reproducing the MSSM gauge coupling unification picture. On the other hand, the presence of these additional gauge backgrounds induces group dependent sub-leading corrections; they essentially come from the F^4 terms in the expansion of (12.53), upon replacing $F^2 \rightarrow \langle F_Y^2 \rangle$ on the 4-cycle, and so are suppressed at large volume. They may, however, be relevant for the detailed comparison of gauge coupling unification in specific models, but their discussion is beyond our scope.

16.1.3 Unification and RCFT type II orientifolds

D-brane models realizing the SM (rather than GUTs) at the string scale have the different gauge factors on different sets of D-branes, and in general there is no gauge coupling unification. On the other hand, in certain cases there are relations between the non-abelian and hypercharge coupling constants, as follows. Consider for instance the class of type IIA models with the SM arising from intersecting D6-brane stacks a, b, c, d, leading to a gauge group $U(3)_a \times U(2)_b \times U(1)_c \times U(1)_d$ or $U(3)_a \times Sp(2)_b \times U(1)_c \times U(1)_d$ as in Sections 10.5 and 10.6. Consider the $SU(3)$ and $SU(2)$ gauge couplings as arbitrary parameters $\alpha_a = \alpha_s$ and $\alpha_b = \alpha_2$, determined by the volume wrapped by the corresponding D-brane. In these models the hypercharge generator is a linear combination

$$Q_Y = \frac{1}{6} Q_a - \frac{1}{2} Q_c - \frac{1}{2} Q_d, \tag{16.23}$$

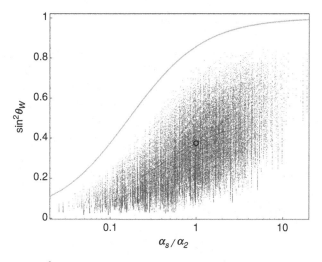

Figure 16.2 Values of $\sin^2 \theta_W$ versus α_s/α_2 for the class of type II RCFT orientifolds studied by Schellekens and collaborators (taken from Dijkstra *et al.* (2005b)).

and so the hypercharge coupling is

$$\frac{1}{\alpha_Y} = \frac{1}{6\alpha_a} + \frac{1}{2\alpha_c} + \frac{1}{2\alpha_d}. \tag{16.24}$$

In many semi-realistic models, e.g. in toroidal compactifications with a potential enhancement to an $SU(4)$, the volumes of the stacks a, d are equal, and so $\alpha_a = \alpha_d$. Some other models, e.g. those with possible enhancements to LR symmetric models, have $\alpha_c = \alpha_b/2$. In these particular cases, the relation is

$$\frac{1}{\alpha_Y} = \frac{2}{3\alpha_s} + \frac{1}{\alpha_2}. \tag{16.25}$$

This is in principle compatible with the canonical $SU(5)$ unification if we further have $\alpha_s = \alpha_2$; even if not, it still provides an interesting partial unification pattern.

It is interesting to consider the status of unification, and of such relations, in the largest available class of MSSM-like models, realized by RCFT Gepner orientifolds as overviewed in Section 10.7.1. In this case the inverse coupling constants for the ath set of D-branes is $g_a^{-2} = \eta R_{0a}/g_s$, with $\eta = 1$ for unitary groups and $\eta = 1/2$ for orthogonal or symplectic groups, and R_{0a} is the boundary coefficient introduced in Section 10.7.1. The plot in Figure 16.2 provides the different possible values of gauge couplings in MSSM-like models with non-unified gauge groups in this RCFT class, in terms of their values for $\sin^2 \theta_W$ and α_s/α_2; the point in the middle corresponds to the canonical $SU(5)$ value 3/8 (the top solid line is an upper bound on $\sin^2 \theta_W$). The general observation is that in arbitrary MSSM-like orientifolds, at least within this class, there are no preferred values for the couplings, which are broadly scattered around the $SU(5)$ value. It is still interesting that a non-negligible fraction, around 10%, satisfy the conditions leading to the partial unification relation (16.25).

In summary, standard MSSM gauge coupling unification may be adjusted in certain classes of string compactifications. On the other hand, it appears naturally in others, namely: (i) strongly coupled heterotic compactifications (with non-standard embedding) based on an underlying GUT symmetry explicitly broken by discrete Wilson lines; (ii) type II orientifolds with an underlying GUT symmetry, broken by either Wilson lines or hypercharge fluxes (in perturbative orientifolds, this can be problematic, since the presence of a large Yukawa coupling for the top quark would require very large instanton corrections); and (iii) local F-theory GUT models with symmetry broken by hypercharge flux.

If MSSM gauge coupling unification is not an accident, those large classes of compactifications would be phenomenologically favored.

16.2 Axions in string theory

In string compactifications there are generically plenty of axion-like scalar fields, namely pseudoscalars a with an effective action invariant under a perturbative shift symmetry $a \to a + c$. These fields appear from (dimensional reduction of) the antisymmetric tensor fields, like the 2-form B_2, or the RR p-form fields C_1, C_3, C_5 in type IIA and C_0, C_2, C_4, C_6 in type IIB; the shift symmetry is a 4d manifestation of the gauge invariance (B.21) of the original p-form fields. These fields are actually the imaginary components of the 4d dilaton and geometric moduli fields, e.g. as described in Section 12.1 (see Section 12.1.4 for the toroidal orientifold case). Since CY manifolds generically have hundreds of Kähler and/or complex structure moduli, these compactifications may potentially contain hundreds of such axion-like fields.

This genericity of axion-like fields is a very attractive property of string theory, as they may play a role in solving the strong CP problem, as described in Section 1.3.2. For this purpose, however, the candidate axion-like fields should couple to QCD gauge bosons, and also satisfy additional properties, like having a decay coupling constant F_a in the allowed window $10^9 \, \text{GeV} \leq F_a \leq 10^{12} \, \text{GeV}$ (however, see the comments at the end of Section 16.2.2). The nature of the QCD axion candidate and its value of F_a depend on the kind of string compactification considered, as we now describe.

16.2.1 Axions in heterotic strings

In perturbative heterotic compactifications the gauge kinetic function has (in the large volume regime) a dilaton and Kähler moduli dependence of the form

$$f = S + \sum_i \epsilon_i T_i, \qquad (16.26)$$

where ϵ_i are small model-dependent one-loop coefficients, like those arising from threshold corrections in orbifolds, recall Sections 9.6.1 and 16.1.1. Ignoring these corrections for simplicity, consider the complex dilaton S as axion candidate. Defining

$$S = \frac{1}{g^2} + \frac{i\eta}{8\pi^2}, \tag{16.27}$$

the gauge kinetic terms read

$$\int d^2\theta \, \frac{S}{4} \, \mathrm{Tr} \, W^\alpha W_\alpha + \text{h.c.} = \frac{1}{4g^2} \, \mathrm{Tr} \, F_{\mu\nu} \, F^{\mu\nu} + \frac{i\eta}{32\pi^2} \, \mathrm{Tr} \, F_{\mu\nu} \, \tilde{F}^{\mu\nu}. \tag{16.28}$$

In standard axion models the canonically normalized axion field is $a = F_a\eta$, with F_a the axion decay constant; then the axion kinetic term is given by

$$\frac{1}{2} \partial_\mu a \, \partial^\mu a = \frac{F_a^2}{2} \partial_\mu \eta \, \partial^\mu \eta, \tag{16.29}$$

and F_a may be extracted from the kinetic term of Im S in the 4d effective theory. This is given by (9.8) in Section 9.1.1, from which

$$\frac{M_p^2}{8\pi (S + S^*)^2} \partial_\mu (\mathrm{Im}\, S) \partial^\mu (\mathrm{Im}\, S) \quad \Longrightarrow \quad F_a^2 = \frac{M_p^2}{4\pi (8\pi^2)^2 (S + S^*)^2}. \tag{16.30}$$

Recalling $(S + S^*) = 1/(2\pi\alpha_G)$, with $\alpha_G \simeq 1/25$ being the QCD coupling at the string scale, one finds

$$F_a = \frac{\alpha_G}{8\pi^{3/2}} M_p \simeq 1.2 \times 10^{16} \mathrm{GeV}. \tag{16.31}$$

This large value for F_a is rather generic in heterotic compactifications. In the alternative possibility of the axion candidate given by the imaginary part of some of the Kähler moduli T_i in (16.26), the F_a is in fact larger, since the coefficients ϵ_i are one-loop; in specific compactifications, its typical value is $F_a \simeq 10^{17}$ GeV.

The situation is similar in the strongly coupled limit of the $E_8 \times E_8$ heterotic string, in terms of the Hořawa–Witten theory on $\mathbf{X}_6$ times an interval $\mathbf{S}^1/\mathbf{Z}_2$ of length ρ. The main modification arises because, as reviewed in Section 7.5, the CY volume changes along the interval, so that the averaged compactification volume can be parametrized as $V_7 = V_6 \pi \rho \, \epsilon$; here $\epsilon = 1$ for small ρ, and $\epsilon = 1/2$ if the CY volume varies linearly between 0 and V_6 along the interval. The axion candidate Im S has then a decay constant

$$F_a = \frac{\alpha_G}{8\pi^{3/2}\epsilon} M_p. \tag{16.32}$$

Note that the result does not show an explicit dependence on ρ, so that the decay constant cannot be lowered by increasing ρ; in fact, it is still $\sim 10^{16}$ GeV, and actually reproduces the perturbative result for $\epsilon = 1$.

As in the perturbative case, there are axionic partners of Kähler moduli arising from the M-theory 3-form C_{ijk} with one leg along the interval, and two along a 2-form on $\mathbf{X}_6$. These fields can also enter the gauge kinetic functions, although with coefficients similar to the perturbative case. In summary, F_a is of order 10^{16}–10^{17} GeV in generic perturbative or non-perturbative heterotic compactifications.

16.2.2 Axions in type II orientifolds

Type II orientifolds have plenty of potential axions coming from the diverse 10d RR and NSNS p-form fields. Moreover, as already described, the localization of the SM sector on D-branes allows to lower the string scale much below the Planck scale, suggesting an extra flexibility in the values of the axion decay constants. We now consider these issues, concentrating on type IIB orientifolds with the SM on D7- or D3-branes; other setups, like type IIA orientifolds, or type IIB D9/D5-brane models, yield similar results.

As a first toy example we consider the case of the SM sector (or at least QCD) on a stack of D7-branes wrapped on a 4-cycle with associated Kähler modulus T, whose imaginary part is the axion candidate. The gauge kinetic function is of the form $2\pi f = T$, and we ignore possible S-dependent terms from worldvolume magnetic fluxes, as they do not change the argument. Assuming a typical Kähler potential $K = -\log(T + T^*)$, the computation of the axion decay constant is as in the above heterotic case, with the replacement $S \to T$, and again yields $F_a \sim 10^{16}$ GeV. Note that this holds even if the string scale is lowered, against the naive expectation for a more flexible range of predicted axion scales.

In fact, the above assumed form for the Kähler potential is very restrictive, and in generic CY compactifications it applies only to the overall Kähler modulus. Let us therefore consider the alternative situation, in which the D7-brane wraps a 4-cycle, whose volume is independent of the overall Kähler modulus. A prototypical example are the "small" 4-cycles in the swiss cheese CYs mentioned in Section 15.3.2; hence we consider the simple example therein, with two moduli and Kähler potential (neglecting the α' corrections, which do not modify the argument)

$$\kappa_4^2 K = -2 \log \left(\tau_b^{3/2} - \tau_s^{3/2} \right) \simeq -3 \log(\tau_b) + 2 \left(\frac{\tau_s}{\tau_b} \right)^{3/2}, \qquad (16.33)$$

where recall that Re $T_b = \tau_b$ determines the overall volume. The SM is assumed to localize on D7-branes wrapped on the small cycle, with modulus T_s, so the gauge kinetic function is $2\pi f = T_s$; the axion candidate is Im T_s, and its kinetic term from (16.33) leads to the axion decay constant

$$F_a = \left(\frac{3}{32\pi} \right)^{1/2} \frac{1}{\left(\tau_s \tau_b^3 \right)^{1/4}} \frac{M_p}{8\pi^2} = \left(\frac{9\alpha_G}{\pi^2 g_s} \right)^{1/4} \frac{M_s}{16\pi^2}. \qquad (16.34)$$

Taking $\alpha_G = 1/25$ (in the convention with tr $T^2 = 1/2$) and $g_s \simeq 0.1$, one obtains $F_a \simeq 5 \times 10^{-3} M_s$. The axion scale is now directly related to the string scale, and may indeed be lowered. So we conclude that axions corresponding to small 4-cycle Kähler moduli may have arbitrarily low F_a, which in particular can be chosen within the allowed window; this conclusion extends to any other model where the SM 7-branes wrap on small 4-cycles.

A similar pattern is found in D3-brane models. For D3-branes at smooth points in the internal space, the gauge kinetic function is $2\pi f = S$, and the Kähler potential

$K = -\log(S + S^*)$ leads to axion decay constant (16.31), by a computation analogous to the heterotic one. On the other hand, realistic models are actually based on D3-branes at singularities, for which the gauge kinetic function is $2\pi f = S + \sum_k \Phi_k$, i.e. depends on blowing-up moduli Φ_k, recall Section 12.3.2. The fields Im Φ_k lead to axion candidates with decay constants controlled by the string scale, which is the only relevant local scale; this fits the interpretation in Section 11.4.3 of D3-branes at singularities as D7-branes wrapped on the 4-cycles collapsed at the singular point.

In summary, string models produce an axion decay constant in the 10^{16}–10^{17} GeV range, unless the SM and the axion candidate are associated to "small" cycles, in which case F_a can be tuned by lowering the string scale. In the former case, the predicted decay constant is much larger than the cosmological upper bound of 10^{12} GeV, coming from overclosure of the universe. This would suggest a phenomenological preference for the latter scenario, with lower string scale and the SM localized on small cycles. However, there are proposals to circumvent the problematic cosmological upper bound (e.g. by assuming the universe contains an atypically small axion density selected on anthropic grounds), which would thus keep alive the possibility of string axions with large F_a.

A final important issue for axions in the string theory setup is their survival at low energies. Note that, in order to solve the strong CP problem, the axion should be extremely light, with the dominant contribution to its potential arising from non-perturbative QCD dynamics. In string compactifications, there are several possible sources of large masses for axion-like scalars, thus preventing them to play as phenomenologically viable axions. For instance, as discussed in Sections 9.5 and 12.4, certain axion-like fields become massive by combining with D-brane $U(1)$ fields, via Stückelberg $B \wedge F$ couplings. A second source of masses are closed string fluxes. In the type IIB compactifications in Section 14.1, the NSNS and RR 3-form fluxes induce masses for the dilaton S and complex structure moduli, including their axionic imaginary parts. These fluxes do not generate masses for the (axionic parts of the) Kähler moduli; however, such masses may appear in the type IIB compactification with non-geometric fluxes in Section 14.4, or through non-perturbative effects from D-brane instantons or gaugino condensation, see Sections 13.1 and 15.3.1. Similarly, in type IIA flux compactifications, stabilization of the dilaton, complex structure, and Kähler moduli leads to mass terms for the corresponding axion-like fields; analogous effects exist also in the heterotic case. The conclusion is that building explicit models of moduli fixing in which some QCD axion-like field remains massless may be challenging in some models, particularly those with full moduli stabilization.

16.3 R-parity and B/L-violation

In SUSY extensions of the SM the existence or not of R-parity, and the possible violation of baryon and lepton numbers are important issues. In string theory, the existence of these symmetries and their properties are very model dependent. We will thus settle for a general view of these issues in the string context, and direct the reader to the Bibliography for more information.

16.3.1 R-parity and (B-L) in string theory

As we discussed in Section 2.6.1, the most general renormalizable superpotential consistent with the SM gauge symmetries contains four types of terms violating baryon or lepton number, denoted by UDD, QDL, LLE, and LH_u in (2.70). In order for the proton to be stable, some or all of these superpotential terms must be absent. The simplest possibility is having an R-parity symmetry forbidding all these couplings. We thus turn to the important question of the existence of R-parity in specific examples of string compactifications.

Most semi-realistic $\mathcal{N} = 1$ string compactifications constructed to date achieve the absence of dimension-4 proton decay exploiting a gauged $U(1)_{B-L}$ symmetry (or a discrete version thereof), at some level. The presence of such gauged $U(1)_{B-L}$ is fairly generic in many compactifications, both in heterotic and type II orientifolds. A heuristic argument for this is as follows: $B - L$ is a non-anomalous global symmetry of the SM with three right-handed neutrinos, so string compactifications resembling closely the SM will tend to contain a $U(1)_{B-L}$ symmetry; however, general arguments in Section 17.1.4 show that consistent theories containing quantum gravity cannot have continuous global symmetries. Thus in string compactifications, the $U(1)_{B-L}$ symmetry, if present, must be a gauge symmetry.

In heterotic orbifold compactifications, the construction of realistic models involves introducing vevs along SM singlet flat directions, as in Section 9.7. Their choice is crucial in the final structure of the observable (SUSY) SM superpotential, and its symmetries; one can then maintain sufficient proton stability by choosing appropriate flat directions. An interesting way to get an effective R-parity, in models with an underlying $SO(10)$ symmetry, is to have vevs for SM singlets with even charges under $U(1)_{B-L}$. There is then an unbroken $\mathbf{Z}_2$ subgroup of $U(1)_{B-L}$, which may be identified with R-parity. However, this mechanism requires choosing very particular directions in the singlet moduli space, so although possible, it may be regarded as non-generic.

In the case of the SM or MSSM-like type IIA intersecting D-brane models in Sections 17.1.4 and 10.6 (and their type IIB mirrors), B and L are gauge symmetries, which remain unbroken at the perturbative level; therefore $B - L$, and hence R-parity, are automatically preserved. In MSSM-like toroidal orientifolds with 4 D-brane stacks, $U(1)_{B+L}$ is anomalous and necessarily becomes massive, remaining at low energies as a perturbative anomalous global symmetry, just like in the SM; on the other hand, $B - L$ is anomaly-free, and may remain or not as a gauge symmetry at low energies, depending on whether it has $B \wedge F$ couplings to RR fields. If it becomes massive, there may be non-perturbative instanton effects generating R-parity violating terms, so they should be adequately suppressed.

Type II orientifold models with an underlying $SU(5)$ GUT symmetry are somewhat problematic in what concerns R-parity, since in general there are allowed couplings $\mathbf{10}_f \cdot \mathbf{\bar{5}}_f \cdot \mathbf{\bar{5}}_f$, containing UDD and QDL terms, like in the $SU(5)$ example at the end of Section 11.2.3. This also holds in the F-theory $SU(5)$ GUTs in Section 11.5.3, in which proton stability must be ensured by assuming that the fermion matter curves do not have intersections which could lead to such couplings.

16.3.2 Proton decay through dimension-5/-6 operators

Even if dimension-4 operators are forbidden, dimension-5 operators like $QQQL$ or $UUDE$ may give rise to fast proton decay. These operators are not forbidden by $U(1)_{B-L}$, and hence are harder to avoid in string compactifications. On the other hand, these higher-dimensional operators have weaker experimental constraints, and in concrete examples, model dependent properties may suppress them sufficiently. This is particularly the case in heterotic constructions, since their string scale is necessarily large.

In MSSM-like D-brane models, the (gauged) B and L symmetries forbid those couplings at the perturbative level; they may still be generated by (model-dependent) non-perturbative D-brane instanton effects, although these are generically expected to be sufficiently suppressed. On the other hand, in D-brane or F-theory models with underlying $SU(5)$ GUT structure, there is no symmetry forbidding those couplings. In the F-theory setting they can be sufficiently suppressed by geometrically localizing the two Higgs multiplets $\mathbf{5}_H$ and $\bar{\mathbf{5}}_H$ on different matter curves; this way the colored triplet higgsino $\tilde{H}_3$ mass term in Figure 2.2 vanishes, since multiplets on different matter curves cannot couple in a mass term. In this case, which is fairly generic, the dangerous dimension-5 operator is absent.

Dimension-6 operators, in analogy with the ordinary SUSY-GUT case, are very much suppressed in SUSY models with a large fundamental scale. This is the case in heterotic string compactifications, whose string scale is necessarily large. In MSSM-like D-brane models, they are again forbidden in perturbation theory by the B and L symmetries. Finally, in D-brane or F-theory models with an underlying $SU(5)$ GUT, the fundamental scale is $\sim 10^{16}$ GeV, sufficient to yield proton lifetimes $\sim 10^{36}$ years, well above experimental bounds of present or forthcoming experiments.

16.3.3 Neutrino masses

Most of the semi-realistic compactifications constructed to date have a number of SM singlet fields which may be considered as right-handed neutrinos N_i, in the sense that they have Yukawa couplings with left-handed leptons. As described in Section 1.2.6, an elegant explanation for the smallness of observed neutrinos is the seesaw mechanism, which involves a large Majorana mass for these singlets. Hence, a natural question is whether these semi-realistic string compactifications can produce such mass terms. Again, the answer is rather model dependent.

Typically in all string models, such Majorana masses for candidate right-handed neutrinos are forbidden by $U(1)$ gauge symmetries. There are, however, two main mechanisms potentially solving this problem. The first, often arising in heterotic orbifold models, is the possible existence of superpotential couplings of the form

$$\frac{h_{ij}}{M^{n-1}} N_i N_j \prod_{a=1}^{n} s_a;$$ (16.35)

here the s_a are SM singlet fields, which acquire large vevs in the one-loop re-stabilization of the vacuum, recall Section 9.7, and M is a large (string or compactification) scale. The second mechanism, explored in D-brane models, consists in non-perturbative instanton effects generating couplings of the form

$$M_s \, e^{-S_{\rm cl.}} h_{ij} \, N_i N_j,$$ (16.36)

as discussed in Section 13.3.1. As explained there, the $U(1)$ transformation of the $N_i N_j$ bilinear is compensated by a shift of the classical instanton action $S_{\rm cl}$.

The first possibility has been mostly studied in heterotic orbifold compactifications. In this case there are typically many singlets N_i (often many more than 3), and the possible existence of the required superpotential couplings and vevs can be studied using the orbifold selection rules in Section 9.3.2; this turns out to be a non-trivial computation, with negative results in many cases, but notably leading to successful Majorana mass terms in a few working examples.

The second mechanism, i.e. instanton induced neutrino masses, is not realized in most heterotic compactifications constructed to date. The reason is that in heterotic compactifications on orbifolds, or on smooth CYs without $U(1)$ gauge backgrounds, the only modulus shifting under $U(1)$ symmetries is the dilaton S; the corresponding instantons euclidean NS5-branes wrapped on the internal 6d space; their strengths are of order $\simeq e^{-1/\alpha_G}$, just like ordinary gauge theory instantons, and thus result in far too small (in fact, negligible) right-handed neutrino masses. On the other hand, in MSSM-like models from type II orientifolds, there is a richer pattern of instanton effects, which can indeed lead to right-handed neutrino Majorana masses of the right magnitude, as already discussed in Section 13.3.1. We recall that a necessary condition is the existence of a Stückelberg mass for the $U(1)_{B-L}$ generator.

16.4 Extra U(1) gauge bosons

In explicit string model building there are often extra $U(1)$ gauge interactions beyond hypercharge. In fact, we already mentioned that in many compactifications there may remain an extra unbroken $U(1)_{B-L}$ gauge symmetry, guaranteeing the absence of dimension-4 B/L-violating operators. We now review several patterns of extra $U(1)$ gauge interactions in string compactifications, and possible experimental signatures of the corresponding Z' gauge bosons.

16.4.1 Extra U(1)'s in heterotic (2,2) compactifications

In the heterotic CY compactifications with standard embedding in Section 7.3.2, the E_6 group may be broken down to the SM by discrete Wilson lines, recall Section 7.3.3. This is a possible source of extra $U(1)$'s, appearing in the branching

$$E_6 \; \rightarrow \; SO(10) \times U(1)_\psi \; \rightarrow \; SU(5) \times U(1)_\chi \times U(1)_\psi.$$ (16.37)

Table 16.1 *SM fields and their exotic partners in the* **27***, with their charges under the extra U(1) symmetries inside* E_6

$SO(10)$	$SU(5)$	Q_χ	Q_ψ
16	**10** : (Q, U, E)	-1	1
	$\bar{\mathbf{5}}$: (D, L)	3	1
	1 : ν_R	-5	1
10	**5** : (D', H_u)	2	-2
	$\bar{\mathbf{5}}$: $(\bar{D}', H_d)$	-2	-2
1	**1**: N	0	4

The quantum numbers of the SM particles and their exotic E_6 partners in the representation **27** are shown in Table 16.1, in $SO(10)$ and $SU(5)$ multiplet notation. The $U(1)_\chi$ symmetry is anomaly free in the SM with three right-handed neutrinos ν_R; it is in fact a linear combination $Q_\chi = 4Q_Y - 5Q_{B-L}$ of hypercharge and $U(1)_{B-L}$. On the other hand, if the $U(1)_\psi$ symmetry remains light upon compactification, cancellation of its anomalies requires three families of massless exotics, transforming as $\mathbf{1} + \mathbf{5} + \bar{\mathbf{5}}$ under $SU(5)$. If those symmetries are unbroken in the compactification, the corresponding Z' gauge bosons get massive only upon electroweak symmetry breaking. Hence they may be produced at colliders, and also contribute to precision measurements of SM radiative corrections. The present Fermilab data imply a lower bound of about 800 GeV on such Z' gauge boson masses, which will soon be much improved by the LHC.

Heterotic string compactifications with non-standard embedding may give rise to other $U(1)$ symmetries surviving at low energies. However, in general they bring along additional chiral exotic fermions, required to cancel their anomalies. It is therefore generically expected that these symmetries are broken by the SM singlet vevs introduced to remove the additional exotic fields, recall Section 9.7. From this perspective, the $U(1)_\chi$ symmetry (or other linear combinations of hypercharge and $U(1)_{B-L}$) is quite special, since it is anomaly free in the SM with the addition of just three singlets, the right-handed neutrinos.

16.4.2 Extra U(1)'s in type II orientifolds

Extra $U(1)$ gauge bosons also appear in most semi-realistic models based on type II orientifold compactifications, as is clear from the various examples in earlier chapters. Most of these $U(1)$ symmetries are anomalous, however, and their gauge bosons get masses of order the string scale – with modulo volume factors as described around (12.69). Again $U(1)_{B-L}$ is special, and appears naturally in many explicit models, as already mentioned in Section 16.3.1.

Table 16.2 *Example of D6-brane wrapping numbers giving rise to a SM spectrum*

N_α	$\left(n_i^1, m_i^1\right)$	$\left(n_i^2, m_i^2\right)$	$\left(n_i^3, m_i^3\right)$
$N_a = 3$	$(1, 0)$	$(2, 1)$	$(1, 1/2)$
$N_b = 2$	$(0, -1)$	$(1, 0)$	$(1, 3/2)$
$N_c = 1$	$(1, 3)$	$(1, 0)$	$(0, 1)$
$N_d = 1$	$(1, 0)$	$(0, -1)$	$(1, 3/2)$

Even if the extra $U(1)$'s have a mass of order the string scale, they may still be accessible to experiment in models with low string scale. To illustrate this possibility, let us consider the pattern of $U(1)$ symmetries in the non-supersymmetric SM-like intersecting D6-brane models in Section 10.5.1. We consider four stacks of D6-branes, labeled a, b, c, d for the baryonic, left, right, and leptonic branes. We focus on one example with wrapping numbers $\left(n_i^\alpha, m_i^\alpha\right)$ on $\mathbf{T}^2 \times \mathbf{T}^2 \times \mathbf{T}^2$ as given in Table 16.2, with α running through the stacks a, b, c, d. The initial gauge group is $U(3)_a \times U(2)_b \times U(1)_c \times U(1)_d$, and the gauge quantum numbers of the massless bi-fundamental fermions are given in Table 10.1; they correspond to three generations of quarks, leptons, and right-handed neutrinos.

There are four $U(1)$ symmetries, with charges Q_a, Q_b, Q_c, and Q_d, with nice interpretations in terms of familiar SM global symmetries. Indeed, $Q_a = 3B$ and $Q_d = -L$, with B and L being the baryon and lepton numbers. Also $Q_c = 2T_R^3$, i.e. twice the third component of the $SU(2)_R$ familiar from left–right symmetric models. Finally, Q_b has mixed $SU(3)$ anomalies and has the properties of a Peccei–Quinn symmetry. The two $U(1)$ symmetries Q_b and $Q_a/3 - Q_d$ have triangle anomalies, canceled by the Green-Schwarz mechanism, as explained in Section 10.3.1. More concretely, there are four 4d RR 2-form fields B_2^k, $k = 0, 1, 2, 3$, and their dual RR scalars a_k. The couplings (10.23) in the toroidal orientifold case become

$$N_\alpha m_1^\alpha m_2^\alpha m_3^\alpha \int_{4d} B_2^{(0)} \wedge \mathrm{tr}\, F_\alpha, \quad n_1^\beta n_2^\beta n_3^\beta \int_{4d} a_0 \, \mathrm{tr}\, F_\beta^2, \tag{16.38}$$

$$N_\alpha n_j^\alpha n_k^\alpha m_i^\alpha \int_{4d} B_2^{(i)} \wedge \mathrm{tr}\, F_\alpha, \quad m_j^\beta m_k^\beta n_i^\beta \int_{4d} a_i \, \mathrm{tr}\, F_\beta^2 \quad i \neq j \neq k, \; i = 1, 2, 3,$$

where the contribution from branes and their orientifold images are included. For the example in Table 16.2, one has

$$
\begin{aligned}
&0, & &2a_0 \, \mathrm{tr}\, F_a^2 \\
&B_2^{(1)} \wedge (-2\, \mathrm{tr}\, F_b), & &\tfrac{1}{2} a_1 \left(\mathrm{tr}\, F_a^2 - 3\mathrm{tr}\, F_d^2 \right) \\
&B_2^{(2)} \wedge (3\mathrm{tr}\, F_a - \mathrm{tr}\, F_d), & &\tfrac{1}{2} a_2 \left(-3\mathrm{tr}\, F_b^2 + 6\mathrm{tr}\, F_c^2 \right) \\
&B_2^{(3)} \wedge (3\mathrm{tr}\, F_a + \mathrm{tr}\, F_c), & &0.
\end{aligned}
\tag{16.39}
$$

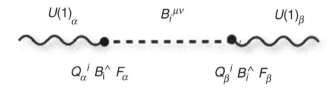

Figure 16.3 Mass term for $U(1)$'s mixing with RR 2-form fields $(B_2)_k$.

It is easy to check that the combination of both kinds of couplings for the RR fields labeled 1 and 2 leads to cancellation of mixed anomalies for Q_b and $3Q_a - Q_d$. In addition, the RR scalar a_0 has no Stückelberg coupling, but couples to the $SU(3)$ D6-brane stack, and so provides a QCD axion candidate, exemplifying the ubiquity of axions in string theory mentioned in Section 16.2.2. Finally, the field $B_2^{(3)}$ couples to (non-anomalous) $3Q_a + Q_c$, and makes its gauge boson massive.

The diagram for the $U(1)$ mass terms is shown in Figure 16.3. From (12.69) the $U(1)$ boson mass matrix is

$$(M^2)_{\alpha\beta} = \frac{M_s^2}{4\pi} g_\alpha g_\beta \sum_{i=0}^{3} Q_\alpha^k Q_\beta^k = \frac{M_s^2}{4\pi} \begin{pmatrix} 18g_a^2 & 0 & 3g_a g_c & -3g_a g_d \\ 0 & 4g_b^2 & 0 & 0 \\ 3g_c g_a & 0 & g_c^2 & 0 \\ -3g_d g_a & 0 & 0 & g_d^2 \end{pmatrix},$$

where $Q_a^0 = m_a^1 m_a^2 m_a^3$, $Q_a^i = -m_a^i n_a^j n_a^k$, $i \neq j \neq k$, and the dependence on the gauge couplings g_α arises from normalizing the gauge fields canonically. To avoid volume factors we have also considered an isotropic torus as explained below (12.69). The vector $(1/(6g_a), 0, -1/(2g_c), 1/(2g_d))$ is the hypercharge generator, and is the only massless eigenstate. Hence, the non-anomalous $U(1)_{B-L}$ gauge boson is massive, a phenomenologically interesting property, as mentioned at several points.

If the string scale is sufficiently low, just above the weak scale, there is the possibility of producing these three massive extra Z' bosons at the LHC, even before actually reaching the string threshold. Note also that in this case, after electroweak symmetry breaking, the four $U(1)$'s mix with the diagonal $SU(2)_L$ generator, producing a 5×5 mass matrix with the photon as only massless eigenstate. The lightest of the four massive $U(1)$ bosons is identified with the observed Z^0, which is essentially the electroweak Z^0 with a slight contamination of the other massive $U(1)$'s. Present precision measurements like the ρ-parameter constrain the masses of the latter to be above 1 TeV; also, the massive $U(1)_{B-L}$ should be heavier than ~ 800 GeV from the Tevatron measurements, and results from the LHC are steadily increasing this lower bound as we write this book.

Let us finally mention that the above toroidal example does not actually allow for a TeV string scale, as there is no dimension simultaneously transverse to all SM D6-branes, as required in large extra dimension models. The same analysis, however, applies to other existing constructions which do allow for low string scale, and have an analogous $U(1)$ structure.

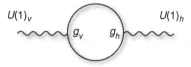

Figure 16.4 One-loop kinetic mixing between visible and hidden $U(1)$ symmetries.

16.4.3 Hidden sector U(1)'s and kinetic mixing

Another interesting piece of $U(1)$ phenomenology in string compactifications is the appearance of additional $U(1)$'s with suppressed or non-existent direct couplings to the SM fields, hence termed *hidden sector $U(1)$'s*. Several possible instances are:

- Type II orientifold compactifications can lead to RR $U(1)$ gauge bosons when $h_{2,1}^+ \neq 0$ (in type IIB) or $h_{1,1}^+ \neq 0$ (in type IIA), recall Section 12.1. These conditions are not realized in the toroidal setup, but are not particularly restrictive in more general CY orientifolds, where they are quite generic. The only states charged under RR $U(1)$'s are very heavy D-brane particle states, so there are no light charged fields, and these symmetries certainly qualify as hidden $U(1)$'s. As shown in Section 12.3.3, in type IIB models their gauge kinetic function depends on complex structure moduli, and not on Kähler moduli, while the converse holds for type IIA orientifolds (as required by mirror symmetry); note that this dependence is opposite to that of gauge kinetic functions for D-brane sectors in the models. The precise form of the gauge kinetic function may be obtained from the underlying $\mathcal{N} = 2$ prepotential before orientifolding, see Section 12.3.3.
- In type II orientifolds there may be additional D-branes carrying $U(1)$ gauge interactions, located away from the SM branes in the compact manifold, i.e. not intersecting them (in type IIA terminology). Such $U(1)$'s are hidden, as the only states charged under both the SM and $U(1)_h$ are massive open strings stretching between the two D-brane stacks.
- In heterotic compactifications with non-standard embedding, there often appear $U(1)$ symmetries from the hidden E_8 sector, thus hidden from the SM sector.

Although hidden, such $U(1)$'s may lead to interesting phenomenology, which we summarize here for the general (not necessarily stringy) case. The main point is that hidden $U(1)$'s may mix with the SM hypercharge at some level. In particular, if there are massive particles charged under a visible $U(1)_v$ and a hidden $U(1)_h$, the one-loop diagram in Figure 16.4 induces a mixing in the kinetic terms, i.e.

$$\mathcal{L} = -\frac{1}{4g_v^2} F_{\mu\nu}^v F_v^{\mu\nu} - \frac{1}{4g_h^2} F_{\mu\nu}^h F_h^{\mu\nu} + \frac{\chi}{2g_v g_h} F_{\mu\nu}^v F_h^{\mu\nu}. \tag{16.40}$$

Since the mixing operator is dimension 4, the coupling χ may be non-negligible even if the particles in the loop are very heavy; there is no mass scale suppression, but rather the mixing parameter is generically of order

$$\chi \simeq \frac{g_v g_h}{16\pi^2}. \tag{16.41}$$

The experimental signatures of a mixing between hypercharge (and hence the photon) and a hidden $U(1)_h$ depend on whether there are hidden sector light particles charged under $U(1)_h$. If no such charged particles exist, the signatures depend on the $U(1)_h$ gauge boson mass M_h. For $M_h \simeq 100\,\text{GeV}$–$1\,\text{TeV}$, the limits come from EW precision data, and yield $\chi \leq 10^{-2}$; for this range of masses, the hidden $U(1)_h$ behaves like a Z' and could be produced à la Drell–Yan at the LHC. For smaller M_h, the best limits come from $\gamma \leftrightarrow \gamma'$ oscillations and from deviations from Coulomb's law; the typical bound is $\chi \leq 10^{-6} - 10^{-7}$ for $M_h \leq 10^{-3}\,\text{eV}$, although these limits are expected to improve by orders of magnitude in future planned experiments.

If the $U(1)_h$ gauge boson is massless ($M_h = 0$), there are no limits on χ, since the kinetic mixing matrix can be diagonalized, leading to just a redefined hypercharge with redefined coupling constant. In the supersymmetric case the situation may be more interesting because the $U(1)_h$ gauginos will in general be massive and can mix with the MSSM neutralinos; this could lead to new signatures at the LHC, if SUSY is found, through different cascade decays depending on the neutralino mass patterns.

The phenomenology changes significantly if there are light particles charged under $U(1)_h$. In this case the mixing of $U(1)_h$ with hypercharge induces an effective electric *minicharge* q_h for those particles, given by

$$q_h = \chi g_h = \epsilon\, e. \tag{16.42}$$

At present there are limits on ϵ depending on the mass m_q of the minicharged particles (which may be easily pair-created, if sufficiently light) and the mass of the $U(1)_h$ (from laser experiments, "shining through a wall" experiments, and others). The typical bound is $\epsilon \lesssim 10^{-5} - 10^{-7}$ for m_q smaller than the electron mass. There are also much stronger cosmological limits, which are however very model-dependent and less reliable.

In string theory this kind of effect could open up a useful window into the structure of hidden sectors and yield important information on the underlying compactification. These signatures could in principle arise from mixing between visible and hidden sectors coming from distant branes (or antibranes) in type II orientifolds, as well as in heterotic compactifications. In both cases there may be massive string states charged under both visible and hidden sectors which could lead to $U(1)$ mixing. The size of the possible mixing is very model dependent with values $\chi \simeq 10^{-4}$ in some D-brane scenarios. For hidden RR $U(1)$ gauge bosons, one-loop kinetic mixing requires the $U(1)_h$-charged non-perturbative D-brane states to couple to the SM sector; although this is in principle possible, no detailed study of such states has been carried out hitherto. In addition, such RR $U(1)$ bosons may become massive through Stückelberg couplings in certain flux compactifications, in which case they cannot be considered as relevant hidden $U(1)$'s of the 4d effective theory.

A final related point is the possible appearance of microgauge (i.e. extremely weakly coupled) interactions in scenarios like, e.g., the large volume IIB compactifications, arising by a different mechanism; namely from D7-branes wrapping the large volume 4-cycle in

Table 16.3 *Values of large extra dimension radii and scale of first KK excitation consistent with a TeV string scale*

n	1	2	3	$\cdots$	7
R	10^8 km	0.1 mm	10^{-6} mm	$\cdots$	10^{-12} mm
R^{-1}	10^{-18} eV	10^{-3} eV	100 eV	$\cdots$	100 MeV

the manifold, leading to e.g. a suppressed coupling $g_v \simeq 10^{-4}$ for an intermediate string scale $M_s \simeq 10^{10}$ GeV.

16.5 Strings at the weak scale

As discussed in Section 16.1.2, the string scale M_s may be arbitrarily close to the EW scale, only constrained by experimental limits. This is based on the large extra dimension scenario of Section 1.4.2, realized by localizing the SM sector on Dp-branes wrapped on $(p-3)$-cycles on the internal space. For instance, for D3-brane models, (16.11) implies that $M_s \simeq 1$ TeV can be achieved for a sufficiently large transverse volume V_6. For general Dp-branes, the string scale may be lowered for sufficiently large transverse dimensions; parametrizing them in terms of a single radius R, we may write $\overline{M}_p^2 = M_*^{2+n} R^n$, where n is the number of large extra dimensions, $\overline{M}_p^2 = 1/\kappa_4^2$ and M_* is of order the string scale M_s. The precise connection between M_* and M_s depends on n and may be easily established using the results in Section 16.1.2; the values of R required to set $M_* \simeq 1$ TeV consistently with the experimental value of M_p are given in Table 16.3, for different values of n. Note that the case $n = 1$ is excluded, since it requires a radius, $R \sim 10^8$ km, producing too large deviations from Newton's law at astronomical scales.

If the string scale is close to the weak scale, there are several possible signatures at colliders. There are also constraints coming from limits on deviations from Newtonian gravity, as well as from astrophysical implications. All these constraints are very sensitive to the number of large transverse dimensions and also to the particular string compactification underlying the SM interactions. Some generic signatures are the following:

- *Deviations from Newton's law*: Unlike the SM fields, the graviton is allowed to propagate on the large extra dimensions, and has a tower of KK replicas with masses $m_k^2 = k^2/R^2$. These KK gravitons are very light for large radii (see Table 16.3), but couple to the SM sector with gravitational strength. Still, they can be exchanged in diagrams like Figure 16.5 and lead to deviations from Newton's law; these can be parametrized by a modified gravitational potential

$$V(r) = -\frac{G_N m_1 m_2}{r}\left(1 + \alpha\, e^{-\frac{r}{\lambda}}\right), \tag{16.43}$$

and may be constrained by tabletop Cavendish experiments. In a simple toroidal setting with a single compactification length scale R, one has $\alpha = 8n/3$ and $\lambda = R$. Present

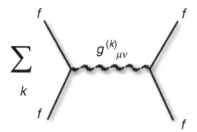

Figure 16.5 Light Kaluza–Klein copies of the graviton exchanged between matter fields produces deviations from Newton's law.

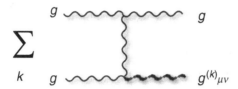

Figure 16.6 Inclusive production of KK gravitons in gluon–gluon collisions.

bounds exclude the case $n = 1$ as mentioned above, and yield $R < 37 \times 10^{-3}$ mm for $n = 2$, which corresponds to $M_* > 3.6$ TeV. The bounds become weaker for larger n.

- *Kaluza–Klein graviton emission*: KK replicas $g_{\mu\nu}^{(k)}$ of the graviton may also be produced at colliders like the LHC, by diagrams as in Figure 16.6. The gravitational strength coupling of each such KK mode suppresses its production rate, but the inclusive cross-section sums over all the states in the KK tower; on dimensional grounds, the cross-section is

$$\sigma \simeq \frac{(E_{\mathrm{CM}})^n}{M_*^{n+2}}. \tag{16.44}$$

These KK gravitons would escape undetected, and would give rise to a missing energy signature at colliders. Present experimental bounds from these processes are relatively weak, but the LHC at 14 TeV would be able to test values up to $M_* = 2, 3$ TeV, for $n = 4$, 3, from this type of signatures.

- *Astrophysical bounds on extra dimensions*: The presence of a large multiplicity of very light KK gravitons may have important astrophysical consequences, which in fact lead to the strongest constraints on the large extra dimensions scenario. For instance, KK graviton emission could profusely occur in supernovae explosion, carrying out a fraction of its available energy; the requirement of not depleting too much the observed neutrino flux from SN1987A, leads to a bound $M_* > 14, 1.6$ TeV for $n = 2, 3$, respectively. Other limits arise from diffuse γ-ray emission from KK gravitons in the halo of neutron stars; limits from the EGRET satellite imply $M_* > 38, 4.1$ TeV for $n = 2, 3$, respectively. Finally, KK gravitons emitted by supernovae or neutron stars, and trapped in their gravitational field, could occasionally decay into photons, giving rise to an overheating of the observed neutron star surface; the measured luminosities of some pulsars imply $M_* > 750, 35\,\text{TeV}$ for $n = 2, 3$, respectively. For larger n the limits are much weaker since the KK states are heavier.

- *Production of string excitations*: It is interesting to consider the possibility of exploring experimentally possible resonances of SM fields localized on the D-branes, in analogy with the KK resonances of the bulk graviton. Actually SM fields have KK excitations along the $(6 - n)$ extra dimensions wrapped by the brane. These dimensions are not large, however, but presumably of order $\simeq 1/M_s$, leading to relatively heavy SM KK excitations. Although these may be accessible to experiment, their spectrum is rather model dependent, making it difficult to single out generic features. This in fact holds for any SM field resonance depending non-trivially on the internal space geometry. Remarkably there is a sector of resonances which is actually model-independent, corresponding to the string excitations obtained by applying (4d spacetime) oscillator operators to the massless string groundstates describing the SM fields. These string resonances share the gauge quantum numbers of the SM groundstate, but have higher spin; they lie along Regge trajectories in which angular momentum is quadratic in the square of the mass,

$$j = j_0 + \alpha' M^2. \tag{16.45}$$

For low enough string scale, these string excitations may be produced or contribute to scattering amplitudes of partons in collider experiments. In models with low string scale and at weak coupling g_s, such events dominate over other exotic processes, e.g. the micro black hole production mentioned below.

It is possible to compute the corrections coming from these excited states to, e.g., four-point scattering amplitudes of gluons $gg \to gg$. The gauge boson amplitudes at tree level are independent of the structure of the compactification, since they involve vertex operators in a universal sector. The computation of the disk diagram leads to the celebrated *Veneziano amplitude*

$$V(s, t, u) = \frac{\Gamma\left(1 - s/M_s^2\right)\Gamma\left(1 - u/M_s^2\right)}{\Gamma\left(1 + t/M_s^2\right)} \simeq -\frac{su}{t}\sum_{n=0}^{\infty}\frac{\alpha'^n}{n!}\frac{1}{s - nM_s^2}\prod_{J=1}^{n}\left(u + M_s^2 J\right),$$

where s, t, u are the Mandelstam variables (with $s + t + u = 0$), and Γ is Euler's Γ-function. The right-hand side expansion exhibits the expected infinite sum of s-channel resonances corresponding to string excitations with masses $\sqrt{n}M_s$, see Figure 16.7. In practice, only the first excitation would be dominant at the energies available at the LHC, and hence the expected signature is essentially a peak in jet–jet mass distributions. The width of such s-channel resonances is of order $100\,(M_s/\text{TeV})\,\text{GeV}$. As we write this book, the LHC has found already limits of order 2.5 TeV for such string resonances.

- *Micro black hole production*: In models with a low fundamental gravity scale $M_* \sim$ TeV, certain quantum gravity effects may be observable at colliders. In particular, there may be creation of mini black holes, with an estimated cross section of order their classical horizon area $\sigma \simeq \pi R^2$. For $M_* \simeq 1\,\text{TeV}$, the production rate at the LHC may be large, with some estimates yielding $\sigma \simeq 10$ pb for 6 TeV black holes with $M_* \simeq 1.5\,\text{TeV}$.

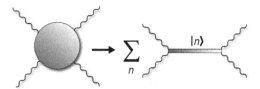

Figure 16.7 Four-point amplitude of massless partons include an infinite sum over string resonances in the s-channel.

Black holes decay with a plausible lifetime of order $10^{-26} - 10^{-27}$ s for $M_* \simeq 1$ TeV, through (the last stages of) Hawking radiation, potentially leading to spectacular multi-particle signatures at the LHC. Nevertheless, the above figures rely on the semiclassical approximation, which is only valid for energies well above the quantum gravity scale, and beyond the actual LHC reach; thus the detailed properties of the mini black hole formation and decay processes could depart from such estimates.

The astrophysical constraints on the size of the quantum gravity/string scale M_* are particularly strong for $n = 2, 3$, which have the lightest KK gravitons. Taken at face value, they would imply that only for $n > 3$ one could expect to discover a low-energy string threshold at the LHC; still, the possibility of finding such effects at the LHC is certainly exciting.

16.6 Strings and cosmology

In the last decade there has been amazing progress in the field of observational cosmology. The measurements of the fluctuations in the cosmic microwave background (CMB) are now in a precision era, which will continue through the next decade. Since string theory is a fundamental theory automatically containing gravity, we expect it to accommodate the cosmological history of the universe, and hopefully address some of the puzzles of the present standard model of cosmology. Conversely, cosmological observational data may provide important constraints on the possible underlying string theory vacua.

The most ambitious approach to achieve a string description of cosmological phenomena would be to study full string theory compactifications with 4d time dependent backgrounds describing, e.g., an expanding universe containing the SM and possibly other matter fields. However, the study of time-dependent backgrounds in string theory, even in far simpler setups, turns out to be a conceptually very hard task. Hence many questions requiring a full string theoretical description, like the nature and resolution of initial cosmological singularities, remain open.

On the other hand, there are many questions in cosmology which can be efficiently addressed using just the 4d effective field theory. The information from the underlying string theory is encoded in the structure of such effective action, and in particular the absence or size of certain UV sensitive terms. The most interesting question of this kind is the realization of cosmological inflation in string theory. A further manifestation of the underlying string theory in cosmology is related to the additional degrees of freedom in the theory, which may lead to potentially problematic or observationally interesting relic densities, for instance of moduli fields, or of cosmic strings. String cosmology is by now a broad area of research in itself, and here we settle for a brief discussion of the above mentioned topics.

16.6.1 Inflation

Let us start with a brief summary of the standard model of cosmology, focusing on properties whose elucidation may relate to particle physics. The most general metric with homogeneous and isotropic spatial slices is the Friedmann–Robertson–Walker (FRW) metric

$$ds^2 = -dt^2 + a^2(t)\, dx_3^2, \tag{16.46}$$

where $a(t)$ is the scale factor which characterizes the (time dependent) spatial length scales, and $H = \dot{a}/a$ gives the instantaneous Hubble parameter. Einstein's equations then give rise to the two coupled non-linear Friedmann equations

$$\left(\frac{\dot{a}}{a}\right)^2 + \frac{\kappa}{a^2} = \frac{\rho}{3M_p^2}, \quad \frac{\ddot{a}}{a} = -\frac{\rho + 3p}{6M_p^2}, \tag{16.47}$$

where ρ and p are the local density and pressure, and $\kappa = 0, 1, -1$ for a flat, positively or negatively curved universe respectively. The energy density of a spatially flat universe is known as the *critical density*, $\rho_c = 3M_p^2 H^2$; present observations are consistent with $\kappa = 0$, so the total density is essentially the critical one, at present times $\rho_{\text{tot}} = 3M_p^2 H_0^2 \simeq 10^{-29}$ g/cm^3, where $H_0 = 71$ (km/s)/Mpc is the Hubble parameter at present. All cosmological data may be described essentially in terms of four contributions ρ_i to this density coming from baryonic matter ρ_B, radiation ρ_{rad}, dark matter ρ_{DM} and dark energy ρ_{DE} (customarily encoded in the density ratios $\Omega_i = \rho_i/\rho_{\text{crit}}$, with $\Omega_{\text{tot}} = \sum_i \Omega_i = 1$). In the standard cosmological model, their equations of state are $p_{\text{rad}} = \rho_{\text{rad}}/3$, $p_B = 0$, $p_{\text{DM}} = 0$ and $p_{\text{DE}} = -\rho_{\text{DE}}$; the latter two describe a non-relativistic (cold) fluid of dark matter, presumably corresponding to massive and very stable neutral particles, whose precise nature is yet unknown, and a dark energy as a (possible effective) vacuum cosmological constant Λ (see Section 1.3.1). The standard cosmological model is hence known as ΛCDM model. Note that (16.47) implies $a(t) = a_0(t/t_0)^\beta$ and $H(t) = \beta/t$ with $\beta = 1/2, 2/3$ for a radiation and matter dominated universe respectively; the comoving Hubble radius $1/(aH)$ increases with time in both cases.

The universe has a rich biography, known as the big bang model. It started out 13.75×10^9 years ago in a hot, relativistic, radiation dominated expansion era. As it cooled down, light nuclei formed in the first few minutes in the primordial nucleosynthesis process. The radiation era ended around $70\,000$ years later leading to a matter dominated era. Electrons and nuclei combined into atoms at the recombination time, around 3.8×10^5 years later, making the universe transparent to photons, whose (suitably redshifted) uniform radiation is observed today as the 2.715 K cosmic microwave background (CMB). The dilution of radiation and matter densities made them subdominant with respect to the constant dark energy density, which around the age of 5×10^9 years started triggering an accelerated expansion of the universe, in which we live today.

Taken at face value, the standard cosmological model has several important naturalness problems – besides the unnaturally small (from the particle physics viewpoint) value of the cosmological constant, already discussed in Section 1.3.1. For instance, present observations, including the very precise measurement of CMB temperature anisotropies, are still consistent with a small but non-zero curvature factor κ/a^2, at the level of a 10%. Dividing the first equation in (16.47) by H^2, we obtain $\Omega_{\text{tot}} - 1 = \kappa/(aH)^2$; since $1/(aH)$ decreases as we go back in time, the observed almost critical value $\Omega = 1$ at present requires a very delicate fine-tuning in the early universe, e.g. with $\Omega - 1 \simeq 3.6 \times 10^{-18}$ at the time of primordial nucleosynthesis. This is known as the *flatness problem*. Another characteristic feature of the present universe is its extreme homogeneity, e.g. with CMB anisotropies at the tiny level of 10^{-5}; this is puzzling, since homogeneity holds even for regions which are causally disconnected, i.e. which are so far apart that did not have time to get into causal contact. This is known as the *horizon problem*.

A natural solution to both problems is provided by *inflation*. In this scenario, the universe is assumed to undergo a period of exponentially accelerated expansion in its very early stages. During this period the evolution of the universe is dominated by a very large and positive energy density $\rho_I \simeq M_I^4$, with M_I typically of order the GUT or other large scale, and with approximate equation of state $\rho = -p$. The Friedman equations then dictate an exponential expansion $a(t) = a(t_0) e^{H_I(t-t_0)}$, while the Hubble scale remains approximately constant, $H_I^2 \simeq M_I^4/\left(3M_p^2\right)$. On the other hand, the spatial curvature piece κ/a^2 is exponentially damped, thus solving the flatness problem for a sufficiently long inflation epoch. This dilution power of inflation over time Δt is encoded in the so-called number of *e-foldings* $N_e \equiv H_I \Delta t$, required to be $N_e \gtrsim 60$ for typical (high-scale) inflation models. The inflationary exponential expansion also solves the horizon problem; correlations between regions initially inside a Hubble radius (and hence in causal contact) before inflation, are stretched by the exponential expansion to scales larger than the Hubble scale, purporting to be causally disconnected at present.

Inflation is assumed to end in a process known as *reheating*, in which the inflationary energy density ρ_I is released into standard degrees of freedom, leaving a bath of matter and radiation at a high temperature T_H. The final stage of the inflation model is a very flat and homogeneous hot universe in a standard expansion phase, which thus provides the initial conditions for big bang cosmology.

In the context of cosmological implications of certain particle physics models, like GUTs, supergravity or string theory, inflation may lead to yet another welcome implication. Some of these models predict a density of heavy relic particles or cosmic defects (e.g. cosmic strings or domain walls) in the early universe, which may even dominate or overclose it; a prototypical example are magnetic monopoles in GUTs. An inflationary period washes away the density of any relic formed before inflation to essentially innocuous values (which is not regenerated at reheating for relics whose mass scale is above T_H).

Most inflation models are based on the introduction of scalar fields with the inflationary energy density corresponding to their potential energy. The general features can be illustrated in the simplest case of a single scalar field ϕ, called the *inflaton*, coupled to gravity in the action

$$S = \int dx^4 \sqrt{-g} \left[\frac{1}{2\kappa_4^2} R - \frac{1}{2} g^{\mu\nu} \partial_\mu \phi \partial_\nu \phi - V(\phi) \right]. \tag{16.48}$$

Assuming a spatial homogeneous profile $\phi(t)$, the inflaton dynamics is governed by

$$\ddot{\phi} + 3H\dot{\phi} + V' = 0, \tag{16.49}$$

where dots denote time derivatives, and primes are ϕ derivatives. The cosmological evolution is controlled by the inflaton energy density and pressure, given by $\rho_\phi = \dot{\phi}^2/2 + V$, $p_\phi = \dot{\phi}^2/2 - V$. Inflation is achieved if there is a slow roll regime in which the potential energy dominates, $\dot{\phi}^2 \ll V$, so that $\rho_\phi \simeq -p_\phi \simeq V(\phi)$. Inflation holds while the inflaton vev rolls down its potential, with inflation scale $M_I \simeq V^{1/4}$, and ends when the slow roll conditions cease to be satisfied; a long enough roll requires $\ddot{\phi} \ll H\dot{\phi}$. The above requirements lead to the *slow roll conditions*

$$\epsilon \equiv \frac{1}{2} \left(\frac{V' M_p}{V} \right)^2 \ll 1; \quad |\eta| \equiv \left| \frac{V'' M_p^2}{V} \right| \ll 1. \tag{16.50}$$

Remarkably, in addition to solving the flatness and horizon problems, inflation provides a simple mechanism to generate the primordial fluctuations eventually giving rise to the observed CMB anisotropies and seeding structure formation. They originate from quantum fluctuations of the inflaton $\delta\phi$ and the metric $\delta g_{\mu\nu}$, which during the inflationary expansion get stretched to large scales and become classical perturbations. The observed CMB anisotropies fit a gaussian distribution with two-point correlations encoded in a power spectrum of the form $P_s(k) \simeq k^{n_s}$, with the so-called spectral index $n_s \simeq 1$. The single inflaton model discussed above leads to a gaussian spectrum of perturbations, since inflaton fluctuations are weakly coupled, and can be shown to predict

$$n_s - 1 = -6\epsilon + 2\eta, \tag{16.51}$$

so that (16.50) indeed imply $n_s \simeq 1$. It also predicts the observed coherent structure of acoustic peaks in the angular power spectrum of the CMB temperature fluctuations, essentially due to the super-horizon scale of the perturbations.

Fluctuations of the metric give rise to primordial gravitational waves, which translate into the so-called tensor modes of the CMB, not yet observed. The single inflaton model leads to a ratio of tensor to scalar fluctuations $r = 16\epsilon$, much below present or forthcoming observational bounds. This ratio is directly linked in single inflaton models to the inflaton field range $\Delta\phi$ covered during inflation, by the Lyth bound

$$\frac{\Delta\phi}{M_p} \simeq \left(\frac{r}{0.01}\right)^{1/2}. \tag{16.52}$$

In general, there are two large varieties of inflationary models, according to whether $\Delta\phi$ remains small in M_p units or is required to be super-planckian. Large r is only attainable in models with super-planckian field ranges, and thus requires ensuring the inflaton potential to be very flat over a large field range, a strong (and UV sensitive) requisite in building successful models of this kind. Large field inflation examples are provided by the *chaotic inflation* models, in which $V(\phi) = \lambda\phi^k/M_p^k$, which satisfy the slow roll conditions in the super-planckian large ϕ regime. A prototype of small field inflation is *hybrid inflation*, in which the slow inflaton roll reaches a regime where a second field becomes tachyonic and condenses, thus ending inflation. There are many other classes of models beyond these; in particular, models with multiple inflaton fields, or higher derivative terms, may give rise to other signatures, like non-gaussianities or isocurvature fluctuations, which might be detected in future experiments.

Let us close this brief summary of inflation with a generic problem in building specific models of inflation, known as the η *problem*. Inflation, and in particular the existence of a slow roll regime in a given theory, is very sensitive to its UV structure. In particular, the scalar potential in the effective theory may generically acquire terms of the form $\left(\mathcal{O}_4/M_p^2\right)\phi^2$, with $\mathcal{O}_4$ some dimension-4 coefficient, which need not be suppressed (in the absence of symmetries forbidding it). For instance, in the context of models with $\mathcal{N}=1$ supersymmetry, such operators appear upon expansion of the supergravity scalar potential (2.50) in powers of the inflaton. Also, once SUSY is broken, the inflaton generically gets mass terms analogous to the soft terms described in Section 2.6.3. Such terms lead to $\eta \simeq 1$ and spoil the slow roll inflation; in most inflation models, the η problem typically requires appropriate fine-tuning of the underlying parameters.

16.6.2 String theory models for inflation

In 4d string theory compactifications with a structure of scales $H_I \ll M_{\text{KK}} \ll M_s$, the physics of inflation lies in a regime which can be studied using the 4d effective field theory; yet the UV sensitivity of inflation may connect measurable quantities with properties of the underlying string model. Thus there are many efforts to explore the nature and physics of inflaton candidates among the different light scalar fields in string compactifications. There are two main classes of string inflation models, according to the nature of the inflaton candidate. One class realizes the inflaton as a closed string Kähler or complex structure modulus; in a second class of models, the inflaton is realized as an open string modulus, describing for instance the location of D-branes in the internal space of type II (orientifold) compactifications. Both classes are in principle quite attractive, since the corresponding scalar potential vanishes perturbatively (at least in the absence of fluxes), and further contributions may be expected to yield interesting scalar potentials for inflation. An important point is that establishing the existence of slow roll requires knowledge of the full scalar potential, to guarantee the absence of steep directions in field space spoiling slow roll; this implies that the construction of inflation models in string theory is necessarily correlated with the issue of moduli stabilization.

Closed string moduli as inflatons

Closed string moduli fields in string compactifications were considered as inflaton candidates already in the early days of heterotic string phenomenology. However, this possibility has only recently become sufficiently rigorous, mainly due to the advances in moduli stabilization. As described in Chapters 14 and 15, these are more developed in the context of type II orientifold compactifications. We consider the setup of type IIB compactifications with NSNS and RR 3-form fluxes, assuming they stabilize the dilaton and complex structure moduli at a high scale; the resulting 4d effective theory contains the Kähler moduli, which remain as flat directions of the flux potential, with a scalar potential arising from non-perturbative effects, as in Section 15.3.1, so it may be expected to be quite flat and well-suited for inflation.

It turns out that such inflating potentials are not realized for the simplest case of a single Kähler modulus; however, they are possible in the generic case of multiple Kähler moduli, as we now illustrate in a two-moduli CY example. Consider a type IIB orientifold on the CY $\mathbf{P}^4_{1,1,1,6,9}$[18], already appeared in Section 15.3.2, which has two Kähler moduli $T_{1,2}$. Assuming suitable D-brane instantons to generate non-perturbative superpotential terms controlled by these moduli, we have

$$W = W_0 + A e^{aT_1} + B e^{bT_2}, \qquad (16.53)$$

where W_0 is the constant value of the flux superpotential at its minimum, assumed to be small as in Section 15.3.1. The model as it stands leads to AdS vacua, and should eventually include an uplifting mechanism to a small and positive cosmological constant, e.g. $\overline{D3}$-branes. In this example, known as "better racetrack inflation," the inflaton is played by a linear combination of the two axion-like fields Im T_1, Im T_2, with slow roll inflation occurring near a saddle point in the scalar potential. It leads to predictions characteristic of small field inflation models, and with suitable choices of parameters produces a correct magnitude of density perturbations with a spectral index $n_s = 0.95$ and negligible r, in agreement with observations. The model is however very sensitive to the choices of parameters, requiring a fine-tuning around or beyond the percent level to achieve the slow roll regime.

Other variants of moduli inflation constructions are based on the large volume moduli stabilization in Section 15.3.2, and are known as "Kähler moduli inflation." The simplest toy model in this class involves at least three Kähler moduli T_i, whose Kähler potential has a swiss cheese structure, i.e.

$$K = -2\log \mathcal{V}; \quad \mathcal{V} = \frac{\alpha}{2\sqrt{2}}\left[(T_b + \bar{T}_b)^{3/2} - \sum_{i=1}^{n}(T_{s,i} + \bar{T}_{s,i})^{3/2} \right], \quad (16.54)$$

where $\mathrm{Re}\, T_b$ controls the overall volume of the CY. In the $n = 2$ case one also assumes there is a non-perturbative superpotential involving the "small" 4-cycle moduli $T_{s,1}$, $T_{s,2}$. The model yields minima at $\mathrm{Re}\, T_b$ much larger than the small moduli, one of which has an exponentially flat potential of the form

$$V \simeq V_0 - C(\mathrm{Re}\, T_c)^{4/3} \exp\left[-c\,(\mathrm{Re}\, T_c)^{4/3} \right], \quad (16.55)$$

where T_c is the modulus with canonical kinetic term. This inflaton candidate again leads to $n_s \simeq 0.96$ and negligible tensor fluctuations. In these models the η problem, mentioned in Section 16.6.1, ameliorates due to the extreme flatness of the potential. Their main short-coming is the difficulty to evaluate modifications of this flatness due to higher-order loop corrections, which can be estimated to be under control, yet are not explicitly computable.

Brane motion inflation

Another class of models is based on realizing the inflaton as a scalar field parametrizing the location of D-branes in the internal space of type II (orientifold) compactifications. We consider the prototypical case of D3-branes, in a type IIB compactification in the presence of $\overline{\text{D3}}$-branes, so that the brane–antibrane attraction leads to a non-trivial potential for the modulus controlling the inter-brane distance, denoted r (not to be confused with the ratio of tensor to scalar fluctuations). Assuming a single D3-$\overline{\text{D3}}$ pair, in the regime $\alpha'^{1/2} \ll r \ll L$, with L a typical length scale of the compact space, the potential is well-described by the 10d gravitational and RR Coulomb attraction; one obtains

$$V(r) = \frac{2\mu_3}{g_s}\left(1 - \frac{\kappa_{10}^2 \mu_3}{2\pi^3 g_s}\frac{1}{r^4} \right), \quad (16.56)$$

where κ_{10}^2 is the 10d gravitational constant and μ_3 is defined in (6.7). Here the first term describes the total brane–antibrane tension, and r should be regarded as the vev of the inter-brane distance modulus, to be identified with the inflaton ϕ. Unfortunately, the above potential yields too large slow roll parameters, in particular

$$\eta = M_p^2 \frac{V''}{V} = -\frac{10}{\pi^3}\left(\frac{L}{r}\right)^6 \simeq -0.3\left(\frac{L}{r}\right)^6, \quad (16.57)$$

where we have set $M_p^2 = L^6/\kappa_{10}^2$. Thus the conditions (16.50) require $r \gg L$, beyond the validity of (16.56); it can be shown that the situation is not significantly improved for anisotropic compactifications, or even using a potential valid for any r (i.e. using the complete gravitational and Coulomb potential in the compact space).

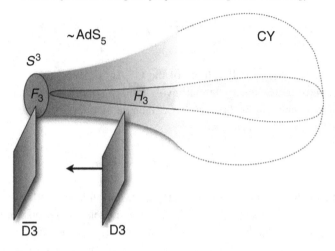

Figure 16.8 Brane inflation model in a warped throat.

This issue is solved in a different implementation of the brane–antibrane inflation idea, in which the objects are located on a warped throat, see Section 14.1.3, which induces additional suppressions due to its gravitational redshift; this setup fits well with the KKLT full moduli stabilization picture in Section 15.3.1, since the 3-form fluxes stabilizing moduli can support warped throats, and the uplifting $\overline{D3}$-branes are naturally driven to their tip. The inflaton candidate is thus the position modulus of a D3-brane, whose scalar potential does not get contributions from the 3-form fluxes, recall Sections 15.4.1 and 15.5.3, and is hence dominated by the interaction with the $\overline{D3}$-brane. Hence the dynamics of D3-branes falling onto $\overline{D3}$-branes at the bottom of a warped throat, sketched in Figure 16.8, can be expected to dominate the last (and only relevant for our observable universe) efoldings of inflation.

The main explicit example in this class, known as KKLMMT model, is based on the conifold throat introduced in Section 14.1.3. The D3-$\overline{D3}$ interaction is approximated by solving the Laplace equation for the gravitational and Coulomb potential in an AdS$_5$ geometry, and gives the result

$$V_0 = \frac{2\mu_3}{g_s} \frac{r_0^4}{R^4} \left(1 - \frac{1}{N} \frac{r_0^4}{r^4} \right), \tag{16.58}$$

where R is the AdS$_5$ length scale, N denotes the number of 5-form flux quanta, r_0 is the position of the end of the throat at which $\overline{D3}$-branes are located, and r is the D3-brane position, which should be regarded as the vev for the inflaton candidate ϕ. The warped geometry produces a redshift factor r_0^4/R^4 on the tension of the brane–antibrane system, and a factor r_0^8/R^4 on the attractive term; these suppression factors would in principle lead to a very flat potential, leading easily to slow roll inflation.

The story is a bit more complicated, however, since the moduli fixing dynamics gives rise to additional contributions to the scalar potential. In particular, as usual in models with an underlying 4d $\mathcal{N} = 1$ supergravity description, there is a new avatar of the above mentioned η problem; in the present setup, it is related to the generic presence of mass contributions for open string moduli after SUSY breaking, recall Section 15.5.3. The scalar potential has a structure

$$V = V_0 + H_0^2 \phi^2 + \delta V(\phi), \qquad (16.59)$$

where the second term is the mass arising from the supergravity scalar potential (2.50), and δV summarizes further corrections, discussed below. The mass term destroys the slow roll unless there are cancellations with extra pieces from δV. There are explicit computations of corrections δV sourced by non-perturbative effects depending non-trivially on the open string modulus ϕ, as in Section 13.1.3. A more systematic classification of possible contributions to δV is based on the use of the AdS/CFT correspondence, see Section 6.4, to map them into operators in the holographic dual field theory description of the throat, in Section 14.1.3; this analysis shows the possible existence of corrections $\delta V = \phi^\Delta$ with $\Delta = 1, 3/2, 2$. Such contributions may cancel the mass term and allow for slow roll inflation, at the cost of a moderate fine tuning in the parameters of the model.

An important general aspect of the brane–antibrane model is that the end of inflation occurs by the annihilation of branes and antibranes, as follows. The sector of open strings stretched between the two objects is very massive for large inter-brane distance, but its groundstate becomes tachyonic as they approach at distances below $\alpha'^{1/2}$, see Section 6.5.2. Condensation of this tachyon leads to brane–antibrane annihilation, and the release of the energy density stored in their tension during inflation. Interestingly, tachyon condensation in D3-$\overline{\text{D3}}$ brane systems can produce topological defects in 4d spacetime, corresponding to leftover fundamental strings (denoted F1s), D1-strings, or (p, q)-strings describing bound states thereof. These may remain as relics of the early universe, as we discuss further in Section 16.6.4. This annihilation does not allow for other stable topological defects like domain walls, a fortunate circumstance as their energy density (generated at reheating and so not diluted by inflation) would overclose the universe.

String inflation, tensor modes and non-gaussianities

The above examples for the realization of inflation in string theory share the property of being small field inflation models, and hence according to (16.52) do not produce sizable tensor perturbations. On the other hand there are already important experimental limits on the values of n_s and r from the WMAP satellite data, see Figure 16.9, and improved sensitivities in upcoming experiments over the next decade will allow to measure a non-negligible ratio r, or lower its bounds by several orders of magnitude. In the string inflation models described above, tensor modes are strongly suppressed (thick line close to the n_s axis in Figure 16.9). This raises the question of whether small r is a general signature of inflation models in string theory, or else which explicit string models lead to non-negligible values of r.

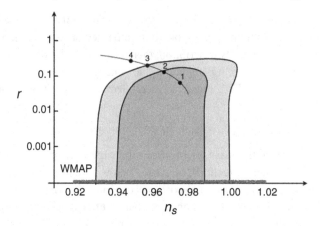

Figure 16.9 Observational limits from the CMB on the parameters r, n_s of slow roll inflation models. The shaded areas correspond to the 68% and 95% CL limits allowed by WMAP data and the thin line to ϕ^k *chaotic inflation* models with general k. D-brane, Kähler, and race-track inflation models correspond to the lowest thin band close to the n_s axis.

Although this is still an open question, its interest motivates us to brush over several ideas put forward to try and construct string models of large (super-planckian) field inflation, thus leading to sizable r. The so-called *N-flation* model considers using a large number N (in the hundreds) of axion-like fields, e.g. the imaginary parts of geometric moduli in CY compactifications with large Hodge numbers; this large number of scalars with field values of order M_p may collectively lead to an effect analogous to a single field with super-planckian range, leading to a sizable r ($k = 2$ in the Figure 16.9). A possible problem is that the large number of scalars induces a large renormalization of Newton's constant which reduces the field range. Another proposed approach, known as *monodromy inflation*, uses multiple circuits of a single periodic open string (axion-like) field. The model leads to a potential linear in the (canonically normalized) inflaton field ($k = 1$ in Figure 16.9), and leads to $r \simeq 0.07$, within reach of upcoming CMB observations.

A further interesting signature beyond the slow roll paradigm would be non-gaussianities in the spectrum of fluctuations; they are typically absent in single field inflaton models, but may appear, e.g., in multiple field inflationary models. Sizable non-gaussianities can also be realized in string theory, e.g. in the so-called *DBI inflation*; it realizes the inflaton as the position modulus of a D3-brane moving at relativistic speeds, so its description requires the full (higher derivative) DBI brane action, and goes beyond the slow roll scenario. We refer the reader to the Bibliography for more information on these and other string inflation scenarios.

Inflation and SM sector in string theory

The above discussion focuses on the realization of the inflaton in string theory, as a question decoupled from the realization of the SM sector. There are, however, certain questions in which the physics of both sectors is relevant, and which lead to further open issues in string inflation. Clearly one such question is reheating, i.e. the efficient transfer of the released inflationary energy density into SM degrees of freedom; this is a very model dependent issue, which certainly deserves further exploration to achieve a more complete realization of inflation in string theory.

But even at a more general level, there is a rather generic problem in the realization of inflation in models with moduli stabilization achieved in the KKLT framework of Section 15.3.1, or its extensions. In most inflationary models, the Hubble scale H_I is large, typically around, or a few orders of magnitude below, the GUT scale. On the other hand, string inflation models in the KKLT-like class lead to $H_I < m_{3/2}$, where $m_{3/2}$ is the gravitino mass, which is typically $m_{3/2} \leq O(1)$ TeV for $\mathcal{N} = 1$ SUSY to address the hierarchy problem. Indeed, raising H_I above $m_{3/2}$ in these string models spoils the moduli stabilization pattern, as follows. Consider the KKLT scenario extended by adding some inflaton in a chiral multiplet Φ. The scalar potential has the structure (setting $\kappa_4 = 1$)

$$V = e^K \left(G^{\Phi\bar{\Phi}} |D_\Phi W|^2 + G^{T\bar{T}} |D_T W|^2 - 3|W|^2 \right) + \frac{D}{(T + T^*)^3}. \qquad (16.60)$$

The height of the potential barrier before the addition of the inflaton piece is of order the AdS$_4$ vacuum energy before uplifting, $V \simeq |V_{\text{AdS}}| \simeq 3e^K |W_0|^2 \simeq m_{3/2}^2$; in order to maintain the structure of the moduli fixing minimum, the inflaton contribution to the scalar potential must be smaller than this barrier, i.e.

$$e^{K/2} G^{\Phi\bar{\Phi}} |D_\Phi W|^2 < m_{3/2}^2. \qquad (16.61)$$

However, this inflaton potential energy precisely determines the scale of inflation, hence we are forced onto $H_I < m_{3/2}$, as announced. The problem is actually rather general, and arises also in other moduli stabilization set-ups like the large volume scenario of Section 15.3.2. The underlying reason is that any scalars, and in particular the moduli, acquire an effective mass $\sim H_I^2$ due to their coupling to the inflationary de Sitter gravitational background, which competes with (and can override) the moduli stabilization mass scale. This clash between the inflation and SUSY breaking scales may be avoided by appropriate fine-tuning, or other proposed mechanisms. Still it is fair to say that there is a tension between large-scale inflation and low-energy SUSY in the moduli-fixing scenarios discussed so far.

16.6.3 The cosmological moduli problem

There are other issues related to the role of string moduli fields in cosmology, as we now review for the so-called cosmological moduli problem. One expects the string moduli (Kähler, complex structure and complex dilaton) to be abundantly produced in the early universe. In compactifications in which the moduli masses are generated upon SUSY breaking, their expected scale (for both scalar moduli and their modulino fermionic partners) is of order the gravitino mass; namely, $m_M \lesssim m_{3/2} \simeq \mathcal{O}(\text{TeV})$ in models with low-energy supersymmetry.

Modulinos in this mass range are actually problematic for cosmology, as follows. If they are stable, their density would overclose the universe, unless their mass is $\lesssim 1$ keV; if they are unstable, they are long-lived due to their gravitational strength coupling to SM fields, with an estimated lifetime

$$\tau_M \simeq \frac{M_p^2}{m_{\tilde{M}}^3} \simeq \frac{M_p^2}{m_{3/2}^3}. \tag{16.62}$$

Taking $m_{3/2} \simeq 1$ TeV, one has $\tau_M \simeq 10^3$ s, somewhat after primordial nucleosynthesis; their decays would produce a large number of high-energy photons which can photodissociate the nuclei, changing the ^{4}He and deuterium abundances, and hence ruining their remarkable agreement with observation. There is an analogous gravitino problem in general $\mathcal{N} = 1$ supergravity models, including $\mathcal{N} = 1$ string models. To avoid these problems, these fields should have masses m_M, $m_{3/2} > \mathcal{O}(10)$ TeV, slightly heavier than expected in low-energy SUSY; however, this bound may be avoided by invoking the dilution of primordial modulino density during inflation.

Scalar moduli lead to an analogous problem, similar to the so-called *Polonyi problem* of general $\mathcal{N} = 1$ supergravity models, and which is slightly more troublesome than for modulinos, as it is not satisfactorily solved by inflation. Indeed, during inflation the scalars ϕ_M are in general not localized around their zero temperature minimum, rather they are displaced by a distance of order $\delta\phi_M \simeq M_p$, given the flatness of the moduli potential. After inflation their oscillations create moduli particles behaving as non-relativistic matter, with energy density dominating over radiation, $\rho_{\phi_M}/\rho_{\text{rad}} \simeq 1/T$, as the universe cools down. As their fermionic partners, these fields are long-lived with $\tau \simeq M_p^2/m_M^3$, and their decay products may spoil nucleosynthesis unless $m_M > \mathcal{O}(10)$ TeV. Since inflation is over, it cannot help in diluting the scalar energy density, and this mass bound is more difficult to avoid. An additional problem of scalar moduli decays is the generation of large amounts of entropy, which can erase any preexisting baryon asymmetry.

The problem of the scalar moduli would disappear if $\delta\phi_M \ll M_p$ during inflation, but this condition is in principle unnatural. An interesting possible realization is the existence of an additional last stage of *thermal inflation*, at energies much lower than the original inflation; in this scheme the finite temperature corrections keep the moduli fields at small values, and give rise to a short inflationary period (of about 10 efoldings) diluting any previously existing moduli energy density, without affecting the density perturbations generated during ordinary inflation. There is another obvious solution to the cosmological moduli problem, by fixing all moduli at a high scale, e.g. using a general enough set of fluxes as in some examples in Chapter 14. Hence, the cosmological moduli problem is actually more severe for models of moduli stabilization at the scale of low-energy SUSY breaking dynamics.

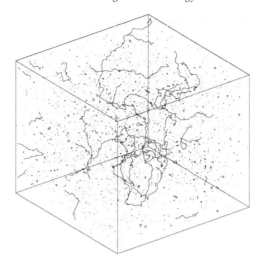

Figure 16.10 Computer simulation of cosmic string network (from Allen & Shellard (1990)).

16.6.4 Cosmic string relics

Cosmic strings and other defects may be present in the early universe, although their primordial density would be diluted during inflation. However, as already mentioned in Section 16.6.2, certain realizations of inflation in string theory lead to cosmic string production during reheating; prototypical examples are brane–antibrane (e.g. D3-$\overline{\text{D3}}$) inflation models ending in annihilation.

A relevant question is the amount of energy density stored in such cosmic strings once they are created. The total energy carried by a string is proportional to its length and hence (assuming stability, as we implicitly do, skipping further discussion) grows with the scale factor $a(t)$. The comoving volume scales like $a(t)^3$ and so the cosmic string density scales like $\rho_s \simeq 1/a(t)^2$. This would dominate over radiation and matter densities, which decrease as $1/a(t)^4$ and $1/a(t)^3$ respectively. However, the density in strings is depleted by a number of processes in the evolution of the cosmic string network. For instance, when two string segments collide they can recombine (intercommute) producing two kinked strings, which subsequently relax by straightening to reduce the total string length; also when a string intersects itself the loop breaks off and eventually decays through gravitational emission; more generally, long excited strings tend to straighten over time. Detailed simulations (see Figure 16.10) of string network evolution show that the ratio of the cosmic string density over radiation and matter densities is approximately constant with $\rho_s/\rho_r \simeq 400\,G_N\mu$ during the radiation dominated era and $\rho_s/\rho_m \simeq 60\,G_N\mu$ in the matter dominated era. Here G_N is Newton's constant and μ is the string tension.

The dimensionless quantity $G_N\mu$, describing the string tension in M_p units, is the relevant parameter for most physical questions on cosmic strings. In particular cosmic strings can create density perturbations, expected on dimensional grounds to be of order $\delta\rho/\rho \simeq$

$G_N\mu$. Explicit computations show that they are scale invariant, although their CMB power-spectrum is rather smooth and structureless, and hence very different from that predicted by inflation, and from the observed one. The constraints from the observed CMB anisotropies require in general $G_N\mu < 10^{-5.5}$; a more detailed fit of the CMB in a scenario including standard cold dark matter plus cosmological constant plus cosmic strings gives an upper limit $G_N\mu < 10^{-6}$.

Different classes of 4d string models lead to different cosmic string candidates with different possible values of $G_N\mu$. In heterotic string models, the string tension is $\mu = 1/(2\pi\alpha')$ and the gravitational constant is $8\kappa_4^2 = \alpha' g^2$, and so $G_N\mu = \alpha_G/(32\pi)$; taking $\alpha_G \simeq 1/24$, we obtain $G_N\mu \simeq 4 \times 10^{-4}$, which is excluded by the above bounds. Similar conclusions follow for the case of type I string compactifications. However, in type II (orientifold) models with lower-dimensional D-branes, it is possible to decrease the string tension μ by considering lower string scales, or D-brane sectors on warped regions of the compactification space. The latter case indeed occurs in the KKLMMT warped throat model, where the effective string tension is $\mu = e^{2A(r_0)}/(2\pi\alpha')$, where r_0 denotes the location of the string; thus, such effective string tension μ is in practice a free parameter. For realistic inflationary parameters in this scenario, the string tensions for F1-, D1-, and (p, q)-strings are

$$G_N\mu_F \simeq 2 \times 10^{-10}\sqrt{g_s}, \quad G_N\mu_D \simeq 2 \times 10^{-10}/\sqrt{g_s}, \quad G_N\mu_{p,q} = \mu_F\sqrt{p^2 + q^2/g_s^2}$$

$$(16.63)$$

(with g_s assumed fixed by moduli stabilization, e.g. 3-form fluxes in the CY bulk). More generally, there are models realizing values in a wide range $10^{-12} < G_N\mu < 10^{-6}$. Therefore, warped compactifications/throats with brane–antibrane annihilation lead to interesting cosmic string candidates with viable values of $G_N\mu$.

There are several ways to detect the existence of cosmic strings in the sky. The main is arguably gravitational lensing, with a very characteristic pattern of images symmetric with respect to an axis rather than a point; despite claims of candidate events in the past, none of them has been confirmed yet. A second detection opportunity is gravitational radiation emission from string cusps, which are generically present at least once in each loop of an oscillating cosmic string; the large energy density concentrated at the tip of the cusp produces an intense and characteristic gravitational wave beam, detectable at LIGO2 or LISA for values as low as $G_N\mu \simeq 10^{-10}$. Finally, observations of millisecond pulsars put strong limits on the cosmic string gravitational background; the gravitational radiation would disturb the timing of the pulses, so the observed pulse stability over more than two decades gives an upper limit $G_N\mu < 1.5 \times 10^{-8}c^{-3/2}$, where c is the average number of cusps on each string loop.

A natural further question is how to discriminate between cosmic strings in purely field theoretical models (like GUTs) and in string models. One difference is that cosmic string intercommutation in (perturbative) field theory generically has a probability $P \simeq 1$, whereas for a fundamental string, it is a quantum effect with probability $P \simeq g_s^2$. This probability plays an important role in the string network evolution, so detailed measurements of its gravitational wave spectrum could in principle allow to discriminate field theoretical and stringy cosmic strings. Another characteristic feature of certain string models is the possibility of complicated network topologies, with triple junctions involving different kinds of (p, q) strings.

In summary, there is a fair chance for forthcoming observations to detect experimentally the presence of cosmic strings from string theory. Such observations would not only give experimental support for string theory, but would also provide important information on the possible underlying string models.

17

The space of string vacua

In the preceding chapters we have displayed the wealth of possibilities to build specific 4d string vacua. In this chapter we take a more general viewpoint and present an aerial view of the space of string vacua (at our present level of knowledge), in particular of those corners with low-energy physics resembling closely the SM of particle physics. In spite of the apparent enormous number of consistent vacua, the space of field theories obtainable as low-energy effective descriptions of string theories can be argued to be strongly constrained, forming a minute subset in the space of all possible field theories. We present some of these general properties of string vacua, and then turn to overviewing the status of realistic vacua searches in different string compactification setups. These explicit models are presumably only the tip of the iceberg in the set of semi-realistic models in string theory, which we term the *flavour landscape*. In addition, the latter may be complemented by additional ingredients like flux degrees of freedom, yielding a large multiplicity of vacua, known as the *flux landscape*. Our outlook summarizes some general reflections on string theory and particle physics, inspired by this general perspective.

17.1 General properties of the massless spectrum in string compactifications

The spectrum and symmetries of light particles in 4d string compactifications is very model dependent. There are, however, several important general properties valid in any compactification, and which can provide non-trivial constraints in model building. These include restrictions on the 4d gauge group and light matter representations, the absence of global symmetries and other quantum gravity constraints, and the possible genericity of fractionally charged exotic states.

17.1.1 The gauge groups

In perturbative type II orientifolds, the gauge factors arise from D-branes, and lead to $U(N)$ or $SO(N)$, $USp(N)$ symmetries, depending on their relative interplay with the O-planes; it is not possible to obtain exceptional groups like E_6 at the perturbative level. The total rank of the gauge group depends on the number of D-branes in the corresponding vacuum, and is strongly constrained by the RR tadpole cancellation conditions. In fact, in specific orientifold models the actual gauge group may be quite small, and even reduce to just the SM. In general, however, the presence of additional hidden (or semi-hidden) sectors is fairly generic.

In perturbative heterotic compactifications there are non-trivial constraints on the gauge group. Its rank has to be necessarily smaller that $16 + 6 = 22$, and can include classical or exceptional Lie groups. The bound on the rank follows from the realization of spacetime gauge symmetries as Kac–Moody (KM) algebras on the left-moving worldsheet CFT, see Section E.3, as follows. The contribution of each gauge factor G, realized as a level k KM algebra, to the central charge is

$$c = \frac{k \dim G}{k + h_G}, \tag{17.1}$$

where h_G is the dual Coxeter number, given by the quadratic Casimir of the adjoint representation, e.g. $h_G = N, (2N - 2), 12, 18, 30$ for $SU(N), SO(2N), E_6, E_7, E_8$ respectively; for $U(1)$ factors, $h_{U(1)} = 0$ and $c = 1$. Since the total internal left-moving central charge must equal 22, there is a bound on the contributions from gauge factors G_i, as

$$c_{\text{KM}} = \sum_i \frac{k_i \dim G_i}{k_i + h_{G_i}} \leq 22. \tag{17.2}$$

For level $k = 1$ KM algebras of simply-laced groups, one obtains $c = \text{rk } G$, and the constraint reproduces the rank-22 bound mentioned above. Higher-level realizations or non-simply laced groups lead to stronger constraints on the gauge group rank. Alternatively, the condition (17.2) gives, for sufficiently large gauge groups, an upper bound on their KM level. For instance, for $SO(10)$, the maximum value for the level is $k = 7$; for smaller groups the bound is less restrictive, and, e.g., the SM group can contribute at most $c = 8 + 3 + 1 = 12$ units at any level, so this bound is useless. However, in practice it is somewhat difficult to construct models with higher-level gauge factors, see Section 9.8.

Non-perturbative heterotic models can be obtained by introducing NS5-branes, or using the Hořava–Witten description, so are easier to discuss using suitable dual descriptions, like type II orientifolds or their F- and M-theory generalizations.

In non-perturbative versions of the above compactifications, some constraints are avoided or relaxed. For instance, and in contrast with perturbative orientifolds, F- and M-theory models allow to realize exceptional gauge symmetries. In general they also lead to a priori more flexible tadpole conditions, and hence larger allowed gauge group ranks. For instance, F-theory compactifications on certain fourfolds with large Euler characteristic allow the introduction of thousands of bulk D3-branes. However, this large rank does not necessarily imply a large non-abelian particle physics group. For instance, gauge enhancement requires coincident branes, which can be argued to be a non-generic situation; also, particle physics models require chiral matter, and bulk D3-branes necessarily lead to non-chiral spectra.

A general lesson is therefore that bounds on *phenomenologically interesting* gauge groups are stronger, due to the constraints in realizing *phenomenologically interesting* matter representations, to which we turn next.

17.1.2 Massless matter representations are small

One very attractive, but often not fully appreciated, feature of string theory is that it predicts that massless particles must transform in small representations of the gauge factors; this is in sharp contrast with standard field theory, where there are no constraints on the gauge representations (beyond anomaly cancellation), and, e.g., the representation **2** of $SU(2)$ is in principle not more fundamental than the representation of dimension 137. In string theory the lightness of a particle constrains its possible gauge quantum numbers, as we now review for the different compactification setups.

The gauge representations in perturbative type II orientifolds, including in particular the massless spectrum, are very constrained, since they arise from the Chan–Paton indices at open string endpoints. Particles must necessarily lie in either bi-fundamental representations or 2-index tensor representations (antisymmetric, symmetric or adjoints). This is very well suited to describe the SM, as quarks and lepton quantum numbers may be accommodated in those representations. On the other hand, it is not possible to obtain massless spinor representations of $SO(N)$ groups, and hence it is not possible to build $SO(10)$ GUT models.

In perturbative heterotic compactifications, the possible gauge representations of light particles are somewhat more flexible, and include e.g. $SO(N)$ spinors or 3-index tensors of $SU(N)$. However, large representations are penalized, as follows. For a fixed level k of the underlying KM algebra, unitarity of the KM algebra representation implies a bound on the complexity of possible gauge representations; namely, the only possible representations are those satisfying

$$\sum_{j=1}^{\text{rk}\,G} n_j m_j \leq k, \tag{17.3}$$

where m_j are positive integers, sometimes called co-marks, known for every simple Lie algebra, and n_j are the Dynkin labels of the highest weight of the representation of the group G – an effective measure of its complexity. In particular, for $SU(N)$ all $m_j = 1$ and the only allowed representations at level $k = 1$ have Dynkin labels $(1, 0, 0, \ldots)$ and permutations. This corresponds to (anti)fundamental and completely antisymmetric representations of $SU(N)$, and would thus explain the role of $SU(2)$ doublet and $SU(3)$ triplet representations in the SM. On the other hand, it precludes the presence of light adjoints **24** of $SU(5)$ (or other GUT groups) at $k = 1$, and thus forces the realization of standard GUTs in string theory to the (more difficult) realm of higher-level KM algebras.

There are further constraints on which of these representations may be light, since the KM/gauge representation R contributes to the left-moving conformal weight of the state, and hence to its spacetime mass via (E.14). Concretely we have

$$\frac{\alpha' M^2}{4} = h_{\text{KM},R} + h_{\text{int}} + N_{\text{osc}} - 1, \quad \text{with } h_{\text{KM},R} = \frac{C(R)}{k + C(A)}. \tag{17.4}$$

Here h_{int} encodes other possible contributions to the conformal weight. We also define $C(R) = T(R)\dim G/\dim R$, where $T(R)$ is the second quadratic Casimir tr $(T_a^{(R)}T_b^{(R)}) = T(R)\delta_{ab}$; for $U(1)$, $h_{KM} = q^2/k_1$, where q is the charge and k_1 is the $U(1)$ normalization factor defined in Section 9.4. Hence lightness of a particle implies the bound $h_{KM} < 1$. This can be used, e.g., to constrain the KM levels using the SM massless representations: for a SM field the $SU(3) \times SU(2) \times U(1)_Y$ contribution to the conformal weight is

$$h_{KM; R_3, R_3, Y} = \frac{C(R_3)}{k_3 + 3} + \frac{C(R_2)}{k_2 + 2} + \frac{Y^2}{k_1} \leq 1; \qquad (17.5)$$

hence lightness of the right-handed electron ($Y = 1$) implies $k_1 \geq 1$. Alternatively, for fixed levels, it constrains the representations which can appear in the massless spectrum. For instance, for the most common case with $k_3 = k_2 = 1$, the massless matter sector can include only bi-fundamentals or antisymmetric representations; thus, e.g., $SU(2)_R$ triplets, often used in left–right symmetric models (to break the symmetry and to induce right-handed neutrino masses), cannot appear at level one. Note also that, for $SO(10)$ at level one, the **16** leads to $h_{KM} = 5/8 < 1$ and hence can (and does) appear in the massless spectrum of heterotic compactifications; on the other hand, the representation **126**, which used in the Higgs sector could be relevant for right-handed neutrino masses, can appear only for $k \geq 4$, making its realization in the massless spectrum very difficult (and in fact it has been shown to be impossible in any heterotic free fermion construction).

Remarkably, gauge representations in massless sectors remain severely constrained even in non-perturbative setups, like F- or M-theory compactifications; there, the massless representations are inherited from adjoint representations of larger gauge groups, through the unfolding in Sections 10.7.2 and 11.5.2. This constrains the possible representations to those available upon decomposition of the adjoint of, e.g., E_6, E_7, E_8, or other large classical groups. This allows the realization of $SO(10)$ spinors, but certainly implies strong restrictions for the possible massless representations.

17.1.3 Fractionally charged particles

String theory contains electric and magnetic charges under any $U(1)$ symmetry, so application of the Dirac quantization argument in Section B.3 implies in particular quantization of electric charge in multiples of some basic unit. The appearance of the correct values for hypercharge, and hence of electric charges, for SM fermions is expected in any SM-like string construction, since they are very much determined by anomaly cancellation. In GUT-like constructions, quantization of electric charge in multiples of the electron charge (for color singlets) applies even to possible particles beyond the SM, since it follows from the GUT multiplet structure. However, in string models without a GUT symmetry, the string spectrum can generically contain (either massless or massive) additional charged fermions beyond the SM, with non-standard hypercharge and hence fractional electric charges.

Fractionally charged states can be classified according to whether they are chiral or not under the SM group. The former are known as chiral exotics, and can only become massive once the EW symmetry is broken, so they are relatively light; on the other hand, non-chiral states can be massive to start with, or get large masses by diverse mechanisms, see below. Fractionally charged states are stable particles, since there is nothing they can decay into. Millikan-type experiments constrain their abundance to less than 10^{-21} fractionally charged particles per ordinary matter nucleon. Fractionally charged particles light enough to have been produced abundantly in the early universe lead to too large relic abundances, ruled out by the above bound. Thus fractionally charged chiral exotics are phenomenologically problematic; on the other hand, superheavy fractionally charged particles may be easily diluted during inflation and may not be problematic.

The origin of fractionally charged states is quite clear in the context of type II orientifolds with a SM structure, where they appear from open strings stretched between SM branes and extra "hidden" branes generically present in the model. They are massless or massive depending on the "distance" between these two sectors. As a simple example, consider the SUSY example in Section 10.6 with massless spectrum in Table 10.6. In Pati–Salam $SU(4) \times SU(2)_L \times SU(2)_R$ terminology, it contains three quark/lepton generations, one Higgs multiplet, and two copies of exotics transforming as $(1, 2, 1) + (1, 1, 2)$. The latter arise from open strings between the $SU(2)_L$ (or $SU(2)_R$) branes and the hidden branes h_1, h_2; they have fractionally charged chiral exotics, with electric charge $\pm 1/2$, and are initially massless. In other models, fractionally charged particles may already be massive, or come in vector-like combinations; many such examples have been constructed in terms of type II RCFT orientifolds.

Fractionally charged states are also rather pervasive in perturbative $(0, 2)$ compactifications of the heterotic string, as manifest by direct inspection of explicit fermionic models or abelian orbifolds, such as the $\mathbf{Z}_3$ example in Section 9.7. There is in fact a theorem ensuring that compactifications containing the SM group, with underlying KM algebras realizing $SU(5)$ normalizations $k_3 = k_2 = 1$, $k_1 = 5/3$, either contain fractionally charged exotics or else lead to a full $SU(5)$ gauge symmetry group. The argument, somewhat similar to the proof of the absence of global symmetries in string theory discussed below, runs as follows.

Given a level k $SU(N)$ KM algebra one can define simple currents (see Section E.5.2) $J_N^{(i)}$, $i = 0, \ldots, N - 1$, with conformal dimension $h_i = k(N - i)/2N$. The monodromy (E.36) of a primary field Φ_R in the $SU(N)$ representation R under $J_N^{(i)}$ is given by $Q_{J_N^{(i)}}(\Phi_R) = it/N$, where t in the N-ality of R; for $U(1)$ KM algebras, there is an $h = k/4$ simple current J (with k the normalization factor introduced in Section 9.4) with monodromy charges given by the $U(1)$ charges themselves. Note also that, from the partition function (E.37), a simple current J under which all fields in the spectrum have integer monodromy, is a current extending the Virasoro algebra, and is in the spectrum itself. Consider now a compactification with SM group, and take the simple current $J = \left(J_3^{(1)}, J_2^{(1)}, J_1 \right)$ corresponding to the three SM gauge factors. The monodromy charge of a field Φ under J is

$$Q_J(\Phi) = \frac{t_3}{3} + \frac{t_2}{2} + Y. \tag{17.6}$$

One can check that the condition $Q_J(\Phi) \in \mathbf{Z}$ is precisely the integrality of its electric charge. Imposing this condition for all fields in the theory implies, by the above argument, that J itself must be in the spectrum, and $h_J = k_3/3 + k_2/4 + k_1/4$ must be integer. In particular, for $k_3 = k_2 = 1$, $k_1 = 5/3$, the conformal dimension is $h_J = 1$ and J produces a massless gauge boson with SM quantum numbers $(\mathbf{3}, \mathbf{2}, 5/6)$. These gauge particles, together with their conjugates, actually enhance the gauge symmetry to $SU(5)$, as announced.

The presence of massless fractionally charged vector-like exotics may not necessarily be a problem. In many heterotic examples, such exotics are present at tree level, but become very massive in the one-loop vacuum re-stabilization triggered by the non-vanishing FI term, as in Section 9.7.

Finally, string models with underlying GUT symmetries broken in the compactification, e.g. like heterotic CY compactifications with discrete Wilson lines, also contain fractionally charged states. They arise from twisted states of the freely-acting twist, (i.e. strings wrapped around the non-trivial homotopy 1-cycles); these are very massive due to the stretching, hence not very problematic.

17.1.4 Absence of continuous global symmetries

It is believed on general grounds that in consistent theories of quantum gravity there are no continuous global symmetries. The argument is based on a thought experiment, in which particles charged under the symmetry are accreted to form a black hole, which eventually evaporates into a thermal bath of Hawking radiation with zero total charge, thus leading to an overall violation of charge conservation.

String theory is a consistent theory of quantum gravity, and hence should not lead to vacua with continuous global symmetries; indeed, any seeming continuous global symmetry turns out to be gauged at some level. Let us illustrate the argument in the case of a $U(1)$ symmetry in the closed bosonic string theory (see Appendix E for background). A global $U(1)$ symmetry must be necessarily realized in terms of a conserved charge on the worldsheet, which may be written as

$$Q = \frac{1}{2\pi i} \oint \left[dz\, j(z) - d\bar{z}\, j(\bar{z}) \right]. \tag{17.7}$$

For this to be conformally invariant, the currents $j(z)$, $j(\bar{z})$ must have conformal dimensions $(1, 0),(0, 1)$; but then they may be used to construct vertex operators

$$j(z)\, \bar{\partial}\, X^\mu e^{ik\cdot X}, \quad j(\bar{z})\, \partial\, X^\mu e^{ik\cdot X}, \tag{17.8}$$

which have conformal weights $(1, 1)$ and thus correspond to spacetime gauge bosons. Hence the symmetry defined by Q actually corresponds to a spacetime gauge symmetry. This argument may be easily generalized to type II strings, heterotic strings, and open string sectors. Due to duality symmetries, one expects its validity also in non-perturbative frameworks like F- and M-theory.

The above derivation allows for a couple of exceptions, which, however, do not contradict the general lore on quantum gravity. First there are continuous axionic shift symmetries $a \rightarrow a + \chi$ of certain moduli, e.g. manifest in the classical actions in most string compactifications; these symmetries have no associated generator Q in the worldsheet action, since there are no worldsheet fields transforming under them, and hence there is no associated gauge symmetry. These are, however, not true continuous global symmetries, as they are explicitly broken to some discrete subgroups by worldsheet or spacetime instantons, as explained in Section 13.1.2 – in fact, they correspond to discrete *gauge* symmetries of the theory. A second exception is the Lorentz symmetry of uncompactified dimensions; in this case, it turns out that, due to the non-compactness of Minkowski space, the corresponding $j(z)$ and $j(\bar{z})$ operators do not allow for the construction of any new gauge boson vertex operator. This symmetry also ends up as a *gauge* symmetry, associated to spacetime coordinate reparametrizations, precisely when taking gravity into account.

At the practical level, there is however a natural source of effectively global *perturbative $U(1)$* symmetries in string vacua. As discussed in Section 12.4, in type II orientifolds the D-branes contain $U(1)$ gauge symmetries, whose gauge bosons become massive by a Stückelberg $B \wedge F$ coupling to a RR 2-form field; the $U(1)$ symmetry remains as a perturbatively exact global symmetry, only broken at the non-perturbative level by D-brane instanton effects. A phenomenologically interesting example is the baryon number, which often arises in SM D-brane configurations, and automatically ensures proton stability, see Section 16.3.

This phenomenon is rather different in heterotic compactifications. In compactifications not involving $U(1)$ gauge backgrounds, such as orbifolds or smooth CYs with $SU(n)$ backgrounds, the only modulus potentially mixing with a $U(1)$ is the 4d dilaton S. The corresponding FI-term is proportional to the string coupling Re S, and cannot be set to zero; it rather must be canceled by large chiral multiplet vevs, which thus break the $U(1)$ symmetry at the perturbative level, leaving no approximate remnant global symmetry. In heterotic vacua with $U(1)$ gauge backgrounds, the discussion is more similar to the type II setup, although some of the $U(1)$ symmetries may be violated by worldsheet instanton effects, which are non-perturbative in α', but perturbative (in fact, tree level) in g_s.

In any event, none of these symmetries represent a violation of the general folk theorem mentioned above. From a fundamental viewpoint, they are gauged; from an effective field theory perspective (below the gauge boson mass), they are only approximate global symmetries, violated by non-perturbative effects.

17.1.5 Further proposed constraints from quantum gravity

There are other general properties, on slightly less rigorous grounds, which have been conjectured to hold in any string vacuum, and whose validity is ultimately connected to the consistency of string theory as a theory of quantum gravity.

- *Finiteness of the volume of the moduli space of scalars*: The volume of the moduli space of scalars turns out to be finite in string vacua. Consider for example the volume of the type IIB complex coupling, given by

$$V_S = \int_{F_0} \frac{d^2 S}{(S + S^*)^2},$$

(17.9)

where the integration is over the $SL(2, \mathbf{Z})$ fundamental domain, so the result is finite thanks to S-duality. Similarly, the volume of the Narain moduli space in type II or heterotic toroidal compactifications is finite, due to T-duality symmetries. Yet another example is the moduli space of D3-brane positions in a CY compactification, whose volume is essentially the volume of the compactification manifold. These examples illustrate that finiteness of the moduli space volume actually holds for fixed finite M_p; indeed, in the limit $M_p \to \infty$ the moduli space volumes are in general no longer finite, e.g. the moduli of a non-compact CY are not bounded, or neither are the D3-brane locations on a non-compact transverse space. Hence, the property can be regarded as a constraint for a consistent coupling to gravity.

- *Finiteness of the number of the massless states*: Indeed there is no known string compactification in which the number of massless states is infinite. This is again not true in the absence of gravity, $M_p \to \infty$, since in non-compact CYs there are infinite sequences of vacua, e.g. an arbitrarily large number of D3-branes on a singularity, with an arbitrarily large number of massless states.

- *Gravity as the weakest force*: The absence of global symmetries in a theory containing quantum gravity, as described in Section 17.1.4, implies the existence of a lower bound on the gauge coupling g of any gauge symmetry, since it becomes effectively a global symmetry in the limit $g \to 0$. In fact, it is possible to use charged black hole thought experiments to argue that, e.g., a $U(1)$ gauge theory with coupling g necessarily has an ultraviolet cutoff at a scale $\Lambda = g M_p$, rather than the naive M_p; hence the limit $g \to 0$ is not smooth, since the cutoff goes to zero in that limit. A manifestation of the cutoff is conjectured to be the existence of at least a charged state with mass $m < g M_p$; this has the important implication that any gauge interaction must obey the bound

$$g > \frac{m}{M_p}$$

(17.10)

with m the mass of the lightest charged particle. This can be stated as gravity being the weakest force in any consistent theory; in string theory, there is no example violating this bound. Note again the bound disappears as $M_p \to \infty$.

The existence of these and other proposed constraints for theories to admit a consistent coupling to quantum gravity, shows that there is a vast space of field theories (sometimes called the *swampland*) which cannot possibly be obtained as a low-energy limit of string theory.

17.2 The flavour landscape

In this section we shift from general properties of string vacua to the space of vacua of potential phenomenological interest, namely with a SM gauge group (or small extensions thereof) and three (net) quark-lepton generations. This is interesting to explore the extent to which the above general properties constrain the structure of string realizations of the SM. Unfortunately the only means to study this problem appears to be the brute force systematic construction of models, via different computer searches, whose present status we overview. This merely scratches the surface of the set of SM-like string models, which we term the *flavour landscape*, and in particular most of the models explicitly constructed are not even expected to be *the* string theory model realized in Nature; yet they can be used to draw lessons on the general properties of particle physics when realized in string theory.

17.2.1 Systematic MSSM-like model building

Most of the efforts have been devoted to the search for supersymmetric MSSM-like (rather than non-supersymmetric SM-like) vacua, as illustrated in the examples in previous chapters. The most systematic searches of realistic 4d compactifications have focused on perturbative heterotic and type II orientifold models, and have led to models with diverse degrees of subsequent phenomenological study, as follows.

Heterotic models

- For $(2, 2)$ and $(0, 2)$ compactifications on smooth CYs, a few examples have been worked out in detail (e.g. see the models in Sections 7.3.3 and 7.4.2).
- In abelian toroidal orbifolds, there are a number of MSSM-like models, mostly based on $\mathbf{Z}_3, \mathbf{Z}_6', \mathbf{Z}_2 \times \mathbf{Z}_2, \mathbf{Z}_7, \mathbf{Z}_{12}$, including also some large scans, although only a few orbifold examples have been analyzed in some detail (e.g. see Sections 8.3.3 and 9.7).
- In the fermionic construction, there are large scans, particularly with $\mathbf{Z}_2 \times \mathbf{Z}_2$ boundary conditions, leading to large classes of models with SM, flipped $SU(5)$ or Pati–Salam groups, with a few examples analyzed in detail (e.g. see Section 8.5.2).
- For $(0, 2)$ Gepner-like models, some 4000 SM-like models have been built, although their detailed phenomenology remains unexplored (see Sections 8.6.5 and 17.2.2).

Type II orientifolds

- In orientifolds of toroidal orbifolds there are a few examples of SM-like models based on $\mathbf{Z}_2 \times \mathbf{Z}_2$, and a large class based on the $\mathbf{Z}_6$ and $\mathbf{Z}_6'$, a few of which have been analyzed in some detail (e.g. see the models in Sections 10.6, 11.4.2, 12.6, and 13.3.3).
- For non-SUSY toroidal models, there is the class of models with just the SM particle content of Section 10.5.1 and generalizations.
- In the class of models of D3-branes at singularities, a few semi-realistic models have been analyzed (e.g. see Sections 11.3.3 and 11.3.4).

- There is a huge class of RCFT orientifolds, with MSSM-like spectra, whose phenomenology remains unexplored (see Sections 10.7.1, 16.1.3, and 17.2.2).
- There are a few examples of local F-theory GUT models with MSSM spectrum (see Section 11.5.3).

In total, there are only a few tens of MSSM-like string compactifications for which a detailed phenomenological analysis, beyond the mere determination of the massless spectrum, has been carried out; by "detailed analysis," we mean, for instance, addressing issues like $U(1)$ anomaly cancellation, vacuum re-stabilization, and flat directions in heterotic models, Yukawa coupling structure, gauge coupling unification, proton stability, neutrino spectra, SUSY-breaking, moduli stabilization, etc. Some of these models get quite close to the observed physics (or rather, its SUSY extension), but it is fair to say that there is no particular model passing all tests *at the same time*. The most prominent bottleneck is the issue of full moduli stabilization, and the realization of realistic Yukawa structures.

17.2.2 Flavour statistics

In most of the above systematic searches of MSSM-like models, the algorithm often requires scanning through a larger class of consistent 4d compactifications, including also models with non-MSSM spectrum, but rather arbitrary gauge groups and chiral multiplet content. This has motivated the statistical analysis of the relative abundance of certain properties, e.g. particular gauge groups, or numbers of families. We summarize some of these results, see the Bibliography for references.

Heterotic models

- A large scan of $(0, 2)$ compactifications, based on the 168 Gepner models with A and E modular invariants, produced a set of 85 000 3-generation models with SM group or extensions thereof (including LR symmetric, Pati–Salam, $SU(5)$ and $SO(10)$ gauge groups) with some 4000 having the SM group and three net generations; beyond the MSSM spectrum, the models include additional Higgs multiplets, vector-like multiplets, and typically chiral or vector-like fractionally charged particles. Models in this class have typically a small number of generations, with a distribution shown in Figure 17.1 (excluding the case of no net generations, which dominates), with one and two generations being most generic.
- The fermionic construction has been used to explore the realization of $\mathbf{Z}_2 \times \mathbf{Z}_2$ orbifold vacua with Pati–Salam gauge group; out of approximately 10^{11} models constructed, a fraction 3×10^{-3} corresponds to 3-generation models.
- Toroidal orbifold compactifications based on $\mathbf{Z}_6'$, with discrete Wilson lines, have allowed the construction of order 10^7 models; some 300 lead to a SM gauge group and three generations (and also reasonable gauge coupling unification and a heavy top quark).

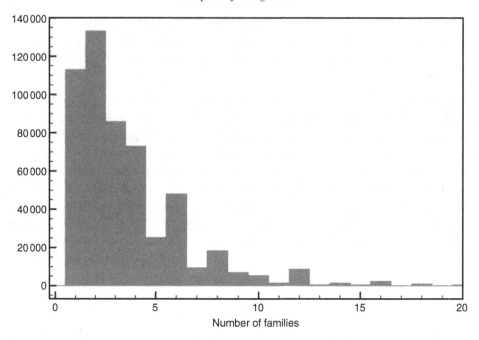

Figure 17.1 Distribution of number of chiral families for heterotic $(0, 2)$ Gepner MSSM-like models (taken from Gato-Rivera and Schellekens (2010a)).

Type II orientifolds

- For RCFT orientifolds based on the 168 Gepner models, and using of order 10^{20} tadpole free combinations of boundary states (i.e. sets of D-branes), there have been constructed some 200 000 explicit 3-generation models with SM group (or LR extensions). The massless spectrum typically contains additional vector-like multiplets and vector-like fractionally charged states. Similar searches for different numbers of families indicate that in this class the ratio of 3-generation to 1/2-generation models is about 10^{-3}, see Figure 17.2.

- In type IIA intersecting D6-brane models in the $\mathbf{Z}_2 \times \mathbf{Z}_2$ toroidal orientifold, a systematic search produced around 10^8 explicit models, with rather general gauge groups and numbers of families. In this search, none of the models actually corresponds to a 3-generation SM-like theory, and in particular the model in Section 10.6 is absent, due to computational limitations. The fraction of 3-family SM theories was statistically estimated to be 10^{-9}.

- Type IIA intersecting D6-brane models in the $\mathbf{Z}_6$ and $\mathbf{Z}_6'$ toroidal orientifolds, have allowed the construction of order 10^{28} $\mathbf{Z}_6$ models, with a fraction of 10^{-22} leading to 3-generation MSSM-like models (all containing chiral exotics), and some 10^{23} $\mathbf{Z}_6'$ models, with a fraction of 10^{-8} corresponding to 3-generation MSSM-like models, out of which a further fraction of 10^{-8} contains no chiral fractionally charged exotics.

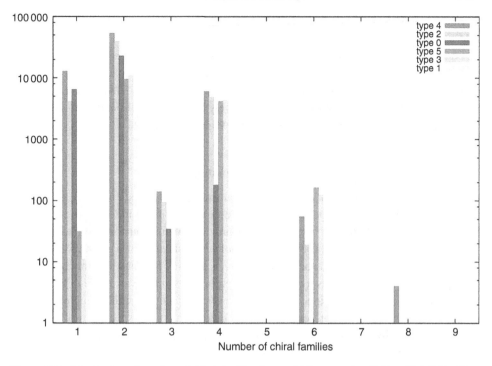

Figure 17.2 Distribution of number of chiral families in type II Gepner orientifolds with MSSM-like structure. The notation for the models is as in Section 10.7.1 (taken from Dijkstra *et al.* (2005b)).

Three-generation models are 10^4 times less frequent than the one-generation case, but 10^3 times more frequent than two generations.

The interpretation of the above information is very much open to discussion. Let us simply emphasize that one must resist the temptation to extract statistical conclusions, for a number of reasons. First, the above scans admittedly explore a minute subset of the complete set of string compactifications, since toroidal orbifold geometries or Gepner models relate to compactifications on very particular CYs, the full general class being considerably much larger. In addition, there are classes of promising realistic setups, like F-theory fourfold compactifications, which are still largely unexplored. Also, the methods to extract statistical information in the above different scans have different biases, associated to computational limitations or to different priors (e.g. the choice of spectra of the initial models among which to search the SM structure). Finally, the above models do not account for moduli stabilization and supersymmetry breaking. Although these ingredients may be restricted to the moduli sector, they have a non-trivial impact on the visible sector through, e.g., flux contributions to the RR tadpole conditions, which may change the above information on correlations or relative abundances of certain properties.

Still, some general lessons may be drawn. For instance, the realization of models with SM group and three generations seems relatively rare, compared to the other possible structures. Still the large set of possible 4d compactifications suggests the existence of a large *flavour landscape* of string vacua with properties very close to those of the MSSM, whose full extension is not yet known. Only some tens of explicit solutions have been analyzed in detail, none of which has been conclusively shown to fulfill *simultaneously all* the phenomenological requirements to describe the observed physics.

17.3 The flux landscape

A realistic string vacuum should include mechanisms for moduli stabilization and supersymmetry breaking, for instance in terms of the fluxes discussed in Chapters 14 and 15; hence any model in the flavour landscape in the previous section must be complemented by suitable fluxes. Since there may be many flux choices consistent with a given visible sector, any model in the flavour landscape potentially leads to many variants differing in their flux configurations. Clearly there is a non-trivial interplay between the structure of the visible sector and the fluxes, as mentioned below. However, the analysis of the combined system is in general too difficult at present, and most descriptions of the space of flux vacua are carried out ignoring the structures required to accommodate a realistic visible sector, or using simple parametrizations of its expected impact in the flux sector. We thus adopt this viewpoint in this section.

To illustrate the large multiplicity of flux vacua, let us concentrate on the simple case of type IIB orientifolds with NSNS and RR 3-form fluxes H_3, F_3. A flux configuration is determined by the flux quanta along the independent 3-cycles, whose number is the third Betti number $b_3(\mathbf{X}_6)$ of the CY $\mathbf{X}_6$; we denote the flux quanta collectively by n_i, $i = 1, \ldots, 2b_3$. The contribution of fluxes to the RR 4-form tadpole is schematically

$$\sum_{i,j} n_i Q_{ij} n_j \leq L, \tag{17.11}$$

with Q defining a suitable bilinear product. The integer L should correspond to the RR tadpole left unsaturated by the D-brane sector, and can be assumed to typically lie in the range 10–100. An estimate of the number of flux configurations in a fixed CY may be obtained by allowing for, e.g., 10 different values for each flux quanta, and taking a typical value for b_3 in the range 100–300, yielding $10^{2b_3} \sim 10^{100}$–10^{600} flux vacua.

This number should be taken with several grains of salt. For starters, and contrary to widespread statements in journalism and popular science, it should not be considered as counting the number of consistent (or realistic) string vacua in any sense; the latter should account (among many other issues) for e.g. the large number of possible choices of compactification space, i.e. of CY manifolds in the simplest setup, which is in fact only *conjectured* to be finite. Taken at face value, the above number is an estimate of the multiplicity of 3-form fluxes in a particular CY. It may also be interpreted as an estimate of the multiplicity of 3-form flux configurations consistent with a given visible (e.g. realistic) D-brane sector, although there are several caveats due to the interplay of fluxes and D-branes. For instance, the D-brane contribution to the RR tadpole, assumed above to be parametrized through L. Also, the possible fluxes that may be added to a given D-brane configuration are restricted by the Freed-Witten constraints described in Section 14.3.1. In addition, a given D-brane configuration is stable in particular regions of moduli space (e.g. due to dynamical brane recombination processes away from it), and hence should be coupled only to fluxes stabilizing moduli on those regions. Furthermore, full moduli stabilization may involve non-perturbative instanton effects as in Section 15.3.1, or generalized fluxes as in Section 14.4, not included in the above discussion.

Still, it may not be a bad approximation to assume that each semi-realistic D-brane sector in the type IIB orientifold flavour landscape can be coupled to a flux sector, with a similarly large multiplicity of flux choices. We may thus assume this picture, and study its possible implications/applications. For each flux choice the scalar potential and the corresponding dynamics change as described by the flux superpotential, whose rough structure is

$$W_{\text{flux}} = \sum_i^{2b_3} n_i \Pi_i(U),$$ (17.12)

where U denotes the dilaton and complex structure moduli, and $\Pi_i(U)$ are computable holomorphic functions. The large multiplicity of flux choices can then lead to vacua in which there are accidental cancellations, resulting in tiny adimensional quantities, which would be regarded as unnatural from other standpoints. For instance, the $\mathcal{O}(10^{2b_3})$ choices of flux parameters can lead to cancellations among the $2b_3$ additive terms in (17.12), resulting in very small values of the superpotential at the minimum, of order $W_0 \simeq 10^{-16}$ in Planck units, as required in the KKLT moduli stabilization proposal in Section 15.3.1.

An interesting application along these lines is to use the large number of flux choices to accommodate the otherwise vexing fine-tuning of the cosmological constant Λ, as already anticipated in Section 1.3.1. The general idea, put forward by Bousso and Polchinski, may be illustrated in a simple toy model. Consider a scalar potential depending on some scalars ϕ and some integer flux quanta n_i, $i = 1, \dots, N$ of the form

$$V = V_0(\phi) + \sum_{i,j} g_{ij}(\phi) n^i n^j$$ (17.13)

where g_{ij} is some positive definite metric in the moduli space, with $(\det g)^{1/N} \ll M_p^4$. The moduli scalars ϕ are ignored in the remainder, and simply assumed to sit at their minima, leading typically to a negative V_0 of order M_p^4 (or smaller), as is often the case in $\mathcal{N} = 1$ supergravity vacua. The number of flux vacua δN_{vac} leading to a cosmological constant Λ, with precision $\delta\Lambda \ll |V_0|$, can then be computed as the number of points in the N-dimensional lattice of flux vectors n_i lying on a hypersphere shell $\mathbf{S}^{N-1}$ of radius $R^2 = |V_0| + \Lambda$ and thickness $\delta R = \delta\Lambda/(2R)$, see Figure 17.3. In the case of physical interest, $\Lambda \ll M_p^4 \simeq |V_0|$, so the sphere is large in lattice spacing units and the number of

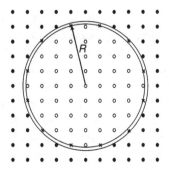

Figure 17.3 The set of flux configurations finely scans through possible values of the cosmological constant.

points in the shell can be estimated in a continuous approximation, as proportional to its volume

$$\delta \text{Vol}(R) = \frac{1}{\sqrt{\det g}} \frac{(\pi R^2)^{\frac{N}{2}}}{(\frac{N}{2})!} \delta R. \tag{17.14}$$

This produces a distribution for Λ of the form

$$\delta N_{\text{vac}}(\Lambda) \simeq \frac{1}{\sqrt{\det g}} \frac{\pi^{\frac{N}{2}} (|V_0| + \Lambda)^{\frac{N}{2}-1}}{(\frac{N}{2}-1)!} \delta\Lambda \simeq \left[\frac{2\pi e (|V_0| + \Lambda)}{\mu^4} \right]^{\frac{N}{2}} \frac{\delta\Lambda}{|V_0| + \Lambda}, \tag{17.15}$$

where $\mu^4 \simeq N (2\pi \det g)^{1/N} \lesssim M_p^4$ gives a measure of the size of the flux piece of the scalar potential; also e is the base of natural logarithms, and the last equation uses Stirling formula for large N. To illustrate the implications of this distribution, consider the number of vacua required for a small cosmological constant $\Lambda \simeq 0$, consistent with cosmological observations. Taking, e.g., $|V_0|/\mu^4 \simeq 10$, one obtains

$$\delta N_{\text{vac}} \simeq 10^{\frac{N}{2}} \frac{\delta\Lambda}{|V_0|}. \tag{17.16}$$

So that for $N \simeq 2b_3 \simeq 100 - 600$, the multiplicity is so large so that it is reasonable to expect the existence of vacua with $\Lambda \sim 10^{-120} M_p$. Although the above is a toy model, such structure has been encountered in more realistic scalar potentials derived from type IIB flux compactifications in explicit CY manifolds. Its realization in the mechanisms of full moduli stabilization in Section 15.3, including the effect of uplifting, is less clear; however, the basic intuition that the large multiplicity of flux vacua allows to finely scan through different values of the cosmological constant seems a lesson valid in more general setups.

The large multiplicity of flux vacua allows for some vacua to display an accidentally small cosmological constant, with no underlying naturalness argument. But more interestingly, it also allows for an environmental (i.e. anthropic) explanation of the observed small cosmological constant, along the lines advanced in Section 1.3.1; any mechanism allowing the universe to populate the set of flux vacua (e.g. a string theory embedding of eternal inflation) would lead to a situation where the cosmological constant measured by any observer is necessarily small, to be consistent with galaxy formation and the very existence of observers. This constitutes a possible solution for the cosmological constant problem in string theory.

This viewpoint does not necessarily imply giving up completely on the traditional search for fundamental explanations of physical properties, but rather complements it. In particular, a natural question in this framework is whether there are other magnitudes scanned over in the space of flux vacua, suggesting an environmental (rather than fundamental) explanation. The natural *suspects* are magnitudes for which small variations can easily spoil properties allowing the existence of complex systems and hence observers. It is certainly difficult to draw a clear borderline between "environmental" and "fundamental" quantities, but there are interesting proposals for both. For instance, the proton lifetime is experimentally above 10^{32} years, thus far longer than the age of the universe or any obvious timescale associated to the existence of observers; it is hence expected to admit a fundamental explanation, arising from properties generic in the space of string vacua, as reviewed in Section 16.3. Other examples may be the QCD θ parameter, the existence of several generations, etc. On the other extreme, the precise up, down and electron masses may be regarded as tuned, since small variations of their values ruin the possibility of complex chemistry; for instance, the d quark is just massive enough to make the proton lighter than the neutron (permitting the existence of stable hydrogen), yet light enough to make the deuteron stable (enabling its role in the formation of heavier nuclei). Hence the u and d quark mass difference may (arguably) be bound to an environmental explanation. Note in this respect that in string compactifications the lightest quark and lepton masses arise from small corrections, e.g. instanton effects or non-renormalizable couplings, depending on moduli vevs, so are likely to be finely scanned over in the set of flux vacua.

17.4 Outlook

Our knowledge of the structure of the space of string vacua is still very limited. It is mostly based on the analysis of the low-energy effective field theory in the weakly coupled and large compactification volume limits, in which semiclassical arguments are reliable. We have a pretty good understanding of vacua like toroidal orbifolds, free fermions, or Gepner models, for which worldsheet CFT methods allow for an explicit computation of many quantities of phenomenological interest, like gauge coupling constants or Yukawa couplings; our knowledge of compactifications on general CY spaces (or their generalizations in the presence of fluxes) is still rather modest, and the study of F-theory and M-theory vacua is still in its infancy. Still, the progress in the last two decades has been impressive and our view of the space of string models, and how they relate to the realization of the properties of the SM of particle physics, is now far richer; new large classes of string compactifications have been found, and there are new available mechanisms to address moduli stabilization and SUSY-breaking.

String theory has a number of attractive features supporting its proposal to describe a unified theory of all interactions:

- String theory implies the very existence of gravity, as the graviton is always present as a massless closed string mode. It is at present the only candidate for a consistent theory of quantum gravity coupled to matter. It also gives a microscopic description of black hole degrees of freedom, and provides an explicit realization of holography through the AdS/CFT correspondence.
- It provides an ultraviolet completion of the SM interactions, thus allowing for the possibility of a unified fundamental (as opposed to an effective) theory of all interactions.
- String theory has no external adimensional parameters; they are all derived from the vevs of scalar moduli fields at the physical minima. All standard dimensionful fundamental constants are derived from reintroduction of the scale-fixing constants c, $\hbar$, and a length scale $(\alpha')^{1/2}$ (or its analog in M-, F-theory).
- String theory provides an understanding of family replication, since generically string SM-like vacua contains more than one quark-lepton generation.
- In string compactifications the massless fermions must necessarily transform in small representations of the gauge group, mostly bi-fundamentals, as in the SM.
- The existence of axion fields coupling to the QCD field strength is built-in in string vacua. These axions may potentially solve the strong CP problem.
- There is a landscape of flavour and flux vacua with sufficient multiplicity to provide for an environmental explanation of the observed cosmological constant.
- There are explicit examples of string compactifications, with an SM or MSSM structure and three generations, which reproduce quite closely the properties of the observed physics (essentially to the level of our computational ability).
- In 3-generation models, there is very often one generation with Yukawa coupling of order the gauge coupling constant g, while the remaining Yukawas arise from subleading instanton or non-renormalizable couplings; also, Yukawa couplings naturally contain CP-violating phases. This qualitative structure is consistent with observations.
- Closed and open string moduli scalars provide natural candidates for the inflaton.
- In the supersymmetric SM context, string vacua often have hidden sectors which can naturally source and transmit SUSY-breaking.
- Euclidean brane instantons provide a new promising source for the origin of large Majorana masses for right-handed neutrinos, not requiring large Higgs multiplets.

At the moment there is no alternative theory with these successful properties. In this connection, ideas like grand unification are at best effective descriptions, which do not address either the issue of quantum gravity, nor other SM questions, like the origin of family replication or the origin of Yukawa couplings. It is remarkable that string theory, introduced in principle as a solution to the first two points above, brings along a nicely packaged bonus with the remaining properties.

There are some experimental opportunities for testing or constraining the structure of realistic string theory vacua, with diverse (and possibly subjective) ranges of plausibility. With no particular ordering, we may mention the following:

- String theory could be amenable to direct experimental test at colliders like the LHC or CLIC, if the string scale is of order a few TeV.
- Deviations from Newton's law may be sensitive to light moduli fields or large extra dimensions arising for low-string-scale models.
- Discovery of SUSY at the LHC would strongly support an underlying string theory description, since SUSY is a fundamental built-in ingredient in the theory – although its preservation down to the TeV scale is not guaranteed.
- Eventual measurement of the SUSY spectrum at future colliders would allow a comparison with specific SUSY breaking mechanisms in string theory, like moduli/dilaton domination, or others, potentially discriminating among different large classes of compactifications.
- Extra gauge $U(1)$'s, in particular $U(1)_{B-L}$, could be a signature of certain classes of string theory vacua.
- The observation of kinetic mixing of hypercharge with hidden sector $U(1)$'s, in "shining through a wall," laser, or other experiments, would support a string theory picture, where such hidden sectors are pervasive. Hidden sector gauginos could also be produced at the LHC.
- The detection of cosmic strings, with intercommutation properties differing from field theoretical cosmic strings, could give a direct test of string theory.
- The discovery of exotic particles with fractional charges would be a signature of MSSM-like (rather than GUT-like) string vacua.
- Cosmological CMB observations including measurement of tensor modes and/or non-gaussianities may yield important information on string models for inflation.
- Eventual detection of axions and/or dark matter particles can strongly restrict classes of string compactifications.

Apart from these opportunities, there remains the challenge of constructing a string vacuum capable of reproducing strictly all experimental information accumulated about the SM. As mentioned, some explicit constructions come close to this goal, reproducing a 3-generation MSSM spectrum with qualitatively correct hierarchical Yukawa coupling textures. They fail, however, to simultaneously implement a compelling description of moduli stabilization, SUSY-breaking, neutrino masses, and other detailed properties. Progress in these directions has been continuous in the last two decades and the prospects are promising. Further improvements will probably require an advanced understanding of moduli stabilization in realistic vacua, and the exploration of new regions in the flavour landscape, like compactifications beyond the CY ansatz, and F- and M-theory vacua. In this challenge, the experimental findings at the LHC, clarifying the nature of particle physics at the TeV scale, are expected to be crucial to guide the search.

Appendix A
Modular functions

There is a rich mathematical literature on modular functions, namely the functions of a complex parameter τ with nice transformation properties under the $SL(2, \mathbf{Z})$ modular group

$$\tau \to \frac{a\tau + b}{c\tau + d} \quad \text{with } a, b, c, d \in \mathbf{Z} \quad \text{and } ad - bc = 1. \tag{A.1}$$

The group is generated by $T: \tau \to \tau + 1$ and $S: \tau \to -1/\tau$. A useful parameter to construct modular functions is

$$q = e^{2\pi i \tau}. \tag{A.2}$$

The Dedekind eta function is defined by

$$\eta(\tau) = q^{1/24} \prod_{n=1}^{\infty} (1 - q^n). \tag{A.3}$$

Under T and S, it transforms as

$$\eta(\tau + 1) = e^{\pi i / 12} \eta(\tau),$$
$$\eta(-1/\tau) = (-i\tau)^{1/2} \eta(\tau). \tag{A.4}$$

The theta function with characteristics θ, ϕ is defined as

$$\vartheta \begin{bmatrix} \theta \\ \phi \end{bmatrix} = \eta \, e^{2\pi i \theta \phi} \, q^{\frac{1}{2}\theta^2 - \frac{1}{24}} \prod_{n=1}^{\infty} \left(1 + q^{n+\theta-1/2} e^{2\pi i \phi}\right)\left(1 + q^{n-\theta-1/2} e^{-2\pi i \phi}\right). \tag{A.5}$$

These functions also have an expression as infinite sums

$$\vartheta \begin{bmatrix} \theta \\ \phi \end{bmatrix}(\tau) = \sum_{n \in \mathbf{Z}} q^{\frac{1}{2}(n+\theta)^2} e^{2\pi i (n+\theta)\phi}. \tag{A.6}$$

The equality of (A.6) and (A.5) is related to bosonization/fermionization, as discussed in Section 4.2.6.

Particular theta functions that appear often are

$$\vartheta\begin{bmatrix} 0 \\ 0 \end{bmatrix} = \prod_{n=1}^{\infty}(1-q^n)(1+q^{n-1/2})^2,$$

$$\vartheta\begin{bmatrix} 0 \\ 1/2 \end{bmatrix} = \prod_{n=1}^{\infty}(1-q^n)(1-q^{n-1/2})^2,$$

$$\vartheta\begin{bmatrix} 1/2 \\ 0 \end{bmatrix} = q^{1/8}\prod_{n=1}^{\infty}(1-q^n)(1+q^n)(1+q^{n-1}) = 2q^{1/8}\prod_{n=1}^{\infty}(1-q^n)(1+q^n)^2,$$

$$\vartheta\begin{bmatrix} 1/2 \\ 1/2 \end{bmatrix} = iq^{1/8}\prod_{n=1}^{\infty}(1-q^n)^2(1-q^{n-1}) = 0. \tag{A.7}$$

They satisfy the so-called abstruse identity

$$\vartheta\begin{bmatrix} 0 \\ 0 \end{bmatrix}^4 - \vartheta\begin{bmatrix} 0 \\ 1/2 \end{bmatrix}^4 - \vartheta\begin{bmatrix} 1/2 \\ 0 \end{bmatrix}^4 = 0. \tag{A.8}$$

A further useful identity is

$$\sum_{\alpha,\beta} \eta_{\alpha,\beta}\, \vartheta\begin{bmatrix} \alpha \\ \beta \end{bmatrix} \prod_{i=1}^{3} \vartheta\begin{bmatrix} \alpha \\ \beta+u_i \end{bmatrix} = 0, \tag{A.9}$$

with $\eta_{0,0} = -\eta_{0,\frac{1}{2}} = -\eta_{\frac{1}{2},0} = 1$ and $u_1 + u_2 + u_3 = 0$.

A useful property for the twisted bosonic partition functions in Section 8.2.5 is

$$\vartheta\begin{bmatrix} 1/2 \\ 1/2+\phi \end{bmatrix} = -2\sin(\pi\phi)\eta q^{\frac{1}{12}}\prod_{n=1}^{\infty}(1-q^n e^{2\pi i\phi})(1-q^n e^{-2\pi i\phi})$$

$$\implies \quad \lim_{\varphi\to 0} \frac{-2\sin\pi\varphi}{\vartheta\begin{bmatrix} \frac{1}{2} \\ \frac{1}{2}+\varphi \end{bmatrix}} = \frac{1}{\eta^3}. \tag{A.10}$$

The theta functions behave nicely under shifts of their characteristics

$$\vartheta\begin{bmatrix} \theta+m \\ \phi+n \end{bmatrix}(\tau) = e^{2\pi i\theta n}\,\vartheta\begin{bmatrix} \theta \\ \phi \end{bmatrix}(\tau). \tag{A.11}$$

Finally, under modular transformations

$$\vartheta\begin{bmatrix} \theta \\ \phi \end{bmatrix}(\tau+1) = e^{-\pi i(\theta^2-\theta)}\,\vartheta\begin{bmatrix} \theta \\ \theta+\phi-1/2 \end{bmatrix}(\tau),$$

$$\vartheta\begin{bmatrix} \theta \\ \phi \end{bmatrix}(-1/\tau) = (-i\tau)^{1/2}\,\vartheta\begin{bmatrix} \phi \\ -\theta \end{bmatrix}(\tau). \tag{A.12}$$

The first is easily shown using (A.6) and noting $e^{\pi in^2} = e^{\pi in}$ (since $n^2 = n$ mod 2). The second is shown from (A.6) by using the Poisson resummation formula

$$\sum_{n\in\mathbf{Z}}\exp[-\pi A(n+\theta)^2 + 2\pi i\,(n+\theta)\phi] = A^{-\frac{1}{2}}\sum_{k\in\mathbf{Z}}\exp[-\pi A^{-1}(k+\phi)^2 - 2\pi i\,k\theta].$$

$$\tag{A.13}$$

This is a particular case of the general Poisson resummation formula, which we describe now. Consider an n-dimensional lattice Λ in $\mathbf{R}^n$, endowed with an internal product such that $v \cdot w \in \mathbf{Z}$ for all $v, w \in \mathbf{Z}$. The dual lattice Λ^* is defined as the set of vectors k in $\mathbf{R}^n$ such that $k \cdot v \in \mathbf{Z}$ for any vector $v \in \Lambda$. Noting that $\Lambda \subseteq \Lambda^*$, we introduce the index $|\Lambda^*/\Lambda|$ as the cardinal of the quotient. We then have

$$\sum_{v \in \Lambda} \exp\left[-\pi(v+\theta) \cdot A \cdot (v+\theta) + 2\pi i(v+\theta) \cdot \phi\right]$$

$$= \frac{1}{|\Lambda^*/\Lambda|\sqrt{\det A}} \sum_{k \in \Lambda^*} \exp\left[-\pi(k+\phi) \cdot A^{-1} \cdot (k+\phi) - 2\pi i k\theta\right]. \qquad (A.14)$$

This general formula can be shown by iterating the one-dimensional version.

In modular invariance of string theory partition functions, a prominent role is played by even and self-dual lattices. An even lattice Λ is defined as satisfying $v \cdot v \in 2\mathbf{Z}$ for all $v \in \Lambda$. Note that, since $(v+w)^2 = v^2 + w^2 + 2v \cdot w$ must be even for any $v, w \in \Lambda$, and so are v^2, w^2, it follows that even lattices satisfy $v \cdot w \in \mathbf{Z}$ for all $v, w \in \Lambda$, and hence $\Lambda \subset \Lambda^*$. A self-dual lattice is one satisfying $\Lambda = \Lambda^*$. Lattice sums in partition functions are invariant under $\tau \to \tau + 1$ and $\tau \to -1/\tau$ for even and self-dual lattices. The main examples are the $E_8 \times E_8$ and $SO(32)$ lattices in the 10d heterotic, Section 4.3.2, and the Narain lattice of left and right momenta in toroidal compactifications, e.g. (3.82) in Section 3.2.3, and (5.12) in Section 5.1.3, and (5.26) in Section 5.2.2.

Appendix B

Some topological tools

B.1 Forms and cycles: cohomology and homology

B.1.1 Differential manifolds, tangent and cotangent space

A *differential manifold* $\mathbf{X}$ of dimension d is defined as a space with local patches $U_{(\alpha)}$ parametrized by local coordinates $(x^1_{(\alpha)}, \ldots, x^d_{(\alpha)})$, glued together with smooth transition functions $f_{(\alpha\beta)}$ on the overlaps $U_{(\alpha)} \bigcap U_{(\beta)}$, schematically as

$$x_{(\alpha)} = f_{(\alpha\beta)}(x_{(\beta)}). \tag{B.1}$$

From now on we drop the patch labels, and the adjective "differential," and simply refer to "manifolds." We skip the description of familiar examples (like the n-sphere $\mathbf{S}^n$ or the n-torus $\mathbf{T}^n$) in this language, and move on to defining new geometric objects on manifolds. We focus on orientable manifolds without boundary, unless stated otherwise.

The *tangent space* $T_P(\mathbf{X})$ of a manifold $\mathbf{X}$ at a point P with coordinates x^m (in an appropriate patch), is defined as the vector space spanned by the *tangent vectors* $\partial_m|_P \equiv \partial/\partial x^m|_P$. The latter are formally defined as objects mapping smooth functions to real numbers, by extracting their x^m-derivative at P

$$\partial_m|_P : f(x) \to \partial_m f(x)|_P. \tag{B.2}$$

The tangent vectors can be extended to vector fields ∂_m by considering the whole set of possible base points $P \in \mathbf{X}$. A *vector field* is a point-dependent linear combination of the tangent vector fields $v(x) = \sum v^m(x)\partial_m$.

The *cotangent space* $T_P^*(\mathbf{X})$ is defined as the space dual to $T_P(\mathbf{X})$. Namely, it is spanned by the basis cotangent vectors $dx^m|_P$, defined as linear maps from $T_P(\mathbf{X})$ to the real numbers, acting on the tangent vectors as $dx^m(\partial_n) = \delta_n^m$. This definition is implicitly local at P, but extends to $\mathbf{X}$, leading to the definition of the cotangent fields dx^m. A general cotangent field, usually referred to as 1-form, is a point-dependent linear combination $\lambda = \lambda_m(x)dx^m$.

In the following the term "field" is implicit in all definitions.

B.1.2 Tensors and differential forms

A *tensor* of type $[k, l]$ is a linear combination

$$T = T^{n_1 \cdots n_k}_{m_1 \cdots m_l} \, dx^{m_1} \otimes \cdots \otimes dx^{m_l} \otimes \partial_{n_1} \otimes \cdots \otimes \partial_{n_k}, \tag{B.3}$$

where we have used tensor products of the basis tangent and cotangent vectors. A familiar example of a tensor is the metric, $g = g_{ij} dx^i \otimes dx^j$.

A differential *p-form* (or p-form for short) C_p is a tensor of type $[0, p]$, which has completely antisymmetric components $C_{m_1 \cdots m_p}$. Equivalently, a linear combination of the basis (with brackets denoting antisymmetrization averaged with $1/p!$)

$$dx^{m_1} \wedge \cdots \wedge dx^{m_p} = dx^{[m_1} \otimes \cdots \otimes dx^{m_p]}. \tag{B.4}$$

We define the component expansion as

$$C_p = \frac{1}{p!} C_{m_1 \cdots m_p}(x) dx^{m_1} \wedge \cdots \wedge dx^{m_p}. \tag{B.5}$$

The *wedge product* of a p-form A_p and a q-form B_q is defined as the $(p+q)$-form

$$A_p \wedge B_q = \frac{1}{p! q!} A_{m_1 \cdots m_p} B_{n_1 \cdots n_q} dx^{m_1} \wedge \cdots \wedge dx^{m_p} \wedge dx^{n_1} \wedge \cdots \wedge dx^{n_q}. \tag{B.6}$$

It satisfies $A_p \wedge B_q = (-1)^{pq} B_q \wedge A_p$. In many p-form equations, wedge products are often assumed, and not explicitly displayed.

The *exterior derivative d* is defined as mapping a p-form to a $(p+1)$-form, by

$$dC_p = \frac{1}{p!} \partial_{m_0} C_{m_1 \cdots m_p} \, dx^{m_0} \wedge dx^{m_1} \wedge \cdots \wedge dx^{m_p}. \tag{B.7}$$

It satisfies $d(A_p \wedge B_q) = dA_p \wedge B_q + (-1)^p A_p \wedge dB_q$. Differential forms and their exterior derivatives are ubiquitous in physics. The electromagnetic gauge potential can be described as a 1-form A_1, with a 2-form field strength $F_2 = dA_1$. Their non-abelian generalization involves Lie algebra valued 1- and 2-forms A_1^a, F_2^a.

A key property of the exterior derivative is $d(dC_p) = 0$ on any p-form C_p. Formally

$$d^2 \equiv 0. \tag{B.8}$$

This underlies the definition of cohomology, to which we turn next.

B.1.3 Cohomology

A p-form A_p satisfying $dA_p = 0$ is said to be *closed*. Also a p-form A_p is said to be *exact* if there exists a globally defined $(p-1)$-form B_{p-1} such that $A_p = dB_{p-1}$. Clearly (B.8) implies that any exact form is closed. The reverse is also true for the trivial case of $\mathbf{R}^d$ (by the so-called Poincaré lemma), but may not be so for a general manifold $\mathbf{X}$. In fact, the existence of closed forms which are not exact provides a characterization of the non-trivial topology of manifolds, through the cohomology groups, as follows.

Recall the definition of $\mathbf{X}$ by glueing local coordinate patches $U_{(\alpha)}$. A closed p-form A_p on a manifold $\mathbf{X}$ defines, by restriction, a closed p-form on each $U_{(\alpha)}$. Since each such patch is locally identical to $\mathbf{R}^d$, the Poincaré lemma implies that on each $U_{(\alpha)}$ it is possible to express $A_p = dB_{p-1}^{(\alpha)}$ for some $(p-1)$-form $B_{p-1}^{(\alpha)}$. However, there is no guarantee that the locally defined forms $B_{p-1}^{(\alpha)}$ patch up to a globally defined $(p-1)$-form in $\mathbf{X}$. Such obstructions clearly depend on the global, rather than local, structure of $\mathbf{X}$, and thus the existence of closed forms which are not exact provides a measure of the topological non-triviality of $\mathbf{X}$.

The (de Rham) *cohomology groups* characterize these global obstructions. To define them, denote by $\mathcal{Z}^p$ the set of closed p-forms on $\mathbf{X}$, and by $\mathcal{B}_p$ the set of exact p-forms. The pth de Rham cohomology group of $\mathbf{X}$ is defined as the quotient

$$H^p(\mathbf{X}, \mathbf{R}) = \frac{\mathcal{Z}^p}{\mathcal{B}^p}. \tag{B.9}$$

Namely, the set of closed p-forms modulo the equivalence relation

$$A_p \simeq A_p + dB_{p-1}. \tag{B.10}$$

Two closed p-forms differing by an exact form thus define the same equivalence class. The cohomology class of a form A_p is denoted $[A_p]$; exact forms, which are trivially closed, are in the so-called trivial class $[0]$. The sets $H^p(\mathbf{X}, \mathbf{R})$ have the structure of finite-dimensional vector spaces; their dimensions $b^p(\mathbf{X}) = \dim H^p(\mathbf{X}, \mathbf{R})$ are topological invariants, known as *Betti numbers*.

To illustrate these ideas, consider the example of $\mathbf{T}^2$, with two periodic coordinates $x \simeq x + 1$, $y \simeq y + 1$. It has $b_1(\mathbf{T}^2) = 2$, because there are two non-trivial 1-forms, which abusing language are usually denoted by their local expressions dx, dy; still they are globally well defined, and closed, but not exact since x, y are not good global coordinates. A second familiar example of cohomologically non-trivial form in a general d-dimensional manifold $\mathbf{X}$ is the volume form, usually described by its local expression

$$d\mathrm{vol}_d = dx^1 \wedge \cdots \wedge dx^d. \tag{B.11}$$

B.1.4 Homology

An alternative way of characterizing the same topological information about a manifold $\mathbf{X}$ is through properties of its submanifolds, in terms of the homology groups, reviewed next.

A p-dimensional submanifold N of $\mathbf{X}$ is a subset of $\mathbf{X}$ with the structure of a p-dimensional differential manifold, i.e. described by local patches glued with smooth transition functions. In general, we also consider submanifolds with boundary, of dimension $(p-1)$, and denoted as ∂N. The symbol ∂ can be regarded as the operator "take the boundary of," formally mapping p-dimensional manifolds to $(p-1)$-dimensional ones. An important property for what follows is that $\partial(\partial N) \equiv 0$, i.e. the boundary of a manifold has no boundary itself.

The set of submanifolds of $\mathbf{X}$ can be endowed with a vector space structure, by introducing *p-chains*, $a_p = c_k N_p^k$, formal linear combinations of p-dimensional submanifolds N_p^k (with real coefficients $c_k \in \mathbf{R}$). The upcoming formal definitions for p-chains can be made more intuitive by replacing "chain" by "submanifold."

A p-chain $a_p \subset \mathbf{X}$ with no boundary, i.e. satisfying $\partial a_p = 0$, is called a *p-cycle*. On the other hand, a p-chain a_p which is the boundary of a $(p+1)$-chain $a_p = \partial b_{p+1}$ is called trivial. Since $\partial^2 \equiv 0$, any trivial p-chain is a p-cycle. The reverse is also true for the trivial case of $\mathbf{R}^d$ (by the dual Poincaré lemma), but may not be so for a general manifold $\mathbf{X}$. In fact, the existence of non-trivial p-cycles characterizes the non-trivial topology of manifolds, through the homology groups.

The *homology groups* characterize non-trivial cycles. To define them, denote $\mathcal{Z}_p$ the sets of p-cycles in $\mathbf{X}$, and $\mathcal{B}_p$ the set of trivial p-chains. The pth homology group of $\mathbf{X}$ is defined as the quotient

$$H_p(\mathbf{X}, \mathbf{R}) = \frac{\mathcal{Z}_p}{\mathcal{B}_p}. \tag{B.12}$$

Namely, two p-cycles are considered equivalent if they differ by a trivial p-chain

$$a_p \simeq a_p + \partial b_{p+1}. \tag{B.13}$$

The homology class of a p-cycle a_p is denoted $[a_p]$, and trivial cycles lie in the trivial class $[0]$. The sets $H_p(\mathbf{X}, \mathbf{R})$ are vector spaces, with dimensions given by the above introduced Betti numbers $b_p(\mathbf{X})$, for reasons explained below.

To illustrate these ideas, consider $\mathbf{T}^2$ defined by two periodic coordinates $x \simeq x + 1$, $y \simeq y + 1$. The two non-trivial 1-cycles, denoted $[a]$ and $[b]$, can be represented as, e.g., the submanifolds defined by $y = 0$ and $x = 0$, respectively.

B.1.5 De Rahm duality of $H^p(X,R)$ and $H_p(X,R)$

There are strong analogies in the definitions of cohomology and homology groups, with closed forms, exact forms, and the exterior derivative d, relating to cycles, trivial chains, and the boundary operator ∂. This actually follows from a duality of the vector spaces of forms and chains, defined by integration operation.

The *integral* of a p-form A_p over a p-dimensional submanifold $N \subseteq \mathbf{X}$ is roughly defined as

$$\int_N A_p \equiv \int d^p x \, A_{1\cdots p}, \tag{B.14}$$

in which the right-hand side integral is in standard calculus. In a more precise version, the integral is defined as a sum of similar expressions for the local patches glued up to form N, carefully avoiding overcounting. The integral of a p-form A_p on a general p-chain

$a_p = \sum_k c_k N_p^k$ is defined by linear extension.

$$\int_{a_p} A_p = \sum_k c_k \int_{N_p^k} A_p. \tag{B.15}$$

An important property of integration is *Stokes theorem*, which states that

$$\int_{a_p} dB_{p-1} = \int_{\partial a_p} B_{p-1}. \tag{B.16}$$

It implies that the integral of a closed p-form A_p over a p-cycle a_p depends only on their cohomology and homology classes, $[A]$ and $[a]$, as follows. Consider different representatives of these classes $A'_p = A_p + dB_{p-1}$, $a'_p = a_p + \partial b_{p+1}$, to obtain

$$\int_{a'} A' = \int_a (A + dB) + \int_{\partial b} (A + dB) = \int_a A + \int_{\partial a} B + \int_b d(A + dB) = \int_a A,$$

where we used (B.16), as well as $\partial a = 0$, $dA = 0$ and $d(dB) = 0$. Hence, integration descends to an operation for cohomology and homology classes, and defines a linear mapping $H^p(\mathbf{X}, \mathbf{R}) \times H_p(\mathbf{X}, \mathbf{R}) \to \mathbf{R}$. Equivalently, $H^p(\mathbf{X}, \mathbf{R})$ and $H_p(\mathbf{X}, \mathbf{R})$ are dual as vector spaces, and in particular have the same dimensions $b_p(\mathbf{X})$, as anticipated. The duality implies that it is always possible to choose basis of cycles $\{(a_p)_i\}$ and forms $\{(A_p)_j\}$, $i, j = 1, \ldots, b_p$, such that $\int_{[a_i]} [A_j] = \delta_{ij}$.

For instance, in the already used $\mathbf{T}^2$ example, we have $\int_{[a]} dx = 1$, $\int_{[a]} dy = 0$, $\int_{[b]} dy = 1$, $\int_{[b]} dx = 0$, and so dx, dy are dual to $[a]$, $[b]$, respectively.

The above and upcoming definitions hold also for forms and cycles defined by linear combinations with integer coefficients. The corresponding cohomology and homology groups are denoted $H^p(\mathbf{X}, \mathbf{Z})$, $H_p(\mathbf{X}, \mathbf{Z})$. More generally, it is sometimes useful to consider forms defined by point-dependent linear combinations whose coefficients are not just functions, but rather carry some additional (e.g. gauge) quantum number. In formal terms, the coefficients are sections of a (e.g. gauge) vector bundle E. The corresponding cohomology groups are denoted $H^p(\mathbf{X}, E)$. A particular application is the computation of the massless scalar from gauge fields in heterotic string compactifications in sections 7.3.2, 7.4.

B.1.6 Poincaré duality

Integration of the wedge product of a p- and a $(d - p)$-form, $\int_\mathbf{X} A_p \wedge B_{d-p}$, defines a map $H^p(\mathbf{X}, \mathbf{R}) \times H^{d-p}(\mathbf{X}, \mathbf{R}) \to \mathbf{R}$. This implies that H^p and H^{d-p} are dual vector spaces, and in particular $b_p(\mathbf{X}) = b_{n-p}(\mathbf{X})$. Combining with de Rahm duality, this induces an identification (isomorphism) of H^{d-p} and H_p. Namely, to each non-trivial p-cycle a_p we may associate a non-trivial $(d - p)$-cohomology class. An intuitive representative of the latter is a "bump" $(d - p)$ form $\delta(a_p)$, with indices along the $(d - p)$ directions transverse to the p-cycle and coefficient given by a Dirac delta function with support on a_p. This is consistent with the defining relation for the isomorphism

$$\int_{a_p} B_p = \int_X B_p \wedge \delta(a_p) \quad \text{for any } B_p. \tag{B.17}$$

The Poincaré dual form allows to extend integrals on p-cycles to integrals on X.

The Poincaré dual forms allow to define the topological intersection number of integer homology classes $[a_p] \in H_p(X, Z)$, $[b_{d-p}] \in H_{d-p}(X, Z)$, as

$$[a_p] \cdot [b_{d-p}] = \int_X \delta(a_p) \wedge \delta(b_{d-p}). \tag{B.18}$$

The bump form representation shows that this computes the number of intersection of the non-trivial cycles, weighted with signs due to orientations. The intersection number is integer for integer homology classes. For instance, in T^2 the basic intersection numbers are $[a] \cdot [a] = [b] \cdot [b] = 0$, $[a] \cdot [b] = -[b] \cdot [a] = 1$. Hence for two 1-cycles $[\pi_1] = n_1[a] + m_1[b]$ and $[\pi_2] = n_2[a] + m_2[b]$ in $H_1(T^2, Z)$, with $n_i, m_i \in Z$, we have

$$[\pi_1] \cdot [\pi_2] = n_1 m_2 - n_2 m_1. \tag{B.19}$$

B.2 Hodge dual

The previous geometric objects were constructed without invoking a metric on X, and are hence topological. We now consider additional structures present for a d-dimensional Riemannian manifold X, i.e. a manifold endowed with a (euclidean or lorentzian signature) metric g.

The *Hodge duality* $*$ maps p-forms to $(d - p)$-forms, acting on the basis as

$$*(dx^{m_1} \cdots dx^{m_p}) = \frac{1}{(d-p)!} |g|^{\frac{1}{2}} g^{m_1 n_1} \cdots g^{m_p n_p} \epsilon_{n_1 \cdots n_p n_{p+1} \cdots n_d} dx^{n_{p+1}} \cdots dx^{n_d},$$

with implicit wedge products and $g = \det(g_{mn})$. It satisfies $* * = (-1)^{p(d-p)}$ or $** = (-1)^{p(d-p)} + 1$ for euclidean and lorentzian signatures, respectively.

The Hodge operator defines a positive-definite inner product between p-forms

$$(A_p, B_p) = \frac{1}{p!} \int_X A_p \wedge *B_p = \int_X |g|^{\frac{1}{2}} A_{m_1 \cdots m_p} B^{m_1 \cdots m_p} dx^1 \cdots dx^d. \tag{B.20}$$

A *harmonic form* C_p is defined by the conditions $dC_p = 0$, $d(*C_p) = 0$. Although we skip the derivation, it can be shown that for each cohomology class there is a unique harmonic representative. The precise expression of harmonic forms depends on the metric, but not their number, which is given by $b_p(X)$.

Many of the objects in this and the previous section admit a refinement when defined on complex manifolds, as described for CY spaces in Section 7.2.2

B.3 Application: *p*-form gauge fields

The antisymmetric tensor fields in string and M-theory are naturally described as differential forms, here denoted collectively as C_{p+1}, with the subindex indicating the degree. They are generalized gauge potentials, with gauge transformations

$$C_{p+1} \to C_{p+1} + d\Lambda_p, \tag{B.21}$$

where Λ_p is a p-form gauge parameter. The field strength tensor $F_{p+2} = dC_{p+1}$ is gauge invariant. The kinetic term for the gauge potential is

$$S_{C_p, \text{kin}} = -\frac{1}{2} \int F_{p+2} \wedge *F_{p+2} = -\frac{1}{2} \int d^d x \, (-G)^{\frac{1}{2}} \frac{1}{(p+2)!} F_{m_1 \cdots m_{p+2}} F^{m_1 \cdots m_{p+2}}$$

$$\equiv -\frac{1}{2} \int d^d x \, (-G)^{\frac{1}{2}} |F_{p+2}|^2, \tag{B.22}$$

where the integral is over 10d or 11d, and $*$ is the corresponding Hodge duality.

In KK compactifications in Section 7.2.4, zero modes of p-form fields correspond to harmonic forms, as follows. Defining the adjoint $d^\dagger \simeq *d*$ of d with respect to the product (B.20) as $(A, dB) = (d^\dagger A, B)$, the kinetic term operator for C_p is $dd^\dagger + d^\dagger d$. Zero modes must satisfy $dC_p = 0$ and $d^\dagger C_p = 0$, hence are harmonic p-forms.

The objects electrically charged under C_{p+1} are p-branes, i.e. extended objects with p spatial dimensions, thus sweeping out a $(p+1)$-dimensional subspace W_{p+1} of spacetime as they evolve in time. The electric coupling reads

$$S_{\text{electr.}} = Q \int_{W_{p+1}} C_{p+1}, \tag{B.23}$$

as present, e.g., in (6.16). From (B.22) and (B.23), the equations of motion are

$$d * F_{p+2} = Q \, \delta(W_{p+1}), \tag{B.24}$$

where $\delta(W_{p+1})$ is the Poincaré dual of W_{p+1}. In a d-dimensional theory, it is useful to define the dual field strength $F_{d-p-2} = *F_{p+2}$, and the dual $(d-p-3)$-form C_{d-p-3}, defined locally as $F_{d-p-2} = dC_{d-p-3}$. A p-brane is said to be electrically charged under C_{p+1} and magnetically under C_{d-p-3} (and vice versa for the "dual" $(d-p-4)$-brane). The gauge potential and its dual gauge potential are defined locally, but in general not globally. The kinetic term (B.22) can be shown to be equivalent to a similar kinetic term involving the dual field strength tensor.

The electric charge carried by a p-brane can be measured using Gauss' theorem, as the integrated flux of F_{d-p-2} around a $(d-p-2)$-sphere surrounding the object in the transverse $(d-p-1)$-dimensional space, as

$$\int_{\mathbf{S}^{d-p-2}} F_{d-p-2} = \int_{\mathbf{B}^{d-p-1}} dF_{d-p-2} = Q \int_{\mathbf{B}^{d-p-1}} \delta(W_{p+1}) = Q, \tag{B.25}$$

with $\mathbf{B}^{d-p-1}$ the interior of the sphere, i.e. $\partial \mathbf{B}^{d-p-1} = \mathbf{S}^{d-p-2}$. The charges Q are quantized, as follows from the general flux quantization condition reviewed next.

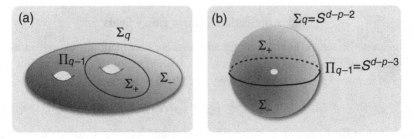

Figure B.1 Dirac quantization of fluxes (a) and charges (b).

B.3.1 Dirac quantization of fluxes and charges

Consider a d-dimensional theory with a q-form field strength tensor F_q, satisfying $dF_q = 0$ on a space supporting non-trivial homology q-cycles. If the theory contains objects, $(q-2)$-branes, charged under the $(q-1)$-form gauge potential C_{q-1}, with minimal charge Q_e, the F_q integrals over any q-cycle Σ_q satisfy the quantization

$$Q_e \int_{\Sigma_q} F_q \in 2\pi \mathbf{Z}. \tag{B.26}$$

Hence, fluxes are classified by integer cohomology classes in $H^q(\mathbf{X}, \mathbf{Z})$. To show it, consider a trivial $(q-1)$-cycle Π_{q-1} in Σ_q, splitting the latter in two parts, the "inside" Σ_+ and the "outside" Σ_-, see Figure B.1. They satisfy $\partial\Sigma_+ = \partial\Sigma_- = \Pi_{q-1}$ and $\Sigma_+ - \Sigma_- = \Sigma_q$, with the latter sign arising from the orientation flip to glue both pieces. Consider the quantum amplitude of a process in which a $(q-2)$-brane of charge Q_e has all its worldvolume dimensions wrapped on Π_{q-1}. From the coupling (B.23), there is a phase contribution involving an integral of C_{q-1}. Since the latter is in general not globally well-defined, the contribution is better expressed in terms of the globally well-defined F_q, using Stokes theorem as

$$\exp\!\left(i Q_e \int_{\Pi_{q-1}} C_{q-1}\right) = \exp\!\left(i Q_e \int_{\Sigma_\pm} F_q\right). \tag{B.27}$$

Actually the two choices $\Sigma_\pm$ in principle differ by a phase

$$Q_e\!\left(\int_{\Sigma_+} F_q - \int_{\Sigma_-} F_q\right) = Q_e \int_{\Sigma_q} F_q. \tag{B.28}$$

The quantization (B.26) guarantees that the two choices define the same amplitude.

As examples, this leads to quantization of worldvolume magnetic fluxes in magnetized brane and F-theory models in Sections 11.4 and 11.5, and in flux compactifications in Chapter 14. Also, it leads to charge quantization in theories with electric and magnetic charges, as follows, see Figure B.1(b). Consider an object (a p-brane) with charge Q'_m under C_{p+1}, leading to the equation of motion (B.24), so the dual field strength F_{d-p-2} satisfies $dF_{d-p-2} = 0$, except at the p-brane volume W_{p+1}. Consider then the d-dimensional spacetime with W_{p+1} removed (more simply, the transverse $\mathbf{R}^{d-p-1}$ with the origin removed); it is topologically non-trivial, and contains a non-trivial $(d - p - 2)$-cycle (represented by a sphere $\mathbf{S}^{d-p-2}$ around the origin in $\mathbf{R}^{d-p-1}$). Recalling that the charge Q'_m is measured in (B.25) as the flux of F_{d-p-2} over this $\mathbf{S}^{d-p-2}$, the quantization condition (B.26) with $q = d - p - 2$, $\Sigma_q = \mathbf{S}^{d-p-2}$ leads to the charge quantization condition

$$Q_e Q'_m \in 2\pi \mathbf{Z}. \tag{B.29}$$

This generalizes easily to theories containing dyonic objects, carrying electric and magnetic charges simultaneously. The general condition requires

$$Q_e Q'_m - Q_m Q'_e \in 2\pi \mathbf{Z}, \tag{B.30}$$

for any pair of dyons with charges (Q_e, Q_m), (Q'_e, Q'_m); this general formula explains *a posteriori* some earlier choices of notation. For string theory and M-theory, this implies the quantization of BPS brane charges. For 1-form gauge potentials in $d = 4$, this implies the quantization of electric and magnetic charges in gauge theories, for instance in 4d $\mathcal{N} = 4$ SYM in Section 2.5.2.

B.3.2 Chern and Pontryagin classes

The above quantization condition for 1-forms has a natural generalization to non-abelian gauge fields, described as 1-forms A taking values in the adjoint representation of the gauge group G, which we take as $SU(N)$. The field strength 2-form is defined by $F^a_c = dA^a_c + A^a_b \wedge A^b_c$, with a, b, c labeling the fundamental of $SU(N)$. There is a quantization condition on the Chern classes, which are defined by the formal expansion of $\mathrm{Tr}\,\exp(\frac{F}{2\pi})$, i.e.

$$c_n = \frac{1}{n!(2\pi)^n} \mathrm{Tr}\, F^n, \qquad \frac{1}{n!(2\pi)^n} \int_{\Sigma_{2n}} \mathrm{Tr}\, F^n \in \mathbf{Z}, \tag{B.31}$$

on any $(2n)$-cycle of $\mathbf{X}$. For $n = 1$ we recover the above flux quantization, while for $n = 2$ we obtain the quantized instanton number (13.3).

The quantization of gauge field strength integrals has an analog for geometric curvatures, when described in the differential form language. Consider a d-dimensional Riemannian manifold and introduce the "tangent space 1-forms" or d-bein e^a_m, defined by $g_{mn} = e^a_m e^b_n \delta_{ab}$, with $a, b = 1, \ldots, d$ labeling the tangent space directions, on which a local $SO(d)$ rotation group acts. We define the 1-forms $e^a = e^a_m dx^m$, and use parallel transport $de^a = \omega^a_b \wedge e^b$ to define ω^a_b, a connection described as a 1-form taking values in the adjoint of the tangent $SO(d)$, very analogous to a non-abelian gauge field. For instance, $R^a_c = d\omega^a_c + \omega^a_b \wedge \omega^b_c$ defines a curvature 2-form, taking values in the adjoint of $SO(d)$. There is a quantization condition on the Pontryagin classes, which are defined by the formal expansion of $\det(1 - \frac{R}{2\pi})$. For instance

$$\frac{1}{8\pi^2} \int_{\Sigma_4} \mathrm{Tr}\, R^2 \in \mathbf{Z}. \tag{B.32}$$

B.4 Homotopy groups

Consider the set $\mathcal{P}_1(\mathbf{X})$ of continuous oriented paths starting and ending at a "base point" P in a manifold $\mathbf{X}$. This has a group structure under composition of paths, with the identity being the trivial path staying at P, and inverses given by orientation reversed paths. Moreover, the resulting set is actually independent of the base point P, thus ignored in the notation. The first homotopy group of $\mathbf{X}$, also known as "fundamental group," and denoted $\Pi_1(\mathbf{X})$, is obtained from $\mathcal{P}_1(\mathbf{X})$ as a quotient, with equivalence relation given by continuous deformation of paths. It thus characterizes topologically inequivalent embeddings of $\mathbf{S}^1$ into $\mathbf{X}$. Spaces with non-trivial $\Pi_1(\mathbf{X})$ are called non-simply connected. A simple example is $\mathbf{S}^1$ with $\Pi_1(\mathbf{S}^1) = \mathbf{Z}$, where the integer characterizes the winding number in maps $\mathbf{S}^1 \to \mathbf{S}^1$. In string compactifications (or other KK compactifications with gauge fields) on $\mathbf{X}$, the group $\Pi_1(\mathbf{X})$ characterizes the possibility to turn on non-trivial Wilson lines.

This definition generalizes to the pth homotopy group $\Pi_p(\mathbf{X})$, by considering embeddings of p-spheres $\mathbf{S}^p$ into $\mathbf{X}$, modulo continuous deformations. Homology and homotopy groups are actually different topological structures, not to be confused. Homology groups are defined by considering arbitrary submanifolds, and imposing an equivalence relation identifying submanifolds differing by a pure boundary; homotopy groups are in contrast defined by considering p-dimensional spheres, and equivalence relation given by continuous deformation.

In physics, the topological classification of non-trivial gauge field configurations in theories with gauge group G can often be recast in terms of homotopy groups $\Pi_p(G)$. The $\mathbf{S}^p$ describes the sphere at infinity in $(p+1)$ dimensions, so $\Pi_p(G)$ characterizes the possible gauge-inequivalent ways in which the gauge field can "wind" around infinity. For instance, $\Pi_3(G)$ characterizes non-trivial 4d gauge instanton configurations by the winding around the $\mathbf{S}^3$ at infinity in (euclidean) 4d space; for G a simple group, $\Pi_3(G) = \mathbf{Z}$, and the integer corresponds to the instanton number (13.3). In 4d gauge theories with spontaneous breaking of G into a subgroup H, $\Pi_2(G/H)$ classifies magnetic monopoles, by describing the winding of the Higgs field around spatial infinity in the three spatial dimensions. Finally, $\Pi_4(G)$ is related to global gauge anomalies in 4d, with the $\mathbf{S}^4$ played by 4d spacetime compactified adding a point at infinity. For instance, $\Pi_4(USp(2k)) = \mathbf{Z}_2$ encoding the global gauge anomaly of $USp(2k)$ gauge theories with an odd number of fermions in the fundamental representation, as mentioned in Section 10.4.2.

Appendix C

Spectrum and charges of a semi-realistic $\mathbf{Z}_3$ heterotic orbifold

In the following table the charges are under the linear combinations of the $E_8 \times E_8$ Cartan subalgebra defined by the following vectors

$$Q_1 = 6\,(1, 1, 1, 0, \ldots, 0) \times (0, \ldots, 0)'$$
$$Q_2 = 6\,(0, 0, 0, 1, -1, 0, 0, 0) \times (0, \ldots, 0)'$$
$$Q_3 = 6\,(0, \ldots, 01, 0, 0) \times (0, \ldots, 0)'$$
$$Q_4 = 6\,(0, \ldots, 0, 0, 1, 0) \times (0, \ldots, 0)'$$
$$Q_5 = 6\,(0, \ldots, 0, 0, 0, 1) \times (0, \ldots, 0)'$$
$$Q_6 = 6\,(0, \ldots, 0) \times (1, 1, 0, 0, \ldots, 0)'$$
$$Q_7 = 6\,(0, \ldots, 0) \times (0, 1, 1, 0, \ldots, 0)'$$
$$X = 6\,(0, \ldots, 0) \times (1, -1, 1, 0, \ldots, 0)'$$
$$Y = \left(\tfrac{1}{3}, \tfrac{1}{3}, \tfrac{1}{3}, -\tfrac{1}{2}, \tfrac{1}{2}, 0, 1, 0\right) \times (0, \ldots, 0)'. \tag{C.1}$$

Table C.1 *Spectrum of the $\mathbf{Z}_3$ example in Section 8.3.3. The indices (m_1, m_3), $m_1, m_3 = 0, \pm 1$ label fixed points in the first and second $\mathbf{T}^2$. All fields have three copies*

(m_1, m_3)	Field	Rep	Q_1	Q_2	Q_3	Q_4	Q_5	Q_6	Q_7	X	$6Y$
Unt	Q_L	$(\mathbf{3}, \mathbf{2})$	6	6	0	0	0	0	0	0	1
	u_1	$(\bar{\mathbf{3}}, \mathbf{1})$	6	0	0	6	0	0	0	0	4
	$\bar{G}_1$	$(\mathbf{1}, \mathbf{2})$	0	6	0	6	0	0	0	0	3
	16'	$\mathbf{1}$	0	0	0	0	0	0	0	9	0
$(0, 0)$	D_1	$(\mathbf{3}, \mathbf{1})$	0	4	0	0	0	4	0	4	-2
	$\bar{G}_2$	$(\mathbf{1}, \mathbf{2})$	6	-2	0	6	0	4	0	4	3
	$\bar{A}_1$	$\mathbf{1}$	-3	-2	-3	-3	-3	4	0	4	-3
	$\bar{A}_2$	$\mathbf{1}$	-3	-2	3	-3	3	4	0	4	-3
	A_1	$\mathbf{1}$	-3	-2	-3	3	3	4	0	4	3
	A_2	$\mathbf{1}$	-3	-2	3	3	-3	4	0	4	3

Table C.1 *(Cont.)*

(m_1, m_3)	Field	Rep	Q_1	Q_2	Q_3	Q_4	Q_5	Q_6	Q_7	X	$6Y$
$(1, 0)$	S_4	**1**	6	4	0	0	−2	6	4	4	0
	S_5	**1**	6	4	0	0	−2	−6	−2	4	0
	S_6	**1**	6	4	0	0	−2	0	−2	−8	0
	$\bar{A}_3$	**1**	−3	−2	−3	−3	1	6	4	4	−3
	$\bar{A}_4$	**1**	−3	−2	−3	−3	1	−6	−2	4	−3
	$\bar{A}_5$	**1**	−3	−2	−3	−3	1	0	−2	−8	−3
	A_3	**1**	−3	−2	3	3	1	6	4	4	3
	A_4	**1**	−3	−2	3	3	1	−6	−2	4	3
	A_5	**1**	−3	−2	3	3	1	0	−2	−8	3
$(−1, 0)$	S_7	**1**	6	4	0	0	2	2	−4	4	0
	S_8	**1**	6	4	0	0	2	2	2	−8	0
	S_9	**1**	6	4	0	0	2	−4	2	4	0
	$\bar{A}_6$	**1**	−3	−2	3	−3	−1	2	−4	4	−3
	$\bar{A}_7$	**1**	−3	−2	3	−3	−1	2	2	−8	−3
	$\bar{A}_8$	**1**	−3	−2	3	−3	−1	−4	2	4	−3
	A_6	**1**	−3	−2	−3	3	−1	2	−4	4	3
	A_7	**1**	−3	−2	−3	3	−1	2	2	−8	3
	A_8	**1**	−3	−2	−3	3	−1	−4	2	4	3
$(0, 1)$	d_1	$(\bar{\mathbf{3}}, \mathbf{1})$	0	0	0	2	2	−4	−4	4	2
	F_1	$(\mathbf{1}, \mathbf{2})$	3	0	−3	−1	−1	−4	−4	4	0
	$\bar{A}_9$	**1**	3	6	3	−1	−1	−4	−4	4	−3
	A_9	**1**	3	−6	3	−1	−1	−4	−4	4	−3
	$\bar{l}_1$	**1**	−6	0	0	−4	2	−4	−4	4	−6
	S_{10}	**1**	−6	0	0	2	−4	−4	−4	4	0
$(1, 1)$	D_2	$(\mathbf{3}, \mathbf{1})$	6	0	0	2	0	−2	0	4	4
	u_2	$(\bar{\mathbf{3}}, \mathbf{1})$	0	0	0	−4	0	−2	0	4	4
	F_2	$(\mathbf{1}, \mathbf{2})$	3	0	3	−1	−3	−2	0	4	0
	F_3	$(\mathbf{1}, \mathbf{2})$	3	0	−3	−1	3	−2	0	4	0
	S_1	**1**	−6	0	0	2	0	4	0	−8	0
	Y_1	**1**	−6	0	0	2	0	−2	0	4	0
	$\bar{A}_{10}$	**1**	3	6	−3	−1	−3	−2	0	4	−3
	$\bar{A}_{11}$	**1**	3	6	3	−1	3	−2	0	4	−3
	A_{10}	**1**	3	−6	3	−1	3	−2	0	4	3
	A_{11}	**1**	3	−6	−3	−1	−3	−2	0	4	−3
$(−1, 1)$	d_2	$(\bar{\mathbf{3}}, \mathbf{1})$	0	0	0	2	−2	0	4	4	2
	F_4	$(\mathbf{1}, \mathbf{2})$	3	0	3	−1	1	0	4	4	0
	$\bar{A}_{12}$	**1**	3	6	−3	−1	1	0	4	4	−3
	A_{12}	**1**	3	−6	−3	−1	1	0	4	4	3
	$\bar{l}_2$	**1**	−6	0	0	−4	−2	0	4	4	−6
	S_{11}	**1**	−6	0	0	2	4	0	4	4	0

Table C.1 *(Cont.)*

(m_1, m_3)	Field	Rep	Q_1	Q_2	Q_3	Q_4	Q_5	Q_6	Q_7	X	$6Y$
$(0,-1)$	$\bar{D}_1$	$(\bar{3}, 1)$	-3	2	-3	1	1	0	-2	4	-1
	D_3	$(3, 1)$	3	2	3	1	1	0	-2	4	1
	$\bar{G}_3$	$(1, 2)$	0	2	0	4	-2	0	-2	4	3
	G_1	$(1, 2)$	0	2	0	-2	4	0	-2	4	-3
	S_2	1	0	-4	0	-2	-2	0	4	-8	0
	Y_2	1	0	-4	0	-2	-2	0	-2	4	0
	l_1	1	0	-4	0	4	4	0	-2	4	6
	$\bar{l}_3$	1	0	8	0	-2	-2	0	-2	4	-6
	$\bar{A}_{13}$	1	-9	2	3	1	1	0	-2	4	-3
	A_{13}	1	9	2	-3	1	1	0	-2	4	3
$(1,-1)$	$\bar{D}_2$	$(\bar{3}, 1)$	-3	2	3	1	-1	2	2	4	-1
	D_4	$(3, 1)$	3	2	-3	1	-1	2	2	4	1
	$\bar{G}_4$	$(1, 2)$	0	2	0	4	2	2	2	4	3
	G_2	$(1, 2)$	0	2	0	-2	-4	2	2	4	-3
	S_3	1	0	-4	0	-2	2	-4	-4	-8	0
	Y_3	1	0	-4	0	-2	2	2	2	4	0
	l_2	1	0	-4	0	4	-4	2	2	4	6
	$\bar{l}_4$	1	0	8	0	-2	2	2	2	4	-6
	$\bar{A}_{14}$	1	-9	2	-3	1	-1	2	2	4	-3
	A_{14}	1	9	2	3	1	-1	2	2	4	3
$(-1,-1)$	G_3	$(1, 2)$	0	2	0	-2	0	-2	-6	4	-3
	G_4	$(1, 2)$	0	2	0	-2	0	4	6	4	-3
	G_5	$(1, 2)$	0	2	0	-2	0	-2	0	-8	-3
	l_3	1	0	-4	0	4	0	4	6	4	6
	l_4	1	0	-4	0	4	0	-2	-6	4	6
	l_5	1	0	-4	0	4	0	-2	0	-8	6

Appendix D
Computation of RR tadpoles

The simplest way to evaluate the total tadpoles is to compute the sum of Klein bottle, Moebius strip, and cylinder diagrams, transform them into the dual channel, and factorize them. As shown in Figure D.1, the result is the square of the disk plus crosscap tadpole (or sums thereof for various fields). In supersymmetric models, the total amplitude vanishes due to a cancellation of NSNS and RR closed string exchanges. Yet the consistency of the models requires the cancellation of the tadpoles for the diverse RR fields, which are extracted from the RR exchange piece in this factorization limit.

D.1 RR tadpoles in type I theory
D.1.1 The cylinder

The annulus/cylinder amplitude of an open string propagating for a time $2T\ell$ and closing onto itself, as in Figure 3.13, is given by

$$Z_\mathcal{C} = \int_0^\infty \frac{dT}{2T} \mathcal{C}(T), \quad \text{with } \mathcal{C}(T) = \frac{1}{2} N^2 \operatorname{Tr}_{\mathcal{H}_{\text{open}}} e^{-2T\ell H_{\text{open}}}. \tag{D.1}$$

The factor of $1/2$ arises from the $(1+\Omega)/2$ projector. Direct computation gives

$$\mathcal{C}(T) = \frac{N^2}{4} \frac{V_{10}}{(8\pi^2\alpha' T)^5} \eta^{-8} \eta^{-4} \left(\vartheta\!\left[\begin{smallmatrix} 0 \\ 0 \end{smallmatrix}\right]^4 - \vartheta\!\left[\begin{smallmatrix} 0 \\ 1/2 \end{smallmatrix}\right]^4 - \vartheta\!\left[\begin{smallmatrix} 1/2 \\ 0 \end{smallmatrix}\right]^4 + \vartheta\!\left[\begin{smallmatrix} 1/2 \\ 1/2 \end{smallmatrix}\right]^4 \right), \tag{D.2}$$

with $q = e^{-2\pi T}$, and the factor $1/4$ arises from the GSO and Ω projections. The diagram in the dual channel describes a closed string propagating between two boundaries for a time $T'\ell$ with $T' = 1/(2T)$. Its amplitude can be obtained by replacing $q = e^{-2\pi T} = e^{-\pi/T'}$ and using modular transformations, to obtain

$$\mathcal{C}(T') = \frac{V_{10} N^2 T'}{2(8\pi^2\alpha')^5} \tilde{\eta}^{-8} \tilde{\eta}^{-4} \left(\tilde{\vartheta}\!\left[\begin{smallmatrix} 0 \\ 0 \end{smallmatrix}\right]^4 - \tilde{\vartheta}\!\left[\begin{smallmatrix} 0 \\ 1/2 \end{smallmatrix}\right]^4 - \tilde{\vartheta}\!\left[\begin{smallmatrix} 1/2 \\ 0 \end{smallmatrix}\right]^4 + \tilde{\vartheta}\!\left[\begin{smallmatrix} 1/2 \\ 1/2 \end{smallmatrix}\right]^4 \right), \tag{D.3}$$

where the tilde indicates that the modular functions have argument $\tilde{q} = e^{-4\pi T'}$. Note that this instance of open–closed duality implies the requirement of GSO projection in the open string sector, as noted in Section 4.4.1.

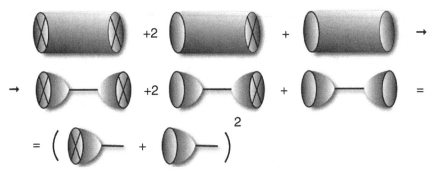

Figure D.1 The sum of the cylinder, Klein bottle, and Moebius strip amplitudes, in the dual channel, factorizes as the square of the total (disk plus crosscap) tadpole.

In the factorization limit $T' \to \infty$, the RR tadpole can be extracted by isolating the corresponding theta functions, namely

$$\mathcal{C}(T')_{\mathrm{RR}} = -\frac{N^2 V_{10} T'}{2(8\pi^2 \alpha')^5} \tilde{\eta}^{-8} \tilde{\eta}^{-4} \left(\tilde{\vartheta} \begin{bmatrix} 1/2 \\ 0 \end{bmatrix}^4 - \tilde{\vartheta} \begin{bmatrix} 1/2 \\ 1/2 \end{bmatrix}^4 \right) \to -\frac{8 V_{10} T'}{(8\pi^2 \alpha')^5} N^2. \qquad (\mathrm{D.4})$$

The non-zero disk tadpole is proportional to N, as advanced in (4.95). The NS–NS tadpole is obtained similarly, and by the identity (A.8), which reflects spacetime supersymmetry, it equals the RR tadpole.

D.1.2 The Klein bottle

As shown in Section 3.4.2, see Figure 3.17(a), a Klein bottle diagram for a closed string propagating for a time $T\ell$ can be regarded in a dual channel as a diagram of closed strings traveling between two crosscaps for a time $T'\ell$ with $T' = 1/(4T)$. The Klein bottle amplitude is given by

$$Z_K = \int_0^\infty \frac{dT}{2T} \mathcal{K}, \quad \mathcal{K} = \frac{1}{2} \mathrm{tr}\,_{\mathcal{H}_{\mathrm{closed}}} (\Omega e^{-T\ell H_{\mathrm{closed}}}), \qquad (\mathrm{D.5})$$

where the factor of $1/2$ arises from the $(1 + \Omega)/2$ projector. The trace runs over the Hilbert space of closed oriented strings, but the Ω insertion implies that only left–right symmetric states contribute to the trace. Hence we can simply sum over the left states and double the energy of each state, and obtain

$$\mathcal{K} = \frac{1}{4} \frac{V_{10}}{(4\pi^2\alpha' T)^5} \tilde{\eta}^{-8}\, \tilde{\eta}^{-4} \left(\tilde{\vartheta} \begin{bmatrix} 0 \\ 0 \end{bmatrix}^4 - \tilde{\vartheta} \begin{bmatrix} 0 \\ 1/2 \end{bmatrix}^4 - \tilde{\vartheta} \begin{bmatrix} 1/2 \\ 0 \end{bmatrix}^4 + \tilde{\vartheta} \begin{bmatrix} 1/2 \\ 1/2 \end{bmatrix}^4 \right),$$

where the tilded functions have modular parameter $\tilde{q} = e^{-4\pi T}$. The dual channel amplitude is obtained by using $T = 1/(4T')$ and modular transformations, giving

$$\mathcal{K}(T') = \frac{V_{10} T'}{(4\pi^2\alpha')^5} 2^4 \tilde{\eta}^{-8}\, \tilde{\eta}^{-4} \left(\tilde{\vartheta} \begin{bmatrix} 0 \\ 0 \end{bmatrix}^4 - \tilde{\vartheta} \begin{bmatrix} 0 \\ 1/2 \end{bmatrix}^4 - \tilde{\vartheta} \begin{bmatrix} 1/2 \\ 0 \end{bmatrix}^4 + \tilde{\vartheta} \begin{bmatrix} 1/2 \\ 1/2 \end{bmatrix}^4 \right),$$

where modular functions have argument $\tilde{q} = e^{-4\pi T'}$. Isolating the RR contribution in the dual channel and taking the $T' \to \infty$ limit

$$\mathcal{K}(T')_{\text{RR}} = -\frac{2^9 V_{10} T'}{(8\pi^2 \alpha')^5} \tilde{\eta}^{-8} \tilde{\eta}^{-4} \left(\tilde{\vartheta} \begin{bmatrix} 1/2 \\ 0 \end{bmatrix}^4 - \tilde{\vartheta} \begin{bmatrix} 1/2 \\ 1/2 \end{bmatrix}^4 \right) \to -\frac{8 V_{10} T'}{(8\pi^2 \alpha')^5} 2^{10}. \quad (\text{D.6})$$

This is proportional to the square of the RR crosscap tadpole, which is therefore non-vanishing as anticipated in (4.101).

D.1.3 The Moebius strip

As shown in Section 3.4.3, see Figure 3.17(b), a Moebius strip diagram for an open string propagating for a time $2T\ell$ can be regarded in a dual channel as a diagram of closed strings traveling between a crosscap and a boundary for a time $T'\ell$ with $T' = 1/(8T)$. The Moebius strip amplitude is given by

$$Z_{\mathcal{M}} = \int_0^\infty \frac{dT}{2T} \mathcal{M}, \quad \mathcal{M} = \text{tr} \left(\gamma_\Omega^{-1} \gamma_\Omega^T \right) \frac{1}{2} \text{tr}\, \mathcal{H}_{\text{open}} \left(\Omega\, e^{-2T\ell H_{\text{open}}} \right), \quad (\text{D.7})$$

with the factor of $1/2$ from the $(1+\Omega)/2$ projector. The trace over the Chan–Paton indices is $\text{tr} \left(\gamma_\Omega^{-1} \gamma_\Omega^T \right) = \pm N$ for the SO or Sp projections. The remaining trace is over the Hilbert space of open oriented strings, properly accounting for the Ω action. For instance, the trace over bosonic degrees of freedom gives

$$\text{Tr}_{\text{bos.}} \left(q^{N_B + E_0^B} \Omega \right) = \left[q^{\frac{1}{24}} \prod_{n=1}^\infty (1 - (-1)^n q^n) \right]^{-8} = \tilde{\eta}^4 \tilde{\vartheta} \begin{bmatrix} 0 \\ 0 \end{bmatrix}^{-4}, \quad (\text{D.8})$$

where the second equality follows upon splitting the product into even and odd moddings. Working similarly with the fermionic sectors, the total amplitude is

$$\mathcal{M} = \pm N \frac{1}{4} \frac{V_{10}}{(8\pi^2 \alpha')^5} (1 - 1) \tilde{\eta}^{-8} \tilde{\eta}^{-4} \tilde{\vartheta} \begin{bmatrix} 0 \\ 0 \end{bmatrix}^{-4} \tilde{\vartheta} \begin{bmatrix} 0 \\ 1/2 \end{bmatrix}^4 \tilde{\vartheta} \begin{bmatrix} 1/2 \\ 0 \end{bmatrix}^4, \quad (\text{D.9})$$

where we have isolated the (equal and opposite) RR and NSNS tadpoles, as indicated by the $(1-1)$ factor. The dual channel amplitude is obtained by using $T = 1/(8T')$ and modular transformations. Isolating the RR piece we have

$$\mathcal{M}(T')_{\text{RR}} = \frac{\pm 2^5 N V_{10} T'}{(8\pi^2 \alpha')^5} \hat{\eta}^{-8} \hat{\eta}^{-4} \hat{\vartheta} \begin{bmatrix} 0 \\ 0 \end{bmatrix}^{-4} \hat{\vartheta} \begin{bmatrix} 0 \\ 1/2 \end{bmatrix}^4 \hat{\vartheta} \begin{bmatrix} 1/2 \\ 0 \end{bmatrix}^4 \to \frac{8 V_{10} T'}{(8\pi^2 \alpha')^5} (\pm 2^6 N),$$
$$(\text{D.10})$$

where the hatted functions have modular parameter $\hat{q} = e^{-8\pi T'}$.

The RR tadpole condition is obtained from the sum of factorized cylinder, Klein bottle and Moebius strip amplitudes. Using (D.4), (D.6), and (D.10), we have

$$-\frac{8 V_{10} T'}{(8\pi^2 \alpha')^5} (N^2 + 2^{10} \mp 2^6 N)^2 \simeq (N \mp 32)^2, \quad (\text{D.11})$$

The RR tadpole cancels for $N = 32$ with SO projection, as used in Section 4.4.3.

D.2 Tadpoles for $\mathbf{T^6/Z_N}$ type IIB orientifolds

We can apply a similar strategy to the computation of RR tadpoles in the toroidal type IIB orientifolds of Section 11.2. The amplitude sums over possible orbifold twisted boundary conditions along the non-trivial cycles in the worldsheet, making the computation technically more involved. For concreteness we focus only on aspects relevant to the $\mathbf{Z}_3$ orbifold in Section 11.2.1 (or other odd order $\mathbf{Z}_N$ orbifolds).

In the Klein bottle amplitude, the non-zero contributions arise only from states mapped to themselves under Ω. This exchanges oppositely twisted sectors, so in odd order orbifolds the only contributions arise from untwisted states (and from $\theta^{N/2}$-twisted states in even order orbifolds, not considered here). We have

$$Z_K = \frac{1}{N} \sum_{k=0}^{N-1} \int_0^\infty \frac{dT}{2T} \mathcal{K}(1,\theta^k), \quad \mathcal{K}(1,\theta^k) = \frac{1}{2} \mathrm{Tr}\,_{\mathcal{H}_{\mathrm{cl.unt}}} \left[\Omega\, \theta^k\, e^{-T\ell(H_L + H_R)} \right].$$

For $k=0$ this produces the type I RR 10-form tadpole, and is ignored in what follows. Since only left–right symmetric states contribute, we may sum over the left sector and double its contribution to the exponent. Direct evaluation gives

$$\mathcal{K}(1,\theta^k) = \frac{V_4}{4(4\pi^2\alpha'T)^2} \sum_{\alpha,\beta=0,\frac{1}{2}} \eta_{\alpha,\beta} \frac{\tilde{\vartheta}\begin{bmatrix}\alpha\\\beta\end{bmatrix}}{\tilde{\eta}^3} \prod_{i=1}^3 \frac{-2\sin 2\pi k v_i}{4\sin^2 \pi k v_i} \frac{\tilde{\vartheta}\begin{bmatrix}\alpha\\\beta+2kv_i\end{bmatrix}}{\tilde{\vartheta}\begin{bmatrix}\frac{1}{2}\\\frac{1}{2}+2kv_i\end{bmatrix}}, \tag{D.12}$$

where $\eta_{0,0} = -\eta_{0,\frac{1}{2}} = -\eta_{\frac{1}{2},0} = 1$, and tilded functions have argument $\tilde{q} = e^{-4\pi T}$. The factor $(4\sin^2 \pi k v_i)^{-1}$ is the trace of θ^k over the center of mass degrees of freedom. The amplitude (D.12) vanishes, by supersymmetry, using (A.9). Hence

$$\mathcal{K}(1,\theta^k) = (1-1)\frac{V_4}{4(4\pi^2\alpha'T)^2} \frac{\tilde{\vartheta}\begin{bmatrix}0\\\frac{1}{2}\end{bmatrix}}{\tilde{\eta}^3} \prod_{i=1}^3 \frac{\cos \pi k v_i}{\sin \pi k v_i} \frac{\tilde{\vartheta}\begin{bmatrix}0\\\frac{1}{2}+2kv_i\end{bmatrix}}{\tilde{\vartheta}\begin{bmatrix}\frac{1}{2}\\\frac{1}{2}+2kv_i\end{bmatrix}}. \tag{D.13}$$

The tadpoles can be read in the factorization limit $T' = 1/(4T) \to \infty$, where

$$\mathcal{K}(1,\theta^k)_{\mathrm{RR}} \to \frac{V_4 T'}{4(8\pi^2\alpha')^2} 32 \prod_{i=1}^3 \frac{\cos \pi k v_i}{\sin \pi k v_i} \quad \text{for } k \neq 0. \tag{D.14}$$

The cylinder amplitude for the D9–D9 open strings is given by

$$Z_C = \frac{V_4}{4N} \frac{1}{N} \sum_{k=0}^{N-1} \int_0^\infty \frac{dT}{2T} \mathcal{C}(\theta^k), \quad \mathcal{Z}_C(\theta^k) = \mathrm{Tr}\,_{\mathcal{H}_{\mathrm{op.}}} \left(\theta^k e^{-2T\ell H_{\mathrm{open}}} \right). \tag{D.15}$$

Direct computation gives

$$\mathcal{C}(\theta^k) = \frac{V_4}{4(8\pi^2\alpha'T)^2} \sum_{\alpha,\beta=0,\frac{1}{2}} \eta_{\alpha,\beta} \frac{\vartheta\begin{bmatrix}\alpha\\\beta\end{bmatrix}}{\eta^3} \prod_{i=1}^{3} \frac{-2\sin\pi k v_i}{4\sin^2\pi k v_i} \frac{\vartheta\begin{bmatrix}\alpha\\\beta+k v_i\end{bmatrix}}{\vartheta\begin{bmatrix}\frac{1}{2}\\\frac{1}{2}+k v_i\end{bmatrix}} (\mathrm{Tr}\,\gamma_{k,9})^2$$

$$= (1-1)\frac{V_4}{4(8\pi^2\alpha'T)^2} \frac{\vartheta\begin{bmatrix}0\\\frac{1}{2}\end{bmatrix}}{\eta^3} \prod_{i=1}^{3} \frac{1}{2\sin\pi k v_i} \frac{\vartheta\begin{bmatrix}0\\\frac{1}{2}+k v_i\end{bmatrix}}{\vartheta\begin{bmatrix}\frac{1}{2}\\\frac{1}{2}+k v_i\end{bmatrix}} (\mathrm{Tr}\,\gamma_{k,9})^2, \qquad \text{(D.16)}$$

where the last equality uses (A.9). The factorization limit $T' = 1/(2T) \to \infty$ gives

$$\mathcal{C}(\theta^k)_{\mathrm{RR}} \to \frac{V_4 T'}{4(8\pi^2\alpha')^2} 2 \prod_{i=1}^{3} \frac{1}{2\sin\pi k v_i} (\mathrm{Tr}\,\gamma_{k,9})^2 \quad \text{for } k \neq 0. \qquad \text{(D.17)}$$

Finally, we consider the Moebius strip amplitudes, whose structure is

$$Z_{\mathcal{M}} = \frac{1}{N} \sum_{k=0}^{N-1} \int_0^{\infty} \frac{dT}{2T} \mathcal{M}(\theta^k), \quad \mathcal{M}(\theta^k) = \mathrm{Tr}_{\mathcal{H}_{\mathrm{op}}}\left(\Omega\theta^k e^{-2T\ell H_{\mathrm{open}}}\right).$$

The main difference between $Z_{\mathcal{M}}$ and $Z_{\mathcal{C}}$ is the insertion of Ω, which introduces extra phases due to its action on oscillators. The amplitude is

$$\mathcal{M}(\theta^k) = (1-1)\frac{-V_4}{4(8\pi^2\alpha'T)^2} \frac{\tilde{\vartheta}\begin{bmatrix}\frac{1}{2}\\0\end{bmatrix}\tilde{\vartheta}\begin{bmatrix}0\\\frac{1}{2}\end{bmatrix}}{\tilde{\eta}^3\tilde{\vartheta}\begin{bmatrix}0\\0\end{bmatrix}} \prod_{i=1}^{3} \frac{-2\sin\pi k v_i}{4\sin^2\pi k v_i} \frac{\tilde{\vartheta}\begin{bmatrix}\frac{1}{2}\\k v_i\end{bmatrix}\tilde{\vartheta}\begin{bmatrix}0\\\frac{1}{2}+k v_i\end{bmatrix}}{\tilde{\vartheta}\begin{bmatrix}\frac{1}{2}\\\frac{1}{2}+k v_i\end{bmatrix}\tilde{\vartheta}\begin{bmatrix}0\\k v_i\end{bmatrix}} \mathrm{Tr}\,\gamma_{2k,9},$$

where again the tilded functions have variable $\tilde{q} = e^{-4\pi T}$, and we have used $\mathrm{Tr}\left(\gamma_{\Omega k,9}^{-1} \gamma_{\Omega k,9}^{T}\right) = \mathrm{Tr}\,\gamma_{2k,9}$. In the factorization limit, the RR contribution is

$$\mathcal{M}(\theta^k)_{\mathrm{RR}} \to \frac{V_4 T'}{4(8\pi^2\alpha')^2} 16 \prod_{i=1}^{3} \frac{-1}{2\sin\pi k v_i} (\mathrm{Tr}\,\gamma_{2k,9}) \quad \text{for } k \neq 0. \qquad \text{(D.18)}$$

Adding (D.17) for $\mathcal{C}(\theta^{2k})_{\mathrm{RR}}$ and (D.14), (D.18) the amplitude is proportional to

$$\sum_k \left[\prod_i \frac{1}{2\sin 2\pi k v_i} (\mathrm{tr}\,\gamma_{2k,9})^2 - 8 \prod_i \frac{1}{2\sin\pi k v_i} \mathrm{tr}\,\gamma_{2k,9} + 16 \prod_i \frac{\cos\pi k v_i}{\sin\pi k v_i} \right]$$

$$= \sum_k \frac{1}{\prod_i 2\sin 2\pi k v_i} \left(\mathrm{tr}\,\gamma_{2k,9} - 32 \prod_i \cos\pi k v_i \right)^2. \qquad \text{(D.19)}$$

This factorizes as a sum of squares, as expected. The RR tadpole conditions are

$$\mathrm{tr}\,\gamma_{2k,9} = 32 \prod_i \cos\pi k v_i. \qquad \text{(D.20)}$$

Appendix E
CFT toolkit

In this appendix we provide some background material on 2d conformal and superconformal field theory (CFT and SCFT, respectively). We merely state the main results, giving up a complete logical flow at certain points in favor of easy reference from the main text.

E.1 Conformal symmetry and conformal fields

We trade the left and right coordinates $t \pm \sigma$ for holomorphic and antiholomorphic coordinates $z = e^{-t+i\sigma/\ell}, \bar{z} = e^{-t-i\sigma/\ell}$. In this picture, time runs radially so hamiltonian quantization is called *radial* quantization, and the hamiltonian relates to the dilatation operator, and thus to the behavior of fields under conformal transformations. Eventually both z and $\bar{z}$ can be extended to independent complex coordinates. For most of the analysis, we focus on the holomorphic sector only, with different combinations of left and right sector corresponding to the different string theories.

As described in Section 3.2.1, a 2d conformal transformation can be described as a holomorphic reparametrization $z \to f(z)$ on the worldsheet. Let us consider an infinitesimal transformation $z \to z + \epsilon(z)$ generated by the 2d energy-momentum tensor $T(z)$. Integrating over the σ direction, we obtain the integrated generator

$$T_\epsilon = \frac{1}{2\pi i} \oint_{C_0} dz\, \epsilon(z)\, T(z), \qquad (E.1)$$

with integration along a contour C_0 around the origin. The transformation of a 2d field $\phi(z)$ can be written

$$\delta_\epsilon \phi(w) = [T_\epsilon, \phi(w)] = \frac{1}{2\pi i} \oint_{C_w} dz\, \epsilon(z) T(z) \phi(w), \qquad (E.2)$$

with implicit radial (i.e. time) ordering in the absolute values $|z|, |w|$. In the last equality the difference of integrals along two z contours at $|z| > |w|$ and $|z| < |w|$ has been recast as an integral along a contour C_z around z, see Figure E.1. The latter expression depends only on the structure of poles in $(z - w)$ of $T(z)\phi(w)$, which is known as the operator product expansion (OPE) and provides an economic encoding of the commutation relations.

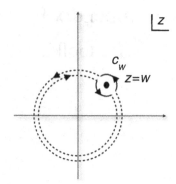

Figure E.1 Radially ordered commutator relations between operators at points z and w can be encoded in the structure of poles in $(z - w)$ in the operator product expansion.

A primary field of conformal dimension h is defined as transforming under $z \to f(z) = z + \epsilon(z)$ as

$$\phi(z) \to \left(\frac{\partial f}{\partial z}\right)^h \phi'(f(z)) \quad \text{namely} \quad \delta\phi(z) = h\partial\epsilon(z)\phi(z) + \epsilon(z)\partial\phi(z), \quad \text{(E.3)}$$

which is reproduced by the OPE

$$T(z)\phi(w) \sim \frac{h\,\phi(w)}{(z-w)^2} + \frac{\partial\phi(w)}{z-w}. \quad \text{(E.4)}$$

In a general CFT the energy-momentum tensor has the OPE structure

$$T(z)T(w) \sim \frac{c/2}{(z-w)^4} + \frac{2T(w)}{(z-w)^2} + \frac{\partial T(w)}{z-w}, \quad \text{(E.5)}$$

and so is essentially a dimension 2 primary operator, except for a possible conformal anomaly controlled by the quantity c, known as central charge of the CFT. The above structure is known as Virasoro algebra.

Simple examples of CFTs can be obtained using free theories. The free massless bosons $X(z)$ and fermions $\psi(z)$ on string worldsheets have

$$\partial X(z)\partial X(w) \sim -\frac{\alpha'}{2(z-w)^2}, \quad T(z) = -\frac{1}{\alpha'}\partial X\partial X, \quad c = 1,$$

$$\psi(z)\psi(w) \sim -\frac{1}{z-w}, \quad T(z) = -\frac{1}{2}(\partial\psi)\psi, \quad c = \frac{1}{2}, \quad \text{(E.6)}$$

where here and in what follows, normal ordering is implicit. The field $X(z)$ has OPE $X(z)X(w) \sim -\frac{\alpha'}{2}\log(z-w)$ and is not a primary field itself, but can be used to build primary fields. For instance, ∂X, with conformal dimension $h = 1$, which is the worldsheet current associated to momentum conservation in the spacetime coordinate X. Also, the operator e^{ikX}, for which useful OPEs are

$$\partial X(z)\, e^{ikX(w)} \sim -\frac{ik\alpha'}{2(z-w)}\, e^{ikX(w)},$$

$$e^{ik_1 X(z)}\, e^{ik_2 X(w)} \sim (z-w)^{k_1 k_2}\, e^{i(k_1+k_2)X(w)}. \tag{E.7}$$

The operator e^{ikX} carries spacetime momentum k along the coordinate X; it has conformal dimension $h = \frac{\alpha' k^2}{4}$.

The fermion ψ has conformal dimension $h(\psi) = 1/2$. The bosonization in Section 4.2.6 can be stated as the equivalence of a free complex fermion $\psi(z), \overline{\psi}(z)$ and a free real boson $H(z)$ with $H(z)H(w) \sim -\log(z-w)$ as

$$\psi(z) \cong e^{iH(z)}, \quad \overline{\psi}(z) \cong e^{-iH(z)}, \quad \text{namely } i\partial H(z) \cong \psi\overline{\psi}(z), \tag{E.8}$$

where "$\cong$" stands for "reproduces the same OPEs." The last expression relates momentum conservation along H (known as H-momentum) and fermion number conservation.

In the covariant quantization of the bosonic string theory, one fixes the conformal gauge, and then introduces a ghost system for the left-over conformal symmetries. The ghosts $b(z), c(z)$ are free anticommuting fields with

$$c(z)b(w) \sim \frac{1}{z-w}, \quad T(z) = 2(\partial c)b + c\partial b, \quad c = -26. \tag{E.9}$$

The total energy-momentum tensor is the sum of contributions from the (b, c) ghost system and D spacetime coordinates $X^M(z)$. The total central charge is $c = D-26$, so cancellation of the conformal anomaly requires the critical dimension $D = 26$.

In the superstring, there are in addition commuting (so-called superconformal) ghosts $\beta(z), \gamma(z)$ with

$$\gamma(z)\beta(w) \sim \frac{1}{z-w}, \quad T(z) = -\frac{1}{2}\gamma\partial\beta - \frac{3}{2}(\partial\gamma)\beta, \quad c = 11. \tag{E.10}$$

The total energy-momentum tensor is the sum of contributions from the (b, c) and (β, γ) ghosts and D sets of free bosons and fermions $X^M(z), \psi^M(z)$, resulting in $c = \frac{3}{2}D - 15$, and hence $D = 10$ as critical dimension.

E.2 Vertex operators and structure of scattering amplitudes

In conformal field theories there exists a *state-operator map*, which associates a so-called *vertex operator* to each state. In radial quantization, this is done as follows. For a given operator $V_{\mathcal{O}}$, the corresponding state $|\mathcal{O}\rangle$ is obtained by performing the CFT path integral over the unit disk $|z| \leq 1$, with an insertion of $\mathcal{O}$ at the origin. In standard hamiltonian quantization on the cylinder, the origin is sent to $t \to -\infty$ and the above procedure defines an incoming asymptotic state, see Figure E.2.

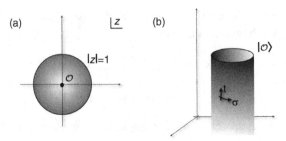

Figure E.2 State-operator map.

In string theory, scattering amplitudes for different asymptotic states are computed as correlators of the corresponding vertex operators in the 2d quantum field theory in a worldsheet Σ

$$\langle V_{\mathcal{O}_1} \cdots V_{\mathcal{O}_n} \rangle_\Sigma. \tag{E.11}$$

A more careful study of vertex operators and the topology of Σ leads to some interesting selection rules, as described below.

In closed bosonic string theory, the vertex operators have a structure

$$V_{\text{int}} = \int_\Sigma d^2 z \, \mathcal{O} \tilde{\mathcal{O}} e^{ikX(z,\bar{z})}, \tag{E.12}$$

where "int" denotes that the operator is integrated over the worldsheet. The integrand must have conformal weights $(h, \tilde{h}) = (1, 1)$ to compensate the transformation of the measure $d^2 z$ and lead to conformal invariant object. This relates the vertex operator conformal weight and the spacetime mass of the corresponding string state, using the mass-shell condition $k^2 + m^2 = 0$, and $M_L^2 = M_R^2 = m^2/2$. Namely

$$\frac{\alpha' k^2}{4} + h_{\mathcal{O}} = 1 \quad \rightarrow \quad \frac{\alpha' M_L^2}{2} = h_{\mathcal{O}} - 1, \tag{E.13}$$

and similarly for $\tilde{h}$. This corresponds to the closed bosonic string mass formula. Indeed, the closed bosonic string tachyon vertex operator has $\mathcal{O} = \tilde{\mathcal{O}} = 1$, hence $m^2 = -4/\alpha'$. Also, the graviton, 2-form and dilaton states, $\alpha_{-1}^M \tilde{\alpha}_{-1}^N |0\rangle$, have vertex operators with $\mathcal{O}\tilde{\mathcal{O}} = \partial X^M \partial \tilde{X}^N$, and we recover $m^2 = 0$.

The existence of a contribution to the spacetime mass from the conformal dimension is a general property. Splitting into left and right sectors, and considering the CFT under discussion one particular factor in the full theory, the contribution of a conformal weight h piece of the vertex operator to the left-moving spacetime mass of the corresponding state is

$$\left(\frac{\alpha' M_L^2}{2} \right)_{\mathcal{O}} = h_{\mathcal{O}}. \tag{E.14}$$

Vertex operators like (E.12) have the correct structure for insertions whose position is integrated over the worldsheet. However, one can use symmetries on the worldsheet to set the position of certain vertex operator insertions to arbitrary fiducial positions. The number of such symmetries, and hence the number of "fixed" operators, depends on the topology of the Riemann surface. For instance, for closed string tree level amplitudes it is possible to use an $SL(2, \mathbf{C})$ symmetry of the sphere to fix the position of three vertices. In one-loop amplitudes, it is possible to fix the position of one vertex, by shifting the origin of $\mathbf{T}^2$.

In the covariant quantization, the above freedom manifests as the existence of zero modes for the ghosts $c, \tilde{c}$. Their number is precisely the number of vertices whose positions can be fixed. In this language a "fixed" vertex operator is

$$V_{\text{fix}} = c\tilde{c}\mathcal{O}\tilde{\mathcal{O}}, \tag{E.15}$$

so that there is saturation of fermion zero modes in the path integral. Hence, e.g., a closed string tree level amplitude must involve three operators in the "picture" (E.15) and the remaining in the picture (E.12).

The left-moving part of the above statement applies to the supersymmetric side of heterotic strings. Similarly, they apply to open strings, which effectively work as the left-moving sector of the corresponding closed string theories.

There is an analogous but more involved story in superstrings. Using 2d superspace language, vertex operators must in general be integrated also over $\theta, \bar{\theta}$. However, there may be the freedom of fixing the $\theta, \bar{\theta}$ positions for a number of vertex operators. This is signaled by zero modes for the superconformal ghost γ (E.10).

The superconformal ghosts can be traded for a boson $\phi(z)$, with $\phi(z)\phi(w) \sim -\log(z - w)$, via

$$\beta(z) \cong e^{-\phi(z)} \dots, \quad \gamma(z) \cong e^{-\phi} \dots, \quad \partial\phi \cong \beta\gamma, \tag{E.16}$$

where the dots denote contributions from an extra CFT sector required to reproduce all OPEs correctly. The field $\phi(z)$ is called the bosonized superconformal ghost, given the analogy of these formulas with (E.8). The existence of γ zero modes on a worldsheet topology Σ translates into an anomaly for the current $\partial\phi$; that is, a background charge which must be canceled by the total superconformal ghost charge from the vertex operator insertions in correlators (E.11) to lead to a non-zero amplitude.

In a left-moving sector of a superstring (or in an open superstring), the vertex operator for the NS state $\psi^M_{-1/2}|0\rangle$ is

$$V_{-1} = e^{-\phi}\psi^M(z), \tag{E.17}$$

where from now on we ignore the momentum factor $\exp(ikX)$. It has superconformal ghost charge -1, as indicated in the subindex. This is known as the (-1)-picture, and provides the analog of the fixed operators (E.15), since it can be shown that $e^{-\phi} \sim \delta(\gamma)$ is effectively a delta function fixing the position along the γ zero modes. The conformal weight of $\exp(-\phi)$ is $1/2$, and so (E.17) has conformal dimension $h = 1$ as required.

It is convenient to introduce the vertex operators in the "integrated" picture with zero ghost charge, known as 0-picture. It is given by

$$V_0 = -\frac{2}{\alpha'}\left(i\partial X^M + \frac{1}{2}\alpha' k \cdot \psi\psi^M\right). \tag{E.18}$$

For R groundstates, the state-operator map gives vertices with ghost charge $-1/2$, in the so-called$(-1/2)$-picture. They have the structure

$$V_{-\frac{1}{2}} = e^{-\phi/2}e^{is_a H_a(z)}, \tag{E.19}$$

where $(s_0, \ldots, s_5)$ denote the $SO(10)$ spinor weights and $H_a(z)$ are the bosonized world-sheet fermions. The conformal dimension of $\exp(-\phi/2)$ is $3/8$, hence (E.19) has conformal dimension $h = 1$, as required. The bosonized language is convenient also for states in the NS sector, which read

$$V_{-1} = e^{-\phi}e^{i\alpha_v \cdot H}, \tag{E.20}$$

where α_v is a weight in the $SO(10)$ vector representation $\alpha_v = (\underline{\pm 1, 0, 0, 0, 0})$, with underlining meaning permutation of entries.

For tree level amplitudes on a sphere, it can be shown that there are two γ (and two $\tilde{\gamma}$) zero modes. Hence, the left-moving vertex operators in any correlator must have super-conformal ghost charges adding up to -2 for a non-vanishing amplitude (and similarly for the right-moving sector). For amplitudes involving two spacetime fermions and an arbitrary number of bosons, one can take the two R vertex operators in the $(-1/2)$-picture, one boson in the (-1)-picture, and the remaining bosons in the 0-picture. Similar results hold for the supersymmetric right-moving sector of the heterotic, and for open superstrings on the disk.

E.3 Kac–Moody algebras

A Kac–Moody (KM) algebra associated to a Lie algebra G with structure constants f_{ABC} are a set of operators $J^A(z)$ with OPE

$$J^A(z)J^B(w) \sim \frac{\tilde{k}}{(z-w)^2}\delta_{AB} + \frac{if_{ABC}}{z-w}J^C(w), \tag{E.21}$$

with $\tilde{k} = k|\rho|^2/2$, where k is a positive integer known as the level of the algebra, and ρ is the length of the highest root. In our conventions, $\rho = 1$ for ADE Lie algebras, on which we focus. KM algebras are closely related to CFTs, since we can construct an energy-momentum tensor satisfying the Virasoro algebra, given by

$$T_{KM} = \frac{1}{k+h_G}\sum_A : J^A(z)J^A(z) :, \tag{E.22}$$

where h_G is the dual Coxeter number of G. This is known as the Sugawara construction. With this definition, the $T_{KM}J^A$ OPE implies that J^A are conserved currents with conformal dimension 1. The $T_{KM}T_{KM}$ OPE shows that the associated sector of the CFT has central charge

$$c_{G,k} = \frac{k\dim G}{k + h_G}.$$

(E.23)

Kac–Moody algebras arise naturally in the realization of gauge symmetries in heterotic string theory. Indeed, if the internal left-moving CFT contains a KM sector, the corresponding dimension 1 currents $J^A(z)$ are associated with left-moving massless states, which can combine with the right-moving NS groundstates to produce massless spacetime gauge bosons.

As a simple realization, consider the 10d heterotic theory, focusing on the 16d internal coordinates X^I. The associated KM currents split into Cartan and non-zero root currents

$$J^{H_I}(z) = i\partial X^I(z), \quad J^{E_P} = e^{iP\cdot X(z)},$$

(E.24)

with P the momentum in the 16d lattice, with $P^2 = 2$, hence corresponding to the non-zero roots of $E_8 \times E_8$ or $SO(32)$. Using the OPEs (E.7) one can check that this is a free-field realization of a KM algebra at level 1. When used as vertex operators, these correspond to the massless states $\alpha^I_{-1}|0\rangle$, $|P\rangle_{P^2=2}$ in (4.76).

Two KM algebras based on the same Lie algebra G, at levels k, k', can be used to realize a KM algebra at level $k + k'$. The latter is defined by the currents $J^A_{\text{diag}}(z) = J^A(z) + J^{A'}(z)$, where J^A, $J^{A'}$ denote the original currents. This can be used to provide a free field realization of higher-level KM algebras. In particular, as exploited in Section 9.8, a set of k level 1 KM algebras with Lie algebra G can be combined into a diagonal level k KM algebra with Lie algebra G. Higher-level KM algebras modify the gauge coupling unification relations in heterotic models, as discussed in Section 9.4.

A further procedure to generate new CFTs is the coset construction. Consider a level k KM algebra based on the Lie algebra G, with a level l KM subalgebra based on the Lie algebra H. It is possible to construct an energy-momentum tensor

$$T_{G/H} = T_{KM,G,k} - T_{KM,H,l},$$

(E.25)

satisfying a Virasoro algebra with central charge $c_{G/H} = c_{G,k} - c_{H,l}$.

A prototypical example of this construction are the $\mathcal{N} = 2$ minimal models, used in the construction of Gepner model compactifications in Sections 8.6 and 10.7.1. They are obtained as a coset $[SU(2)_k \times SO(2)]/U(1)$, where the $SU(2)$ is realized at level k, so the central charge is $c = 3k/(k + 2)$, see Section 8.6.1 for further properties.

E.4 $\mathcal{N}=2$ superconformal field theories

The 2d $\mathcal{N}=2$ superconformal algebra is an extension of the Virasoro algebra by two fermionic (super)currents $G^{\pm}(z)$ and a $U(1)$ (superconformal) current $J(z)$, with OPEs

$$T(z)G^{\pm}(0) \sim \frac{3/2}{z^2} G^{\pm}(0) + \frac{1}{z}\partial G^{\pm}(0),$$

$$T(z)J(0) \sim \frac{1}{z^2} J(0) + \frac{1}{z}\partial J(0),$$

$$G^{+}(z)G^{-}(z) \sim \frac{2c/3}{z^3} + \frac{2}{z^2} J(0) + \frac{2}{z} T(0) + \frac{1}{z}\partial J(0),$$

$$J(z)G^{\pm}(0) \sim \pm\frac{1}{z}G^{\pm}(0),$$

$$J(z)J(0) \sim \frac{c/3}{z^2}. \tag{E.26}$$

In particular, the superconformal current J has $h=1$, and the supercurrents $G^{\pm}(z)$ have $h=3/2$ and superconformal charge ± 1.

There is a close relationship between $\mathcal{N}=2$ worldsheet supersymmetry and spacetime supersymmetry. This is due to the so-called *spectral flow*, an isomorphism between representations of the algebra by superconformal fields differing in the periodicity of their realizations of $G^{\pm}$. In particular, spectral flow relates the spectrum of states in the NS and R sectors, and manifests as spacetime supersymmetry. Type II compactifications to 4d with an internal $\mathcal{N}=2$ SCFT on both left and right sectors lead to two sets of spacetime supercharges (from left and right sectors), i.e. 4d $\mathcal{N}=2$ supersymmetry; this corresponds to CY compactifications in geometric compactifications. Heterotic CY compactifications with standard embedding have gauge backgrounds completely determined in terms of the CY geometry. Their supersymmetric right-moving sector has an internal $\mathcal{N}=2$ SCFT, while the left-moving sector is related to a $\mathcal{N}=2$ SCFT by simple modifications, hence the models are known as $(2,2)$ heterotic compactifications. Heterotic CY compactifications with non-standard embedding are known as $(0,2)$ compactifications.

E.5 Rational conformal field theory and simple currents
E.5.1 Virasoro representations and characters

The Virasoro algebra is generated by the conserved currents L_n of $T(z)$

$$L_n = \frac{1}{2\pi i}\oint dz\, z^{n+1} T(z), \quad T(z) = \sum_n z^{-n-2} L_n, \tag{E.27}$$

which have commutation relations (equivalent to the OPEs (E.5))

$$[L_n, L_m] = (n-m)L_{m+n} + \frac{c}{12}n(n^2-1)\delta_{n,-m}, \tag{E.28}$$

with c the central charge.

A (unitary) highest weight representation is a representation containing a state $|h\rangle$ with a smallest value of L_0, denoted h. This implies that $L_n|h\rangle = 0$ for $n \geq 1$. The representation is spanned by the state $|h\rangle$, called *primary state*, and the so-called *descendants*, whose form (at level n) is

$$L_{-n_1} \cdots L_{-n_k}|0\rangle, \quad \text{with } \sum_i n_i = n. \tag{E.29}$$

The vacuum of the theory is defined by $L_n|0\rangle = 0$ for $n \geq -1$. For a conformal field $\phi(z)$ of conformal weight h, the state $|h\rangle = \phi(0)|0\rangle$ is a primary state with weight h. Either fields or states or representations are often referred to as *primaries*.

The partition function of the theory can be recast as a sum over representations

$$Z(\tau, \bar{\tau}) = \text{Tr}_{\mathcal{H}_{\text{CFT}}} q^{L_0 - c/24} \bar{q}^{\bar{L}_0 - \tilde{c}/24} = \sum_{i,j} Z_{ij} \chi_i(\tau) \chi_j(\tau)^*, \tag{E.30}$$

where i, j run over all highest weight representations, $L_0 - c/24$ is the hamiltonian, and we have defined the (Virasoro) character of the representation i

$$\chi_i(\tau) = \text{Tr}_i q^{L_0 - c/24}. \tag{E.31}$$

The multiplicities Z_{ij} are non-negative integers. The case $i = 0$ is the representation containing the CFT vacuum state, and uniqueness of the latter implies $Z_{00} = 1$. Modular transformation of characters defines the S and T "modular matrices"

$$\chi_i(\tau + 1) = \sum_j T_{ij} \chi_j(\tau), \quad \chi_i(-1/\tau) = \sum_j S_{ij} \chi_j(\tau), \tag{E.32}$$

with actually $T_{ij} = \exp[2\pi i(h_i - c/24)]\delta_{ij}$. Modular invariance of (E.30) amounts to $[S, Z] = [T, Z] = 0$. A trivial solution is the diagonal invariant $Z_{ij} = \delta_{ij}$. For example, the Hilbert space of a free 2d boson theory falls, for each value p of the (target) momentum, in a single Virasoro representation, with character $\chi_p = q^{p^2/2}\eta^{-1}$. The closed bosonic string is constructed using the diagonal modular invariant.

The selection rules implied by the Virasoro algebra on 3-point functions are encoded in the so-called *fusion rules* for the corresponding primaries, written

$$[i] \times [j] = \sum_k N_{ijk}[k]. \tag{E.33}$$

The N_{ijk} is a completely symmetric tensor of non-negative integers. Values $N_{ij}^k > 1$ indicate that there exists more than one way of coupling the fields. A more explicit definition is in terms of the Verlinde formula, relating them to modular matrices

$$N_{ijk} = \sum_n \frac{S_{in} S_{jn} S_{kn}}{S_{0n}}. \tag{E.34}$$

In most applications, worldsheet theories have extended symmetries beyond the Virasoro algebra. In analogy with the currents $T(z)$ generating the Virasoro algebra, extended algebras are generated by additional currents, which for our purposes are of integer or half-integer spin. There are different types of extension currents, including the already appeared

Kac–Moody currents of Section E.3, and the superconformal currents of Section E.4. Other examples are currents generating GSO projections (or generalizations thereof), as studied in the next section. There is a natural generalization of the definitions of representations, primaries, characters, fusion rules, etc., to extended algebras.

Certain theories, known as rational conformal field theories or (RCFTs), have a finite number of primaries with respect to their extended algebras. This simplification allows the construction of solvable interacting CFTs, which can be used in string compactifications, like Gepner models.

E.5.2 Simple currents and modular invariant partition functions

Most known examples of RCFTs can be understood in terms of extensions of the Virasoro algebra by *simple currents*. A simple current is a primary that upon fusion with any other field produces just one field

$$[J] \times [i] = [i'], \quad \forall [i] \neq [J]. \tag{E.35}$$

From now on, the fusion rule product is implicit, and we drop square brackets for primaries. A simple current J thus organizes fields into orbits, generically of N elements, on which it acts as a $\mathbf{Z}_N$ group. The number N is called the order of the current. Shorter orbits are associated to the existence of *fixed points*, i.e. primaries i satisfying $J^k i = i$ for some $k < N$, so J^k generates a proper subgroup of $\mathbf{Z}_N$.

We define the (*monodromy*) *charge* of a primary i under a simple current J by

$$Q_J(i) = h(i) + h(J) - h(Ji) \bmod 1. \tag{E.36}$$

It measures the violation of conformal weight additivity in the fusion product (E.35).

Simple currents of integer spin $h(J) \in \mathbf{Z}$ extending the algebra can be used to build non-diagonal partition functions, with the structure

$$Z(\tau, \bar{\tau}) = \sum_{i \,|\, Q_J(i)=0} \left| \sum_{k=0}^{N} \chi_{i, J^k i} \right|^2. \tag{E.37}$$

Namely, the representations of the extended algebra are the $\mathbf{Z}_N$ orbits of the zero monodromy charge primaries of the unextended algebra. The conformal weight additivity properties for zero charge primaries imply the modular invariance of the non-diagonal partition function. Most non-diagonal partition functions (in particular, those involving GSO projections, orbifold constructions, etc.) can be understood in terms of simple current extensions.

The above construction can be generalized to construct a modular invariant partition function for the extension by a set of simple currents, generating a group $\mathcal{H}$ and satisfying

$h_J N_J \in \mathbf{Z}$ for any $J \in \mathcal{H}$. In general $\mathcal{H} = \prod_a \mathbf{Z}_{N_\alpha}$, and we denote J_α the generator of $\mathbf{Z}_{N_\alpha}$. We introduce a symmetric matrix $X_{\alpha\beta}$ satisfying

$$2X_{\alpha\beta} = Q_{J_\alpha}(J_\beta) \bmod 1 \quad \text{for } \alpha \neq \beta,$$
$$X_{\alpha\alpha} = -h_{J_\alpha},$$
$$N_{J_\alpha} X_{\alpha\beta} \in \mathbf{Z} \quad \forall \alpha, \beta, \tag{E.38}$$

and for simple currents $J = \prod_\alpha J_\alpha^{n_\alpha}$, $J' = \prod_\beta J_\beta^{N_\beta}$, we define

$$X(J, J') = \prod_{\alpha\beta} n_\alpha m_\beta X_{\alpha\beta}. \tag{E.39}$$

Then the partition function is

$$Z(\tau, \bar{\tau}) = \sum_{ij} Z_{ij} \chi_i \chi_j^*, \tag{E.40}$$

where now Z_{ij} is the number of currents $L \in \mathcal{H}$ such that

$$j = Li,$$
$$Q_M(i) + X(M, L) = 0 \bmod 1 \quad \forall M \in \mathcal{H}. \tag{E.41}$$

E.5.3 Gepner models and simple currents

Gepner models are based on a tensor product of the $\mathcal{N} = 2$ minimal models described in Section 8.6.1, tensored with the 4d spacetime one

$$\mathcal{A}_{\text{tensor}} = \mathcal{A}_{4d} \otimes_{i=1}^r \mathcal{A}_{k_i}, \tag{E.42}$$

with $\sum_i c_{k_i} = 9$. As described in Section 8.6.1, actual 4d string compactifications require further projections, which here we describe in terms of simple current extensions. These are easily obtained from certain simple currents in the individual factors in (E.42). In particular, each factor contains the $\mathcal{N} = 2$ supercurrents, which are simple currents denoted v, and two Ramond simple currents denoted s and c.

In order for the tensor product theory to have overall $\mathcal{N} = 2$ worldsheet supersymmetry, we need to extend the algebra by the *fermion alignment* simple currents $(v, \underline{v, 0, 0, \ldots, 0})$, where underlining means permutation of entries. As a result of this extension, all primaries are either in the Ramond (R) or Neveu–Schwarz (NS) sector; this clearly corresponds to the β_i projections (8.109). A further extension by the *spectral flow simple current* $(s, s, s, s, \ldots)$, that relates the R and NS sector, implements spacetime supersymmetry, and is clearly equivalent to the β_0 projection (8.107). The description of Gepner models as simple current RCFTs is useful in the description of its orientifold quotients in Section 10.7.1.

Bibliography

Chapter 1

For an introduction to Grand Unified Theories see Langacker (1981) and Ross (1981, 1984). The idea of grand unification first appeared in Pati and Salam (1973) and Georgi and Glashow (1974), where the minimal $SU(5)$ model was first constructed. The $SO(10)$ GUT was constructed by H. Georgi (unpublished) and in Fritzsch and Minkowski (1975), see also Georgi and Nanopoulos (1979). The E_6 gauge group was first considered in Gursey et al. (1976) and flipped $SU(5)$ in Barr (1982, 1989), see also Derendinger et al. (1984). The left–right symmetric model was considered in Mohapatra and Pati (1975a,b). The joining of the SM couplings according to the renormalization group was pointed out in Georgi et al. (1974) and the renormalization of quark/lepton mass ratios in GUTs in Buras et al. (1978). The see-saw mechanism for generation of neutrino masses was put forward in Minkowski (1977), Gell-Mann et al. (1979) and Yanagida et al. (1979). Baryon number generation through leptogenesis was proposed in Fukugita and Yanagida (1986), see Buchmüller et al. (2005) for a review.

For reviews about the cosmological constant problem see Weinberg (1989), Carroll (2001), and Polchinski (2006). Weinberg's first proposal for an anthropic solution of the cosmological constant problem appeared in Weinberg (1987). For other possible anthropic fundamental constants see Donoghue (2007), Hogan (2000), and references therein. The axion solution to the strong CP problem of the SM was presented in Peccei and Quinn (1977a,b), see also Weinberg (1978) and Wilczek (1978). The invisible axion idea was put forward in Kim (1979), Shifman et al. (1980), and Dine et al. (1981a), see Kim and Carosi (2010) for a review. The Technicolor solution to the hierarchy problem was proposed in Weinberg (1976, 1979) and Susskind (1979) and extended to include the generation of fermion masses in Dimopoulos and Susskind (1979) and Eichten and Lane (1980), see Lane (2002) for a review.

For recent reviews on extra dimension models see Gabadadze (2003), Cheng (2010), Krippendorf et al. (2010a), and Shifman (2010). For phenomenological aspects see Giudice and Wells (2008), Kong et al. (2010), and Rizzo (2010). The possibility of a fifth dimension was proposed in Kaluza (1921) and physically reinterpreted in Klein (1926). The possibility of confining the SM on a lower-dimensional brane appeared in Rubakov and Shaposhnikov (1983). Large extra-dimensional models were introduced in Arkani-Hamed *et al.* (1998, 1999) and in Antoniadis *et al.* (1998). Hierarchies from warped dimensions were first discussed in Randall and Sundrum (1999a). Warping as an alternative to compactification was proposed in Randall and Sundrum (1999b).

Chapter 2

The idea of supersymmetry appeared for the first time in Ramond (1971) in the context of the two-dimensional string theory worldsheet. In four dimensions it was first discussed in Golfand and Likhtman (1971), Volkov and Akulov (1973), and Wess and Zumino (1974). There are several standard textbooks on the topic, e.g. Wess and Bagger (1992) (whose conventions we follow), Drees *et al.* (2004), Baer and Tata (2006), Binétruy (2006), Dine (2007), Terning (2006), and Weinberg (2000). For shorter reports, see Krippendorf *et al.* (2010a), Martin (1997) and Lykken (1996a). $\mathcal{N} = 1$ supergravity was introduced in Freedman *et al.* (1976) and Deser and Zumino (1976), see Nilles (1984) for its applications in particle physics. The complete $\mathcal{N} = 1$ supergravity action was written in Cremmer *et al.* (1979, 1983b). SUSY breaking mechanisms discussed in Section 2.2 were introduced in O'Raifeartaigh (1975) and Fayet and Iliopoulos (1974), and the Polonyi model of SUSY breaking in supergravity was proposed in Polonyi (1977).

For the multiplet structure in extended supersymmetry and supergravity, see Strathdee (1987), see also Ortin (2004) and references therein for further information. The action of $N = 8$ supergravity was worked out in Cremmer and Julia (1978, 1979).

Non-perturbative aspects of $\mathcal{N} = 1$ supersymmetric theories are discussed in Affleck *et al.* (1984, 1985) and Seiberg (1994) see Intriligator and Seiberg (1996, 2007) and Terning (2006) for reviews. Duality in $\mathcal{N} = 2$ supersymmetric field theories was discussed in Seiberg and Witten (1994a,b). Duality between weak and strong coupling in gauge theories was conjectured in Montonen and Olive (1977) and S-duality in $\mathcal{N} = 4$ was conjectured in Font *et al.* (1990b), see Harvey (1996) for a review.

Pioneer supersymmetric versions of the SM were constructed in Fayet (1977), Fayet (1978), and Fayet (1979), see also Farrar and Fayet (1978, 1980), Dine *et al.* (1981b), and Dimopoulos and Raby (1981). Supersymmetric versions of $SU(5)$ were first considered in Witten (1981), Dimopoulos and Georgi (1981), and Sakai (1981), and the computation of gauge coupling unification in the MSSM was worked out in Dimopoulos *et al.* (1981), Dimopoulos and Georgi (1981), and Ibáñez and Ross (1981). Proton decay in supersymmetric unified theories was studied in Weinberg (1982), Dimopoulos *et al.* (1982), Sakai and Yanagida (1982), and Ellis *et al.* (1982). Softly broken supersymmetric theories were first discussed in Girardello and Grisaru (1982), and first applied to the SUSY SM in Dimopoulos and Georgi (1981). Radiative EW symmetry breaking induced by the top quark Yukawa was first discussed in Ibáñez and Ross (1982). The renormalization group improved version of the mechanism was described in Inoue *et al.* (1982), and in the supergravity context in Ibáñez (1983), Ibáñez and López (1983), Ibáñez *et al.* (1985),

Alvarez-Gaumé *et al.* (1983), Ellis *et al.* (1983), and Derendinger and Savoy (1984). The one-loop positive contribution raising the upper bound of the lightest Higgs mass in the MSSM was computed in Haber and Hempfling (1991), Ellis *et al.* (1991), and Okada *et al.* (1991).

The generation of soft terms via supergravity mediation was first discussed in Chamseddine *et al.* (1982), Ibáñez (1982), and Barbieri *et al.* (1982). This was further systematized in Hall *et al.* (1983) and Soni and Weldon (1983). In the text we follow the review of Brignole *et al.* (1997). The μ-problem was first discussed in Kim and Nilles (1984) and the Giudice–Masiero proposed solution appeared in Giudice and Masiero (1988). Gauge mediated supersymmetry breaking was first proposed in Alvarez-Gaumé *et al.* (1982), Dine and Fischler (1982), and Dimopoulos and Raby (1983), see Giudice and Rattazzi (1999) and Luty (2005) for reviews of more recent developments, and Meade *et al.* (2009) for a general parametrization. Anomaly mediation of supersymmetry breaking was introduced in Randall and Sundrum (1999c) and Giudice *et al.* (1998). Split supersymmetry was discussed in Arkani-Hamed and Dimopoulos (2005), see also Giudice and Romanino (2004).

Chapters 3, 4, 5, and 6

There are several standard textbooks introducing string theory and covering the topics in these chapters, for instance Green *et al.* (1987a,b), Polchinski (1998a,b), also Zwiebach (2004), Becker *et al.* (2007), Kiritsis (2007), Johnson (2003), Lüst and Theisen (1989), and Ortin (2004). There are also several reviews, e.g. Uranga (2001a), Polchinski (2010), Tong (2009), Ooguri and Yin (1996), Kiritsis (1997), Szabo (2002), see also Gómez and Hernández (2002), Alvarez and Meessen (1999), Alvarez-Gaumé and Vázquez-Mozo (1992).

The miraculous cancellation of anomalies in 10d type IIB theory was derived in Alvarez-Gaumé and Witten (1984). The choices of gauge groups potentially leading to anomaly-free 10d $\mathcal{N} = 1$ theories were obtained in Green and Schwarz (1984). The 10d Green–Schwarz anomaly cancellation mechanism was uncovered (in type I) in Green and Schwarz (1985a,b). The inconsistency of the gauge groups $E_8 \times U(1)^{248}$ and $U(1)^{496}$ has been recently shown in Adams *et al.* (2010). For reviews on anomalies, see Alvarez-Gaumé (1985) and Harvey (2005).

T-duality in toroidal string compactifications was first observed in Kikkawa and Yamasaki (1984). The transformation of background fields under T-duality transformations was analyzed in Buscher (1987). For reviews on toroidal compactification and T-duality, see Giveon *et al.* (1994) and Alvarez *et al.* (1995). The appearance of D-branes in T-duals of type I was already noticed in Dai *et al.* (1989). Their role was latent, until diverse evidence for new non-perturbative dynamics in string theory was presented, e.g. Shenker (1995) and Strominger (1995), and crystallized in the proposal in Polchinski (1995) of D-branes as non-perturbative states. For reviews on D-branes see Johnson (2000), see also Polchinski (1996), Polchinski *et al.* (1996), and Bachas (1998).

For reviews covering general dualities in string theory see, e.g., Sen (1998) and Schwarz (1997). The notion of S-duality in string theory was first conjectured in Font *et al.* (1990b) for the toroidal compactification of the heterotic string and was further explored in Sen (1994). Pioneer work underlying string dualities included Duff *et al.* (1987), Bergshoeff *et al.* (1987), and Strominger (1990). The S-duality of type IIB string theory was proposed in Hull and Townsend (1995) and the proposal that type IIA at strong coupling becomes 11-dimensional appeared in Townsend (1995). The action of 11-dimensional supergravity was obtained in Cremmer *et al.* (1978). The structure of the strong coupling limit of the 10d type IIB, type IIA and $SO(32)$ heterotic and type I string theories was further completed in Witten (1995), which also conjectured the existence of 11d M-theory. Its compactification on $\mathbf{S}^1/\mathbf{Z}_2$, and its duality to the $E_8 \times E_8$ heterotic, was revealed in Hořava and Witten (1996a,b). Further evidence for the S-duality between type I and the $SO(32)$ heterotic string was presented in Polchinski and Witten (1996). The heuristic argument using the lowest dimensional D-brane as guiding principle appeared in Hull (1996). The relation between M-theory on $\mathbf{T}^2$ and type IIB was introduced in Schwarz (1995) and Aspinwall (1996). For the relation of Hořava–Witten theory with type I, see Hořava and Witten (1996b); also Polchinski and Witten (1996), Kachru and Silverstein (1997). The M-theory lift of type IIA D6-branes as Taub-NUT geometries was introduced in Townsend (1995), while the appearance of enhanced gauge symmetries from M-theory singularities follow from Section 4.6 in Witten (1995).

The AdS/CFT correspondence was proposed in Maldacena (1998), and further clarified in Witten (1998a) and Gubser *et al.* (1998). For reviews on AdS/CFT, see e.g. Aharony *et al.* (2000), Maldacena (2003), Klebanov (2000), and D'Hoker and Freedman (2002). The conifold throat and its field theory dual were constructed in Klebanov and Strassler (2000).

For reviews on antibranes and non-BPS states, see Sen (1999) and Lerda and Russo (2000); the more mathematical K-theory perspective was introduced in Witten (1998b), see Olsen and Szabo (1999) for a review.

Chapter 7

For some reviews and books on CY manifolds and compactification see Greene (1996), Candelas (1987), Hubsch (1992), and Green *et al.* (1987b). Heterotic string compactifications on CY manifolds were first introduced in Candelas *et al.* (1985). Gauge symmetry breaking through discrete Wilson lines was proposed in Witten (1985c), Breit *et al.* (1985), and Sen (1985), inspired by work in Hosotani (1983). Some general phenomenological properties of CY compactifications were described in Witten (1986) and Dine *et al.* (1985b). CY compactifications as subspaces of weighted projective spaces were constructed in Candelas *et al.* (1988a,b, 1990). For results and a review on toric CY compactifications see Kreuzer and Skarke (2002); see also Kreuzer (2010a). For recent results on CY manifolds with small Hodge numbers see Candelas and Davies (2010) and Braun *et al.* (2010). Mirror symmetry was discovered in Candelas *et al.* (1990) and Greene and Plesser (1990), see also Candelas *et al.* (1991, 1994a).

For a nice presentation of the quintic CY example see Green *et al.* (1987b). The Tian–Yau 3-generation manifold was presented in Tian *et al.* (1985). It was phenomenologically analyzed in Greene *et al.* (1986a,b, 1987a,b, 1989b), see Ross (1988) for a review. The stability conditions for vector bundles in CY manifolds were put forward in Donaldson (1985) and Uhlenbeck and Yau (1986). For techniques to build holomorphic stable bundles, see e.g. Friedman *et al.* (1997a,b), Donagi (1997), and Anderson *et al.* (2007, 2008, 2010). Heterotic $(0, 2)$ CY compactifications were studied in, e.g., Distler and Greene (1988), Distler and Kachru (1994), and Kachru (1995). The 3-generation $SO(10)$ example in Section 7.4.2 was introduced in Braun *et al.* (2005, 2006a,b) and the $SU(5)$ example in Bouchard and Donagi (2006). Supersymmetric heterotic vacua with $U(1)$ bundles are explored in Andreas and Hernández Ruiperez (2005) and Blumenhagen *et al.* (2006, 2007c). The structure of CY compactifications of the Hořava–Witten theory was first discussed in Witten (1996c), see also Lukas *et al.* (1999), Donagi *et al.* (2002), and references for Chapter 9.

Chapter 8

Orbifold compactifications were introduced in Dixon *et al.* (1985, 1986), see also Ibáñez *et al.* (1988). Reviews in orbifolds include Font and Theisen (2005), Choi and Kim (2006), Ibáñez (1987), and Ramos-Sánchez (2009). The effect of Wilson lines on the orbifold spectrum was analyzed in Ibáñez *et al.* (1987b) and the first 3-generation models based on the Z_3 orbifold were constructed in Ibáñez *et al.* (1987a), see also Bailin *et al.* (1987) and Chamseddine and Quirós (1989). The Z_N orbifold partition functions were constructed in Minahan (1988), Senda and Sugamoto (1988), and Ibáñez *et al.* (1988). In the text we follow the last reference in which the projector on massless states was constructed, other relevant references include Markushevich *et al.* (1987), Vafa (1986), Font *et al.* (1989b, 1988a), Katsuki *et al.* (1990), and Erler and Klemm (1993). For the analysis of $U(1)$ charges and anomalies in orbifold models see Casas *et al.* (1989). The 3-generation Z_3 model discussed in Section 8.3.3 was constructed in Ibáñez *et al.* (1987a) and analyzed in Casas and Muñoz (1988b,a) and Font *et al.* (1988b, 1990d). For recent orbifold model building see Kobayashi *et al.* (2004, 2005), Buchmüller *et al.* (2006, 2007b), Forste *et al.* (2004), and Ramos-Sánchez (2009). In particular the 3-generation Z'_6 model in Section 8.3.3. was constructed in Buchmüller *et al.* (2006).

Asymmetric heterotic orbifolds were first introduced in Narain *et al.* (1987, 1991) and Mueller and Witten (1986), see also Ibáñez *et al.* (1988) for some explicit models. For some model building see Kakushadze *et al.* (1998) and references therein. Four-dimensional strings using the fermionic construction were introduced in Kawai *et al.* (1986, 1987, 1988), Antoniadis *et al.* (1987), and Antoniadis and Bachas (1988), see Lykken (1995) and Cleaver (2007) for reviews. Section 8.5 follows the presentation in the latter. Model building includes flipped $SU(5)$ models (see Antoniadis *et al.* (1989) and references therein) as well SM constructions, see Faraggi *et al.* (1990) and Cleaver *et al.* (1999), see also

Chaudhuri *et al.* (1996). The three-generation example in Section 8.5.2 was introduced in Faraggi *et al.* (1990) and further analyzed in Cleaver *et al.* (1999). The covariant lattice approach to 4d string model building using the bosonic formulation was introduced in Lerche *et al.* (1987a), see also Lerche *et al.* (1989).

Gepner models were introduced in Gepner (1987, 1988a,b), for reviews on this and related RCFT constructions see Gepner (1989), Greene (1996), and Schellekens (1990a, 1996). The 168 Gepner models based on A and E modular invariants were constructed in Lutken and Ross (1988) and Lynker and Schimmrigk (1988), see also Fuchs *et al.* (1990). The larger class of models including twists and discrete torsion described in Section 8.6.4 was discussed in Font *et al.* (1989a, 1990e). Mirror symmetry and its realization by twists in Gepner models were first discussed in Greene and Plesser (1990). The general construction of RCFT models in terms of simple currents was introduced in Schellekens and Yankielowicz (1990a, 1989, 1990b). See Kreuzer (2010b) for a recent discussion of (0, 2) Gepner constructions. The connection between Landau–Ginzburg models and Gepner models is discussed in Greene *et al.* (1989a), Martinec (1989), and Witten (1993), see Greene (1996) for a nice review of the main ideas. For a phenomenological analysis of the Gepner counterpart of the Tian–Yau 3-generation model see Ross (1988) and Greene *et al.* (1989b). Recently a large class of (0, 2) RCFT heterotic models with MSSM-like spectrum was constructed in Gato-Rivera and Schellekens (2010a,b,c).

Chapter 9

The dimensional reduction described in Section 9.1.1 was carried out in Witten (1985b), see also Derendinger *et al.* (1986), Burgess *et al.* (1986), and Ferrara *et al.* (1986). In its presentation we follow Polchinski (1998b). Additional general aspects of heterotic string CY compactifications are discussed in Witten (1986) and Strominger and Witten (1985), see also Candelas (1988) for the Yukawa couplings. The Kahler potential with no-scale structure was first found in the supergavity context in Cremmer *et al.* (1983a). The Kähler metrics for matter fields in orbifold compactifications were obtained in Dixon *et al.* (1990), see also Cvetič *et al.* (1988, 1989) and Ibáñez and Lüst (1992). In our presentation we follow this latter reference. Yukawa couplings in orbifold compactifications were computed in Dixon *et al.* (1987) and Hamidi and Vafa (1987), see also Burwick *et al.* (1991) and Erler *et al.* (1993), which we follow in our presentation. For phenomenological applications of these orbifold Yukawas see Ibáñez (1986), Casas and Muñoz (1990), Casas *et al.* (1993, 1992), and Kobayashi and Ohtsubo (1990, 1994). For general selection rules for couplings in orbifold models see Cvetič (1987b,a) and Font *et al.* (1988a, 1990d). For analogous results for models obtained from the fermionic construction see Kalara *et al.* (1990, 1991).

The detailed relationship among the string scale, Planck mass and the gauge coupling is worked out in Ginsparg (1987) and Kaplunovsky (1985, 1988). Cancellation of $U(1)$ anomalies through the Green–Schwarz mechanism in heterotic 4d vacua is discussed in Lerche *et al.* (1987b, 1988), see Casas *et al.* (1989) for an explicit application to orbifold constructions. The associated Fayet–Illiopoulos terms discussed in Section 9.5.3 were computed in Dine *et al.* (1987a,b) and Atick *et al.* (1987). The computation of the weak angle in terms of anomaly coefficients was discussed in Ibáñez (1993a), and phenomenological fermion mass models exploiting anomalous $U(1)$'s were introduced in Ibáñez and Ross (1994), see also Binétruy and Ramond (1995).

The T-duality of heterotic effective actions was remarked in Ferrara *et al.* (1989a,b), see also Shapere and Wilczek (1989) and Ferrara *et al.* (1991). The effect of T-duality transformations on orbifold effective actions was further studied in Lauer *et al.* (1989, 1991), Chun *et al.* (1989), and Erler *et al.* (1992). The computation of string threshold corrections to gauge couplings in orbifold models was worked out in Dixon *et al.* (1991) and Antoniadis *et al.* (1991, 1992), see also Bailin *et al.* (1994). The general field dependence of gauge couplings in the effective actions from string and orbifold compactifications is discussed in Kaplunovsky and Louis (1994, 1995). T-duality anomalies are discussed in Derendinger *et al.* (1992) and Lopes Cardoso and Ovrut (1992, 1993), we follow Ibáñez and Lüst (1992) in our presentation.

The explicit $\mathbf{Z}_3$ 3-generation orbifold model in Section 9.7 is described in Font *et al.* (1988b, 1990d), see also Casas and Muñoz (1988b,a). Heterotic compactifications with higher Kac–Moody levels were first discussed in Lewellen (1990). For a discussion of general phenomenological constraints on KM level model building see Font *et al.* (1990a). For string GUT models see Aldazabal *et al.* (1995, 1996), Finnell (1996), Chaudhuri *et al.* (1994a,b), and Dienes and March-Russell (1996). For GUT models from asymmetric orbifolds see Erler (1996), Kakushadze and Tye (1997), and Kakushadze *et al.* (1998). The general structure of 4d compactifications of the Hořava–Witten theory was discussed in Witten (1996c). For more details on the structure of the Kähler potential and superpotential see Banks and Dine (1996), Lukas *et al.* (1998, 1999), Dudas and Grojean (1997), and Lalak and Thomas (1998).

Chapter 10

For compactification of type II strings on CY manifolds and mirror symmetry, see the general reviews of CYs in the references for Chapter 7. There are several additional reviews and books centered on mirror symmetry, for instance Hori (2002), Hosono *et al.* (1994), and Hori *et al.* (2003), see also the compilations Yau (1998) and Greene and Yau (1997). The realization of mirror symmetry as T-duality was proposed in Strominger *et al.* (1996).

For reviews about particle physics models from intersecting branes in type IIA and type IIB orientifolds see, e.g., Marchesano (2007), Uranga (2003, 2007, 2005), Blumenhagen *et al.* (2007a, 2005b), and Marchesano (2003). Configurations of intersecting D-branes in 10d flat space were originally considered in Berkooz *et al.* (1996b). Compact non-chiral models of intersecting D-branes on top of O-planes were constructed in Blumenhagen *et al.* (2000b,c) and Forste *et al.* (2001b). Non-supersymmetric intersecting D6-brane models of particle physics were first constructed in Blumenhagen *et al.* (2000a) and Aldazabal *et al.*

(2001a,b). Models with just the SM spectrum were constructed in Ibáñez *et al.* (2001), see also Cremades *et al.* (2002a,c). For supersymmetric models in the $Z_2 \times Z_2$ orientifold, see Cvetič *et al.* (2001a,c) and Marchesano and Shiu (2004, 2005). The MSSM-like model described in Section 10.6 was constructed in Cremades *et al.* (2003) and a tadpole-free completion was constructed in Marchesano and Shiu (2004). Pati–Salam models were explored, e.g., in Cvetič *et al.* (2004). Non-supersymmetric models with a low-energy string scale were considered in, e.g., Cremades *et al.* (2002b) and Kokorelis (2004). For other abelian orbifold orientifolds see Forste *et al.* (2001a), Blumenhagen *et al.* (2003), Honecker (2003), and Honecker and Ott (2004).

The general BPS conditions for branes wrapping special lagrangian 3-cycles appeared in Becker *et al.* (1995), see also Ooguri *et al.* (1996), Mariño *et al.* (2000), and Douglas (2001). These can be described in terms of the calibrated geometries introduced in Harvey and Lawson (1982). For examples of intersecting D6-brane models in general CY geometries see Uranga (2002) and Blumenhagen *et al.* (2002a,b). The K-theory constraints sketched in Section 10.4.2 were obtained in Uranga (2001b). Chiral supersymmetric orientifolds with frozen open string moduli were constructed in Blumenhagen *et al.* (2005a).

Type II Gepner 4d orientifolds were constructed in Blumenhagen and Wisskirchen (1998), Blumenhagen (2003), Blumenhagen and Weigand (2004), and Aldazabal *et al.* (2003, 2004). More systematic searches of MSSM-like models in RCFT orientifolds were worked out in Dijkstra *et al.* (2005b,a) and Anastasopoulos *et al.* (2006), based on earlier work, see e.g. Fuchs *et al.* (2000), Schellekens (2000), and references therein. We follow these references in our presentation. For detailed information on the particle spectra in this large class of models visit www.nikhef.nl/ t58/filtersols.php.

Mathematical aspects of G_2 manifolds are covered in Joyce (2000). For a review of M-theory on G_2 manifolds, see Acharya and Gukov (2004), Gubser (2002), and references therein. For G_2 from lifts of SUSY type IIA D6-brane models, see Cvetič *et al.* (2001b). A direct analysis of G_2 singularities in M-theory was carried out in Atiyah and Witten (2003), and the unfolding trick was introduced in the present context in Acharya and Witten (2001).

Chapter 11

For general reviews on type IIB orientifolds see Angelantonj and Sagnotti (2002), also those already mentioned in the bibliography of Chapter 10. Type IIB orientifolds in $D = 6$ were first constructed in Pradisi and Sagnotti (1989), Bianchi and Sagnotti (1990), Gimon and Polchinski (1996), Gimon and Johnson (1996), and Dabholkar and Park (1996). Concerning 4d $\mathcal{N} = 1$ models, the non-chiral $Z_2 \times Z_2$ toroidal orientifold was constructed in Berkooz and Leigh (1997), while the chiral Z_3 orientifolds first appeared in Angelantonj *et al.* (1996). See Aldazabal *et al.* (1998) for the constructions of more general Z_N orientifolds, also Kakushadze and Shiu (1998, 1997), Zwart (1998), and references therein. The notion of vector structure in orientifolds was described in Berkooz *et al.* (1996a) and

Witten (1998c). For $U(1)$ anomaly cancellation in $\mathbf{Z}_N$ orientifolds see Ibáñez *et al.* (1999a) and for the related induced FI term Poppitz (1999).

The $SU(5)$ example in Section 11.2.3 was constructed in Lykken *et al.* (1999). Non-supersymmetric orientifold models with antibranes were considered in Sugimoto (1999) in 10d, and in Antoniadis *et al.* (1999a), Aldazabal and Uranga (1999), and Angelantonj *et al.* (2000b) in the 4d context; see Aldazabal *et al.* (2000c,a) for particle physics models.

The spectrum and couplings of D-branes at singularities were studied in Douglas and Moore (1996) and Douglas *et al.* (1997), see also Kachru and Silverstein (1998) and Hanany and Uranga (1998). The bottom-up philosophy for the SM embedding based on branes at singularities was put forward in Aldazabal *et al.* (2000b), including several non-compact and compact explicit models. Further model building work with branes at singularities is contained in Berenstein *et al.* (2002) and Verlinde and Wijnholt (2007), see Malyshev and Verlinde (2007) for a review. Dimer diagrams as a tool to study D-branes at more general toric singularities were introduced in Hanany and Kennaway (2005), see Kennaway (2007), Franco *et al.* (2006), and García-Etxebarría *et al.* (2006b) for reviews, and Conlon *et al.* (2009) and Krippendorf *et al.* (2010b) for model building applications.

Magnetized type I compactifications were considered in Bachas (1995), Angelantonj *et al.* (2000a), and Blumenhagen *et al.* (2001), see also Cremades *et al.* (2004). Magnetized coisotropic D8-branes in type IIA toroidal compactifications were discussed in Font *et al.* (2006), see Antoniadis and Maillard (2005) and Antoniadis *et al.* (2007) for related type IIB constructions.

F-theory was introduced in Vafa (1996) and Morrison and Vafa (1996b,a). For the appearance of matter fields in F-theory see Bershadsky *et al.* (1996) and Katz and Vafa (1997). The orientifold limit in a class of F-theory compactifications was revealed in Sen (1996, 1997). For reviews introducing model building with F-theory see Weigand (2010) and Denef (2008). F-theory GUT model building was initiated in Donagi and Wijnholt (2008b), Beasley *et al.* (2009a,b), and Donagi and Wijnholt (2008a), for reviews see Heckman and Vafa (2008) and Heckman (2010). References about Yukawa couplings in F-theory local models are given in the bibliography to Chapter 12. The construction of global compact F-theory models is addressed in, e.g., Marsano *et al.* (2009b), Blumenhagen *et al.* (2010a), Grimm *et al.* (2010), and Cordova (2009).

Chapter 12

The moduli Kähler potential and the gauge kinetic couplings in general type II orientifolds were worked out in detail in Jockers and Louis (2005) and Grimm and Louis (2004, 2005), see Grimm (2005) for a review. For applications to the toroidal case see Cremades *et al.* (2002a,c), Blumenhagen *et al.* (2003a), and Font and Ibáñez (2005). For reviews on different aspects of the effective action of type II orientifolds see Marchesano (2003), Blumenhagen *et al.* (2005b, 2007a), and Marchesano (2007). Some aspects of the effective action of moduli in M-theory compactifications on G_2 manifolds were described in

Harvey and Moore (1999), Gutowski and Papadopoulos (2001), and Beasley and Witten (2002), see Grimm (2005) for a review. The Kähler metrics of matter fields in type IIB toroidal orientifolds were worked out in Ibáñez *et al.* (1999b) for unmagnetized branes using duality arguments, see also Körs and Nath (2004b). A full computation including fluxes was performed in Lüst *et al.* (2004), Lüst *et al.* (2005a) extracting the metrics from string scattering amplitudes, see also Di Vecchia *et al.* (2009).

Yukawa couplings in type IIA toroidal intersecting D6-brane models were computed in Cremades *et al.* (2003) (semiclassical piece) and in Cvetič and Papadimitriou (2003) and Abel and Owen (2003) (quantum piece). The T-dual type IIB computation with magnetized D-branes was worked out in Cremades *et al.* (2004), see Conlon *et al.* (2008) for a non-toroidal example. The basic setting for Yukawa couplings in F-theory unified theories was put forward in Beasley *et al.* (2009a,b), see also Font and Ibáñez (2009b,a), Heckman and Vafa (2010), Hayashi *et al.* (2009, 2010), Conlon and Palti (2010), Dudas and Palti (2010), and King *et al.* (2010). Corrections to Yukawa couplings from fluxes were obtained in Cecotti *et al.* (2009) using a non-conmutative geometry. Analogous corrections coming from instanton effects were obtained in Marchesano and Martucci (2010).

Chapter 13

For a recollection of classical articles on field theory instantons, see Shifman (1994), which includes also Coleman's lectures, see Coleman (1979). More recent reviews are Bianchi *et al.* (2008) and Tong (2005).

Worldsheet instanton effects were first discussed in Dine *et al.* (1986, 1987c), and have been profusely studied in relation to mirror symmetry, see references for Chapter 10. Brane and D-brane instantons were discussed in, e.g., Becker *et al.* (1995), Witten (2000), and Harvey and Moore (1999). The conditions for an instanton to generate a non-perturbative superpotential in F-theory were spelled out in Witten (1996a), see Donagi *et al.* (1996), also Denef *et al.* (2004, 2005) for applications. D-brane instantons as gauge instantons have been discussed in Douglas (1995) and Witten (1996b), see also Billò *et al.* (2003) and Petersson (2008) for the $U(1)$ gauge theory case.

String instanton-induced superpotentials involving charged fields were first studied in Blumenhagen *et al.* (2007b), Ibáñez and Uranga (2007), and Florea *et al.* (2007), for reviews see Blumenhagen *et al.* (2009a), Akerblom *et al.* (2008), and Cvetič *et al.* (2007b), also Bianchi and Samsonyan (2009). The structure of zero modes required for superpotentials to be generated was discussed in Argurio *et al.* (2007b) and Ibáñez *et al.* (2007), see also Bianchi and Kiritsis (2007), Argurio *et al.* (2007c), and Bianchi *et al.* (2007).

Neutrino masses from string instanton effects were discussed in Blumenhagen *et al.* (2007b) and Ibáñez and Uranga (2007), see also Antusch *et al.* (2007), Cvetič *et al.* (2007a), and Cvetič and Langacker (2008). For instanton corrections to Yukawa couplings see Blumenhagen *et al.* (2008b), Abel and Goodsell (2007), and Ibáñez and Richter (2009). For other phenomenological applications, see also Aharony *et al.* (2007), Ibáñez and Uranga

(2008), Cvetič and Weigand (2008), and Cvetič *et al.* (2009). For instanton effects in local F-theory GUTs, see Blumenhagen *et al.* (2010b), also Heckman *et al.* (2008) and Marsano *et al.* (2010).

Chapter 14

For reviews about string vacua in the presence of fluxes see Graña (2006a,b), Denef (2008), Denef *et al.* (2007), and Douglas and Kachru (2007). String vacua with fluxes were considered in Polchinski and Strominger (1996) and Becker and Becker (1996). The case of type IIB compactifications on CY orientifolds with 3-form fluxes (and its generalization to M/F-theory on CY fourfolds) was described in Dasgupta *et al.* (1999), Giddings *et al.* (2002), and Becker and Becker (2001), see also Frey and Polchinski (2002) and Kachru *et al.* (2003b,c). The flux superpotential in Section 14.1.2 was derived in Gukov *et al.* (2000) for M-theory on CY fourfolds, see Taylor and Vafa (2000) for the type IIB version that we follow. The Klebanov–Strassler solution was obtained in Klebanov and Strassler (2000); particle physics model building applications of analogous warped throats may be found, e.g., in Cascales *et al.* (2004, 2005), the latter describing the example in page 467, see also Franco *et al.* (2009). For the effective action of warped compactifications, see e.g. DeWolfe and Giddings (2003), Shiu *et al.* (2008), Douglas and Torroba (2009), and Frey *et al.* (2009). For references on the effect of fluxes on D-branes see the bibliography for Chapter 15.

Fluxes in type IIB toroidal orientifolds were discussed in Kachru *et al.* (2003b). For an analogous discussion in type IIA see Derendinger *et al.* (2005), Villadoro and Zwirner (2005), DeWolfe *et al.* (2005), Camara *et al.* (2005b), and Aldazabal *et al.* (2006). Type IIA orientifolds with all moduli fixed in AdS were discussed in DeWolfe *et al.* (2005) and Camara *et al.* (2005b); we follow this last reference in the discussion in Section 14.2. Metric fluxes were discussed in Kaloper and Myers (1999), and its interplay with mirror symmetry/T-duality was appreciated in Gurrieri *et al.* (2003); the Scherk–Schwarz symmetry breaking mechanism was introduced in Scherk and Schwarz (1979a,b).

Freed–Witten anomalies were discussed in Freed and Witten (1999), see also Maldacena *et al.* (2001), and were applied to D-brane instantons in Kashani-Poor and Tomasiello (2005). Lifting of brane instanton fermion zero modes by fluxes has been discussed, e.g., in Bergshoeff *et al.* (2005), Tripathy and Trivedi (2005), Uranga (2009), and Billò *et al.* (2008a,b); we quote results from the latter. Semi-realistic orientifold models with fluxes were first constructed in Blumenhagen *et al.* (2003b) and Cascales and Uranga (2003). The $\mathbf{Z}_2 \times \mathbf{Z}_2$ toroidal orientifold with fluxes was analyzed in Marchesano and Shiu (2004) and Marchesano and Shiu (2005).

Non-geometric fluxes are explored in Shelton *et al.* (2005), Hellerman *et al.* (2004), Dabholkar and Hull (2003), and Aldazabal *et al.* (2006), see Wecht (2007) for a review. Early work on fluxes in heterotic compactifications was carried out in Strominger (1986), see e.g. Becker *et al.* (2003, 2004), Lopes Cardoso *et al.* (2003, 2004), and Gurrieri *et al.* (2004) for more recent developments.

Chapter 15

Gaugino condensation inducing SUSY breaking in supergravity theories was pointed out in Ferrara *et al.* (1983). In the context of heterotic compactifications it was proposed in Derendinger *et al.* (1985) and Dine *et al.* (1985a). Multiple gaugino condensation was first considered in Krasnikov (1987), see also Dixon (1990) and Casas *et al.* (1990). Modular invariant gaugino condensation was discussed in Font *et al.* (1990c) (which we follow in the text), Ferrara *et al.* (1990), Nilles and Olechowski (1990), and Binétruy and Gaillard (1991), see also Binétruy *et al.* (1997). The effect of matter fields transforming under the condensing group was analyzed in Lüst and Taylor (1991) and de Carlos *et al.* (1993c). The problem of the *runaway* behavior of the dilaton/moduli scalar potential in string compactifications was remarked in Dine and Seiberg (1985). For more recent developments about SUSY breaking in heterotic vacua see, e.g., Gaillard and Nelson (2007), de Carlos *et al.* (2006), and references therein. Soft terms in Hořawa–Witten heterotic compactifications were discussed, e.g., in Choi *et al.* (1998).

The KKLT scenario for full moduli fixing was introduced in Kachru *et al.* (2003a), see Denef *et al.* (2005, 2004) for specific CY examples of moduli fixing in AdS, and Denef *et al.* (2007) and Denef (2008) for reviews. For alternative ways to obtain the *uplifting* of the KKLT scalar potential see Burgess *et al.* (2003) and Saltman and Silverstein (2004). The structure of symmetry breaking in the observable sector of KKLT compactifications was explored in Choi *et al.* (2004, 2005), Falkowski *et al.* (2005), and Choi and Nilles (2007). Ways to address full moduli fixing in M-theory compactified in G_2 manifolds were proposed in Acharya *et al.* (2007), see Acharya *et al.* (2008) for phenomenological consequences. The large volume scenario was presented in Balasubramanian *et al.* (2005), Conlon *et al.* (2005), and Cicoli *et al.* (2008). The CY example in the text with two Kähler moduli was first studied in Candelas *et al.* (1994b,a). For further aspects of large volume models see Conlon *et al.* (2007b,c) and Blumenhagen *et al.* (2008a, 2009b).

The computation of soft terms from the DBI+CS actions in the presence of fluxes was studied in Graña (2003) and Camara *et al.* (2004) for D3-branes, see also Graña *et al.* (2004). For D7/D3 systems it was worked out in Camara *et al.* (2005a) and Lüst *et al.* (2005a). Flux induced soft-terms in the $Z_2 \times Z_2$ MSSM-like orientifold in Section 11.4.2 were computed in Lüst *et al.* (2005b) and Font and Ibáñez (2005).

Some aspects of SUSY breaking soft terms in heterotic string models were first addressed in Cvetič *et al.* (1991) and Ibáñez and Lüst (1992). A systematic analysis was presented in Brignole *et al.* (1994) and Kaplunovsky and Louis (1993), we follow the review Brignole *et al.* (1997) in the text. The low-energy sparticle spectrum from dilaton domination soft terms was analyzed among others in Barbieri *et al.* (1993), Casas *et al.* (1996), de Carlos *et al.* (1993b), and Abel *et al.* (2000).

For SUSY breaking terms in large volume type IIB scenarios see Conlon *et al.* (2007c) and Blumenhagen *et al.* (2009b), and in other type II orientifold models see, e.g., Allanach *et al.* (2005) and Kane *et al.* (2005). The computation of the Kähler metrics for matter fields located at the intersection of local sets of D7-brane was given in Conlon *et al.* (2007a).

Flavour independence of soft terms in type II orientifolds is discussed in Conlon (2008). The phenomenology of soft terms from modulus dominance in local sets of D7-branes is worked out in detail in Aparicio *et al.* (2008). We follow this reference in Sections 15.5.3 and 15.6. Signatures at LHC in some string scenarios have been worked out in Kane *et al.* (2007, 2008), Baer *et al.* (2007), and Heckman *et al.* (2009).

SUSY breaking at meta-stable minima was proposed in Dimopoulos *et al.* (1998), and implemented in SQCD in Intriligator *et al.* (2006), see Argurio *et al.* (2007b) for the realization in string theory sketched in the text. For other discussions of gauge mediated SUSY breaking in string theory, see e.g. Diaconescu *et al.* (2006), García-Etxebarría *et al.* (2006a), Argurio *et al.* (2007a), and Marsano *et al.* (2009a). The possibility of sequestering in string theory has been explored in Kachru *et al.* (2007) and Berg *et al.* (2010).

Chapter 16

Coupling unification in heterotic compactifications was described in Ginsparg (1987), Kaplunovsky (1985, 1988, 1992), see Dienes (1997) for a review of coupling unification in heterotic vacua. The effect of threshold corrections on coupling unification was explored in Ibáñez *et al.* (1991), Nilles and Stieberger (1996), and Hebecker and Trapletti (2005). The effect of smaller than canonical k_1 values was explored in Ibáñez (1993b), Dienes and Faraggi (1995), Dienes *et al.* (1996), and Chaudhuri *et al.* (1996).

Lowering the string scale to match with the coupling unification scale in strongly coupled heterotic compactifications was proposed in Witten (1996c), see also Lykken (1996b). Gauge coupling constants in type I and type II orientifolds were studied in, e.g., Ibáñez *et al.* (1999b), Antoniadis *et al.* (2000), Aldazabal *et al.* (2001b), Cremades *et al.* (2002a), and Blumenhagen *et al.* (2003a). For gauge couplings for D3-branes at singularities see Aldazabal *et al.* (2000b) and Verlinde and Wijnholt (2007). Unification in the left–right symmetric model in Section 16.1.2 was worked out in Aldazabal *et al.* (2000a). Coupling constants in RCFT type II orientifolds were discussed in Dijkstra *et al.* (2005b), from which Figure 16.2 is taken. Threshold corrections to gauge couplings in type II orientifolds were computed in Bachas and Fabre (1996), Antoniadis *et al.* (1999b), and Lüst and Stieberger (2007). In F-theory threshold effects were estimated in Donagi and Wijnholt (2008a) and Blumenhagen (2009). General threshold effects in extra dimension models are discussed in Dienes *et al.* (1998, 1999) and Ghilencea and Ross (1998).

The generic presence of axions in string theory was first pointed out in Witten (1984), see also Choi and Kim (1985a,b) and Kim and Carosi (2010) for a recent review. For more recent references see Banks and Dine (1997), Banks *et al.* (2003), and Fox *et al.* (2004), whose heterotic discussion we follow. Svrcek and Witten (2006) contains a discussion of the value of the axion decay constant in a variety of string compactifications. The possibility of relaxing the upper cosmological bound on this constant has been considered in Linde (1988), Tegmark *et al.* (2006), and Hertzberg *et al.* (2008). Astrophysical consequences of a large multiplicity of axion strings have been explored in Arvanitaki *et al.* (2010a).

Heterotic orbifold models with preserved R-parity have been studied in Lebedev *et al.*
(2008b). Proton decay through dimension 6 operators in intersecting brane models has
been explored in Klebanov and Witten (2003) and Cvetič and Richter (2007), see Nath
and Fileviez Pérez (2007) for a review of proton stability in string models. The generation
of Majorana neutrino masses in heterotic compactifications has been analyzed, e.g., in
Faraggi and Halyo (1993), Giedt *et al.* (2005), Lebedev *et al.* (2008b), and Buchmüller
et al. (2007a). Neutrino masses in type II orientifold models coming from instanton effects
have been discussed in Ibáñez and Uranga (2007), Blumenhagen *et al.* (2007b), Antusch
et al. (2007), and Cvetič and Langacker (2008).

For the phenomenology of new Z' bosons from E_6-inspired heterotic compactifications
see the reviews Hewett and Rizzo (1989), Langacker (2009), and Brooijmans (2010). Extra
$U(1)$'s from low string scale orientifold models were explored in Ghilencea *et al.* (2002)
(which we follow in Section 16.4.2), see also Körs and Nath (2004a), Coriano *et al.* (2006),
Feldman *et al.* (2007), and Anastasopoulos *et al.* (2008). One loop kinetic mixing among
two $U(1)$'s was first pointed out in Holdom (1986). For kinetic mixing between hypercharge and hidden $U(1)$'s in string compactifications see Dienes *et al.* (1997), Abel and
Schofield (2004), Abel *et al.* (2008), and Goodsell and Ringwald (2010). For the possibility of production of hidden sector photini at LHC see Ibarra *et al.* (2009) and Arvanitaki
et al. (2010b).

For different experimental and astrophysical limits on the size of extra dimensions see
Giudice and Wells (2008), see also Hewett and Spiropulu (2002). Cross sections for production of KK gravitons were obtained in Giudice *et al.* (1999), Mirabelli *et al.* (1999), Han
et al. (1999), and Cullen *et al.* (2000). For the LHC phenomenology of string excitations
in low scale string models see Anchordoqui *et al.* (2009a,b), Lüst (2009), and references
therein; the string scattering amplitude in the text is essentially as in Veneziano (1968),
which historically provided the starting point of string theory. For a review on micro black
hole production at colliders see Giddings (2007).

For reviews on string cosmology see Baumann (2009), Burgess (2007), McAllister and
Silverstein (2008), Linde (2008), Cline (2006), Quevedo (2002), and Carroll (1999). Models with Kähler moduli inflatons discussed in Section 16.6.2 are presented in Blanco-
Pillado *et al.* (2004, 2006) and Conlon and Quevedo (2006). D-brane inflation was first
proposed in Dvali and Tye (1999). The KKLMMT model was described in Kachru *et al.*
(2003d) and further elaborated in Baumann *et al.* (2006, 2008, 2010). The *N-flation* idea
was presented in Dimopoulos *et al.* (2008), *monodromy* inflation in Silverstein and Westphal (2008), and DBI inflation in Alishahiha *et al.* (2004), see also Cicoli *et al.* (2009) for
a different model with measurable non-gaussianities. The CMB bounds in Figure 16.9 are
taken from Komatsu (2009). The possible tension between inflation and SUSY breaking
scales in KKLT-like models was pointed out in Kallosh and Linde (2004). The cosmological moduli problem was discussed in de Carlos *et al.* (1993a) and Banks *et al.* (1994), see
also Coughlan *et al.* (1983). A possible solution via thermal inflation was pointed out in
Lyth and Stewart (1996). The possible role of fundamental strings as cosmic strings was
first pointed out in Witten (1985a). For reviews on string theoretical cosmic strings see

Copeland and Kibble (2010) and Polchinski (2004), which we follow in the text, see also Copeland *et al.* (2004). The computer simulation in Figure 16.10 is taken from Allen and Shellard (1990).

Chapter 17

For constraints on the gauge group and matter content in the heterotic string see Lewellen (1990) and Font *et al.* (1990a). The generic presence of massive fractionally charged particles in heterotic compactifications was first pointed out in Wen and Witten (1985), see also Athanasiu *et al.* (1988). The theorem on charge quantization described in Section 17.1.3 appeared in Schellekens (1990b), and the absence of continuous global symmetries in Banks and Dixon (1988), see also Burgess *et al.* (2008). Other constraints from quantum gravity arguments were discussed in Vafa (2005) and Ooguri and Vafa (2007). The conjecture of gravity as the weakest force was put forward in Arkani-Hamed *et al.* (2007).

Statistical results on the construction of MSSM-like models (or extensions thereof) using (0, 2) RCFT heterotic models were presented in Gato-Rivera and Schellekens (2010a,b,c). Statistical distributions for Pati–Salam models in the heterotic fermionic formalism were given in Assel *et al.* (2010). A landscape of MSSM-like models from the heterotic Z'_6 orbifold was analyzed in Lebedev *et al.* (2007, 2008a). Other statistical analyses of heterotic vacua were discussed in Dienes (2006). In the type II orientifold class of vacua, statistical distributions of a large class of MSSM-like models constructed using RCFT techniques were presented in Dijkstra *et al.* (2005b), see also Anastasopoulos *et al.* (2006) and Gato-Rivera and Schellekens (2009). Statistical properties of a class of toroidal $Z_2 \times Z_2$ orientifolds with MSSM-like spectra were studied in Gmeiner *et al.* (2006) and Douglas and Taylor (2007). A landscape of Z_6 orientifolds was described in Gmeiner *et al.* (2007) and that of Z'_6 orientifolds in Gmeiner and Honecker (2007, 2008).

For reviews on the flux landscape see Denef (2008), Douglas and Kachru (2007), Denef *et al.* (2007), Polchinski (2006), and Graña (2006b). The general idea discussed in Section 17.3 appeared in Bousso and Polchinski (2000), we follow Denef *et al.* (2007) in the presentation. Examples in type IIB orientifold vacua are presented in Denef and Douglas (2004, 2005), see Ashok and Douglas (2004) for analytic results of flux vacua distributions. For the anthropic solution to the cosmological constant problem, see the bibliography for Chapter 1. For different viewpoints on the string theory landscape, see Susskind (2003), Schellekens (2006, 2008), Acharya and Douglas (2006), and Douglas (2006), and for criticisms of the idea see Banks *et al.* (2004) and Banks (2004).

Appendices

For Appendix A, a comprehensive reference is Mumford (2006). Useful reviews for Appendix B are Candelas (1987) and Eguchi *et al.* (1980), as well as standard textbooks such as Nakahara (2003). The model in Appendix C is presented in Font *et al.* (1988b, 1990d), see also Casas and Muñoz (1988b,a).

The computation of RR tadpoles in Appendix D follows Aldazabal *et al.* (1998), see also Gimon and Johnson (1996), Zwart (1998), and O'Driscoll (1998). For Appendix E, some general CFT references are, e.g., Ginsparg (1988), Schellekens (1996), Gaberdiel (2000), and Chapters 8–10 of Gómez *et al.* (1996). The applications to string theory are profusely present in the diverse textbooks in the references for Chapters 3–6.

References

Abel, S. A. and Goodsell, M. D. (2007). Realistic Yukawa couplings through instantons in intersecting brane worlds. *JHEP*, **0710**, 034.

Abel, S. A. and Owen, A. W. (2003). Interactions in intersecting brane models. *Nucl. Phys.*, **B663**, 197–214.

Abel, S. A., and Schofield, B. W. (2004). Brane-antibrane kinetic mixing, millicharged particles and SUSY breaking. *Nucl. Phys.*, **B685**, 150–70.

Abel, S. A., Allanach, B. C., Quevedo, F., Ibáñez, L. E., and Klein, M. (2000). Soft SUSY breaking, dilaton domination and intermediate scale string models. *JHEP*, **0012**, 026.

Abel, S. A., Goodsell, M. D., Jaeckel, J., Khoze, V. V., and Ringwald, A. (2008). Kinetic mixing of the photon with hidden $U(1)$'s in string phenomenology. *JHEP*, **0807**, 124.

Acharya, B. S. and Douglas, M. R. (2006). A finite landscape? [arXiv:hep-th/0606212].

Acharya, B. S. and Gukov, S. (2004). M theory and singularities of exceptional holonomy manifolds. *Phys. Rept.*, **392**, 121–89.

Acharya, B. S. and Witten, E. (2001). Chiral fermions from manifolds of G_2 holonomy. [arXiv:hep-th/0109152].

Acharya, B. S., Bobkov, K., Kane, G. L., Kumar, P., and Shao, J. (2007). Explaining the electroweak scale and stabilizing moduli in M-theory. *Phys. Rev.*, **D76**, 126010.

Acharya, B. S., Bobkov, K., Kane, G. L., Shao, J., and Kumar, P. (2008). The G_2-MSSM: an M-theory motivated model of particle physics. *Phys. Rev.*, **D78**, 065038.

Adams, A., DeWolfe, O., and Taylor, W. (2010). String universality in ten dimensions. *Phys. Rev. Lett.*, **105**, 071601.

Affleck, I., Dine, M., and Seiberg, N. (1984). Dynamical supersymmetry breaking in supersymmetric QCD. *Nucl. Phys.*, **B241**, 493–534.

Affleck, I., Dine, M., and Seiberg, N. (1985). Dynamical supersymmetry breaking in four-dimensions and its phenomenological implications. *Nucl. Phys.*, **B256**, 557.

Aharony, O., Gubser, S. S., Maldacena, J. M., Ooguri, H., and Oz, Y. (2000). Large N field theories, string theory and gravity. *Phys. Rept.*, **323**, 183–386.

Aharony, O., Kachru, S., and Silverstein, E. (2007). Simple stringy dynamical SUSY breaking. *Phys. Rev.*, **D76**, 126009.

Akerblom, N., Blumenhagen, R., Lüst, D., and Schmidt-Sommerfeld, M. (2008). D-brane instantons in 4D supersymmetric string vacua. *Fortsch. Phys.*, **56**, 313–23.

Aldazabal, G. and Uranga, A. M. (1999). Tachyon free nonsupersymmetric type IIB orientifolds via brane–antibrane systems. *JHEP*, **9910**, 024.

Aldazabal, G., Font, A., Ibáñez, L. E., and Uranga, A. M. (1995). String GUTs. *Nucl. Phys.*, **B452**, 3–44.

Aldazabal, G., Font, A., Ibáñez, L. E., and Uranga, A. M. (1996). Building GUTs from strings. *Nucl. Phys.*, **B465**, 34–70.

Aldazabal, G., Font, A., Ibáñez, L. E., and Violero, G. (1998). D = 4, $\mathcal{N} = 1$, type IIB orientifolds. *Nucl. Phys.*, **B536**, 29–68.

Aldazabal, G., Ibáñez, L. E., and Quevedo, F. (2000a). A D-brane alternative to the MSSM. *JHEP*, **0002**, 015.

Aldazabal, G., Ibáñez, L. E., Quevedo, F., and Uranga, A. M. (2000b). D-branes at singularities: a bottom up approach to the string embedding of the standard model. *JHEP*, **0008**, 002.

Aldazabal, G., Ibáñez, L. E., and Quevedo, F. (2000c). Standard-like models with broken supersymmetry from type I string vacua. *JHEP*, **0001**, 031.

Aldazabal, G., Franco, S., Ibáñez, L. E., Rabadán, R., and Uranga, A. M. (2001a). D = 4 chiral string compactifications from intersecting branes. *J. Math. Phys.*, **42**, 3103–26.

Aldazabal, G., Franco, S., Ibáñez, L. E., Rabadán, R., and Uranga, A. M. (2001b). Intersecting brane worlds. *JHEP*, **0102**, 047.

Aldazabal, G., Andrés, E. C., Leston, M., and Nuñez, C. (2003). Type IIB orientifolds on Gepner points. *JHEP*, **0309**, 067.

Aldazabal, G., Andrés, E. C., and Juknevich, J. E. (2004). Particle models from orientifolds at Gepner orbifold points. *JHEP*, **0405**, 054.

Aldazabal, G., Camara, P. G., Font, A., and Ibáñez, L. E. 2006. More dual fluxes and moduli fixing. *JHEP*, **0605**, 070.

Alishahiha, M., Silverstein, E., and Tong, D. (2004). DBI in the sky. *Phys. Rev.*, **D70**, 123505.

Allanach, B. C., Brignole, A., and Ibáñez, L. E. (2005). Phenomenology of a fluxed MSSM. *JHEP*, **0505**, 030.

Allen, B., and Shellard, E. P. S. (1990). Cosmic string evolution: a numerical simulation. *Phys. Rev. Lett.*, **64**, 119–22.

Alvarez, E. and Meessen, P. (1999). String primer. *JHEP*, **9902**, 015.

Alvarez, E., Alvarez-Gaumé, L., and Lozano, Y. (1995). An introduction to T duality in string theory. *Nucl. Phys. Proc. Suppl.*, **41**, 1–20.

Alvarez-Gaumé, L. (1985). An antroduction to anomalies, *Erice School Proc., Math. Phys.* 1985:0093 (QCD161:I569:1985).

Alvarez-Gaumé, L. and Witten, E. (1984). Gravitational anomalies. *Nucl. Phys.*, **B234**, 269.

Alvarez-Gaumé, L., Claudson, M., and Wise, M. (1982). Low-energy supersymmetry. *Nucl. Phys.*, **B207**, 96.

Alvarez-Gaumé, L., Polchinski, J., and Wise, M. (1983). Minimal low-energy supergravity. *Nucl. Phys.*, **B221**, 495.

Alvarez-Gaumé, L. and Vázquez-Mozo, M. A. (1992). Topics in string theory and quantum gravity. [arXiv:hep-th/9212006].

Anastasopoulos, P., Dijkstra, T. P. T., Kiritsis, E., and Schellekens, A. N. (2006). Orientifolds, hypercharge embeddings and the Standard Model. *Nucl. Phys.*, **B759**, 83–146.

Anastasopoulos, P., Fucito, F., Lionetto, A., Pradisi, G., Racioppi, A., and Stanev, Y. S. (2008). Minimal anomalous $U(1)'$ extension of the MSSM. *Phys. Rev.*, **D78**, 085014.

Anchordoqui, L. A., Goldberg, H., Lüst, D., Nawata, S., Stieberger, S., and Taylor, T. R. (2009a). LHC phenomenology for string hunters. *Nucl. Phys.*, **B821**, 181–96.

Anchordoqui, L. A., Goldberg, H., Lüst, D., Stieberger, S., and Taylor, T. R. (2009b). String phenomenology at the LHC. *Mod. Phys. Lett.*, **A24**, 2481–90.

Anderson, L. B., He, Y.-H., and Lukas, A. (2007). Heterotic compactification, an algorithmic approach. *JHEP*, **0707**, 049.

Anderson, L. B., He, Y.-H., and Lukas, A. (2008). Monad bundles in heterotic string compactifications. *JHEP*, **0807**, 104.

Anderson, L. B., Gray, J., He, Y.-H., and Lukas, A. (2010). Exploring positive monad bundles and a new heterotic Standard Model. *JHEP*, **1002**, 054.

Andreas, B. and Hernández Ruiperez, D. (2005). $U(n)$ vector bundles on Calabi-Yau threefolds for string theory compactifications. *Adv. Theor. Math. Phys.*, **9**, 253–84.

Angelantonj, C. and Sagnotti, A. (2002). Open strings. *Phys. Rept.*, **371**, 1–150.

Angelantonj, C., Bianchi, M., Pradisi, G., Sagnotti, A., and Stanev, Y. S. (1996). Chiral asymmetry in four-dimensional open string vacua. *Phys. Lett.*, **B385**, 96–102.

Angelantonj, C., Antoniadis, I., Dudas, E., and Sagnotti, A. (2000a). Type I strings on magnetized orbifolds and brane transmutation. *Phys. Lett.*, **B489**, 223–32.

Angelantonj, C., Antoniadis, I., D'Appollonio, G., Dudas, E., and Sagnotti, A. (2000b). Type I vacua with brane supersymmetry breaking. *Nucl. Phys.*, **B572**, 36–70.

Antoniadis, I. and Bachas, C. (1988). 4-D fermionic superstrings with arbitrary twists. *Nucl. Phys.*, **B298**, 586.

Antoniadis, I. and Maillard, T. (2005). Moduli stabilization from magnetic fluxes in type I string theory. *Nucl. Phys.*, **B716**, 3–32.

Antoniadis, I., Bachas, C., and Kounnas, C. (1987). Four-dimensional superstrings. *Nucl. Phys.*, **B289**, 87.

Antoniadis, I., Ellis, J., Hagelin, J., and Nanopoulos, D. V. (1989). The flipped $SU(5) \times U(1)$ string model revamped. *Phys. Lett.*, **B231**, 65.

Antoniadis, I., Narain, K. S., and Taylor, T. R. (1991). Higher genus string corrections to gauge couplings. *Phys. Lett.*, **B267**, 37–45.

Antoniadis, I., Gava, E., and Narain, K. S. (1992). Moduli corrections to gauge and gravitational couplings in four-dimensional superstrings. *Nucl. Phys.*, **B383**, 93–109.

Antoniadis, I., Arkani-Hamed, N., Dimopoulos, S., and Dvali, G. (1998). New dimensions at a millimeter to a fermi and superstrings at a TeV. *Phys. Lett.*, **B436**, 257–63.

Antoniadis, I., Dudas, E., and Sagnotti, A. (1999a). Brane supersymmetry breaking. *Phys. Lett.*, **B464**, 38–45.

Antoniadis, I., Bachas, C., and Dudas, E. (1999b). Gauge couplings in four-dimensional type I string orbifolds. *Nucl. Phys.*, **B560**, 93–134.

Antoniadis, I., Kiritsis, E., and Tomaras, T. N. (2000). A D-brane alternative to unification. *Phys. Lett.*, **B486**, 186–93.

Antoniadis, I., Kumar, A., and Maillard, T. (2007). Magnetic fluxes and moduli stabilization. *Nucl. Phys.*, **B767**, 139–62.

Antusch, S., Ibáñez, L. E., and Macri, T. (2007). Neutrino masses and mixings from string theory instantons. *JHEP*, **0709**, 087.

Aparicio, L., Cerdeño, D. G., and Ibáñez, L. E. (2008). Modulus-dominated SUSY-breaking soft terms in F-theory and their test at LHC. *JHEP*, **0807**, 099.

Argurio, R., Bertolini, M., Franco, S., and Kachru, S. (2007a). Gauge/gravity duality and meta-stable dynamical supersymmetry breaking. *JHEP*, **0701**, 083.

Argurio, R., Bertolini, M., Franco, S., and Kachru, S. (2007b). Metastable vacua and D-branes at the conifold. *JHEP*, **0706**, 017.

Argurio, R., Bertolini, M., Ferretti, G., Lerda, A., and Petersson, C. (2007c). Stringy instantons at orbifold singularities. *JHEP*, **0706**, 067.

Arkani-Hamed, N. and Dimopoulos, S. (2005). Supersymmetric unification without low energy supersymmetry and signatures for fine-tuning at the LHC. *JHEP*, **0506**, 073.

Arkani-Hamed, N., Dimopoulos, S., and Dvali, G. 1998. The Hierarchy problem and new dimensions at a millimeter. *Phys. Lett.*, **B429**, 263–72.

Arkani-Hamed, N., Dimopoulos, S., and Dvali, G. (1999). Phenomenology, astrophysics and cosmology of theories with submillimeter dimensions and TeV scale quantum gravity. *Phys. Rev.*, **D59**, 086004.

Arkani-Hamed, N., Motl, L., Nicolis, A., and Vafa, C. (2007). The String landscape, black holes and gravity as the weakest force. *JHEP*, **0706**, 060.

Arvanitaki, A., Dimopoulos, S., Dubovsky, S., Kaloper, N., and March-Russell, J. (2010a). String Axiverse. *Phys. Rev.*, **D81**, 123530.

Arvanitaki, A., Craig, N., Dimopoulos, S., Dubovsky, S., and March-Russell, J. (2010b). String photini at the LHC. *Phys. Rev.*, **D81**, 075018.

Ashok, S. and Douglas, M. R. (2004). Counting flux vacua. *JHEP*, **0401**, 060.

Aspinwall, P. S. (1996). Some relationships between dualities in string theory. *Nucl. Phys. Proc. Suppl.*, **46**, 30–8.

Assel, B., Christodoulides, K., Faraggi, A. E., Kounnas, C., and Rizos, J. (2010). Classification of heterotic Pati–Salam models. [arXiv:hep-th/1007.2268].

Athanasiu, G. G., Atick, J. J., Dine, M., and Fischler, W. (1988). Remarks on Wilson lines, modular invariance and possible string relics in Calabi–Yau compactifications. *Phys. Lett.*, **B214**, 55.

Atick, J., Dixon, L. J., and Sen, A. (1987). String calculation of Fayet–Iliopoulos D-terms in arbitrary supersymmetric compactifications. *Nucl. Phys.*, **B292**, 109–49.

Atiyah, M. and Witten, E. (2003). M theory dynamics on a manifold of G_2 holonomy. *Adv. Theor. Math. Phys.*, **6**, 1–106.

Bachas, C. (1995). A way to break supersymmetry. [arXiv:hep-th/9503030].

Bachas, C. (1998). Lectures on D-branes. [arXiv:hep-th/9806199].

Bachas, C. and Fabre, C. (1996). Threshold effects in open string theory. *Nucl. Phys.*, **B476**, 418–36.

Baer, H. and Tata, X. (2006). *Weak Scale Supersymmetry: From Superfields to Scattering Events*. Cambridge: Cambridge University Press.

Baer, H., Park, E.-K., Tata, X., and Wang, T. T. (2007). Collider and dark matter phenomenology of models with mirage unification. *JHEP*, **0706**, 033.

Bailin, D., Love, A., and Thomas, S. (1987). A three generation orbifold compactified superstring model with realistic gauge group. *Phys. Lett.*, **B194**, 385.

Bailin, D., Love, A., Sabra, W. A., and Thomas, S. (1994). String loop threshold corrections for Z_N Coxeter orbifolds. *Mod. Phys. Lett.*, **A9**, 67–80.

Balasubramanian, V., Berglund, P., Conlon, J. P., and Quevedo, F. (2005). Systematics of moduli stabilisation in Calabi–Yau flux compactifications. *JHEP*, **0503**, 007.

Banks, T. (2004). Landskepticism or why effective potentials don't count string models. [arXiv:hep-th/0412129].

Banks, T. and Dine, M. (1996). Couplings and scales in strongly coupled heterotic string theory. *Nucl. Phys.*, **B479**, 173–96.

Banks, T. and Dine, M. (1997). The cosmology of string theoretic axions. *Nucl. Phys.*, **B505**, 445–60.

Banks, T. and Dixon, L. J. (1988). Constraints on string vacua with space-time supersymmetry. *Nucl. Phys.*, **B307**, 93–108.

Banks, T., Kaplan, D. B., and Nelson, A. E. (1994). Cosmological implications of dynamical supersymmetry breaking. *Phys. Rev.*, **D49**, 779–87.

Banks, T., Dine, M., Fox, P., and Gorbatov, E. (2003). On the possibility of large axion decay constants. *JCAP*, **0306**, 001.

Banks, T., Dine, M., and Gorbatov, E. (2004). Is there a string theory landscape? *JHEP*, **0408**, 058.

Barbieri, R., Ferrara, S., and Savoy, C. (1982). Gauge models with spontaneously broken local supersymmetry. *Phys. Lett.*, **B119**, 343.

Barbieri, R., Louis, J., and Moretti, M. (1993). Phenomenological implications of supersymmetry breaking by the dilaton. *Phys. Lett.*, **B312**, 451–60.

Barr, S. M. (1982). A new symmetry breaking pattern for $SO(10)$ and proton decay. *Phys. Lett.*, **B112**, 219.

Barr, S. M. (1989). Some comments on flipped $SU(5) \times U(1)$ and flipped unification in general. *Phys. Rev.*, **D40**, 2457.

Baumann, D. (2009). TASI lectures on inflation. [arXiv:hep-th0907.5424].

Baumann, D., Dymarsky, A., Klebanov, I. R. Maldacena, J., McAllister, L., and Murugan, A. (2006). On D3-brane potentials in compactifications with fluxes and wrapped D-branes. *JHEP*, **0611**, 031.

Baumann, D., Dymarsky, A., Klebanov, I. R., and McAllister, L. (2008). Towards an explicit model of D-brane inflation. *JCAP*, **0801**, 024.

Baumann, D., Dymarsky, A., Kachru, S., Klebanov, I. R., and McAllister, L. (2010). D3-brane potentials from fluxes in AdS/CFT. *JHEP*, **1006**, 072.

Beasley, C. and Witten, E. (2002). A note on fluxes and superpotentials in M theory compactifications on manifolds of G_2 holonomy. *JHEP*, **0207**, 046.

Beasley, C., Heckman, J. J., and Vafa, C. (2009a). GUTs and exceptional branes in F-theory–I. *JHEP*, **0901**, 058.

Beasley, C., Heckman, J. J., and Vafa, C. (2009b). GUTs and exceptional branes in F-theory–II: experimental predictions. *JHEP*, **0901**, 059.

Becker, K. and Becker, M. (1996). M theory on eight manifolds. *Nucl. Phys.*, **B477**, 155–67.

Becker, K. and Becker, M. (2001). Supersymmetry breaking, M theory and fluxes. *JHEP*, **0107**, 038.

Becker, K., Becker, M., and Strominger, A. (1995). Five-branes, membranes and nonperturbative string theory. *Nucl. Phys.*, **B456**, 130–52.

Becker, K., Becker, M., Dasgupta, K., and Green, P. S. (2003). Compactifications of heterotic theory on non-Kahler complex manifolds. 1. *JHEP*, **0304**, 007.

Becker, K., Becker, M., Green, P. S., Dasgupta, K., and Sharpe, E. (2004). Compactifications of heterotic strings on non-Kahler complex manifolds. 2. *Nucl. Phys.*, **B678**, 19–100.

Becker, K., Becker, M., and Schwarz, J. H. (2007). *String Theory and M-theory: A Modern Introduction.* Cambridge: Cambridge University Press.

Berenstein, D., Jejjala, V., and Leigh, R. G. (2002). The standard model on a D-brane. *Phys. Rev. Lett.*, **88**, 071602.

Berg, M., Marsh, D., McAllister, L., and Pajer, E. (2010). Sequestering in string compactifications. [arXiv:hep-th/1012.1858].

Bergshoeff, E., Sezgin, E., and Townsend, P. K. (1987). Supermembranes and eleven-dimensional supergravity. *Phys. Lett.*, **B189**, 75–8.

Bergshoeff, E., Kallosh, R., Kashani-Poor, A.-K., Sorokin, D., and Tomasiello, A. (2005). An index for the Dirac operator on D3 branes with background fluxes. *JHEP*, **0510**, 102.

Berkooz, M. and Leigh, R. G. (1997). A D $= 4$, $\mathcal{N} = 1$ orbifold of type I strings. *Nucl. Phys.*, **B483**, 187–208.

Berkooz, M. Leigh, R. G., Polchinski, J., Schwarz, J. H., Seiberg, N., and Witten, E. (1996a). Anomalies, dualities, and topology of D = 6$\mathcal{N}$ = 1 superstring vacua. *Nucl. Phys.*, **B475**, 115–48.

Berkooz, M., Douglas, M. R., and Leigh, R. G. (1996b). Branes intersecting at angles. *Nucl. Phys.*, **B480**, 265–78.

Bershadsky, M., Vafa, C., and Sadov, V. (1996). D-branes and topological field theories. *Nucl. Phys.*, **B463**, 420–34.

Bianchi, M. and Kiritsis, E. (2007). Non-perturbative and flux superpotentials for Type I strings on the Z_3 orbifold. *Nucl. Phys.*, **B782**, 26–50.

Bianchi, M. and Sagnotti, A. (1990). On the systematics of open string theories. *Phys. Lett.*, **B247**, 517–24.

Bianchi, M. and Samsonyan, M. (2009). Notes on unoriented D-brane instantons. *Int. J. Mod. Phys.*, **A24**, 5737–63.

Bianchi, M., Fucito, F., and Morales, J. F. (2007). D-brane instantons on the T^6/Z_3 orientifold. *JHEP*, **0707**, 038.

Bianchi, M., Kovacs, S., and Rossi, G. (2008). Instantons and supersymmetry. *Lect. Notes Phys.*, **737**, 303–470.

Billò, M., Frau, M., Pesando, I., Fucito, F., Lerda, A., and Liccardo, A. (2003). Classical gauge instantons from open strings. *JHEP*, **0302**, 045.

Billò, M., Ferro, L., Frau, M., Fucito, F., Lerda, A., and Morales, J. F. (2008a). Flux interactions on D-branes and instantons. *JHEP*, **0810**, 112.

Billò, M., Ferro, L., Frau, M., Fucito, F., Lerda, A., and Morales, J. F. (2008b). Non-perturbative effective interactions from fluxes. *JHEP*, **0812**, 102.

Binétruy, P. (2006). *Supersymmetry: Theory, Experiment and Cosmology*. Oxford: Oxford University Press.

Binétruy, P. and Gaillard, M. K. (1991). A modular invariant formulation of gaugino condensation with a positive semidefinite potential. *Phys. Lett.*, **B253**, 119–28.

Binétruy, P. and Ramond, P. (1995). Yukawa textures and anomalies. *Phys. Lett.*, **B350**, 49–57.

Binétruy, P., Gaillard, M. K., and Wu, Y.-Y. (1997). Modular invariant formulation of multi – gaugino and matter condensation. *Nucl. Phys.*, **B493**, 27–55.

Blanco-Pillado, J. J., Burgess, C. P., Cline, J. M., *et al.* (2004). Racetrack inflation. *JHEP*, **0411**, 063.

Blanco-Pillado, J. J., Burgess, C. P., Cline, J. M., *et al.* (2006). Inflating in a better racetrack. *JHEP*, **0609**, 002.

Blumenhagen, R. (2003). Supersymmetric orientifolds of Gepner models. *JHEP*, **0311**, 055.

Blumenhagen, R. (2009). Gauge coupling unification in F-theory Grand Unified Theories. *Phys. Rev. Lett.*, **102**, 071601.

Blumenhagen, R. and Weigand, T. (2004). Chiral supersymmetric Gepner model orientifolds. *JHEP*, **0402**, 041.

Blumenhagen, R. and Wisskirchen, A. (1998). Spectra of 4-D, $\mathcal{N}$ = 1 type I string vacua on nontoroidal CY threefolds. *Phys. Lett.*, **B438**, 52–60.

Blumenhagen, R., Görlich, L., Körs, B., and Lüst, D. (2000a). Non-commutative compactifications of type I strings on tori with magnetic background flux. *JHEP*, **0010**, 006.

Blumenhagen, R., Görlich, L., and Körs, B. (2000b). Supersymmetric 4-D orientifolds of type IIA with D6-branes at angles. *JHEP*, **0001**, 040.

Blumenhagen, R., Görlich, L., and Körs, B. (2000c). Supersymmetric orientifolds in 6-D with D-branes at angles. *Nucl. Phys.*, **B569**, 209–28.

Blumenhagen, R., Körs, B., and Lüst, D. (2001). Type I strings with F flux and B flux. *JHEP*, **0102**, 030.

Blumenhagen, R., Braun, V., Körs, B., and Lüst, D. (2002a). Orientifolds of K3 and Calabi–Yau manifolds with intersecting D-branes. *JHEP*, **0207**, 026.

Blumenhagen, R., Braun, V., Körs, B., and Lüst, D. (2002b). The standard model on the quintic. [arXiv:hep-th/0210083].

Blumenhagen, R., Lüst, D., and Stieberger, S. (2003a). Gauge unification in supersymmetric intersecting brane worlds. *JHEP*, **0307**, 036.

Blumenhagen, R., Lüst, D., and Taylor, T. R. (2003b). Moduli stabilization in chiral type IIB orientifold models with fluxes. *Nucl. Phys.*, **B663**, 319–42.

Blumenhagen, R., Cvetič, M., Marchesano, F., and Shiu, G. (2005a). Chiral D-brane models with frozen open string moduli. *JHEP*, **0503**, 050.

Blumenhagen, R., Cvetič, M., Langacker, P., and Shiu, G. (2005b). Toward realistic intersecting D-brane models. *Ann. Rev. Nucl. Part. Sci.*, **55**, 71–139.

Blumenhagen, R., Moster, S., and Weigand, T. (2006). Heterotic GUT and standard model vacua from simply connected Calabi–Yau manifolds. *Nucl. Phys.*, **B751**, 186–221.

Blumenhagen, R., Körs, B., Lüst, D., and Stieberger, S. (2007a). Four-dimensional string compactifications with D-branes, orientifolds and fluxes. *Phys. Rept.*, **445**, 1–193.

Blumenhagen, R., Cvetič, M., and Weigand, T. (2007b). Spacetime instanton corrections in 4D string vacua: The Seesaw mechanism for D-brane models. *Nucl. Phys.*, **B771**, 113–42.

Blumenhagen, R., Moster, S., and Plauschinn, E. (2008a). Moduli stabilisation versus chirality for MSSM like type IIB orientifolds. *JHEP*, **0801**, 058.

Blumenhagen, R., Cvetič, M., Lüst, D., Richter, R., and Weigand, T. (2008b). Non-perturbative Yukawa couplings from string instantons. *Phys. Rev. Lett.*, **100**, 061602.

Blumenhagen, R., Cvetič, M., Kachru, S., and Weigand, T. (2009a). D-brane instantons in type II orientifolds. *Ann. Rev. Nucl. Part. Sci.*, **59**, 269–96.

Blumenhagen, R., Conlon, J. P., Krippendorf, S., Moster, S., and Quevedo, F. (2009b). SUSY breaking in local string/F-theory models. *JHEP*, **0909**, 007.

Blumenhagen, R., Grimm, T. W., Jurke, B., and Weigand, T. (2010a). Global F-theory GUTs. *Nucl. Phys.*, **B829**, 325–69.

Blumenhagen, R., Collinucci, A., and Jurke, B. (2010b). On instanton effects in F-theory. *JHEP*, **1008**, 079.

Blumenhagen, R., Moster, S., Reinbacher, R., and Weigand, T. (2007c). Massless spectra of three generation $U(N)$ heterotic string vacua. *JHEP*, **0705**, 041.

Blumenhagen, R., Görlich, L., and Ott, T. (2003). Supersymmetric intersecting branes on the type IIA T^6/Z_4 orientifold. *JHEP*, **0301**, 021.

Bouchard, V. and Donagi, R. (2006). An $SU(5)$ heterotic standard model. *Phys. Lett.*, **B633**, 783–91.

Bousso, R. and Polchinski, J. (2000). Quantization of four form fluxes and dynamical neutralization of the cosmological constant. *JHEP*, **0006**, 006.

Braun, V., He, Y.-H., Ovrut, B., and Pantev, T. (2005). A heterotic standard model. *Phys. Lett.*, **B618**, 252–8.

Braun, V., He, Y.-H., Ovrut, B., and Pantev, T. (2006a). The exact MSSM spectrum from string theory. *JHEP*, **0605**, 043.

Braun, V., He, Y.-H., and Ovrut, B. (2006b). Yukawa couplings in heterotic standard models. *JHEP*, **0604**, 019.

Braun, V., Candelas, P., and Davies, R. (2010). A three-generation Calabi–Yau manifold with small Hodge numbers. *Fortsch. Phys.*, **58**, 467–502.

Breit, J., Ovrut, B., and Segre, G. (1985). E_6 symmetry breaking in the superstring theory. *Phys. Lett.*, **B158**, 33.

Brignole, A., Ibáñez, L. E., and Muñoz, C. (1994). Towards a theory of soft terms for the supersymmetric standard model. *Nucl. Phys.*, **B422**, 125–71.

Brignole, A., Ibáñez, L. E., and Muñoz, C. (1997). Soft supersymmetry-breaking terms from supergravity and superstring models. In Kane, G. L. (ed.) *Perspectives on Supersymmetry*. Singapore: World Scientific, pp. 125-48. [arXiv:hep-ph/9707209].

Brooijmans, *et al.* (2010). New physics at the LHC. [arXiv:hep-ph/1005.1229].

Buchmüller, W., Di Bari, P., and Plumacher, M. (2005). Leptogenesis for pedestrians. *Annals Phys.*, **315**, 305–51.

Buchmüller, W., Hamaguchi, K., Lebedev, O., and Ratz, M. (2006). Supersymmetric standard model from the heterotic string. *Phys. Rev. Lett.*, **96**, 121602.

Buchmüller, W., Hamaguchi, K., Lebedev, O., Ramos-Sánchez, S., and Ratz, M. (2007a). Seesaw neutrinos from the heterotic string. *Phys. Rev. Lett.*, **99**, 021601.

Buchmüller, W., Hamaguchi, K., Lebedev, O., and Ratz, M. (2007b). Supersymmetric standard model from the heterotic string. II. *Nucl. Phys.*, **B785**, 149–209.

Buras, A., Ellis, J., Gaillard, M. K., and Nanopoulos, D. V. (1978). Aspects of the Grand Unification of strong, weak and electromagnetic interactions. *Nucl. Phys.*, **B135**, 66–92.

Burgess, C. P. (2007). Lectures on cosmic inflation and its potential stringy realizations. *Class. Quant. Grav.*, **24**, S795.

Burgess, C. P., Font, A., and Quevedo, F. (1986). Low-energy effective action for the superstring. *Nucl. Phys.*, **B272**, 661.

Burgess, C. P., Kallosh, R., and Quevedo, F. (2003). De Sitter string vacua from supersymmetric D terms. *JHEP*, **0310**, 056.

Burgess, C. P., Conlon, J. P., Hung, L.-Y., Kom, C. H., Maharana, A., and Quevedo, F. (2008). Continuous global symmetries and hyperweak interactions in string compactifications. *JHEP*, **0807**, 073.

Burwick, T. T., Kaiser, R. K., and Muller, H. F. (1991). General Yukawa couplings of strings on Z_N orbifolds. *Nucl. Phys.*, **B355**, 689–711.

Buscher, T. H. (1987). A symmetry of the string background field equations. *Phys. Lett.*, **B194**, 59.

Camara, P. G., Ibáñez, L. E., and Uranga, A. M. (2004). Flux induced SUSY breaking soft terms. *Nucl. Phys.*, **B689**, 195–242.

Camara, P. G., Ibáñez, L. E., and Uranga, A. M. (2005a). Flux-induced SUSY-breaking soft terms on D7-D3 brane systems. *Nucl. Phys.*, **B708**, 268–316.

Camara, P. G., Font, A., and Ibáñez, L. E. (2005b). Fluxes, moduli fixing and MSSM-like vacua in a simple IIA orientifold. *JHEP*, **0509**, 013.

Candelas, P. (1987). *Lectures on Complex Manifolds, Proceedings of Superstrings 87*, Trieste, p. 1.

Candelas, P. (1988). Yukawa couplings between (2,1) forms. *Nucl. Phys.*, **B298**, 458.

Candelas, P. and Davies, R. (2010). New Calabi–Yau manifolds with small Hodge numbers. *Fortsch. Phys.*, **58**, 383–466.

Candelas, P., Horowitz, G., Strominger, A., and Witten, E. (1985). Vacuum configurations for superstrings. *Nucl. Phys.*, **B258**, 46–74.

Candelas, P., Dale, A. M., Lutken, C. A., and Schimmrigk, R. (1988a). Complete intersection Calabi–Yau manifolds. *Nucl. Phys.*, **B298**, 493.

Candelas, P., Lutken, C. A., and Schimmrigk, R. (1988b). Complete intersection Calabi–Yau manifolds 2: three generation manifolds. *Nucl. Phys.*, **B306**, 113.

Candelas, P., Lynker, M., and Schimmrigk, R. (1990). Calabi–Yau manifolds in weighted P_4. *Nucl. Phys.*, **B341**, 383–402.

Candelas, P., De La Ossa, X., Green, P., and Parkes, L. (1991). A pair of Calabi–Yau manifolds as an exactly soluble superconformal theory. *Nucl. Phys.*, **B359**, 21–74.

Candelas, P., De La Ossa, X., Font, A., Katz, S., and Morrison, D. (1994a). Mirror symmetry for two parameter models. 1. *Nucl. Phys.*, **B416**, 481–538.

Candelas, P., Font, A., Katz, S. H., and Morrison, D. R. (1994b). Mirror symmetry for two parameter models. 2. *Nucl. Phys.*, **B429**, 626–74.

Carroll, S. M. (1999). TASI lectures: cosmology for string theorists. [arXiv:hep-th/0011110].

Carroll, S. M. (2001). The cosmological constant. *Living Rev. Rel.*, **4**, 1.

Casas, J. A. and Muñoz, C. (1988a). Three generation $SU(3) \times SU(2) \times U(1)_Y$ models from orbifolds. *Phys. Lett.*, **B214**, 63.

Casas, J. A. and Muñoz, C. (1988b). Three generation $SU(3) \times SU(2) \times U(1)_Y \times U(1)$ orbifold models through Fayet–Iliopoulos terms. *Phys. Lett.*, **B209**, 214.

Casas, J. A. and Muñoz, C. (1990). Fermion masses and mixing angles: a test for string vacua. *Nucl. Phys.*, **B332**, 189.

Casas, J. A., Katehou, E. K., and Muñoz, C. (1989). $U(1)$ Charges in orbifolds: anomaly cancellation and phenomenological consequences. *Nucl. Phys.*, **B317**, 171.

Casas, J. A., Gómez, F., and Muñoz, C. (1992). Fitting the quark and lepton masses in string theories. *Phys. Lett.*, **B292**, 42–54.

Casas, J. A., Gómez, F., and Muñoz, C. (1993). Complete structure of Z_N Yukawa couplings. *Int. J. Mod. Phys.*, **A8**, 455–506.

Casas, J. A., Lalak, Z., Muñoz, C., and Ross, G. G. (1990). Hierarchical supersymmetry breaking and dynamical determination of compactification parameters by non-perturbative effects. *Nucl. Phys.*, **B347**, 243–69.

Casas, J. A., Lleyda, A., and Muñoz, C. (1996). Problems for supersymmetry breaking by the dilaton in strings from charge and color breaking. *Phys. Lett.*, **B380**, 59–67.

Cascales, J. F. G. and Uranga, A. M. (2003). Chiral 4d string vacua with D branes and NSNS and RR fluxes. *JHEP*, **0305**, 011.

Cascales, J. F. G., Garcia del Moral, M. P., Quevedo, F., and Uranga, A. M. (2004). Realistic D-brane models on warped throats: fluxes, hierarchies and moduli stabilization. *JHEP*, **0402**, 031.

Cascales, J. F. G., Saad, F., and Uranga, A. M. (2005). Holographic dual of the standard model on the throat. *JHEP*, **0511**, 047.

Cecotti, S., Cheng, M. C. N., Heckman, J. J., and Vafa, C. (2009). Yukawa couplings in F-theory and non-commutative geometry. [arXiv:hep-th/0910.0477].

Chamseddine, A. and Quirós, M. (1989). Three generation four-dimensional $SU(3) \times SU(2) \times U(1)^5$ string models. *Nucl. Phys.*, **B316**, 101.

Chamseddine, A., Arnowitt, R., and Nath, P. (1982). Locally supersymmetric Grand Unification. *Phys. Rev. Lett.*, **49**, 970.

Chaudhuri, S., Chung, S.-W., and Lykken, J. D. (1994a). Fermion masses from superstring models with adjoint scalars. [arXiv:hep-ph/9405374].

Chaudhuri, S., Chung, S. W., Hockney, G., and Lykken, J. D. (1994b). String models for locally supersymmetric grand unification. [arXiv:hep-th/9409151].

Chaudhuri, S., Hockney, G., and Lykken, J. D. (1996). Three generations in the fermionic construction. *Nucl. Phys.*, **B469**, 357–86.

Cheng, H.-C. (2010). TASI lectures: introduction to extra dimensions. [arXiv:hep/1003.1162].

Choi, K. and Kim, J. E. (1985a). Compactification and axions in $E_8 \times E_8$ superstring models. *Phys. Lett.*, **B165**, 71.

Choi, K., and Kim, J. E. (1985b). Harmful axions in superstring models. *Phys. Lett.*, **B154**, 393.

Choi, K. and Nilles, H. P. (2007). The gaugino code. *JHEP*, **0704**, 006.

Choi, K., Kim, H. B., and Muñoz, C. (1998). Four-dimensional effective supergravity and soft terms in M theory. *Phys. Rev.*, **D57**, 7521–8.

Choi, K., Falkowski, A., Nilles, H. P., Olechowski, M., and Pokorski, S. 2004. Stability of flux compactifications and the pattern of supersymmetry breaking. *JHEP*, **0411**, 076.

Choi, K., Falkowski, A., Nilles, H. P., and Olechowski, M. (2005). Soft supersymmetry breaking in KKLT flux compactification. *Nucl. Phys.*, **B718**, 113–33.

Choi, K.-S., and Kim, J. E. (2006). Quarks and leptons from orbifolded superstring. *Lect. Notes Phys.*, **696**, 1–406.

Chun, E. J., Mas, J., Lauer, J., and Nilles, H. P. (1989). Duality and Landau–Ginzburg models. *Phys. Lett.*, **B233**, 141.

Cicoli, M., Conlon, J. P., and Quevedo, F. (2008). Systematics of string loop corrections in type IIB Calabi–Yau flux compactifications. *JHEP*, **0801**, 052.

Cicoli, M., Burgess, C. P., and Quevedo, F. (2009). Fibre inflation: observable gravity waves from IIB string compactifications. *JCAP*, **0903**, 013.

Cleaver, G. B. (2007). In search of the (minimal supersymmetric) standard model string. [arXiv:hep-ph/0703027].

Cleaver, G. B., Faraggi, A. E., and Nanopoulos, D. V. (1999). String derived MSSM and M theory unification. *Phys. Lett.*, **B455**, 135–46.

Cline, J. M. (2006). String cosmology. [arXiv:hep-th/0612129].

Coleman, S. R. (1979). The uses of instantons. *Subnucl. Ser.*, **15**, 805.

Conlon, J. P. (2008). Mirror mediation. *JHEP*, **0803**, 025.

Conlon, J. P. and Palti, E. (2010). Aspects of flavour and supersymmetry in F-theory GUTs. *JHEP*, **1001**, 029.

Conlon, J. P. and Quevedo, F. (2006). Kahler moduli inflation. *JHEP*, **0601**, 146.

Conlon, J. P., Quevedo, F., and Suruliz, K. (2005). Large-volume flux compactifications: moduli spectrum and D3/D7 soft supersymmetry breaking. *JHEP*, **0508**, 007.

Conlon, J. P., Cremades, D., and Quevedo, F. (2007a). Kahler potentials of chiral matter fields for Calabi–Yau string compactifications. *JHEP*, **0701**, 022.

Conlon, J. P., Abdussalam, S. S., Quevedo, F., and Suruliz, K. (2007b). Soft SUSY breaking terms for chiral matter in IIB string compactifications. *JHEP*, **0701**, 032.

Conlon, J. P., Kom, C. H., Suruliz, K., Allanach, B. C., and Quevedo, F. (2007c). Sparticle spectra and LHC signatures for large volume string compactifications. *JHEP*, **0708**, 061.

Conlon, J. P., Maharana, A., and Quevedo, F. (2008). Wave functions and Yukawa couplings in local string compactifications. *JHEP*, **0809**, 104.

Conlon, J. P., Maharana, A., and Quevedo, F. (2009). Towards realistic string vacua. *JHEP*, **0905**, 109.

Copeland, E. J., and Kibble, T. W. B. (2010). Cosmic strings and superstrings. *Proc. Roy. Soc. Lond.*, **A466**, 623–57.

Copeland, E. J., Myers, R. C., and Polchinski, J. (2004). Cosmic F and D strings. *JHEP*, **0406**, 013.

Cordova, C. (2009). Decoupling gravity in F-theory. [arXiv:hep-th/0910.2955].

Coriano, C., Irges, N., and Kiritsis, E. (2006). On the effective theory of low scale orientifold string vacua. *Nucl. Phys.*, **B746**, 77–135.

Coughlan, G. D., Fischler, W., Kolb, E. W., Raby, S., and Ross, G. G. (1983). Cosmological problems for the Polonyi potential. *Phys. Lett.*, **B131**, 59.

Cremades, D., Ibáñez, L. E., and Marchesano, F. (2002a). Intersecting brane models of particle physics and the Higgs mechanism. *JHEP*, **0207**, 022.

Cremades, D., Ibáñez, L. E., and Marchesano, F. (2002b). Standard model at intersecting D5-branes: lowering the string scale. *Nucl. Phys.*, **B643**, 93–130.

Cremades, D., Ibáñez, L. E., and Marchesano, F. (2002c). SUSY quivers, intersecting branes and the modest hierarchy problem. *JHEP*, **0207**, 009.

Cremades, D., Ibáñez, L. E., and Marchesano, F. (2003). Yukawa couplings in intersecting D-brane models. *JHEP*, **0307**, 038.

Cremades, D., Ibáñez, L. E., and Marchesano, F. (2004). Computing Yukawa couplings from magnetized extra dimensions. *JHEP*, **0405**, 079.

Cremmer, E. and Julia, B. (1978). The N = 8 supergravity theory. 1. The Lagrangian. *Phys. Lett.*, **B80**, 48.

Cremmer, E. and Julia, B. (1979). The SO(8) supergravity. *Nucl. Phys.*, **B159**, 141.

Cremmer, E., Julia, B., and Scherk, J. (1978). Supergravity theory in eleven-dimensions. *Phys. Lett.*, **B76**, 409–12.

Cremmer, E., Julia, B., Scherk, J., Ferrara, S., Girardello, L., and v. Nieuwenhuizen, P. (1979). Spontaneous symmetry breaking and Higgs effect in supergravity without cosmological constant. *Nucl. Phys.*, **B147**, 105.

Cremmer, E., Ferrara, S., Kounnas, C., and Nanopoulos, D. V. (1983a). Naturally vanishing cosmological constant in $\mathcal{N} = 1$ supergravity. *Phys. Lett.*, **B133**, 61.

Cremmer, E., Ferrara, S., Girardello, L., and Van Proeyen, A. (1983b). Yang–Mills theories with local supersymmetry: Lagrangian, transformation laws and super-Higgs effect. *Nucl. Phys.*, **B212**, 413.

Cullen, S., Perelstein, M., and Peskin, M. E. (2000). TeV strings and collider probes of large extra dimensions. *Phys. Rev.*, **D62**, 055012.

Cvetič, M. (1987a). Exact construction of (0, 2) Calabi–Yau manifolds. *Phys. Rev. Lett.*, **59**, 2829.

Cvetič, M. (1987b). Supression of nonrenormalizable terms in the effective superpotential for (blown-up) orbifold compactification. *Phys. Rev. Lett.*, **59**, 1795.

Cvetič, M. and Langacker, P. (2008). D-instanton generated Dirac neutrino masses. *Phys. Rev.*, **D78**, 066012.

Cvetič, M. and Papadimitriou, I. (2003). Conformal field theory couplings for intersecting D-branes on orientifolds. *Phys. Rev.*, **D68**, 046001.

Cvetič, M. and Richter, R. (2007). Proton decay via dimension-six operators in intersecting D6-brane models. *Nucl. Phys.*, **B762**, 112–47.

Cvetič, M. and Weigand, T. (2008). Hierarchies from D-brane instantons in globally defined Calabi–Yau orientifolds. *Phys. Rev. Lett.*, **100**, 251601.

Cvetič, M., Louis, J., and Ovrut, B. (1988). A string calculation of the Kahler potentials for moduli of Z_N orbifolds. *Phys. Lett.*, **B206**, 227.

Cvetič, M., Molera, J., and Ovrut, B. (1989). Kahler potentials for matter scalars and moduli of Z_N orbifolds. *Phys. Rev.*, **D40**, 1140.

Cvetič, M., Font, A., Ibáñez, L.E., Lüst, D., and Quevedo, F. (1991). Target space duality, supersymmetry breaking and the stability of classical string vacua. *Nucl. Phys.*, **B361**, 194–232.

Cvetič, M., Shiu, G., and Uranga, A. M. (2001a). Chiral four-dimensional $\mathcal{N} = 1$ supersymmetric type IIA orientifolds from intersecting D6 branes. *Nucl. Phys.*, **B615**, 3–32.

Cvetič, M., Shiu, G., and Uranga, A. M. (2001b). Chiral type II orientifold constructions as M theory on G_2 holonomy spaces. [arXiv:hep-th/0111179].

Cvetič, M., Shiu, G., and Uranga, A. M. (2001c). Three family supersymmetric standard-like models from intersecting brane worlds. *Phys. Rev. Lett.*, **87**, 201801.

Cvetič, M., Li, T., and Liu, T. (2004). Supersymmetric Pati–Salam models from intersecting D6-branes: a road to the standard model. *Nucl. Phys.*, **B698**, 163–201.

Cvetič, M., Richter, R., and Weigand, T. (2007a). Computation of D-brane instanton induced superpotential couplings: Majorana masses from string theory. *Phys. Rev.*, **D76**, 086002.

Cvetič, M., Richter, R., and Weigand, T. (2007b). D-brane instanton effects in type II orientifolds: Local and global issues. [arXiv:hep-th/0712.2845].

Cvetič, M., Halverson, J., and Richter, R. (2009). Realistic Yukawa structures from orientifold compactifications. *JHEP*, **0912**, 063.

Dabholkar, A. and Hull, C. (2003). Duality twists, orbifolds, and fluxes. *JHEP*, **0309**, 054.

Dabholkar, A. and Park, J. (1996). An orientifold of type IIB theory on K3. *Nucl. Phys.*, **B472**, 207–20.

Dai, Jin, Leigh, R. G., and Polchinski, J. (1989). New connections between string theories. *Mod. Phys. Lett.*, **A4**, 2073–83.

Dasgupta, K., Rajesh, G., and Sethi, S. (1999). M theory, orientifolds and G-flux. *JHEP*, **9908**, 023.

de Carlos, B., Casas, J. A., Quevedo, F., and Roulet, E. (1993a). Model independent properties and cosmological implications of the dilaton and moduli sectors of 4-d strings. *Phys. Lett.*, **B318**, 447–56.

de Carlos, B., Casas, J. A., and Muñoz, C. (1993b). Soft SUSY breaking terms in stringy scenarios: computation and phenomenological viability. *Phys. Lett.*, **B299**, 234–46.

de Carlos, B., Casas, J. A., and Muñoz, C. (1993c). Supersymmetry breaking and determination of the unification gauge coupling constant in string theories. *Nucl. Phys.*, **B399**, 623–53.

de Carlos, B., Gurrieri, S., Lukas, A., and Micu, A. (2006). Moduli stabilisation in heterotic string compactifications. *JHEP*, **0603**, 005.

Denef, F. (2008). Les Houches lectures on constructing string vacua. [arXiv:hep-th/0803.1194].

Denef, F. and Douglas, M. R. (2004). Distributions of flux vacua. *JHEP*, **0405**, 072.

Denef, F. and Douglas, M. R. (2005). Distributions of nonsupersymmetric flux vacua. *JHEP*, **0503**, 061.

Denef, F., Douglas, M. R., and Florea, B. (2004). Building a better racetrack. *JHEP*, **0406**, 034.

Denef, F., Douglas, M. R., Florea, B., Grassi, A., and Kachru, S. (2005). Fixing all moduli in a simple F-theory compactification. *Adv. Theor. Math. Phys.*, **9**, 861–929.

Denef, F., Douglas, M. R., and Kachru, S. (2007). Physics of string flux compactifications. *Ann. Rev. Nucl. Part. Sci.*, **57**, 119–44.

Derendinger, J. P., and Savoy, C. (1984). Quantum effects and $SU(2) \times U(1)$ breaking in supergravity gauge theories. *Nucl. Phys.*, **B237**, 307.

Derendinger, J. P., Ibáñez, L. E., and Nilles, H. P. (1985). On the low-energy d = 4, $\mathcal{N} = 1$ supergravity theory extracted from the d = 10, $\mathcal{N} = 1$ superstring. *Phys. Lett.*, **B155**, 65.

Derendinger, J. P., Ibáñez, L. E., and Nilles, H. P. (1986). On the low-energy limit of superstring theories. *Nucl. Phys.*, **B267**, 365.

Derendinger, J. P., Ferrara, S., Kounnas, C., and Zwirner, F. (1992). On loop corrections to string effective field theories: field dependent gauge couplings and sigma model anomalies. *Nucl. Phys.*, **B372**, 145–88.

Derendinger, J.-P., Kounnas, C., Petropoulos, P. M., and Zwirner, F. (2005). Superpotentials in IIA compactifications with general fluxes. *Nucl. Phys.*, **B715**, 211–33.

Derendinger, J. P., Kim, J. E., and Nanopoulos, D. V. (1984). Anti-SU(5). *Phys. Lett.*, **B139**, 170.

Deser, S. and Zumino, B. (1976). Consistent supergravity. *Phys. Lett.*, **B62**, 335.

DeWolfe, O. and Giddings, S. B. (2003). Scales and hierarchies in warped compactifications and brane worlds. *Phys. Rev.*, **D67**, 066008.

DeWolfe, O., Giryavets, A., Kachru, S., and Taylor, W. (2005). Type IIA moduli stabilization. *JHEP*, **0507**, 066.

D'Hoker, E. and Freedman, D. Z. (2002). Supersymmetric gauge theories and the AdS/CFT correspondence. [arXiv-hep-th/0201253].

Di Vecchia, P., Liccardo, A., Marotta, R., and Pezzella, F. (2009). Kahler metrics and Yukawa couplings in magnetized brane models. *JHEP*, **0903**, 029.

Diaconescu, D.-E., Florea, B., Kachru, S., and Svrcek, P. (2006). Gauge: mediated supersymmetry breaking in string compactifications. *JHEP*, **0602**, 020.

Dienes, K. R. (1997). String theory and the path to unification: a review of recent developments. *Phys. Rept.*, **287**, 447–525.

Dienes, K. R. (2006). Statistics on the heterotic landscape: gauge groups and cosmological constants of four-dimensional heterotic strings. *Phys. Rev.*, **D73**, 106010.

Dienes, K. R. and Faraggi, A. E. (1995). Gauge coupling unification in realistic free fermionic string models. *Nucl. Phys.*, **B457**, 409–83.

Dienes, K. R. and March-Russell, J. (1996). Realizing higher level gauge symmetries in string theory: new embeddings for string GUTs. *Nucl. Phys.*, **B479**, 113–72.

Dienes, K. R., Faraggi, A. E., and March-Russell, J. (1996). String unification, higher level gauge symmetries, and exotic hypercharge normalizations. *Nucl. Phys.*, **B467**, 44–99.

Dienes, K. R., Kolda, C. F., and March-Russell, J. (1997). Kinetic mixing and the supersymmetric gauge hierarchy. *Nucl. Phys.*, **B492**, 104–18.

Dienes, K. R., Dudas, E., and Gherghetta, T. (1998). Extra space-time dimensions and unification. *Phys. Lett.*, **B436**, 55–65.

Dienes, K. R., Dudas, E., and Gherghetta, T. (1999). Grand unification at intermediate mass scales through extra dimensions. *Nucl. Phys.*, **B537**, 47–108.

Dijkstra, T. P. T., Huiszoon, L. R., and Schellekens, A. N. (2005a). Chiral supersymmetric standard model spectra from orientifolds of Gepner models. *Phys. Lett.*, **B609**, 408–17.

Dijkstra, T. P. T., Huiszoon, L. R., and Schellekens, A. N. (2005b). Supersymmetric standard model spectra from RCFT orientifolds. *Nucl. Phys.*, **B710**, 3–57.

Dimopoulos, S., and Georgi, H. (1981). Softly broken supersymmetry and SU(5). *Nucl. Phys.*, **B193**, 150.

Dimopoulos, S. and Raby, S. (1981). Supercolor. *Nucl. Phys.*, **B192**, 353.

Dimopoulos, S. and Raby, S. (1983). Geometric hierarchy. *Nucl. Phys.*, **B219**, 479.

Dimopoulos, S. and Susskind, L. (1979). Mass without scalars. *Nucl. Phys.*, **B155**, 237–52.

Dimopoulos, S., Raby, S., and Wilczek, F. (1981). Supersymmetry and the scale of unification. *Phys. Rev.*, **D24**, 1681–3.

Dimopoulos, S., Raby, S., and Wilczek, F. (1982). Proton decay in supersymmetric models. *Phys. Lett.*, **B112**, 133.

Dimopoulos, S., Dvali, G. R., Rattazzi, R., and Giudice, G. F. (1998). Dynamical soft terms with unbroken supersymmetry. *Nucl. Phys.*, **B510**, 12–38.

Dimopoulos, S., Kachru, S., McGreevy, J., and Wacker, J. G. (2008). N-flation. *JCAP*, **0808**, 003.

Dine, M. (2007). *Supersymmetry and String Theory: Beyond the Standard Model*. Cambridge: Cambridge University Press.

Dine, M. and Fischler, W. (1982). A phenomenological model of particle physics based on supersymmetry. *Phys. Lett.*, **B110**, 227.

Dine, M. and Seiberg, N. (1985). Is the superstring weakly coupled? *Phys. Lett.*, **B162**, 299.

Dine, M., Fischler, W., and Srednicki, M. (1981a). A simple solution to the strong CP problem with a harmless axion. *Phys. Lett.*, **B104**, 199.

Dine, M., Fischler, W., and Srednicki, M. (1981b). Supersymmetric technicolor. *Nucl. Phys.*, **B189**, 575–93.

Dine, M., Rohm, R., Seiberg, N., and Witten, E. (1985a). Gluino condensation in superstring models. *Phys. Lett.*, **B156**, 55.

Dine, M., Kaplunovsky, V., Mangano, M., Nappi, C., and Seiberg, N. (1985b). Superstring model building. *Nucl. Phys.*, **B259**, 549–71.

Dine, M., Seiberg, N., Wen, X. G., and Witten, E. (1986). Nonperturbative effects on the string world sheet. *Nucl. Phys.*, **B278**, 769.

Dine, M., Ichinose, I., and Seiberg, N. (1987a). F-terms and D-terms in string theory. *Nucl. Phys.*, **B293**, 253.

Dine, M., Seiberg, N., and Witten, E. (1987b). Fayet–Iliopoulos terms in string theory. *Nucl. Phys.*, **B289**, 589.

Dine, M., Seiberg, N., Wen, X. G., and Witten, E. (1987c). Nonperturbative effects on the string world sheet. 2. *Nucl. Phys.*, **B289**, 319.

Distler, J., and Greene, B. R. (1988). Aspects of (2,0) string compactifications. *Nucl. Phys.*, **B304**, 1.

Distler, J. and Kachru, S. (1994). (0,2) Landau–Ginzburg theory. *Nucl. Phys.*, **B413**, 213–43.

Dixon, L. J. (1990). Supersymmetry breaking in string theory. In *Proceedings of the APS, DPF 1990*, p. 811.

Dixon, L. J., Harvey, J., Vafa, C., and Witten, E. (1985). Strings on orbifolds. *Nucl. Phys.*, **B261**, 678–86.

Dixon, L. J., Harvey, J., Vafa, C., and Witten, E. (1986). Strings on orbifolds. 2. *Nucl. Phys.*, **B274**, 285–314.

Dixon, L. J., Friedan, D., Martinec, E. J., and Shenker, S. (1987). The conformal field theory of orbifolds. *Nucl. Phys.*, **B282**, 13–73.

Dixon, L. J., Kaplunovsky, V., and Louis, J. (1990). On effective field theories describing (2, 2) vacua of the heterotic string. *Nucl. Phys.*, **B329**, 27–82.

Dixon, L. J., Kaplunovsky, V., and Louis, J. (1991). Moduli dependence of string loop corrections to gauge coupling constants. *Nucl. Phys.*, **B355**, 649–88.

Donagi, R. (1997). Principal bundles on elliptic fibrations. *Asian J. Math.*, **1**, 214–23.

Donagi, R. and Wijnholt, M. (2008a). Breaking GUT groups in F-theory. [arXiv:hep-th/0808.2223].

Donagi, R. and Wijnholt, M. (2008b). Model building with F-theory. [arXiv:hep-th/0802.2969].

Donagi, R., Grassi, A., and Witten, E. (1996). A nonperturbative superpotential with E_8 symmetry. *Mod. Phys. Lett.*, **A11**, 2199–212.

Donagi, R., Ovrut, B., Pantev, T., and Waldram, D. (2002). Standard models from heterotic M theory. *Adv. Theor. Math. Phys.*, **5**, 93–137.

Donaldson, S. K. (1985). Anti self-dual Yang–Mills connections over complex algebraic surfaces and stable vector bundles. *Proc. Lond. Math. Soc.*, **50**, 1–26.

Donoghue, J. (2007). The fine-tuning problems of particle physics and anthropic mechanisms. In, Carr, B. (ed.) *Universe or Multiverse*. Cambridge: Cambridge University Press, pp. 231–46. [arXiv:hep-ph/0710.4080].

Douglas, M. R. (1995). Branes within branes. [arXiv:hep-th/9512077].

Douglas, M. R. (2001). D-branes, categories and $\mathcal{N} = 1$ supersymmetry. *J. Math. Phys.*, **42**, 2818–43.

Douglas, M. R. (2006). Understanding the landscape. [arXiv:hep-th/0602266].

Douglas, M. R. and Kachru, S. (2007). Flux compactification. *Rev. Mod. Phys.*, **79**, 733–96.

Douglas, M. R. and Moore, G. W. (1996). D-branes, quivers, and ALE instantons. [arXiv:hep-th/9603167].

Douglas, M. R. and Taylor, W. (2007). The landscape of intersecting brane models. *JHEP*, **0701**, 031.

Douglas, M. R. and Torroba, G. (2009). Kinetic terms in warped compactifications. *JHEP*, **0905**, 013.

Douglas, M. R., Greene, B. R., and Morrison, D. R. (1997). Orbifold resolution by D-branes. *Nucl. Phys.*, **B506**, 84–106.

Drees, M., Godbole, R., and Roy, P. (2004). *Theory and Phenomenology of Sparticles*. Singapore: World Scientific.

Dudas, E. and Grojean, C. (1997). Four-dimensional M theory and supersymmetry breaking. *Nucl. Phys.*, **B507**, 553–70.

Dudas, E. and Palti, E. (2010). Froggatt–Nielsen models from E_8 in F-theory GUTs. *JHEP*, **1001**, 127.

Duff, M. J., Howe, Paul S., Inami, T., and Stelle, K. S. (1987). Superstrings in D = 10 from supermembranes in D = 11. *Phys. Lett.*, **B191**, 70.

Dvali, G. and Tye, S. H. H. (1999). Brane inflation. *Phys. Lett.*, **B450**, 72–82.

Eguchi, T., Gilkey, P. B., and Hanson, A. J. (1980). Gravitation, gauge theories and differential geometry. *Phys. Rept.*, **66**, 213.

Eichten, E. and Lane, K. (1980). Dynamical breaking of weak interaction symmetries. *Phys. Lett.*, **B90**, 125–130.

Ellis, J., Nanopoulos, D. V., and Rudaz, S. (1982). GUTs 3: SUSY GUTs 2. *Nucl. Phys.*, **B202**, 43.

Ellis, J., Hagelin, J., Nanopoulos, D. V., and Tamvakis, K. (1983). Weak symmetry breaking by radiative corrections in broken supergravity. *Phys. Lett.*, **B125**, 275.

Ellis, J., Ridolfi, G., and Zwirner, F. (1991). Radiative corrections to the masses of supersymmetric Higgs bosons. *Phys. Lett.*, **B257**, 83–91.

Erler, J. (1996). Asymmetric orbifolds and higher level models. *Nucl. Phys.*, **B475**, 597–626.

Erler, J. and Klemm, A. (1993). Comment on the generation number in orbifold compactifications. *Commun. Math. Phys.*, **153**, 579–604.

Erler, J., Jungnickel, D., and Nilles, H. P. (1992). Space duality and quantized Wilson lines. *Phys. Lett.*, **B276**, 303–310.

Erler, J., Jungnickel, D., Spalinski, M., and Stieberger, S. (1993). Higher twisted sector couplings of Z_N orbifolds. *Nucl. Phys.*, **B397**, 379–416.

Falkowski, A., Lebedev, O., and Mambrini, Y. (2005). SUSY phenomenology of KKLT flux compactifications. *JHEP*, **0511**, 034.

Faraggi, A. E., and Halyo, E. (1993). Neutrino masses in superstring derived standard-like models. *Phys. Lett.*, **B307**, 311–17.

Faraggi, A. E., Nanopoulos, D. V., and Yuan, K. (1990). A standard like model in the 4D free fermionic string formulation. *Nucl. Phys.*, **B335**, 347.

Farrar, G. and Fayet, P. (1978). Bounds on R-hadron production from calorimetry experiments. *Phys. Lett.*, **B79**, 442.

Farrar, G. and Fayet, P. (1980). Searching for the spin 0 leptons of supersymmetry. *Phys. Lett.*, **B89**, 191.

Fayet, P. (1977). Spontaneously broken supersymmetric theories of weak, electromagnetic and strong interactions. *Phys. Lett.*, **B69**, 489.

Fayet, P. (1978). Massive gluinos. *Phys. Lett.*, **B78**, 417.

Fayet, P. (1979). Relations between the masses of the superpartners of leptons and quarks, the Goldstino couplings and the neutral currents. *Phys. Lett.*, **B84**, 416.

Fayet, P. and Iliopoulos, J. (1974). Spontaneously broken supergauge symmetries and Goldstone spinors. *Phys. Lett.*, **B51**, 461–4.

Feldman, D., Liu, Z., and Nath, P. (2007). The Stueckelberg Z' extension with kinetic mixing and milli-charged dark matter from the hidden sector. *Phys. Rev.*, **D75**, 115001.

Ferrara, S., Girardello, L., and Nilles, H. P. (1983). Breakdown of local supersymmetry through gauge fermion condensates. *Phys. Lett.*, **B125**, 457.

Ferrara, S., Kounnas, C., and Porrati, M. (1986). General dimensional reduction of ten-dimensional supergravity and superstring. *Phys. Lett.*, **B181**, 263.

Ferrara, S., Lüst, D., Shapere, A., and Theisen, S. (1989a). Modular invariance in supersymmetric field theories. *Phys. Lett.*, **B225**, 363.

Ferrara, S., Lüst, .D., and Theisen, S. (1989b). Target space modular invariance and low-energy couplings in orbifold compactifications. *Phys. Lett.*, **B233**, 147.

Ferrara, S., Magnoli, N., Taylor, T. R., and Veneziano, G. (1990). Duality and supersymmetry breaking in string theory. *Phys. Lett.*, **B245**, 409–16.

Ferrara, S., Kounnas, C., Lüst, D., and Zwirner, F. (1991). Duality invariant partition functions and automorphic superpotentials for (2,2) string compactifications. *Nucl. Phys.*, **B365**, 431–66.

Finnell, D. (1996). Grand unification with three generations in free fermionic string models. *Phys. Rev.*, **D53**, 5781–9.

Florea, B., Kachru, S., McGreevy, J., and Saulina, N. (2007). Stringy instantons and quiver gauge theories. *JHEP*, **0705**, 024.

Font, A. and Ibáñez, L. E. (2005). SUSY-breaking soft terms in a MSSM magnetized D7-brane model. *JHEP*, **0503**, 040.

Font, A. and Ibáñez, L.E. (2009a). Matter wave functions and Yukawa couplings in F-theory Grand Unification. *JHEP*, **0909**, 036.

Font, A. and Ibáñez, L. E. (2009b). Yukawa structure from $U(1)$ fluxes in F-theory Grand Unification. *JHEP*, **0902**, 016.

Font, A. and Theisen, S. (2005). Introduction to string compactification. *Lect. Notes Phys.*, **668**, 101–81.

Font, A., Ibáñez, L. E., Nilles, H. P., and Quevedo, F. (1988a). Degenerate orbifolds. *Nucl. Phys.*, **B307**, 109.

Font, A., Ibáñez, L. E., Nilles, H. P., and Quevedo, F. (1988b). Yukawa couplings in degenerate orbifolds: towards a realistic $SU(3) \times SU(2) \times U(1)$ superstring. *Phys. Lett.*, **210B**, 101.

Font, A., Ibáñez, L. E., Mondragon, M., Quevedo, F., and Ross, G. G. (1989a). (0, 2) Heterotic string compactifications from $\mathcal{N} = 2$ superconformal theories. *Phys. Lett.*, **B227**, 34.

Font, A., Ibáñez, L. E., and Quevedo, F. (1989b). $Z_N \times Z_M$ orbifolds and discrete torsion. *Phys. Lett.*, **B217**, 272.

Font, A., Ibáñez, L. E., and Quevedo, F. (1990a). Higher level Kac–Moody string models and their phenomenological implications. *Nucl. Phys.*, **B345**, 389–430.

Font, A., Ibáñez, L. E., Lüst, D., and Quevedo, F. (1990b). Strong–weak coupling duality and nonperturbative effects in string theory. *Phys. Lett.*, **B249**, 35–43.

Font, A., Ibáñez, L. E., Lüst, D., and Quevedo, F. (1990c). Supersymmetry breaking from duality invariant gaugino condensation. *Phys. Lett.*, **B245**, 401–408.

Font, A., Ibáñez, L. E., Quevedo, F., and Sierra, A. (1990d). The construction of "realistic" four-dimensional strings through orbifolds. *Nucl. Phys.*, **B331**, 421–74.

Font, A., Ibáñez, L. E., Quevedo, F., and Sierra, A. (1990e). Twisted $\mathcal{N} = 2$ coset models, discrete torsion and asymmetric heterotic string compactifications. *Nucl. Phys.*, **B337**, 119.

Font, A., Ibáñez, L. E., and Marchesano, F. (2006). Coisotropic D8-branes and model-building. *JHEP*, **0609**, 080.

Forste, S., Honecker, G., and Schreyer, R. (2001a). Orientifolds with branes at angles. *JHEP*, **06**, 004.

Forste, S., Honecker, G., and Schreyer, R. (2001b). Supersymmetric $Z_N \times Z_M$ orientifolds in 4-D with D branes at angles. *Nucl. Phys.*, **B593**, 127–154.

Forste, S., Nilles, H. P., Vaudrevange, P., and Wingerter, A. (2004). Heterotic brane world. *Phys. Rev.*, **D70**, 106008.

Fox, P., Pierce, A., and Thomas, S. (2004). Probing a QCD string axion with precision cosmological measurements. [arXiv:hep-th/0409059].

Franco, S., Hanany, A., Kennaway, K. D., Vegh, D., and Wecht, B. (2006). Brane dimers and quiver gauge theories. *JHEP*, **0601**, 096.

Franco, S., Rodríguez-Gomez, D., and Verlinde, H. (2009). N-ification of forces: a holographic perspective on D-brane model building. *JHEP*, **0906**, 030.

Freed, D. S. and Witten, E. (1999). Anomalies in string theory with D-branes. [arXiv:hep-th/9907189].

Freedman, D., van Nieuwenhuizen, P., and Ferrara, S. (1976). Progress toward a theory of supergravity. *Phys. Rev.*, **D13**, 3214–18.

Frey, A. R. and Polchinski, J. (2002). N = 3 warped compactifications. *Phys. Rev.*, **D65**, 126009.

Frey, A. R., Torroba, G., Underwood, B., and Douglas, M. R. (2009). The universal Kähler modulus in warped compactifications. *JHEP*, **0901**, 036.

Friedman, R., Morgan, J., and Witten, E. (1997a). Vector bundles and F theory. *Commun. Math. Phys.*, **187**, 679–743.

Friedman, R., Morgan, J., and Witten, E. (1997b). Vector bundles over elliptic fibrations. [arXiv:alg-geom/9709029].

Fritzsch, H. and Minkowski, P. (1975). Unified interactions of leptons and hadrons. *Annals Phys.*, **93**, 193–266.

Fuchs, J., Klemm, A., Scheich, C., and Schmidt, M. G. (1990). Spectra and symmetries of Gepner models compared to Calabi–Yau compactifications. *Annals Phys.*, **204**, 1–51.

Fuchs, J., Huiszoon, L. R., Schellekens, A. N., Schweigert, C., and Walcher, J. (2000). Boundaries, crosscaps and simple currents. *Phys. Lett.*, **B495**, 427–34.

Fukugita, M. and Yanagida, T. (1986). Baryogenesis without Grand Unification. *Phys. Lett.*, **B174**, 45.

Gabadadze, G. (2003). ICTP lectures on large extra dimensions. [arXiv:hep-ph/0308112].

Gaberdiel, M. R. (2000). An introduction to conformal field theory. *Rept. Prog. Phys.*, **63**, 607–67.

Gaillard, M. K. and Nelson, B. D. (2007). Kahler stabilized, modular invariant heterotic string models. *Int. J. Mod. Phys.*, **A22**, 1451–588.

García-Etxebarría, I., Saad, F., and Uranga, A. M. (2006a). Local models of gauge mediated supersymmetry breaking in string theory. *JHEP*, **0608**, 069.

García-Etxebarría, I., Saad, F., and Uranga, A. M. (2006b). Quiver gauge theories at resolved and deformed singularities using dimers. *JHEP*, **0606**, 055.

Gato-Rivera, B. and Schellekens, A. N. (2009). Non-supersymmetric orientifolds of Gepner models. *Phys. Lett.*, **B671**, 105–10.

Gato-Rivera, B. and Schellekens, A. N. (2010a). Asymmetric Gepner models II: heterotic weight lifting. [arXiv:hep-th/1009.1320].

Gato-Rivera, B., and Schellekens, A. N. (2010b). Asymmetric Gepner models III: B-L lifting. [arXiv:hep-th/1012.0796].

Gato-Rivera, B. and Schellekens, A. N. (2010c). Asymmetric Gepner models: revisited. *Nucl. Phys.*, **B841**, 100–129.

Gell-Mann, M., Ramond, P., Slansky, R. (1979) Complex spinors and unified theories. In P. v. Nieuwenhuizen and D. Friedman (eds.) *Supergravity*. Amsterdam: North Holland, p. 315.

Georgi, H. and Glashow, S. L. (1974). Unity of all elementary particle forces. *Phys. Rev. Lett.*, **32**, 438–41.

Georgi, H. and Nanopoulos, D. V. (1979). Ordinary predictions from grand principles: t-quark mass in O(10). *Nucl. Phys.*, **B155**, 52.

Georgi, H., Quinn, H., and Weinberg, S. (1974). Hierarchy of interactions in unified gauge theories. *Phys. Rev. Lett.*, **33**, 451–4.

Gepner, D. (1987). Exactly solvable string compactifications on manifolds of $SU(N)$ holonomy. *Phys. Lett.*, **B199**, 380–8.

Gepner, D. (1988a). Space-time supersymmetry in compactified string theory and super-conformal models. *Nucl. Phys.*, **B296**, 757.

Gepner, D. (1988b). Yukawa couplings for Calabi–Yau string compactification. *Nucl. Phys.*, **B311**, 191.

Gepner, D. (1989). Lectures on $\mathcal{N} = 2$ string theory. *Lectures at Spring School on Superstrings*, Trieste, Italy, April 3–14.

Ghilencea, D. and Ross, G. G. (1998). Unification and extra space-time dimensions. *Phys. Lett.*, **B442**, 165–72.

Ghilencea, D. M., Ibáñez, L. E., Irges, N., and Quevedo, F. (2002). TeV scale Z' bosons from D-branes. *JHEP*, **0208**, 016.

Giddings, S. B. (2007). High-energy black hole production. *AIP Conf. Proc.*, **957**, 69–78.

Giddings, S. B., Kachru, S., and Polchinski, J. (2002). Hierarchies from fluxes in string compactifications. *Phys. Rev.*, **D66**, 106006.

Giedt, J., Kane, G. L., Langacker, P., and Nelson, B. D. (2005). Massive neutrinos and (heterotic) string theory. *Phys. Rev.*, **D71**, 115013.

Gimon, E. G. and Johnson, C. V. (1996). K3 orientifolds. *Nucl. Phys.*, **B477**, 715–45.

Gimon, E. G. and Polchinski, J. (1996). Consistency conditions for orientifolds and D manifolds. *Phys. Rev.*, **D54**, 1667–76.

Ginsparg, P. (1987). Gauge and gravitational couplings in four-dimensional string theories. *Phys. Lett.*, **B197**, 139.

Ginsparg, P. (1988). Applied conformal field theory. [arXiv:hep-th/9108028].

Girardello, L. and Grisaru, M. (1982). Soft breaking of supersymmetry. *Nucl. Phys.*, **B194**, 65.

Giudice, G. and Romanino, A. (2004). Split supersymmetry. *Nucl. Phys.*, **B699**, 65–89.

Giudice, G. F. and Masiero, A. (1988). A natural solution to the μ problem in supergravity theories. *Phys. Lett.*, **B206**, 480–4.

Giudice, G. F. and Rattazzi, R. (1999). Theories with gauge mediated supersymmetry breaking. *Phys. Rept.*, **322**, 419–99.

Giudice, G. F. and Wells, J. D. (2008). Extra dimensions. In *Review of Particle Properties, Phys. Lett.*, **B667**,1.

Giudice, G. F., Luty, M., Murayama, H., and Rattazzi, R. (1998). Gaugino mass without singlets. *JHEP*, **9812**, 027.

Giudice, G. F., Rattazzi, R., and Wells, J. D. (1999). Quantum gravity and extra dimensions at high-energy colliders. *Nucl. Phys.*, **B544**, 3–38.

Giveon, A., Porrati, M., and Rabinovici, E. (1994). Target space duality in string theory. *Phys. Rept.*, **244**, 77–202.

Gmeiner, F. and Honecker, G. (2007). Mapping an island in the landscape. *JHEP*, **0709**, 128.

Gmeiner, F. and Honecker, G. (2008). Millions of standard models on Z_6'? *JHEP*, **0807**, 052.

Gmeiner, F., Blumenhagen, R., Honecker, G., Lüst, D., and Weigand, T. (2006). One in a billion: MSSM-like D-brane statistics. *JHEP*, **0601**, 004.

Gmeiner, F., Lüst, D., and Stein, M. (2007). Statistics of intersecting D-brane models on T^6/Z_6. *JHEP*, **0705**, 018.

Golfand, Yu. A. and Likhtman, E. P. (1971). Extension of the algebra of Poincare group generators and violation of P invariance. *JETP Lett.*, **13**, 323–6.

Gómez, C. and Hernández, R. (2002). Fields, strings and branes. *Lect. Notes Math.*, **1776**, 39–191.

Gómez, C., Sierra, G., and Ruiz Altaba, M. (1996). *Quantum Groups in Two-Dimensional Physics*. Cambridge: Cambridge University Press.

Goodsell, M. and Ringwald, A. (2010). Light hidden-sector $U(1)$'s in string compactifications. *Fortsch. Phys.*, **58**, 716–20.

Graña, M. (2003). MSSM parameters from supergravity backgrounds. *Phys. Rev.*, **D67**, 066006.

Graña, M. (2006a). Flux compactifications and generalized geometries. *Class. Quant. Grav.*, **23**, S883–926.

Graña, M. (2006b). Flux compactifications in string theory: a comprehensive review. *Phys. Rept.*, **423**, 91–158.

Graña, M., Grimm, T. W., Jockers, H., and Louis, J. (2004). Soft supersymmetry breaking in Calabi–Yau orientifolds with D-branes and fluxes. *Nucl. Phys.*, **B690**, 21–61.

Green, M. B., and Schwarz, J. H. (1984). Anomaly cancellation in supersymmetric $D = 10$ gauge theory and superstring theory. *Phys. Lett.*, **B149**, 117–22.

Green, M. B. and Schwarz, J. H. (1985a). Infinity cancellations in SO(32) superstring theory. *Phys. Lett.*, **B151**, 21–5.

Green, M. B. and Schwarz, J. H. (1985b). The hexagon gauge anomaly in type I superstring theory. *Nucl. Phys.*, **B255**, 93–114.

Green, M. B., Schwarz, J. H., and Witten, E. (1987a). *Superstring Theory, Vol 1: Introduction*. Cambridge: Cambridge University Press.

Green, M. B., Schwarz, J. H., and Witten, E. (1987b). *Superstring Theory, Vol 2: Loop Amplitudes, Anomalies and Phenomenology*. Cambridge: Cambridge University Press.

Greene, B. R. (1996). String theory on Calabi–Yau manifolds. [arXiv:hep-th/9702155].

Greene, B. R. and Plesser, M. R. (1990). Duality in Calabi–Yau moduli space. *Nucl. Phys.*, **B338**, 15–37.

Greene, B. R. and Yau, S.-T. (1997). *Mirror Symmetry II*. Ann Arbor, MI: American Mathematical Society/International Press.

Greene, B. R., Kirklin, K., Miron, P., and Ross, G. G. (1986a). A superstring inspired Standard Model. *Phys. Lett.*, **B180**, 69.

Greene, B. R., Kirklin, K., Miron, P., and Ross, G. G. (1986b). A three generation superstring model. 1. Compactification and discrete symmetries. *Nucl. Phys.*, **B278**, 667.

Greene, B. R., Kirklin, K., Miron, P., and Ross, G. G. (1987a). 27^3 Yukawa couplings for a three generation superstring model. *Phys. Lett.*, **B192**, 111.

Greene, B. R., Kirklin, K., Miron, P., and Ross, G. G. (1987b). A three generation superstring model. 2. Symmetry breaking and the low-energy theory. *Nucl. Phys.*, **B292**, 606.

Greene, B. R., Vafa, C., and Warner, N. P. (1989a). Calabi–Yau manifolds and renormalization group flows. *Nucl. Phys.*, **B324**, 371.

Greene, B. R., Lutken, C. A., and Ross, G. G. (1989b). Couplings in the heterotic superconformal three generation model. *Nucl. Phys.*, **B325**, 101.

Grimm, T. W. (2005). The effective action of type II Calabi–Yau orientifolds. *Fortsch. Phys.*, **53**, 1179–271.

Grimm, T. W. and Louis, J. (2004). The effective action of $\mathcal{N} = 1$ Calabi–Yau orientifolds. *Nucl. Phys.*, **B699**, 387–426.

Grimm, T. W. and Louis, J. (2005). The effective action of type IIA Calabi–Yau orientifolds. *Nucl. Phys.*, **B718**, 153–202.

Grimm, T. W., Krause, S., and Weigand, T. (2010). F-theory GUT vacua on compact Calabi–Yau fourfolds. *JHEP*, **1007**, 037.

Gubser, S. S. (2002). TASI lectures: special holonomy in string theory and M theory. [arXiv:hep-th/0201114], pp. 197–233.

Gubser, S. S., Klebanov, I. R., and Polyakov, A. M. (1998). Gauge theory correlators from noncritical string theory. *Phys. Lett.*, **B428**, 105–14.

Gukov, S., Vafa, C., and Witten, E. (2000). CFT's from Calabi–Yau four folds. *Nucl. Phys.*, **B584**, 69–108.

Gurrieri, S., Louis, J., Micu, A., and Waldram, D. (2003). Mirror symmetry in generalized Calabi–Yau compactifications. *Nucl. Phys.*, **B654**, 61–113.

Gurrieri, S., Lukas, A., and Micu, A. (2004). Heterotic on half-flat. *Phys. Rev.*, **D70**, 126009.

Gursey, F., Ramond, P., and Sikivie, P. (1976). A universal gauge theory model based on E_6. *Phys. Lett.*, **B60**, 177.

Gutowski, J. and Papadopoulos, G. (2001). Moduli spaces and brane solitons for M theory compactifications on holonomy G_2 manifolds. *Nucl. Phys.*, **B615**, 237–65.

Haber, H. E. and Hempfling, R. (1991). Can the mass of the lightest Higgs boson of the minimal supersymmetric model be larger than m(Z)? *Phys. Rev. Lett.*, **66**, 1815–18.

Hall, L., Lykken, J., and Weinberg, S. (1983). Supergravity as the messenger of supersymmetry breaking. *Phys. Rev.*, **D27**, 2359–78.

Hamidi, S. and Vafa, C. (1987). Interactions on orbifolds. *Nucl. Phys.*, **B279**, 465.

Han, T., Lykken, J. D., and Zhang, R.-J. (1999). On Kaluza–Klein states from large extra dimensions. *Phys. Rev.*, **D59**, 105006.

Hanany, A. and Kennaway, K. D. (2005). Dimer models and toric diagrams, [arXiv:hep-th/0503149].

Hanany, A. and Uranga, A. M. (1998). Brane boxes and branes on singularities. *JHEP*, **9805**, 013.

Harvey, J. (2005). TASI 2003 lectures on anomalies. [arXiv:hep-th/0509097].

Harvey, J. and Moore, G. W. (1999). Superpotentials and membrane instantons. [arXiv:hep-th/9907026].

Harvey, J. A. (1996). Magnetic monopoles, duality and supersymmetry. [arXiv:hep-th/9603086].

Harvey, R. and Lawson, H. B. (1982). Calibrated geometries. *Acta Math.*, **148**, 47.

Hayashi, H., Kawano, T., Tatar, R., and Watari, T. (2009). Codimension-3 singularities and Yukawa couplings in F-theory. *Nucl. Phys.*, **B823**, 47–115.

Hayashi, H., Kawano, T., Tsuchiya, Y., and Watari, T. (2010). Flavour structure in F-theory compactifications. *JHEP*, **1008**, 036.

Hebecker, A. and Trapletti, M. (2005). Gauge unification in highly anisotropic string compactifications. *Nucl. Phys.*, **B713**, 173–203.

Heckman, J. J. (2010). Particle physics implications of F-theory. [arXiv:hep-th/1001.0577].

Heckman, J. J. and Vafa, C. (2008). From F-theory GUTs to the LHC. [arXiv:hep-ph/0809.3452].

Heckman, J. J. and Vafa, C. (2010). Flavour hierarchy from F-theory. *Nucl. Phys.*, **B837**, 137–51.

Heckman, J. J., Marsano, J., Saulina, N., Schafer-Nameki, S., and Vafa, C. (2008). Instantons and SUSY breaking in F-theory. [arXiv:hep-th/0808.1286].

Heckman, J. J., Kane, G. L., Shao, J., and Vafa, C. (2009). The footprint of F-theory at the LHC. *JHEP*, **0910**, 039.

Hellerman, S., McGreevy, J., and Williams, B. (2004). Geometric constructions of nongeometric string theories. *JHEP*, **0401**, 024.

Hertzberg, M., Tegmark, M., and Wilczek, F. (2008). Axion cosmology and the energy scale of inflation. *Phys. Rev.*, **D78**, 083507.

Hewett, J. L. and Rizzo, T. G. (1989). Low-energy phenomenology of superstring inspired E_6 models. *Phys. Rept.*, **183**, 193.

Hewett, J. L., and Spiropulu, M. (2002). Particle physics probes of extra space-time dimensions. *Ann. Rev. Nucl. Part. Sci.*, **52**, 397–424.

Hogan, C. J. (2000). Why the universe is just so. *Rev. Mod. Phys.*, **72**, 1149–61.

Holdom, B. (1986). Two $U(1)$'s and epsilon charge shifts. *Phys. Lett.*, **B166**, 196.

Honecker, G. (2003). Chiral supersymmetric models on an orientifold of $Z_4 \times Z_2$ with intersecting D6-branes. *Nucl. Phys.*, **B666**, 175–96.

Honecker, G. and Ott, T. (2004). Getting just the supersymmetric standard model at intersecting branes on the Z_6 orientifold. *Phys. Rev.*, **D70**, 126010.

Hořava, P. and Witten, E. (1996a). Eleven-dimensional supergravity on a manifold with boundary. *Nucl. Phys.*, **B475**, 94–114.

Hořava, P. and Witten, E. (1996b). Heterotic and type I string dynamics from eleven-dimensions. *Nucl. Phys.*, **B460**, 506–24.

Hori, K. (2002). Trieste lectures on mirror symmetry. Available at http://users.ictp.it/~pub_off/lectures/lns013/Hori/Hori.pdf.

Hori, K., Katz, S., Klemm, A., *et al.* (2003). *Mirror Symmetry*. Ann Arbor, MI American Mathematical Society/International Press.

Hosono, S., Klemm, A., and Theisen, S. (1994). Lectures on mirror symmetry. [arXiv:hep-th/9403096].

Hosotani, Y. (1983). Dynamical gauge symmetry breaking as the casimir effect. *Phys. Lett.*, **B129**, 193.

Hubsch, T. (1992). *Calabi–Yau Manifolds: A Bestiary for Physicists*. Singapore: World Scientific.

Hull, C. M. (1996). String dynamics at strong coupling. *Nucl. Phys.*, **B468**, 113–54.

Hull, C. M. and Townsend, P. K. (1995). Unity of superstring dualities. *Nucl. Phys.*, **B438**, 109–37.

Ibáñez, L. E. (1982). Locally supersymmetric SU(5) Grand Unification. *Phys. Lett.*, **B118**, 73.

Ibáñez, L. E. (1983). Grand Unification with local supersymmetry. *Nucl. Phys.*, **B218**, 514.

Ibáñez, L. E. (1986). Hierarchy of quark-lepton masses in orbifold superstring compactification. *Phys. Lett.*, **B181**, 269.

Ibáñez, L. E. (1987). The search for a Standard Model $SU(3) \times SU(2) \times U(1)$ superstring: a introduction to orbifold constructions. In J. E. Kim (ed.) *Strings and Superstrings*. Singapore: World Scientific.

Ibáñez, L. E. (1993a). Computing the weak mixing angle from anomaly cancellation. *Phys. Lett.*, **B303**, 55–62.

Ibáñez, L. E. (1993b). Gauge coupling unification: strings versus SUSY GUTs. *Phys. Lett.*, **B318**, 73–6.

Ibáñez, L. E. and López, C. (1983). $\mathcal{N} = 1$ supergravity, the breaking of $SU(2) \times U(1)$ and the top quark mass. *Phys. Lett.*, **B126**, 54.

Ibáñez, L. E., and Lüst, D. (1992). Duality anomaly cancellation, minimal string unification and the effective low-energy Lagrangian of 4-D strings. *Nucl. Phys.*, **B382**, 305–64.

Ibáñez, L. E. and Richter, R. (2009). Stringy instantons and Yukawa couplings in MSSM-like orientifold models. *JHEP*, **0903**, 090.

Ibáñez, L. E. and Ross, G. G. (1981). Low-energy predictions in supersymmetric Grand Unified Theories. *Phys. Lett.*, **B105**, 439.

Ibáñez, L. E. and Ross, G. G. (1982). $SU(2)_L \times U(1)$ symmetry breaking as a radiative effect of supersymmetry breaking in GUT's. *Phys. Lett.*, **B110**, 215–220.

Ibáñez, L. E. and Ross, G. G. (1994). Fermion masses and mixing angles from gauge symmetries. *Phys. Lett.*, **B332**, 100–10.

Ibáñez, L. E. and Uranga, A. M. (2007). Neutrino Majorana masses from string theory instanton effects. *JHEP*, **0703**, 052.

Ibáñez, L. E. and Uranga, A. M. (2008). Instanton induced open string superpotentials and branes at singularities. *JHEP*, **0802**, 103.

Ibáñez, L. E., López, C., and Muñoz, C. (1985). The low-energy supersymmetric spectrum according to $\mathcal{N} = 1$ supergravity GUT's. *Nucl. Phys.*, **B256**, 218–52.

Ibáñez, L. E., Kim, J. E., Nilles, H. P., and Quevedo, F. (1987a). Orbifold compactifications with three families of $SU(3) \times SU(2) \times U(1)^n$. *Phys. Lett.*, **B191**, 282–6.

Ibáñez, L. E., Nilles, H. P., and Quevedo, F. (1987b). Orbifolds and Wilson lines. *Phys. Lett.*, **B187**, 25–32.

Ibáñez, L. E., Mas, J., Nilles, H. P., and Quevedo, F. (1988). Heterotic strings in symmetric and asymmetric orbifold backgrounds. *Nucl. Phys.*, **B301**, 157.

Ibáñez, L. E., Lüst, D., and Ross, G. G. (1991). Gauge coupling running in minimal $SU(3) \times SU(2) \times U(1)$ superstring unification. *Phys. Lett.*, **B272**, 251–60.

Ibáñez, L. E., Rabadán, R., and Uranga, A. M. (1999a). Anomalous $U(1)$'s in type I and type IIB D $= 4$, $\mathcal{N} = 1$ string vacua. *Nucl. Phys.*, **B542**, 112–38.

Ibáñez, L. E., Muñoz, C., and Rigolin, S. (1999b). Aspects of type I string phenomenology. *Nucl. Phys.*, **B553**, 43–80.

Ibáñez, L. E., Marchesano, F., and Rabadán, R. (2001). Getting just the standard model at intersecting branes. *JHEP*, **0111**, 002.

Ibáñez, L. E., Schellekens, A. N., and Uranga, A. M. (2007). Instanton induced neutrino majorana masses in CFT orientifolds with MSSM-like spectra. *JHEP*, **0706**, 011.

Ibarra, A., Ringwald, A., and Weniger, C. (2009). Hidden gauginos of an unbroken $U(1)$: Cosmological constraints and phenomenological prospects. *JCAP*, **0901**, 003.

Inoue, K., Kakuto, A., Komatsu, H., and Takeshita, S. (1982). Aspects of grand unified models with softly broken supersymmetry. *Prog. Theor. Phys.*, **68**, 927.

Intriligator, K. and Seiberg, N. (1996). Lectures on supersymmetric gauge theories and electric–magnetic duality. *Nucl. Phys. Proc. Suppl.*, **45BC**, 1–28. [arXiv:hep-th/9509066].

Intriligator, K. and Seiberg, N. (2007). Lectures on supersymmetry breaking. *Class. Quant. Grav.*, **24**, S741–72. [arXiv:hep-ph/0702069].

Intriligator, K. A., Seiberg, N., and Shih, D. (2006). Dynamical SUSY breaking in meta-stable vacua. *JHEP*, **0604**, 021.

Jockers, H. and Louis, J. (2005). The effective action of D7-branes in N $= 1$ Calabi–Yau orientifolds. *Nucl. Phys.*, **B705**, 167–211.

Johnson, C. V. (2000). D-brane primer, lectures at ICTP, TASI, and BUSSTEPP. [arXiv:hep-th/0007170].

Johnson, C. V. (2003). *D-branes*. Cambridge: Cambridge University Press.

Joyce, D. (2000). *Compact Manifolds with Special Holonomy*. Oxford: Oxford University Press.

Kachru, S. (1995). Some three generation (0,2) Calabi–Yau models. *Phys. Lett.*, **B349**, 76–82.

Kachru, S. and Silverstein, E. (1997). On gauge bosons in the matrix model approach to M theory. *Phys. Lett.*, **B396**, 70–6.

Kachru, S. and Silverstein, E. (1998). 4-D conformal theories and strings on orbifolds. *Phys. Rev. Lett.*, **80**, 4855–8.

Kachru, S., Kallosh, R., Linde, A. D., and Trivedi, S. P. (2003a). De Sitter vacua in string theory. *Phys. Rev.*, **D68**, 046005.

Kachru, S., Schulz, M. B., and Trivedi, S. (2003b). Moduli stabilization from fluxes in a simple IIB orientifold. *JHEP*, **0310**, 007.

Kachru, S., Schulz, M. B., Tripathy, P. K., and Trivedi, S. P. (2003c). New supersymmetric string compactifications. *JHEP*, **0303**, 061.

Kachru, S., Kallosh, R., Linde, A. D., Maldacena, J., McAllister, L., and Trivedi, S. P. (2003d). Towards inflation in string theory. *JCAP*, **0310**, 013.

Kachru, S., McAllister, L., and Sundrum, R. (2007). Sequestering in string theory. *JHEP*, **0710**, 013.

Kakushadze, Z. and Shiu, G. (1997). A chiral $\mathcal{N} = 1$ type I vacuum in four-dimensions and its heterotic dual. *Phys. Rev.*, **D56**, 3686–97.

Kakushadze, Z. and Shiu, G. (1998). 4-D chiral $\mathcal{N} = 1$ type one vacua with and without D5-branes. *Nucl. Phys.*, **B520**, 75–92.

Kakushadze, Z. and Tye, S. H. H. (1997). Three family $SU(5)$ grand unification in string theory. *Phys. Lett.*, **B392**, 335–42.

Kakushadze, Z., Shiu, G., Tye, S. H. H., and Vtorov-Karevsky, Y. (1998). A review of three-family grand unified string models. *Int. J. Mod. Phys.*, **A13**, 2551–98.

Kalara, S., López, J. L., and Nanopoulos, D. V. (1990). Nonrenormalizable terms in the free fermionic formulation of 4D strings. *Phys. Lett.*, **B245**, 421–8.

Kalara, S., López, J. L., and Nanopoulos, D. V. (1991). Calculable nonrenormalizable terms in string theory: a guide for the practitioner. *Nucl. Phys.*, **B353**, 650–82.

Kallosh, R. and Linde, A. D. (2004). Landscape, the scale of SUSY breaking, and inflation. *JHEP*, **0412**, 004.

Kaloper, N. and Myers, R. C. (1999). The $O(d, d)$ story of massive supergravity. *JHEP*, **9905**, 010.

Kaluza, T. (1921). On the problem of unity in physics. *Sitzungsber. Preuss. Akad. Wiss. Berlin (Math. Phys.)*, **1921**, 966–72.

Kane, G. L., Kumar, P., Lykken, J. D., and Wang, T. T. (2005). Some phenomenology of intersecting D-brane models. *Phys. Rev.*, **D71**, 115017.

Kane, G. L., Kumar, P., and Shao, J. (2007). LHC string phenomenology. *J. Phys. G*, **G34**, 1993–2036.

Kane, G. L., Kumar, P., and Shao, J. (2008). Unravelling strings at the CERN LHC. *Phys. Rev.*, **D77**, 116005.

Kaplunovsky, V. (1985). Mass scales of the string unification. *Phys. Rev. Lett.*, **55**, 1036.

Kaplunovsky, V. (1988). One loop threshold effects in string unification. *Nucl. Phys.*, **B307**, 145.

Kaplunovsky, V. S. (1992). One loop threshold effects in string unification. [arXiv:hep-th/9205070].

Kaplunovsky, V. and Louis, J. (1993). Model independent analysis of soft terms in effective supergravity and in string theory. *Phys. Lett.*, **B306**, 269–75.

Kaplunovsky, V. and Louis, J. (1994). Field dependent gauge couplings in locally supersymmetric effective quantum field theories. *Nucl. Phys.*, **B422**, 57–124.

Kaplunovsky, V. and Louis, J. (1995). On gauge couplings in string theory. *Nucl. Phys.*, **B444**, 191–244.

Kashani-Poor, A.-K., and Tomasiello, A. (2005). A stringy test of flux-induced isometry gauging. *Nucl. Phys.*, **B728**, 135–47.

Katsuki, Y., Kawamura, Y., Kobayashi, T., Ohtsubo, N., Ono, Y., and Tanioka, K. (1990). Z_N orbifold models. *Nucl. Phys.*, **B341**, 611–40.

Katz, S. H. and Vafa, C. (1997). Matter from geometry. *Nucl. Phys.*, **B497**, 146–54.

Kawai, H., Lewellen, D. C., and Tye, S. H. H. (1986). Construction of four-dimensional fermionic string models. *Phys. Rev. Lett.*, **57**, 1832.

Kawai, H., Lewellen, D. C., and Tye, S. H. H. (1987). Construction of fermionic string models in four dimensions. *Nucl. Phys.*, **B288**, 1.

Kawai, H., Lewellen, D. C., Schwartz, J. A., and Tye, S. H. H. (1988). The spin structure construction of string models and multi-loop modular invariance. *Nucl. Phys.*, **B299**, 431.

Kennaway, K. D. (2007). Brane tilings. *Int. J. Mod. Phys.*, **A22**, 2977–3038.

Kikkawa, K. and Yamasaki, M. (1984). Casimir effects in superstring theories. *Phys. Lett.*, **B149**, 357.

Kim, J. E. (1979). Weak interaction singlet and strong CP invariance. *Phys. Rev. Lett.*, **43**, 103.

Kim, J. E. and Carosi, G. (2010). Axions and the strong CP problem. *Rev. Mod. Phys.*, **82**, 557–602.

Kim, J. E. and Nilles, H. P. (1984). The mu problem and the strong CP problem. *Phys. Lett.*, **B138**, 150.

King, S. F., Leontaris, G. K., and Ross, G. G. (2010). Family symmetries in F-theory GUTs. *Nucl. Phys.*, **B838**, 119–35.

Kiritsis, E. (1997). Introduction to superstring theory. *Lectures presented at the Catholic University of Leuven and University of Padova, 1996–1997.* [arXiv:hep-th/9709062].

Kiritsis, E. (2007). *String Theory in a Nutshell.* Princeton, NJ: Princeton University Press.

Klebanov, I. R. (2000). TASI lectures: introduction to the AdS/CFT correspondence. [arXiv:hep-th/0009139].

Klebanov, I. R. and Strassler, M. J. (2000). Supergravity and a confining gauge theory: Duality cascades and chi SB resolution of naked singularities. *JHEP*, **0008**, 052.

Klebanov, I. R. and Witten, E. (2003). Proton decay in intersecting D-brane models. *Nucl. Phys.*, **B664**, 3–20.

Klein, O. (1926). Quantum theory and five-dimensional theory of relativity. *Z. Phys.*, **37**, 895–906.

Kobayashi, T. and Ohtsubo, N. (1990). Yukawa coupling condition of Z_N orbifold models. *Phys. Lett.*, **B245**, 441–6.

Kobayashi, T. and Ohtsubo, N. (1994). Geometrical aspects of Z_N orbifold phenomenology. *Int. J. Mod. Phys.*, **A9**, 87–126.

Kobayashi, T., Raby, S., and Zhang, R.-J. (2004). Constructing 5d orbifold grand unified theories from heterotic strings. *Phys. Lett.*, **B593**, 262–70.

Kobayashi, T., Raby, S., and Zhang, R.-J. (2005). Searching for realistic 4d string models with a Pati–Salam symmetry: orbifold grand unified theories from heterotic string compactification on a Z_6 orbifold. *Nucl. Phys.*, **B704**, 3–55.

Kokorelis, C. (2004). Exact standard model structures from intersecting D5-branes. *Nucl. Phys.*, **B677**, 115–63.

Komatsu, E. *et al.* (2009). Five-year Wilkinson Microwave Anisotropy Probe (WMAP) observations: cosmological interpretation. *Astrophys. J. Suppl.*, **180**, 330–76.

Kong, K., Matchev, K., and Servant, G. (2010). Extra dimensions at the LHC. [arXiv:hep-ph/1001.4801].

Körs, B. and Nath, P. (2004a). A Stueckelberg extension of the standard model. *Phys. Lett.*, **B586**, 366–72.

Körs, B. and Nath, P. (2004b). Effective action and soft supersymmetry breaking for intersecting D-brane models. *Nucl. Phys.*, **B681**, 77–119.

Krasnikov, N. V. (1987). On supersymmetry breaking in superstring theories. *Phys. Lett.*, **B193**, 37–40.

Kreuzer, M. (2010a). Toric geometry and Calabi–Yau compactifications. *Ukr. J. Phys.*, **55**, 613.

Kreuzer, M. (2010b). Heterotic (0,2) Gepner models and related geometries. In Grumiller, D., Rebhan, A., and Vassilevich, D. (eds.) *Fundamental Interactions: A Memorial Volume for Wolfgang Kummer.* Singapore: World Scientific, pp. 335–62. [arXiv:hep-th/0904.4467].

Kreuzer, M. and Skarke, H. (2002). Complete classification of reflexive polyhedra in four-dimensions. *Adv. Theor. Math. Phys.*, **4**, 1209–30.

Krippendorf, S., Quevedo, F., and Schlotterer, O. (2010a). Cambridge lectures on supersymmetry and extra dimensions. [arXiv:hep-th/1011.1491].

Krippendorf, S., Dolan, M. J., Maharana, A., and Quevedo, F. (2010b). D-branes at toric singularities: model building, Yukawa couplings and flavour physics. *JHEP*, **1006**, 092.

Lalak, Z. and Thomas, S. (1998). Gaugino condensation, moduli potential and supersymmetry breaking in M theory models. *Nucl. Phys.*, **B515**, 55–72.

Lane, K. (2002). Two lectures on technicolor. [arXiv:hep-ph/0202255].

Langacker, P. (1981). Grand Unified Theories and proton decay. *Phys. Rept.*, **72**, 185.

Langacker, P. (2009). The physics of heavy Z' gauge bosons. *Rev. Mod. Phys.*, **81**, 1199–228.

Lauer, J., Mas, J., and Nilles, H. P. (1989). Duality and the role of non-perturbative effects on the worldsheet. *Phys. Lett.*, **B226**, 251.

Lauer, J., Mas, J., and Nilles, H. P. (1991). Twisted sector representations of discrete background symmetries for two-dimensional orbifolds. *Nucl. Phys.*, **B351**, 353–424.

Lebedev, O., Nilles, H. P., Raby, S., *et al.* (2007). A mini-landscape of exact MSSM spectra in heterotic orbifolds. *Phys. Lett.*, **B645**, 88–94.

Lebedev, O., Nilles, H. P., Ramos-Sánchez, S., Ratz, M., and Vaudrevange, P. K. S. (2008a). Heterotic mini-landscape (II): completing the search for MSSM vacua in a Z_6 orbifold. *Phys. Lett.*, **B668**, 331–5.

Lebedev, O., Nilles, H. P., Raby, S., *et al.* (2008b). The heterotic road to the MSSM with R parity. *Phys. Rev.*, **D77**, 046013.

Lerche, W., Lüst, D., and Schellekens, A. N. (1987a). Chiral four-dimensional heterotic strings from selfdual lattices. *Nucl. Phys.*, **B287**, 477.

Lerche, W., Nilsson, B. E. W., and Schellekens, A. N. (1987b). Heterotic string loop calculation of the anomaly canceling term. *Nucl. Phys.*, **B289**, 609.

Lerche, W., Nilsson, B. E. W., Schellekens, A. N., and Warner, N. P. (1988). Anomaly canceling terms from the elliptic genus. *Nucl. Phys.*, **B299**, 91.

Lerche, W., Schellekens, A. N., and Warner, N. P. (1989). Lattices and strings. *Phys. Rept.*, **177**, 1.

Lerda, A., and Russo, R. (2000). Stable non-BPS states in string theory: a pedagogical review. *Int. J. Mod. Phys.*, **A15**, 771–820.

Lewellen, D. C. (1990). Embedding higher level Kac–Moody algebras in heterotic string models. *Nucl. Phys.*, **B337**, 61.

Linde, A. D. (1988). Inflation and axion cosmology. *Phys. Lett.*, **B201**, 437.

Linde, A. D. (2008). Inflationary cosmology. *Lect. Notes Phys.*, **738**, 1–54.

Lopes Cardoso, G. and Ovrut, B. (1992). A Green–Schwarz mechanism for $D = 4$, $\mathcal{N} = 1$ supergravity anomalies. *Nucl. Phys.*, **B369**, 351–72.

Lopes Cardoso, G. and Ovrut, B. (1993). Coordinate and Kahler sigma model anomalies and their cancellation in string effective field theories. *Nucl. Phys.*, **B392**, 315–44.

Lopes Cardoso, G., Curio, G., Dall'Agata, G., and Lüst, D. (2003). BPS action and superpotential for heterotic string compactifications with fluxes. *JHEP*, **0310**, 004.

Lopes Cardoso, G., Curio, G., Dall'Agata, G., and Lüst, D. (2004). Heterotic string theory on non-Kahler manifolds with H flux and gaugino condensate. *Fortsch.Phys.*, **52**, 483–8.

Lukas, A., Ovrut, B., and Waldram, D. (1998). On the four-dimensional effective action of strongly coupled heterotic string theory. *Nucl. Phys.*, **B532**, 43–82.

Lukas, A., Ovrut, B., Stelle, K. S., and Waldram, D. (1999). Heterotic M theory in five-dimensions. *Nucl. Phys.*, **B552**, 246–90.

Lüst, D. (2009). Seeing through the string landscape – a string hunter's companion in particle physics and cosmology. *JHEP*, **0903**, 149.

Lüst, D. and Stieberger, S. (2007). Gauge threshold corrections in intersecting brane world models. *Fortsch. Phys.*, **55**, 427–65.

Lüst, D. and Taylor, T. R. (1991). Hidden sectors with hidden matter. *Phys. Lett.*, **B253**, 335–41.

Lüst, D. and Theisen, S. (1989). Lectures on string theory. *Lect. Notes Phys.*, **346**, 1–346.

Lüst, D., Mayr, P., Richter, R., and Stieberger, S. (2004). Scattering of gauge, matter, and moduli fields from intersecting branes. *Nucl. Phys.*, **B696**, 205–50.

Lüst, D., Reffert, S., and Stieberger, S. (2005a). Flux-induced soft supersymmetry breaking in chiral type IIB orientifolds with D3/D7-branes. *Nucl. Phys.*, **B706**, 3–52.

Lüst, D., Reffert, S., and Stieberger, S. (2005b). MSSM with soft SUSY breaking terms from D7-branes with fluxes. *Nucl. Phys.*, **B727**, 264–300.

Lutken, C. A. and Ross, G. G. (1988). Taxonomy of heterotic superconformal field theories. *Phys. Lett.*, **B213**, 152.

Luty, M. A. (2005). TASI lectures on supersymmetry breaking. [arXiv:hep-th/0509029].

Lykken, J. D. (1995). Four-dimensional superstring models. [arXiv:hep-ph/9511456].

Lykken, J. D. (1996a). Introduction to supersymmetry. [arXiv:hep-th/9612114].

Lykken, J. D. (1996b). Weak scale superstrings. *Phys. Rev.*, **D54**, 3693–7.

Lykken, J. D., Poppitz, E., and Trivedi, S. P. (1999). Branes with GUTs and supersymmetry breaking. *Nucl. Phys.*, **B543**, 105–21.

Lynker, M. and Schimmrigk, R. (1988). On the spectrum of (2, 2) compactifications of the heterotic string on conformal field theories. *Phys. Lett.*, **B215**, 681.

Lyth, D. H., and Stewart, E. D. (1996). Thermal inflation and the moduli problem. *Phys. Rev.*, **D53**, 1784–98.

Maldacena, J., Moore, G. W., and Seiberg, N. (2001). D-brane instantons and K theory charges. *JHEP*, **0111**, 062.

Maldacena, J. M. (1998). The large N limit of superconformal field theories and supergravity. *Adv. Theor. Math. Phys.*, **2**, 231–52.

Maldacena, J. M. (2003). TASI 2003 lectures on AdS/CFT. [arXiv:hep-th/0309246].

Malyshev, D. and Verlinde, H. (2007). D-branes at singularities and string phenomenology. *Nucl. Phys. Proc. Suppl.*, **171**, 139–63.

Marchesano, F. (2003). Intersecting D-brane models. Ph.D. thesis. [arXiv:hep-th/0307252].

Marchesano, F. (2007). Progress in D-brane model building. *Fortsch. Phys.*, **55**, 491–518.

Marchesano, F. and Martucci, L. (2010). Non-perturbative effects on seven-brane Yukawa couplings. *Phys. Rev. Lett.*, **104**, 231601.

Marchesano, F. and Shiu, G. (2004). Building MSSM flux vacua. *JHEP*, **0411**, 041.

Marchesano, F. and Shiu, G. (2005). MSSM vacua from flux compactifications. *Phys. Rev.*, **D71**, 011701.

Mariño, M., Minasian, R., Moore, G. W., and Strominger, A. (2000). Nonlinear instantons from supersymmetric p-branes. *JHEP*, **0001**, 005.

Markushevich, D. G., Olshanetsky, M. A., and Perelomov, A. M. (1987). Description of a class of superstring compactifications related to semisimple Lie algebras. *Commun. Math. Phys.*, **111**, 247.

Marsano, J., Saulina, N., and Schafer-Nameki, S. (2009a). Gauge mediation in F-theory GUT models. *Phys. Rev.*, **D80**, 046006.

Marsano, J., Saulina, N., and Schafer-Nameki, S. (2009b). Monodromies, fluxes, and compact three-generation F-theory GUTs. *JHEP*, **0908**, 046.

Marsano, J., Saulina, N., and Schafer-Nameki, S. (2010). An instanton toolbox for F-Theory model building. *JHEP*, **1001**, 128.

Martin, S. P. (1997). A supersymmetry primer. [arXiv:hep-ph/9709356].

Martinec, E. J. (1989). Algebraic geometry and effective lagrangians. *Phys. Lett.*, **B217**, 431.

McAllister, L. and Silverstein, E. (2008). String cosmology: a review. *Gen. Rel. Grav.*, **40**, 565–605.

Meade, P., Seiberg, N., and Shih, D. (2009). General gauge mediation. *Prog. Theor. Phys. Suppl.*, **177**, 143–58.

Minahan, J. A. (1988). One loop amplitudes on orbifolds and the renormalization of coupling constants. *Nucl. Phys.*, **B298**, 36.

Minkowski, P. (1977). $\mu \to e\gamma$ at a rate of one out of 1-billion muon decays? *Phys. Lett.*, **B67**, 421.

Mirabelli, E. A., Perelstein, M., and Peskin, M. E. (1999). Collider signatures of new large space dimensions. *Phys. Rev. Lett.*, **82**, 2236–9.

Mohapatra, R. and Pati, J. (1975a). A natural left–right symmetry. *Phys. Rev.*, **D11**, 2558.

Mohapatra, R. and Pati, J. (1975b). Left–right gauge symmetry and an isoconjugate model of CP violation. *Phys. Rev.*, **D11**, 566–71.

Montonen, C. and Olive, D. (1977). Magnetic monopoles as gauge particles? *Phys. Lett.*, **B72**, 117.

Morrison, D. R. and Vafa, C. (1996a). Compactifications of F theory on Calabi–Yau threefolds. 1. *Nucl. Phys.*, **B473**, 74–92.

Morrison, D. R. and Vafa, C. (1996b). Compactifications of F theory on Calabi–Yau threefolds. 2. *Nucl. Phys.*, **B476**, 437–69.

Mueller, M. T. and Witten, E. (1986). Twisting toroidally compactified heterotic strings with enlarged symmetry groups. *Phys. Lett.*, **B182**, 28.

Mumford, D. (2006). *Tata Lectures on Theta*. Berlin: Birkhäuser.

Nakahara, M. (2003). *Geometry, Topology and Physics* 2nd edn. Bristol: Institute of Physics.

Narain, K. S., Sarmadi, M. H., and Vafa, C. (1987). Asymmetric orbifolds. *Nucl. Phys.*, **B288**, 551.

Narain, K. S., Sarmadi, M. H., and Vafa, C. (1991). Asymmetric orbifolds: path integral and operator formulations. *Nucl. Phys.*, **B356**, 163–207.

Nath, P. and Fileviez Pérez, P. (2007). Proton stability in grand unified theories, in strings and in branes. *Phys. Rept.*, **441**, 191–317.

Nilles, H. P. (1984). Supersymmetry, supergravity and particle physics. *Phys. Rept.*, **110**, 1–162.

Nilles, H. P. and Olechowski, M. (1990). Gaugino condensation and duality invariance. *Phys. Lett.*, **B248**, 268–72.

Nilles, H. P. and Stieberger, S. (1996). How to reach the correct $sin^2\theta_W$ and α_s in string theory. *Phys. Lett.*, **B367**, 126–33.

O'Driscoll, D. (1998). General Abelian orientifold models and one loop amplitudes. [arXiv:hep-th/9801114].

Okada, Y., Yamaguchi, M., and Yanagida, T. (1991). Upper bound of the lightest Higgs boson mass in the minimal supersymmetric standard model. *Prog. Theor. Phys.*, **85**, 1–6.

Olsen, K. and Szabo, R. J. (1999). Constructing D-branes from K theory. *Adv. Theor. Math. Phys.*, **3**, 889–1025.

Ooguri, H. and Yin, Z. (1996). TASI lectures on perturbative string theories. [arXiv:hep-th/9612254].

Ooguri, H., Oz, Y., and Yin, Z. (1996). D-branes on Calabi–Yau spaces and their mirrors. *Nucl. Phys.*, **B477**, 407–30.

Ooguri, Hi. and Vafa, C. (2007). On the geometry of the string landscape and the Swampland. *Nucl. Phys.*, **B766**, 21–33.

O'Raifeartaigh, L. (1975). Spontaneous symmetry breaking for chiral scalar superfields. *Nucl. Phys.*, **B96**, 331.

Ortin, T. (2004). *Gravity and Strings*. Cambridge: Cambridge University Press.

Pati, J. and Salam, A. (1973). Unified lepton-hadron symmetry and a gauge theory of the basic interactions. *Phys. Rev.*, **D8**, 1240–51.

Peccei, R. and Quinn, H. (1977a). Constraints imposed by CP conservation in the presence of instantons. *Phys. Rev.*, **D16**, 1791–97.

Peccei, R. and Quinn, H. (1977b). CP conservation in the presence of instantons. *Phys. Rev. Lett.*, **38**, 1440–3.

Petersson, C. (2008). Superpotentials from stringy instantons without orientifolds. *JHEP*, **0805**, 078.

Polchinski, J. (1995). Dirichlet branes and Ramond–Ramond charges. *Phys. Rev. Lett.*, **75**, 4724–7.

Polchinski, J. (1996). TASI lectures on D-branes. [arXiv:hep-th/9611050].

Polchinski, J. (1998a). *String Theory. Vol. 1: An Introduction to the Bosonic String.* Cambridge: Cambridge University Press.

Polchinski, J. (1998b). *String Theory. Vol. 2: Superstring Theory and Beyond.* Cambridge: Cambridge University Press.

Polchinski, J. (2004). Introduction to cosmic F- and D-strings. [arXiv:hep-th/0412244].

Polchinski, J. (2006). The cosmological constant and the string landscape. [arXiv:hep-th/0603249].

Polchinski, J. (2010). *Joe's Little Book of String*. Available at www.kitp.ucsb.edu/joep/JLBS.pdf.

Polchinski, J. and Strominger, A. (1996). New vacua for type II string theory. *Phys. Lett.*, **B388**, 736–42.

Polchinski, J. and Witten, E. (1996). Evidence for heterotic–type I string duality. *Nucl. Phys.*, **B460**, 525–40.

Polchinski, J., Chaudhuri, S., and Johnson, C. V. (1996). Notes on D-branes. [arXiv:hep-th/9602052].

Polonyi, J. (1977). Generalization of the massive scalar multiplet coupling to the supergravity. Report KFKI-77-93, Hungary Central Institute of Research.

Poppitz, E. (1999). On the one loop Fayet–Iliopoulos term in chiral four-dimensional type I orbifolds. *Nucl. Phys.*, **B542**, 31–44.

Pradisi, G. and Sagnotti, A. (1989). Open string orbifolds. *Phys. Lett.*, **B216**, 59.

Quevedo, F. (2002). Lectures on string/brane cosmology. *Class. Quant. Grav.*, **19**, 5721–79.

Ramond, P. (1971). Dual theory for free fermions. *Phys. Rev.*, **D3**, 2415–18.

Ramos-Sánchez, S. (2009). Towards low energy physics from the heterotic string. *Fortsch. Phys.*, **10**, 907–1036.

Randall, L. and Sundrum, R. (1999a). A large mass hierarchy from a small extra dimension. *Phys. Rev. Lett.*, **83**, 3370–73.

Randall, L. and Sundrum, R. (1999b). An alternative to compactification. *Phys. Rev. Lett.*, **83**, 4690–3.

Randall, L. and Sundrum, R. (1999c). Out of this world supersymmetry breaking. *Nucl. Phys.*, **B557**, 79–118.

Rizzo, T. G. (2010). Introduction to extra dimensions. *AIP Conf. Proc.*, **1256**, 27–50.

Ross, G. G. (1981). Unified Field Theories. *Rept. Prog. Phys.*, **44**, 655–718.

Ross, G. G. (1984). *Grand Unified Theories*. Frontiers in Physics, 60. Menlo Park, CA: Benjamin/Cummings.

Ross, G. G. (1988). Models and phenomenology of superstring theories. CERN preprint CERN-TH.5109/88.

Rubakov, V. A. and Shaposhnikov, M. E. (1983). Do we live inside a domain wall? *Phys. Lett.*, **B125**, 136–8.

Sakai, N. (1981). Naturalness in supersymmetric GUT's. *Z. Phys.*, **C11**, 153.

Sakai, N. and Yanagida, T. (1982). Proton decay in a class of supersymmetric grand unified models. *Nucl. Phys.*, **B197**, 533.

Saltman, A. and Silverstein, E. (2004). The scaling of the no scale potential and de Sitter model building. *JHEP*, **0411**, 066.

Schellekens, A. N. (1990a). Conformal field theory for four-dimensional strings. *Nucl. Phys. Proc. Suppl.*, **15**, 3–34.

Schellekens, A. N. (1990b). Electric charge quantization in string theory. *Phys. Lett.*, **B237**, 363.

Schellekens, A. N. (1996). Introduction to conformal field theory. *Fortsch. Phys.*, **44**, 605–705.

Schellekens, A. N. (2000). Open strings, simple currents and fixed points. [arXiv:hep-th/0001198].

Schellekens, A. N. (2006). The Landscape "avant la lettre." [arXiv:physics/0604134].

Schellekens, A. N. (2008). The emperor's last clothes? Overlooking the string theory landscape. *Rept. Prog. Phys.*, **71**, 072201.

Schellekens, A. N. and Yankielowicz, S. (1989). Extended chiral algebras and modular invariant partition functions. *Nucl. Phys.*, **B327**, 673.

Schellekens, A. N. and Yankielowicz, S. (1990a). New modular invariants for $\mathcal{N} = 2$ tensor products and four-dimensional strings. *Nucl. Phys.*, **B330**, 103.

Schellekens, A. N. and Yankielowicz, S. (1990b). Simple currents, modular invariants and fixed points. *Int. J. Mod. Phys.*, **A5**, 2903–52.

Scherk, J. and Schwarz, J. H. (1979a). How to get masses from extra dimensions. *Nucl. Phys.*, **B153**, 61–88.

Scherk, J. and Schwarz, J. H. (1979b). Spontaneous breaking of supersymmetry through dimensional reduction. *Phys. Lett.*, **B82**, 60.

Schwarz, J. H. (1997). Lectures on superstring and M theory dualities. *Nucl. Phys. Proc. Suppl.*, **55B**, 1–32.

Schwarz, J. H. (1995). An SL(2,Z) multiplet of type IIB superstrings. *Phys. Lett.*, **B360**, 13–18.

Seiberg, N. (1994). Exact results on the space of vacua of four-dimensional SUSY gauge theories. *Phys. Rev.*, **D49**, 6857–63.

Seiberg, N. and Witten, E. (1994a). Electric–magnetic duality, monopole condensation, and confinement in $\mathcal{N} = 2$ supersymmetric Yang–Mills theory. *Nucl. Phys.*, **B426**, 19–52.

Seiberg, N. and Witten, E. (1994b). Monopoles, duality and chiral symmetry breaking in $\mathcal{N} = 2$ supersymmetric QCD. *Nucl. Phys.*, **B431**, 484–550.

Sen, A. (1985). Naturally light Higgs doublet in supersymmetric E_6 Grand Unified Theory. *Phys. Rev. Lett.*, **55**, 33.

Sen, A. (1994). Strong–weak coupling duality in four-dimensional string theory. *Int. J. Mod. Phys.*, **A9**, 3707–50.

Sen, A. (1996). F theory and orientifolds. *Nucl. Phys.*, **B475**, 562–78.

Sen, A. (1997). Orientifold limit of F theory vacua. *Phys. Rev.*, **D55**, 7345–49.

Sen, A. (1998). An introduction to nonperturbative string theory. [arXiv:hep-th/9802051].

Sen, A. (1999). Non BPS states and branes in string theory. [arXiv:hep-th/9904207].

Senda, I. and Sugamoto, A. (1988). Orbifold models and modular transformation. *Nucl. Phys.*, **B302**, 291.

Shapere, A. and Wilczek, F. (1989). Selfdual models with theta terms. *Nucl. Phys.*, **B320**, 669.

Shelton, J., Taylor, W., and Wecht, B. (2005). Nongeometric flux compactifications. *JHEP*, **0510**, 085.

Shenker, S. H. (1995). Another length scale in string theory? [arXiv:hep-th/9509132].

Shifman, M. (1994). *Instantons in Gauge Theories*. Singapore: World Scientific.

Shifman, M. (2010). Large extra dimensions: becoming acquainted with an alternative paradigm. *Int. J. Mod. Phys.*, **A25**, 199–225.

Shifman, M., Vainshtein, A., and Zakharov, V. (1980). Can confinement ensure natural CP invariance of strong interactions? *Nucl. Phys.*, **B166**, 493.

Shiu, G., Torroba, G., Underwood, B., and Douglas, M. R. (2008). Dynamics of warped flux compactifications. *JHEP*, **0806**, 024.

Silverstein, E. and Westphal, A. (2008). Monodromy in the CMB: gravity waves and string inflation. *Phys. Rev.*, **D78**, 106003.

Soni, S. K. and Weldon, H. A. (1983). Analysis of the supersymmetry breaking induced by $\mathcal{N} = 1$ supergravity theories. *Phys. Lett.*, **B126**, 215.

Strathdee, J. A. (1987). Extended poincare supersymmetry. *Int. J. Mod. Phys.*, **A2**, 273.

Strominger, A. (1986). Superstrings with torsion. *Nucl. Phys.*, **B274**, 253.

Strominger, A. (1990). Heterotic solitons. *Nucl. Phys.*, **B343**, 167–84.

Strominger, A. (1995). Massless black holes and conifolds in string theory. *Nucl. Phys.*, **B451**, 96–108.

Strominger, A. and Witten, E. (1985). New manifolds for superstring compactification. *Commun. Math. Phys.*, **101**, 341.

Strominger, A., Yau, S.-T., and Zaslow, E. (1996). Mirror symmetry is T duality. *Nucl. Phys.*, **B479**, 243–59.

Sugimoto, S. (1999). Anomaly cancellation in type I D9-anti-D9 system and the USp(32) string theory. *Prog. Theor. Phys.*, **102**, 685–99.

Susskind, L. (1979). Dynamics of spontaneous symmetry breaking in the Weinberg–Salam theory. *Phys. Rev.*, **D20**, 2619–25.

Susskind, L. (2003). The anthropic landscape of string theory. [arXiv:hep-th/0302219].

Svrcek, P. and Witten, E. (2006). Axions in string theory. *JHEP*, **0606**, 051.

Szabo, R. J. (2002). BUSSTEPP lectures on string theory: an introduction to string theory and D-brane dynamics. [arXiv:hep-th/0207142].

Taylor, T. R. and Vafa, C. (2000). RR flux on Calabi–Yau and partial supersymmetry breaking. *Phys. Lett.*, **B474**, 130–7.

Tegmark, M., Aguirre, A., Rees, M., and Wilczek, F. (2006). Dimensionless constants, cosmology and other dark matters. *Phys. Rev.*, **D73**, 023505.

Terning, J. (2006). *Modern Supersymmetry: Dynamics and Duality*. Oxford: Clarendon.

Tian, G. and Yau, S. T. (1985). Examples of three dimensional Ricci-Flat algebraic manifolds with Euler number -6. In *Proc. of Argonne Symposium on Anomalies, Geometry, and Topology*. Singapore: World Scientific.

Tong, D. (2005). TASI lectures on solitons: instantons, monopoles, vortices and kinks. [arXiv:hep-th/0509216].

Tong, D. (2009). String theory. [arXiv:hep-th/0908.0333]. Also available at www.damtp. cam.ac.uk/user/tong/string.html.

Townsend, P. K. (1995). The eleven-dimensional supermembrane revisited. *Phys. Lett.*, **B350**, 184–7.

Tripathy, P. K. and Trivedi, S. P. (2005). D3 brane action and fermion zero modes in presence of background flux. *JHEP*, **0506**, 066.

Uhlenbeck, K. and Yau, S.-T. (1986). On the existence of hermitian–Yang–Mills connections in stable vector bundles. *Comm. Pure Appl. Math.*, **39**, 257–93.

Uranga, A. M. (2001a). Introduction to string theory. Available at www.ift.uam.es/ paginaspersonales/angeluranga/firstpage.html.

Uranga, A. M. (2001b). D-brane probes, RR tadpole cancellation and K theory charge. *Nucl. Phys.*, **B598**, 225–46.

Uranga, A. M. (2002). Local models for intersecting brane worlds. *JHEP*, **0212**, 058.

Uranga, A. M. (2003). Chiral four-dimensional string compactifications with intersecting D-branes. *Class. Quant. Grav.*, **20**, S373–94.

Uranga, A. M. (2005). Intersecting brane worlds. *Class. Quant. Grav.*, **22**, S41–76.

Uranga, A. M. (2007). The standard model in string theory from D-branes. *Nucl. Phys. Proc. Suppl.*, **171**, 119–38.

Uranga, A. M. (2009). D-brane instantons and the effective field theory of flux compactifications. *JHEP*, **0901**, 048.

Vafa, C. (1986). Modular invariance and discrete torsion on orbifolds. *Nucl. Phys.*, **B273**, 592.

Vafa, C. (1996). Evidence for F theory. *Nucl. Phys.*, **B469**, 403–18.

Vafa, C. (2005). The string landscape and the swampland. [arXiv:hep-th/0509212].

Veneziano, G. (1968). Construction of a crossing–symmetric, Regge behaved amplitude for linearly rising trajectories. *Nuovo Cim.*, **A57**, 190–7.

Verlinde, H. and Wijnholt, M. (2007). Building the standard model on a D3-brane. *JHEP*, **0701**, 106.

Villadoro, G. and Zwirner, F. (2005). $\mathcal{N} = 1$ effective potential from dual type-IIA D6/O6 orientifolds with general fluxes. *JHEP*, **0506**, 047.

Volkov, D. V. and Akulov, V. P. (1973). Is the neutrino a Goldstone particle? *Phys. Lett.*, **B46**, 109–10.

Wecht, B. (2007). Lectures on nongeometric flux compactifications. *Class. Quant. Grav.*, **24**, S773–94.

Weigand, T. (2010). Lectures on F-theory compactifications and model building. *Class. Quant. Grav.*, **27**, 214004. [arXiv:hep-th/1009.3497].

Weinberg, S. (1976). Implications of dynamical symmetry breaking. *Phys. Rev.*, **D13**, 974–96.

Weinberg, S. (1978). A new light boson? *Phys. Rev. Lett.*, **40**, 223–6.

Weinberg, S. (1979). Implications of dynamical symmetry breaking: an addendum. *Phys. Rev.*, **D19**, 1277–80.

Weinberg, S. (1982). Supersymmetry at ordinary energies, 1: masses and conservation laws. *Phys. Rev.*, **D26**, 287.

Weinberg, S. (1987). Anthropic bound on the cosmological constant. *Phys. Rev. Lett.*, **59**, 2607.

Weinberg, S. (1989). The cosmological constant problem. *Rev. Mod. Phys.*, **61**, 1–23.

Weinberg, S. (2000). *The Quantum Theory of Fields, Vol. 3: Supersymmetry*. Cambridge: Cambridge University Press.

Wen, X.-G. and Witten, E. (1985). Electric and magnetic charges in superstring models. *Nucl. Phys.*, **B261**, 651.

Wess, J. and Bagger, J. (1992). *Supersymmetry and Supergravity*. Princeton, NJ: Princeton University Press.

Wess, J. and Zumino, B. (1974). Supergauge transformations in four-dimensions. *Nucl. Phys.*, **B70**, 39–50.

Wilczek, F. (1978). Problem of strong P and T invariance in the presence of instantons. *Phys. Rev. Lett.*, **40**, 279–82.

Witten, E. (1981). Dynamical breaking of supersymmetry. *Nucl. Phys.*, **B188**, 513.

Witten, E. (1984). Some properties of O(32) superstrings. *Phys. Lett.*, **B149**, 351–6.

Witten, E. (1985a). Cosmic superstrings. *Phys. Lett.*, **B153**, 243.

Witten, E. (1985b). Dimensional reduction of superstring models. *Phys. Lett.*, **B155**, 151.

Witten, E. (1985c). Symmetry breaking patterns in superstring models. *Nucl. Phys.*, **B258**, 75.

Witten, E. (1986). New issues in manifolds of $SU(3)$ holonomy. *Nucl. Phys.*, **B268**, 79.

Witten, E. (1993). Phases of $\mathcal{N} = 2$ theories in two-dimensions. *Nucl. Phys.*, **B403**, 159–222.

Witten, E. (1995). String theory dynamics in various dimensions. *Nucl. Phys.*, **B443**, 85–126.

Witten, E. (1996a). Nonperturbative superpotentials in string theory. *Nucl. Phys.*, **B474**, 343–60.

Witten, E. (1996b). Small instantons in string theory. *Nucl. Phys.*, **B460**, 541–59.

Witten, E. (1996c). Strong coupling expansion of Calabi–Yau compactification. *Nucl. Phys.*, **B471**, 135–58.

Witten, E. (1998a). Anti-de Sitter space and holography. *Adv. Theor. Math. Phys.*, **2**, 253–91.

Witten, E. (1998b). D-branes and K theory. *JHEP*, **9812**, 019.

Witten, E. (1998c). Toroidal compactification without vector structure. *JHEP*, **9802**, 006.

Witten, E. (2000). World sheet corrections via D instantons. *JHEP*, **0002**, 030.

Yanagida, T. (1979). Horizontal symmetry and masses of neutrious. In *Proc. of the Workshop of Unified Theories and Baryon Number in the Universe*, O. Sawada and A. Sugamoto (eds.). Tuskuba, Japan: KEK.

Yau, S.-T. (1998). *Mirror Symmetry I*. New York: American Mathematical Society/International Press.

Zwart, G. (1998). Four-dimensional $\mathcal{N} = 1$ $Z_N \times Z_M$ orientifolds. *Nucl. Phys.*, **B526**, 378–92.

Zwiebach, B. (2004). *A First Course in String Theory*. Cambridge: Cambridge University Press.

Index

Printed in the United States
By Bookmasters